Progress in Mathematics

Volume 260

Arithmetic and Geometry Around Hypergeometric Functions

Lecture Notes of a CIMPA Summer School held at Galatasaray University, Istanbul, 2005

Rolf-Peter Holzapfel
A. Muhammed Uludağ
Masaaki Yoshida
Editors

Birkhäuser
Basel · Boston · Berlin

Editors:

Rolf-Peter Holzapfel
Institut für Mathematik
Humboldt-Universität zu Berlin
Unter den Linden 6
D-10099 Berlin
e-mail: holzapfl@mathematik.hu-berlin.de

Masaaki Yoshida
Department of Mathematics
Kyushu University
Fukuoka 810-8560
Japan
e-mail: myoshida@math.kyushu-u.ac.jp

A. Muhammed Uludağ
Department of Mathematics
Galatasaray University
34357 Besiktas, Istanbul
Turkey
e-mail: muhammed.uludag@gmail.com

2000 Mathematics Subject Classification 14J10, 14J28, 14J15, 11F06, 11S80, 33C65, 22E40, 11F55, 11G15,11K22, 33C70, 32Q30, 33C05

Library of Congress Control Number : 2006939568

A CIP catalogue record for this book is available from the Library of Congress, Washington D.C., USA

Bibliographic information published by Die Deutsche Bibliothek
Die Deutsche Bibliothek lists this publication in the Deutsche Nationalbibliografie; detailed bibliographic data is available in the Internet at <http://dnb.ddb.de>.

ISBN 978-3-7643-8283-4 Birkhäuser Verlag AG, Basel – Boston – Berlin

Part of Springer Science+Business Media
Printed on acid-free paper produced of chlorine-free pulp. TCF ∞
Printed in Germany
ISBN-10: 3-7643-8283-X
ISBN-13: 978-3-7643-8283-4
e-ISBN-10: 3-7643-7449-7
e-ISBN-13: 978-3-7643-8284-1

9 8 7 6 5 4 3 2 1

www.birkhauser.ch

Preface

This volume comprises the Lecture Notes of the CIMPA Summer School *Arithmetic and Geometry around Hypergeometric Functions* held at Galatasaray University, Istanbul during June 13-25, 2005. In the Summer School there were fifteen lectures forming an impressive group of mathematicians covering a wide range of topics related to hypergeometric functions. The full schedule of talks from the workshop appears on the next page. In addition to the lecture notes submitted by its lecturers, this volume contains several research articles.

A group of fourty graduate students and young researchers attended the school. Among the participants there were 2 Algerian, 3 American, 1 Armenian, 1 Bulgarian, 1 Canadian, 3 Dutch, 2 Georgian, 7 German, 1 Indian, 2 Iraqi, 1 Iranian, 1 Italian, 1 Russian, 5 Japanese, 23 Turkish and 1 Ukrainian mathematicians, including the lecturers.

We would like to thank the *Centre des Mathématiques Pures et Appliquées*, for their financial support and Professor Michel Jambu for organizational help. We could support participants from across the region thanks to the generous financial help provided by the *International Center for Theoretical Physics* (ICTP) and the *International Mathematical Union* (IMU). The local participants has been supported by the *Scientific and Technological Research Councel of Turkey* (TÜBİTAK).

This summer school has been realized not only by financial support from its sponsors but also thanks to the generousity of its lecturers, who all agreed to finance their travel from their own personal grants. Some of them did so also for the accomodation.

The proposal for the AGAHF Summer School was submitted to CIMPA in February 2004. During the long preparatory process and during the summer school, Ayşegül Ulus, Özgür Ceyhan, and Özgür Kişisel contributed at various levels to the organization. We are grateful to them.

Sabine Buchmann is a French artist living in Istanbul, who likes to draw Ottoman-style miniatures of the boats serving across the bosphorus; these boats are an inseparable part of the city panorama. When asked, she liked the idea of a boat full of mathematicians and drew it for the conference poster — with the names of all the lecturers hidden inside, written in minute letters. Her miniature helped us much in attracting the audience of the summer school.

We are thankful to the student team hired by the university comprising Anet İzmitli, Egemen Kırant, Günce Orman, Haris Saybaşılı and Eylem Şentürk for turning this summer school into a pleasant experience.

Finally we would like to thank warmly Prof. Dr. Duygun Yarsuvat, the rector of the Galatasaray University for offering us the great location and financial support of the university.

The second named editor was supported by TÜBİTAK grant Kariyer 103T136 during the summer school and during the preperation of this volume.

Rolf-Peter Holzapfel, A. Muhammed Uludağ and Masaaki Yoshida, Editors

PROGRAM

Daniel Allcock: Real hyperbolic geometry in moduli problems

Igor Dolgachev: Moduli spaces as ball quotients (followed by Kondo's lectures)

Rolf Peter Holzapfel: Orbital Varieties and Invariants

Michel Jambu: Arrangements of Hyperplanes

A. Kochubei: Hypergeometric functions and Carlitz differential equations over function fields

Shigeyuki Kondo: Complex ball uniformizations of the moduli spaces of del Pezzo surfaces

Edward Looijenga: (first week) Introduction to Deligne-Mostow theory

Edward Looijenga: (second week) Hypergeometric functions associated to arrangements

Keiji Matsumoto: Invariant functions with respect to the Whitehead link

Hironori Shiga: Hypergeometric functions and arithmetic geometric means (followed by Wolfart's lectures)

Jan Stienstra: Gel'fand-Kapranov-Zelevinsky hypergeometric systems and their role in mirror symmetry and in string theory

Toshiaki Terada: Hypergeometric representation of the group of pure braids.

A. Muhammed Uludağ: Geometry of Complex Orbifolds

Alexander Varchenko: Special functions, KZ type equations, and representation theory

Jürgen Wolfart: Arithmetic of Schwarz maps (preceded by Shiga's lectures)

Masaaki Yoshida: Schwarz maps (general introduction)

Contents

Progress in Mathematics, Vol. 260, 1–22

Hyperbolic Geometry and the Moduli Space of Real Binary Sextics

Daniel Allcock, James A. Carlson and Domingo Toledo

Abstract. The moduli space of real 6-tuples in $\mathbb{C}P^1$ is modeled on a quotient of hyperbolic 3-space by a nonarithmetic lattice in $\operatorname{Isom} H^3$. This is partly an expository note; the first part of it is an introduction to orbifolds and hyperbolic reflection groups.

Keywords. Complex hyperbolic geometry, hyperbolic reflection groups, orbifolds, moduli spaces, ball quotients.

These notes are an exposition of the key ideas behind our result that the moduli space $\mathcal{M}_s$ of stable real binary sextics is the quotient of real hyperbolic 3-space H^3 by a certain Coxeter group (together with its diagram automorphism). We hope they can serve as an aid in understanding our work [3] on moduli of real cubic surfaces, since exactly the same ideas are used, but the computations are easier and the results can be visualized.

These notes derive from the first author's lectures at the summer school "Algebra and Geometry around Hypergeometric Functions", held at Galatasary University in Istanbul in July 2005. He is grateful to the organizers, fellow speakers and students for making the workshop very rewarding. To keep the flavor of lecture notes, not much has been added beyond the original content of the lectures; some additional material appears in an appendix. The pictures are hand-drawn to encourage readers to draw their own.

Lecture 1

Hyperbolic space H^3 is a Riemannian manifold for which one can write down an explicit metric, but for us the following model will be more useful; it is called the upper half-space model. Its underlying set is the set of points in $\mathbb{R}^3$ with

First author partly supported by NSF grant DMS 0231585. Second and third authors partly supported by NSF grants DMS 9900543 and DMS 0200877.

positive vertical coordinate, and geodesics appear either as vertical half-lines, or as semicircles with both ends resting on the bounding $\mathbb{R}^2$:

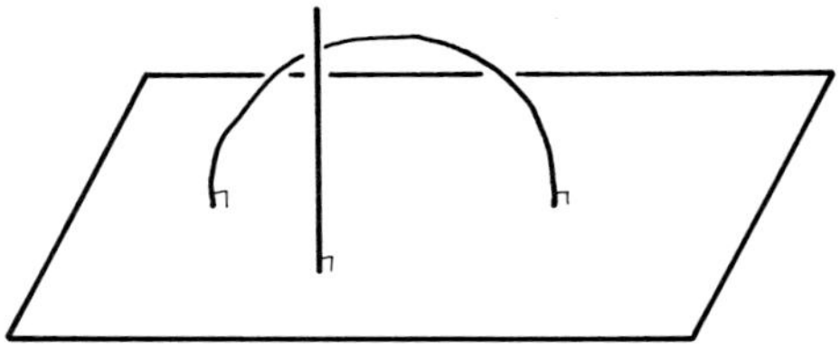

Note that the 'endpoints' of these geodesics lie in the boundary of H^3, not in H^3 itself. Planes appear either as vertical half-planes, or as hemispheres resting on $\mathbb{R}^2$:

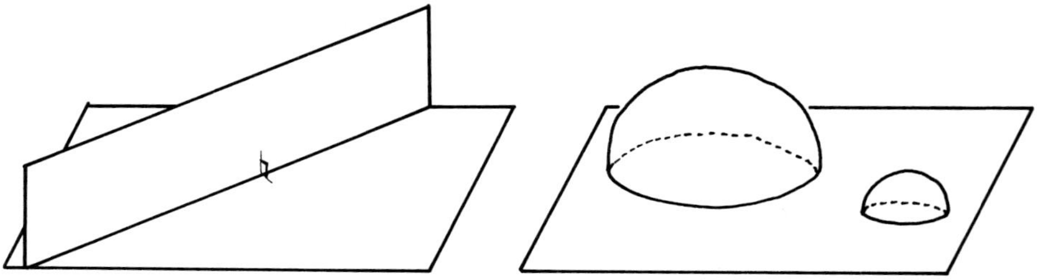

If two planes meet then their intersection is a geodesic. The most important property of the upper half-space model is that it is conformal, meaning that an angle between planes under the hyperbolic metric equals the Euclidean angle between the half-planes and/or hemispheres. For example, the following angle θ looks like a $\pi/4$ angle, so it *is* a $\pi/4$ angle:

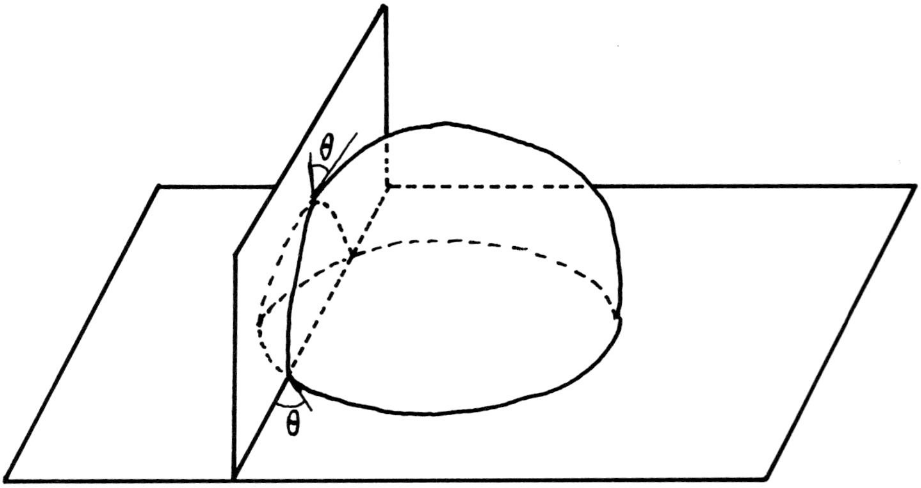

This lets us build hyperbolic polyhedra with specified angles by pushing planes around. For example, the diagram

 P_0 (1)

describes a polyhedron P_0 with four walls, corresponding to the nodes, with the interior angle between two walls being $\pi/2$, $\pi/3$ or $\pi/4$ according to whether the nodes are joined by no edge, a single edge or a double edge. For now, ignore the colors of the nodes; they play no role until Theorem 2. We can build a concrete model of P_0 by observing that the first three nodes describe a Euclidean $(\pi/2, \pi/4, \pi/4)$ triangle, so the first three walls should be arranged to appear as vertical halfplanes. Sometimes pictures like this can be easier to understand if you also draw the view down from vertical infinity; here are both pictures:

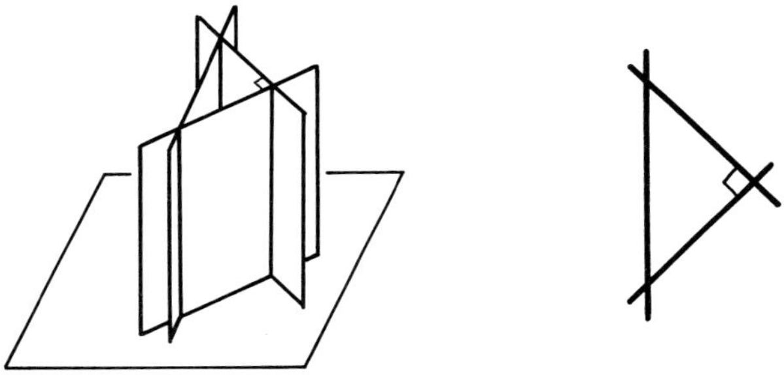

How to fit in the fourth plane? After playing with it one discovers that it cannot appear as a vertical halfplane, so we look for a suitable hemisphere. It must be orthogonal to two of our three walls, so it is centered at the foot of one of the halflines of intersection. The radius of the hemisphere is forced to be 2 because of the angle it makes with the remaining wall (namely $\pi/3$). We have drawn the picture so that the hemisphere is centered at the foot of the back edge. The figure should continue to vertical infinity, but we cut it off because seeing the cross-section makes the polyhedron easier to understand. We've also drawn the view from above; the boundary circle of the hemisphere strictly contains the triangle, corresponding to the fact that P_0 does not descend all the way to the boundary $\mathbb{R}^2$.

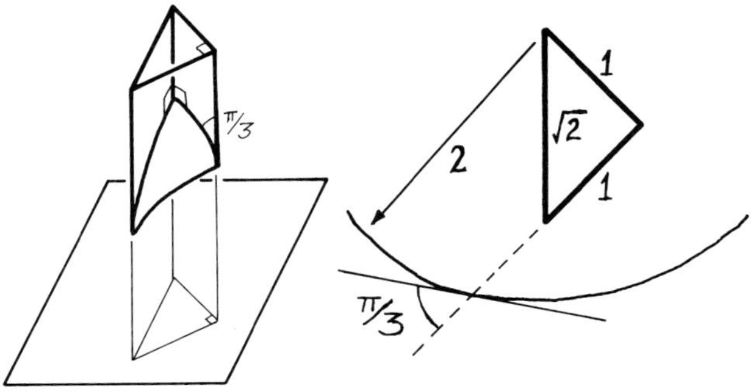

(2)

We think of P_0 as an infinitely tall triangular chimney with its bottom bitten off by a hemisphere. The dimensions we have drawn on the overhead view refer to Euclidean distances, not hyperbolic ones. The "radius" of a hemisphere has no intrinsic meaning in hyperbolic geometry; indeed, the isometry group of H^3 acts transitively on planes.

Readers may enjoy trying their hands at this by drawing polyhedra for the diagrams

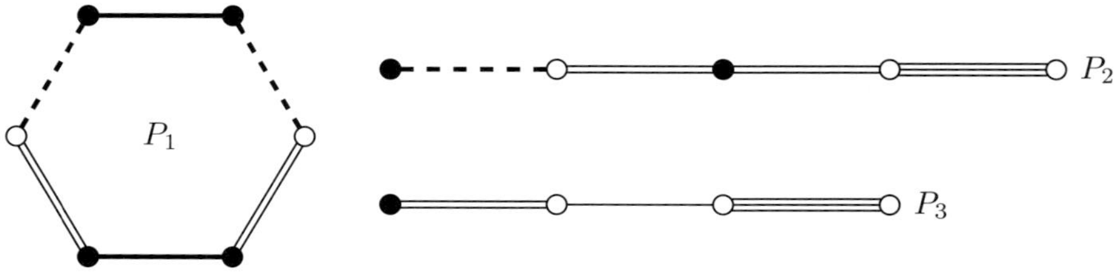

(3)

where the absent, single and double bonds mean the same as before, a triple bond indicates a $\pi/6$ angle, a heavy bond means parallel walls and a dashed bond means ultraparallel walls. In the last two cases we describe the meaning by pictures: Parallelism means

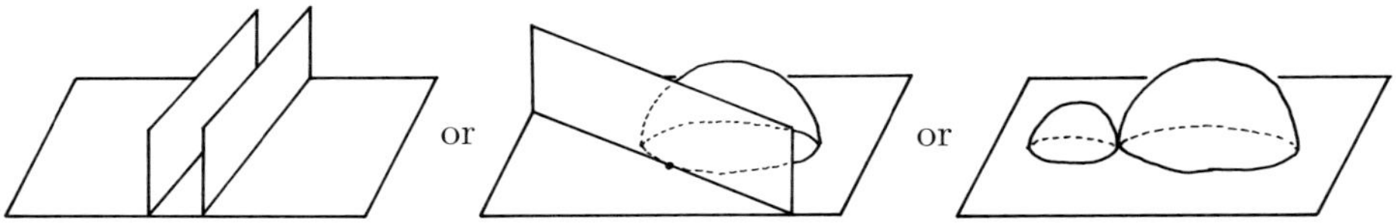

and ultraparallelism means

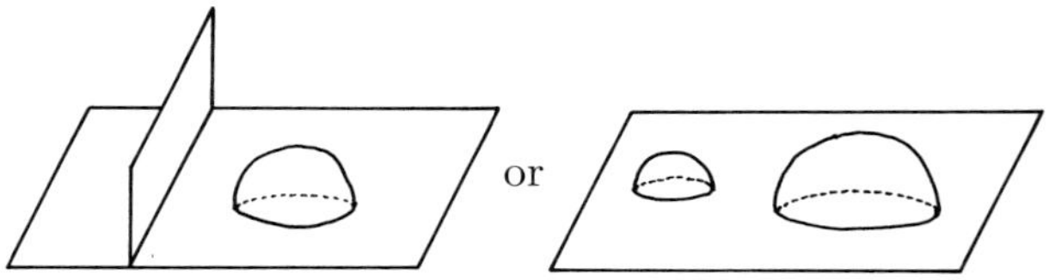

That is, when two planes do not meet in H^3, we call them parallel if they meet at the boundary of H^3, and ultraparallel if they do not meet even there.

Diagrams like (1) and (3) are called Coxeter diagrams after H. S. M. Coxeter, who introduced them to classify the finite groups generated by reflections. Given a random diagram, there is no guarantee that one can find a hyperbolic polyhedron with those angles, but if there is one then it describes a discrete group acting on H^3:

Theorem 1 (Poincaré Polyhedron Theorem). *Suppose $P \subseteq H^3$ is a polyhedron (i.e., the intersection of a finite number of closed half-spaces) with every dihedral angle of the form $\pi/$(an integer). Let Γ be the group generated by the reflections across the walls of P. Then Γ is discrete in* $\operatorname{Isom} H^3$ *and P is a fundamental* *domain for* Γ *in the strong sense: every point of* H^3 *is* Γ*-equivalent to exactly one point of* P.

The proof is a very pretty covering space argument; see [5] for this and for a nice introduction to Coxeter groups in general. A reflection across a plane means the unique isometry of H^3 that fixes the plane pointwise and exchanges the components of its complement. A reflection across a vertical half-plane looks like an ordinary Euclidean reflection, and a reflection across a hemisphere means an inversion in it; here are before-and-after pictures of an inversion.

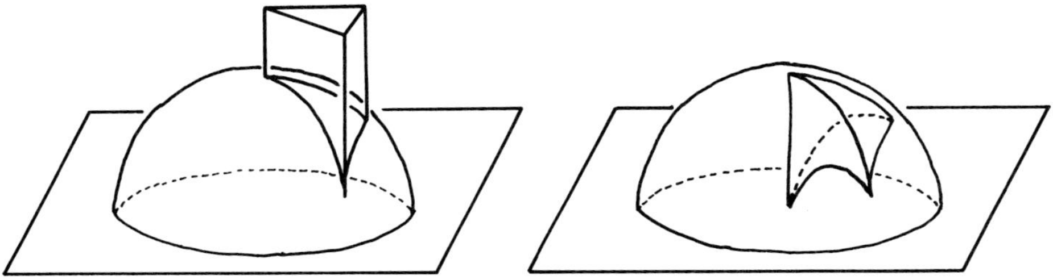

An inversion exchanges vertical infinity with the point of $\mathbb{R}^2$ "at the center" of the hemisphere.

The data of a group Γ acting discretely on H^3 is encoded by an object called an orbifold. As a topological space it is H^3/Γ, but the orbifold has more structure. An orbifold chart on a topological space X is a continuous map from an open subset U of $\mathbb{R}^n$ to X, that factors as

$$U \to U/\Gamma_U \to X,$$

where Γ_U is a finite group acting on U and the second map is a homeomorphism onto its image. Our H^3/Γ has lots of such charts, because if $x \in H^3$ has stabilizer Γ_x and U is a sufficiently small open ball around x, then

$$U \to U/\Gamma_x \to H^3/\Gamma$$

is an orbifold chart. An orbifold is a space locally modeled on a manifold modulo finite groups. Formally, an orbifold X is a hausdorff space covered by such charts, with the compatibility condition that if $x \in X$ lies in the image of charts $U \to U/\Gamma_U \to X$ and $U' \to U'/\Gamma_{U'} \to X$ then there are preimages v and v' of x in U and U' with neighborhoods V and V' preserved by $\Gamma_{U,v}$ and $\Gamma_{U',v'}$, an isomorphism $\Gamma_{U,v} \cong \Gamma_{U',v'}$ and an equivariant isomorphism $\tau_{V,V'}$ between V and V' identifying v with v'. The group $\Gamma_{U,v}$ is called the local group at x, and the nature of the isomorphisms $\tau_{V,V'}$ determines the nature of the orbifold. That is, if all the $\tau_{V,V'}$ are homeomorphisms then X is a topological orbifold, if all are real-analytic diffeomorphisms then X is a real-analytic orbifold, if all are hyperbolic isometries then X is a hyperbolic orbifold, and so on. So H^3/Γ is a hyperbolic orbifold. There is a notion of orbifold universal cover which allows one to reconstruct H^3 and its Γ-action from the orbifold H^3/Γ.

Only in two dimensions is it easy to draw pictures of orbifold charts; here they are for the quotient of the upper half-plane H^2 by the group Γ generated by reflections across the edges of the famous $(\pi/2, \pi/3, \pi/\infty)$ triangle.

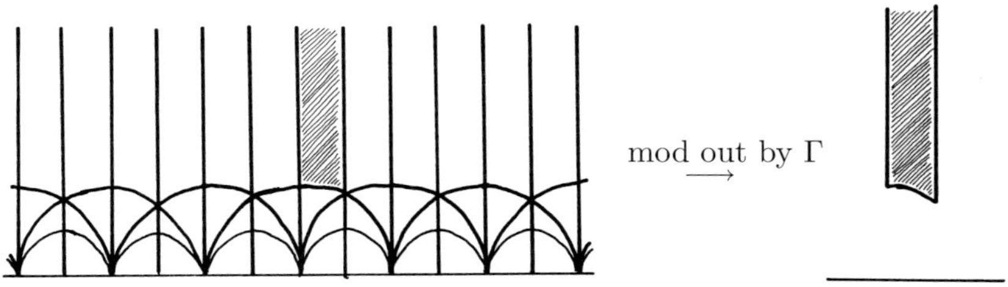

Here are local orbifold charts around various points of H^3/Γ:

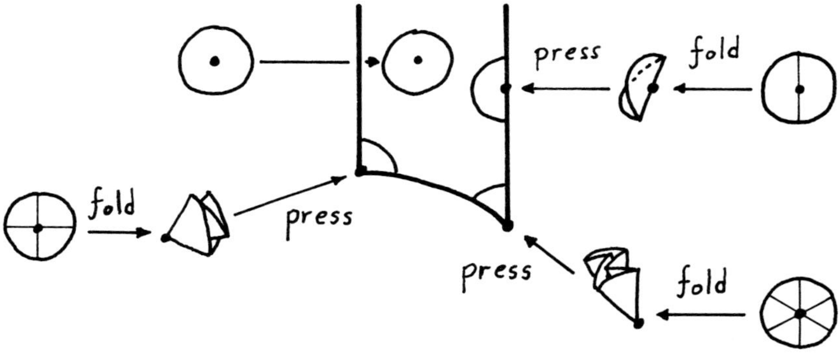

For three-dimensional Coxeter groups essentially the same thing happens: the local chart at a generic point of a wall is the quotient of a 3-ball by a reflection, and along an edge it is the quotient of a 3-ball by a dihedral group. One needs to understand the finite Coxeter groups in dimension 3 in order to understand the folding at the vertices, but this is not necessary here.

We care about hyperbolic orbifolds because it turns out that moduli spaces arising in algebraic geometry are usually orbifolds, and it happens sometimes that such a moduli space happens to coincide with a quotient of hyperbolic space (or complex hyperbolic space or one of the other symmetric spaces). So we can sometimes gain insight into the algebraic geometry by manipulating simple objects like tilings of hyperbolic space.

Suppose a Lie group G acts properly on a smooth manifold X, with finite stabilizers. (Properly means that each compact set K in X meets only "compactly many" of its translates—that is, there exists a compact set in G such that if $g \in G$ lies outside it, then $K \cap gK = \emptyset$. This is needed for the quotient space to be Hausdorff.) Because G acts on the left, we write $G\backslash X$ for the quotient, which is an orbifold by the following construction. For $x \in X$ one can find a small transversal T to the orbit $G.x$, which is preserved by the stabilizer G_x. Then $T \to G_x\backslash T \to G\backslash X$ gives an orbifold chart. In particular, the local group at the image of x in $G\backslash X$

is G_x. If X is real-analytic and G acts real-analytically then $G\backslash X$ is a real-analytic orbifold.

Now we come to the case which concerns us. Let $\mathcal{C}$ be the set of binary sextics, i.e., nonzero 2-variable homogeneous complex polynomials of degree 6, modulo scalars, so $\mathcal{C} = \mathbb{C}P^6$. Let $\mathcal{C}^{\mathbb{R}}$ be the subset given by those with real coefficients, $\mathcal{C}_0$ the smooth sextics (those with 6 distinct roots), and $\mathcal{C}_0^{\mathbb{R}}$ the intersection. Then $G = \mathrm{PGL}_2\mathbb{C}$ acts on $\mathcal{C}$ and $\mathcal{C}_0$ and $G^{\mathbb{R}} = \mathrm{PGL}_2\mathbb{R}$ acts on $\mathcal{C}^{\mathbb{R}}$ and $\mathcal{C}_0^{\mathbb{R}}$. The moduli space $\mathcal{M}_0$ of smooth binary sextics is $G\backslash\mathcal{C}_0$, of 3 complex dimensions. The real moduli space $\mathcal{M}_0^{\mathbb{R}} = G^{\mathbb{R}}\backslash\mathcal{C}_0^{\mathbb{R}}$ is *not* the moduli space of 6-tuples in $\mathbb{R}P^1$; rather it is the moduli space of nonsingular 6-tuples in $\mathbb{C}P^1$ which are preserved by complex conjugation. This set has 4 components, $\mathcal{M}_{0,j}^{\mathbb{R}}$ being $G^{\mathbb{R}}\backslash\mathcal{C}_{0,j}^{\mathbb{R}}$, where j indicates the number of pairs of conjugate roots. It turns out that G acts properly on $\mathcal{C}_0$, and since the point stabilizers are compact algebraic subgroups of G they are finite; therefore $\mathcal{M}_0$ is a complex-analytic orbifold and the $\mathcal{M}_{0,j}^{\mathbb{R}}$ are real-analytic orbifolds. The relation with hyperbolic geometry begins with the following theorem:

Theorem 2. *Let Γ_j be the group generated by the Coxeter group of P_j from* (1) *or* (3)*, together with the diagram automorphism when $j = 1$. Then $\mathcal{M}_{0,j}^{\mathbb{R}}$ is the orbifold H^3/Γ_j, minus the image therein of the walls corresponding to the blackened nodes and the edges corresponding to triple bonds. Here, 'is' means an isomorphism of real-analytic orbifolds.*

In the second lecture we will see that the faces of the P_j corresponding to blackened nodes and triple bonds are very interesting; we will glue the P_j together to obtain a real-hyperbolic description of the entire moduli space.

References. The canonical references for hyperbolic geometry and an introduction to orbifolds are Thurston's notes [15] and book [16]. The book is a highly polished treatment of a subset of the material in the notes, which inspired a great deal of supplementary material, e.g., [4]. For other applications of hyperbolic geometry to real algebraic geometry, see Nikulin's papers [12] and [13], which among other things describe moduli spaces of various sorts of K3 surfaces as quotients of H^n.

Lecture 2

We will not really provide a proof of Theorem 2; instead we will develop the ideas behind it just enough to motivate the main construction leading to Theorem 4 below. Although Theorem 2 concerns smooth sextics, it turns out to be better to consider mildly singular sextics as well. Namely, let $\mathcal{C}_s$ be the set of binary sextics with no point of multiplicity 3 or higher, and let $\Delta \subseteq \mathcal{C}_s$ be the discriminant, so $\mathcal{C}_0 = \mathcal{C}_s - \Delta$. (For those who have seen geometric invariant theory, $\mathcal{C}_s$ is the set of stable sextics, hence the subscript s.) It is easy to see that Δ is a normal crossing divisor in $\mathcal{C}_s$. (In the space of *ordered* 6-tuples in $\mathbb{C}P^1$ this is clear; to get the picture in $\mathcal{C}_s$ one mods out by permutations.) Now let $\mathcal{F}_s$ be the universal branched cover of $\mathcal{C}_s$, with ramification of order 6 along each component of the preimage of Δ. $\mathcal{F}_s$ turns out to be smooth and the preimage of Δ a normal crossing divisor. More precisely, in a neighborhood of a point of $\mathcal{F}_s$ describing a sextic with k double points, the map to $\mathcal{C}_s$ is given locally by

$$(z_1, \ldots, z_6) \mapsto (z_1^6, \ldots, z_k^6, z_{k+1}, \ldots, z_6) ,$$

where the branch locus is the union of the hypersurfaces $z_1 = 0, \ldots, z_k = 0$. Let $\mathcal{F}_0$ be the preimage of $\mathcal{C}_0$ and let Γ be the deck group of $\mathcal{F}_s$ over $\mathcal{C}_s$. We call an element of $\mathcal{F}_s$ (resp. $\mathcal{F}_0$) a framed stable (resp. smooth) binary sextic. Geometric invariant theory implies that G acts properly on $\mathcal{C}_s$, and one can show that this G-action lifts to one on $\mathcal{F}_s$ which is not only proper but free, so $G\backslash\mathcal{F}_s$ is a complex manifold. The reason we use 6-fold branching rather than some other sort of branching is that in this case $G\backslash\mathcal{F}_s$ has a nice description, given by the following theorem. See the appendix for a sketch of the Hodge theory involved in the proof.

Theorem 3 (Deligne–Mostow [6]). *There is a properly discontinuous action of Γ on complex hyperbolic 3-space $\mathbb{C}H^3$ and a Γ-equivariant complex-manifold diffeomorphism $g : G\backslash\mathcal{F}_s \to \mathbb{C}H^3$, identifying $G\backslash\mathcal{F}_0$ with the complement of a hyperplane arrangement $\mathcal{H}$ in $\mathbb{C}H^3$.*

Complex hyperbolic space is like ordinary hyperbolic space except that it has 3 complex dimensions, and hyperplanes have complex codimension 1. There is an upper-half space model analogous to the real case, but the most common model for it is the (open) complex ball. This is analogous to the Poincaré ball model for real hyperbolic space; we don't need the ball model except to see that complex conjugation of $\mathbb{C}H^3$, thought of as the complex 3-ball, has fixed-point set the real 3-ball, which is H^3.

Given a framed stable sextic $\tilde{S}$, Theorem 3 gives us a point $g(\tilde{S})$ of $\mathbb{C}H^3$. If $\tilde{S}$ lies in $\mathcal{F}_0^{\mathbb{R}}$ (the preimage of $\mathcal{C}_0^{\mathbb{R}}$), say over $S \in \mathcal{C}_0^{\mathbb{R}}$, then we can do better, obtaining not just a point of $\mathbb{C}H^3$ but also a copy of H^3 containing it. The idea is that complex conjugation κ of $\mathcal{C}_0$ preserves S and lifts to an antiholomorphic involution (briefly, an anti-involution) $\tilde{\kappa}$ of $\mathcal{F}_0$ that fixes $\tilde{S}$. This uses the facts that $\mathcal{F}_0 \to \mathcal{C}_0$ is a covering space and that $\pi_1(\mathcal{F}_0) \subseteq \pi_1(\mathcal{C}_0)$ is preserved by κ. Riemann extension extends $\tilde{\kappa}$ to an anti-involution of $\mathcal{F}_s$. Since κ normalizes G's action on $\mathcal{C}_s$, $\tilde{\kappa}$ normalizes G's action on $\mathcal{F}_s$, so $\tilde{\kappa}$ descends to an anti-involution κ' of $\mathbb{C}H^3 = G\backslash\mathcal{F}_s$. Each anti-involution of $\mathbb{C}H^3$ has a copy of H^3 as its fixed-point set, so we have defined a map $g^{\mathbb{R}}$ from $\mathcal{F}_0^{\mathbb{R}}$ to the set of pairs

$$(x \in \mathbb{C}H^3, \text{ a copy of } H^3 \text{ containing } x). \tag{4}$$

Note that $\tilde{\kappa}$ fixes every point of $\mathcal{F}_0^{\mathbb{R}}$ sufficiently near $\tilde{S}$, so all nearby framed real sextics determine the same anti-involution κ' of $\mathbb{C}H^3$. Together with the G-invariance of g, this proves that $g^{\mathbb{R}}$ is invariant under the identity component of $G^{\mathbb{R}}$. A closer study of $g^{\mathbb{R}}$ shows that it is actually invariant under all of $G^{\mathbb{R}}$. We write K_0 for the set of pairs (4) in the image $g^{\mathbb{R}}(\mathcal{F}_0^{\mathbb{R}})$. An argument relating points of $\mathcal{C}_s$ preserved by anti-involutions in $G \rtimes (\mathbb{Z}/2)$ to points of $\mathbb{C}H^3$ preserved by anti-involutions in $\Gamma \rtimes (\mathbb{Z}/2)$ shows that if $x \in \mathcal{F}_0^{\mathbb{R}}$ has image $(g(x), H)$, then every pair $(y \in H - \mathcal{H}, H)$

also lies in K_0. That is, K_0 is the disjoint union of a bunch of H^3's, minus their intersections with $\mathcal{H}$. The theoretical content of Theorem 2 is that $g^{\mathbb{R}} : G^{\mathbb{R}} \backslash \mathcal{F}_0^{\mathbb{R}} \to K_0$ is a diffeomorphism.

The computational part of Theorem 2 is the explicit description of K_0, in enough detail to understand $\mathcal{M}_0 = G\backslash\mathcal{F}_0^{\mathbb{R}}/\Gamma = K_0/\Gamma$ concretely. It turns out that Γ, $\mathcal{H}$ and the anti-involutions can all be described cleanly in terms of a certain lattice Λ over the Eisenstein integers $\mathcal{E} = \mathbb{Z}[\omega{=}e^{2\pi i/3}]$. Namely, Λ is a rank 4 free $\mathcal{E}$-module with Hermitian form

$$\langle a|a\rangle = a_0\bar{a}_0 - a_1\bar{a}_1 - a_2\bar{a}_2 - a_3\bar{a}_3\ . \tag{5}$$

The set of positive lines in $P(\mathbb{C}^{1,3} = \Lambda \otimes_{\mathcal{E}} \mathbb{C})$ is a complex 3-ball (i.e., $\mathbb{C}H^3$), $\Gamma = P\mathrm{Aut}\,\Lambda$, $\mathcal{H}$ is the union of the hyperplanes orthogonal to norm -1 elements of Λ, and the anti-involutions of $\mathbb{C}H^3$ corresponding to the elements of K_0 are exactly

$$\begin{aligned}
\kappa_0 &: (x_0,x_1,x_2,x_3) \mapsto (\bar{x}_0,\ \ \bar{x}_1,\ \ \bar{x}_2,\ \ \bar{x}_3)\\
\kappa_1 &: (x_0,x_1,x_2,x_3) \mapsto (\bar{x}_0,\ \ \bar{x}_1,\ \ \bar{x}_2,-\bar{x}_3)\\
\kappa_2 &: (x_0,x_1,x_2,x_3) \mapsto (\bar{x}_0,\ \ \bar{x}_1,-\bar{x}_2,-\bar{x}_3)\\
\kappa_3 &: (x_0,x_1,x_2,x_3) \mapsto (\bar{x}_0,-\bar{x}_1,-\bar{x}_2,-\bar{x}_3)
\end{aligned} \tag{6}$$

and their conjugates by Γ. We write H^3_j for the fixed-point set of κ_j.

Since $H^3_0,\dots,H^3_3$ form a complete set of representatives for the H^3's comprising K_0, we have

$$\mathcal{M}_0^{\mathbb{R}} = K_0/\Gamma = \coprod_{j=0}^{3} \ (H^3_j - \mathcal{H})\Big/(\text{its stabilizer } \Gamma_j \text{ in } \Gamma)$$

Understanding the stabilizers Γ_j required a little luck. Vinberg devised an algorithm for searching for a fundamental domain for a discrete group acting on H^n that is generated by reflections [18]. It is not guaranteed to terminate, but if it does then it gives a fundamental domain. We were lucky and it did terminate; the reflection subgroup of Γ_j turns out to be the Coxeter group of the polyhedron P_j.

One can obtain our polyhedra by applying his algorithm to the $\mathbb{Z}$-sublattices of Λ fixed by each κ_j. For example, an element of the κ_2-invariant part of Λ has the form $(a_0, a_1, a_2\sqrt{-3}, a_3\sqrt{-3})$ with $a_0,\dots,a_3 \in \mathbb{Z}$, of norm $a_0^2 - a_1^2 - 3a_2^2 - 3a_3^2$. Similar analysis leads to the norm forms

$$\begin{aligned}
\langle a|a\rangle &= a_0^2 - \ \ a_1^2 - \ \ a_2^2 - \ \ a_3^2\\
\langle a|a\rangle &= a_0^2 - \ \ a_1^2 - \ \ a_2^2 - 3a_3^2\\
\langle a|a\rangle &= a_0^2 - \ \ a_1^2 - 3a_2^2 - 3a_3^2\\
\langle a|a\rangle &= a_0^2 - 3a_1^2 - 3a_2^2 - 3a_3^2
\end{aligned}$$

in the four cases of (6). Now, Γ_j lies between its reflection subgroup and the semidirect product of this subgroup by its diagram automorphisms. After checking that the diagram automorphism of P_1 lies in Γ_1, the identification of the Γ_j is complete.

The final part of Theorem 2 boils down to considering how the H^3's comprising K_0 meet the hyperplanes comprising $\mathcal{H}$. There is no big idea here; one just works out the answer and writes it down. There are essentially two ways that the H^3 fixed by an anti-involution κ of Λ can meet a hyperplane $r^\perp$, where $r \in \Lambda$ has norm -1. It might happen that $\kappa(r)$ is proportional to r, in which case $H^3 \cap r^\perp$ is a copy of H^2; this accounts for the deleted walls of the P_j. It can also happen that $\kappa(r) \perp r$, in which case $H^3 \cap r^\perp$ is a copy of H^1; this accounts for the deleted edges.

Now, the deleted faces are very interesting, and the next step in our discussion is to add them back in. By Theorem 3 we know that points of $\mathcal{H}$ represent singular sextics, which occur along the boundary between two components of $\mathcal{C}_0^{\mathbb{R}}$. For example,

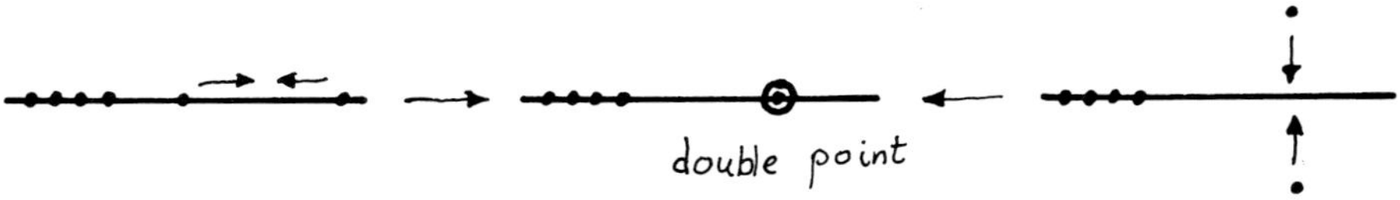

Varying the remaining four points gives a family of singular sextics which lie in the closures of both $\mathcal{C}_{0,0}^{\mathbb{R}}$ and $\mathcal{C}_{0,1}^{\mathbb{R}}$. This suggests reinstating the deleted walls of P_0 and P_1 and gluing the reinstated wall of P_0 to one of the reinstated walls of P_1. Which walls, and by what identification? There is really no choice here, because H_0^3 and H_1^3 meet along an H^2 that lies in $\mathcal{H}$, namely the locus

$$\{(a_0, a_1, a_2, a_3) \in \mathbb{C}^{1,3} \mid a_0, a_1, a_2 \in \mathbb{R} \text{ and } a_3 = 0\}\ .$$

This gives a rule for identifying the points of P_0 and P_1 that lie in this H^2.

Carrying out the gluing visually is quite satisfying; we will draw the pictures first and then worry about what they mean. We have indicated why P_0 and P_1 are glued; in a similar way, P_1 and P_2 are glued, as are P_2 and P_3. This uses up all the gluing walls of the various H_j^3/Γ_j because each has only two, except for P_0 and P_3 which have one each. The $j = 1$ case is interesting because P_1 has four gluing walls, but H_1^3/Γ_1 has only two because the diagram automorphism of P_1 exchanges them in pairs. So the gluing pattern is

$$P_0 \;—\; P_1/(\mathbb{Z}/2) \;—\; P_2 \;——\; P_3 \qquad (7)$$

Working with polyhedra is so much simpler than working with quotients of them by isometries that we will carry out the gluing by assembling P_1 and two copies each of P_0, P_2 and P_3, according to

$$\begin{array}{ccccc} P_0 & & & P_2 \;——\; P_3 \\ & \searrow & & \nearrow & \\ & & P_1 & & \\ & \nearrow & & \searrow & \\ P_0 & & & P_2 \;——\; P_3 \end{array} \qquad (8)$$

and take the quotient of the result by the diagram automorphism.

We begin by assembling P_1 and the copies of P_0 and P_2. This requires pictures of the polyhedra. P_0 appears in (2), and for the others we draw both 3-dimensional and an overhead views.

P_2

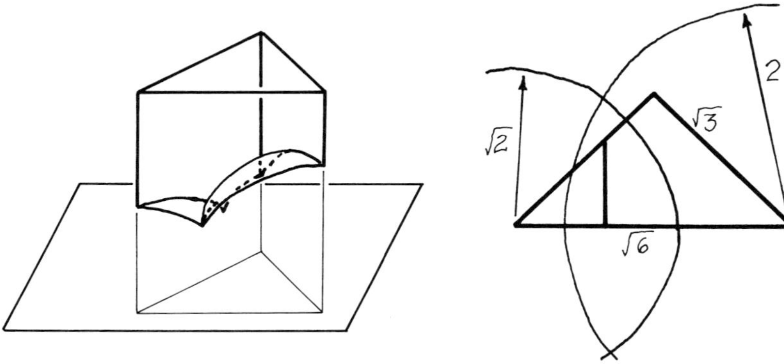

P_1

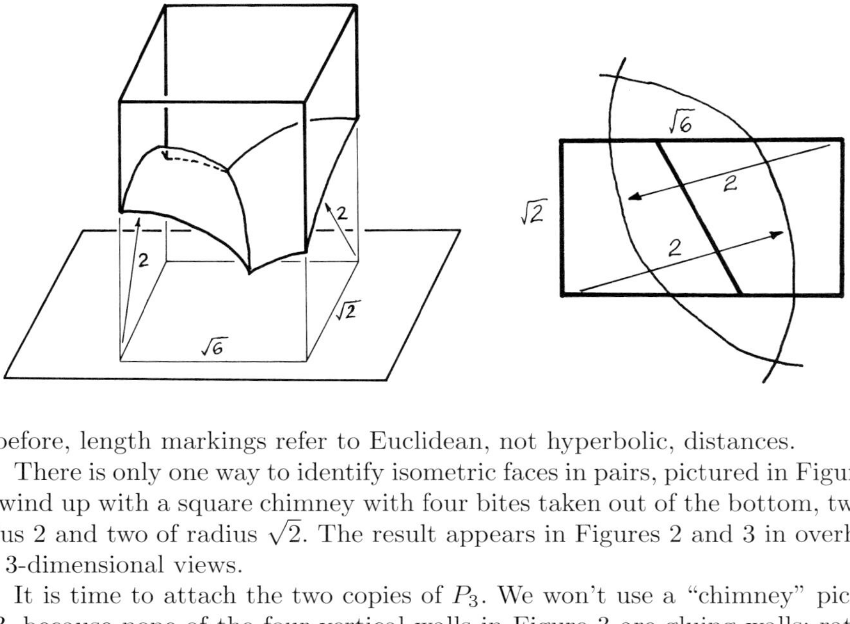

As before, length markings refer to Euclidean, not hyperbolic, distances.

There is only one way to identify isometric faces in pairs, pictured in Figure 1. We wind up with a square chimney with four bites taken out of the bottom, two of radius 2 and two of radius $\sqrt{2}$. The result appears in Figures 2 and 3 in overhead and 3-dimensional views.

It is time to attach the two copies of P_3. We won't use a "chimney" picture of P_3 because none of the four vertical walls in Figure 3 are gluing walls; rather, the two gluing walls are the two small faces on the bottom. Happily, the region bounded by one of these walls and the extensions across it of the three walls it meets is a copy of P_3. That is, P_3 may be described as the interior of a hemisphere of radius $\sqrt{2}$, intersected with a half-space bounded by a vertical half-plane and

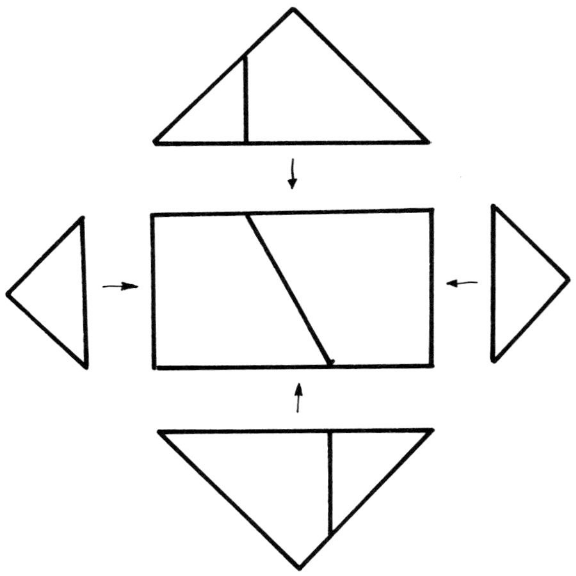

FIGURE 1. Overhead view of instructions for gluing P_1 to two copies of P_0 and two copies of P_2.

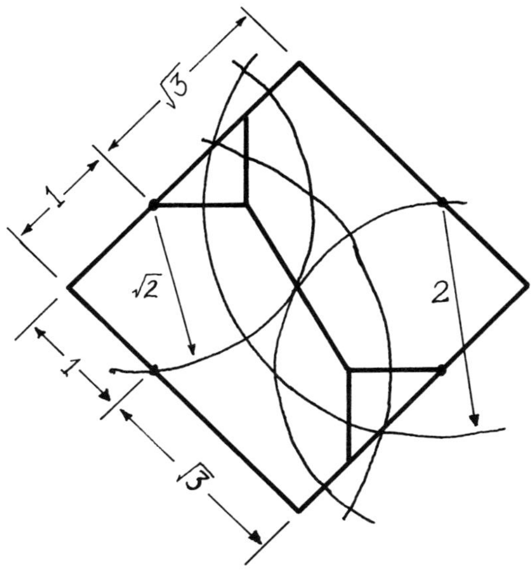

FIGURE 2. Overhead view of the result of gluing P_1 to two copies of P_0 and two copies of P_2.

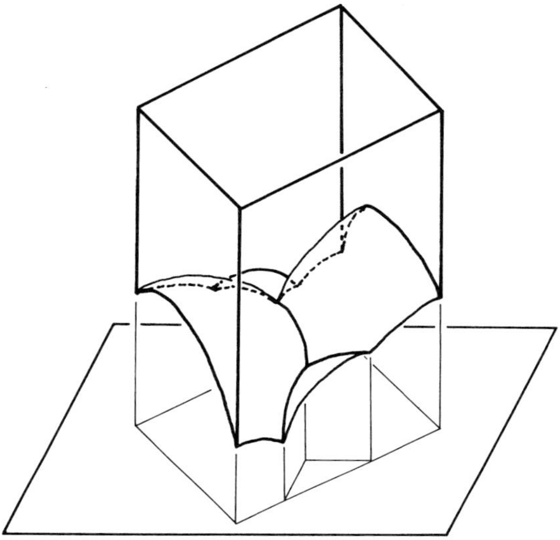

FIGURE 3. Three-dimensional view of the result of gluing P_1 to two copies of P_0 and two copies of P_2.

the *exteriors* of two hemispheres of radius 2:

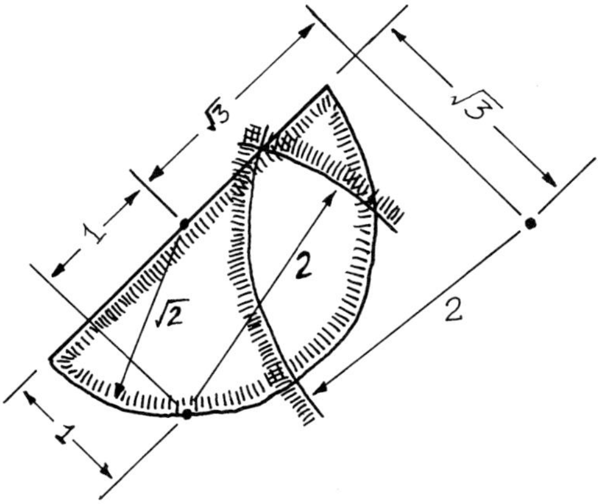

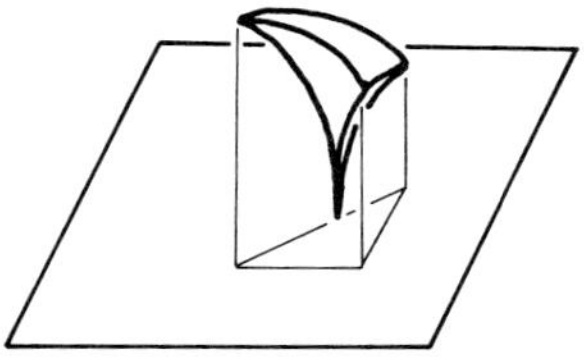

The 3-dimensional picture shows a copy of P_3 that fits neatly beneath one of the bottom walls of Figure 3 (the back one). Adjoining it, and another copy of P_3 in the symmetrical way, completes the gluing described in (8). The result appears in Figures 4 and 5.

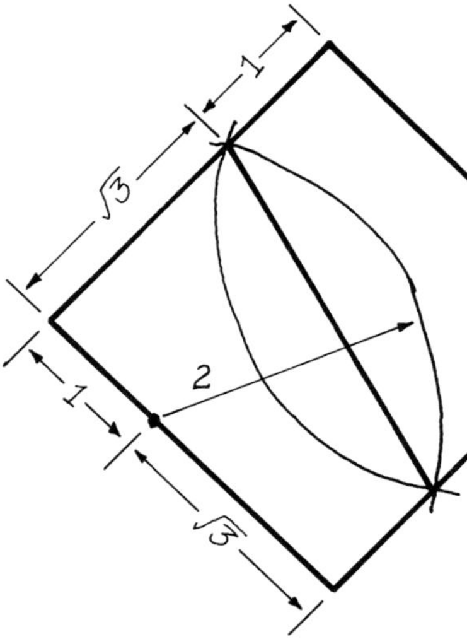

FIGURE 4. Overhead view of the final result of gluing the polyhedra according to (8).

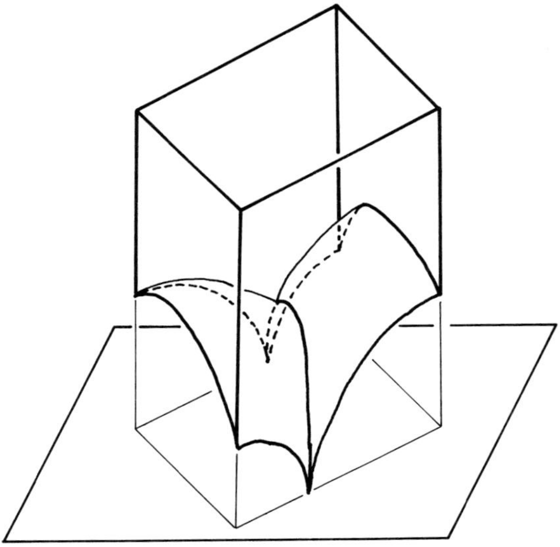

FIGURE 5. Three-dimensional view of the final result of gluing the polyhedra according to (8).

One can find its dihedral angles from our pictures; it is a Coxeter polyhedron with diagram

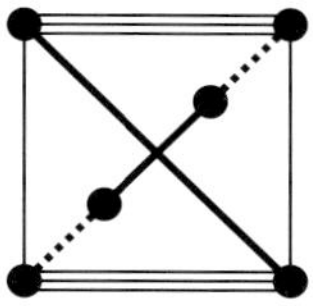

This leads to our main result; we write $\Gamma^{\mathbb{R}}$ for the group generated by this Coxeter group and its diagram automorphism, and Q for $H^3/\Gamma^{\mathbb{R}}$.

Theorem 4. *We have* $\mathcal{M}_s^{\mathbb{R}} \cong Q = H^3/\Gamma^{\mathbb{R}}$*, where "*$\cong$*" means the following:*

(i) $\mathcal{M}_s^{\mathbb{R}} \to Q$ *is a homeomorphism;*

(ii) $\mathcal{M}_s^{\mathbb{R}} \to Q$ *is an isomorphism of topological orbifolds if the orbifold structure of* Q *is changed along the edges associated to triple bonds, by replacing the dihedral group* D_6 *of order* 12 *by* $\mathbb{Z}/2$ *(see below);*

(iii) $\mathcal{M}_s^{\mathbb{R}} \to Q$ *is an isomorphism of real-analytic orbifolds if* Q *is altered as in* (ii) *and also along the loci where the* P_j *are glued together.*

For the rest of the lecture we will focus on the perhaps-surprising subtlety regarding the orbifold structures of $\mathcal{M}_s^{\mathbb{R}}$ and Q. We take $\mathcal{F}_s^{\mathbb{R}}$ to be the preimage of $\mathcal{C}_s^{\mathbb{R}}$, or equivalently the closure of $\mathcal{F}_0^{\mathbb{R}}$. Now, $\mathcal{F}_s^{\mathbb{R}}$ is not a manifold because of the branching of the cover $\mathcal{F}_s \to \mathcal{C}_s$. One example occurs at $\tilde{S} \in \mathcal{F}_s^{\mathbb{R}}$ lying over a sextic $S \in \mathcal{C}_s^{\mathbb{R}}$ with a single double point, necessarily real. In a neighborhood U of S, $\mathcal{C}_s^{\mathbb{R}}$ is a real 6-manifold meeting the discriminant (a complex 5-manifold) along a real 5-manifold. A neighborhood of $\tilde{S}$ is got by taking a 6-fold cover of U, branched along Δ. Therefore near $\tilde{S}$, $\mathcal{F}_s^{\mathbb{R}}$ is modeled on 12 half-balls of dimension 6 meeting along their common 5-ball boundary. Here are pictures of the relevant parts of $\mathcal{C}_s^{\mathbb{R}}$ and $\mathcal{F}_s^{\mathbb{R}}$:

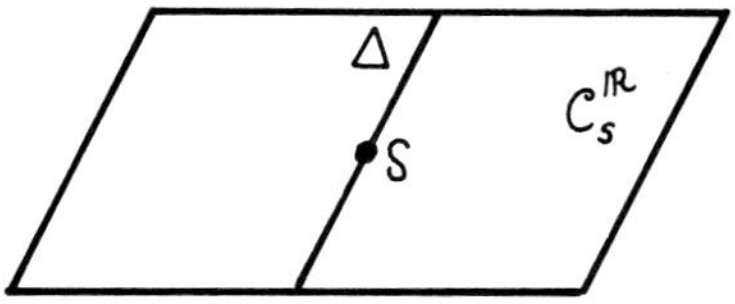

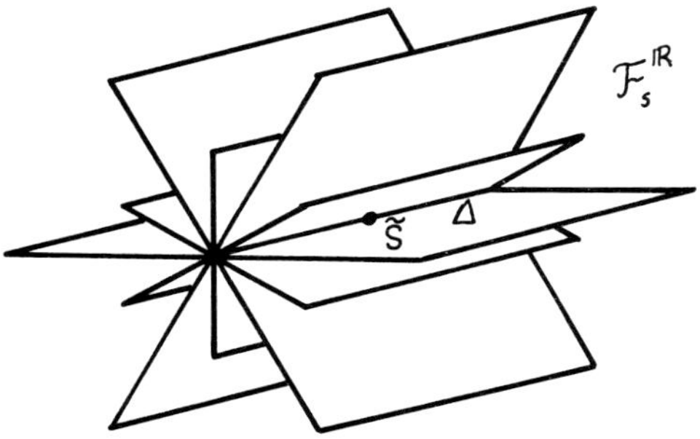

To get an orbifold chart around the image of S in $\mathcal{M}_s^{\mathbb{R}}$, we take a small transversal to $G^{\mathbb{R}}.S$ and mod out by the stabilizer of S in $G^{\mathbb{R}}$, as explained in lecture 1. To get an orbifold chart around the image of $\tilde{S}$ in $G^{\mathbb{R}}\backslash\mathcal{F}_s^{\mathbb{R}}/\Gamma$ we do the following, necessarily more complicated than before because $\mathcal{F}_s^{\mathbb{R}}$ isn't a manifold. We choose a transversal to $G^{\mathbb{R}}.\tilde{S}$, which is identified under g with a neighborhood of $g(\tilde{S})$ in the union X of six H^3's meeting along an H^2. We take the quotient of X by the stabilizer $\mathbb{Z}/6$ of $\tilde{S}$ in Γ. The result is isometric to H^3, and we take an open set in *this* H^3 as the domain for the orbifold chart, mapping to $G^{\mathbb{R}}\backslash\mathcal{F}_s^{\mathbb{R}}/\Gamma$ by taking the quotient of it by

$$(\text{the simultaneous stabilizer of } g(\tilde{S}) \text{ and } X \text{ in } \Gamma) \ / \ (\mathbb{Z}/6) \ ,$$

which is exactly the stabilizer of S in $G^{\mathbb{R}}$. Identifying $\mathcal{M}_s^{\mathbb{R}}$ with $G^{\mathbb{R}}\backslash\mathcal{F}_s^{\mathbb{R}}/\Gamma$ leads to two orbifold charts around the same point. One can check that these charts define the same topological orbifold structure but different real-analytic structures. This leads to (iii) in Theorem 4.

A slightly different phenomenon leads to (ii). The second possibility for how Δ meets $\mathcal{C}_s^{\mathbb{R}}$ is at a sextic S with two complex conjugate double points. Then in a neighborhood U of S, Δ has two branches through S, meeting transversely. The real 6-manifold $\mathcal{C}_s^{\mathbb{R}}$ meets Δ along a real 4-manifold lying in the intersection of these two branches. Since Δ has two branches through S, there is not 6-to-1 but 36-to-1 branching near $\tilde{S} \in \mathcal{F}_s^{\mathbb{R}}$ lying over S. It turns out that a neighborhood $\tilde{U}$ of $\tilde{S}$ in $\mathcal{F}_s^{\mathbb{R}}$ may be taken to be the union of six real 6-balls meeting along a common 4-ball, with each of the 6-balls mapping to U as a 6-to-1 cover branched over the 4-ball. We get an orbifold chart around the image of $\tilde{S}$ in $G^{\mathbb{R}}\backslash\mathcal{F}_s^{\mathbb{R}}/\Gamma$ as follows. Choose a transversal to $\Gamma^{\mathbb{R}}.\tilde{S}$, which maps bijectively to its image in $\mathbb{C}H^3$, which can be described as a neighborhood of $g(\tilde{S})$ in the union of six H^3's meeting along an H^1. Choose *one* of these H^3's and take the quotient of it by the subgroup of Γ which carries both it and $g(\tilde{S})$ to themselves. Generically this subgroup is D_6, because of the $\mathbb{Z}/6$ coming from the branching and the fact that S has a $\mathbb{Z}/2$ symmetry exchanging its double points. This gives an orbifold chart $U \to U/D_6 \to G^{\mathbb{R}}\backslash\mathcal{F}_s^{\mathbb{R}}/\Gamma$. (The idea also applies if S has more symmetry than the generic $\mathbb{Z}/2$.)

Now, this *cannot* be a valid description of the orbifold $\mathcal{M}_s^{\mathbb{R}}$, because the symmetry group of S is $\mathbb{Z}/2$ and so the local group at the image of S in $\mathcal{M}_s^{\mathbb{R}}$ should be $\mathbb{Z}/2$ not D_6. The problem is that the $\mathbb{Z}/6$ coming from the branching is an artifact of our construction. To eliminate it, we take the quotient of the chart by the $\mathbb{Z}/6$, obtaining a topological ball, and use this ball rather than the original one as the domain for the orbifold chart, with local group $D_6/(\mathbb{Z}/6) = \mathbb{Z}/2$. The effect of this operation is to replace the orbifold chart

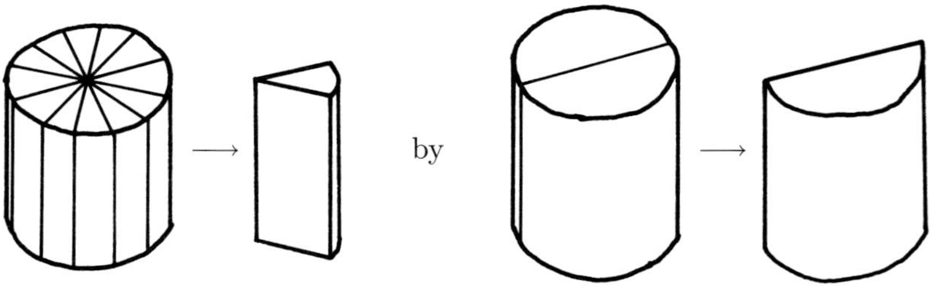

We may picture this as a smoothing of the crease:

Therefore $\mathcal{M}_s^{\mathbb{R}}$'s topological orbifold structure can be completely visualized by taking the hyperbolic polyhedron in Figure 5 and smoothing two of its edges in this manner.

Appendix

We will give a sketch of the Hodge theory behind Theorem 3 and then make a few remarks.

Theorem 3 is due to Deligne and Mostow [6], building on ideas of Picard; our approach is more explicitly Hodge-theoretic, along the lines of our treatment of moduli of cubic surfaces in [1]. Let $S \in \mathcal{C}_0$ be a smooth binary sextic, defined by $F(x_0, x_1) = 0$, and let C be the 6-fold cyclic cover of $\mathbb{C}P^1$ defined in $\mathbb{C}P^2$ by $F(x_0, x_1) + x_2^6 = 0$, which is a smooth curve of genus 10. It has a 6-fold symmetry $\sigma : x_2 \to -\omega x_2$, where ω is our fixed cube root of unity. Now, σ^* acts on $H^1(C; \mathbb{C})$, and its eigenspaces refine the Hodge decomposition because σ acts holomorphically. One finds $H^1_\omega(C; \mathbb{C}) = H^{1,0}_\omega(C) \oplus H^{0,1}_\omega(C)$, the summands having dimensions 1 and 3 respectively. In fact, $H^{1,0}_\omega(C)$ is generated by the residue of the rational differential

$$\frac{(x_0\, dx_1 \wedge dx_2 + x_1\, dx_2 \wedge dx_0 + x_2\, dx_0 \wedge dx_1)\, x_2^3}{F(x_0, x_1) + x_2^6}\ .$$

We remark that our construction really only uses the 3-fold cover of $\mathbb{C}P^1$ rather than the 6-fold cover, because we are working with the ω-eigenspace. We have used the 6-fold cover because the residue calculus is less fussy in projective space than in weighted projective space. See remark 9 for a comparison of the approaches using the 3-fold and 6-fold covers.

The Hermitian form

$$\langle \alpha | \beta \rangle = i\sqrt{3} \int_C \alpha \wedge \bar{\beta} \tag{9}$$

on $H^1(C; \mathbb{C})$ is positive-definite on $H^{1,0}$ and negative-definite on $H^{0,1}$. Therefore $H^{1,0}_\omega(C) \hookrightarrow H^1_\omega(C; \mathbb{C})$ is an inclusion of a positive line into a Hermitian vector space of signature $(1, 3)$, i.e., a point of the complex 3-ball consisting of all such lines in $P(H^1_\omega(C; \mathbb{C}))$. The $\sqrt{3}$ in (9) is not very important; it makes the map Z defined below be an isometry.

To identify this ball with a single fixed complex 3-ball we need an additional structure, namely a choice of basis for the relevant part of $H^1(C;\mathbb{Z})$, that is suitably compatible with σ. Let $\Lambda(C)$ be the sublattice of $H^1(C;\mathbb{Z})$ where σ^* has order 3, together with the 0 element. Then $\Lambda(C)$ is an $\mathcal{E}$-module, with ω acting as σ^*. The eigenspace projection

$$Z : \Lambda(C) \otimes_{\mathcal{E}} \mathbb{C} = \Lambda(C) \otimes_{\mathbb{Z}} \mathbb{R} \to H^1_\omega(C;\mathbb{C})$$

is an isomorphism of complex vector spaces. The $\mathcal{E}$-module structure and the intersection pairing Ω together define a Hermitian form on $\Lambda(C)$, namely

$$\langle x|y\rangle = -\frac{\Omega(\theta x, y) + \theta\Omega(x,y)}{2},$$

where $\theta = \omega - \bar{\omega}$. This turns out to be a copy of Λ, the lattice from (5). A framing of S is a choice of isometry $\phi : \Lambda(C) \to \Lambda$, taken modulo scalars. (The term 'marking' is already taken, usually indicating an ordering of the six points of S.) It turns out that Z is an isometry, so together with ϕ it identifies the ball in $P(H^1_\omega(C;\mathbb{C}))$ with the standard one, i.e., the one in $P(\mathbb{C}^{1,3} = \Lambda \otimes_{\mathcal{E}} \mathbb{C})$. This defines a holomorphic map $g : \mathcal{F}_0 \to B^3$. One constructs an extension of the covering space $\mathcal{F}_0 \to \mathcal{C}_0$ to a branched covering $\mathcal{F}_s \to \mathcal{C}_s$ and extends g to $\mathcal{F}_s$; g is then the isomorphism of Theorem 3. One can show (see, e.g., [1, Lemma 7.12]) that the monodromy homomorphism $\pi_1(\mathcal{C}_0, S) \to P\mathrm{Aut}\,\Lambda(C)$ is surjective, and it follows that $\mathcal{F}_0$ and $\mathcal{F}_s$ are connected, with deck group $\Gamma = P\mathrm{Aut}\,\Lambda$, and that g is Γ-equivariant.

The reason that $\mathcal{F}_s \to \mathcal{C}_s$ has 6-fold branching along each component of the preimage of Δ is that one can use [14] to work out the monodromy in $P\mathrm{Aut}\,\Lambda$ of a small loop encircling Δ at a general point of Δ; it turns out to have order 6.

We close with some remarks relevant but not central to the lectures.

Remark 1. We have treated moduli of unordered real 6-tuples in $\mathbb{C}P^1$, which at first might sound like only a slight departure from the considerable literature on the hyperbolic structure on the moduli space of ordered 6-tuples in $\mathbb{R}P^1$. Briefly, Thurston [17, pp. 515–517] developed his own approach to Theorem 3, and described a component of $G^{\mathbb{R}}\backslash\big((\mathbb{R}P^1)^6 - \Delta\big)$ as the interior of a certain polyhedron in H^3. Using hypergeometric functions, Yoshida [19] obtained essentially the same result, described the tessellation of $G^{\mathbb{R}}\backslash\big((\mathbb{R}P^1)^6 - \Delta\big)$ by translates of this open polyhedron, and discussed the degenerations of 6-tuples corresponding to the boundaries of the components. See also [8] and [10]. The relation to our work is the following: the space $G^{\mathbb{R}}\backslash\big((\mathbb{R}P^1)^6 - \Delta\big)$ is the quotient of $H^3_0 - \mathcal{H}$ by the level 3 principal congruence subgroup $\Gamma_{0,3}$ of Γ_0. A component C of $H^3_0 - \mathcal{H}$ is a copy of Thurston's open polyhedron, its stabilizer in Γ_0 is $S_3 \times \mathbb{Z}/2$, and the quotient of C by this group is the Coxeter orbifold P_0, minus the wall corresponding to the blackened node of the Coxeter diagram. There are $|S_6|/|S_3 \times \mathbb{Z}/2| = 60$ components of $G^{\mathbb{R}}\backslash\big((\mathbb{R}P^1)^6 - \Delta\big)$, permuted by S_6. The S_6 action is visible because the κ_0-invariant part of Λ is $\mathbb{Z}^{1,3}$, and $\Gamma_0/\Gamma_{0,3}$ acts on the $\mathbb{F}_3$-vector space $\mathbb{Z}^{1,3}/3\mathbb{Z}^{1,3}$. Reducing inner products of lattice vectors modulo 3 gives a quadratic form on this vector space, and S_6 happens to be isomorphic to the corresponding projective orthogonal group.

In a similar way, one could consider the moduli space of ordered 6-tuples of distinct points in $\mathbb{C}P^1$ such that (say) points 1 and 2 are conjugate and points $3, \ldots, 6$ are all real. This moduli space is a quotient of $H_1^3 - \mathcal{H}$ by a subgroup of Γ_1. Other configurations of points give quotients by subgroups of the other Γ_j. It is only by considering *unordered* 6-tuples that one sees all four types of 6-tuples occurring together, leading to our gluing construction. One way that our results differ from earlier ones is that the gluing leads to a nonarithmetic group (see Remark 5 below), whereas the constructions using ordered 6-tuples lead to arithmetic groups.

Remark 2. Γ has a single cusp in $\mathbb{C}H^3$, corresponding to the 6-tuple consisting of two triple points; this is the unique minimal strictly semistable orbit in $\mathcal{C}$ (in the sense of geometric invariant theory). The two cusps of $\Gamma^{\mathbb{R}}$ correspond to the two possible real structures on such a 6-tuple—the triple points can be conjugate, or can both be real.

Remark 3. Part (ii) of Theorem 4 lets us write down the orbifold fundamental group $\pi_1^{\text{orb}}(\mathcal{M}_s^{\mathbb{R}})$. The theory of Coxeter groups shows that the reflection subgroup R of $\Gamma^{\mathbb{R}}$ is defined as an abstract group by the relations that the six generating reflections are involutions, and that the product of two has order n when the corresponding walls meet at angle π/n. The modification of orbifold structures amounts to setting two of the generators equal if their walls meet at angle $\pi/6$. This reduces R to $D_\infty \times \mathbb{Z}/2$ where D_∞ denotes the infinite dihedral group. Adjoining the diagram automorphism gives $\pi_1^{\text{orb}}(\mathcal{M}_s^{\mathbb{R}}) \cong (D_\infty \times \mathbb{Z}/2) \rtimes (\mathbb{Z}/2)$, where the $\mathbb{Z}/2$ acts on $D_\infty \times \mathbb{Z}/2$ by exchanging the involutions generating D_∞. This larger group is also isomorphic to $D_\infty \times \mathbb{Z}/2$, so we conclude $\pi_1^{\text{orb}}(\mathcal{M}_s^{\mathbb{R}}) \cong D_\infty \times \mathbb{Z}/2$. This implies that $\mathcal{M}_s^{\mathbb{R}}$ is not a good orbifold in the sense of Thurston [16].

Remark 4. One can work out the volumes of the P_j by dissecting them into suitable simplices, whose volumes can be expressed in terms of the Lobachevsky function $\Lambda(z)$. For background see [9] and [11]. The results are

j	**covolume(Γ_j)**			**fraction of total**
0	$\Lambda(\pi/4)/6$	=	.07633...	$\sim$ 8.66 %
1	$15\Lambda(\pi/3)/16$	=	.31716...	$\sim$ 36.01 %
2	$5\Lambda(\pi/4)/6$	=	.38165...	$\sim$ 43.33 %
3	$5\Lambda(\pi/3)/16$	=	.10572...	$\sim$ 12.00 %

These results suggest that Γ_0 and Γ_2 are commensurable, that Γ_1 and Γ_3 are commensurable, and that these two commensurability classes are distinct. We have verified these statements.

Remark 5. The group $\Gamma^{\mathbb{R}}$ is nonarithmetic; this is suggested by the fact that we built it by gluing together noncommensurable arithmetic groups in the spirit of Gromov and Piatetski-Shapiro's construction of nonarithmetic lattices in $O(n, 1)$.

(See [7].) Their results do not directly imply the nonarithmeticity of $\Gamma^{\mathbb{R}}$, so we used 12.2.8 of [6]. That is, we computed the trace field of $\Gamma^{\mathbb{R}}$, which turns out to be $\mathbb{Q}(\sqrt{3})$, showed that $\Gamma^{\mathbb{R}}$ is a subgroup of the isometry group of the quadratic form diag$[-1,+1,+1,+1]$ over $\mathbb{Z}[\sqrt{3}]$, and observed that the Galois conjugate of this group is noncompact over $\mathbb{R}$.

Remark 6. The anti-involutions (6) and their Γ-conjugates do not account for all the anti-involutions of $\mathbb{C}H^3$ in $\Gamma \rtimes (\mathbb{Z}/2)$: there is exactly one more conjugacy class. Pick a representative κ_4 of this class and write H^3_4 for its fixed-point set. The points of H^3_4 correspond to 6-tuples in $\mathbb{C}P^1$ invariant under the non-standard anti-involution of $\mathbb{C}P^1$, which can be visualized as the antipodal map on the sphere S^2. A generic such 6-tuple cannot be defined by a sextic polynomial with real coefficients, so it corresponds to no point of $\mathcal{M}_0^{\mathbb{R}}$, but it does represent a real point of $\mathcal{M}_0$. One can show that the stabilizer Γ_4 of H^3_4 in Γ is the Coxeter group

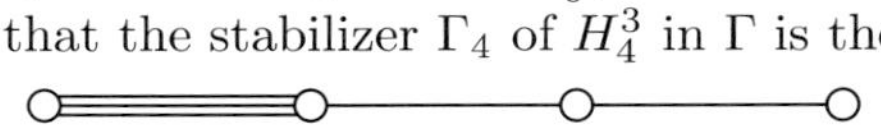

and that the moduli space of such 6-tuples is H^3_4/Γ_4, with the edge corresponding to the triple bond playing exactly the same role as before.

Remark 7. When discussing the gluing patterns (7) and (8) we did not specify information such as which gluing wall of P_2 is glued to the gluing wall of P_3. It turns out that there is no ambiguity because the only isometries between walls of the P_j are the ones we used. But for the sake of explicitness, here are the identifications. The gluing wall of P_0 is glued to one of the top gluing walls of P_1, the gluing wall of P_3 is glued to the left gluing wall of P_2, and the other gluing wall of P_2 is glued to one of the bottom gluing walls of P_1. The words 'left', 'right', 'top' and 'bottom' refer to the Coxeter diagrams (1) and (3), not to the pictures of the polyhedra.

Remark 8. In these notes we work projectively, while in [3] we do not. This means that our space $\mathcal{C}$ is analogous to the $\mathbb{C}P^{19}$ of cubic surfaces in $\mathbb{C}P^3$, which is the projectivization of the space called $\mathcal{C}$ in [3], and similarly for the various versions of $\mathcal{F}$. The group in [3] analogous to G here is the projectivization of the group called G there, and similarly for $G^{\mathbb{R}}$, Γ, Γ_j and $\Gamma^{\mathbb{R}}$.

Remark 9. As mentioned above, our treatment of the Hodge theory uses the 6-fold cover when the 3-fold cover would do; in [2] we used just the 3-fold cover, and the translation between the approaches deserves some comment. There, a 6-tuple is described by a sextic polynomial $f(x,y)$, X_f is the 3-fold cover of $\mathbb{C}P^1$ branched over it, namely the zero-locus of $f(x,y)+z^3=0$ in the weighted projective space $\mathbb{C}P^{1,1,2}$, σ acts by $z \mapsto \omega z$, and the period map is defined by the inclusion $H^{1,0}_{\bar\omega}(X_f) \to H^1_{\bar\omega}(X_f;\mathbb{C})$. Because we use σ in a different way, we write $\sigma_{[2]}$ for the σ of [2]. Now, the pairs $(X_f, \sigma_{[2]})$ and $(C/\langle\sigma^3\rangle, \sigma^{-1})$ are isomorphic, where the last σ indicates the induced action of σ on $C/\langle\sigma^3\rangle$. The isomorphism is just $x = x_0$, $y = x_1$, $z = x_2^2$. This identification gives

$$H^1_{\sigma_{[2]}=\bar\omega}(X_f;\mathbb{C}) = H^1_{\sigma=\omega}(C/\langle\sigma^3\rangle;\mathbb{C}) = H^1_{\sigma=\omega}(C;\mathbb{C}) \ .$$

References

[1] D. Allcock, J. Carlson, and D. Toledo, The complex hyperbolic geometry of the moduli space of cubic surfaces, *J. Alg. Geom.* **11** (2002), 659–724.

[2] D. Allcock, J. Carlson, and D. Toledo, Non-arithmetic uniformization of some real moduli spaces, to appear in *Geom. Dedicata.*

[3] D. Allcock, J. Carlson and D. Toledo, Hyperbolic geometry and moduli of real cubic surfaces, in preparation.

[4] R. D. Canary, D. B. A. Epstein and P. Green, Notes on notes of Thurston, in *Analytical and geometric aspects of hyperbolic space (Coventry/Durham, 1984)* 3–92, London Math. Soc. Lecture Note Ser., 111, Cambridge Univ. Press, Cambridge, 1987.

[5] P. de la Harpe, An invitation to Coxeter groups, in *Group theory from a geometrical viewpoint (Trieste, 1990)*, 193–253, World Sci. Publishing, River Edge, NJ, 1991.

[6] P. Deligne and G. D. Mostow, Monodromy of hypergeometric functions and non-lattice integral monodromy, *Publ. Math. IHES* **63** (1986), 5–89.

[7] M. Gromov and I. Piatetski-Shapiro, Nonarithmetic groups in Lobachevsky spaces, *Publ. Math. I.H.E.S.* **66** (1988), 93–103.

[8] M. Kapovich and J. Millson, On the moduli spaces of polygons in the Euclidean plane", *J. Diff. Geometry.* **42** (1995), 133–164.

[9] R. Kellerhals, On the volume of hyperbolic polyhedra, *Math. Ann.* **285** (1989), 541–569.

[10] S. Kojima, H. Nishi and Y. Yamashita, Configuration spaces of points on the circle and hyperbolic Dehn fillings, *Topology* **38** (1999), 497–516.

[11] J. Milnor, How to compute volume in hyperbolic space, in *Collected Works, Volume 1*, 189–212. Publish or Perish, Houston, 1994.

[12] V. V. Nikulin, Involutions of integral quadratic forms and their applications to real algebraic geometry, *Math. USSR Izvestiya* **22** (1984), 99–172.

[13] V. V. Nikulin, Discrete reflection groups in Lobachevsky spaces and algebraic surfaces, *Proc. I.C.M. Berkeley 1986*, 654–671.

[14] M. Sebastiani and R. Thom, Un résultat sur la monodromie, *Invent. Math.* **13** (1971), 90–96.

[15] W. Thurston, *The Geometry and Topology of Three-Manifolds*, Princeton University notes, electronic version http://www.msri.org/publications/books/gt3m.

[16] W. Thurston, *Three-dimensional geometry and topology,* Princeton Mathematical Series, 35. Princeton University Press, Princeton, NJ, 1997.

[17] W. Thurston, Shapes of polyhedra and triangulations of the sphere, *Geom. Top. Monographs* **1**, 511–549.

[18] E. Vinberg, Some arithmetical discrete groups in Lobacevskii spaces. In *Discrete Subgroups of Lie Groups and Applications to Moduli*, 328–348. Oxford, 1975.

[19] M. Yoshida, The real loci of the configuration space of six points on the projective line and a Picard modular 3-fold, *Kumamoto J. Math.* **11** (1998), 43–67.

Daniel Allcock
Department of Mathematics
University of Texas at Austin
Austin, TX 78712
USA
e-mail: allcock@math.utexas.edu
URL: http://www.math.utexas.edu/~allcock

James A. Carlson
Department of Mathematics
University of Utah
Salt Lake City, UT 84112
USA
and
Clay Mathematics Institute
One Bow Street
Cambridge, Massachusetts 02138
USA
e-mail: carlson@math.utah.edu; carlson@claymath.org
URL: http://www.math.utah.edu/~carlson

Domingo Toledo
Department of Mathematics
University of Utah
Salt Lake City, UT 84112
USA
e-mail: toledo@math.utah.edu
URL: http://www.math.utah.edu/~toledo

Progress in Mathematics, Vol. 260, 23–42

Gauss' Hypergeometric Function

Frits Beukers

Abstract. We give a basic introduction to the properties of Gauss' hypergeometric functions, with an emphasis on the determination of the monodromy group of the Gaussian hypergeometric equation.

Keywords. Gauss hypergeometric function, monodromy, triangle group.

1. Definition, first properties

Let $a, b, c \in \mathbf{R}$ and $c \notin \mathbf{Z}_{\leq 0}$. Define *Gauss' hypergeometric function* by

$$F(a,b,c|z) = \sum \frac{(a)_n (b)_n}{(c)_n n!} z^n. \tag{1}$$

The *Pochhammer symbol* $(x)_n$ is defined by $(x)_0 = 1$ and $(x)_n = x(x+1)\cdots \cdot (x+n-1)$. The radius of convergence of (1) is 1 unless a or b is a non-positive integer, in which cases we have a polynomial.

Examples.

$$\begin{aligned}
(1-z)^{-a} &= F(a,1,1|z)\\
\log\frac{1+z}{1-z} &= 2zF(1/2,1,3/2|z^2)\\
\arcsin z &= zF(1/2,1/2,3/2|z^2)\\
K(z) &= \frac{\pi}{2}F(1/2,1/2,1,z^2)\\
P_n(z) &= 2^n F(-n,n+1,1|(1+z)/2)\\
T_n(z) &= (-1)^n F(-n,n,1/2|(1+z)/2).
\end{aligned}$$

Here $K(z)$ is the Jacobi's elliptic integral of the first kind given by

$$K(z) = \int_0^1 \frac{dx}{\sqrt{(1-x^2)(1-z^2x^2)}}.$$

The polynomials P_n, T_n given by $P_n = (1/n!)(d/dz)^n(1-z^2)^n$ and $T_n(\cos z) = \cos(nz)$ are known as the *Legendre* and *Chebyshev* polynomials respectively. They are examples of orthogonal polynomials.

One easily verifies that (1) satisfies the linear differential equation

$$z(D+a)(D+b)F = D(D+c-1)F, \qquad D = z\frac{d}{dz}.$$

Written more explicitly,

$$z(z-1)F'' + ((a+b+1)z-c)F' + abF = 0. \tag{2}$$

There exist various ways to study the analytic continuation of (1), via *Euler integrals*, *Kummer's solutions* and *Riemann's* approach. The latter will be discussed in later sections. The Euler integral reads

$$F(a,b,c|z) = \frac{\Gamma(c)}{\Gamma(b)\Gamma(c-b)} \int_0^1 t^{b-1}(1-t)^{c-b-1}(1-tz)^{-a}dt \quad (c > b > 0)$$

and allows choices of z with $|z| > 1$. The restriction $c > b > 0$ is included to ensure convergence of the integral at 0 and 1. We can drop this condition if we take the Pochhammer contour γ given by

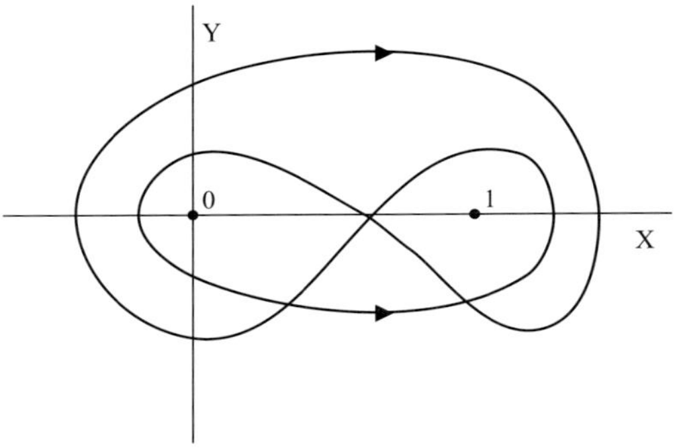

as integration path. Notice that the integrand acquires the same value after analytic continuation along γ.

It is a straightforward exercise to show that for any $b, c-b \notin \mathbf{Z}$ we have

$$F(a,b,c|z) = \frac{\Gamma(c)}{\Gamma(b)\Gamma(c-b)} \frac{1}{(1-e^{2\pi ib})(1-e^{2\pi i(c-b)})} \int_\gamma t^{b-1}(1-t)^{c-b-1}(1-tz)^{-a}dt.$$

Kummer gave the following 24 solutions to (2):

$$\begin{aligned} & F(a,b,c|z) \\ = \; & (1-z)^{c-a-b}F(c-a,c-b,c|z) \\ = \; & (1-z)^{-a}F(a,c-b,c|z/(z-1)) \\ = \; & (1-z)^{-b}F(a-c,b,c|z/(z-1)), \end{aligned}$$

$$\begin{aligned}
&z^{1-c}F(a-c+1,b-c+1,2-c|z)\\
=\;&z^{1-c}(1-z)^{c-a-b}F(1-a,1-b,2-c|z)\\
=\;&z^{1-c}(1-z)^{c-a-1}F(a-c+1,1-b,2-c|z/(z-1))\\
=\;&z^{1-c}(1-z)^{c-b-1}F(1-a,b-c+1,2-c|z/(z-1)),
\end{aligned}$$

$$\begin{aligned}
&F(a,b,a+b-c+1|1-z)\\
=\;&x^{1-c}F(a-c+1,b-c+1,a+b-c+1|1-z)\\
=\;&z^{-a}F(a,a-c+1,a+b-c+1|1-1/z)\\
=\;&z^{-b}F(b-c+1,b,a+b-c+1|1-1/z),
\end{aligned}$$

$$\begin{aligned}
&(1-z)^{c-a-b}F(c-a,c-b,c-a-b+1|1-z)\\
=\;&(1-z)^{c-a-b}z^{1-c}F(1-a,1-b,c-a-b+1|1-z)\\
=\;&(1-z)^{c-a-b}z^{a-c}F(1-a,c-a,c-a-b+1|1-1/z)\\
=\;&(1-z)^{c-a-b}z^{b-c}F(c-b,1-b,c-a-b+1|1-1/z),
\end{aligned}$$

$$\begin{aligned}
&z^{-a}F(a,a-c+1,a-b+1|1/z)\\
=\;&z^{-a}(1-1/z)^{c-a-b}F(1-b,c-b,a-b+1|1/z)\\
=\;&z^{-a}(1-1/z)^{c-a-1}F(a-c+1,1-b,2-c|1/(1-z))\\
=\;&z^{-a}(1-1/z)^{-a}F(a,c-b,a-b+1|1/(1-z)),
\end{aligned}$$

$$\begin{aligned}
&z^{-b}F(b,b-c+1,b-a+1|1/z)\\
=\;&z^{-b}(1-1/z)^{c-a-b}F(1-a,c-a,b-a+1|1/z)\\
=\;&z^{-b}(1-1/z)^{c-b-1}F(b-c+1,1-a,2-c|1/(1-z))\\
=\;&z^{-b}(1-1/z)^{-b}F(b,c-a,b-a+1|1/(1-z)).
\end{aligned}$$

Strictly speaking, the above six 4-tuples of functions are only distinct when c, $c-a-b$, $a-b \notin \mathbf{Z}$. If one of these numbers is an integer we find that there are other solutions containing logarithms. For example, when $c=1$ we find that z^{1-c} becomes $\log z$ and a second solution near $z=0$ reads

$$(\log z)F(a,b,1|z)+\sum_{n=1}^{\infty}\frac{(a)_n(b_n)}{(n!)^2}z^n\left[\sum_{k=1}^{n}\left(\frac{1}{a+k-1}+\frac{1}{b+k-1}-\frac{2}{k}\right)\right].$$

Notice that this solution can be obtained by taking the difference of solutions $z^{1-c}F(a-c+1,b-c+1,2-c|z)-F(a,b,c|z)$, divide it by $c-1$ and take the limit as $c\to 1$.

Later it will turn out that Riemann's approach to hypergeometric functions gives a remarkably transparent insight into these formulas as well as the quadratic transformations of Kummer and Goursat.

Examples of such transformations are

$$F(a,b,a+b+1/2|4z-4z^2)=F(2a,2b,a+b+1/2|z)$$

and

$$F(a,b,a+b+1/2|z^2/(4z-4)) = (1-z)^a F(2a,a+b,2a+2b|z).$$

Finally we mention the 6 *contiguous* functions

$$F(a\pm 1,b,c|z), \qquad F(a,b\pm 1,c|z), \qquad F(a,b,c\pm 1|z).$$

Gauss found that $F(a,b,c|z)$ and any two contiguous functions satisfy a linear relation with coefficients which are linear polynomials in z or constants, for example,

$$(c-a)F(a-1,b,c|z)+(2a-c-az+bz)F(a,b,c|z)+a(z-1)F(a+1,b,c|z)=0.$$

Notice also that $F'(a,b,c|z) = (ab/c)F(a+1,b+1,c+1|z)$. These observations are part of the following theorem.

Theorem 1.1. *Suppose $a,b \not\equiv 0, c \pmod{\mathbf{Z}}$ and $c \notin \mathbf{Z}$. Then any function $F(a+k, b+l, c+m|z)$ with $k,l,m \in \mathbf{Z}$ equals a linear combination of F, F' with rational functions as coefficients.*

Proof. One easily verifies that

$$\begin{aligned} F(a+1,b,c|z) &= \frac{1}{a}(z\frac{d}{dz}+a)F(a,b,c|z) \\ F(a-1,b,c|z) &= \frac{1}{c-a}(z(1-z)\frac{d}{dz}-bz+c-a)F(a,b,c|z) \end{aligned}$$

and similarly for $F(a,b+1,c|z), F(a,b-1,c|z)$. Furthermore,

$$\begin{aligned} F(a,b,c+1|z) &= \frac{c}{(c-a)(c-b)}(z(1-z)\frac{d}{dz}+c-a-b)F(a,b,c|z) \\ F(a,b,c-1|z) &= \frac{1}{c-1}(z\frac{d}{dz}+c-1)F(a,b,c|z). \end{aligned}$$

Hence there exists a linear differential operator $\mathcal{L}_{k,l,m} \in \mathbf{C}(z)[\frac{d}{dz}]$ such that $F(a+k,b+l,c+m|z) = \mathcal{L}_{k,l,m}F(a,b,c|z)$. Since F satisfies a second order linear differential equation, $\mathcal{L}_{k,l,m}F$ can be written as a $\mathbf{C}(z)$-linear combination of F and F'. □

In general we shall call any function $F(a+k,b+l,c+m|z)$ with $k,l,m \in \mathbf{Z}$ contiguous with $F(a,b,c|z)$. Thus we see that, under the assumptions of Theorem 1.1, any three contiguous functions satisfy a $\mathbf{C}(z)$-linear relation.

For many more identities and formulas we refer to [AS] and [E].

2. Ordinary linear differential equations, local theory

Consider the linear differential equation of order n,

$$y^{(n)} + p_1(z)y^{(n-1)} + \cdots + p_{n-1}(z)y' + p_n(z)y = 0, \tag{3}$$

where the p_i are analytic in a neighbourhood of $z=0$, except for a possible pole at 0. In this section we recall, without proof, a number of facts from the local theory of ordinary linear differential equations. Most of it can be found in standard text books such as Poole, Ince, Hille.

Lemma 2.1 (Wronski). *Let $f_1, \ldots, f_m$ be meromorphic functions on some open subset $G \subset \mathbf{C}$. There exists a $\mathbf{C}$-linear relation between these function if and only if $W(f_1, \ldots, f_m) = 0$, where*

$$W(f_1, \ldots, f_m) = \begin{vmatrix} f_1 & \cdots & f_m \\ f_1' & \cdots & f_m' \\ \vdots & & \vdots \\ f_1^{(m-1)} & \cdots & f_m^{(m-1)} \end{vmatrix}$$

is the Wronskian determinant of $f_1, \ldots, f_m$.

If $z = 0$ is not a pole of any p_i, it is called a *regular point* of (3), otherwise it is called a *singular point* of (3). The point $z = 0$ is called a *regular singularity* if p_i has a pole of order at most i for $i = 1, \ldots, n$.

Theorem 2.2 (Cauchy). *Suppose 0 is a regular point of (3). Then the vector space of solutions of (3) is spanned by n $\mathbf{C}$-linear independent Taylor series solutions $f_1, \ldots, f_n$ in z with positive radius of convergence.*
Moreover, the f_i can be chosen such that $f_i(z)/z^{i-1}$ has a non-zero limit as $z \to 0$ for $i = 1, 2, \ldots, n$.
Finally, the Wronskian determinant $W(f_1, \ldots, f_n)$ satisfies the equation $W' = -p_1(z)W$.

As an important remark we note that it may happen that there is a basis of holomorphic solutions near $z = 0$ but 0 may still be a singular point. In that case we call 0 an *apparent singularity*. An example is given by the differential equation $(D-1)(D-3)y = 0$ which obviously has the holomorphic solutions z, z^3. However, working $(D-1)(D-3)y = 0$ out, we find $y'' - \frac{3}{z}y' + \frac{3}{z^2}y = 0$, hence $z = 0$ is a singularity. However, we do have the following theorem which we shall repeatedly apply.

Theorem 2.3. *Suppose there exists a basis of power series solutions $f_1, \ldots, f_n$ such that f_i/z^{i-1} has a non-zero limit as $z \to 0$ for $i = 1, \ldots, n$. Then $z = 0$ is a regular point.*

Suppose that $z = 0$ is regular or a regular singularity. We can rewrite (3) by multiplication with z^n and using the rule $z^r(d/dz)^r = D(D-1)\cdots(D-r+1)$ where $D = z\frac{d}{dz}$. We obtain

$$D^n y + q_1(z)D^{n-1}y + \cdots + q_{n-1}(z)Dy + q_n(z)y = 0. \tag{4}$$

The condition of regular singularity sees to it that the functions $q_i(z)$ are holomorphic near $z = 0$. The *indicial equation* of (3) at $z = 0$ is defined as

$$X^n + q_1(0)X^{n-1} + \cdots + q_{n-1}(0)X + q_n(0) = 0.$$

Suppose we introduce a local parameter t at 0 given by $z = c_1t + c_2t^2 + c_3t^3 + \cdots$ with $c_1 \neq 0$. The differential equation can be rewritten in the new variable t. We obtain, writing $D_t = t\frac{d}{dt}$,

$$D_t^n y + \tilde{q}_1(t)D_t^{n-1}y + \cdots + \tilde{q}_{n-1}(t)D_t y + \tilde{q}_n(t)y = 0,$$

with new functions $\tilde{q}_i(t)$ holomorphic at $t = 0$.

One can show that $\tilde{q}_i(0) = q_i(0)$ for $i = 1, \ldots, n$, hence the indicial equation does not depend on the choice of local parameter at 0. The roots of the indicial equation are called the *local exponents* at $z = 0$.

Remark 2.4. Notice that if we replace y by $z^\mu w$, the differential equation for w reads

$$(D+\mu)^n w + q_1(z)(D+\mu)^{n-1}w + \cdots + q_{n-1}(z)(D+\mu)w + q_n(z)w = 0.$$

In particular, the local exponents all decreased by μ.

Remark 2.5. Show that the local exponents at a regular point read $0, 1, \ldots, n-1$. Theorem 2.3 can be rephrased by saying that if there is a basis of holomorphic solutions around $z = 0$, and if the local exponents are $0, 1, \ldots, n-1$, then $z = 0$ is a regular point of (3).

In the following theorem we shall consider expressions of the form z^A where A is a constant $n \times n$ matrix. This is short hand for

$$z^A = \exp(A \log z) = \sum_{k \geq 0} \frac{1}{k!} A^k (\log z)^k.$$

In particular z^A is an $n \times n$ matrix of multivalued functions around $z = 0$. Examples are,

$$z^{\begin{pmatrix} 1/2 & 0 \\ 0 & -1/2 \end{pmatrix}} = \begin{pmatrix} z^{1/2} & 0 \\ 0 & z^{-1/2} \end{pmatrix}, \qquad z^{\begin{pmatrix} 0 & 1 \\ 0 & 0 \end{pmatrix}} = \begin{pmatrix} 1 & \log z \\ 0 & 1 \end{pmatrix}.$$

Theorem 2.6 (Fuchs). *Let $z = 0$ be a regular singularity of (3). Let ρ be a local exponent at 0 such that none of the numbers $\rho+1, \rho+2, \ldots$ is a local exponent. Then there exists a holomorphic power series $g(z)$ with non-zero constant term such that $z^\rho g(z)$ is a solution of (3).*

Let $\rho_1, \ldots, \rho_n$ be the set of local exponents ordered in such a way that exponents which differ by an integer occur in decreasing order. Then there exists a nilpotent $n \times n$ matrix N, and functions $g_1, \ldots, g_n$, analytic near $z = 0$ with $g_i(0) \neq 0$, such that $(z^{\rho_1}g_1, \ldots, z^{\rho_n}g_n)z^N$ is a basis of solutions of (3). Moreover, $N_{ij} \neq 0$ implies $i \neq j$ and $\rho_i - \rho_j \in \mathbf{Z}_{\geq 0}$.

Example 2.7. Consider the linear differential equation

$$(z^3 + 11z^2 - z)y'' + (3z^2 + 22z - 1)y' + (z+3)y = 0.$$

The local exponents at $z = 0$ are $0, 0$, and a basis for the local solutions is given by

$$f_1(z) = 1 + 3z + 19z^2 + 147z^3 + 1251z^4 + \cdots$$

$$f_2(z) = f_1(z)\log z + 5z + \frac{75}{2}z^2 + \frac{1855}{6}z^3 + \frac{10875}{4}z^4 + \cdots.$$

3. Fuchsian linear differential equations

Consider the linear differential equation

$$y^{(n)} + p_1(z)y^{(n-1)} + \cdots + p_{n-1}(z)y' + p_n(z)y = 0, \qquad p_i(z) \in \mathbf{C}(z). \tag{5}$$

To study this differential equation near any point $P \in \mathbf{P}^1$ we choose a local parameter $t \in \mathbf{C}(z)$ at this point (usually $t = z - P$ if $P \in \mathbf{C}$ and $t = 1/z$ if $P = \infty$), and rewrite the equation with respect to the new variable t. We call the point P a regular point or a regular singularity if this is so for the equation in t at $t = 0$. It is not difficult to verify that a point $P \in \mathbf{C}$ is regular if and only if the p_i have no pole at P. It is a regular point or a regular singularity if and only if $\lim_{z\to P}(z-P)^i p_i(z)$ exists for $i = 1, \ldots, n$. The point ∞ is regular or a regular singularity if and only if $\lim_{z\to\infty} z^i p_i(z)$ exists for $i = 1, \ldots, n$.

Let $P \in \mathbf{P}^1$ be any point which is regular or a regular singularity. Let t be a local parameter around this point and rewrite the equation with respect to the variable t. The corresponding indicial equation will be called the indicial equation of (5) at P. The roots of the indicial equation at P are called the *local exponents* of (5) at P.

As a shortcut to compute indicial equations we use the following lemma.

Lemma 3.1. *Let $P \in \mathbf{C}$ be a regular point or regular singularity of* (5)*. Let*

$$a_i = \lim_{z\to P}(z-P)^i p_i(z)$$

for $i = 1, \ldots, n$. The indicial equation at P is given by

$$X(X-1)\cdots(X-n+1) + a_1X(X-1)\cdots(X-n+2) + \cdots + a_{n-1}X + a_n = 0.$$

When ∞ is regular or a regular singularity, let $a_i = \lim_{z\to\infty} z^i p_i(z)$ for $i = 1, \ldots, n$. The indicial equation at ∞ is given by

$$X(X+1)\cdots(X+n-1) - a_1X(X+1)\cdots(X+n-2) + \cdots$$
$$+(-1)^{n-1}a_{n-1}X + (-1)^n a_n = 0.$$

Proof. Exercise. □

From Cauchy's theorem of the previous section follows automatically

Theorem 3.2 (Cauchy). *Suppose $P \in \mathbf{C}$ is a regular point of* (5)*. Then there exist n $\mathbf{C}$-linear independent Taylor series solutions $f_1, \ldots, f_n$ in $z - P$ with positive radius of convergence. Moreover, any Taylor series solution of* (5) *is a $\mathbf{C}$-linear combination of $f_1, \ldots, f_n$.*

Corollary 3.3. *Any analytic solution of* (5) *near a regular point can be continued analytically along any path in $\mathbf{C}$ not meeting any singularity.*

Let S be the set of singularities of (5) and let $z_0 \in \mathbf{P}^1 \setminus S$. Let $f_1, \ldots, f_n$ be an independent set of analytic solutions around z_0. Let $\gamma \in \pi_1(\mathbf{P}^1 \setminus S, z_0)$. After analytic continuation of $f_1, \ldots, f_n$ along γ we obtain continuations $\tilde{f}_1, \ldots, \tilde{f}_n$, which are again solutions of our equation. Hence there exists a square matrix $M(\gamma) \in GL(n, \mathbf{C})$ such that $(\tilde{f}_1, \ldots, \tilde{f}_n)^t = M(\gamma)(f_1, \ldots, f_n)^t$. The map $\rho : \pi_1(\mathbf{P}^1 \setminus S) \to GL(n, \mathbf{C})$ given by $\rho : \gamma \mapsto M(\gamma)$ is a group homomorphism and its image is called the *monodromy group* of (3). Notice also that after analytic continuation along γ we have $W(f_1, \ldots, f_n) \to \det(M(\gamma))W(f_1, \ldots, f_n)$.

Definition 3.4. *The equation* (5) *is called Fuchsian if all points on* $\mathbf{P}^1$ *are regular or a regular singularity.*

Theorem 3.5 (Fuchs' relation). *Suppose* (5) *is a Fuchsian equation. Let* $\rho_1(P), \ldots, \rho_n(P)$ *the set of local exponents at any* $P \in \mathbf{P}^1$. *Then,*

$$\sum_{P \in \mathbf{P}^1} (\rho_1(P) + \cdots + \rho_n(P) - \binom{n}{2}) = -2\binom{n}{2}.$$

Since the local exponents at a regular point are always $0, 1, \ldots, n-1$, the terms in the summation are zero when P is a regular point. So, in fact, the summation in this theorem is a finite sum.

Proof. From the explicit shape of the indicial equations, given in the Lemma above, we infer that for $P \in \mathbf{C}$,

$$\rho_1(P) + \cdots + \rho_n(P) = \binom{n}{2} - \mathrm{res}_P(p_1(z)dz)$$

and

$$\rho_1(\infty) + \cdots + \rho_n(\infty) = -\binom{n}{2} - \mathrm{res}_\infty(p_1(z)dz).$$

Substract $\binom{n}{2}$ on both sides and add over all $P \in \mathbf{P}^1$. Using the fact that $\sum_{P \in \mathbf{P}^1} \mathrm{res}_P(p_1(z)dz) = 0$ yields our theorem. □

The hypergeometric equation (1) is an example of a Fuchsian equation. Its singularities are $0, 1, \infty$, and the local exponents are given by the following scheme (Riemann scheme):

0	1	∞
0	0	a
$1-c$	$c-a-b$	b

It also turns out that Fuchsian equations with three singular points can be characterised easily.

Theorem 3.6. *Let* $A, B, C \in \mathbf{P}^1$ *be distinct points. Let* $\alpha, \alpha', \beta, \beta', \gamma, \gamma'$ *be any complex numbers which satisfy* $\alpha + \alpha' + \beta + \beta' + \gamma + \gamma' = 1$. *Then there exists a unique Fuchsian equation of order two with rational function coefficients, no* *singularities other than A, B, C and having local exponents given by the following Riemann scheme:*

$$\begin{array}{ccc} A & B & C \\ \hline \alpha & \beta & \gamma \\ \alpha' & \beta' & \gamma' \end{array}$$

Proof. Exercise. □

The solutions of this Fuchsian equation are related to the hypergeometric function as follows. Via a Möbius transformation we can map A, B, C to any three distinct points of $\mathbf{P}^1$. Let us take the mapping $A, B, C \to 0, 1, \infty$. So we have to deal with the Fuchsian equation having Riemann scheme

$$\begin{array}{ccc} 0 & 1 & \infty \\ \hline \alpha & \beta & \gamma \\ \alpha' & \beta' & \gamma' \end{array}$$

If we multiply the solutions of the latter equation by z^μ we obtain a set of functions that satisfy the Fuchsian equation with Riemann scheme

$$\begin{array}{ccc} 0 & 1 & \infty \\ \hline \alpha+\mu & \beta & \gamma-\mu \\ \alpha'+\mu & \beta' & \gamma'-\mu \end{array}$$

A fortiori, after multiplication of the solutions with $z^{-\alpha'}(1-z)^{-\beta'}$ we obtain a Fuchsian equation with a scheme of the form

$$\begin{array}{ccc} 0 & 1 & \infty \\ \hline \alpha'' & \beta'' & \gamma'' \\ 0 & 0 & 1-\alpha''-\beta''-\gamma'' \end{array}$$

This scheme corresponds to a hypergeometric equation with suitable parameters. The 24 solutions of Kummer can now be characterised very easily. Suppose we apply the above procedure to the hypergeometric equation itself. There exist 6 ways to map the set $\{0, 1, \infty\}$ to itself. Having chosen such a map, there exist four ways to multiply by $z^{-\lambda}(1-z)^{-\mu}$ since there are four choices for the pair (λ,μ) of local exponents at 0 and 1. Choose the hypergeometric function (with suitable parameters) as a solution of the final equation, then we obtain the $4 \times 6 = 24$ solutions given by Kummer.

It is also very simple to prove for example the quadratic relation

$$F(a, b, a+b+1/2|t^2/(4t-4)) = (1-t)^a F(2a, a+b, 2a+2b|t).$$

Substitute $z = t^2/(4t-4)$ in the hypergeometric equation with parameters $a, b, a+b+1/2$, then we obtain a new Fuchsian equation. The map $t \to z = t^2/(4t-4)$ ramifies above $0, 1$ in $t = 0, 2$ respectively. Above $z = 1$ we have the point $t = 2$, above $z = 0$ the point $t = 0$ and above $z = \infty$ the two points $t = 1, \infty$. Notice that our equation has local exponents $0, 1/2$ in $z = 1$. Hence the new equation has local exponents $0, 1$ in $t = 2$, with regular solutions, and $t = 2$ turns out to be a regular point. At $t = 0$ we get the local exponents $0, 2(1/2 - a - b)$ and in $t = 1, \infty$, the points above $z = \infty$, we have the local exponents a, b and a, b. Thus our equation in t has again three singular points and Riemann scheme

0	1	∞
0	a	a
$1-2a-2b$	b	b

By the method sketched above, one easily sees that $(1-t)^a F(2a, a+b, 2a+2b|t)$ is a solution of this equation. Moreover, this is the unique (up to a constant factor) solution holomorphic near $t = 0$. At the same time $F(a, b, a+b+1/2|t^2/(4t-4))$ is a solution, and by the uniquess equality follows.

Example 3.7. In a similar way one can show the equality

$$F(a, b, a+b+1/2|4z-4z^2) = F(2a, 2b, a+b+1/2|z).$$

3.8. Monodromy of the hypergeometric function

Let us now turn to the monodromy of the hypergeometric equation. Consider the three loops g_0, g_1, g_∞ which satisfy the relation $g_0 g_1 g_\infty = 1$.

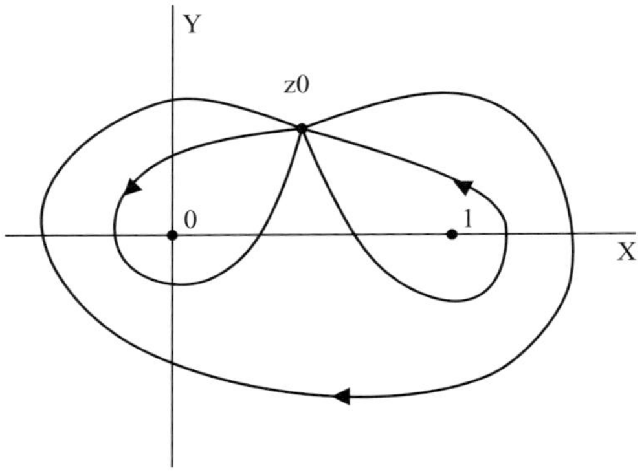

We denote the corresponding monodromy matrices by M_0, M_1, M_∞. They also satisfy $M_0 M_1 M_\infty = 1$, and M_0, M_∞ generate the monodromy group. Since the local exponents at $0, 1, \infty$ are $0, 1-c$, $0, c-a-b$ and a, b respectively, the eigenvalues of the matrices M_0, M_1 and M_∞ are $1, \exp(2\pi i(1-c))$, $1, \exp(2\pi i(c-a-b))$ and $\exp(2\pi i a), \exp(2\pi i b)$ respectively. The monodromy group can be considered as being generated by M_0, M_∞, and we know that $M_\infty M_0 = M_1^{-1}$ has eigenvalue 1. This scant information already suffices to draw some important conclusions.

Lemma 3.9. *Let $A, B \in GL(2, \mathbf{C})$. Suppose that AB^{-1} has eigenvalue 1. Then there exists a common eigenvector of A, B if and only if A, B have a common eigenvalue.*

Proof. Notice that $\ker(A-B)$ has dimension at least 1. If the dimension were 2 we would have $A = B$ and our lemma would be trivial. So we can assume $\dim(\ker(A-B)) = 1$. In this proof we let $v \in \ker(A-B)$, $v \neq 0$.

Suppose there exists a common eigenvector, w say, of A, B with eigenvalues λ_A, λ_B. If these eigenvalues are equal, we are done. Suppose they are not equal. Then w, v span $\mathbf{C}^2$. Choose α, β such that $Av = \alpha v + \beta w$. Since $Av = Bv$ we also have $Bv = \alpha v + \beta w$. Hence with respect to the basis v, w the matrices of A, B read

$$\begin{pmatrix} \alpha & \beta \\ 0 & \lambda_A \end{pmatrix} \qquad \begin{pmatrix} \alpha & \beta \\ 0 & \lambda_B \end{pmatrix}$$

Hence they have the common eigenvalue α.

Suppose A, B have a common eigenvalue λ. If v is an eigenvectore of A, we are done, since $Av = Bv$ implies that it is also an eigenvector of B. So suppose v is not an eigenvector of A. Consider the vector $w = (A - \lambda)v$. Since $A - \lambda$ has non-trivial kernel, we have $< w >_{\mathbf{C}} = (A - \lambda)\mathbf{C}^2$. In particular, $(A - \lambda)w$ is a scalar multiple of w, i.e., w is an eigenvector of A. We also have $w = (B - \lambda)v$ and a similar argument shows that w is an eigenvector of B. Hence A, B have a common eigenvector. □

Corollary 3.10. *The monodromy group of* (2) *acts reducibly on the space of solutions if and only if at least one of the numbers $a, b, c - a, c - b$ is integral.*

Proof. This follows by application of the previous lemma to the case $A = M_\infty, B = M_0^{-1}$. Since $M_1^{-1} = M_\infty M_0$, the condition that AB^{-1} has eigenvalue 1 is fullfilled. Knowing the eigenvalues of M_0, M_∞ one easily checks that equality of eigenvalues comes down to the non-empty intersection of the sets $\{0, c\}$ and $\{a, b\}$ considered modulo $\mathbf{Z}$. □

Definition 3.11. *A hypergeometric equation is called reducible if its monodromy group is reducible. A hypergeometric equation is called abelian if its monodromy group is abelian.*

Typical examples of abelian equations are (2) with $a = c = 0$ having solutions $1, (1 - z)^{-(b+1)}$ and $a = b = 1, c = 2$ having solutions $1/z, \log(1 - z)/z$. Here is a simple necessary condition for abelian equations, which has the pleasant property that it depends only on $a, b, c (\mathrm{mod}\ \mathbf{Z})$:

Lemma 3.12. *If* (2) *is abelian, then at least two of the numbers $a, b, c - a, c - b$ are integral.*

Proof. Abelian monodromy implies reducibility of the monodromy, hence at least one of the four numbers is integral. Let us say $a \in \mathbf{Z}$, the other cases can be dealt with similarly. It suffices to show that in at least one of the points $0, 1, \infty$ the local exponent difference of (2) is integral. Then clearly, $1 - c \in \mathbf{Z}$ implies $c - a \in \mathbf{Z}$, $c - a - b \in \mathbf{Z}$ implies $c - b \in \mathbf{Z}$ and $a - b \in \mathbf{Z}$ implies $b \in \mathbf{Z}$.

Suppose that all local exponent differences are non-integral. In particular the eigenvalues of each of the generating monodromy elements M_0, M_1, M_∞ are distinct. Then abelian monodromy implies that the monodromy group acts on the solution space in a completely reducible way as a sum of two one-dimensional representations. In particular the generators of these representations are functions of the form

$$z^{\lambda}(1-z)^{\mu}q(z) \qquad z^{\lambda'}(1-z)^{\mu'}p(z)$$

where $p(z), q(z)$ are polynomials with the property that they do not vanish at $z = 0$ or 1. The local exponents can be read off immediately, λ, λ' at 0, μ, μ' at 1 and $-\lambda - \mu - \deg(q), -\lambda' - \mu' - \deg(p)$ at ∞. The sum of the local exponents must be 1, hence $-\deg(p) - \deg(q) = 1$. Clearly this is a contradiction. □

Lemma 3.13. *Suppose that $A, B \in GL(2, \mathbf{C})$ have disjoint sets of eigenvalues and suppose that AB^{-1} has eigenvalue 1. Then, letting $X^2+a_1X+a_2$ and $X^2+b_1X+b_2$ be the characteristic polynomials of A, B, we have up to common conjugation,*

$$A = \begin{pmatrix} 0 & -a_2 \\ 1 & -a_1 \end{pmatrix}, \qquad B = \begin{pmatrix} 0 & -b_2 \\ 1 & -b_1 \end{pmatrix}.$$

Proof. Choose $v \in \ker(A - B)$ and $w = Av = Bv$. Since A, B have disjoint eigenvalue sets, v is not an eigenvector of A and B. Hence w, v form a basis of $\mathbf{C}^2$. With respect to this basis A, B automatically obtain the form given in our Lemma. □

Corollary 3.14. *Suppose that (2) is irreducible. Then, up to conjugation, the monodromy group depends only on the values of a, b, c modulo $\mathbf{Z}$.*

Let us now assume that $a, b, c \in \mathbf{R}$, which is the case most frequently studied. The eigenvalues of M_0, M_1, M_∞ then lie on the unit circle.

Definition 3.15. *Let R, S be two disjoint finite subsets of the unit circle of equal cardinality. The sets R, S are said to interlace if every segment on the unit circle, connecting two points of R, contains a point of S.*

Lemma 3.16. *Let A, B be non-commuting elements of $GL(2, \mathbf{C})$. Suppose that the eigenvalues of A, B have absolute value 1 and that AB^{-1} has eigenvalue 1. Let G be the group generated by A, B. Then there exists a unique (up to a constant factor) non-trivial hermitian form F on $\mathbf{C}^2$ such that $F(g(x), g(y)) = F(x, y)$ for every $g \in G$ and every pair $x, y \in \mathbf{C}^2$. Moreover,*

$$F \text{ degenerate} \iff A, B \text{ have common eigenvalues.}$$

Supposing A, B have disjoint eigenvalue sets, we have, in addition,

$$F \text{ definite} \iff \text{eigenvalues of } A, B \text{ interlace}$$

$$F \text{ indefinite} \iff \text{eigenvalues of } A, B \text{ do not interlace.}$$

We call these three cases the euclidean, spherical and hyperbolic case respectively.

Proof. Let $v \in \ker(A - B)$ and $w = Av$. Suppose first that v, w form a basis of $\mathbf{C}^2$. Of course, with respect to this basis A and B have the form given in the previous lemma. In particular we see that A, B cannot have the same characteristic equation, since this would imply that $A = B$.

We have to find a hermitean form F such that

$$\begin{aligned} F(gv,gv) &= F(v,v), \qquad F(gv,gw) = F(v,w), \\ F(gw,gv) &= F(w,v), \qquad F(gw,gw) = F(w,w), \end{aligned}$$

for every $g \in G$. It suffices to take $g = A, B$. Let $X^2 + a_1 X + a_2$ and $X^2 + b_1 X + b_2$ be the characteristic polynomials of A, B. Since the roots are on the unit circle we have $a_2\bar{a}_2 = 1, a_2\bar{a}_1 = a_1$ and similarly for b_1, b_2.

Let us first take $g = A$. Then $F(Av, Av) = F(v, v)$ implies

$$F(w,w) = F(v,v).$$

The conditions $F(Av, Aw) = F(v, w)$ and $F(Aw, Av) = F(w, v)$ imply $F(w, A^2v) = F(v, w)$ and $F(A^2v, w) = F(w, v)$. Hence, using $A^2 = -a_1 A - a_2$,

$$-\bar{a}_1 F(w,w) - \bar{a}_2 F(w,v) = F(v,w) \tag{6}$$

$$-a_1 F(w,w) - a_2 F(v,w) = F(w,v). \tag{7}$$

Because of the relations $a_2 = \bar{a}_2^{-1}$ and $a_2\bar{a}_1 = a_1$ these equations are actually the same. The condition $F(Aw, Aw) = F(w, w)$ yields $F(A^2v, A^2v) = F(w, w)$ and hence

$$|a_1|^2 F(w,w) + a_1\bar{a}_2 F(w,v) + \bar{a}_1 a_2 F(v,w) + |a_2|^2 F(v,v) = F(w,w).$$

Using $|a_2|^2 = 1$, $a_2\bar{a}_1 = a_1$ and $F(w, w) = F(v, v)$ this is equivalent to

$$a_1\bar{a}_1 F(w,w) + a_1\bar{a}_2 F(w,v) + a_1 F(v,w) = 0,$$

which is precisely (6) times a_1. Hence A-invariance of F is equivalent to

$$F(v,v) = F(w,w), \qquad F(w,v) + a_1 F(w,w) + a_2 F(v,w) = 0.$$

Invariance of F with respect to B yields the additional condition

$$F(w,v) + b_1 F(w,w) + b_2 F(v,w) = 0.$$

Since A and B do not have the same characteristic equation the solutionspace for F is one-dimensional. When $a_2 = b_2$, a solution is given by

$$F(w,w) = F(v,v) = 0, \qquad F(w,v) = (-a_2)^{1/2}, \qquad F(v,w) = (-a_2)^{-1/2},$$

when $a_2 \neq b_2$, a solution is given by

$$F(w,w) = F(v,v) = 1, \qquad F(w,v) = \epsilon, \qquad F(v,w) = \bar{\epsilon}, \qquad \epsilon = \frac{a_1 - b_1}{b_2 - a_2}.$$

We formally take $\epsilon = \infty$ if $a_2 = b_2$. In both cases cases we see that F is definite, degenerate, indefinite according to the conditions $|\epsilon| < 1$, $|\epsilon| = 1$, $|\epsilon| > 1$ respectively. It now a straightforward excercise to see that these inequalities correspond to interlacing, coinciding or non-interlacing of the eigenvalues of A and B.

We are left with the case when v is an eigenvector of A and B. Let α be the eigenvalue. If both A and B have only eigenvalues α, they automatically commute, which case is excluded. So either A or B has an eigenvalue different from α. Let us say that A has the distinct eigenvalues α, α'. Let w be an eigenvector corresponding to α'. Then, with respect to v, w the matrix of B must have the form

$$\begin{pmatrix} \alpha & b_{12} \\ 0 & \beta \end{pmatrix}.$$

with $b_{12} \neq 0$. It is now straightforward to verify that $\begin{pmatrix} 0 & 0 \\ 0 & 1 \end{pmatrix}$ is the unique invariant hermitean matrix. Moreover it is degenerate, which it should be as A, B have a common eigenvector. □

Definition 3.17. *With the assumptions as in the previous lemma let G be the group generated by A and B. Then G is called hyperbolic, euclidean, spheric if F is indefinite, degenerate, definite respectively.*

Corollary 3.18. *Let $\{x\}$ denote the fractional part of x (x minus largest integer $\le x$). Suppose that (2) is irreducible. Let F be the invariant hermitean form for the monodromy group. In particular, the sets $\{\{a\}, \{b\}\}$ and $\{0, \{c\}\}$ are disjoint. If $\{c\}$ is between $\{a\}$ and $\{b\}$, then F is positive definite (spherical case). If $\{c\}$ is not between $\{a\}$ and $\{b\}$, then F is indefinite (hyperbolic case).*

The most pittoresque way to describe the monodromy group is by using *Schwarz' triangles*.

First a little geometry.

Definition 3.19. *A curvilinear triangle is a connected open subset of $\mathbf{C} \cup \infty = \mathbf{P}^1$ whose boundary is the union of three open segments of a circle or straight line and three points. The segments are called the edges of the triangles, the points are called the vertices.*

It is an exercise to prove that, given the vertices and the corresponding angles ($< \pi$), a curvilinear triangle exists and is uniquely determined This can be seen best by taking the vertices to be $0, 1, \infty$. Then the edges connected to ∞ are actually straight lines.

More generally, a curvilinear triangle in $\mathbf{C} \cup \infty = \mathbf{P}^1$ is determined by its angles (in clockwise ordering) up to a Möbius transformation.

Let z_0 be a point in the upper half plane $\mathcal{H} = \{z \in \mathbf{C} | \Im(z) > 0\}$ and let f, g be two independent solutions of the hypergeometric equation near z_0. The quotient $D(z) = f/g$, considered as a map from $\mathcal{H}$ to $\mathbf{P}^1$, is called the *Schwarz map* and we have the following theorem and picture.

Theorem 3.20 (Schwarz). *Let $\lambda = |1-c|$, $\mu = |c-a-b|$, $\nu = |a-b|$ and suppose $0 \le \lambda, \mu, \nu < 1$. Then the map $D(z) = f/g$ maps $\mathcal{H} \cup \mathbf{R}$ one-to-one onto a curvilinear triangle. The vertices correspond to the points $D(0), D(1), D(\infty)$ and the corresponding angles are $\lambda\pi, \mu\pi, \nu\pi$.*

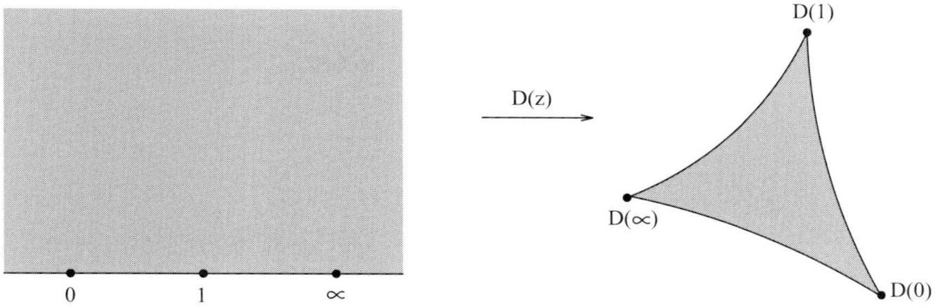

As to the proof of Schwarz' theorem, the following three ingredients are important.

- The map $D(z)$ is locally bijective in every point of $\mathcal{H}$. Notice that $D'(z) = (f'g - fg')/g^2$. The determinant $f'g - fg'$ is the *Wronskian determinant* of our equation and equals $z^{-c}(1-z)^{c-a-b-1}$. In particular it is non-zero in $\mathcal{H}$. When g has a zero at some point z_1 we simply consider $1/D(z)$ instead. Since f and g cannot vanish at the same time in a regular point, we have $f(z_1) \neq 0$.
- The map $D(z)$ maps the segments $(\infty, 0), (0,1), (1,\infty)$ to segments of circles or straight lines. For example, since $a, b, c \in \mathbf{R}$ we have two real solutions on $(0,1)$ (see Kummer's solutions). Call them $\tilde{f}, \tilde{g}$. Clearly, the function $\tilde{D}(z) = \tilde{f}/\tilde{g}$ maps $(0,1)$ on a segment of $\mathbf{R}$. Since f, g are $\mathbf{C}$-linear combinations of $\tilde{f}, \tilde{g}$ we see that $D(z)$ is a Möbius transform of $\tilde{D}(z)$. Hence $D(z)$ maps $(0,1)$ to a segment of a circle or a straight line.
- The map $D(z)$ maps a small neighbourhood of 0 to a sector with angle $|1-c| = \lambda$ and similarly for $1, \infty$. This follows from the fact that near $z = 0$ the functions f, g are $\mathbf{C}$-linear combinations of $F(a,b,c|z)$ and $z^{1-c}F(a-c+1, b-c+1, 2-c|z)$.

For the exact determination of the image of the Schwarz map we need the following additional result.

Proposition 3.21 (Gauss). *Suppose that $a, b, c \in \mathbf{R}$, $c \notin \mathbf{Z}_{\leq 0}$ and $c > a + b$. Then*

$$F(a,b,c|1) = \frac{\Gamma(c-a)\Gamma(c-b)}{\Gamma(c)\Gamma(c-a-b)}.$$

This can be proven by evaluation of Euler's integral using the Euler Beta-function.

To study the analytic continuation of $D(z)$ we use *Schwarz' reflection principle*. Hopefully, the following picture illustrates how this works.

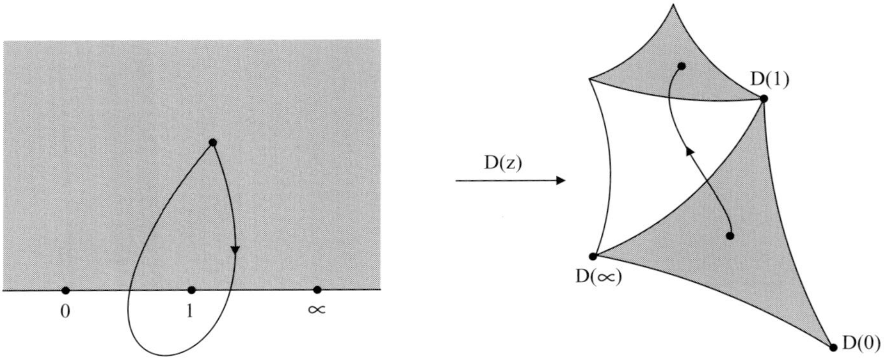

The monodromy group modulo scalars arises as follows. Let W be the group generated by the reflections in the edges of the curvilinear triangle. The monodromy group is the subgroup of W consisting of all elements which are product of an even number of reflections. In the following section we shall study precisely such groups.

3.22. Triangle groups

In this section we let S be either the Poincaré disk $\{z \in \mathbf{C} | \ |z| < 1\}$, $\mathbf{C}$ or $\mathbf{P}^1$, equipped with the hyperbolic, euclidean and spherical metric respectively.

Definition 3.23. *A (geodesic) triangle is an connected open subset of S, of finite volume, whose boundary in S is a union of three open segments of a geodesic and at most three points. The segments are called the edges of the triangles, the points are called the vertices.*

We first point out that under very mild conditions any curvilinear triangle can be thought of as a geodesic triangle.

Lemma 3.24. *Let λ, μ, ν be real numbers in the interval $[0, 1)$. There exists a geodesic triangle with angles $\lambda\pi, \mu\pi, \nu\pi$ if and only if $\lambda + \mu + \nu < 1 + 2\min(\lambda, \mu, \nu)$.*

Proof. Suppose first that $\lambda + \mu + \nu < 1$. Our condition is then trivially satisfied. For any such curvilinear triangle we can take the common orthogonal circle of the three edges, which will become the boundary of a Poincaré disk. The edges are then automatically geodesics.

Suppose that $\lambda + \mu + \nu = 1$. Our condition is equivalent to saying that all angles are positive. In this case geodesic triangles are planar triangles in the euclidean geometry with finite area. The latter property is equivalent to positivity of all angles.

Suppose that $\lambda+\mu+\nu > 1$. From spherical geometry it follows that a spherical triangle exists if and only if our condition is satisfied. □

We let $W(\Delta)$ be the group of isometries of S generated by the 3 reflections through the edges of a geodesic triangle Δ. First we look at subgroups generated by reflection in two intersecting geodesics.

Lemma 3.25. *Let ρ, σ be two geodesics intersecting in a point P with an angle $\pi\lambda$. Let r, s be the reflections in ρ, σ respectively. Then the group D generated by r, s is a dihedral group consisting of rotations $(rs)^n$ around P with angles $2n\pi\lambda$, $n \in \mathbf{Z}$ and reflections in the lines $(rs)^n(\rho), (rs)^n(\sigma)$. In particular D is finite of order $2m$ if and only if $\lambda = q/m$ for some $q \in \mathbf{Z}$ with $q \neq 0$ and $\gcd(m, q) = 1$. Furthermore, D is discrete if and only if λ is either zero or a rational number.*

Theorem 3.26. *For any geodesic triangle Δ we have $S = \bigcup_{\gamma \in W(\Delta)} \gamma(\overline{\Delta})$, where $\overline{\Delta}$ denotes the closure of Δ in S.*

Proof. First of all we note that there exists a positive d_0 with the following property. For any point P whose distance to Δ is less than d_0 there exists $\gamma \in W(\Delta)$ such that $P \in \gamma(\overline{\Delta})$. For γ we can simply take a suitable element from one of the dihedral reflection groups around the vertices.

A fortiori, any point P with distance less than d_0 from $\bigcup_{\gamma \in W(\Delta)} \gamma(\overline{\Delta})$ belongs to this set.

As a consequence the set $\bigcup_{\gamma \in W(\Delta)} \gamma(\overline{\Delta})$ is open and closed in S, hence our theorem follows. □

Definition 3.27. *An elementary triangle is a geodesic triangle whose vertex angles are all of the form π/n, $n \in \mathbf{Z}_{\geq 2} \cup \{\infty\}$.*

Theorem 3.28. *Let Δ be an elementary triangle. Then, for any $\gamma \in W(\Delta), \gamma \neq \mathrm{Id}$ we have $\gamma(\Delta) \cap \Delta = \emptyset$.*

Proof. This is a special case of the theorem of Coxeter–Tits on representations of Coxeter groups. See Humphreys book on Reflection groups and Coxeter groups [H]. □

A group G of isometries acting on S is said to act discretely if there exists a point $P \in S$ and a positive d_0 such that distance$(P, g(P)) > d_0$ whenever $g \neq \mathrm{Id}$. In particular it follows from the previous theorem that triangle groups generated by elementary triangles act discretely. The following theorem characterises all groups $W(\Delta)$ which act discretely on the symmetric space S.

Theorem 3.29. *Suppose $W = W(\Delta)$ acts discretely. Then there exists an elementary triangle Δ_{el} such that $W(\Delta) = W(\Delta_{\mathrm{el}})$. Moreover, $\overline{\Delta}$ is a finite union of copies of $\overline{\Delta_{\mathrm{el}}}$ under elements of W.*

Proof. First of all note that the vertex angles must be either 0 or rational multiples of π, otherwise the corresponding dihedral group is not discrete.

We shall show that if Δ is not elementary, then there exists a geodesic triangle Δ' such that $W(\Delta) = W(\Delta')$ and $\mathrm{Vol}(\Delta') \leq \mathrm{Vol}(\Delta)/2$. If Δ' is not elementary we repeat the process and so on. However, there is a limit to these processes since, by discreteness, there is a positive lower bound to $\mathrm{Vol}(\Delta'')$ for any Δ'' satisfying $W(\Delta) = W(\Delta'')$. Hence we must hit upon an elementary triangle Δ_{el} such that $W(\Delta) = W(\Delta_{\mathrm{el}})$.

Let α, β, γ be the edges of Δ and $r_\alpha, r_\beta, r_\gamma$ the corresponding reflections. Suppose that the vertex angle between α and β is of the form $m\pi/n$ with $\gcd(m,n) = 1$, but $m > 1$. Let δ be the geodesic between α and β whose angle with α is π/n. Let r_δ be the reflection in δ. Then the dihedral group generated by r_α and r_β is the same as the one generated by r_α and r_δ. Let Δ' be the triangle with edges α, δ, γ. Then, clearly, $W(\Delta) = W(\Delta')$. If the volume of Δ' is larger than half the volume of Δ we simply perform the above construction with α and β interchanged. □

Below we give a list of non-elementary triangles $\Delta = (\lambda, \mu, \nu)$ with vertex angles $\lambda\pi, \mu\pi, \nu\pi$ which allow a dissection with elementary triangles Δ_{el} such that $W(\Delta) = W(\Delta_{\mathrm{el}})$. In the spherical case discreteness of $W(\Delta)$ implies finiteness. The list of spherical cases was already found by H.A. Schwarz and F. Klein (see [Kl]). In the following table N denotes the number of congruent elementary triangles needed to cover Δ.

λ	μ	ν	N		elementary
$2/n$	$1/m$	$1/m$	2	×	$(1/2, 1/n, 1/m)$ n odd
$1/2$	$2/n$	$1/n$	3	×	$(1/2, 1/3, 1/n)$ n odd
$1/3$	$3/n$	$1/n$	4	×	$(1/2, 1/3, 1/n)$ $n \not\equiv 0 \bmod 3$
$2/n$	$2/n$	$2/n$	6	×	$(1/2, 1/3, 1/n)$ n odd
$4/n$	$1/n$	$1/n$	6	×	$(1/2, 1/3, 1/n)$ n odd
$2/3$	$1/3$	$1/5$	6	×	$(1/2, 1/3, 1/5)$
$1/2$	$2/3$	$1/5$	7	×	$(1/2, 1/3, 1/5)$
$3/5$	$2/5$	$1/3$	10	×	$(1/2, 1/3, 1/5)$
$1/3$	$2/7$	$1/7$	10	×	$(1/2, 1/3, 1/7)$

As an application we construct a hypergeometric function which is algebraic over $\mathbf{C}(z)$. Take the triangle $(4/5, 1/5, 1/5)$, which is spherical. Corresponding values for a, b, c can be taken to be $1/10, -1/10, 1/5$. Hence the quotient of any two solutions f, g of the corresponding hypergeometric is algebraic. Its derivative $(f'g - fg')/g^2$ is algebraic and so is the Wronskian determinant $f'g - fg' = z^{-c}(1-z)^{c-a-b-1}$. Hence g and, a fortiori, f are algebraic. In particular, $F(1/10, -1/10, 1/5|z)$ is an algebraic function.

In many cases it is also possible to find elementary triangles Δ_{el} which can be dissected into isometric copies of a smaller elementary triangle Δ'_{el}. Hence $W(\Delta_{\mathrm{el}}) \subset W(\Delta'_{\mathrm{el}})$. The most spectacular example is the dissection of the triangle $(1/7, 1/7, 1/7)$ into 24 copies of $(1/2, 1/3, 1/7)$. As a corollary of this dissection we find the remarkable identity

$$ {}_2F_1\left(\frac{2}{7}, \frac{3}{7}, \frac{6}{7}\middle| z\right) = b(z)^{-1/28}\, {}_2F_1\left(\frac{1}{84}, \frac{29}{84}, \frac{6}{7}\middle| 12^3 \frac{z(z-1)(z^3 - 8z^2 + 5 * z + 1)}{b(z)^3}\right) $$

where $b(z) = 1 - 236z + 1666z^2 - 3360z^3 + 3395z^4 - 1736z^5 + 42z^6 + 228z^7 + z^8$. For a complete list of such dissections and the corresponding identities we refer to [V].

3.30. Some loose ends

In the Schwarz map we have assumed that the parameters a, b, c are such that $\lambda = |1-c|$, $\mu = |c-a-b|$, $\nu = |a-b|$ are all less than 1. It turns out that in the irreducible case this is no restriction, since we can shift a, b, c by integers without affecting the monodromy group. In fact:

Lemma 3.31. *Assume that none of the numbers $a, b, c-a, c-b$ is integral. There exist $a' \in a(\mathrm{mod}\ \mathbf{Z})$, $b' \in b(\mathrm{mod}\ \mathbf{Z})$, $c' \in c(\mathrm{mod}\ \mathbf{Z})$ such that*

$$0 \le \lambda, \mu, \nu < 1 \qquad \lambda + \mu + \nu < 1 + 2\min(\lambda, \mu, \nu),$$

where $\lambda = |1-c'|$, $\mu = |c'-a'-b'|$, $\nu = |a'-b'|$. In the case $\lambda+\mu+\nu < 1$ there exists only one choice for a', b', c' and in the case $\lambda+\mu+\nu > 1$ there exist four possible choices.

Proof. First of all let us suppose that $0 \le a, b, c < 1$. Without loss of generality we can assume that $a \le b$. We consider the following cases.

Case i) $0 < a < c < b < 1$. We take $a' = a, b' = b, c' = c$. Then, $\lambda = 1-c, \mu = a+b-c, \nu = b-a$ and the inequalities are satisfied. Moreover, $\lambda+\mu+\nu = 1+2b-2c > 1$.

Case ii) $0 < a \le b < c < 1$. We take $a' = a, b' = b, c' = c$. When $c \ge a+b$ we get $\lambda = 1-c, \mu = c-a-b, \nu = b-a$ and the inequalities hold. Moreover, $\lambda+\mu+\nu = 1-2a < 1$. When $c \le a+b$ we get $\lambda = 1-c, \mu = a+b-c, \nu = b-a$ and the inequalities hold. Moreover, $\lambda+\mu+\nu = 1+2b-2c < 1$.

Case iii) $0 \le c < a \le b < 1$. We take $a' = a, b' = b, c' = c+1$. Then, $\lambda = c, \mu = c+1-a-b, \nu = b-a$ and the inequalities are readily verified. Moreover, $\lambda+\mu+\nu = 1+2c-2a < 1$.

As to uniqueness we note that an integral shift in the a, b, c such that the corresponding values of λ, μ, ν stay below 1 necessarily gives the substitutions of the form $\lambda \to 1-\lambda, \mu \to 1-\mu, \nu \to \nu$ and similar ones where two of the parameters are replaced by 1 minus their value. In this case the condition $\lambda+\mu+\nu < 1+2\min(\lambda,\mu,\nu)$ is violated by such a substitution. For example, $\lambda+\mu+\nu \le 1$ implies $1-\lambda+1-\mu+\nu = 2-(\lambda+\mu+\nu)+2\nu \ge 1+2\nu$. In the spherical case the condition is not violated. □

When we have obtained a geodesic Schwarz triangle in our construction we automatically have a metric which is invariant under the projective monodromy group. This closely reflects the nature of the natural hermitian form on the monodromy group itself.

Theorem 3.32. *Let $a, b, c \in \mathbf{R}$ be such that*

$$0 \le \lambda, \mu, \nu < 1, \qquad \lambda + \mu + \nu < 1 + 2\min(\lambda, \mu, \nu),$$

where $\lambda = |1-c|, \mu = |c-a-b|, \nu = |a-b|$. Let M be the monodromy group of (2). Then,

$$M \text{ is spheric} \iff \lambda+\mu+\nu > 1$$

$$M \text{ is euclidean} \iff \lambda+\mu+\nu = 1$$

$$M \text{ is hyperbolic} \iff \lambda + \mu + \nu < 1.$$

Proof. In the case when none of the numbers $a, b, c - a, c - b$ is integral, this statement can already be inferred from the proof of the previous lemma (we get only the hyperbolic and spheric case). It remains to show that if one of the numbers $a, b, c-a, c-b$ is integral, we have $\lambda+\mu+\nu = 1$. Let us suppose for example that $a \in \mathbf{Z}$. Notice that $|a-b| < 1$ and $|a+b| < |c|+1 < 3$. Hence $|a| \leq |a-b|/2+|a+b|/2 < 2$. So, $a = 0, \pm 1$. A case-by-case analysis using the inequalities for λ, μ, ν yields our statement. □

References

[AS] M. Abramowitz, I. Stegun, *Handbook of Mathematical Functions*, Dover, 1970.

[E] A. Erdélyi, *Higher transcendental functions, Vol. I*, Bateman Manuscript Project, New York, McGraw-Hill, 1953.

[H] J.E. Humphreys, *Reflection groups and Coxeter groups*, Cambridge University Press, 1992.

[Kl] F. Klein, *Vorlesungen über das Ikosaeder*, New edition with a preface by P. Slodowy, Birkhäuser-Teubner, 1993.

[V] R. Vidunas, Transformations of Gauss hypergeometric functions, *J. Computational and Applied Math.* **178** (2005), 473–487, arXiv: math.CA/0310436.

Frits Beukers
Department of Mathematics
Utrecht University
P.O. Box 80.010
Utrecht
The Netherlands
e-mail: `beukers@math.uu.nl`

Progress in Mathematics, Vol. 260, 43–100

Moduli of K3 Surfaces and Complex Ball Quotients

Igor V. Dolgachev and Shigeyuki Kondō

Abstract. These notes are based on a series of talks given by the authors at the CIMPA Summer School on Algebraic Geometry and Hypergeometric Functions held in Istanbul in Summer of 2005. They provide an introduction to recent work on the complex ball uniformization of the moduli spaces of del Pezzo surfaces, K3 surfaces and algebraic curves of lower genus. We discuss the relationship of these constructions with the Deligne–Mostow theory of periods of hypergeometric differential forms. For convenience to a non-expert reader we include an introduction to the theory of periods of integrals on algebraic varieties with emphasis on abelian varieties and K3 surfaces.

Mathematics Subject Classification (2000). Primary 14J10; Secondary 14J28, 14H15.

Keywords. Hodge structure, periods, moduli, Abelian varieties, arrangements of hyperplanes, $K3$ surfaces, complex ball.

1. Introduction

These notes are based on a series of talks at the CIMPA Summer School on Algebraic Geometry and Hypergeometric Functions held in Istanbul in Summer of 2005. The topic of the talks was an introduction to recent work of various people on the complex ball uniformization of the moduli spaces of del Pezzo surfaces and algebraic curves of lower genus ([ACT], [Vo], [K1]–[K4], [HL]). Keeping in mind the diverse background of the audience we include a general introduction to the theory of Hodge structures and period domains with more emphasis on abelian varieties and K3 surfaces. So, an expert may start reading the notes from Section 6.

Research of the first author is partially supported by NSF grant 0245203.
Research of the second author is partially supported by Grant-in-Aid for Scientific Research A-14204001, Japan.

It has been known for more than a century that a complex structure on a Riemann surface of genus g is determined up to isomorphism by the period matrix $\Pi = (\int_{\gamma_j} \omega_i)$, where $(\gamma_1, \ldots, \gamma_{2g})$ is a basis of 1-homology and $(\omega_1, \ldots, \omega_g)$ is a basis of holomorphic 1-forms. It is possible to choose the bases in a such a way that the matrix Π has the form $(Z\ I_g)$, where Z is a symmetric complex matrix of size g with positive definite imaginary part. All such matrices are parametrized by a complex domain $\mathcal{Z}_g$ in $\mathbb{C}^{g(g+1)/2}$ which is homogeneous with respect to the group $\mathrm{Sp}(2g, \mathbb{R})$. In fact, it represents an example of a Hermitian symmetric space of non-compact type, a Siegel half-plane of degree g. A different choice of bases with the above property of the period matrix corresponds to a natural action of the group $\Gamma_g = \mathrm{Sp}(2g, \mathbb{Z})$ on $\mathcal{Z}_g$. In this way the moduli space $\mathcal{M}_g$ of complex structures on Riemann surfaces of genus g admits a holomorphic map to the orbit space $\Gamma_g \backslash \mathcal{Z}_g$ which is called the period map. The fundamental fact is the Torelli Theorem which asserts that this map is an isomorphism onto its image. This gives a moduli theoretical interpretation of some points of the orbit space. All points can be interpreted as the period matrices of principally polarized abelian varieties, i.e., g-dimensional complex tori equipped with an ample line bundle whose space of holomorphic sections is one-dimensional. In this way the orbit space $\Gamma_g \backslash \mathcal{Z}_g$ becomes isomorphic to the moduli space $\mathcal{A}_g$ of such complex tori. The Siegel half-plane $\mathcal{Z}_g$ is a Hermitian symmetric space of non-compact type III in Cartan's classification. The development of the general theory of periods of integrals on algebraic varieties in the 1960s due to P. Griffiths raised a natural question on moduli-theoretical interpretation of other Hermitian symmetric spaces and their arithmetic quotients and the analogs of the Torelli Theorem. For the spaces of classical types I–IV, this can be achieved by embedding any type of space into a Siegel half-plane, and introducing the moduli space of abelian varieties with some additional structure (like a level, complex multiplication or some tensor form on cohomology). All of this becomes a part of the fancy theory of Shimura varieties. However, a more explicit interpretation remained to be searched for. For the type IV Hermitian symmetric space of dimension ≤ 19 such an interpretation had been found in terms of moduli of complex algebraic surfaces of type K3. The fundamental result of I. Piatetsky-Shapiro and I. Shafarevich in the 1970s gives an analog of the Torelli Theorem for polarized algebraic K3 surfaces. Although this gives only a realization of the 19-dimensional type IV space by a very special arithmetic group depending only on the degree of the polarization, a modified version of the polarization structure due to V. Nikulin allows one to extend this construction to type IV space of any dimension $d \leq 19$, with a variety of arithmetic groups realized as certain orthogonal groups of integral quadratic forms of signature $(2, d)$. Recently, some arithmetic quotients of type IV domain of dimension 20 have been realized as periods of holomorphic symplectic manifolds of dimension 4.

A complex ball is an example of a Hermitian symmetric space of type I. Some of its arithmetic quotients have a realization as periods of hypergeometric differential forms via the Deligne–Mostow theory. We refer for the details to Looijenga's article in the same volume [Lo]. A hypergeometric differential form has an interpretation as a holomorphic 1-form on a certain algebraic curve on which a cyclic group acts by automorphisms and the form is transformed according to a character of this group. In Section 6 we discuss in a more general setting the theory of what we call eigenperiods of algebraic varieties.

The periods of hypergeometric functions allows one to realize some complex ball quotients as the moduli space of weighted semi-stable ordered point sets in projective line modulo projective equivalence. In some cases these moduli spaces are isomorphic to moduli spaces of other structures. For example, via the period map of algebraic curves, the hypergeometric complex ball quotients are mapped onto a subvariety of $\mathcal{A}_g$ parametrizing principally polarized abelian varieties with a certain cyclic group action. For example, the moduli space of 6 points with equal weights defines a curve of genus 4 admitting a cyclic triple cover of the projective line ramified at the 6 points. Its Jacobian variety is an abelian variety of dimension 4 with a cyclic group of order 3 acting by automorphism of certain type. The moduli space of such abelian varieties is a complex ball quotient. Another example is the moduli space of equally marked sets of 5 points which leads to the moduli space of marked del Pezzo surfaces of degree 4 [K3]. Other examples relating the arithmetic complex ball quotients arising in the Deligne–Mostow theory to moduli spaces of Del Pezzo surfaces were found in [MT], [HL]. It turns out that all of these examples are intimately related to the moduli space of K3 surfaces with special structure of its Picard group of algebraic cycles and an action of a cyclic group. In Section 10 we develop a general theory of such moduli spaces. In Section 11 we briefly discuss all the known examples of moduli spaces of Del Pezzo surfaces and curves of low genus which are isomorphic to the moduli space of such structures on K3 surfaces, and via this isomorphism admit a complex ball uniformization by an arithmetic group. Some of these examples arise from the Deligne–Mostow theory. We conjecture that all Deligne–Mostow arithmetic complex ball quotients are moduli spaces of K3 surfaces.

We would like to thank the organizers of the Summer School, and especially Professor Uludag for the hospitality and for providing a stimulating and pleasant audience for our lectures.

2. Introduction to Hodge theory

In this and the next four sections we give a brief introduction to the Hodge theory of periods of integrals on algebraic varieties. We refer for details to [GH] or [Vo].

Let M be a smooth compact oriented connected manifold of even dimension $2n$. Its cohomology $H^*(M,\mathbb{Z})$ is a graded algebra over $\mathbb{Z}$ with multiplication defined by the cup-product

$$\cup : H^k(M,\mathbb{Z}) \times H^l(M,\mathbb{Z}) \to H^{k+l}(M,\mathbb{Z})$$

satisfying $x \cup y = (-1)^{kl} y \cup x$. In particular, the restriction of $\cup$ to the middle-dimensional cohomology $H^n(M, \mathbb{Z})$ is a $\mathbb{Z}$-bilinear form

$$b_M : H^n(M, \mathbb{Z}) \times H^n(M, \mathbb{Z}) \to H^{2n}(M, \mathbb{Z}) \cong \mathbb{Z},$$

where the latter isomorphism is defined by using the fundamental class $[M]$ of M. The Poincaré duality asserts that this bilinear form is a perfect pairing, modulo torsion. It is also symmetric if n is even, and skew-symmetric if n is odd.

Recall that the cohomology $H^*(M, \mathbb{R}) = H^*(M, \mathbb{Z}) \otimes \mathbb{R}$ can be computed by using the De Rham theorem:

$$H^*(M, \mathbb{R}) \cong H^*(0 \to A^0(M) \xrightarrow{d} A^1(M) \xrightarrow{d} A^2(M) \to \ldots),$$

where $A^k(M)$ is the space of smooth differential k-forms on M. The cup-product is defined by

$$[\alpha] \cup [\beta](\gamma) = \int_\gamma \alpha \wedge \beta, \ \gamma \in H_*(M, \mathbb{R}),$$

where we consider cohomology as linear functions on homology. In particular, the bilinear form b_M is an inner product on $H^n(M, \mathbb{R})$ defined by the formula

$$b_M([\alpha], [\beta]) = \int_M \alpha \wedge \beta. \tag{2.1}$$

The same is true for the cohomology $H^*(M, \mathbb{C}) = H^*(M, \mathbb{R}) \otimes \mathbb{C}$ if we replace $A^k(M)$ with complex valued smooth k-forms. Now suppose M is the underlying differentiable structure of a complex manifold X. Then local coordinates $t_1, \ldots, t_{2n}$ can be expressed in terms of complex coordinates $z_1, \ldots, z_n$ and its conjugates $\bar{z}_1, \ldots, \bar{z}_n$. This allows us to express locally a smooth k-form as a sum of forms of type (p, q):

$$\omega = \sum a_{i_1 \ldots i_p; j_1 \ldots j_q}(z, \bar{z}) dz_{i_1} \wedge \ldots \wedge dz_{i_p} \wedge d\bar{z}_{j_1} \wedge \ldots \wedge d\bar{z}_q, \quad p + q = k, \tag{2.2}$$

where $a_{i_1 \ldots i_p; j_1 \ldots j_q}(z, \bar{z})$ are smooth complex-valued functions. This gives a decomposition

$$A^k(X) = \bigoplus_{p+q=k} A^{pq}(X)$$

and $d = d' + d''$, where d' (resp. d'') is the derivation operator with respect to the variables z_i (resp. $\bar{z}_i$)). The Dolbeaut Theorem gives an isomorphism

$$H^q(X, \Omega_X^p) \cong H^q(0 \xrightarrow{d''} A^{p,0} \xrightarrow{d''} A^{p,1} \xrightarrow{d''} \ldots),$$

where Ω_X^p is the sheaf of holomorphic p-forms, i.e., forms from (2.2) of type $(p, 0)$, where the coefficients are holomorphic functions. The Dolbeaut theorem generalizes a well-known fact that a smooth function $f(z, \bar{z})$ of a complex variable is holomorphic if and only if it satisfies the equation $\frac{\partial f(z)}{\partial \bar{z}} = 0$.

A structure of a complex manifold X on a smooth manifold gives a decomposition of the complexified tangent bundle $T_M \otimes \mathbb{C}$ into a holomorphic and anti-holomorphic part with local basis $(\frac{\partial}{\partial z_i})$ and $(\frac{\partial}{\partial \bar{z}_i})$, respectively. We denote the holomorphic part by T_X. An additional structure of a hermitian complex manifold on X is given by a choice of a holomorphically varying structures of a hermitian vector space on tangent spaces $T_{X,x}$ defined by a tensor

$$ds^2 = \sum h_{ij}(z)dz_i \otimes d\bar{z}_j. \tag{2.3}$$

It allows one to define the adjoint operator δ'' of d'', the Laplace operator $\Delta'' = \delta''d'' + d''\delta''$, and the notion of a *harmonic form* of type (p,q) (an element of the kernel of the Laplace operator). One shows that each d''-closed form of type (p,q) is d''-cohomologous to a unique harmonic form of type (p,q). In particular, there is a canonical isomorphism of vector spaces

$$\mathcal{H}^{pq} \cong H^q(X, \Omega_X^p),$$

where $\mathcal{H}^{pq}$ is the space of harmonic forms of type (p,q). On the other hand, a hermitian form defines a structure of a Riemannian manifold on M. The latter defines the adjoint operator δ of d, the Laplace operator $\Delta = \delta d + d\delta$, and the space of harmonic forms $\mathcal{H}^k$. Each cohomology class has a unique representative by a harmonic form, i.e., there is a canonical isomorphism of vector spaces

$$H^k(M, \mathbb{R}) = \mathcal{H}^k.$$

A fundamental fact proved by Hodge asserts that if a hermitian metric on X is a *Kähler metric*, i.e., the $(1,1)$-form

$$\omega = \frac{i}{2}\sum h_{ij}(z)dz_i \wedge d\bar{z}_j \tag{2.4}$$

is closed (in this case the 2-form is called the *Kähler form*), the Laplace operators Δ'' and Δ extended to $H^*(X, \mathbb{C})$ coincide up to a constant factor. This shows that

$$\sum_{p+q=k} \mathcal{H}^{pq} = \mathcal{H}^k,$$

and we have a *Hodge decomposition*

$$H^k(X, \mathbb{C}) = \bigoplus_{p+q=k} H^{pq}(X), \tag{2.5}$$

where

$$H^{pq}(X) = \mathcal{H}^{pq} \cong H^q(X, \Omega_X^p).$$

The Hodge decomposition satisfies the following properties:

(HD1) The decomposition (2.5) does not depend on the choice of a Kähler metric;

(HD2) $H^{pq} = \overline{H^{qp}}$, where the bar denotes the complex conjugation. In particular $h^{pq}(X) = h^{qp}(X)$, where

$$h^{pq}(X) = \dim_{\mathbb{C}} H^{pq}(X);$$

(HD3) The bilinear form $Q : H^k(X, \mathbb{R}) \times H^k(X, \mathbb{R}) \to \mathbb{R}$ defined by the cup-product $(\phi, \psi) \mapsto \phi \cup \psi \cup [\omega]^{n-k}$ is symmetric when k is even and skew-symmetric otherwise. It satisfies

$$Q(x, y) = 0, \quad x \in H^{pq}, y \in H^{p'q'}, p \neq q'.$$

(HD4) Let $H^k_{prim}(X,\mathbb{R})$ denote the orthogonal complement of $[\omega]^{n-\frac{k}{2}}$ with respect to Q (for k odd $H^k(X,\mathbb{R})_{prim} := H^k(X,\mathbb{R})$). For any nonzero $x \in H^{pq}_{prim} := H^{pq} \cap H^k_{prim}(X,\mathbb{R})$,

$$i^{p-q}(-1)^{k(k-1)/2}Q(x,\bar{x}) > 0.$$

Using property (HD3) one can compute the signature $I(M)$ of the cup-product on $H^n(M,\mathbb{R})$ if n is even and M admits a Kähler structure:

$$I(M) = \sum_{p\equiv q \mathrm{mod} 2} (-1)^p h_1^{pq}, \tag{2.6}$$

where $h_1^{pq} = h^{pq}$ if $p \neq q$ and $h_1^{pq} = h^{pq} - 1$ otherwise. In particular, the Sylvester signature (t_+, t_-) (recall that, by definition, $I(M) = t_+ - t_-$) of the quadratic form on $H^n(M,\mathbb{R})$ when n is even, is given by

$$t_+ = \frac{1}{2}(b_n(M) + I(M) + 1), \quad t_- = \frac{1}{2}(b_n(M) - I(M) + 1), \tag{2.7}$$

where $b_i(M) = \dim_{\mathbb{R}} H^i(M,\mathbb{R})$ are the Betti numbers of M.

Define the subspace F^p of $H^k(X,\mathbb{C})$ by

$$F^p = \sum_{p'\geq p} H^{p'q}(X),$$

so that $F^0 = H^k(X,\mathbb{C}), F^k = H^{k,0}, F^p = \{0\}, p > k$. This defines a flag (F^p) of linear subspaces

$$0 \subset F^k \subset F^{k-1} \subset \ldots \subset F^0 = H^k(X,\mathbb{C}).$$

Note that Hodge decomposition can be reconstructed from this flag by using property (HD2),

$$H^{pq}(X) = \{x \in F^p : Q(x,\bar{y}) = 0, \ \forall y \in F^{p+1}\}. \tag{2.8}$$

Assume that the Kähler form (2.4) is a *Hodge form*, i.e., its cohomology class $[\omega]$ belongs to $H^2(X,\mathbb{Z})$. By a theorem of Kodaira this implies that X is isomorphic to a complex projective algebraic variety. It admits a projective embedding such that the cohomology class of a hyperplane section is equal to some positive multiple of $[\omega]$. The Hodge decomposition in this case has an additional property that the form Q takes integer values on the image of $H^k(X,\mathbb{Z})$ in $H^k(X,\mathbb{R})$.

The flag (F^p) defined by a Hodge decomposition is an invariant of a complex structure on M, and a bold conjecture (not true in general) is that it completely determines the complex structure on M up to isomorphism. In fact, this was shown in the 19th century for Riemann surfaces, i.e., complex manifolds of dimension 1.

One can define an *abstract Hodge structure of weight* k (AHS) on a real vector space V to be a decomposition into direct sum of complex subspaces

$$V_{\mathbb{C}} = \bigoplus_{p+q=k} V^{pq}$$

such that $\overline{V^{pq}} = V^{qp}$. A *polarization* of AHS on V is a a non-degenerate bilinear form Q on V which is symmetric if k is even, and skew-symmetric otherwise. It satisfies the conditions (HD3) and (HD4) from above, where H^{pq}_{prim} is replaced with V^{pq}.

An *integral structure* of a AHS is a free abelian subgroup $\Lambda \subset V$ of rank equal to $\dim V$ (a *lattice*) such that $Q(\Lambda \times \Lambda) \subset \mathbb{Z}$. One can always find an integral structure by taking Λ to be the $\mathbb{Z}$-span of a *standard basis* of Q for which the matrix of Q is equal to a matrix

$$I(a, n-a) = \begin{pmatrix} I_a & 0_{a,b} \\ 0_{b,a} & -I_b \end{pmatrix} \tag{2.9}$$

if k is even, and the matrix

$$J = \begin{pmatrix} 0_m & I_m \\ -I_m & 0_m \end{pmatrix} \tag{2.10}$$

if k is odd.

A Hodge structure on cohomology $V = H^k_{prim}(X, \mathbb{R})$ is an example of an AHS of weight k. If $[\omega]$ is a Hodge class, then the cohomology admits an integral structure with respect to the lattice Λ equal to the intersection of the image of $H^k(X, \mathbb{Z})$ in $H^k(X, \mathbb{R})$ with $H^k_{prim}(X, \mathbb{R})$.

3. The period map

Let (F^p) be the flag of subspaces of $V_{\mathbb{C}}$ defined by a polarized AHS of weight k on a vector space V. Let

$$f_p = \sum_{p' \geq p} h^{p'q}(X) = \dim F^p, \quad \mathbf{f} = (f_0, \ldots, f_k).$$

Let $\mathrm{Fl}(\mathbf{f}, V_{\mathbb{C}})$ be the variety of flags of linear subspaces F^p of dimensions f_p. It is a closed algebraic subvariety of the product of the Grassmann varieties $G(f_p, V_{\mathbb{C}})$. A polarized AHS of weight k defines a point (F^p) in $\mathrm{Fl}(\mathbf{f}, V_{\mathbb{C}})$. It satisfies the following conditions:

(i) $V_{\mathbb{C}} = F^p \oplus \overline{F^{k-p+1}}$;
(ii) $Q(F^p, F^{k-p+1}) = 0$;
(iii) $(-1)^{k(k-1)/2} Q(Cx, \bar{x}) > 0$, where C acts on H^{pq} as multiplication by i^{p-q}.

The subset of flags in $\mathrm{Fl}(\mathbf{f}, V_{\mathbb{C}})$ satisfying the previous conditions is denoted by $\mathcal{D}_{\mathbf{f}}(V, Q)$ and is called the *period space* of (V, Q) of type $\mathbf{f}$.

Fix a standard basis in V with respect to Q to identify V with the space $\mathbb{R}^r$, where $r = f_0$. Let F^p be the column space of a complex matrix Π_p of size $r \times f_p$. We assume that the first f_{p+1} columns of Π_p form the matrix Π_{p+1}. Then any flag in $\mathcal{D}_{\mathbf{f}} := \mathcal{D}_{\mathbf{f}}(\mathbb{R}^r, A)$ is described by a set of matrices Π_p satisfying the following conditions:

(PM1) $\det(\Pi_p|\overline{\Pi_{k-p+1}}) \neq 0$;
(PM2) ${}^t\Pi_p \cdot A \cdot \Pi_{k-p+1} = 0$;
(PM3) $(-1)^{k(k-1)/2}{}^t\Pi_p \cdot A \cdot \overline{\Pi_{k-p+1}} \cdot K > 0$, where K is a diagonal matrix with $\pm i$ at the diagonal representing the operator C.

Note that condition (ii) is a closed algebraic condition and defines a closed algebraic subvariety of $\mathrm{Fl}(\mathbf{f}, V_\mathbb{C})$. Other conditions are open conditions in complex topology. Thus the collection of matrices (Π_p) has a natural structure of a complex manifold. Two collections of matrices (Π_p) and (Π'_p) define the same point in the period space if and only if there exists an invertible complex matrix X of size r such that $\Pi'_0 = \Pi_0 \cdot X$. The matrix X obviously belongs to a subgroup $P_\mathbf{f}$ of $\mathrm{GL}(r, \mathbb{C})$ preserving the flag of the subspaces F^p generated by the first f_p unit vectors $\mathbf{e}_i$. The flag variety $\mathrm{Fl}(\mathbf{f}, V_\mathbb{C})$ is isomorphic to the homogeneous space

$$\mathrm{Fl}(\mathbf{f}, V_\mathbb{C}) \cong \mathrm{GL}(r, \mathbb{C})/P_\mathbf{f}.$$

The period space $\mathcal{D}_\mathbf{f}$ is an open subset (in complex topology) of a closed algebraic subvariety $\check{\mathcal{D}}_\mathbf{f}$ of $\mathrm{Fl}(\mathbf{f}, V_\mathbb{C})$ defined by condition (ii). It is known that the group $G_\mathbb{C} = \mathrm{Aut}(\mathbb{C}^r, Q_0)$ acts transitively on $\check{\mathcal{D}}_\mathbf{f}$ with isotropy subgroup $P = G_\mathbb{C} \cap P_\mathbf{f}$ so that

$$\check{\mathcal{D}}_\mathbf{f} \cong G_\mathbb{C}/P$$

is a projective homogeneous variety. The group $G_\mathbb{R} = \mathrm{Aut}(\mathbb{R}^r, Q_0)$ acts transitively on $\mathcal{D}_\mathbf{f}$ with a compact isotropy subgroup K so that

$$\mathcal{D}_\mathbf{f} \cong G_\mathbb{R}/K$$

is a complex non-compact homogeneous space.

In the case when the AHS is the Hodge structure on cohomology $H^k(X, \mathbb{R})$ of a Kähler manifold X, the matrices Π_p are called the *period matrices*. If $(\gamma_1, \ldots, \gamma_{f_0})$ is a basis in $H_k(X, \mathbb{Z})/\mathrm{Tors}$ such that the dual basis $(\gamma_1^*, \ldots, \gamma_{f_0}^*)$ is a standard basis of the polarization form defined by a choice of a Kähler structure, then

$$\Pi_p = (\int_{\gamma_i} \omega_j),$$

where $(\omega_1, \ldots, \omega_{k-p})$ is a basis of F^p represented by differential k-forms.

Now, suppose we have a *family* of compact connected complex manifolds. It is a holomorphic smooth map $f : \mathcal{X} \to T$ of complex manifolds with connected base T such that its fibre $X_t = f^{-1}(t)$. For any point $t \in T$ we have the real vector space $V_t = H^k(X_t, \mathbb{R})$ equipped with a Hodge structure. We also fix a Kähler class $[\omega]$ on $\mathcal{X}$ whose restriction to each X_t defines a polarization Q_t of the Hodge structure on V_t. One can prove (and this is not trivial!) that the Hodge numbers $h^{pq}(X_t)$ do not depend on t. Fix an isomorphism

$$\phi_t : (V, Q_0) \to (H^k(X_t, \mathbb{R}), \cup) \tag{3.1}$$

called a *kth marking* of X_t. Then pre-image of the Hodge flag (F_t^p) is a Q_0-polarized AHS of weight k on (V, Q_0). Let $\mathcal{D}_\mathbf{f}$ be the period space of (V, Q_0) of type $\mathbf{f}$, where

$f_p = \dim F_t^p$. We have a "multi-valued" map

$$\phi : T \to \mathcal{D}, \quad t \mapsto \big(\phi^{-1}(F_t^p)\big).$$

called the *multi-valued period map* associated to f. According to a theorem of Griffiths this map is a multi-valued holomorphic map.

To make the period map one-valued we need to fix a standard basis in (V_t, Q_t) for each t which depends holomorphically on t. This is not possible in general. The vector spaces $(V_t)_{t\in T}$ form a real *local coefficient system* $\mathcal{V}$ on T, i.e., a real vector bundle of rank f_0 over T whose transition functions are matrices with constant entries. As any local coefficient system it is determined by its *monodromy representation*. If we fix a point $t_0 \in T$ and let $V = V_{t_0}$, then for any continuous loop $\gamma : [0,1] \to T$ with $\gamma(0) = \gamma(1) = t_0$, the pull-back $\gamma^*(\mathcal{V})$ of $\mathcal{V}$ to $[0,1]$ is a trivial local coefficient system which defines a linear self-map $a_\gamma : V = \gamma^*(\mathcal{V})_0 \to V = \gamma^*(\mathcal{V})_1$. It depends only on the homotopy class $[\gamma]$ of γ. The map $[\gamma] \mapsto a_\gamma$ defines a homomorphism of groups

$$a : \pi_1(T, t_0) \to \mathrm{GL}(V),$$

called the *monodromy representation*. The pull-back of $\mathcal{V}$ to the universal covering $\tilde{T}$ is isomorphic to the trivial local coefficient system $\tilde{T} \times V$ and

$$\mathcal{V} \cong \tilde{T} \times V/\pi_1(T, t_0),$$

where $\pi_1(T, t_0)$ acts by the formula $[\gamma] : (z, v) = ([\gamma]\cdot z, a([\gamma])(v))$. Here the action of $\pi_1(T, t_0)$ on $\tilde{T}$ is the usual action by deck transformations.

One can show that the monodromy representation preserves the polarization form Q_{t_0} on V so that the image $\Gamma(f)$ of the monodromy representation lies in $\mathrm{Aut}(V, Q)$. This image is called the *monodromy group* of f. Of course, if T happens to be simply-connected, say by restriction to a small neighborhood of a point t_0, the monodromy representation is trivial, and we can define a one-valued period map.

Let $\pi : \tilde{T} \to T$ be the universal covering map. The second projection of the fibred product $\tilde{\mathcal{X}} = \mathcal{X} \times_T \tilde{T}$ of complex manifolds defines a holomorphic map $\tilde{f} : \tilde{\mathcal{X}} \to \tilde{T}$. Its fibre over a point $\tilde{t}$ is isomorphic to the fibre X_t over the point $t = \pi(\tilde{t})$. Now we can define the one-valued holomorphic map

$$\tilde{\phi} : \tilde{T} \to \mathcal{D}_{\mathbf{f}}. \tag{3.2}$$

Assume that the monodromy group $\Gamma(f)$ is a discrete subgroup of $\mathrm{Aut}(V, Q)$ with respect to its topology of a real Lie group. Since $\mathrm{Aut}(V, Q)$ acts on $\mathcal{D}_{\mathbf{f}}$ with compact isotropy subgroups, the isotropy subgroups of $\Gamma(f)$ are finite. This implies that the orbit space $\Gamma(f)\backslash\mathcal{D}_{\mathbf{f}}$ is a complex variety with only quotient singularities and the natural projection to the orbit space is a holomorphic map. The period map $\tilde{\phi}$ descends to a one-valued holomorphic map

$$\bar{\phi} : T \to \Gamma(f)\backslash\mathcal{D}_{\mathbf{f}}. \tag{3.3}$$

The discreteness condition for the monodromy group is always satisfied if $f : \mathcal{X} \to T$ is a family of projective algebraic varieties and the Kähler form on each X_t defining the polarization Q_t is a Hodge form. In this case the monodromy group is a subgroup of $G_\Lambda = \mathrm{Aut}(\Lambda, Q|\Lambda)$, where Λ is the image of $H^k(X_{t_0}, \mathbb{Z})$ in V. Thus for any such family there is a holomorphic map

$$\phi : T \to G_\Lambda \backslash \mathcal{D}_{\mathbf{f}}. \tag{3.4}$$

One says that a Kähler manifold X satisfies an *Infinitesimal Torelli Theorem* if for any family of varieties $f : \mathcal{X} \to T$ as above with $X \cong X_{t_0}$ for some $t_0 \in T$ the period map (3.2) is an isomorphic embedding of some analytic neighborhood of t_0 (or, equivalently, the differential of the map at t_0 is injective). We say that X satisfies a *Global Torelli Theorem* if for any two points $t, t' \in T$ with the same image in $\Gamma(f)\backslash\mathcal{D}_{\mathbf{f}}$ there is an isomorphism of complex manifolds $\phi : X_t \to X_{t'}$ such that $f^*([\omega_{t'}]) = [\omega_t]$.

Fix a smooth manifold M underlying some complex manifold. Consider the set of isomorphism classes of complex structures on M. A *moduli problem* is the problem of putting on this set a structure of an analytic space (or a complex variety) $\mathcal{M}$ such that for any holomorphic map $f : Y \to T$ of analytic spaces whose fibres are complex manifolds diffeomorphic to M, the map $T \to \mathcal{M}$ which assigns to $t \in T$ the isomorphism class of $f^{-1}(t)$ defines a holomorphic map $\phi : T \to \mathcal{M}$. Or, equivalently, for any $f : Y \to T$ there exists a holomorphic map $\phi : T \to \mathcal{M}$ such that the fibres which are mapped to the same point must be isomorphic complex manifolds. We require also that for any $\mathcal{M}'$ with the same property there exists a holomorphic map $s : \mathcal{M} \to \mathcal{M}'$ such that $\phi' = s \circ \phi$. If $\mathcal{M}$ exists it is called a *coarse moduli space* of complex structures on M.

If additionally, there exists a holomorphic map $u : \mathcal{X} \to \mathcal{M}$ such that any f as above is obtained via a unique map $\phi : T \to \mathcal{M}$ by taking the fibred product $\mathcal{X} \times_{\mathcal{M}} T \to T$. In this case $\mathcal{M}$ is called a *fine moduli space*, and u the *universal family*. Note that, for any point $m \in \mathcal{M}$, the isomorphism class of the fibre $u^{-1}(m)$ is equal to m. Fine moduli spaces rarely exist unless we put some additional data, for example a marking on cohomology as in (3.1).

Similar definitions one can give for the moduli space of structures of an algebraic variety, or for polarized Kähler manifolds or polarized projective algebraic varieties. For the latter, one fixes a cohomology class in $H^2(M, \mathbb{R})$ and consider the set of complex structures on M such that this cohomology class can be represented by a Kähler (or a Hodge) form.

Suppose a coarse moduli of polarized algebraic manifolds of diffeomorphism type M exists and the Global Torelli Theorem holds for the corresponding complex manifolds. Then, by definition of a coarse moduli space there must be a holomorphic map

$$\mathrm{per} : \mathcal{M} \to G_\Lambda \backslash \mathcal{D}_{\mathbf{f}}.$$

A *Local Torelli Theorem* is the assertion that this map is a local isomorphism which, together with Global Torelli Theorem, will assert that the map is an embedding of complex varieties. Note that the Local Torelli Theorem implies the Infinitesimal Torelli Theorem but the converse is not true in general.

4. Hodge structures of weight 1

An abstract Hodge structure of weight 1 on a real vector space V is a decomposition into direct sum of complex linear subspaces

$$V_{\mathbb{C}} = V^{10} \oplus V^{01}. \tag{4.1}$$

such that

(i) $\overline{V^{10}} = V^{01}$.

A polarization on AHS of weight 1 is a non-degenerate skew-symmetric form Q on V (a symplectic form) such that

(ii) $Q|V^{10} = 0$, $Q|V^{01} = 0$;
(iii) $iQ(x,\bar{x}) > 0, \quad \forall x \in V^{10} \setminus \{0\}$.

Here, as always, we extend Q to the complexification by linearity.

Consider a Hodge structure of weight 1 on V. Let $W = V^{01}$. The composition $V \to W$ of the natural inclusion map $V \hookrightarrow V_{\mathbb{C}}$ and the projection to W with respect to the decomposition (4.1) is an $\mathbb{R}$-linear isomorphism. This allows us to transfer the structure of a complex space on W to V. Recall that a complex structure on a real vector space V is defined by a linear operator I satisfying $I^2 = -1_V$ by setting

$$(a+bi)\cdot v = av + bI(v).$$

We have

$$V = \{w + \bar{w}, w \in W\}$$

and, for any $v = w + \bar{w}$ we define $I(v)$ to be the unique vector in V such that

$$I(v) = iw - i\bar{w}.$$

In particular, the $\mathbb{R}$-linear isomorphism $V \to W$ is defined by $v \mapsto \frac{1}{2}(v - iI(v))$. We will often identify V with W by means of this isomorphism. Conversely, a complex structure I on V defines a decomposition (4.1), where W (resp. $\overline{W}$) is the eigensubspace of I extended to $V_{\mathbb{C}}$ by linearity with eigenvalue $i = \sqrt{-1}$ (resp. $-i$). Thus we obtain a bijective correspondence between AHS of weight 1 on V and complex structures on V. Note that replacing the Hodge structure by switching V^{10} with V^{01} changes the complex structure to the conjugate one defined by the operator $-I$.

Now let us see the meaning of a polarized AHS of weight 1. The polarization form Q makes the pair (V,Q) a real symplectic vector space.

A complex structure I on V is called a *positive complex structure* with respect to Q if $Q(v, I(v'))$ defines a symmetric positive definite bilinear form on V. It follows from the symmetry condition that the operator I is an isometry of the symplectic space (V,Q).

Consider a polarized Hodge structure on V of weight 1 and let I be the complex structure I on V determined by the subspace $W = V^{01}$. Extending by linearity, we will consider Q as a skew-symmetric form on $V_{\mathbb{C}}$. Let $w = v - iI(v)$, $w' = v' - iI(v') \in W$. We have

$$-iQ(w, \bar{w'}) = Q(v, I(v')) + iQ(v, v'). \tag{4.2}$$

By definition of a polarized AHS, $Q(v, I(v)) = -iQ(w, \bar{w}) > 0$ for any $w \neq 0$. This shows that the complex structure I is positive with respect to Q.

Conversely, suppose I is a positive complex structure on (V, Q). Let V^{01} be the i-eigensubspace of $V_{\mathbb{C}}$ of the operator I. Since $Q(w, I(w')) = iQ(w, w')$ is a symmetric and also a skew-symmetric bilinear form on W, it must be zero. Hence W (and, similarly, $V^{10} = \overline{W}$) is an isotropic subspace of Q. This checks property (ii) of the Hodge structure. We have $0 < -iQ(v - iI(v), v + iI(v)) = 2Q(v, v)$. This checks property (iii). Thus we have proved the following.

Lemma 4.1. *There is a natural bijection between the set of Hodge structures of weight 1 on V with polarization form Q and the set of positive complex structures on V with respect to Q.*

Example. Let $V = \mathbb{R}^{2g}$ with standard basis formed by the unit vectors $\mathbf{e}_k$. Define the complex structure I by

$$I(\mathbf{e}_k) = \mathbf{e}_{k+g}, \; I(\mathbf{e}_{g+k}) = -\mathbf{e}_k, \quad k = 1, \ldots, g.$$

The space $V_{\mathbb{C}} = \mathbb{C}^{2g}$ decomposes into the direct sum of the $\pm i$-eigensubspaces $V_\pm$ of I, where $V_\pm$ is spanned by the vectors $\mathbf{e}_k \mp i\mathbf{e}_{k+i}$. Let Q be a skew-symmetric bilinear form defined by the condition

$$Q(\mathbf{e}_k, \mathbf{e}_{k+g}) = 1, \; Q(\mathbf{e}_k, \mathbf{e}_{k'}) = 0 \text{ if } |k - k'| \neq g$$

(the *standard symplectic form* on $\mathbb{R}^{2g}$). Then the matrix of $Q(v, I(v'))$ in the standard basis is the identity matrix, so $Q(v, I(v'))$ is symmetric and positive definite.

Recall that a *hermitian form* on a complex vector space E is a $\mathbb{R}$-bilinear form $H : E \times E \to \mathbb{C}$ which is $\mathbb{C}$-linear in the first argument and satisfies

$$H(x, y) = \overline{H(y, x)}. \tag{4.3}$$

One can view a hermitian form as a $\mathbb{C}$-bilinear map $E \times \bar{E} \to \mathbb{C}$ satisfying (4.3). Here $\bar{E}$ is the same real space as E with conjugate complex structure. The restriction of H to the diagonal is a real-valued quadratic form on the real space E, and the signature of H is the signature of this quadratic form. In particular, we can speak about positive definite hermitian forms.

Lemma 4.2. *The formula*

$$H(x, y) = -iQ(x, \bar{y}) \tag{4.4}$$

defines a hermitian form on $V_{\mathbb{C}}$ of signature (g, g).

Proof. Obviously, formula (4.3) defines a bilinear form on $V_{\mathbb{C}} \times \overline{V_{\mathbb{C}}}$. Write $x, y \in V_{\mathbb{C}}$ in the form $x = v + iw$, $y = v' + iw'$ for some $v, w, v', w' \in V$. Then

$$\begin{aligned} H(x,y) &= iQ(v+iw, v'-iw') = i(Q(v,v') + Q(w,w')) + Q(w,v') - Q(v,w') \\ &= \overline{H(y,x)}. \end{aligned}$$

Let $e_1, \ldots, e_{2g}$ be a standard symplectic basis in V. Let $f_k = e_k - ie_{k+g}$, $\bar{f}_k = e_k + ie_{k+i}$, $k = 1, \ldots, g$. These vectors form a basis of $V_{\mathbb{C}}$. We have

$$\begin{aligned} H(f_k, f_l) &= -iQ(f_k, \bar{f}_l) = -iQ(e_k - ie_{k+g}, e_l + ie_{l+g}) = Q(e_k, e_{l+g}) - Q(e_{k+g}, e_l) \\ &= 1 \end{aligned}$$

if $k = i$ and 0 otherwise. This shows that for any nonzero $x = \sum a_k f_k$, we have

$$H(x,x) = 2\sum a_k \bar{a}_k > 0.$$

Thus the restriction of H to the span W of the f_k's is positive definite. Similarly, we check that the restriction of H to the span of the $\bar{f}_k$'s is negative definite and two subspaces W and $\overline{W}$ are orthogonal with respect to H. This shows that H is of signature (g, g). □

Note that the skew-symmetric form Q on $V_{\mathbb{C}}$ is reconstructed from H by the formula

$$Q(x,y) = iH(x, \bar{y}). \tag{4.5}$$

Let $G(g, V_{\mathbb{C}})$ be the Grassmann variety of g-dimensional subspaces of $V_{\mathbb{C}}$. Let H be a hermitian form on $V_{\mathbb{C}}$ of signature (g, g) and Q be a skew-symmetric form on $V_{\mathbb{C}}$ associated to H by the formula (4.5). Set

$$G(g, V_{\mathbb{C}})_H = \{W \in G(g, V_{\mathbb{C}} : Q|W = 0, H|W > 0\}. \tag{4.6}$$

We see that a Hodge structure of weight 1 on V defines a point $W = V^{01}$ in $G(g, V_{\mathbb{C}})_H$, where H is defined by the formula (4.3). Conversely, for any $W \in G(g, V_{\mathbb{C}})_H$, set $V^{01} = W, V^{10} = \overline{W}$. Since $Q|W = 0$, formula (4.5) implies that W and $\overline{W}$ are orthogonal to each other, hence we have a direct sum decomposition (4.1). We take Q to be defined by (4.5). Properties (ii) and (iii) are obviously satisfied.

The hermitian form H defined in Lemma 4.2 will be called the hermitian form *associated* to Q.

We have proved the following.

Theorem 4.3. *Let (V, Q) be a real symplectic space. There is a natural bijection between Hodge structures on V of weight 1 with polarization form Q and points in $G(g, V_{\mathbb{C}})_H$, where H is the associated hermitian form of Q.*

By choosing a standard symplectic basis in V, $G(g, V_{\mathbb{C}})_H$ can be described as a set of complex $2g \times g$-matrices satisfying conditions (PM1)–(PM3) with $p = 1$:

$${}^t\Pi \cdot J \cdot \Pi = 0 \tag{4.7}$$

$$-i{}^t\Pi \cdot J \cdot \bar{\Pi} > 0. \tag{4.8}$$

Two such matrices Π' and Π define the same AHS if and only if there exists an invertible matrix X such that $\Pi' = \Pi \cdot X$. Recall that W defines a complex structure I on V such that $v \mapsto v - iI(v)$ is an isomorphism of complex vector spaces $(V, I) \to W$. Let E be the real subspace of V spanned by the last g vectors of the standard symplectic basis. We have $E \cap I(E) = \{0\}$ since $w = I(v) \in E$ for some $v \in E$ implies $Q(w, I(w)) = Q(w, I^2(v)) = Q(v, w) = 0$ contradicting the positivity condition of a complex structure unless $v = 0$. This shows that the vectors $v - iI(v), v \in E$ span W as a complex space. Thus we can find a unique basis of W such that the last g rows of the matrix Π form the identity matrix. In other words, we can always assume that

$$\Pi = \begin{pmatrix} Z \\ I_g \end{pmatrix}, \tag{4.9}$$

for a unique square complex matrix Z of size g. The conditions (4.7) and (4.8) are equivalent to the conditions

$${}^tZ = Z, \quad \mathrm{Im}(Z) = \frac{1}{2i}(Z - \bar{Z}) > 0. \tag{4.10}$$

We obtain that the period space $\mathcal{D}_{(2g,g)}$ parametrizing polarized AHS of weight 1 is isomorphic to the complex manifold

$$\mathcal{Z}_g := \{Z \in \mathrm{Mat}_g(\mathbb{C}) : {}^tZ = Z, \mathrm{Im}(Z) > 0\}, \tag{4.11}$$

called the *Siegel half-plane* of degree g. Its dimension is equal to $g(g+1)/2$. We know that $\mathcal{D}_{(2g,g)}$ is a complex homogeneous space of the form $G_{\mathbb{R}}/K$, where $G_{\mathbb{R}} = \mathrm{Aut}(\mathbb{R}^{2g}, Q)$ and K is a compact subgroup of $G_{\mathbb{R}}$. Explicitly, $G_{\mathbb{R}}$ can be identified with the group of matrices (the *symplectic group*)

$$\mathrm{Sp}(2g, \mathbb{R}) = \{M \in \mathrm{GL}(2g, \mathbb{R}) : {}^tM \cdot J \cdot M = J\}.$$

The group $\mathrm{Sp}(2g, \mathbb{R})$ acts on $\mathcal{Z}_g$ by its natural action on isotropic subspaces W of $G(g, V_{\mathbb{C}})$ or on matrices $Z \in M_g(\mathbb{C})$ by the formula

$$M \cdot Z = (A \cdot Z + B) \cdot (C \cdot Z + D)^{-1}, \tag{4.12}$$

where M is written as a block-matrix $M = \begin{pmatrix} A & B \\ C & D \end{pmatrix}$. Let $Z = X + iY \in \mathcal{Z}_g$. Since Y is a symmetric positive definite matrix, it can be written as the product $A \cdot {}^tA$ for some invertible square matrix A. The matrix $M = \begin{pmatrix} A & X{}^tA^{-1} \\ 0 & {}^tA^{-1} \end{pmatrix}$ belongs to $\mathrm{Sp}(2g, \mathbb{R})$ and $M \cdot iI_g = (iA + X{}^tA^{-1}) \cdot A^t = X + iY$. This checks that $\mathrm{Sp}(2g, \mathbb{R})$ acts transitively on $\mathcal{Z}_g$. If we take Z to be the matrix iI_g, then the stabilizer $\mathrm{Sp}(2g, \mathbb{R})_Z$ of Z consists of matrices M with $iA + B = i(iC + D)$, i.e., satisfying $A = D, B = -C$. The map $M \to X = A + iB$ is an isomorphism of $\mathrm{Sp}(2g, \mathbb{R})_Z$ onto a subgroup of complex $g \times g$-matrices. Also the condition ${}^tM \cdot J \cdot M = J$ translates into the condition ${}^t\bar{X} \cdot X = I_g$. Thus a compact subgroup K could be taken to be a subgroup of $\mathrm{Sp}(2g, \mathbb{R})$ isomorphic to the unitary group $U(g)$. So we get

$$\mathcal{Z}_g \cong \mathrm{Sp}(2g, \mathbb{R})/U(g). \tag{4.13}$$

A Siegel upper-half plane is an irreducible hermitian symmetric space of type III in Cartan's classification.

Recall that *hermitian symmetric space* is a connected complex manifold X equipped with a hermitian metric such that each point $x \in X$ is an isolated fixed point of a holomorphic involution of X which preserves the metric. One can show that the metric satisfying this condition must be a Kähler metric. The group of holomorphic isometries acts transitively on a hermitian symmetric space X and its connected component of the identity is a connected Lie group $G(X)$. The isotropy subgroup of a point is a maximal compact subgroup K of $G(X)$ which contains a central subgroup isomorphic to $U(1)$. An element I of this subgroup satisfying $I^2 = -1$ defines a complex structure on the tangent space of each point of X. Any hermitian symmetric space is a symmetric space with respect to the canonical structure of a Riemannian manifold. A hermitian symmetric space of non-compact type is a homogeneous space G/K as above with G a semi-simple Lie group. It is called irreducible if the Lie group is a simple Lie group.

The Siegel half-plane $\mathcal{Z}_1$ is just the upper half-plane

$$\mathcal{H} = \{z = x + iy \in \mathbb{C} : y > 0\}.$$

The action of $\mathrm{Sp}(2g, \mathbb{R})$ on $\mathcal{Z}_g$ is analogous of the Moebius transformation of the upper half-plane $z \mapsto \frac{az+b}{cd+d}$.

It is known that $\mathcal{H}$ is holomorphically isomorphic to the unit 1-ball

$$U = \{z \in \mathbb{C} : |z| < 1\}.$$

Similarly, the Siegel half-plane is isomorphic as a complex manifold to a bounded domain in $\mathbb{C}^{\frac{1}{2}g(g+1)}$ defined by

$$\{Z \in \mathrm{Mat}_g(\mathbb{C}) : {}^tZ = Z, I_g - \bar{Z} \cdot Z > 0\} \tag{4.14}$$

The isomorphism is defined by replacing matrix (4.9) satisfying (4.7) and (4.8) with the matrix

$$\Pi' = \begin{pmatrix} I_g & I_g \\ iI_g & -iI_g \end{pmatrix} \cdot \Pi$$

and then reducing it back to a form (4.9).

There is a natural definition of the direct sum of abstract polarized Hodge structures which we leave to the reader. For AHS of weight 1 this of course corresponds to the operation of the direct sum of symplectic spaces and the direct sum of complex structures. It is easy to see that a polarized AHS of weight 1 decomposes into a direct sum if and only if the corresponding matrix $Z \in \mathcal{Z}_g$ is the direct product of block-matrices of smaller size.

5. Abelian varieties

Recall that a torus T is a smooth manifold V/Λ, where V is a real space of dimension n and Λ is a *lattice* in V. The additive group structure of V defines a structure of a group manifold on T, the space V acts on T by translations with kernel equal to Λ. Obviously, T is diffeomorphic to $\mathbb{R}^n/\mathbb{Z}^n = (S^1)^n$, the product of n circles.

The space V is the universal covering of T, and the group Λ is identified with the fundamental group of T. Since it is abelian, we obtain a canonical isomorphism

$$\Lambda \cong H_1(T, \mathbb{Z}).$$

By the Künneth formula,

$$H_k(T, \mathbb{Z}) \cong \bigwedge^k \Lambda, \ \ H^k(T, \mathbb{Z}) \cong \bigwedge^k \mathrm{Hom}(\Lambda, \mathbb{Z}).$$

There is a natural bijective correspondence between tensor forms on the smooth manifold V/Λ invariant with respect to translations and tensors on V. Tensoring with $\mathbb{R}$ and $\mathbb{C}$ we can identify $H^k(T, \mathbb{R})$ with the space k-multilinear forms on $V = \Lambda \otimes \mathbb{R}$ and $H^k(T, \mathbb{C})$ with the space of k-multilinear forms on the complexification $V_{\mathbb{C}}$ of V.

Now assume $n = 2g$. A complex structure on T invariant with respect to translations (making T a complex Lie group) is defined by a complex structure I on the space V. The holomorphic tangent bundle of T becomes isomorphic to the trivial bundle with fibre $W = (V, I)$. A translation invariant structure of a hermitian manifold on T is defined by a positive definite hermitian form h on W. It is easy to see that

$$h(v, w) = Q(I(v), w) + iQ(v, w), \tag{5.1}$$

where Q is a skew-symmetric form on V and $g(v, w) = Q(I(v), w)$ is a positive definite symmetric form. Since $h(I(v), I(w)) = h(v, w)$, we see that $g(v, w)$ is a Riemannian metric on T invariant with respect to the complex structure and also I is an isometry of the symplectic space (V, Q). It follows from (5.1) that the conjugate complex structure $-I$ on V is positive with respect to Q. Thus a translation invariant structure of a hermitian manifold on T defines a Hodge structure of weight 1 on V with polarization Q and $W = V^{10}$. As is easy to see the converse is true.

Extend h to a hermitian form H on $V_{\mathbb{C}}$ by requiring that the subspaces W and $\overline{W}$ are orthogonal with respect to H and $H|W = h$, $H(\bar{w}, \bar{w}') = -H(w, w')$. Write $w = \frac{1}{2}(v - iI(v)), w' = \frac{1}{2}(v' - iI(v'))$ for some $v, v' \in V$. Since

$$2iQ(w, \bar{w}') = 2iQ(\frac{1}{2}(v - iI(v)), \frac{1}{2}(v' + iI(v')) = Q(v, I(v')) + iQ(v, v') = h(v, v'),$$

we see that H is defined by

$$H(x, y) = 2iQ(x, \bar{y})$$

and hence H and $\Omega = -2Q$ are associated hermitian and skew-symmetric forms on $V_{\mathbb{C}}$ as is defined in the previous section. Let z_α be the coordinate functions on W with respect to some basis $e_1, \ldots, e_g$ and $\bar{z}_\alpha$ be the coordinate functions on $\overline{W}$

with respect to the conjugate basis $\bar{e}_1, \ldots, \bar{e}_g$. Then $z_\alpha, \bar{z}_\alpha$ are coordinate functions on $V_\mathbb{C}$ satisfying

$$\overline{z_\alpha(x)} = \bar{z}_\alpha(\bar{x}).$$

Let $h_{\alpha\beta} = H(e_\alpha, e_\beta)$ so that, for any $x, y \in W$,

$$h(x,y) = \sum h_{\alpha\beta} z_\alpha(x) \bar{z}_\beta(y).$$

Then $\Omega(e_\alpha, e_\beta) = \Omega(\bar{e}_\alpha, \bar{e}_\beta) = 0$ and

$$\Omega(e_\alpha, \bar{e}_\beta) = ih(e_\alpha, e_\beta) = ih_{\alpha\beta}.$$

This shows that, for any $x, y \in V_\mathbb{C}$,

$$\Omega(x,y) = i \sum h_{\alpha\beta} z_\alpha(x) \wedge \bar{z}_\beta(y).$$

Comparing this with (2.3), we see that Ω defines the Kähler form on T associated to the hermitian metric h. Since it has constant coefficients, its closedness is obvious.

Assume that the symplectic form Q satisfies the integrality condition with respect to the lattice Λ, then the Kähler form ω associated to the hermitian metric is a Hodge form, and hence the complex torus $X = W/\Lambda$ admits an embedding in a projective space such that the cohomology class in $H^2(T, \mathbb{Z})$ which is dual (with respect to the Poincaré duality) to the homology class in $H_{2n-2}(T, \mathbb{Z})$ of a hyperplane in $\mathbb{P}^n$ is equal to some positive multiple of the cohomology class $[\omega]$.

Conversely, let $T = W/\Lambda$ be a complex torus. We identify the real vector space V underlying W with $\Lambda \otimes \mathbb{R}$ and $V_\mathbb{C}$ with $\Lambda \otimes \mathbb{C}$. A choice of a symplectic form Q on V such that the complex structure W on V is positive with respect to $-Q$ is called a *polarization* on T. Two polarizations are called equivalent if the symplectic forms differ by a constant factor equal to a positive rational number. A polarization defines a hermitian form on W and a structure of a hermitian complex manifold on T. A pair (T, Q) is called a *polarized complex torus.* If additionally $Q(\Lambda \times \Lambda) \subset \mathbb{Z}$, then the polarized torus is called a *polarized abelian variety.* Recall that an abelian variety (over complex numbers) is a projective algebraic variety isomorphic, as a complex manifold, to a complex torus. A projective embedding of an abelian variety X defines a structure of a polarized abelian variety on X, the Hodge class being the cohomology class of a hyperplane section.

Not every complex torus W/Λ is an abelian variety. A polarization Q is defined by an integral matrix A in a basis of Λ. If we identify $V_\mathbb{C}$ with $\mathbb{C}^{2g}$ by means of this basis and the space $\bar{W}$ with the column space of a matrix Π (called the *coperiod matrix* of the torus), then the necessary and sufficient conditions of positivity of the complex structure of $\bar{W}$ with respect to Q are the conditions

$$^t\Pi \cdot A \cdot \Pi = 0, \quad -i^t\Pi \cdot A \cdot \bar{\Pi} > 0. \tag{5.2}$$

These conditions can be combined by introducing the square matrix $\tilde{\Pi} = (\Pi|\bar{\Pi})$ (corresponding to the matrix Π_0 in (PM1)–(PM3)). The condition is

$$-i^t\tilde{\Pi} \cdot A \cdot \bar{\tilde{\Pi}} = \begin{pmatrix} M & 0_g \\ 0_g & -{}^tM \end{pmatrix}, \tag{5.3}$$

where M is a positive definite hermitian matrix.

Since $V_{\mathbb{C}} = W \oplus \overline{W}$, the matrix $\tilde{\Pi}^{-1}$ is invertible. Write

$$\tilde{P} = \tilde{\Pi}^{-1} = (P|\bar{P})$$

for some $2g \times g$-matrix. The jth column of P gives an expression of the jth basis vector of Λ as a linear combination of a basis of W formed by the columns of Π. It is called the *period matrix* of T. One can restate the condition (5.3) in terms of the period matrix by

$$i^t\overline{\tilde{P}} \cdot A^{-1} \cdot {}^t\tilde{P} = \begin{pmatrix} M' & 0_g \\ 0_g & -{}^tM' \end{pmatrix}, \tag{5.4}$$

where $M' = M^{-1}$ is a positive definite hermitian matrix. These are the so-called *Riemann–Frobenius conditions.*

Example. Any one-dimensional torus is an abelian variety (an elliptic curve). In fact let T be isomorphic to the complex torus $\mathbb{C}/\mathbb{Z}\gamma_1 + \mathbb{Z}\gamma_2$. Under the multiplication by γ_1, the torus is isomorphic to the torus $\mathbb{C}/1 \cdot \mathbb{Z} + \tau\mathbb{Z}$, where $\tau = \gamma_2/\gamma_1$. Replacing τ by $-\tau$, if necessary we may assume that $\mathrm{Im}(\tau) > 0$. Thus the period matrix of T with respect to the basis 1 of $\mathbb{C}$ and the basis $1, \tau$ of Λ is equal to $P = (1, \tau)$. The Riemann–Frobenius condition is the existence of an integer r such that

$$i \begin{pmatrix} 1 & \bar{\tau} \\ 1 & \tau \end{pmatrix} \cdot \begin{pmatrix} 0 & 1/r \\ -1/r & 0 \end{pmatrix} \cdot \begin{pmatrix} 1 & 1 \\ \tau & \bar{\tau} \end{pmatrix} = \begin{pmatrix} a & 0 \\ 0 & -a \end{pmatrix},$$

where a is a positive real number. Computing the product we obtain that the condition is equivalent to the condition $ir^{-1}(\bar{\tau} - \tau)) \in \mathbb{R}_{>0}$ which is satisfied for any negative integer r.

We leave to the reader to check that the complex torus $T = \mathbb{C}^2/P \cdot \mathbb{Z}^2$, where

$$P = \begin{pmatrix} \sqrt{-2} & \sqrt{-3} \\ \sqrt{-5} & \sqrt{-7} \end{pmatrix}$$

does not satisfy condition (5.4) for any integral skew-symmetric matrix A. Thus it is not a projective algebraic variety.

Let $(W/\Lambda, Q)$ be a polarized abelian variety. It is known from linear algebra that Λ admits a basis such that the matrix of $Q|\Lambda$ is equal to

$$J_D = \begin{pmatrix} 0_g & D \\ -D & 0_g \end{pmatrix}, \tag{5.5}$$

where $D = \mathrm{diag}[d_1, \dots, d_g]$ with $0 < d_1|d_2| \dots |d_g$. The vector $(d_1, \dots, d_g)$ is an invariant of an integral valued skew-symmetric form on Λ. The equivalence class of D with respect to the equivalence relation defined by $D \sim D'$ if $D' = aD$ for some positive rational number a is called the *type* of the polarization of $(W/\Lambda, Q)$. We will always represent it by a primitive D (i.e., non-positive multiple of any other D). A polarization of type $(1, \dots, 1)$ is called a *principal polarization.*

Two polarized abelian varieties $(W/\Lambda, Q)$ and $(W'/\Lambda', Q')$ are called isomorphic if there exists an isomorphism of complex spaces $f : W \to W'$ such that

$f(\Lambda) = \Lambda'$ and $Q' \circ (f \times f) : \Lambda \times \Lambda \to \mathbb{Z}$ is equivalent to Q. Clearly, isomorphic polarized varieties have the same type of polarization. Any polarized abelian variety of type $(d_1, \ldots, d_g)$ is isomorphic to the polarized variety $\mathbb{C}^g / P \cdot \mathbb{Z}^{2g}$, where P is the period matrix such that the matrix of Q with respect to the basis of Λ formed by the columns of P is the matrix (5.5). Then P satisfies (5.4). The matrix

$$\Pi' = \begin{pmatrix} 1_g & 0_g \\ 0_g & D \end{pmatrix} \cdot {}^tP$$

satisfies (4.7) and (4.8). As we have seen, there exists an invertible complex matrix X such that $\Pi \cdot X$ is of the form (4.9), where $Z \in \mathcal{Z}_g$. This implies that our torus is isomorphic to the torus $\mathbb{C}^g / P \cdot \mathbb{Z}^{2g}$, where $P = (Z|D)$ and $Z \in \mathcal{Z}_g$. The period matrix of this form is called a *normalized period matrix*. Two polarized abelian varieties with normalized period matrices $P = (Z|D)$ and $P' = (Z'|D)$ are isomorphic if and only if there exists a complex matrix X defining a map $\mathbb{C}^g \to \mathbb{C}^g$ such that

$$X \cdot (Z|D) = (Z'|D) \cdot M,$$

where M is an invertible integral $2g \times 2g$-matrix satisfying

$${}^tM \cdot J_D \cdot M = J_D.$$

Let $\mathrm{Sp}(2g, \mathbb{Z})_D \subset \mathrm{GL}(2g, \mathbb{Z})$ be the group of such matrices. If $D = I_g$, i.e., $J_D = J$, the group $\mathrm{Sp}(2g, \mathbb{Z})_D$ is the *Siegel modular group*

$$\Gamma_g = \mathrm{Sp}(2g, \mathbb{Z})$$

of symplectic integral matrices. For any $M \in \mathrm{Sp}(2g, \mathbb{Z})_D$, the conjugate matrix

$$N = \begin{pmatrix} 1_g & 0_g \\ 0_g & D \end{pmatrix} \cdot M \cdot \begin{pmatrix} 1_g & 0_g \\ 0_g & D \end{pmatrix}^{-1}$$

belongs to $\mathrm{Sp}(2g, \mathbb{Q})$ and leaves invariant the lattice Λ_D in $\mathbb{R}^{2g}$ with period matrix $\begin{pmatrix} 1_g & 0_g \\ 0_g & D \end{pmatrix}$. It is easy to see that any such matrix arises in this way from a matrix from $\mathrm{Sp}(2g, \mathbb{Z})_D$. Denote the group of such matrices by Γ_D. We see that

$$\Gamma_D = \begin{pmatrix} 1_g & 0_g \\ 0_g & D \end{pmatrix} \cdot \mathrm{Sp}(2g, \mathbb{Z})_D \cdot \begin{pmatrix} 1_g & 0_g \\ 0_g & D \end{pmatrix}^{-1}.$$

Assume that $P = (Z|D)$, $P' = (Z'|D)$ are normalized. The corresponding tori are isomorphic as polarized abelian varieties if and only if there exists a matrix $M \in \mathrm{Sp}(2g, \mathbb{Z})_D$ and a matrix $X \in \mathrm{GL}(g, \mathbb{C})$ such that

$$X \cdot (Z'|D) = (Z|D) \cdot M. \tag{5.6}$$

Write M in the form

$$M = \begin{pmatrix} 1_g & 0_g \\ 0_g & D^{-1} \end{pmatrix} \cdot N \cdot \begin{pmatrix} 1_g & 0_g \\ 0_g & D \end{pmatrix}, \tag{5.7}$$

where

$$N = \begin{pmatrix} N_1 & N_2 \\ N_3 & N_4 \end{pmatrix} \in \Gamma_g(D),$$

and the blocks are all of size g. Then (5.6) is equivalent to

$$X = Z \cdot N_3 + N_4, \ X \cdot Z' = Z \cdot N_1 + N_2.$$

This shows that X is determined by N and

$$Z' = (Z \cdot N_3 + N_4)^{-1} \cdot (Z \cdot N_1 + N_2). \tag{5.8}$$

Transposing these matrices and using that ${}^tZ = Z$ because $Z \in \mathcal{Z}_g$, we obtain

$$Z' = ({}^tN_1 \cdot Z + {}^tN_2) \cdot ({}^tN_3 \cdot Z + {}^tN_4)^{-1}. \tag{5.9}$$

The group $\mathrm{Sp}(2g, \mathbb{Q})$ is invariant with respect to the transpose operation, let ${}^t\Gamma_D$ be its subgroup of matrices ${}^tM$, where $M \in \Gamma_D$. Then Γ_D acts on $\mathcal{Z}_g$ by the formula (5.8) or (5.9) and ${}^t\Gamma_D$ acts on $\mathcal{Z}_g$ by the restriction of the action (4.12) of $\mathrm{Sp}(2g, \mathbb{R})$ on $\mathcal{Z}_g$.

Summarizing, we obtain the following.

Theorem 5.1. *The natural map assigning to a polarized abelian variety its normalized period matrix defines a bijection between isomorphism classes of abelian varieties with polarization of type $D = (d_1, \ldots, d_g)$ and the orbit space*

$$\mathcal{A}_D = \mathcal{Z}_g / \Gamma_D.$$

In fact one can show more, the coarse moduli space of abelian varieties with polarization of type D exists and is isomorphic to $\mathcal{Z}_g / \Gamma_D$.

6. Picard and Albanese varieties

Let X be a compact Kähler manifold. It defines a Hodge structure of weight 1

$$V_{\mathbb{C}} = H^1(X, \mathbb{C}) = H^{10}(X) \oplus H^{01}(X)$$

on the real space $V = H^1(X, \mathbb{R})$ of dimension $b_1(X) = 2q$. The map

$$\alpha : H_1(X, \mathbb{Z}) \to H^{01}(X)^*, \quad \alpha(\gamma)(\omega) = \int_\gamma \omega,$$

defines an isomorphism from $H_1(X, \mathbb{Z})/\mathrm{Tors}$ onto a lattice Λ in $H^{10}(X)^* = H^0(X, \Omega^1_X)^*$. The complex torus

$$\mathrm{Alb}(X) = H^{10}(X)^* / \Lambda$$

is called the *Albanese torus* of X. Its period matrix with respect to a basis $\omega_1, \ldots, \omega_q$ in $H^{10}(X)$ and a basis $(\alpha(\gamma_1), \ldots, \alpha(\gamma_{2q}))$ of Λ is the matrix

$$P = (\int_{\gamma_j} \omega_i). \tag{6.1}$$

Fixing a point $x_0 \in X$ defines a natural map (the *Albanese map*)

$$\mathrm{alb}_{x_0} : X \to \mathrm{Alb}(X), \quad x \to (\int_{x_0}^x \omega_1, \ldots, \int_{x_0}^x \omega_g) \bmod \Lambda,$$

where the integration is taken with respect to any 1-chain originated from x_0 and ending at x. Since the difference of two such chains is a 1-cycle, the point in the torus does not depend on a choice of a path. The Albanese map satisfies the following universal property: for any holomorphic map $\phi : X \to T$ to a complex torus T there exists a unique (up to translation) map of complex tori $f : \mathrm{Alb}(X) \to T$ such that $f \circ \mathrm{alb} = \phi$.

One can define another complex torus associated to X. Consider the projection Λ' of $H^1(X,\mathbb{Z}) \subset V$ to $H^{01}(X)$. It is a lattice in the complex space $H^{01}(X)$. The complex torus

$$\mathrm{Pic}^0(X) = H^{01}(X)/\Lambda' \tag{6.2}$$

is called the *Picard torus* of X. Recall that $H^{01}(X) \cong H^1(X, \mathcal{O}_X)$, where $\mathcal{O}_X$ is the sheaf of holomorphic functions on X. The exponential exact sequence

$$0 \to \mathbb{Z} \to \mathbb{C} \overset{z \mapsto e^{2\pi i z}}{\longrightarrow} \mathbb{C}^* \to 0$$

defines an exact sequence of sheaves

$$0 \to \mathbb{Z} \to \mathcal{O}_X \overset{\exp}{\longrightarrow} \mathcal{O}_X^* \to 0$$

and the corresponding exact cohomology sequence

$$H^1(X,\mathbb{Z}) \to H^1(X,\mathcal{O}) \overset{c}{\to} H^2(X,\mathbb{Z}).$$

One can show that the first map coincides with the composition of the maps $H^1(X,\mathbb{Z}) \to H^1(X,\mathbb{C}) \to H^{01}(X)$ so that

$$\mathrm{Pic}^0(X) \cong \mathrm{Ker}(H^1(X,\mathcal{O}_X^*) \overset{c}{\to} H^2(X,\mathbb{Z})).$$

The group $H^1(X,\mathcal{O}_X^*)$ is the *Picard group* of X of isomorphism classes of holomorphic line bundles over X. The coboundary map c is interpreted as taking the first Chern class of a line bundle. Thus the Picard variety is a complex torus whose set of points is naturally bijective to the set of isomorphism classes of holomorphic line bundles L with trivial first Chern class $c_1(L)$.

Of course, one can also consider the complex tori $\mathrm{Alb}(X)'$ and $\mathrm{Pic}^0(X)'$, interchanging $H^{10}(X)$ with H^{01}. This just replaces the complex structure of the torus to the conjugate one.

The complex tori $\mathrm{Alb}(X)$ and $\mathrm{Pic}^0(X)$ are examples of *dual complex tori* in the sense of the following definition. Let $T = V/\Lambda$ be a torus and $W = (V, I)$ be a complex structure on V. Recall that we identify W with the complex subspace of $V_\mathbb{C}$ of vectors $v - iI(v)$. For any $\mathbb{R}$-linear function $\phi \in V^*$ and $w = v + iI(v) \in \overline{W}$ set

$$\tilde{\phi}(w) = \phi(I(v)) + i\phi(v).$$

It is immediately checked that $\tilde{\phi}$ is a $\mathbb{C}$-linear function on $\overline{W}$ with imaginary part equal to ϕ (when we identify $\overline{W}$ with $(V, -I)$). In this way we identify the real spaces V^* and $\overline{W}^*$ and hence consider $\overline{W}$ as a complex structure on V^*. Let

$$\Lambda^* = \{f \in \overline{W}^* : \mathrm{Im}(f)(\Lambda) \subset \mathbb{Z}\}.$$

The dual complex torus is

$$T^* := \overline{W}^* / \Lambda^*.$$

In our case, $H^{10}(X)^* = \overline{H^{01}(X)}^*$. Since $H^1(X,\mathbb{Z}) = \{\phi \in H^1(X,\mathbb{R}) : \phi(H_1(X,\mathbb{Z})) \subset \mathbb{Z}\}$, we see that $\Lambda' = \Lambda^*$.

The dual complex tori correspond to the dual Hodge structure on V^*

$$V^*_{\mathbb{C}} = (V^*)^{10} \oplus (V^*)^{01},$$

where

$$(V^*)^{10} = (V^{01})^* = (V_{\mathbb{C}}/V^{10})^* = (V^{10})^{\perp},$$
$$(V^*)^{01} = (V^{10})^* = (V_{\mathbb{C}}/V^{01})^* = (V^{01})^{\perp}.$$

Now let Q be a polarization of the Hodge structure on V with hermitian form H. It defines a dual symplectic form Q^* on V^* by viewing Q as a bijective map $V \to V^*$ and setting $Q^* = -Q^{-1} : V^* \to V$. Since the subspaces V^{10}, V^{01} are isotropic with respect to Q, $Q(V^{10}) = (V^{10})^{\perp} = (V^{01})^*$, $Q(V^{01}) = (V^{10})^*$. This shows that $(V^{10})^*$ and $(V^{01})^*$ are isotropic with respect to Q^*. Let V^{10} define a complex structure I on V. Then the complex structure on $(V^{10})^*$ is defined by the operator I^* which is adjoint to I with respect to the natural pairing between V and V^*. The symmetric bilinear form $Q(v, I(v'))$ can be considered as the composition of the map $V \to V^*$ and $I^* : V^* \to V^*$. The symmetric form $Q^*(I^*(\alpha), \beta)$ defines the inverse map. This shows that I is positive with respect to Q if and only if Q^* is positive with respect to I^*. Thus Q^* defines a polarization of the dual of a polarized AHS of weight 1. It is called the *dual polarization.*

An example of dual Hodge structures is a Hodge structure on odd-dimensional cohomology $H^k(X,\mathbb{R})$ and on $H^{2n-k}(X,\mathbb{R})$, where X is a Kähler manifold of dimension n. The duality is defined by the Poincaré duality. It is also a duality of polarized Hodge structures.

It is clear that a polarized Hodge structure on $H^1(X,\mathbb{R})$ defines a polarization on $\mathrm{Pic}^0(X)$ and the dual polarization on $\mathrm{Alb}(X)$. An integral polarization defined by a Hodge class makes these tori polarized abelian varieties.

Suppose Λ is an integral structure of Q. Choose a basis (v_i) in Λ such that the matrix of Q is equal to a matrix J_D for some D. Let (v_i^*) be the dual basis in V^*. The map $Q : V \to V^*$ sends v_i to $d_i v_{i+g}^*$ and v_{i+g} to $-d_i v_i^*$ for $i = 1, \ldots, g$. Thus the matrix of Q^* with respect to the basis $v_{1+g}^*, \ldots, v_{2g}^*, v_1^*, \ldots, v_g^*$ is equal to J_D^{-1}. Multiplying Q^* by d_g we get an integral structure of the dual AHS of type $D^* = (1, d_g/d_{g-1}, \ldots, d/d_1)$ with respect to the dual lattice Λ^*.

In particular, we see that the dual of a polarized abelian variety of type D is a polarized abelian variety of type D^*.

Now let Q be a polarization on T with the integral structure defined by Λ. Let H be the associated hermitian form on $V_{\mathbb{C}}$ and h be its restriction to W. It defines a $\mathbb{C}$-linear map $h : W \to \overline{W}^*, x \mapsto H(x,)$. Since $\mathrm{Im}(h)(x,y) = Q(x,y)$, we see that $h(\Lambda) \subset \Lambda^*$. Thus we have a holomorphic map of complex tori

$$h : T \to T^* = \mathrm{Pic}^0(T).$$

Obviously the map is surjective, and its kernel is isomorphic to the finite abelian group of order $\det J_D = (d_1 \cdots d_g)^2$ isomorphic to the cokernel of the map $Q : \Lambda \to \Lambda^*$. In particular, h is an isomorphism if $D = 1_g$.

Let P be the period matrix of T with respect to the chosen bases of Λ and W. It defines a map $V_{\mathbb{C}} = \Lambda_{\mathbb{C}} \to W$, the direct sum of the transpose map $W^* \to V_C^*$ and its conjugate map $\overline{W}^* \to V_C^*$ define the dual Hodge structure. Thus the matrix ${}^t\bar{P}$ is the coperiod matrix Π^* of the dual torus T^* with respect to the bases (v_i^*) and $(\bar{w}_i)$. The period matrix P^* of T^* is reconstructed from the equation

$$\begin{pmatrix} P^* \\ \bar{P}^* \end{pmatrix} = (\Pi^* | \bar{\Pi}^*)^{-1} = ({}^t\bar{P} | {}^tP)^{-1}.$$

Example. Let $T = W/\Lambda$ be a complex torus. The space $T \times W$ is identified with the trivial holomorphic tangent bundle of T. Thus the dual space W^* is the space of holomorphic differentials $H^{10}(X)$. Thus

$$T \cong \mathrm{Alb}(T).$$

Hence

$$\mathrm{Pic}^0(T) = T^*.$$

Also we see that a polarization on T defines the dual polarization of $\mathrm{Pic}^0(T)$. In particular, if T is an abelian variety with a principal polarization, then $T \cong \mathrm{Pic}^0(T)$ as polarized abelian varieties.

So far we have defined complex tori associated to a polarized AHS of weight 1 with integral structure $\Lambda \subset V$. One can associate complex tori to an AHS of any odd weight $k = 2m + 1$. In fact, set

$$V^{10} = \bigoplus_{s=0}^{m} H^{m+1-s,m-s}(X), \quad V^{01} = \bigoplus_{s=0}^{m} H^{s,k-s}(X), \tag{6.3}$$

then $V_{\mathbb{C}} = V^{10} \oplus V^{01}$ is an AHS of weight 1 which is polarized with respect to the polarization form Q of the original AHS. This defines two dual complex tori

$$J = V^{10}/\Lambda, \ \check{J} = (V^{01})^*/\Lambda^*$$

where Λ is the projection of Λ to V^{10}. The torus J is called the *Griffiths complex torus* associated to AHS.

In the case when the Hodge structure is the Hodge structure on odd-dimensional cohomology $H^k(X, \mathbb{R})$ of a n-dimensional Kähler variety with integral structure defined by $\Lambda = H^k(X, \mathbb{Z})$ we obtain the definition of the *kth Griffiths intermediate Jacobian* $J = J^k(X)$. We have

$$\check{J}^k(X) \cong J^{2n-k}(X).$$

In particular, they coincide when $k = n = 2m+1$. In this case $J^n(X)$ is called the *intermediate Jacobian* of X. Clearly,

$$J^1(X) = \mathrm{Pic}^0(X), \quad J^{n-1}(X) = \mathrm{Alb}(X).$$

Note that, in general, Q does not define a polarization of the AHS (6.3) since the hermitian form $iQ(x, \bar{x})$ is not positive on V^{10}; it changes signs on the direct summands. Thus the Griffiths tori are not abelian varieties in general. However, in one special case they are. This is the case when all summands except one in V^{10} are equal to zero. We call such AHS an *exceptional AHS.* In this case J is a polarized abelian variety. It is a principally polarized variety if Q is unimodular on Λ. For example, the intermediate Jacobian $J^n(X)$ of an odd-dimensional Kähler manifold with exceptional Hodge structure on the middle cohomology is a principally polarized abelian variety. For example, any Kähler 3-fold with $h^{30} = 0$ (it must be an algebraic variety in this case) defines a principally polarized abelian variety $J^3(X)$ of dimension h^{21}. Another example is a rigid Calabi–Yau 3-fold for which $h^{30} = 1, h^{21} = 0$.

Let T be a connected complex variety. A *family of polarized abelian varieties* with base T is a closed subvariety Y of $\mathbb{P}^N \times T$ such that each fibre X_t of the second projection $f : Y \to T$ is an abelian variety of dimension g. Under the natural map $X_t \to \mathbb{P}^N \times \{t\}$, each fibre is isomorphic to a closed projective subvariety of $\mathbb{P}^N$ and hence acquires a structure of a polarized abelian variety of some type D independent of t. The period map is defined and is a holomorphic map

$$\bar{\phi} : T \to \mathcal{Z}_g/\Gamma_D.$$

If we identify the orbit space with the coarse moduli space $\mathcal{A}_g(D)$ of abelian varieties with polarization of type D, then the map is $X_t \mapsto \mathrm{Pic}^0(X)$. The abelian variety X can be uniquely reconstructed from $\mathrm{Pic}^0(X)$ as the dual polarized abelian variety. This proves the Global Torelli Theorem for polarized abelian varieties. It gives an isomorphism

$$\mathrm{per} : \mathcal{A}_g(D) \to \mathcal{Z}_g/\Gamma_{D^*} \cong \mathcal{A}_{D^*}.$$

Example. Let M be a compact smooth oriented 2-manifold. Then $b_1(M) = 2g$, where g is the genus of M. We choose a complex structure on M which makes it a compact Riemann surface X, or a projective nonsingular curve. It is known that the group $H_1(M, \mathbb{Z})$ is a free abelian group of rank $2g$ which admits a standard symplectic basis $\gamma_1, \ldots, \gamma_{2g}$ with respect to the cup-product

$$Q : H^1(M, \mathbb{Z}) \times H^1(M, \mathbb{Z}) \to H^2(M, \mathbb{Z}) \cong \mathbb{Z}.$$

This defines an integral polarization of the Hodge structure corresponding the the Kähler class generating $H^2(X, \mathbb{Z}) = \mathbb{Z}$. It gives a principal polarization on $\mathrm{Pic}^0(X)$. Hence $\mathrm{Alb}(X) \cong \mathrm{Pic}(X)$ are isomorphic as principally polarized abelian varieties and coincide, by definition, with the *Jacobian variety* $\mathrm{Jac}(X)$ of X. We know that there exists a basis $\omega_1, \ldots, \omega_g$ in the space of holomorphic 1-forms $H^{10}(X)$ such that the period matrix of $\mathrm{Jac}(X)$

$$\Pi = (\pi_{ij}), \quad \pi_{ij} = \int_{\gamma_i} \omega_j$$

is normalized, i.e., has the form ${}^t(Z|I_g)$, where $Z \in \mathcal{Z}_g$.

A *marked Riemann surface* is a Riemann surface X together with a choice of a symplectic basis in $H^1(X,\mathbb{Z})$. An isomorphism of marked Riemann surfaces $(X,(\gamma_i)) \to (X',(\gamma_i'))$ is an isomorphism $f : X \to X'$ of complex varieties such that $f^*(\gamma_i') = \gamma_i$. The *Torelli Theorem* for Riemann surfaces asserts that two marked Riemann surfaces are isomorphic if and only if their normalized period matrices with respect to the symplectic matrices are equal. In other words they define the same point in the Siegel half-plane $\mathcal{Z}_g$. Two different markings on the same surface define the points in $\mathcal{Z}_g$ equivalent with respect to the action of the Siegel modular group $\Gamma_g = \mathrm{Sp}(2g,\mathbb{Z})$. Thus we see that two Riemann surfaces are isomorphic if and only if the corresponding Jacobian varieties are isomorphic as principal abelian varieties, i.e., define the same point in $\mathcal{A}_g = \mathcal{Z}_g/\Gamma_g$. The coarse moduli space $\mathcal{M}_g$ of Riemann surfaces of genus g exists and is an algebraic variety of dimension $3g-3$. The period map

$$\mathrm{per} : \mathcal{M}_g \to \mathcal{Z}_g/\Gamma_g \cong \mathcal{A}_g$$

is just the map $X \mapsto \mathrm{Jac}(X)$. Thus the Torelli Theorem for Riemann surfaces is the statement that two Riemann surfaces are isomorphic if and only if their Jacobian varieties are isomorphic as principally polarized abelian varieties. Clearly, this also establishes the Global Torelli Theorem for Hodge structures on one-dimensional cohomology of Riemann surfaces.

In case $g=1$, we have an isomorphism $\mathcal{A}_1 = \mathcal{H}/\mathrm{SL}(2,\mathbb{Z}) \cong \mathbb{C}$. It is given by the *absolute invariant* function $j(\tau)$. The Weierstrass function

$$\wp(z) = z^{-2} + \sum_{(m,n)\in\mathbb{Z}^2\setminus\{0\}} \left(\frac{1}{(z+m+n\tau)^2} - \frac{1}{(m+n\tau)^2}\right)$$

defines a holomorphic map

$$\mathbb{C}\setminus(\mathbb{Z}+\tau\mathbb{Z}) \to \mathbb{C}^2, \quad \mathbb{Z} \mapsto (\wp(z),\wp'(z))$$

which can be extended to an isomorphism from the elliptic curve $E_\tau = \mathbb{C}/\mathbb{Z}+\tau\mathbb{Z}$ to the plane cubic curve in $\mathbb{P}^2$ defined by

$$x_2^2x_0 - 4x_1^3 + g_2x_1x_0^2 + g_3x_0^3 = 0.$$

We have

$$j(\tau) = 1728\frac{g_2^3}{4g_2^3+27g_3^2}$$

and $E_\tau \cong E_{\tau'}$ if and only if $j(\tau) = j(\tau')$.

7. Eigenperiods

Let $V_\mathbb{C} = \oplus V^{pq}$ be a polarized AHS of weight k on a vector space (V,Q). A *Hodge isometry* of $V_\mathbb{C}$ is a $\mathbb{R}$-linear automorphism of (V,Q) such that its linear extension to $V_\mathbb{C}$ preserves the Hodge decomposition. Let A be a finite abelian group acting on (V, Q) by Hodge isometries via a linear representation $\rho : A \to \mathrm{Aut}(V, Q)$. The vector space $V_{\mathbb{C}}$ splits into a direct sum of eigensubspaces

$$V_{\mathbb{C}}(\chi) = \{x \in V_{\mathbb{C}} : g(x) = \chi(g)x, \forall g \in A\},$$

corresponding to different characters $\chi \in \hat{A} := \mathrm{Hom}(A, \mathbb{C}^*)$. The reality condition of the representation ρ on $V_{\mathbb{C}}$ is equivalent to the condition

$$\overline{V_{\chi}} = V_{\bar{\chi}}. \tag{7.1}$$

We have a decomposition

$$V_{\mathbb{C}}(\chi) = \bigoplus_{p+q=k} V_{\chi}^{pq}, \quad V_{\chi}^{pq} = V_{\mathbb{C}}(\chi) \cap V^{pq}. \tag{7.2}$$

It is not a Hodge decomposition in general because

$$\overline{V_{\chi}^{pq}} = V_{\bar{\chi}}^{qp}$$

and also because the complex vector space $V_{\mathbb{C}}(\chi)$ does not have a natural identification with a complexification of a real vector space. However it satisfies properties (HD2) and (HD3) of a polarized AHS of weight k. Since, for any $x \in V_{\mathbb{C}}(\chi)$, $y \in V_{\mathbb{C}}(\chi')$,

$$Q(x, y) = Q(g(x), g(y)) = Q(\chi(g)x, \chi'(g)y) = \chi(g)\chi'(g)Q(x, y), \tag{7.3}$$

we obtain that the eigensubspaces $V_{\mathbb{C}}(\chi)$ and $V_{\mathbb{C}}(\chi')$ are orthogonal if $\chi \neq \overline{\chi}'$.

Denote by (F_{χ}^p) the flag of subspaces

$$F_{\chi}^p = \oplus_{p' \geq p} V_{\chi}^{p'q}.$$

It satisfies the properties

(i) $F_{\chi}^p \cap \overline{F_{\chi}^{k-p+1}} = \{0\}$;

(ii) $(-1)^{\frac{k(k-1)}{2}} Q(Cx, \bar{x}) > 0$, where C acts on V_{χ}^{pq} as multiplication by i^{p-q}.

Note that similar to (2.8) the spaces V_{χ}^{pq} can be reconstructed from the flag (F_{χ}^p)

$$V_{\chi}^{pq} = \{x \in F_{\chi}^p : Q(x, \bar{y}) = 0, \ \forall y \in F_{\chi}^{p+1}\}. \tag{7.4}$$

Let

$$f_p(\chi) = \dim F_{\chi}^p, \quad \mathbf{f}(\chi) = (f_0(\chi), \ldots, f_k(\chi)).$$

Let $H(x, y) = Q(x, \bar{y})$ be the hermitian form on $V_{\mathbb{C}}(\chi)$. If χ is real, i.e., $\chi(g) = \pm 1$, we define $\mathrm{Fl}(\mathbf{f}(\chi), V_{\mathbb{C}}(\chi))^0$ to be the open subset of the flag variety $\mathrm{Fl}(\mathbf{f}(\chi), V_{\mathbb{C}}(\chi))$ which consists of flags (F^p) satisfying conditions (i) and (ii). If χ is real, we define $\mathrm{Fl}(\mathbf{f}(\chi), V_{\mathbb{C}}(\chi))^0$ to be the period space $\mathcal{D}_{\mathbf{f}(\chi)}(V(\chi), Q|V(\chi))$. Note that in this case the decomposition (7.2) is a polarized Hodge structure on the real eigenspace $V(\chi)$.

Let $\mathrm{Wt}(\rho)$ denote the subset of $\chi \in \hat{A}$ such that $(V_{\mathbb{C}})(\chi) \neq 0$. Let $\mathrm{Wt}_0(\rho) \subset \hat{A}$ be the subset of real characters. Then $\mathrm{Wt}(\rho)$ can be written as a disjoint union of three subsets

$$\mathrm{Wt}(\rho) = \mathrm{Wt}_+(\rho) \coprod \mathrm{Wt}_-(\rho) \coprod \mathrm{Wt}_0(\rho), \tag{7.5}$$

such that $\overline{\mathrm{Wt}_\pm(\rho)} = \mathrm{Wt}_\mp(\rho)$. The conjugation map of $V_\mathbb{C}$ sends the set $\mathrm{Fl}(\mathbf{f}(\chi), V_\mathbb{C}(\chi))$ to the set $\mathrm{Fl}(\mathbf{f}(\bar\chi), V_\mathbb{C}(\bar\chi))$. The group A acts naturally on the period space $\mathcal{D}_\mathbf{f}(V, Q)$ via its representation ρ. Let

$$\mathcal{D}_\mathbf{f}(V, Q)^\rho = \{x \in \mathcal{D}_\mathbf{f}(V, Q) : \rho(a)(x) = x, \forall a \in A\}.$$

We have a natural map

$$\Phi : \mathcal{D}_\mathbf{f}(V, Q)^\rho \longrightarrow \prod_{\chi \in \mathrm{Wt}_+(\rho) \cup \mathrm{Wt}_0(\rho)} \mathrm{Fl}(\mathbf{f}(\chi), V_\mathbb{C}(\chi))^0. \tag{7.6}$$

It is obviously injective since one can reconstruct the Hodge structure from the image using (7.4) applied to F^p_χ and $\overline{F^p_{\bar\chi}}$.

For any Kähler manifold X together with a Kähler class $[\omega]$ let $\mathrm{Aut}(X, [\omega])$ denote the group of holomorphic automorphisms leaving $[\omega]$ invariant. Fix a real vector space V together with a non-degenerate bilinear form Q_0, symmetric (resp. skew-symmetric) if k is even (resp. odd) and a linear representation $\rho_0 : A \to \mathrm{GL}(V, Q_0)$. A *$\rho$-marking* of a polarized Kähler manifold X on which A acts via a homomorphism $\sigma : A \to \mathrm{Aut}(X, [\omega])$ is an isomorphism

$$\phi : (V, Q_0) \to (H^k(X, \mathbb{R}), Q)$$

such that, for any $g \in A$, $\rho \circ \rho_0(g) \circ \phi^{-1} = \sigma^*(g)$.

Let $f : Y \to T$ be a family of polarized Kähler manifolds. Suppose that the group A acts holomorphically on Y preserving the Kähler form $[\omega]$ on Y which defines the polarization form $[\omega]_t$ on each fibre X_t. Also we assume that $g(X_t) = X_t$ for all $t \in T$. Let $V = H^k(X_{t_0}, \mathbb{R})$ with polarization form Q_{t_0}. Fix an isomorphism $(H^k(X_{t_0}, \mathbb{R}), Q_{t_0}) \to (V, Q)$ and define $\rho : A \to \mathrm{Aut}(V, Q)$ as the representation $g \mapsto g^*$ of A on $H^k(X_{t_0}, \mathbb{R})$. One can show that the monodromy representation $\pi(T, t_0) \to \mathrm{Aut}(V, Q)$ commutes with the representation ρ. Let $\Gamma_\rho(f)$ be the centralizer of the group $\rho(A)$ in $\mathrm{Aut}(V, Q)$. Since $\Gamma_\rho(f)$ preserves a lattice in V, the image of $H^k(X, \mathbb{Z})$, the group $\Gamma_\rho(f)$ is a discrete subgroup of $\mathrm{Aut}(V, Q)$. So the period map

$$\phi : T \to \Gamma_\rho(f) \backslash \mathcal{D}_\mathbf{f}(V, Q_0)^\rho \tag{7.7}$$

is a holomorphic map of complex varieties. Consider the composition of the multi-valued period map $T \to \mathcal{D}_\mathbf{f}(V, Q)^\rho$ with the projection to $\mathrm{Fl}(\mathbf{f}(\chi), V_\mathbb{C}(\chi))^0$. The group $\Gamma_\rho(f)$ leaves each subspace $V_\mathbb{C}(\chi)$ invariant and preserves the flags (F^p_χ). Since, in general, it is not a discrete subgroup of $\mathrm{Aut}(V_\mathbb{C}(\chi), Q)$, only the multi-valued period map is defined as a holomorphic map. By passing to the universal cover $\tilde T$ we have a one-valued holomorphic map

$$\phi_\lambda : \tilde T \to \mathrm{Fl}(\mathbf{f}(\chi), V_\mathbb{C}(\chi))^0. \tag{7.8}$$

We call this map the *eigenperiod map* of the family f.

Consider the case when $k = 1$. A Hodge isometry of V is an automorphism of V which is an automorphism of (V, I), where I is a positive complex structure defined by a Q-polarized Hodge structure on V. Conversely any complex automorphism of $V_{\mathbb{C}}$ preserving the Hodge decomposition and Q arises from a $\mathbb{R}$-linear automorphism of V by extension of scalars.

Given a representation $A \to \mathrm{Sp}(V,Q) := \mathrm{Aut}(V,Q)$ we want to describe the set of positive complex structures I on V such that any $g \in A$ commutes with I. In other words, the group A acts naturally on the Siegel half-plane $G(g, V_{\mathbb{C}})_H$ via its action on $V_{\mathbb{C}}$ and we want to describe the subset

$$G(g, V_{\mathbb{C}})_H^{\rho} = \{W : g(W) \subset W, \forall g \in A\}$$

of fixed points of A.

Let $W_\chi = V_\chi \cap W$. Then

$$V_{\mathbb{C}}(\chi) = W_\chi \oplus \overline{W}_\chi.$$

We know that $H|W_\chi > 0$ and $H|\overline{W}_\chi < 0$. Also W and $\overline{W}$ are orthogonal with respect to H. Therefore the signature of $H|V_{\mathbb{C}}(\chi)$ is equal to (p_λ, q_λ), where $p_\lambda = \dim W_\chi, q = \dim \overline{W}_\chi$.

For any hermitian complex vector space (E,h) of signature (p,q) let us denote by $G(p,(E,h))$ the open subset of the Grassmannian $G(p,E)$ which consists of p-dimensional subspaces L such that $h|L > 0$. It is known that E admits a basis $e_1, \ldots, e_{p+q}$ such that the matrix of E is equal to the matrix $I(p,q)$ (2.9). This shows that the group $U(E,h)$ of isometries of (E,h) is isomorphic to the group

$$U(p,q) = \{A \in \mathrm{GL}(p+q,\mathbb{C}) : {}^tA \cdot I(p,q) \cdot \bar{A} = I(p,q)\}. \tag{7.9}$$

By Witt's theorem, the group $U(E,h)$ acts transitively on $G(p,(E,h))$ with the isotropy subgroup of L equal to $U(L,h|L) \times U(L^\perp, h|L^\perp)$. Thus

$$G(p,(E,h)) \cong U(p,q)/U(p) \times U(q).$$

It has a structure of a hermitian symmetric space of non-compact type I in Cartan's classification.

Identifying E with $\mathbb{C}^g$ and h with the matrix $I(p,q)$, we identify a subspace from $G(p,(E,h))$ with the column space of a matrix Π of size $(p+q)\times p$. Writing the matrix in the form $\Pi = \begin{pmatrix} A \\ B \end{pmatrix}$, where A is a square matrix of size p, the positivity condition is expressed by the condition

$$\begin{pmatrix} {}^tA & {}^tB \end{pmatrix} \cdot \begin{pmatrix} I_p & 0_{pq} \\ 0_{qp} & -I_q \end{pmatrix} \cdot \begin{pmatrix} \bar{A} \\ \bar{B} \end{pmatrix} = {}^tA \cdot \bar{A} - {}^tB \cdot \bar{B} > 0.$$

One can show that $|A| \neq 0$ so, replacing Π with $\Pi \cdot A^{-1}$ we may assume that Π is of the form $\begin{pmatrix} I_p \\ Z \end{pmatrix}$ for some complex $(q \times p)$-matrix. The condition on Z is

$$I_p - {}^tZ \cdot \bar{Z} > 0. \tag{7.10}$$

This shows that $G(p,(E,h))$ is isomorphic to a bounded domain $\mathcal{I}_{pq}$ in $\mathbb{C}^{pq}$ defined by the previous inequality. Any hermitian symmetric space of non-compact type I is isomorphic to such a domain.

Taking $p = 1$ we see that $\mathcal{I}_{1q}$ is isomorphic as a complex manifold to the unit complex ball of dimension q

$$\mathbb{B}_q = \{z \in \mathbb{C}^q : |z_1|^2 + \ldots + |z_q|^2 < 1\}. \tag{7.11}$$

In particular, we we have

$$U(1, q)/U(1) \times U(q) \cong \mathbb{B}_q.$$

There is a natural embedding of $\mathcal{I}_{pq}$ into the Siegel half-plane $\mathcal{Z}_{p+q}$ for which we choose a boundary model from (7.10). It is given by assigning to a matrix Z of size $p \times q$ satisfying (4.14) the symmetric square matrix of size $p + q$

$$Z' = \begin{pmatrix} 0_p & {}^tZ \\ Z & 0_q \end{pmatrix} \tag{7.12}$$

satisfying (7.10). Similarly, we have an embedding

$$\prod_{i=1}^{k} \mathcal{I}_{p_i q_i} \hookrightarrow \mathcal{Z}_g, \tag{7.13}$$

where $g = p_1 + \ldots + p_k = q_1 + \ldots + q_k$. Also we see that the Siegel half-plane $\mathcal{Z}_g$ is isomorphic to a closed subvariety of $\mathcal{I}_{gg}$ defined by the equation ${}^tZ = Z$.

Fix a partition of $\mathrm{Wt}(\rho)$. Let $H_\chi = H|V_{\mathbb{C}}(\chi)$. For any $\chi \in \mathrm{Wt}(\rho)^+$, let (p_χ, q_χ) be the signature of H_χ. By (7.1), the signature of $H_{\bar{\chi}} = (q_\chi, p_\chi)$. The set of pairs of numbers (p_χ, q_χ), well-defined up to permutation, is called the *type of the representation* ρ.

We construct the inverse of the map (7.6)

$$\prod_{\chi \in \mathrm{Wt}(\rho)^+} G(p_\chi, (V_{\mathbb{C}}(\chi), H_\chi)) \to G(g, V_{\mathbb{C}})^\rho_H. \tag{7.14}$$

It is defined by assigning to a collection of subspaces (E_χ) from the left-hand side the direct sum

$$W = \bigoplus_{\chi \in \mathrm{Wt}(\rho)^+} E_\chi \oplus \overline{(E_\chi)^\perp_{H_\chi}}.$$

It is clear that $\overline{(E_\chi)^\perp_{H_\chi}} \subset V_{\bar{\chi}}$ and the restriction of H to this subspace is positive. Thus $H|W > 0$ and

$$\dim W = \sum_{\chi \in \mathrm{Wt}(\rho)^+} p_\chi + \sum_{\chi \in \mathrm{Wt}(\rho)^-} q_\chi = g.$$

Also, using (7.3) and the fact that $H(E_\chi, (E_\chi)^\perp_{H_\chi}) = 0$ implies $Q(E_\chi, \overline{(E_\chi)^\perp_{H_\chi}}) = 0$, we see that $Q|W = 0$. Thus $W \in G(g, V_{\mathbb{C}})_H$. As a sum of eigensubspaces it is obviously ρ-invariant, so the map is well defined and is easy to see coincides with the inverse of the map (7.6).

Thus for any linear representation $\rho : A \to \mathrm{Sp}(V, Q)$ with $\mathrm{Wt}(V_{\mathbb{C}}) = \{\chi_i, \bar{\chi}_i, i = 1, \ldots, k\}$ and $(p_i, q_i) = (p_{\chi_i}, q_{\chi_i})$, then we obtain an embedding of hermitian symmetric domains

$$\prod_{i=1}^{k} \mathcal{I}_{p_i,q_i} \hookrightarrow \mathcal{Z}_g.$$

Example. Let $A = (g)$ be a cyclic group of order 4. Its generator g satisfies $g^2 = -1$. Therefore its action on V is equivalent to a complex structure I on V and hence defines a subspace $L \subset V_{\mathbb{C}}$ on which g acts with eigenvalue i. Thus

$$L = (V_{\mathbb{C}})_{\chi}, \quad \bar{L} = (V_{\mathbb{C}})_{\bar{\chi}},$$

where $\chi(g) = i, \bar{\chi}(g) = -i$. Let Q be a symplectic form on V such that $Q(v, I(v'))$ is a symmetric form of signature (p, q) and $H = iQ(x, \bar{x})$ be the associated hermitian form. Its restriction h to L is of signature (p, q). Let E be a positive subspace of L of dimension p. Let $E_h^{\perp}$ be its orthogonal complement in L with respect to h. The restriction of h to $E_h^{\perp}$ is negative. Its conjugate subspace $E' = \overline{E_h^{\perp}}$ belongs to $\overline{W}$. Since $H(x, \bar{x}) = -H(\bar{x}, x)$, the restriction of H to E' is positive. Thus $W = E \oplus E'$ defines a positive complex structure on V with respect to Q. The operator I acts on a p-dimensional subspace E of W as $i1_E$ and acts on the orthogonal q-dimensional subspace E' as $-i1_{E'}$. The decomposition $V_{\mathbb{C}} = W \oplus \overline{W}$ is a Q-polarized Hodge structure invariant with respect to the representation of A in V.

In coordinates, choose a standard symplectic basis e_i in V for a symplectic form Q, and define I by the formula

$$I(e_k) = e_{k+g}, \quad k = 1, \ldots, p, \quad I(e_{p+k}) = -e_{p+g+k}, \; k = 1, \ldots, q.$$

Then $Q(v, I(v'))$ is a symmetric quadratic form of signature (p, q). Then I preserves the lattice in V spanned by the vectors v_i and acts on W preserving the projection Λ of this lattice in W. Thus g acts an automorphism of the principally polarized abelian variety $A = W/\Lambda$. This automorphism is of type (p, q), i.e., W decomposes into a direct sum of eigensubspaces of dimension p and q.

We can find a basis in $V_{\mathbb{C}}$ which is a sum of a basis in L and $\bar{L}$ such that $W = E \oplus E'$ is represented by a matrix

$$\begin{pmatrix} I_p & 0_{pq} \\ Z & 0_q \\ 0_q & I_q \\ 0_{qp} & {}^tZ \end{pmatrix}$$

where $Z \in \mathcal{I}_{pq}$. After an obvious change of a basis this matrix becomes a matrix (7.12) defining the embedding of $\mathcal{I}_{pq}$ in $\mathcal{Z}_g$.

This shows that the isomorphism classes of principally polarized abelian varieties of dimension g admitting an automorphism of order 4 of type (p, q) is isomorphic to the space of orbits

$$\Gamma_{pq} \backslash \mathcal{I}_{pq} \subset \mathcal{A}_g = \mathrm{Sp}(2g, \mathbb{Z}) \backslash \mathcal{Z}_g,$$

where Γ_{pq} is the subgroup of matrices $M \in \mathrm{Sp}(2g,\mathbb{Z})$ which commute with the matrix $\begin{pmatrix} 0_g & I_{p,q} \\ -I_{p,q} & 0_g \end{pmatrix}$ defining the operator I.

8. Arrangements of hyperplanes

Let $H_1,\dots,H_m$ be hyperplanes in $\mathbb{P}^n(\mathbb{C})$ defined by linear forms

$$f_i(t_0,\dots,t_n) = \sum_{i=0}^{n} a_{ij}t_j, \quad i = 1,\dots,m.$$

We assume that the hyperplanes are in a general position. This means that all maximal minors of the matrix

$$M = (a_{ij}) \tag{8.1}$$

are not equal to zero. Geometrically this means that the intersection of any $n+1$ hyperplanes is a point. Let

$$U = \mathbb{P}^n \setminus \bigcup_{i=1}^{m} H_i$$

be the complementary set. Choose a set of rational numbers

$$\boldsymbol{\mu} = (\mu_1,\dots,\mu_m),$$

satisfying

$$0 < \mu_i < 1; \tag{8.2}$$

$$|\boldsymbol{\mu}| := \sum_{i=1}^{m} \mu_i \in \mathbb{Z}. \tag{8.3}$$

It is well-known that the fundamental group $\pi_1(U,u_0)$ is a free group generated by the homotopy classes g_i of loops which are defined as follows. Choose a line ℓ in $\mathbb{P}^n$ which does not intersect any codimension 2 subspace $H_i \cap H_j$ and contains u_0. Let γ_i be a path in ℓ which connects u_0 with a point on a small circle around the point $p_i = \ell \cap H_i$, goes around the circle in a counterclockwise way, making the full circle, and then returns in the opposite direction to the point u_0. The homotopy classes $g_i = [\gamma_i]$ generate the group $\pi_1(U,u_0)$ with the defining equation $g_1\cdots g_m = 1$.

Let $\mathcal{L}_{\boldsymbol{\mu}}$ be the complex local coefficient system on U defined by the homomorphism

$$\chi : \pi_1(U,u_0) \to \mathbb{C}^*, \ g_i \mapsto e^{-2\pi i\mu_i}. \tag{8.4}$$

Let d be their common denominator of the numbers μ_i and

$$A_d(m) = (\mathbb{Z}/d)^m/\Delta(\mathbb{Z}/d),$$

where Δ denotes the diagonal map. Consider the finite cover U' of U corresponding to the homomorphism $\pi_1(U,u_0) \to A_d(m)$ defined by sending g_1 to the vector $e_1 = (1,0,\dots,0)$, and so on. Let X be the normalization of $\mathbb{P}^1$ in the field of rational functions on U'. This is a nonsingular algebraic variety which can be explicitly described as follows. Consider the map

$$f : \mathbb{P}^n \to \mathbb{P}^{m-1}, \quad (t_0, t_1, \ldots, t_n) \mapsto (f_1(t), \ldots, f_m(t)).$$

Its image is a linear n-dimensional subspace L of $\mathbb{P}^{m-1}$. Let $\phi : \mathbb{P}^{m-1} \to \mathbb{P}^{m-1}$ be the cover defined by

$$(z_0, \ldots, z_{m-1}) \to (z_0^d, \ldots, z_{m-1}^d).$$

Then X is isomorphic to the pre-image of L under the cover ϕ. If we write the linear forms $f_i, i > n+1$, as linear combinations $\alpha_{i0} f_1 + \ldots + \alpha_{in} f_{n+1}$, then X becomes isomorphic to the subvariety in $\mathbb{P}^{m-1}$ given by the equations

$$-x_i^d + \alpha_{i0} x_0^d + \ldots + \alpha_{in} x_n^d = 0, \quad i = n+1, \ldots, m.$$

The group $A_d(m)$ acts naturally on X via multiplication of the coordinates x_i by dth roots of unity. The quotient space X/A is isomorphic to $\mathbb{P}^n$. The group $A_d(m)$ acts on $V = H^n(X, \mathbb{R})$ and on its complexification $V_{\mathbb{C}} = H^n(X, \mathbb{C})$.

Lemma 8.1. *Let $\chi \in \hat{A}_d(m)$ be defined by (8.4). Then*

$$H^j(U, \mathcal{L}_{\boldsymbol{\mu}}) = 0, j \neq n, \quad H^n(U, \mathcal{L}_{\boldsymbol{\mu}}) \cong H^n(X, \mathbb{C})(\chi).$$

Proof. Let

$$U' = X \setminus \bigcup_{i=0}^{m-1} \{x_i = 0\}$$

and let $\pi : U' \to U$ be the natural projection. This is a unramified Galois covering with the Galois group $A_d(m)$. The direct image of the constant coefficient system $\mathbb{C}_{U'}$ in U decomposes into the direct sum of local coefficients systems $\mathcal{L}_\chi$ corresponding to different characters $\chi \in \hat{A}_d(m)$

$$\pi_*(\mathbb{C}_{U'}) = \bigoplus_{\chi \in \hat{A}} \mathcal{L}_\chi.$$

We have an isomorphism of cohomology with compact support

$$H_c^j(U, \pi_*(\mathbb{C}_{U'})) \cong H_c^j(U', \mathbb{C})$$

which is compatible with the action of A and gives an isomorphism

$$H_c^j(U', \mathbb{C})(\chi) \cong H_c^j(U, \mathcal{L}_{\boldsymbol{\mu}}).$$

One can show (see [DM], p. 20) that for any character χ determined by the numbers μ_i satisfying (8.2) and (8.3), $H_c^j(U, \mathcal{L}_\chi) \cong H^j(U, \mathcal{L}_\chi)$. Let $Y = X \setminus U'$. Now consider the long exact sequence

$$\ldots \to H_c^j(U', \mathbb{C}) \to H^j(X, \mathbb{C}) \to H^j(Y, \mathbb{C}) \to H_c^{j+1}(U, \mathbb{C}) \to \ldots.$$

Since U' is affine, it has homotopy type of a CW-complex of dimension n, and hence $H_c^j(U', \mathbb{C}) = 0$ for $j > n$. $H^{n-1}(Y, \mathbb{C})$ is generated by the fundamental classes of the irreducible components $Y_i = Y \cap \{x_i = 0\}$. They are fixed under the action of $A_d(m)$, and hence $H^{n-1}(Y, \mathbb{C})_\chi = 0$. Since Y_i and X are complete intersections in projective space, by Lefschetz's Theorem on a hyperplane section, the natural homomorphism $H^j(\mathbb{P}^{m-1},\mathbb{C}) \to H^j(X,\mathbb{C})$, is an isomorphism for $j < n$. Similarly, $H^j(\mathbb{P}^{m-2},\mathbb{C}) \to H^j(Y_i,\mathbb{C})$, is an isomorphism for $j < n-1$. Since the group $A_d(m)$ acts trivially on cohomology of the projective space, we obtain that

$$H^j(X,\mathbb{C})(\chi) = \{0\}, \quad j \neq n.$$

Also using the Mayer–Vietoris sequence, one shows that

$$H^j(Y,\mathbb{C})(\chi) = 0, \quad j \neq n-1.$$

The long exact sequence now gives

$$H^j(U,\mathcal{L}_{\boldsymbol{\mu}}) \cong H^j_c(U',\mathbb{C})(\chi) = 0, \quad j \neq n,$$

$$H^n(U,\mathcal{L}_{\boldsymbol{\mu}}) \cong H^n_c(U',\mathbb{C})(\chi) \cong H^n(X,\mathbb{C})(\chi).$$ □

Lemma 8.2. *For any character χ defined by a collection $\boldsymbol{\mu}$,*

$$\dim H^{pq}_\chi(X) = \binom{|\boldsymbol{\mu}|-1}{p}\binom{m-1-|\boldsymbol{\mu}|}{q},$$

$$\dim H^n(X,\mathbb{C})_\chi = \binom{m-2}{n}.$$

Proof. We only sketch the proof. Another proof was given by P. Deligne [De]. One can compute explicitly the Hodge decomposition of a nonsingular complete intersection n-dimensional subvariety X in projective space $\mathbb{P}^N$. Let $F_1 = \ldots = F_{N-n} = 0$ be homogeneous equations of X in variables $x_0,\ldots,x_N$. Let $y_1,\ldots,y_{N-n}$ be new variables and

$$F = y_1F_1 + \ldots + y_{N-n}F_{N-n}.$$

Define the *jacobian algebra* of F as the quotient algebra

$$R(F) = \mathbb{C}[x_0,\ldots,x_N,y_1,\ldots,y_{N-n}]/J_F,$$

where J_F is the ideal generated by partial derivatives of F in each variable. It has a natural bigrading defined by $\deg x_i = (1,0)$, $\deg y_j = (0,1)$. There is a natural isomorphism (see [Ter])

$$H^{pq}_{prim}(X) \cong R(F)_{q,((N+1+q)d-n-1)},$$

where d is the sum of the degrees of the polynomials F_i. If G is a group of automorphisms of X induced by linear transformations of $\mathbb{P}^N$, then its action on the cohomology is compatible with the action on the ring $R(F)$ (the representation on this ring must be twisted by the one-dimensional determinant representation). In our case the equations f_i are very simple, and the action of the group $A_d(m)$ can be explicitly computed.

The second formula follows from the first by using the combinatorics of binomial coefficients. □

Let χ be the character corresponding to $\boldsymbol{\mu}$. Using the isomorphism from Lemma 8.1, we can define

$$H^0(U, \mathcal{L}_{\boldsymbol{\mu}})^{pq} = H_\chi^{pq}(X). \tag{8.5}$$

One has the following description of this space (see [DM], [De]).

$$H^0(U, \mathcal{L}_{\boldsymbol{\mu}})^{pq} = H^q(\mathbb{P}^n, \Omega^p_{\mathbb{P}^n}(\log \cup H_i) \otimes \mathcal{L}_{\boldsymbol{\mu}}),$$

where $\Omega^p_{\mathbb{P}^n}(\log \cup H_i)$ is the sheaf of meromorphic differential p-forms with at most simple poles on the hyperplanes H_i. These forms can be considered as multi-valued forms on $\mathbb{P}^n$ with local branches defined by the local coefficient system.

Consider the family $\mathcal{X}$ parametrized by the set of all possible ordered sets of m hyperplanes in general linear position. Any such collection is defined uniquely by the matrix M (8.1) of coefficients of linear forms defining the hyperplanes. Two matrices define the same collection if and only if their columns are proportional. The set of equivalence classes is an algebraic variety $\mathcal{X}_{n,m}$ of dimension mn. Let $\tilde{\mathcal{X}}_{n,m}$ be the universal covering of $\mathcal{X}_{n,m}$. We have the eigenspace period map

$$\phi : \tilde{\mathcal{X}}_{n,m} \to \mathrm{Fl}(\mathbf{f}(\chi), V_{\mathbb{C}}(\chi))^0,$$

where $V = H^n(X_{t_0}, \mathbb{R})$ for some $t_0 \in \mathcal{X}_{n,n}$ and

$$f_p(\chi) = \sum_{p' \geq p} \binom{|\boldsymbol{\mu}| - 1}{p'} \binom{m - 1 - |\boldsymbol{\mu}|}{n - p'}, \quad p = 0, \ldots, n.$$

The group $\mathrm{GL}(n+1, \mathbb{C})$ acts naturally on $\mathcal{X}_{n,m}$ by left-multiplications. Geometrically this means a projective transformation sending a collection of hyperplanes to a projectively equivalent collection. On can show that the orbit space $\mathcal{P}_{n,m} = \mathcal{X}_{n,m}/\mathrm{GL}(n+1, \mathbb{C})$ is a quasi-projective algebraic variety of dimension $(m - n - 1)m$. Two X_t and $X_{t'}$ corresponding to points in the same orbit are $A_d(m)$-equivariantly isomorphic. This shows that the eigenperiod map is defined on the universal covering $\tilde{\mathcal{P}}_{n,m}$.

The following result is a theorem of A. Varchenko [Va].

Theorem 8.3. *Assume that χ corresponds to collection of numbers μ_i such that*

$$|\boldsymbol{\mu}| = \mu_1 + \ldots + \mu_m = n + 1,$$

or, equivalently, $h_\chi^{n0}(X) = 1$. Then the eigenperiod map

$$\tilde{\mathcal{P}}_{n,m} \to Fl(\mathbf{f}(\chi), V_{\mathbb{C}}(\chi))^0$$

is a local isomorphism onto its image.

Example. We take $n = 1$. In this case $U = \mathbb{P}^1 - \{p_1, \ldots, p_m\}$. Without loss of generality we may assume that p_m is the infinity point and other p_i's have affine coordinates z_i.

Consider a multi-valued form

$$\omega = g(z)(z - z_1)^{-\mu_1} \ldots (z - z_{m-1})^{-\mu_{m-1}} dz,$$

where $g(z)$ is a polynomial of degree $|\boldsymbol{\mu}|-2$ (zero if $|\boldsymbol{\mu}|<2$). Under the monodromy transformation the power $(z-z_i)^{-\mu_i}$ is transformed in the way prescribed by the local coefficient system. At the infinity we have to make the variable change $z=1/u$ to transform ω to the form

$$\omega = h(u)(1-z_1u)^{-\mu_1}\dots(1-z_{m-1}u)^{-\mu_{m-1}}u^{-\mu_m}du.$$

Again we see that the local monodromy $u\mapsto ue^{2\pi i\phi}$ agrees with the monodromy of the local coefficient system. This shows that the forms ω span a subspace of dimension $|\boldsymbol{\mu}|-1$. Comparing with the formula from Lemma 8.2 we see that they span the whole space $H^1(U,\mathcal{L}_{\boldsymbol{\mu}})^{10}=H^{10}_\chi(X)$.

Assume

$$|\boldsymbol{\mu}| = \mu_1+\dots+\mu_m = 2, \tag{8.6}$$

or, equivalently, $h^{10}_\chi(X)=1$. In this case, the space $F^1_\chi = H^1(U,\mathcal{L}_{\boldsymbol{\mu}})^{10}$ is generated by the form

$$\omega_{\boldsymbol{\mu}} = (z-z_1)^{-\mu_1}\dots(z-z_{m-1})^{-\mu_{m-1}}dz.$$

The form $\omega_{\boldsymbol{\mu}}$ is called the hypergeometric form with exponents $(\mu_1,\dots,\mu_{m-1})$. The space $V=H^1(X,\mathbb{R})(\chi)$ is of dimension $m-2$ and $\mathrm{Fl}(\mathbf{f}(\chi),V_{\mathbb{C}}(\chi))^0$ is the open subspace of the projective space of lines in $V_{\mathbb{C}}$ such that restriction of the hermitian form $iQ(x,\bar{x})$ to F^1 is positive definite. Thus $\mathrm{Fl}(\mathbf{f}(\chi),V_{\mathbb{C}}(\chi))^0$ is isomorphic to a complex ball $\mathbb{B}_{m-3}$ and we have the eigenperiod map

$$\tilde{\mathcal{P}}_{1,m}\to\mathbb{B}_{m-3}. \tag{8.7}$$

Note that $\dim\mathcal{P}_{1,m}=m-3=\dim\mathbb{B}_{m-3}$. By Theorem 8.3, this map is a local isomorphism (in the case $n=1$ this is an earlier result which can be found in [DM].)

Now recall some constructions from Geometric Invariant Theory. Consider the space $X=(\mathbb{P}^n)^m$ parametrizing m-tuples of points in $\mathbb{P}^n$, not necessary distinct. Consider a map $f:(\mathbb{P}^n)^m\to\mathbb{P}^N$ given by all multi-homogeneous monomials of multi-degrees $\mathbf{k}=(k_1,\dots,k_m),\ k_i>0$. The group $\mathrm{SL}(n+1)$ acts naturally on $\mathbb{P}^n$ by projective transformations and on the product by the diagonal action. Let S be the projective coordinate ring of the image of the product in $\mathbb{P}^N$. The group $G=SL(n+1,\mathbb{C})$ acts on S via its linear action on $\mathbb{P}^N$. Let S^G be the subring of invariant elements. By a theorem of Hilbert, the subring S^G is a finitely generated graded algebra over $\mathbb{C}$. By composing the map f with the Veronese map $\mathbb{P}^N\to\mathbb{P}^M$ defined by all homogeneous monomials of some sufficiently large degree one may assume that S^G is generated by elements of degree 1. A standard construction realizes this ring as the projective coordinate ring of some projective variety. This projective variety (which is uniquely defined up to isomorphism) is denoted by $X//_{\mathbf{k}}G$ and is called the GIT-quotient of X by G with respect to the linearization defined by numbers $\mathbf{k}$. Assume $n=1$. Let U^{s} (resp. U^{ss}) be the open Zariski subset of $(\mathbb{P}^1)^m$ parametrizing collections of points $(p_1,\dots,p_m)$ such that for any $p\in\mathbb{P}^1$

the following conditions is satisfied:

$$2\sum_{p_i=p} k_i < \sum_{i=1}^m k_i \text{ (resp. } 2\sum_{p_i=p} k_i \le \sum_{i=1}^m k_i). \tag{8.8}$$

One shows that there is a canonical surjective map $U^{\mathrm{ss}} \to X//_{\mathbf{k}}G$ whose restriction to U^{s} is isomorphic to the natural projection of the orbit space $U \to U/G$. Two point sets in the set $U^{\mathrm{ss}} \setminus U^s$ are mapped to the same point if and only if they have the same subset of points p_j, $j \in J$, satisfying $2\sum_{j\in J} k_i = \sum_{i=1}^m k_i$ or its complementary set.

Let $P^{\mathrm{ss}}_{1,\mathbf{k}}$ (resp. $P^{\mathrm{s}}_{1,\mathbf{k}}$) denote the GIT-quotient $(\mathbb{P}^1)^m//_{\mathbf{k}}\mathrm{SL}(2,\mathbb{C})$ (resp. its open subset isomorphic to the orbit space $U^{\mathrm{s}}/\mathrm{SL}(2)$). Our variety $P_{1,m}$ is isomorphic to an open subset of $P^{\mathrm{s}}_{1,\mathbf{k}}$ for any $\mathbf{k} = (k_1, \dots, k_m)$.

The following is a main result of Deligne–Mostow's paper [DM].

Theorem 8.4. *Let $\boldsymbol{\mu}$ satisfy condition (8.6). Write $\mu_i = k_i/d$, where d is a common least denominator of the μ_i's. Assume that the following condition is satisfied:*

$$(1-\mu_i-\mu_j)^{-1} \in \mathbb{Z}, \quad \text{for any } i \neq j \text{ such that } \mu_i+\mu_j < 1. \tag{8.9}$$

Then the monodromy group $\Gamma_{\boldsymbol{\mu}}$ is a discrete group of holomorphic automorphisms of $\mathbb{B}_{m-3}$ and the eigenperiod map (8.7) extends to an isomorphism

$$\Phi : P^{\mathrm{s}}_{1,\mathbf{k}} \to \Gamma_{\boldsymbol{\mu}}\backslash\mathbb{B}_{m-3}.$$

Moreover, Φ can be extended a holomorphic isomorphism

$$\overline{\Phi} : P^{\mathrm{s}}_{1,\mathbf{k}} \to \overline{\Gamma_{\boldsymbol{\mu}}\backslash\mathbb{B}_{m-3}},$$

where the target space is a certain compactification of the ball quotient obtained by adding a finite set of points.

The list of collections of numbers $(\mu_1, \dots, \mu_m)$ satisfying (8.9) is finite for $m > 4$. It consists of 27 collections with $m = 5$, of 7 collections with $m = 6$ and one collection with $m = 7$ and $m = 8$. One can extend the theorem to the case when $\boldsymbol{\mu}$ satisfies a weaker condition:

$$(1-\mu_i-\mu_j)^{-1} \in \mathbb{Z} \quad \text{for any } i \neq j \text{ such that } \mu_i+\mu_j < 1,\ \mu_i \neq \mu_j \tag{8.10}$$
$$2(1-\mu_i-\mu_j)^{-1} \in \mathbb{Z} \quad \text{for any } i \neq j \text{ such that } \mu_i+\mu_j < 1,\ \mu_i = \mu_j.$$

This gives additional cases, with largest m equal to 12.

Write $\mu_i = k_i/d$ as in Theorem 8.4. Recall that, by Lemma 8.1 we have an isomorphism

$$H^1(U, \mathcal{L}_{\boldsymbol{\mu}}) \cong H^1(X, \mathbb{C})(\chi),$$

where X is the curve in $\mathbb{P}^{m-1}$ defined by equations

$$-x_i^d + a_{i0}x_0^d + a_{i1}x_1^d = 0,\ i = 2, \dots, m-1.$$

Let $H = \mathrm{Ker}(\chi)$ and $X_{\boldsymbol{\mu}} = X/H$. One can show that the curve $X_{\boldsymbol{\mu}}$ is isomorphic to a nonsingular model of the affine algebraic curve with equation

$$y^d = (x-z_1)^{k_1}\dots(x-z_{m-1})^{k_{m-1}}. \tag{8.11}$$

The cyclic group $C_d = \mathbb{Z}/d\mathbb{Z}$ acts naturally on this curve by acting on the variable y. One proves that

$$H^1(U, \mathcal{L}_{\boldsymbol{\mu}}) \cong H^1(X_{\boldsymbol{\mu}}, \mathbb{C})(\chi).$$

For example, take $m = 6$, $d = 3$, $\mu_1 = \ldots = \mu_6 = 1$. The curve $X_{\boldsymbol{\mu}}$ is the curve of genus 4 with affine equation $y^3 = (x - z_1) \cdots (x - z_6)$. We have $\dim H^{10}(X_{\boldsymbol{\mu}})(\chi) = 1$. Thus the action of C_3 on the Jacobian variety is of type $(1, 3)$. As we see in Section 11, the locus of principally polarized abelian 4-folds admitting an automorphism of order 3 of type $(1, 3)$ is isomorphic to a 3-dimensional ball quotient. This agrees with the theory of Deligne–Mostow.

9. K3 surfaces

This is our second example. This time we consider compact orientable simply-connected 4-manifolds M. Recall that by a theorem of M. Friedman, the homeomorphism type of M is determined by the cup-product on $H_2(M, \mathbb{Z}) \cong \mathbb{Z}^{b_2(M)}$. It is a symmetric bilinear form whose matrix in any $\mathbb{Z}$-basis has determinant ± 1. The corresponding integral quadratic form is called in such case a *unimodular quadratic form*. The group $H_2(M, \mathbb{Z})$ equipped with the cup-product is an example of a *lattice*, a free abelian group equipped with a symmetric bilinear form, or, equivalently, with a $\mathbb{Z}$-valued quadratic form. We will denote the values of the bilinear form by (x, y) and use x^2 to denote (x, x). We apply the terminology of real quadratic forms to a lattice whose quadratic form is obtained by extension of scalars to $\mathbb{R}$. There are two types of indefinite lattices. The *even type* is when the quadratic form takes only even values, and the *odd type* when it takes some odd values. A unimodular indefinite lattice of odd type can be given in some $\mathbb{Z}$-basis by a matrix $I(p, q)$ (2.9). Its Sylvester signature is (p, q). A unimodular indefinite quadratic form S of even type of signature ≤ 0 is isomorphic to the orthogonal sum $U^{\oplus p} \oplus E_8^{\oplus q}$, where U is given by the matrix

$$\begin{pmatrix} 0 & 1 \\ 1 & 0 \end{pmatrix}$$

and E_8 is a unique (up to isomorphism) even negative definite unimodular lattice of rank 8. The index of S is equal to $-8q$ and the Sylvester signature is equal to $(p, p + 8q)$. The lattice E_8 is defined by the Dynkin diagram of type E_8. Its vertices correspond to vectors with $(x, x) = -2$ and $(x_i, x_j) = 1$ or 0 dependent on whether the corresponding vertices are joined by an edge. Note that changing the orientation changes the index $I(M)$ to its negative. Also by a theorem of Donaldson, $p \neq 0$ for the even cup-product on a simply-connected 4-manifold.

A 4-manifold M is called a *K3-manifold* if it realizes the form $L_{K3} = U^3 \oplus E_8^2$, where the direct sum is considered to be an orthogonal sum.

Theorem 9.1. *Each K3-manifold is homeomorphic to a compact simply connected complex Kähler surface X with trivial first Chern class (i.e., the second exterior power of the holomorphic tangent bundle is the trivial line bundle).*

Proof. It is enough to construct such a complex surface realizing the quadratic form L_{K3}. Let X be a nonsingular quartic surface in $\mathbb{P}^3(\mathbb{C})$. The standard exact sequence

$$0 \to T_X \to T_{\mathbb{P}^3}|X \to N_{X/\mathbb{P}^3} \to 0$$

allows one to compute the Chern classes of X. We have

$$1 + c_1 + c_2 = (1+h)^4/(1+4h) = 1 + 6h^2,$$

where h is the class of a plane section. Thus $c_1 = 0$ and $c_2 = 6 \cdot 4 = 24$. Since $c_1 = 0$, the canonical class $K_X = -c_1 = 0$. Thus $H^{20}(X) = H^0(X, \Omega_X^2) = \mathbb{C}$. Since $24 = c_2 = \chi(X) = \sum(-1)^i b_i(X)$ and X is simply-connected, we get $b_2(X) = 22$. Thus $b_2 = h^{20} + h^{11} + h^{02}$ implies $h^{11} = 20$. By the Index formula (2.7) we obtain that $I(M) = -16$, hence the Sylvester signature is $(3, 19)$. One also uses Wu's Theorem which asserts that the cup-product is even if $c_1(X)$ is divisible by 2 in $H^2(X, \mathbb{Z})$. Thus $H^2(X, \mathbb{Z})$ is an even lattice. This implies that the cup-product $H^2(X, \mathbb{Z})$ is isomorphic to the lattice L_{K3}. □

A non-trivial theorem which is a corollary of the theory of periods of K3 surfaces is the following.

Theorem 9.2. *All complex K3 surfaces are diffeomorphic to a nonsingular quartic surface in* $\mathbb{P}^3(\mathbb{C})$.

Let X be an algebraic K3 surface (e.g., a quartic surface). We have already observed that $(h^{20}, h^{11}, h^{02}) = (1, 20, 1)$. The Hodge flag is

$$0 \subset F^2 = H^{20}(X) \subset F^1 = H^{20}(X) + H^{11}(X) \subset F^0 = H^2(X, \mathbb{C}).$$

The Hodge structure on cohomology $V = H^2_{prim}(X, \mathbb{R})$ is a AHS of weight 2 of type $(1, 19, 1)$, where we take the polarization defined by a choice of a Hodge form $h \in H^2(X, \mathbb{Z})$. The polarization form Q admits an integral structure with respect to the lattice $H^2(X, \mathbb{Z})$. Note that the flag (F^p) is completely determined by F^2 (because $F^1 = (F^2)^\perp$). This shows that the period space $D_{\mathbf{f}}(V, Q)$ is isomorphic, as a complex manifold, to

$$\mathcal{D}_h(V) = \{\mathbb{C} \cdot v \in \mathbb{P}(V_{\mathbb{C}}) : Q(v, v) = 0, Q(v, \bar{v}) > 0\}. \tag{9.1}$$

A *marking* of a K3 surface is an isomorphism of lattices $\phi : L_{K3} \to H^2(X, \mathbb{Z})$. After tensoring with $\mathbb{R}$ or $\mathbb{C}$, it defines an isomorphism $\phi_{\mathbb{R}} : (L_{K3})_{\mathbb{R}} \to H^2(X, \mathbb{R})$, $\phi_{\mathbb{C}} : (L_{K3})_{\mathbb{C}} \to H^2(X, \mathbb{C})$. Choose a marking. Let $l = \phi^{-1}(h)$. Let $l^\perp$ denote the orthogonal complement of $\mathbb{Z}l$ in L_{K3}. It is a sublattice of signature $(2, 19)$. The period space $\mathcal{D}_h(V)$ becomes isomorphic to the 19-dimensional complex variety

$$\mathcal{D}_l = \{\mathbb{C}v \in \mathbb{P}((l^\perp)_{\mathbb{C}}) : (v, v) = 0, (v, \bar{v}) > 0\}.$$

Note that, in coordinates, the period space $\mathcal{D}_l$ is isomorphic to a subset of lines in $\mathbb{P}^{20}$ whose complex coordinates $(t_0, \ldots, t_{19})$ satisfy

$$t_0^2 + t_1^2 - t_3^2 - \ldots - t_{19}^2 = 0,$$
$$|t_0|^2 + |t_1|^2 - |t_3|^2 - \ldots - |t_{19}|^2 > 0.$$

One can give another model of the period space as follows. For any line $\mathbb{C}v \in \mathcal{D}_l$ write $v = x + iy$, where $x, y \in (l^\perp)_\mathbb{R}$ and consider the plane P in $V = (l^\perp)_\mathbb{R}$ spanned by x, y. We have $(x+iy, x-iy) = x^2 + y^2 > 0$. Also P carries a canonical orientation defined by the orientation class of the basis formed by x and y, in this order. This defines a map

$$\mathcal{D}_l \to G(2,V)_Q^+ = \{P \in G(2,V)^0 : Q|P > 0\},$$

where $G(2,V)^0$ is the Grassmannian of oriented planes in V. It is easy to construct the inverse map. For any $P \in G(2,V)_Q^+$ with a basis (x,y) we assign a line in $V_\mathbb{C}$ spanned by $x + iy$. A different basis in the same orientation class changes the complex vector by a complex multiple. This gives a well-defined map to the projective space $\mathbb{P}(V_\mathbb{C})$ and, as is easy to see, the image is equal to $\mathcal{D}_l$. The changing of orientation in the plane P corresponds to replacing $x + iy$ with $x - iy$, i.e., replacing v with $\bar{v}$. If v arises as the period of a K3 surface, this switching is achieved by the changing the complex structure of the surface to the conjugate one.

The real Grassmannian model shows that the group $\mathrm{O}(V) \cong O(2,19)$ acts transitively on $\mathcal{D}_l$ with the isotropy group of some point isomorphic to the subgroup $\mathrm{SO}(2)\times\mathrm{O}(19)$. Here $\mathrm{O}(p,q)$ denotes the orthogonal group of the real quadratic form on $\mathbb{R}^n$ defined by the diagonal matrix $I(p,q)$, and $\mathrm{SO}(p,q)$ denotes its subgroup of matrices with determinant 1. So, we obtain

$$\mathcal{D}_l \cong \mathrm{O}(2,19)/\mathrm{SO}(2) \times \mathrm{O}(19).$$

This description shows that the period space is not connected. It consists of two disconnected copies of a Hermitian symmetric space of non-compact type IV isomorphic to the orbit space $\mathrm{SO}(2,19)/\mathrm{SO}(2) \times \mathrm{SO}(19)$. In coordinate description of $\mathcal{D}_l$ the connected components are distinguished by the sign of $\mathrm{Im}(\frac{t_1}{t_0})$.

It is known that the K3-lattice L_{K3} represents any even number $2d$, i.e., the set of elements $x \in L_{K3}$ with $x^2 = 2d$ is not empty. This is easy to see, for example, considering the sublattice of L_{K3} isomorphic to U. Less trivial is the fact that any two vectors x and y with $x^2 = y^2$ differ by an isometry of L_{K3} provided that none of them is divisible by an integer $\neq \pm 1$ (i.e., x, y are *primitive elements* of the lattice). Let (X,h) be a polarized algebraic K3 surface of degree $h^2 = 2d$. As always in this case we assume that the polarization class h belongs to $H^2(X,\mathbb{Z})$, i.e., is a Hodge class. We can always choose h to be a primitive element in $H^2(X,\mathbb{Z})$. For each positive integer d fix a primitive element $l \in L_{K3}$ with $l^2 = 2d$. Let $\phi : L_{K3} \to H^2(X,\mathbb{Z})$ be a marking of (X,h). Then $\sigma(\phi^{-1}(h)) = l$ for some $\sigma \in \mathrm{O}(L_{K3})$. This shows that (X,h) always admits a marking such that $\phi^{-1}(h) = l$. We call it a *marking of a polarized K3 surface.* Its *period*

$$p_{X,\phi} = \phi_\mathbb{C}^{-1}(H^{20}(X))$$

belongs to $\mathcal{D}_l$. Note that the polarization h is determined by the marking since, by definition, $\phi(l) = h$, so there is no need to use the notation $p_{X,h,\phi}$.

We say that two marked polarized K3 surfaces (X, ϕ) and (X', ϕ') are isomorphic if there exists an isomorphism of algebraic varieties $f : X' \to X$ such that $\phi' = f^* \circ \phi$ and $f^*(h') = h$. Obviously this implies that $p_{X,\phi} = p_{X',\phi'}$. Conversely, suppose this happens. Then $\psi = \phi \circ \phi^{-1} : H^2(X', \mathbb{Z}) \to H^2(X, \mathbb{Z})$ is an isomorphism of lattices such that $\psi(H^{20}(X')) = H^{20}(X)$ and $\psi(h') = h$.

The following is a fundamental result due to I.R. Shafarevich and I.I. Piatetsky-Shapiro [SPS], called the *Global Torelli Theorem for polarized algebraic K3 surfaces.*

Theorem 9.3. *Let (X, h) and (X', h') be two polarized algebraic K3 surfaces. Suppose there is an isometry of lattices $\phi : H^2(X', \mathbb{Z}) \to H^2(X, \mathbb{Z})$ such that $\phi(h') = h$ and $\phi^*_{\mathbb{C}}(H^{20}(X')) = H^{20}(X)$. Then there exists a unique isomorphism of algebraic varieties $f : X \to X'$ such that $f^* = \phi$.*

Let
$$\Gamma_l = \{\sigma \in \mathrm{O}(L_{K3}) : \sigma(l) = l\}.$$
For any marked polarized K3 surface (X, ϕ) with $p_{X,\phi} \in \mathcal{D}_l$ and any $\sigma \in \Gamma_l$ we have $p_{X,\phi\circ\sigma} \in \mathcal{D}_l$. So the orbit $\Gamma_l \cdot p_{X,\phi}$ depends only on isomorphism class of (X, h). The group Γ_l is isomorphic to a discrete subgroup of $\mathrm{O}(2, 19)$ which acts transitively on $\mathcal{D}_l$ with compact stabilizers isomorphic to $\mathrm{SO}(2) \times \mathrm{O}(19)$. This shows that Γ_l is a discrete subgroup of the automorphism group of $\mathcal{D}_l$ (i.e., its stabilizers are finite subgroups). Thus the orbit space $\Gamma_l \backslash \mathcal{D}_l$ has a structure of a complex variety (with quotient singularities). In fact, it has a structure of a quasi-projective algebraic variety. Note that, although the period space $\mathcal{D}_l$ is not connected, the orbit space is an irreducible algebraic variety. The reason is that the group Γ_l contains an element which switches the two connected components. In fact, it is easy to see that there is an isomorphism of lattices
$$l^\perp \cong U^2 \oplus E_8^2 \oplus < -2d >,$$
where $< m >$ denotes a lattice of rank 1 defined by the matrix (m). The isometry σ of $l^\perp$ defined by being the minus-identity on one copy of U and the identity on all other direct summands belongs to Γ_l. Since it switches the orientation of a positive definite 2-plane spanned by a vector $x \in U$ and a vector y from another copy of U such that $(x, x) > 0, (y, y) > 0$, the isometry σ switches the two connected components of $\mathcal{D}_l$.

We see that the isomorphism class of (X, h) defines a point
$$p_{X,h} \in \Gamma_l \backslash \mathcal{D}_l$$
and the Global Torelli Theorem from above asserts that $p(X, h) = p(X', h')$ if and only if (X, h) and (X', h') are isomorphic polarized surfaces. This is truly one of the deepest results in mathematics generalizing the Torelli theorem for curves and polarized abelian varieties.

The Global Torelli Theorem can be restated in terms of the period map for a family $\pi : \mathcal{X} \to T$ of polarized K3 surfaces, where we always assume that T is connected. The cohomology groups of fibres $H^2(X_t, \mathbb{Z})$ form a sheaf $\mathcal{H}^2(\pi)$ of

$\mathbb{Z}$-modules (the sheaf $R^2\pi_*(\mathbb{Z}_Y)$). A marking of the family is an isomorphism of sheaves $\phi : (L_{K3})_T \to \mathcal{H}^2(\pi)$, where $(L_{K3})_T$ is the constant sheaf with fibre L_{K3}. For any $t \in T$ the map of fibres defines a marking ϕ_t of X_t. The polarization classes h_t of fibres define a section h of $\mathcal{H}^2(\pi)$ and we require that $\phi^{-1}(h) = l$. For any $t \in T$, the map of fibres $\phi_{\mathbb{C}} : (L_{K3})_{\mathbb{C}} \to H^2(X_t, \mathbb{C})$ defines the period point $p_{X_t,\phi_t} \in \mathcal{D}_l$. This gives a map

$$p_{\pi,\phi} : T \to \mathcal{D}_l. \tag{9.2}$$

By Griffiths' theorem it is a holomorphic map of complex varieties. In general, a family does not admit a marking because the sheaf $\mathcal{H}^2(\pi)$ need not be constant. However, it is isomorphic to a constant sheaf when the base T is simply connected. So replacing T by the universal covering $\tilde{T}$ we can define the period map (9.2). Now for any two points $\tilde{t}$ and $\tilde{t}'$ in $\tilde{T}$ over the same point in T, the fibres of the family $\tilde{f} : Y \times_T \tilde{T} \to \tilde{T}$ over these points are isomorphic polarized surfaces. Therefore their images in $\mathcal{D}_l$ lie in the same orbit of Γ_l. This defines a holomorphic map

$$p_\pi : T \to \Gamma_l \backslash \mathcal{D}_l$$

called the *period map of a family*. The Global Torelli Theorem asserts that the points in the same fibre of the period map are isomorphic polarized K3 surfaces.

One can construct a coarse moduli space of polarized K3 surfaces. Let X be an algebraic K3 surface which contains a Hodge class h with $h^2 = 2d$. One can show that a line bundle L with $c_1(L) = 3h$ defines an isomorphism from X to a closed subvariety of $\mathbb{P}^{9d+1}$. Two different embeddings defined in this way differ by a projective transformation of $\mathbb{P}^{9d+1}$. A construction from algebraic geometry based on the notion of a *Hilbert scheme* allows one to parametrize all closed subvarieties of $\mathbb{P}^{9d+1}$ isomorphic to an embedded K3 surface as above by some quasi-projective algebraic variety H_{2d}. The orbit space $H_{2d}/\mathrm{PGL}(9d+2, \mathbb{C})$ exists and represents a coarse moduli space of K3 surfaces (see [SPS]). We denote it by $\mathcal{M}_{2d}$. Then we have the period map

$$\mathrm{per} : \mathcal{M}_{2d} \to \Gamma_l \backslash \mathcal{D}_l. \tag{9.3}$$

The Global Torelli Theorem implies that this map is injective. A further result, the *Local Torelli Theorem* asserts that this map is a local isomorphism of complex varieties, and hence the period map is an isomorphism with its image.

Before we state a result describing the image of the period map let us remind some terminology from algebraic geometry. Let $\mathrm{Pic}(X)$ be the Picard group of X. It is a subgroup of $H^2(X, \mathbb{Z})$ spanned by the fundamental classes of irreducible 1-dimensional complex subvarieties (curves) on X. An integral linear combination $\sum n_i[\Gamma_i]$ of such classes is cohomologous to zero if and only if the divisor $D = \sum n_i\Gamma_i$ is linearly equivalent to zero (i.e., the divisor of a meromorphic function). So $\mathrm{Pic}(X)$ can be identified with the group of divisor classes on X. Via the well-known relation between divisor classer and isomorphism classes of algebraic line bundles over X (defined by the first Chern class map), we see another description of $\mathrm{Pic}(X)$ as the group of isomorphism classes of line bundles, or invertible sheaves of $\mathcal{O}_X$-modules. We identify $\mathrm{Pic}(X)$ with a sublattice S_X of the unimodular lattice $H^2(X,\mathbb{Z})$ with respect to the cup-product (the *Picard lattice*).

Recall that a line bundle L is called *ample* if sections of some positive tensor power $L^{\otimes m}$ define an embedding of X in projective space. In this case the line bundle becomes isomorphic to the restriction of the line bundle corresponding to a hyperplane. A line bundle L is ample if and only if its first Chern class $c_1(L)$ is the cohomology class of some Hodge form. A line bundle L is ample if and only if

$$c_1(L)^2 > 0, \ (c_1(L),[C]) > 0,$$

for any curve C. We say that L is *pseudo-ample* if the first condition holds but in the second one the equality is possible. It is known that the restriction of the cup-product Q to $\mathrm{Pic}(X)$ has signature $(1, r-1)$, where $r = \mathrm{rank}\,\mathrm{Pic}(X)$. This is because, by the Lefschetz Theorem,

$$\mathrm{Pic}(X) = H^{11}(X) \cap H^2(X,\mathbb{Z}). \tag{9.4}$$

By the property of signature of the intersection form on H^{11}, any irreducible curve C with $(c_1(L),[C]) = 0$ must satisfy $[C]^2 < 0$. Since $K_X = 0$, the adjunction formula implies that $C \cong \mathbb{P}^1$ and $C^2 = -2$ (we will omit writing the brackets $[C]$ indicating the divisor class). Such a curve is called a (-2)-*curve.* Thus if X has no (-2)-curves, any divisor class D with $D \cdot D > 0$ is ample. It follows from the Riemann–Roch Theorem on algebraic surfaces that any divisor class D with $D^2 = -2$ is either effective, i.e., the class of some curve, maybe reducible, or $-D$ is effective. It is known that a line bundle L is pseudo-ample if and only if sections of some positive tensor power of L define a map from X to a surface in X' in some projective space such that all (-2)-curves R with $(R, c_1(L)) = 0$ are blown down to singular points of X' and outside of the union of such curves the map is an isomorphism.

The next result describes the image of the period map (9.3). Let $\delta \in L_{K3}$ satisfy $(\delta, l) = 0$, $\delta^2 = -2$. Let $\mathcal{H}_\delta$ denotes the intersection of $\mathcal{D}_l$ with the hyperplane $\{\mathbb{C}x \in \mathbb{P}((l^\perp)_\mathbb{C}) : (x,\delta) = 0\}$. Let

$$\Delta_l = \bigcup_{\delta^2=-2,(\delta,l)=0} \mathcal{H}_\delta.$$

It is a union of countably many closed hypersurfaces in $\mathcal{D}_l$. We call it the *discriminant locus* of $\mathcal{D}_l$. Note that for any marked polarized K3 surface (X,ϕ) of degree $2d$ its period $p_{X,\phi}$ does not belong to any hyperplane $\mathcal{H}_\delta$. In fact, otherwise $0 = (\delta, p_{X,\phi}) = (\phi(\delta), H^{20}(X))$ implies, by (9.4) that $\phi(\delta) \in \mathrm{Pic}(X)$ and $\phi(\delta)$ or its negative is an effective divisor class. However $0 = (\delta, l) = (\phi(\delta), h)$ contradicts the assumption that $h = c_1(L)$ for some ample line bundle. So we see that the image of the period map is contained in the complement of the discriminant locus. A deep result, called the *Surjectivity Theorem* for the periods of polarized K3 surfaces asserts that the image is exactly the complement of Δ_l. In fact, it gives more.

By weakening the condition on the polarization one can consider pseudo-polarized K3 surfaces as pairs (X, h), where h is a primitive divisor class corresponding to a pseudo-ample line bundle. One can extend the notion of a marked pseudo-polarized K3 surface (X, ϕ) and define its period point $p_{X,\phi}$ and extend the definition of the period map to families of *pseudo-polarized* K3 surfaces. Then one proves the following.

Theorem 9.4. *Any point in $\mathcal{D}_l$ is realized as the period point $p_{X,\phi}$ of a marked pseudo-polarized K3 surface (X, h). The image of the set $\{\delta : p_{X,\phi} \in \mathcal{H}_\delta\}$ under the map ϕ is equal to the set $\Delta(X, h)$ of divisor classes of curves R such that $(R, h) = 0$. In particular, the periods of marked polarized K3 surfaces is equal to the set $\mathcal{D}_l \setminus \Delta_l$. The orbit space $\Gamma_l \backslash \mathcal{D}_l$ is a coarse moduli space for pseudo-polarized K3 surfaces of degree $h^2 = 2d$.*

Note that the set $\Delta(X, h)$ is finite because it is contained in the negative definite lattice equal to the orthogonal complement of h in S_X. This shows that the set of hyperplanes $\mathcal{H}_\delta$ is locally finite, i.e., each point lies in the intersection of only finite many hyperplanes. Let $\mathcal{R}(X, h)$ be the sublattice of S_X generated by the subset of $\Delta(X, h)$. It is negative definite (because it is a sublattice of the hyperbolic lattice S_X orthogonal to an element h with $h^2 > 0$) and is spanned by divisor classes R with $R^2 = -2$. Let $[R] \in \Delta(X, h)$, replacing R with $-R$ we may assume that R is a curve on X. Writing it as a sum of its irreducible components R_i we see that $(R_i, h) = 0$, so $R_i \in \Delta(X, h)$. This shows that $\Delta(X, h)$ is spanned by the classes of (-2)-curves from $\Delta(X, h)$. All negative definite lattices generated by elements x with $x^2 = 2$ lattices have been classified by E. Cartan. They coincide (after multiplying the value of the quadratic form by -1) with root lattices of semi-simple Lie algebras of types A_n, D_n, E_n and their direct sums and described by the corresponding Cartan matrices and their direct sums. Recall that the pseudo-polarization h defines a birational model of X as a surface X' with singular points whose minimal resolution is isomorphic to X. The singular points are rational double points of types corresponding to the irreducible components of the lattice $\mathcal{R}_X$. The simplest case is when $\Delta(X, h)$ consists of one element. In this case, X' has one ordinary singularity (of type A_1). All possible lattices $\mathcal{R}(X, h)$ which may occur have been classified only in the cases $h^2 = 2$ or 4.

10. Lattice polarized K3 surfaces

Let M be any abstract even lattice of signature $(1, r-1)$ (if $r > 1$ it is called a *hyperbolic lattice*). It is known that the cone

$$\mathcal{V}_M = \{x \in M_{\mathbb{R}} : x^2 \geq 0\} \tag{10.1}$$

after deleting the zero vector consists of two connected components. In coordinates, the cone is linearly isomorphic to

$$\{x = (x_1, \ldots, x_r) \in \mathbb{R}^r : x_1^2 - x_2^2 - \ldots - x_r^2 \geq 0\}.$$

The components are distinguished by the sign of x_1.

Let $\mathcal{V}_M^0$ be one of the two "halves" of the cone $\mathcal{V}_M$, and set

$$M_{-2} = \{x \in M : x^2 = -2\}.$$

Fix a connected component $\mathcal{C}(M)^+$ of

$$\mathcal{V}_M^0 \setminus \bigcup_{\delta \in M_{-2}} \delta^{\perp}.$$

Let

$$M_{-2}^+ = \{\delta \in M_{-2} : (m, \delta) > 0, \forall m \in \mathcal{C}(M)^+\}.$$

This gives a decomposition

$$M_{-2} = M_{-2}^+ \coprod M_{-2}^-,$$

where $M_{-2}^- = \{-\delta : \delta \in M_{-2}^+\}$. We have

$$\mathcal{C}(M)^+ = \{x \in \mathcal{V}_M^0 : (x, \delta) > 0, \forall \delta \in M_{-2}^+\}.$$

Take the Picard lattice S_X as M. Then we have a canonical decomposition of

$$(S_X)_{-2} = (S_X)_{-2}^+ \coprod (S_X)_{-2}^-$$

where $(S_X)_{-2}^+$ consists of the classes of effective divisors. The corresponding connected component is denoted by $\mathcal{K}_X^0$ and is called *ample cone*. We denote by $\mathcal{K}_X$ the closure of $\mathcal{K}_X^0$ in $\mathcal{V}_{S_X}^0$.

A *M-polarized K3 surface* is a pair (X, j) consisting of an algebraic K3 surface X and a primitive embedding of lattices $j : M \to \mathrm{Pic}(X)$ (where primitive means that the cokernel is a free abelian group) such that $j(\mathcal{C}(M)^+) \cap \mathcal{K}_X \neq \emptyset$. A M-marking is called *ample* if $j(\mathcal{C}(M)^+) \cap \mathcal{K}_X^0 \neq \emptyset$ (see [Do]).

We say that two M-polarized K3 surfaces (X, j) and (X', j') are isomorphic if there exists an isomorphism of algebraic surfaces $f : X \to X'$ such that $j = f^* \circ j'$.

When M is of rank 1 and is generated by an element x with $x^2 = 2d$, an M-polarized (resp. ample M-polarized) K3 surface is a pseudo-ample (resp. ample) polarized K3 surface of degree $2d$.

It turns out that the isomorphism classes of M-polarized K3 surfaces can be parametrized by a quasi-projective algebraic variety $\mathcal{M}_M$. The construction is based on the period mapping.

First we fix a primitive embedding of M in L_{K_3}. If it does not exist, $\mathcal{M}_M = \emptyset$. We shall identify M with its image. Then we consider a *marking* of (X, j). It is an isomorphism of lattices $\phi : L_{K3} \to H^2(X, \mathbb{Z})$ such that $\phi | M = j$. An isomorphism of marked M-polarized surfaces is an isomorphism of surfaces $f : X \to X'$ such that $f^* \circ \phi' = \phi$.

Let $N = M^{\perp}$ be the orthogonal complement of M in L_{K3}. Set

$$\mathcal{D}_M = \{\mathbb{C}v \in \mathbb{P}(N_{\mathbb{C}}) : v^2 = 0, (v, \bar{v}) > 0\} \tag{10.2}$$

In the case $M = \mathbb{Z}l$ is generated by one element, we see that $D_M \cong \mathcal{D}_l$. Since M is of signature $(1, r-1)$, the signature of the lattice N is $(2, 20-r)$. As before we see that

$$\mathcal{D}_M \cong G(2, N_{\mathbb{R}})^+ \cong \mathrm{O}(2, 20-r)/\mathrm{SO}(2) \times \mathrm{O}(20-r).$$

It is a disjoint union of two copies of a hermitian symmetric space of non-compact type IV of dimension $20-r$.

Let $j(M) \subset S_X$, its orthogonal complement $j(M)^\perp$ in $H^2(X, \mathbb{Z})$ is equal to $\phi(N)$ and does not depend on the choice of ϕ. Set

$$V^{20} = H^{20}(X), \quad V^{11} = H^{11}(X) \cap \phi(N)_{\mathbb{C}}, \quad V^{02} = H^{02}(X). \tag{10.3}$$

This defines an AHS of weight 2 on the vector space $V = \phi(N)_{\mathbb{R}}$ with the period flag of type $\mathbf{f} = (1, 21-r, 22-r)$. The polarization form is obtained from the symmetric bilinear form defining the lattice structure on N. The space $\mathcal{D}_M$ is isomorphic to the period space of such abstract Hodge structures.

Since $\phi(M) \subset \mathrm{Pic}(X) \subset H^{11}(X)$, $\phi^{-1}(H^{20}(X)) \subset N_{\mathbb{C}}$. This shows that

$$p_{(X,\phi)} = \phi^{-1}(H^{20}(X)) \in \mathcal{D}_M.$$

Let

$$N_{-2} = \{\delta \in N : \delta^2 = -2\}.$$

For any $\delta \in N_{-2}$ let $\delta^\perp$ denote the hyperplane of vectors in $\mathbb{P}(N_{\mathbb{C}})$ orthogonal to δ. Let $\mathcal{H}_\delta = \delta^\perp \cap \mathcal{D}_M$, and define

$$\Delta_M = \bigcup_{\delta \in N_{-2}} \mathcal{H}_\delta.$$

Theorem 10.1. *Let (X, j) be a M-polarized K3 surface. Then:*

(i) *The M-polarization is ample if and only if for any choice of a marking $p_{X,\phi} \in \mathcal{D}_M \setminus \Delta_M$.*

(ii) *If (X, ϕ) and (X', ϕ') are two marked M-polarized K3 surfaces, then $p_{X,\phi} = p_{X',\phi'}$ if and only if there exists an isomorphism of marked M-polarized surfaces $(X, \phi) \cong (X', \phi' \circ \alpha)$, where α is a product of reflections with respect to vectors $\delta \in N_{-2}$ such that $p_{X,\phi} \in \mathcal{H}_\delta$.*

(iii) *Any point in $\mathcal{D}_M$ is realized as the point $p_{X,\phi}$ for some M-polarized K3 surface.*

Proof. Suppose (X, j) is an ample M-polarized K3 surface and $\ell = p_{X,\phi} \in \mathcal{H}_\delta$ for some $\delta \in N_{-2}$. This implies that $(\phi(\delta), H^{20}(X)) = 0$ and hence $\phi(\delta)$ is the class of a divisor R (maybe reducible) with $R^2 = -2$. Replacing R with $-R$ if needed, we may assume that R is a curve. Since $j(M)$ contains a polarization class and $[R] \in j(M)^\perp$ we get a contradiction.

Conversely, suppose $p_{X,\phi} \notin \Delta_M$ but (X, j) is not ample. This means that $j(\mathcal{C}(M)^+) \cap \mathcal{K}_X \neq \emptyset$ but $j(\mathcal{C}(M)^+) \cap \mathcal{K}_X^0 = \emptyset$. Equivalently, $j(\mathcal{C}(M)^+)$ is contained in the boundary of $\mathcal{K}_X$ defined by the equations $(x, [R]) = 0$, where R is a (-2)-curve. Since $\mathcal{C}(M)^+$ spans M, this implies that there exists a (-2)-curve

R such that $\phi^{-1}([R]) \in N_{-2} \cap \phi^{-1}(H^{20}(X)^{\perp})$. Therefore $p_{X,\phi} = \phi^{-1}(H^{20}(X)) \in \mathcal{H}_\delta$,where $\delta = \phi^{-1}([R])$, contradicting the assumption.

The last two assertions follow from the Global Torelli Theorem of Burns–Rappoport for non-polarized Kähler surfaces. We refer for the proof of these assertions to [Do]. Note that fixing $\mathcal{C}(M)^+$ in the definition of a M-polarized surface is used for the proof of assertion (ii). □

Now let us see how the period point $p_{X,\phi}$ depends on the marking ϕ. For any isometry σ of the lattice L_{K3} which is identical on M we obtain another marking $\sigma \circ \phi$ of the M-polarized surface (X, j). Let Γ_M denote the group of such isometries. We have a natural injective homomorphism $\Gamma_M \to \mathrm{O}(N)$ and one can easily determine the image. Let $D(N) = N^*/N$ be the discriminant group of N. The group $\mathrm{O}(N)$ acts naturally on N^* leaving N invariant. This defines a homomorphism $\mathrm{O}(N) \to \mathrm{Aut}(D(N))$ and we have

$$\Gamma_M \cong \mathrm{O}(N)^* := \mathrm{Ker}\big(\mathrm{O}(N) \to \mathrm{Aut}(D(N))\big).$$

This is because every isometry from the right-hand side can be extended to an isometry of L acting identically on $M = N^{\perp}$. The group Γ_M is a discrete subgroup of the group of automorphisms of $\mathcal{D}_M$ and the orbit space $\Gamma_M\backslash\mathcal{D}_M$ is a quasi-projective algebraic variety. The image of $p_{X,\phi}$ in the orbit space $\Gamma_M\backslash\mathcal{D}_M$ depends only on (X, j). It is denoted by $p_{X,j}$ and is called the *period of M-polarized surface X*.

Let $\pi : Y \to T$ be a family of K3 surfaces parametrized by some algebraic variety T. The cohomology groups $H^2(X_t, \mathbb{Z})$, $t \in T$, form a local coefficient system $\mathcal{H}^2$ on T. Assume that there exists an embedding $j : M_T \hookrightarrow \mathcal{H}^2$ of the constant coefficient system M_T such that for each $t \in T$ the embedding $j_t : M \to H^2(X_t, \mathbb{Z})$ defines an M-polarization of X_t. We say then that π is a *family of M-polarized K3 surfaces*. We leave to the reader to extend the notion of the period map of polarized K3 surfaces to the case of M-polarized K3 surfaces

$$p_\pi : T \to \Gamma_M\backslash\mathcal{D}_M.$$

This is called the *period map* of the family $\pi : \mathcal{X} \to T$.

Note that the group Γ_M contains reflections $s_\delta : x \mapsto (x, \delta)\delta$ for any $\delta \in N_{-2}$. Applying Theorem 10.1, we obtain that for any M-polarized K3 surfaces (X, j) and (X', j')

$$p(X, j) = p(X', j') \Longleftrightarrow (X, j) \cong (X', j'). \tag{10.4}$$

This implies that $\Gamma_M\backslash\mathcal{D}_M$ (resp. $\Gamma_M\backslash\mathcal{D}_M^0$) can be taken as a coarse moduli space $\mathcal{M}_{M,K3}$ (resp. $\mathcal{M}^a_{M,K3}$) of M-polarized (resp. ample M-polarized) K3 surfaces.

Notice that there are only finitely many orbits of Γ_M on the set N_{-2}. So,

$$\Gamma_M\backslash\Delta_M$$

is a closed subset in the Zariski topology of $\Gamma_M\backslash\mathcal{D}_M$. We know that in the case when M is of rank 1, the orbit space is irreducible because Γ_M contains an element switching two connected components of $\mathcal{D}_M$. The same is true (and the same proof works) if the lattice N contains a direct summand isomorphic to U [Do].

Let (X, j) be a M-polarized K3 surface. Note that in general $j(M) \neq S_X$. As we explained, a surface (X, j) is ample M-polarized if and only if $j(M)^\perp \cap S_X$ contains no curves R with $R^2 = -2$. As in the case of a pseudo-polarized K3 surface we can define the *degeneracy lattice* $\mathcal{R}(X, j)$ as the span of (-2)-curves in $j(M)^\perp$.

11. Eigenperiods of algebraic K3 surfaces

First we recall the fundamental result due to Nikulin [N2].

Theorem 11.1. *Let X be an algebraic K3 surface and let ω_X be a nowhere vanishing holomorphic 2-form on X. Let*

$$\alpha : \mathrm{Aut}(X) \to \mathbb{C}^*$$

be a homomorphism defined by $g^*(\omega_X) = \alpha(g) \cdot \omega_X$ for $g \in \mathrm{Aut}(X)$. Then:

1. *The image of α is a cyclic group of a finite order m.*
2. *Assume that $\alpha(g)$ is a primitive m-th root of 1. Then g^* has no non-zero fixed vectors in $T_X \otimes \mathbb{Q}$.*

Let C_m be a cyclic group of order m acting on a M-polarized K3 surface (X, j) as automorphisms effectively. We assume that the restriction of g^* to $j(M)$ is the identity map for any $g \in C_m$. Fixing a marking $\phi : L_{K3} \to H^2(X, \mathbb{Z})$ such that $\phi|M = j$ we obtain a homomorphism

$$\rho_\phi : C_m \to \mathrm{O}(L_{K3}), \quad g \mapsto \phi^{-1} \circ g^{-1*} \circ \phi.$$

Fix a homomorphism $\rho : C_m \to \mathrm{O}(L_{K3})$ such that

$$M = L_{K3}^\rho := \{x \in L_{K3} : \rho(a)(x) = x, \ \forall a \in C_m\}. \tag{11.1}$$

Define a (ρ, M)-marking of (X, j) as a marking $\phi : L_{K3} \to H^2(X, \mathbb{Z})$ with $\phi|M = j$ such that $\rho = \rho_\phi$.

Let $\chi \in \hat{C}_m$ be the unique character of C_m such that

$$H^{20}(X) \subset H^2(X, \mathbb{C})(\chi).$$

Then

$$p_{X,\phi} = \phi^{-1}(H^{20}(X)) \subset N_{\mathbb{C}}(\chi)$$

where $N = M^\perp$.

Assume that χ is not a real character. Then $N_{\mathbb{C}}(\chi)$ is an isotropic subspace of $N_{\mathbb{C}}$ with respect to the quadratic form Q defined by the lattice N. The restriction of the hermitian form $Q(x, \bar{y})$ to $p_{X,\phi}$ is positive. Let

$$\mathcal{D}_M^{\rho,\chi} = \{\mathbb{C}x \in N_{\mathbb{C}}(\chi) : Q(x, \bar{x}) > 0\}.$$

We know that

$$\mathcal{D}_M^{\rho,\chi} \cong \mathbb{B}_{d(\chi)},$$

where $d(\chi) = \dim N_{\mathbb{C}}(\chi) - 1$.

If χ is a real character, then $N_{\mathbb{C}}(\chi) = N_{\mathbb{R}}(\chi) \otimes \mathbb{C}$ and we set

$$\mathcal{D}_M^{\rho,\chi} = \{\mathbb{C}x \in N_{\mathbb{C}}(\chi) : Q(x,x) = 0, Q(x,\bar{x}) > 0.\}.$$

This is nothing but a type IV Hermitian symmetric space of dimension $d(\chi) - 1$.

There is a natural inclusion $\mathcal{D}_M^{\rho,\chi} \subset \mathcal{D}_M$ corresponding to the inclusion of $N_{\mathbb{C}}(\chi) \subset N_C$.

The group C_m acts on $\mathcal{D}_M$ via the restriction of the representation ρ to the homomorphism $\rho' : C_m \to \mathrm{O}(N)$. We denote by $\mathcal{D}_M^\rho$ the set of points which are fixed under all $\rho(a)$, $a \in C_m$. It is clear that

$$\mathcal{D}_M^\rho = \coprod_{\chi \in \hat{C}_m} \mathcal{D}_M^{\rho,\chi}.$$

Theorem 11.2. *Let (X, ϕ) be a (ρ, M)-marked ample M-polarized K3 surface such that $H^{20}(X) \subset H^2(X,\mathbb{C})(\chi)$. Then $p_{X,\phi} \in \mathcal{D}_M^{\rho,\chi} \setminus \Delta_M$. Conversely, any point in $\mathcal{D}_M^{\rho,\chi} \setminus \Delta_M$ is equal to $p_{X,\phi}$ for some (ρ, M)-marked ample M-polarized K3 surface (X, ϕ).*

Proof. The first part was already explained. Let ℓ be a point in $\mathcal{D}_M^{\rho,\chi} \setminus \Delta_M$. It represents a line in $N_{\mathbb{C}}$ which is fixed under $\rho(a)$, $a \in C_m$. Let $\phi : L_{K3} \to H^2(X,\mathbb{Z})$ be an ample M-marking of X such that $\phi(\ell) = H^{20}(X)$. Then $\phi^{-1} \circ \rho(a) \circ \phi$ acts identically on $j(M)$ and preserves $H^{20}(X)$. Since (X, j) is ample M-polarized, $j(M)$ contains an ample class which is preserved under $\rho(a)$. By the Global Torelli Theorem $\rho(a) = g^*$ for a unique $g \in \mathrm{Aut}(X)$. This defines an action of C_m on X and a (ρ, M)-marking of X. Obviously, $H^{20}(X)$ and $p_{X,\phi}$ are eigenspaces corresponding to the same character of C_m. □

Let

$$\Gamma_{M,\rho} = \{\sigma \in \Gamma_M : \sigma \circ \rho(a) = \rho(a) \circ \sigma, \forall a \in C_m\}. \tag{11.2}$$

Obviously, the group $\Gamma_{M,\rho}$ leaves the connected components $\mathcal{D}_M^{\rho,\chi}$ of $\mathcal{D}_M^\rho$ invariant. Applying Theorem 11.2, we obtain

Theorem 11.3. *The orbit space $\Gamma_{M,\rho} \backslash (\mathcal{D}_M^{\rho,\chi} \backslash \Delta_M)$ parametrizes isomorphism classes of ample (ρ, M)-polarized K3 surfaces such such that $H^{20}(X) \subset H^2(X,\mathbb{C})(\chi)$.*

Remark 11.4. Let

$$\tilde{\Gamma}_{M,\rho} = \{\sigma \in \mathrm{O}(N) : \sigma \circ \rho(a) = \rho(a) \circ \sigma, \forall a \in C_m\}. \tag{11.3}$$

If the canonical map $\tilde{\Gamma}_{M,\rho} \to \mathrm{O}(D(N))$ is surjective, the quotient $\tilde{\Gamma}_{M,\rho}/\Gamma_{M,\rho}$ is isomorphic to $\mathrm{O}(D(N)) \cong \mathrm{O}(D(M))$. The orbit space $\tilde{\Gamma}_{M,\rho} \backslash (\mathcal{D}_M^{\rho,\chi} \setminus \Delta_M)$ (resp. $\Gamma_{M,\rho} \backslash (\mathcal{D}_M^{\rho,\chi} \setminus \Delta_M)$) sometimes parametrizes the moduli space of some varieties (resp. some varieties with a marking). See Section 12.

Lemma 11.5. *Let $p_{X,\phi} \in \mathcal{D}_M^\rho \cap \Delta_M$. Then $\rho = \rho_\phi$ is not represented by any automorphisms of X.*

Proof. Assume that $p_{X,\phi} \in \mathcal{D}^{\rho}_M \cap \Delta_M$ and $\rho = \rho_\phi$ is a representation of C_m acting on X as automorphisms. Since C_m is finite, by averaging, we can find a polarization class h such that $g^*(h) = h$ for all $g \in C_m$. By condition (11.1), $h \in j(M)$ which contradicts the assumption that (X, j) is not ample. □

In the following we study the locus $\mathcal{D}^{\rho}_M \cap \Delta_M$ under the assumption that m is an odd prime number p. Recall that C_p acts trivially on M and acts effectively on $N = M^\perp$. This implies that $D(M) \cong D(N)$ are p-elementary abelian groups. Consider the degeneracy lattice $\mathcal{R}(X, j)$ of (X, j). If $r \in \mathcal{R}(X, j)$ is a (-2)-vector, then $\langle r, \omega_X\rangle = 0$ implies that $\langle \rho^i(r), \omega_X\rangle = 0$. Hence for any marking ϕ of (X, j) its pre-image $R = \phi^{-1}(\mathcal{R}(X, j))$ is an invariant sublattice of L_{K3} under ρ orthogonal to M. We consider the generic case, that is, R is generated by $r, \rho(r), \dots, \rho^{p-2}(r)$.

Lemma 11.6. *Let $r \in \mathcal{R}(X, j)$ with $r^2 = -2$. Let R be the lattice generated by $r, \rho(r), \dots, \rho^{p-2}(r)$. Then R is ρ-invariant and isomorphic to the root lattice A_{p-1}. Moreover ρ acts trivially on $D(R)$.*

Proof. Since R is generated by (-2)-vectors and negative definite, R is a root lattice of rank $p-1$. The ρ-invariance is obvious. Let $m_i = \langle r, \rho^i(r)\rangle$. Then the definiteness implies that $|m_i| \le 1$. Obviously $m_1 = m_{p-1}, m_2 = m_{p-2}, \dots, m_{(p-1)/2} = m_{(p+1)/2}$. Since $-2 = r^2 = \langle r, -\sum_{i=1}^{p-1} \rho^i(r)\rangle = -2m_1 - \cdots - 2m_{(p-1)/2}$, $m_i = 1$ for some $1 \le i \le (p-1)/2$. If $m_i = m_j = 1$ for $1 \le i \ne j \le (p-1)/2$, then $r, \rho^i(r), \rho^j(r), \rho^{p-i}(r), \rho^{p-j}(r)$ generate a degenerate lattice. This contradicts the definiteness of R. Thus there exists a unique i such that $m_i = 1$ and $m_j = 0$ for $1 \le i \le (p-1)/2$. By changing ρ by ρ^i, we may assume that $m_1 = 1$. Then $\langle \rho^i(r), \rho^{i+1}(r)\rangle = 1$ and hence $r, \rho(r), \dots, \rho^{p-2}(r)$ generate a root lattice A_{p-1}. The last assertion follows from the fact that ρ fixes a generator $(r + 2\rho(r) + \cdots + (p-1)\rho^{p-2}(r))/p$ of $D(A_{p-1})$. □

Let R be as in Lemma 11.6. Let M' be the smallest primitive sublattice of L_{K3} containing $M \oplus R$. We have a chain of lattices

$$0 \subset M \oplus R \subset M' \subset M'^* \subset (M \oplus R)^*.$$

Since any $\rho(a)$ acts identically on $D(M \oplus R) = D(M) \oplus D(R)$ (Lemma 11.6), it acts identically on $D(M')$. Let $N' = M'^\perp$. Since N' is the orthogonal complement of M' in a unimodular lattice, the discriminant groups $D(N')$ and $D(M')$ are isomorphic and the action of C_m on $D(N')$ is also identical. This implies that $(1_{M'} \oplus \rho(a)|N')$ can be extended to an isometry $\rho'(a)$ of L_{K3}. This defines a homomorphism $\rho' : C_m \to \mathrm{O}(L_{K3})$ with the sublattice of fixed vectors $L^{\rho'}_{K3}$ equal to M'. Let $j' = \phi|M' : M' \to \mathrm{Pic}(X)$ and let $M'_{-2} = {M'_{-2}}^+ \coprod {M'_{-2}}^-$ be the partition of M'_{-2} which extends the partition of M_{-2} and is defined by the condition that $j'({M'_{-2}}^+)$ is contained in the set of effective divisor classes. Then (X, j') is a M'-polarized K3 surface. The polarization is ample if $p(X, j') \notin \mathcal{H}_\delta$ for any $\delta \in N'_{-2}$. If this condition holds, the pair (X, j') is a (ρ', M')-polarized K3 surface. By induction, we can prove the following result.

Theorem 11.7. *Assume $m = p$. For any point $x \in \mathcal{D}_M^{\rho,\chi}$, let $R(x)$ denote the sublattice of N generated by the vectors $\delta \in N_{-2}$ such that $x \in \mathcal{H}_\delta$. Then $R(x)$ is a p-elementary root lattice, that is, $R(x)^*/R(x)$ is a p-elementary abelian group. Let M' be the smallest primitive sublattice of L_{K3} generated by M and $R(x)$. Then $x = p_{X,\phi}$ for some marked ample (ρ', M')-polarized K3 surface X, where ρ' is a representation of C_p on L_{K3} with $L_{K3}^{\rho'} = M'$. The lattice $R(x)$ is isomorphic to the degeneracy lattice of the M-polarized K3 surface (X, j), where $j = \phi|M$.*

12. Examples

In this section, we shall give examples of eigenperiods of K3 surfaces. For a lattice L and an integer m, we denote by $L(m)$ the lattice over the same $\mathbb{Z}$-module with the symmetric bilinear form multiplied by m.

Example. Let $C_2 = (g)$ be a cyclic group of order 2. Write L_{K3} as a direct sum of two lattices $M = E_8 \oplus U$ and $N = E_8 \oplus U \oplus U$ and define $\rho : C_2 \to \mathrm{O}(L_{K3})$ by $\rho(g)$ to be the identity on the first summand and the minus identity on the second one. Let $f : X \to \mathbb{P}^1$ be an elliptic K3 surface with a section s (taking as the zero in the Mordell–Weil group of sections) and assume that one of the fibres is a reducible fibre of type II* in Kodaira's notation. Choose an M-marking ϕ of X by fixing an isomorphism from E_8 to the subgroup of S_X generated by components of the reducible fibre not intersecting s and an isomorphism from U to the subgroup of S_X to be the pre-image under ϕ generated by any fibre and the section s. The partition of M_{-2} is defined by taking M_{-2}^+ to be the pe-image under ϕ of the set of divisor classes of irreducible components of the reducible fibre and the section. Now the coarse moduli space of isomorphism classes of M-polarized K3 surfaces consists of elliptic K3-surfaces with a section and a fibre of type II^*. Each such surface admits an automorphism of order 2 defined by the inversion automorphism $x \mapsto -x$ on a general fibre with respect to the group law defined by the choice of a section. The M-polarization is ample if and only if the elliptic fibration has only one irreducible fibre. Let $\chi \in \hat{C}_2$ be defined by $\chi(g) = -1$. Then $\mathcal{D}_M^{\rho,\chi} \cong \mathcal{D}_M$ and $\Gamma_{M,\rho} = \Gamma_M$. An ample M-polarized K3 surface is automatically (ρ, M)-polarized K3 surface. The degeneracy lattice $\mathcal{R}(X, j)$ is the sublattice of $\mathrm{Pic}(X)$ generated by components of new reducible fibres not intersecting the section. Using Theorem 11.7 we see that any M-polarized K3 surface admits a structure of a (ρ', M')-polarized K3 surface, where M' is isomorphic to the sublattice of S_X generated by components of fibres and a section and ρ' acts identically on M' and the minus-identity on its orthogonal complement.

Example. (6 points on $\mathbb{P}^1$) Let X' be a surface in weighted projective space $\mathbb{P}(1,1,2,2)$ given by an equation $f_6(x_0, x_1) + x_2^3 + x_3^3 = 0$, where f_6 is a homogeneous form of degree 6 without multiple roots. The surface X' has 3 ordinary nodes $(0,0,1,a)$, where $a^3 = -1$. A minimal resolution of these nodes is a K3 surface X on which the cyclic group $C_3 = \mu_3$ of 3rd roots of unity acts by $(x_0, x_1, x_2, x_3) \mapsto (x_0, x_1, \alpha x_2, \alpha x_3)$. Another way to define this surface is to consider a quadratic cone in $\mathbb{P}^3$, take its transversal intersection with a cubic surface, and then to define X' to be the triple cover of the cone branched along the intersection curve C. The pre-image of the ruling of the cone defines an elliptic fibration $f : X \to \mathbb{P}^1$. Since C has 6 ramification points of order 3 over the roots of f_6, the fibration has 6 reducible fibres of Kodaira type III. The three exceptional curves of the resolution $X \to X'$ form a group of sections isomorphic to $\mathbb{Z}/3\mathbb{Z}$. An explicit computation of the Hodge structure on hypersurfaces in a weighted projective space (see [Do2]) shows that $H^{20}(X) \subset H^2(X, \mathbb{C})(\chi)$, where $\chi(\alpha) = \alpha^2$ and

$$\dim H^{11}(X)(\chi) = 3, \dim H^{11}(X)^{C_3} = 12.$$

Let M be the primitive sublattice of S_X generated by the classes of irreducible components of fibres, and sections. Standard arguments from the theory of elliptic surfaces yield that $M \cong U \oplus E_6 \oplus A_2^3$ and its orthogonal complement $N \cong A_2(-1) \oplus A_2^3$. We see that

$$\mathcal{D}_M^{\rho,\chi} \cong \mathbb{B}_3.$$

The lattice N has a natural structure of a module over the ring of Eisenstein numbers $\mathbb{Z}[e^{2\pi i/3}]$ equipped with a hermitian form over this ring. One shows that

$$\Gamma_{M,\rho} \cong \mathrm{SU}(4, \mathbb{Z}[e^{2\pi i/3}]).$$

Fix a primitive embedding of M in L_{K3} and choose $\mathcal{C}(M)^+$ to be the nef cone in $\mathrm{Pic}(X_0)$ for a fixed X as above. Let ρ be the representation of C_3 on L_{K3} which acts as the identity on $M = U \oplus E_6 \oplus A_2^3$ and acts on N as follows. Choose a standard basis (r_1, r_2) of A_2 corresponding to the vertices of the Dynkin diagram. We want to define an action ρ of the group C_3 on N such that $N^{C_3} = \{0\}$ and C_3 acts trivially on the discriminant group of N (because it acts trivially on M). This easily implies that each direct summand must be invariant, and the action on it has no fixed vectors and the generator of $D(A_2)$ equal to $(r_1 + 2r_2)/3$ is invariant. It is easy to see that the representation on A_2 with this property must be isomorphic to the one given by $\rho(e^{2\pi i/3})(r_1, r_2) = (r_2, -r_1 - r_2)$. Now let $\rho(e^{2\pi i/3})$ act on N as the direct sum of the previous action on each direct summand. In this way we obtain that the coarse moduli space of (M, ρ)-polarized K3 surfaces is isomorphic to the ball quotient

$$\mathcal{D}_M^{\rho,\chi}/\Gamma_{M,\rho} \cong \mathbb{B}_3/\mathrm{SU}(4, \mathbb{Z}[e^{2\pi i/3}]).$$

Note that, using the hypergeometric functions with $\boldsymbol{\mu} = (1/3, 1/3, 1/3, 1/3, 1/3, 1/3)$ the same quotient is isomorphic to the space $\mathcal{P}_{1,\mathbf{k}}$ of ordered 6 points on $\mathbb{P}^1$, where $\mathbf{k} = (1, 1, 1, 1, 1, 1)$. It is also isomorphic to the moduli space of principally polarized abelian varieties of dimension 4 with action of a cyclic group of order 3 of type $(1, 3)$.

Example. (Del Pezzo surfaces of degree 2) Let R be a smooth del Pezzo surface of degree 2. Recall that the anti-canonical model of a smooth del Pezzo surface of degree 2 is a double cover of $\mathbb{P}^2$ branched along a smooth plane quartic curve C.

Let $f_4(x,y,z)$ be the homogeneous polynomial of degree 4 defining C. Let X be a quartic surface in $\mathbb{P}^3$ defined by

$$t^4 = f_4(x,y,z).$$

Then X is a K3 surface which is the 4-cyclic cover of $\mathbb{P}^2$ branched along C. Obviously X admits an automorphism g of order 4. If C is generic, then the Picard lattice S_X (resp. the transcendental lattice T_X) is isomorphic to $M = U(2) \oplus A_1^{\oplus 6}$ (resp. $N = U(2) \oplus U(2) \oplus D_8 \oplus A_1^{\oplus 2}$). In this case, $(g^*)^2$ acts trivially on M, but g^* does not. The above correspondence gives an isomorphism between the moduli space of smooth del Pezzo surfaces of degree 2 and $(\mathbb{B}_6 \setminus \Delta_M)/\tilde{\Gamma}_{M,\rho}$. The quotient $(\mathbb{B}_6 \setminus \Delta_M)/\Gamma_{M,\rho}$ is isomorphic to the moduli space of *marked* smooth del Pezzo surfaces of degree 2. The natural map

$$\mathbb{B}_6/\Gamma_{M,\rho} \to \mathbb{B}_6/\tilde{\Gamma}_{M,\rho}$$

is a Galois covering with $\mathrm{O}(D(M)) \cong \mathrm{Sp}(6,\mathbb{F}_2)$ as its Galois group. We remark that this correspondence gives a uniformization of the moduli space of non-hyperelliptic curves of genus 3. For more details we refer the reader to the paper [K1].

Example. (8 points on $\mathbb{P}^1$) This case is a degenerate case of the previous example since 8 points on $\mathbb{P}^1$ correspond to a hyperelliptic curve of genus 3. Let $\{(\lambda_i : 1)\}$ be a set of distinct 8 points on the projective line. Let $(x_0 : x_1, y_0 : y_1)$ be the bi-homogeneous coordinates on $\mathbb{P}^1 \times \mathbb{P}^1$. Consider a smooth divisor C in $\mathbb{P}^1 \times \mathbb{P}^1$ of bidegree $(4,2)$ given by

$$y_0^2 \cdot \prod_{i=1}^{4}(x_0 - \lambda_i x_1) + y_1^2 \cdot \prod_{i=5}^{8}(x_0 - \lambda_i x_1) = 0. \tag{12.1}$$

Let L_0 (resp. L_1) be the divisor defined by $y_0 = 0$ (resp. $y_1 = 0$). Let ι be an involution of $\mathbb{P}^1 \times \mathbb{P}^1$ given by

$$(x_0 : x_1, y_0 : y_1) \longrightarrow (x_0 : x_1, y_0 : -y_1) \tag{12.2}$$

which preserves C and L_0, L_1. Note that the double cover of $\mathbb{P}^1 \times \mathbb{P}^1$ branched along $C + L_0 + L_1$ has 8 rational double points of type A_1 and its minimal resolution X is a $K3$ surface. The involution ι lifts to an automorphism σ of order 4. The projection

$$(x_0 : x_1, y_0 : y_1) \longrightarrow (x_0 : x_1)$$

from $\mathbb{P}^1 \times \mathbb{P}^1$ to $\mathbb{P}^1$ induces an elliptic fibration

$$\pi : X \longrightarrow \mathbb{P}^1$$

which has 8 singular fibers of type III and two sections. In this case, $M \simeq U(2) \oplus D_4 \oplus D_4$ and $N \simeq U \oplus U(2) \oplus D_4 \oplus D_4$. Thus we have an isomorphism between the moduli space of *ordered* (distinct) 8 points on $\mathbb{P}^1$ and $(\mathbb{B}_5 \setminus \Delta)/\Gamma_{M,\rho}$. This ball quotient appeared in Deligne–Mostow's list [DM]. The group $\mathrm{O}(D(M)) \cong S_8$ naturally acts on the both spaces. Taking the quotient by S_8, we have an isomorphism between the moduli space of (distinct) 8 points on $\mathbb{P}^1$ and $(\mathbb{B}_5 \setminus \Delta)/\tilde{\Gamma}_{M,\rho}$. For more details we refer the reader to the paper [K3].

Example. (Del Pezzo surfaces of degree 3) Let S be a smooth del Pezzo surface. The anti-canonical model of S is a smooth cubic surface. Let l be a line on S. Consider the conic bundle on S defined by hyperplanes through l. The line l is a 2-section of this pencil. For generic S, this pencil has 5 degenerate members. Moreover there are two fibers tangent to l. These give homogeneous polynomials $f_5(x, y)$, $f_2(x, y)$ of degree 5 and 2 in two variables. Now consider the plane sextic curve C defined by

$$z(f_5(x, y) + z^3 f_2(x, y)) = 0.$$

Take the double cover of $\mathbb{P}^2$ branched along C whose minimal resolution X is a K3 surface. The multiplication of z by $e^{2\pi\sqrt{-1}/3}$ induces an automorphism g of X of order 3. In this case, $M \simeq U \oplus A_2^{\oplus 5}$ and $N \simeq A_2(-1) \oplus A_2^{\oplus 4}$. The non-trivial fact is that the K3 surface X is independent of the choice of a line l. Thus we have an isomorphism between the moduli space of smooth cubic surfaces and $(\mathbb{B}_4 \setminus \Delta)/\tilde{\Gamma}_{M,\rho}$. The group $\mathrm{O}(D(M)) \cong W(E_6)$ appears as the Galois group of the covering

$$\mathbb{B}_4/\Gamma_{M,\rho} \to \mathbb{B}_5/\tilde{\Gamma}_{M,\rho}.$$

We remark that the pencil of conics on S induces an elliptic fibration on X with five singular fibers of type IV and two singular fibers of type II. This gives a set of 7 points on $\mathbb{P}^1$. The ball quotient $\mathbb{B}_4/\Gamma_{M,\rho}$ appeared in Deligne–Mostow's list [DM]. For more details we refer the reader to the paper [Vo]. Note that another approach (see [ACT]) consists of attaching to a cubic surface with equation $f(x_0, x_1, x_2, x_3) = 0$, the cubic hypersurface with equation $f(x_0, x_1, x_2, x_3) + x_4^3 = 0$, and considering its intermediate Jacobian variety. It is a principally polarized abelian variety of dimension 5 with cyclic group of order 3 acting on it with type $(1, 4)$. As we know from Section 11, such varieties are parametrized by a 4-dimensional ball.

Example. (Del Pezzo surfaces of degree 4) Let S be a smooth del Pezzo surface of degree 4. Its anti-canonical model is the complete intersection of two quadrics Q_1, Q_2 in $\mathbb{P}^4$. It is known that Q_1 and Q_2 can be diagonalized simultaneously, that is, we may assume

$$Q_1 = \{\sum z_i^2 = 0\}, \quad Q_2 = \{\sum \lambda_i z_i^2 = 0\}.$$

The discriminant of the pencil of quadrics $\{t_1 Q_1 + t_2 Q_2\}_{(t_1:t_2)}$ is distinct 5 points $\{(\lambda_i : 1)\}$ on $\mathbb{P}^1$. Conversely distinct 5 points on $\mathbb{P}^1$ give the intersection of two quadrics. Thus the moduli space of smooth del Pezzo surfaces of degree 4 is isomorphic to the moduli of distinct 5 points on $\mathbb{P}^1$. Next we construct a K3 surface from distinct 5 points $\{(\lambda_i : 1)\}$ on $\mathbb{P}^1$. Let C be a plane sextic curve defined by

$$x_0^6 = x_0 \prod_{i=1}^{5} (x_1 - \lambda_i x_2).$$

Obviously C is invariant under a projective transformation of order 5 by the multiplication of x_0 by a primitive 5-th root of 1. Take the double cover of $\mathbb{P}^2$ branched along C whose minimal resolution X is a K3 surface. The X has an automorphism of order 5 induced from the above projective transformation. In this case M is of rank 10 and $D(M) = (\mathbb{Z}/5\mathbb{Z})^3$, and

$$N \cong \begin{pmatrix} 0 & 1 \\ 1 & 0 \end{pmatrix} \oplus \begin{pmatrix} 2 & 1 \\ 1 & -2 \end{pmatrix} \oplus A_4 \oplus A_4.$$

Thus we have an isomorphism between the moduli space of *ordered* (distinct) 5 points on $\mathbb{P}^1$ and $(\mathbb{B}_2 \setminus \Delta)/\Gamma_{M,\rho}$. The ball quotient $(\mathbb{B}_2 \setminus \Delta)/\Gamma_{M,\rho}$ appeared in Deligne–Mostow [DM]. The group $\mathrm{O}(D(M)) \cong S_5$ naturally acts on the both spaces and the quotients give an isomorphism between the moduli space of smooth del Pezzo surfaces of degree 4 and $(\mathbb{B}_2 \setminus \Delta)/\tilde{\Gamma}_{M,\rho}$. For more details we refer the reader to the paper [K4].

Example. (Curves of genus 4) Let C be a non-hyperelliptic curve of genus 4. The canonical model of C is the intersection of a smooth quadric Q and a cubic surface S in $\mathbb{P}^3$. By taking the triple cover of Q branched along C, we have a smooth K3 surface X with an automorphism of order 3. In this case $M = U(3)$ and $N = U \oplus U(3) \oplus E_8 \oplus E_8$. Thus we have an isomorphism between the moduli space of non-hyperelliptic curves of genus 4 and $(\mathbb{B}_9 \setminus \Delta)/\tilde{\Gamma}_{M,\rho}$. We remark that a ruling of Q induces an elliptic fibration on X with 12 singular fibers of type II. This gives a set of 12 points on $\mathbb{P}^1$. The arithmetic subgroup $\Gamma_{M,\rho}$ is isogeny to the complex reflection group associated to 12 points on $\mathbb{P}^1$ that appeared in Mostow [Mo]. For more details we refer the reader to the paper [K2].

Example. (Del Pezzo surfaces of degree 1) This case is a degenerate case of the previous example. Let S be a smooth del Pezzo surface of degree 1. The anti-bi-canonical model of S is a double cover of a quadric cone Q_0 in $\mathbb{P}^3$ branched along the vertex of Q_0 and a smooth curve C of genus 4. It is known that C is an intersection of Q_0 and a cubic surface. By taking the triple cover of Q_0 branched along C and then taking its minimal resolution, we have a K3 surface X with an automorphism of order 3. In this case, $M = U \oplus A_2(2)$ and $N = U \oplus U \oplus E_8 \oplus D_4 \oplus A_2$. Thus we have an isomorphism between the moduli space of smooth del Pezzo surfaces of degree 1 and $(\mathbb{B}_8 \setminus \Delta)/\tilde{\Gamma}_{M,\rho}$. For more details we refer the reader to the paper [K2].

We summarize these examples in the Table 1.

13. Half-twists of Hodge structures

Here we discuss a version of a construction due to Bert van Geemen [vG].

Let

$$V_{\mathbb{C}} = \bigoplus_{p+q=k} V^{pq}$$

be a polarized AHS of weight k on a real vector space V. Let $\rho : A \to \mathrm{GL}(V)$ be an action of a cyclic group of order d on V by Hodge isometries. Choose a generator g

	Deligne – Mostow	M	N
Curves of genus 4	$(\frac16\frac16\frac16\frac16\frac16\frac16\frac16\frac16\frac16\frac16\frac16\frac16)$	$U(3)$	$U\oplus U(3)\oplus E_8\oplus E_8$
del Pezzo of degree 1	a subball quotient	$U\oplus A_2(2)$	$U\oplus U\oplus E_8\oplus D_4\oplus A_2$
del Pezzo of degree 2	not appear	$U(2)\oplus A_1^{\oplus 6}$	$U(2)\oplus U(2)\oplus D_8\oplus A_1^{\oplus 2}$
del Pezzo of degree 3	$(\frac26\frac26\frac26\frac26\frac26\frac16\frac16)$	$U\oplus A_2^{\oplus 5}$	$A_2(-1)\oplus A_2^{\oplus 4}$
del Pezzo of degree 4	$(\frac25\frac25\frac25\frac25\frac25)$	$U\oplus D_8\oplus D_8$	$U(2)\oplus U(2)$
6 points on $\mathbb{P}^1$	$(\frac13\frac13\frac13\frac13\frac13\frac13)$	$U\oplus E_6\oplus A_2^{\oplus 3}$	$A_2(-1)\oplus A_2^{\oplus 3}$
8 points on $\mathbb{P}^1$	$(\frac14\frac14\frac14\frac14\frac14\frac14\frac14\frac14)$	$U(2)\oplus D_4\oplus D_4$	$U\oplus U(2)\oplus D_4\oplus D_4$

TABLE 1

of A and a subset Σ of $\mathrm{Wt}(\rho)$ which does not contain real characters and satisfies $\mathrm{Im}(\chi(g))>0$ for any $\chi\in\Sigma$. The vector space

$$\bigoplus_{\chi\in\Sigma\cup\overline{\Sigma}} V_{\mathbb{C}}(\chi)\cong W_{\mathbb{C}}$$

for some vector subspace $W\subset V$ (because it is invariant with respect to the conjugation of $V_{\mathbb{C}}$). Write

$$V_{\Sigma}^{pq}=\bigoplus_{\chi\in\Sigma}V_{\chi}^{pq},\quad V_{\overline{\Sigma}}^{pq}=\bigoplus_{\chi\in\overline{\Sigma}}V_{\chi}^{pq}. \tag{13.1}$$

Define the *negative half twist* of V to be the decomposition

$$W_{\mathbb{C}}=\bigoplus_{r+s=k+1}W^{rs},$$

where

$$W^{rs}=V_{\Sigma}^{r-1s}\oplus V_{\overline{\Sigma}}^{rs-1}.$$

Obviously, the decomposition (13.1) satisfies property (HD1) of AHS. Let us define a polarization form on W by changing the polarization form Q on V with

$$Q'(x,y)=Q(x,g(y))-Q(x,g^{-1}(y)).$$

Using that each $V_{\mathbb{C}}(\chi)$ and $V_{\mathbb{C}}(\chi')$ are orthogonal with respect to Q unless $\bar\chi=\chi'$, and V^{ab} orthogonal to $V^{a'b'}$ unless $a=b'$, we check property (HD2). Let $x\in V_{\chi}^{r-1s}$, $\chi\in\Sigma$, write $\chi(g)-\bar\chi(g)=bi$, where $b>0$. We have

$$i^{r-s}Q'(x,\bar x)=i^{r-s}Q(x,g(\bar x)-g^{-1}(\bar x))$$
$$i^{r-s}Q(x,(\bar\chi(g)-\chi(g))\bar x)=i^{r-s}(\bar\chi(g)-\chi(g))Q(x,\bar x)=bi^{r-s-1}Q(x,\bar x)>0.$$

Similarly, we check that $i^{r-s}Q'(x,\bar x)>0$ if $x\in V_{\chi}^{rs-1}$, where $\chi\in\overline{\Sigma}$. This checks property (HD3).

The situation with integral structure is more complicated. Let F be the extension of $\mathbb{Q}$ obtained by joining a dth root of unity. Suppose our AHS has an integral structure with respect to some lattice Λ in V and ρ is obtained from a representation in $\mathrm{GL}(\Lambda)$. Then Λ acquires a structure of a module over the ring of integers $\mathcal{O}$ in the field $F = \mathbb{Q}(e^{2\pi i/d})$ by setting $e^{2\pi i/d} \cdot v = \rho(g)(v)$. The vector space W becomes a vector space over F and Λ is a lattice in V such that Q' can be obtained from a bilinear form on Λ taking its values in $\mathcal{O}$.

Example. Assume $V_{\mathbb{C}} = V^{10} \oplus V^{01}$ is a polarized AHS of weight 1. Then

$$W^{20} = V_{\Sigma}^{10}, \quad W^{11} = V_{\Sigma}^{01} \oplus V_{\bar{\Sigma}}^{10}, \quad W^{02} = V_{\bar{\Sigma}}^{01}.$$

Suppose $\dim V_{\Sigma}^{10} = 1$, then we obtain a Hodge structure of the same type as the Hodge structure arising from a K3 surface. It does not need to be a Hodge structure of a K3 surfaces if, say $\dim W^{11} > 19$.

Let $\boldsymbol{\mu} = (\mu_1, \ldots, \mu_m) = (\frac{a_1}{d}, \ldots, \frac{a_m}{d})$ as in Section 8. We assume that the condition (8.6) holds and $m > 4$. Consider the curve X isomorphic to a nonsingular projective model of the affine curve (8.11). Let C_d be a cyclic group of order d and $\chi \in \hat{A}$ be the character corresponding to $\boldsymbol{\mu}$. We know that $\dim H^{10}(X, \mathbb{C})(\chi) = 1$ and $\dim H^{10}(X, \mathbb{C})(\bar{\chi}) = m - 3$. Take $\Sigma = \{\chi, \bar{\chi}\}$. Then we obtain that the half-twist W is AHS with Hodge numbers $(1, 2m-6, 1)$ and admitting a Hodge isometry g. Assume that W contains a lattice N such that the AHS has an integral structure with respect to N. It is easy to see that the signature of N is equal to $(2, 2m-6)$. Assume that N can be primitively embedded in L_{K3} and let $M = N^{\perp}$. Then our AHS corresponds to a point in $\mathcal{D}_M^{\rho,\chi} \cong \mathbb{B}_{m-3}$. By surjectivity of the period map it corresponds to a (M, ρ)-polarized K3 surface. We conjecture that the needed condition on $\boldsymbol{mu}$ are always satisfied if the monodromy group Γ_μ is an arithmetic subgroup of $\mathrm{Aut}(\mathbb{B}_{m-3})$. For $d = 3, 4, 5, 6$, the conjecture holds (see Section 12).

References

[ACT] D. Allcock, J. A. Carlson, D. Toledo, *The Complex Hyperbolic Geometry of the Moduli Space of Cubic Surfaces*, J. Algebraic Geometry, **11** (2002), 659–724.

[De] P. Deligne, a letter to I. M. Gelfand.

[DM] P. Deligne, G. W. Mostow, *Monodromy of hypergeometric functions and non-lattice integral monodromy*, Publ. Math. IHES, **63** (1986), 5–89.

[Do] I. Dolgachev, *Mirror symmetry for lattice polarized K3-surfaces*, J. Math. Sciences **81** (1996), 2599–2630.

[Do2] I. Dolgachev, *Weighted projective varieties*, in "Group actions on algebraic varieties", Lect. Notes in Math. **956** (1982), 34–71.

[DGK] I. Dolgachev, B. van Geemen, S. Kondō, *A complex ball uniformization of the moduli space of cubic surfaces via periods of K3 surfaces*, math.AG/0310342, J. reine angew. Math. (to appear).

[Ge] *Géometrie des surfaces K3: modules et périodes*, Astérisque, vol. 126, Soc. Math. France, 1985.

[vG] B. van Geemen, *Half twists of Hodge structure of CM-type*, J. Math. Soc. Japan, **53** (2001), 813–833.

[GH] Ph. Griffiths, J. Harris, *Principles of Algebraic Geometry*, Wiley and Sons, 1994.

[HL] G. Heckman, E. Looijenga, *The moduli space of rational elliptic surfaces*, Adv. Studies Pure Math., **36** (2002), Algebraic Geometry 2000, Azumino, 185–248.

[K1] S. Kondō, *A complex hyperbolic structure of the moduli space of curves of genus three*, J. reine angew. Math., **525** (2000), 219–232.

[K2] S. Kondō, *The moduli space of curves of genus* 4 *and Deligne-Mostow's complex reflection groups*, Adv. Studies Pure Math., **36** (2002), Algebraic Geometry 2000, Azumino, 383–400.

[K3] S. Kondō, *The moduli space of* 8 *points on* $\mathbb{P}^1$ *and automorphic forms*, to appear in the Proceedings of the Conference "Algebraic Geometry in the honor of Professor Igor Dolgachev".

[K4] S. Kondō, *The moduli space of* 5 *points on* $\mathbb{P}^1$ *and K3 surfaces*, math.AG/0507006, this volume.

[Lo] E. Looijenga, *Uniformization by Lauricella functions-an overview of the theory of Deligne–Mostow*, math.CV/050753, this volume.

[MT] K. Matsumoto, T. Terasoma, *Theta constants associated to cubic threefolds*, J. Alg. Geom., **12** (2003), 741-755.

[Mo] G. W. Mostow, *Generalized Picard lattices arising from half-integral conditions*, Publ. Math. IHES, **63** (1986), 91–106.

[N1] V. V. Nikulin, *Integral symmetric bilinear forms and its applications*, Math. USSR Izv., **14** (1980), 103–167.

[N2] V. V. Nikulin, *Finite automorphism groups of Kähler* $K3$ *surfaces*, Trans. Moscow Math. Soc., **38** (1980), 71–135.

[N3] V. V. Nikulin, *Factor groups of groups of automorphisms of hyperbolic forms with respect to subgroups generated by* 2*-reflections*, J. Soviet Math., **22** (1983), 1401–1475.

[SPS] I. Piatetski-Shapiro, I. R. Shafarevich, *A Torelli theorem for algebraic surfaces of type* $K3$, Math. USSR Izv., **5** (1971), 547–587.

[S] G. Shimura, *On purely transcendental fields of automorphic functions of several variables*, Osaka J. Math., **1** (1964), 1–14.

[Te] T. Terada, *Fonction hypergéométriques* F_1 *et fonctions automorphes*, J. Math. Soc. Japan **35** (1983), 451–475; II, ibid., **37** (1985), 173–185.

[Ter] T. Terasoma, *Infinitesimal variation of Hodge structures and the weak global Torelli theorem for complete intersections*, Ann. Math., **132** (1990), **37** (1985), 213–235.

[Th] W. P. Thurston, *Shape of polyhedra and triangulations of the sphere*, Geometry & Topology Monograph, **1** (1998), 511–549.

[Va] A. Varchenko, *Hodge filtration of hypergeometric integrals associated with an affine configuration of hyperplanes and a local Torelli theorem*, I. M. Gelfand Seminar, Adv. Soviet Math., **16**, Part 2, Amer. Math. Soc., Providence, 1993, J. Math. Soc. Japan **35** (1983), pp. 167–177.

[Vo] C. Voisin, *Hodge Theory and Complex Geometry*, Cambridge Studies in Adv. Math., vols. 76, 77, Cambridge Univ. Press, 2003.

Igor V. Dolgachev
Department of Mathematics
University of Michigan
Ann Arbor, MI 48109
USA
e-mail: idolga@umich.edu

Shigeyuki Kondō
Graduate School of Mathematics
Nagoya University
Nagoya 464-8602
Japan
e-mail: kondo@math.nagoya-u.ac.jp

Progress in Mathematics, Vol. 260, 101–108

Macbeaths infinite series of Hurwitz groups

Amir Džambić

Abstract. In the present paper we will construct an infinite series of so-called *Hurwitz groups*. One possible way to describe Hurwitz groups is to define them as finite homomorphic images of the Fuchsian triangle group with the signature $(2, 3, 7)$. A reason why Hurwitz groups are interesting lies in the fact, that precisely these groups occur as the automorphism groups of compact Riemann surfaces of genus $g > 1$, which attain the upper bound $84(g-1)$ for the order of the automorphism group. For a long time the only known Hurwitz group was the special linear group $\mathrm{PSL}_2(\mathbb{F}_7)$, with 168 elements, discovered by F. Klein in 1879, which is the automorphism group of the famous *Kleinian quartic*. In 1967 Macbeath found an infinite series of Hurwitz groups using group theoretic methods. In this paper we will give an alternative arithmetic construction of this series.

Mathematics Subject Classification (2000). 11F06, 14H37, 30F10, 30F35.

Keywords. Hurwitz groups, arithmetic Fuchsian groups, Fuchsian triangle groups.

1. Theorem of Hurwitz

We consider a compact Riemann surface X of genus $g > 1$. As it is well-known, X can be written as a quotient $X = \Gamma\backslash\mathcal{H}$, where $\mathcal{H}$ denotes the upper halfplane $\mathcal{H} := \{z \in \mathbb{C} \mid \Im(z) > 0\}$ and Γ is a cocompact Fuchsian group (discrete subgroup of $\mathrm{PSL}_2(\mathbb{R})$ with compact fundamental domain in $\mathcal{H}$), without elliptic elements. Using this description of X the automorphism group

$$\mathrm{Aut}(X) := \{f : X \to X \mid f \text{ is biholomorphic}\}$$

is isomorphic to the factor group

$$\mathrm{Aut}(X) \cong N(\Gamma)/\Gamma \tag{1.1}$$

where $N(\Gamma)$ is the normalizer of Γ in $\mathrm{PSL}_2(\mathbb{R})$. In 1893 Hurwitz proved the following

Theorem 1.1. *Let $X = \Gamma\backslash\mathcal{H}$ be a compact Riemann surface of genus $g > 1$. For the order $|Aut(X)|$ of the automorphism group the following statements hold:*

1. $|\operatorname{Aut}(X)| \leq 84(g-1)$,
2. $|\operatorname{Aut}(X)| = 84(g-1)$ *if and only if* Γ *is a torsion free normal subgroup of finite index in a Fuchsian triangle group with signature* $(2,3,7)$.

Thereby a **Fuchsian triangle group** is defined as a subgroup in $\mathrm{PSL}_2(\mathbb{R})$ generated by three elliptic elements δ_1,δ_2,δ_3, with orders p, q, r respectively. In order to be a discrete subgroup in $\mathrm{PSL}_2(\mathbb{R})$ the relation $\frac{1}{p} + \frac{1}{q} + \frac{1}{r} < 1$ has to be satisfied. The triple (p, q, r) is called the **signature** of the triangle group. The signature determines the triangle group uniquely up to conjugation in $\mathrm{PSL}_2(\mathbb{R})$. Therefore one often identifies the triangle group with its signature.

Definition 1.2. A finite group which can be realized as automorphism group $\operatorname{Aut}(X)$ in the case 2. of Theorem 1.1 is called **Hurwitz group**.

Now one can ask the following questions:

Are there some examples of Hurwitz groups?
And if there are any, are there infinitely many of them?

2. Groups of norm 1 elements in orders of quaternion algebras as Fuchsian groups

2.1. Definitions and notations

Let K denote a totally real algebraic number field (i.e., a number field whose Galois group consists only of elements leading to embeddings of K into $\mathbb{R}$) and $\mathcal{O}_K$ its ring of integers. Let $A = A/K = (\frac{a,b}{K})$ be a quaternion algebra over K, i.e., a central, simple algebra of dimension 4 over K with a basis $\{1, i, j, ij\}$, satisfying the following equations:

$$i^2 = a \in K^*, \quad j^2 = b \in K^*, \quad ij = -ji.$$

Let $\mathcal{O}$ denote a maximal order in A/K (a maximal $\mathcal{O}_K$−module in A such that $\mathcal{O} \otimes_{\mathcal{O}_K} K \cong A$). In A exists an involution $\bar{\ }: A \to A$ sending

$$x := x_0 + x_1 i + x_2 j + x_3 ij$$

to

$$\bar{x} := x_0 - x_1 i - x_2 j - x_3 ij.$$

We define the **reduced norm** $\mathrm{Nrd}(x) := x\bar{x}$ and the **reduced trace** $\mathrm{Trd}(x) := x + \bar{x}$. The **group of elements of reduced norm 1 in** $\mathcal{O}$ is defined as

$$\mathcal{O}^1 := \{x \in \mathcal{O} \mid \mathrm{Nrd}(x) = 1\}.$$

We put

$$\Gamma_{A,\mathcal{O}} := \mathcal{O}^1/\{\pm 1\}.$$

The localizations K_v of K at places v are fields containing K and complete with respect to the valuations v. If v is a finite (i.e., non-archimedian) place, the completion R_v of $\mathcal{O}_K$ in K_v is a discrete valuation ring with precisely one maximal ideal, which is generated by one element, the **uniformizer** of R_v. We define localizations A_v of A as $A_v := A \otimes_K K_v$. A_v is then the completion of A under the embedding $K \to K_v$ and A is dense in A_v. $\mathcal{O}_v$, $\mathcal{O}_v^1$ and $\Gamma_{A,\mathcal{O}}^{(v)}$ should denote the complete closures of $\mathcal{O}$,$\mathcal{O}^1$ and $\Gamma_{A,\mathcal{O}}$ in A_v, respectively.

2.2. Theorem of Borel and Harish-Chandra

From the theory of quadratic forms we have

Lemma 2.1. *For every place v of K A_v is either isomorphic to the matrix algebra $M_2(K_v)$ or isomorphic to the unique division quaternion algebra D_{K_v} over K_v.*

For example, if v is an infinite place, then the two posibilities for A_v are $A_v \cong M_2(\mathbb{R})$ or $A_v \cong \mathbb{H}$, the skew-field of Hamiltonian quaternions.

We say that A is **ramified** at v if $A_v \cong D_{K_v}$ and that A is **unramified** otherwise. We put

$$Ram(A) := \{v \mid A \,\text{is ramified at}\, v\},$$

$$Ram_\infty(A) := \{v \in Ram(A) \mid v \,\text{is infinite place of}\, K\}.$$

Now one observes that, if $|Ram_\infty(A)| < [K : \mathbb{Q}]$, i.e., if A is at least unramified at one infinite place, then we obtain an embedding $\varphi : A \to M_2(\mathbb{R})$. Noting that $\mathrm{Nrd} = \det \circ \varphi$, we have an embedding of $\mathcal{O}^1$ into $\mathrm{SL}_2(\mathbb{R})$. Now we are asking: In which cases is $\varphi(\mathcal{O}^1)$ discrete in $\mathrm{SL}_2(\mathbb{R})$? Or equivalently: In which cases is $\Gamma_{A,\mathcal{O}}$ a Fuchsian group? Borel and Harish-Chandra gave the answer in a general case (see [2], Section 12). It follows from their result:

Theorem 2.2 (Borel and Harish-Chandra). *The group $\Gamma_{A,\mathcal{O}}$ is a Fuchsian group if and only if A is unramified exactly at one infinite place (without loss of generality we may assume that A is unramified at the identity $\sigma_0 : K \hookrightarrow \mathbb{R}$, $\sigma_0 : x \mapsto x$).*

Groups $\Gamma_{A,\mathcal{O}}$, which satisfy the condition of Theorem 2.2 are called **Fuchsian groups derived from the quaternion algebra** A. For the rest of the paper $\Gamma_{A,\mathcal{O}}$ will denote such a group, where $\mathcal{O}$ is a maximal order in A.

Remark 2.3. The discreteness of $\mathcal{O}^1$ does not depend on the choice of $\mathcal{O}$, nevertheless in the following we will restrict our considerations to the maximal orders in A.

3. Principal congruence subgroups in $\Gamma_{A,\mathcal{O}}$

3.1. Definition

Let $\mathcal{O} \subset A$ be a maximal order and I an (two-sided) ideal in $\mathcal{O}$. Let κ be the canonical projection

$$\kappa : \mathcal{O} \longrightarrow \mathcal{O}/I,$$

and κ_1 the restriction of κ to the group $\Gamma_{A,\mathcal{O}}$: $\kappa_1 = \kappa\mid_{\Gamma_{A,\mathcal{O}}}$. We define the **principal congruence subgroup modulo** I, denoted by $\Gamma_{A,\mathcal{O}}(I)$, as the kernel of κ_1:

$$\Gamma_{A,\mathcal{O}}(I) := \ker(\kappa_1) = \{\gamma \in \Gamma_{A,\mathcal{O}} \mid \gamma \equiv 1 \bmod I\}.$$

3.2. The quotients

We are interested in special ideals I, namely lifted prime ideals $\mathfrak{p}\mathcal{O} \subset \mathcal{O}$, where $\mathfrak{p}$ is a prime ideal in $\mathcal{O}_K$. Now we can state our main theorem.

Theorem 3.1. *Let $\mathfrak{p}$ be a prime ideal in $\mathcal{O}_K$, such that A is unramified at the place $\mathfrak{p}$ (here we use the $1:1$ correspondence between the finite places of K and prime ideals in $\mathcal{O}_K$). Then*

$$\Gamma_{A,\mathcal{O}}/\Gamma_{A,\mathcal{O}}(\mathfrak{p}\mathcal{O}) \cong \mathrm{PSL}_2(\mathcal{O}_K/\mathfrak{p}).$$

Proof. Instead of $\Gamma_{A,\mathcal{O}}/\Gamma_{A,\mathcal{O}}(\mathfrak{p}\mathcal{O})$ we look at its image under the diagonal embedding φ of $\Gamma_{A,\mathcal{O}}$ in the product

$$G := \prod_{v\in S, v\notin Ram(A)} \Gamma^{(v)}_{A,\mathcal{O}},$$

where S is a finite set of places containing all infinite places of K. The reasons for considering exactly such a product are the following facts (see [8]):

- The image $\varphi(\Gamma_{A,\mathcal{O}})$ is isomorphic to $\Gamma_{A,\mathcal{O}}$.
- $\varphi(\Gamma_{A,\mathcal{O}})$ is discrete subgroup in G.
- Let S' be a proper subset of S, containing only finite places. Then the projection $\varphi'(\Gamma_{A,\mathcal{O}})$ of $\varphi(\Gamma_{A,\mathcal{O}})$ on every non trivial partial product

$$\{\mathrm{id}\} \neq G' = \prod_{v\in S', v\notin Ram(A)} \Gamma^{(v)}_{A,\mathcal{O}} < G$$

  is dense in G'.
- $\varphi'(\Gamma_{A,\mathcal{O}}) \cong \Gamma_{A,\mathcal{O}}$.

Therefore our strategy will be to compute the local quotients $\Gamma^{(v)}_{A,\mathcal{O}}/\Gamma^{(v)}_{A,\mathcal{O}}(\mathfrak{p}\mathcal{O})$, where v is not an element of $Ram(A)$.

i. At the only one unramified infinite place, which is, as remarked, assumed to be the place corresponding to σ_0 we obtain $A_{\sigma_0} = M_2(\mathbb{R})$. The image of $\mathcal{O}$ under σ_0 is clearly $\mathcal{O}$ itself, which is closed subset in $M_2(\mathbb{R})$ and so $\mathcal{O}_{\sigma_0} = \mathcal{O}$. Consequently $\Gamma^{(\sigma_0)}_{A,\mathcal{O}} = \Gamma_{A,\mathcal{O}}$, $\Gamma^{(\sigma_0)}_{A,\mathcal{O}}(\mathfrak{p}\mathcal{O}) = \Gamma_{A,\mathcal{O}}(\mathfrak{p}\mathcal{O})$ and $\Gamma^{(\sigma_0)}_{A,\mathcal{O}}/\Gamma^{(\sigma_0)}_{A,\mathcal{O}}(\mathfrak{p}\mathcal{O}) = \Gamma_{A,\mathcal{O}}/\Gamma_{A,\mathcal{O}}(\mathfrak{p}\mathcal{O})$. Thus we have here no new information.

ii. Let $v = \mathfrak{p}$. According to assumption A is unramified at $\mathfrak{p}$, thus $A_\mathfrak{p} \cong M_2(K_\mathfrak{p})$. $\mathcal{O}$ is a maximal order in A and therefore $\mathcal{O}_\mathfrak{p}$ is a maximal order in $M_2(K_\mathfrak{p})$. But every maximal order in $M_2(K_v)$ where v is a finite place is conjugate to $M_2(R_v)$ (see [8]). Thus we have $\mathcal{O}_\mathfrak{p} \cong M_2(R_\mathfrak{p})$. It follows that $\Gamma^{(\mathfrak{p})}_{A,\mathcal{O}} \cong \mathrm{PSL}_2(R_\mathfrak{p})$.

   How does $\Gamma^{(\mathfrak{p})}_{A,\mathcal{O}}(\mathfrak{p}\mathcal{O})$ look like?

   The complete closure of $\mathfrak{p}$ in $R_\mathfrak{p}$ is the maximal ideal $\langle\pi\rangle$, where π is an uniformizer of $R_{\mathfrak{p}}$. Hence $\mathfrak{p}\mathcal{O}$ is embedded in $\pi M_2(R_{\mathfrak{p}})$. The group $\Gamma^{(\mathfrak{p})}_{A,\mathcal{O}}(\mathfrak{p}\mathcal{O})$ is the kernel of the restriction $\psi\mid_{\mathrm{PSL}_2(R_{\mathfrak{p}})}$ of

$$\psi : M_2(R_{\mathfrak{p}}) \to M_2(R_{\mathfrak{p}})/\pi M_2(R_{\mathfrak{p}}) \cong M_2(R_{\mathfrak{p}}/\pi R_{\mathfrak{p}}).$$

Then we have

$$\psi\mid_{\mathrm{PSL}_2(R_{\mathfrak{p}})}: \mathrm{PSL}_2(R_{\mathfrak{p}}) \longrightarrow \mathrm{PSL}_2(R_{\mathfrak{p}}/\pi R_{\mathfrak{p}}).$$

This homomorphism is surjective (for the proof see [1]); because of the isomorphy $R_{\mathfrak{p}}/\pi R_{\mathfrak{p}} \cong \mathcal{O}_K/\mathfrak{p}$ we obtain

$$\Gamma^{(\mathfrak{p})}_{A,\mathcal{O}}/\Gamma^{(\mathfrak{p})}_{A,\mathcal{O}}(\mathfrak{p}\mathcal{O}) \cong \mathrm{PSL}_2(\mathcal{O}_K/\mathfrak{p}).$$

iii. Now let $v = \mathfrak{q} \neq \mathfrak{p}$ be a finite place, such that A is not ramified at $\mathfrak{q}$. With the same arguments as in ii. we have $A_{\mathfrak{q}} \cong M_2(K_{\mathfrak{q}})$, $\mathcal{O}_{\mathfrak{q}} \cong M_2(R_{\mathfrak{q}})$ and $\Gamma^{(\mathfrak{q})}_{A,\mathcal{O}} \cong \mathrm{PSL}_2(R_{\mathfrak{q}})$. Now we want to compute $\Gamma^{(\mathfrak{q})}_{A,\mathcal{O}}(\mathfrak{p}\mathcal{O})$. The closure of $\mathfrak{p}\mathcal{O}$ is an ideal in $M_2(R_{\mathfrak{q}})$ and so conjugate to $q^n M_2(R_{\mathfrak{q}})$, where q is an uniformizer of $R_{\mathfrak{q}}$ and n is the $\mathfrak{q}$-valuation of the ideal $\mathfrak{p}$ (see [8]). But $\mathfrak{p}$ and $\mathfrak{q}$ are two distinct prime ideals and therefore is $n = 0$. Hence the closure of $\mathfrak{p}\mathcal{O}$ is $M_2(R_{\mathfrak{q}})$. Then the canonical projection ψ is

$$\psi : M_2(R_{\mathfrak{q}}) \longrightarrow M_2(R_{\mathfrak{q}})/M_2(R_{\mathfrak{q}}) \cong \{0\},$$

and therefore the multiplicative version is

$$\psi\mid_{\mathrm{PSL}_2(R_{\mathfrak{q}})}: \mathrm{PSL}_2(R_{\mathfrak{q}}) \longrightarrow \{1\}.$$

Then the isomorphism theorem says that

$$\Gamma^{(\mathfrak{q})}_{A,\mathcal{O}}/\Gamma^{(\mathfrak{q})}_{A,\mathcal{O}}(\mathfrak{p}\mathcal{O}) \cong \{1\}.$$

It follows, that only at $\mathfrak{p}$ the factor group $\Gamma_{A,\mathcal{O}}/\Gamma_{A,\mathcal{O}}(\mathfrak{p}\mathcal{O})$ has a non-trivial image. It should be a subgroup in $\Gamma^{(\mathfrak{p})}_{A,\mathcal{O}}/\Gamma^{(\mathfrak{p})}_{A,\mathcal{O}}(\mathfrak{p}\mathcal{O}) \cong \mathrm{PSL}_2(\mathcal{O}_K/\mathfrak{p})$. Now our assertion is:

$$\Gamma_{A,\mathcal{O}}/\Gamma_{A,\mathcal{O}}(\mathfrak{p}\mathcal{O}) \cong \Gamma^{(\mathfrak{p})}_{A,\mathcal{O}}/\Gamma^{(\mathfrak{p})}_{A,\mathcal{O}}(\mathfrak{p}\mathcal{O}).$$

To see this we define a map

$$h : \Gamma_{A,\mathcal{O}} \longrightarrow \Gamma^{(\mathfrak{p})}_{A,\mathcal{O}}/\Gamma^{(\mathfrak{p})}_{A,\mathcal{O}}(\mathfrak{p}\mathcal{O}),$$

by

$$h(\gamma) := \gamma\Gamma^{(\mathfrak{p})}_{A,\mathcal{O}}(\mathfrak{p}\mathcal{O}).$$

h is a group homomorphism with the kernel $\Gamma_{A,\mathcal{O}}(\mathfrak{p}\mathcal{O})$. If we are able to show that h is surjective, then the assertion is verified. And indeed:
$\Gamma_{A,\mathcal{O}}$ is dense in $\Gamma^{(\mathfrak{p})}_{A,\mathcal{O}}$, i.e., in every neighbourhood of $\hat{\gamma} \in \Gamma^{(\mathfrak{p})}_{A,\mathcal{O}}$ there is $\gamma \in \Gamma_{A,\mathcal{O}}$. But every coset $\hat{\gamma}\Gamma^{(\mathfrak{p})}_{A,\mathcal{O}}(\mathfrak{p}\mathcal{O})$ is a neighbourhood of $\hat{\gamma}$. Therefore there is $\gamma \in \Gamma_{A,\mathcal{O}}$, such that $\gamma \equiv \hat{\gamma} \bmod \Gamma^{(\mathfrak{p})}_{A,\mathcal{O}}(\mathfrak{p}\mathcal{O})$. So every $\hat{\gamma}$ is an image of h. This implies the surjectivity and completes the proof. □

4. The group of signature $(2,3,7)$

Let $(2,3,7)$ denote a triangle group of this signature. We will keep in mind that the group is unique up to conjugation, as remarked in Section 1.

4.1. $(2,3,7)$ is derived from a quaternion algebra

Shimura showed that $(2,3,7)$ is a Fuchsian group derived from a quaternion algebra (see [5] or more generally [7]). We know from there that $(2,3,7)$ is a group $\Gamma_{A,\mathcal{O}}$, where A is a quaternion algebra over the totally real number field $\mathbb{Q}(c) := \mathbb{Q}(\zeta + \zeta^{-1})$ and $\zeta := \exp(\frac{2\pi i}{7})$ and $\mathcal{O}$ is a maximal order in A (in this case unique up to conjugation in A). A is unramified exactly at one infinite place and unramified at every finite place. In this section let $\Gamma_{A,\mathcal{O}}$ denote the group $(2,3,7)$.

4.2. Congruence subgroups

Now we will apply Theorem 3.1 to the congruence subgroups $\Gamma_{A,\mathcal{O}}(\mathfrak{p}\mathcal{O})$ in $\Gamma_{A,\mathcal{O}}$. First one has to compute the residue fields $\mathcal{O}_{\mathbb{Q}(c)}/\mathfrak{p}$. The following lemma, which is proven by standard arguments in algebraic number theory (see [9]), gives the answer and relates the primes in $\mathcal{O}_{\mathbb{Q}(c)}$ with primes in $\mathbb{Z}$. Note that $\mathcal{O}_{\mathbb{Q}(c)} = \mathbb{Z}[c]$.

Lemma 4.1. *Let p be a prime in $\mathbb{Z}$. Then p has the following prime ideal decomposition in $\mathbb{Z}[c]$:*

1. $p = 7$*:* $p\mathbb{Z}[c] = \mathfrak{p}^3$ *with residue field* $\mathbb{Z}[c]/\mathfrak{p} \cong \mathbb{F}_7$*,*
2. $p \equiv \pm 1 \bmod 7$*:* $p\mathbb{Z}[c] = \mathfrak{p}_1\mathfrak{p}_2\mathfrak{p}_3$ *with* $\mathbb{Z}[c]/\mathfrak{p}_i \cong \mathbb{F}_p$ *for* $i = 1,2,3$*,*
3. $p \equiv \pm 2, 3 \bmod 7$*:* $p\mathbb{Z}[c] = \mathfrak{p}$ *with* $\mathbb{Z}[c]/\mathfrak{p} \cong \mathbb{F}_{p^3}$*.*

Applying Theorem 3.1 to the quotients $\Gamma_{A,\mathcal{O}}/\Gamma_{A,\mathcal{O}}(\mathfrak{p}\mathcal{O})$ we have

Corollary 4.2. *Let $\mathfrak{p}$ be a prime ideal in $\mathbb{Z}[c]$ and p a rational prime.*

1. *If* $\mathfrak{p}|7$*, then* $\Gamma_{A,\mathcal{O}}/\Gamma_{A,\mathcal{O}}(\mathfrak{p}\mathcal{O}) \cong \mathrm{PSL}_2(\mathbb{F}_7)$*.*
2. *If* $\mathfrak{p}|p \equiv \pm 1 \bmod 7$*, then* $\Gamma_{A,\mathcal{O}}/\Gamma_{A,\mathcal{O}}(\mathfrak{p}\mathcal{O}) \cong \mathrm{PSL}_2(\mathbb{F}_p)$*.*
3. *If* $\mathfrak{p}|p \equiv \pm 2, 3 \bmod 7$*, then* $\Gamma_{A,\mathcal{O}}/\Gamma_{A,\mathcal{O}}(\mathfrak{p}\mathcal{O}) \cong \mathrm{PSL}_2(\mathbb{F}_{p^3})$*.*

Let us go back to our starting point, to the Hurwitz groups. Using the definition at the begining, Hurwitz groups are the automorphism groups of Riemann surfaces $X = \Gamma\backslash\mathcal{H}$, where $\Gamma \lhd \Gamma_{A,\mathcal{O}}$ torsion free and $[\Gamma_{A,\mathcal{O}} : \Gamma] < \infty$. According to (1.1) they are of the form $N(\Gamma)/\Gamma$. But we know

Lemma 4.3. *The normalizer of every normal subgroup Γ of $\Gamma_{A,\mathcal{O}}$ is already $\Gamma_{A,\mathcal{O}}$ itself.*

Proof. See, say [6], where it is shown, that $(2,3,7)$ is contained in no other Fuchsian group. □

So the Hurwitz groups are quotients $\Gamma_{A,\mathcal{O}}/\Gamma$ and putting $\Gamma = \Gamma_{A,\mathcal{O}}(\mathfrak{p}\mathcal{O})$ in (1.1) the above lemma gives us good reasons for assumption that the quotients in corollary 4.2 are Hurwitz groups. We have only to check that $\Gamma_{A,\mathcal{O}}(\mathfrak{p}\mathcal{O})$ are torsion

free. This can be done by the following argumentation:
The isomorphisms in corollary 4.2 lead to the surjective homomorphisms

$$h : \Gamma_{A,\mathcal{O}} \longrightarrow \mathrm{PSL}_2(\mathbb{Z}[c]/\mathfrak{p}),$$

with kernels

$$\ker(h) = \Gamma_{A,\mathcal{O}}(\mathfrak{p}\mathcal{O}).$$

Then it is impossible that $\ker(h)$ contains torsion elements. Assume the opposite, i.e., there is $\gamma \in \ker(h)$ with $\gamma^n = 1$. It follows from the theory of Fuchsian groups that γ, as an elliptic element in $\Gamma_{A,\mathcal{O}}$ has to be conjugated to a power of a generator of $\Gamma_{A,\mathcal{O}} = \langle \gamma_0, \gamma_1, \gamma_2 \mid \gamma_0^2 = \gamma_1^3 = \gamma_2^7 = \gamma_0\gamma_1\gamma_2 = 1 \rangle$. To see this let $w_\gamma \in \mathcal{H}$ be the fixed point of γ. Then w_γ lies on the boundary of a fundamental domain $F^{'}$ for $\Gamma_{A,\mathcal{O}}$. Actually w_γ is a vertex of $F^{'}$ (if we enlarge the notion of vertex also to the fixed points of order two elements). Let F denote the fundamental domain with fixed points of γ_0, γ_1, γ_2 as vertices. Now, there is an element of $\Gamma_{A,\mathcal{O}}$ which sends F to $F^{'}$, it automatically sends vertices of F to vertices of $F^{'}$. Via this transformation the stabilizer groups $\mathrm{Stab}_{\Gamma_{A,\mathcal{O}}}(w_\gamma)$ and $\mathrm{Stab}_{\Gamma_{A,\mathcal{O}}}(w_{\gamma_j})$ are conjugated for some $j \in \{0, 1, 2\}$. Since γ_j generates $\mathrm{Stab}_{\Gamma_{A,\mathcal{O}}}(w_{\gamma_j})$ γ is conjugated to a power of γ_j.

So $\gamma = M\gamma_i^{k_i}M^{-1}$ with $M \in \Gamma_{A,\mathcal{O}}$ and $k_i \in \mathbb{N}$ $i = 0, 1, 2$. As a homomorphic image $\mathrm{PSL}_2(\mathbb{Z}[c]/\mathfrak{p})$ is generated by three elements x_0, x_1, x_2 which satisfy same relations as γ_i, but satisfying also some additional relations. Since $h(\gamma) = h(M\gamma_i^{k_i}M^{-1}) = h(M)x_i^{k_i}h(M)^{-1} = 1$ we have $x_i^{k_i} = 1$. Now the relation

$$x_0^2 = x_1^3 = x_2^7 = x_0x_1x_2 = 1$$

forces k_i to be zero or a multiple of 2,3 or 7 since we have three different primes. In any case $\gamma = id$ and this proves

Lemma 4.4. *Principal congruence subgroups* $\Gamma_{A,\mathcal{O}}(\mathfrak{p}\mathcal{O})$ *are torsion free.*

Altogether we obtain

Corollary 4.5. *The groups*

- $\mathrm{PSL}_2(\mathbb{F}_7)$,
- $\mathrm{PSL}_2(\mathbb{F}_p)$, *if* p *is a prime* $p \equiv \pm 1 \bmod 7$,
- $\mathrm{PSL}_2(\mathbb{F}_{p^3})$, *if* p *is a prime* $p \equiv \pm 2, 3 \bmod 7$,

are Hurwitz groups.

This infinite series was found by Macbeath in 1967, who used some group theoretic methods to construct it (see [4]). So he gave a positive answer to the question, if there are infinitely many Hurwitz groups. We should remark that Macbeath proves even a stronger statement: The groups $\mathrm{PSL}_2(\mathbb{F}_q)$ are Hurwitz groups if and only if q takes one of the values $q = 7$, $q = p \equiv \pm 1 \bmod 7$, $q = p^3$ for $p \equiv \pm 2, \pm 3 \bmod 7$, given in corollary 4.5. In contrast to Macbeath's result we get also uniformizing groups of Riemann surfaces with Hurwitz automorphism groups of corollary 4.5, namely the congruence subgroups $\Gamma_{A,\mathcal{O}}(\mathfrak{p}\mathcal{O})$.

Nowadays we know much more infinite series of Hurwitz groups, such as alternating groups A_n (for all but finitely many n), symplectic groups and unitary groups over finite fields (if their dimension is sufficiently large). One famous Hurwitz group is the sporadic simple group big monster with $\approx 8 \cdot 10^{53}$ elements. For details and references on this subject we cite the article by M. Conder [3].

References

[1] H. Bass, K-Theory and stable algebra. *Publ. Math. IHES* **22** (1964).

[2] A. Borel, Harish-Chandra, Arithmetic Subgroups of Algebraic Groups. *Ann. of Math.* **75**, no. 3 (1962).

[3] M. Conder, Hurwitz groups: A brief survey. *Bull. Amer. Math. Soc.* **23**, no. 2 (1990).

[4] A.M. Macbeath, Generators of linear fractional groups. Number Theory. Proc. Symp. Pure. Math. 1969

[5] G. Shimura, Construction of class fields and zeta functions of algebraic curves. *Ann. Math.* **85** (1967).

[6] D. Singerman, Finetely maximal Fuchsian groups. *Jour. Lond. Math. Soc. Ser 2* **6** (1972/73).

[7] K. Takeuchi, Commensurability classes of arithmetic triangle groups. *Journ. Fac. Sci. Tokyo* **24**, no. 1 (1977).

[8] M.F. Vignéras, Arithmétique des Algèbre de Quaternions. Lecture Notes in Mathematics 800, Springer Verlag, Berlin, 1980.

[9] L.C. Washington, Introduction to Cyclotomic Fields. Graduate Texts in Mathematics 83, Springer Verlag, New York, 1982.

Acknowledgment

I am grateful to J. Wolfart for his advice and support during the work on the subject and R. P. Holzapfel for his interest in this topic.

Amir Džambić
Humboldt-University of Berlin
Department of Mathematics
Rudower Chaussee 25
D-10099 Berlin
Germany
e-mail: `dzambic@mathematik.hu-berlin.de`

Progress in Mathematics, Vol. 260, 109–162

Relative Proportionality on Picard and Hilbert Modular Surfaces

Rolf-Peter Holzapfel

Abstract. We introduce “orbital categories”. The background objects are compactified quotient varieties of bounded symmetric domains $\mathbb{B}$ by lattice subgroups of the complex automorphism group of $\mathbb{B}$. Additionally, we endow some subvarieties of a given compact complex normal variety V with a natural weight > 1, imitating ramifications. They define an “orbital cycle” $\mathbf{Z}$. The pairs $\mathbf{V} = (V, \mathbf{Z})$ are orbital varieties. These objects — also understood as an explicit approach to stacks — allow to introduce “orbital invariants” in a functorial manner. Typical are the orbital categories of Hilbert and Picard modular spaces. From the finite orbital data (e.g. the orbital Apollonius cycle on $\mathbb{P}^2$) we read off “orbital Heegner series” as orbital invariants with the help of “orbital intersection theory”. We demonstrate for Hilbert and Picard surface F how their Fourier coefficients can be used to count Shimura curves of given norm on F. On recently discovered orbital projective planes the Shimura curves are joined with well-known classical elliptic modular forms.

Mathematics Subject Classification (2000). 11F06, 11F27, 11F30, 11F41, 11F55, 11G18, 14C17, 14C20, 14D22, 14E20, 14G35, 14G35, 14H30, 14H45, 14J17, 14J25, 20H05, 20H10, 32M15, 32S25, 32S45.

Keywords. Orbital varieties, arithmetic groups, unit ball, Picard modular surfaces, Hilbert modular surfaces, Shimura curves, modular curves, modular forms, surface singularities, rational intersections, theta functions.

Contents

1. Preface

We start with a simple example of an Picard orbiface. The **Apollonius configuration** consists of a quadric together with three tangent lines on the complex projective plane $\mathbb{P}^2$. Explicitly, for instance, we can take the projective curve described by the equation

$$XYZ(X^2+Y^2+Z^2-2XY-2XZ-2YZ)=0:$$

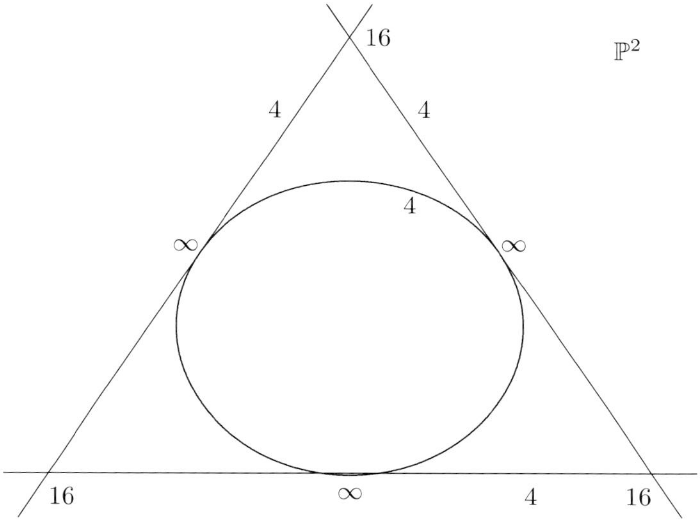

(1)

We endow the points of the plane with weights

- ∞ for the 3 touch points;
- 16 for the 3 intersection points of lines;
- all other points on the 4 curves get weight 4;
- the remaining points of the plane have **trivial weight** 1.

In the mean time it is known that the picture represents an orbital Picard modular plane together with weighted branch locus of a hyperbolic (complex ball) uniformization. A **Picard modular plane** is a Baily–Borel compactified Picard modular surface, which is a projective plane. Without weights, but in connection with ball uniformization, I saw the Apollonius configuration first in the paper [Y-S]. The corresponding Picard modular group acting on the complex 2-ball has been determined precisely first in the HU-preprint [HPV] as congruence subgroup of the full Picard modular group of the imaginary quadratic field $\mathbb{Q}(\sqrt{-1})$ of Gauß numbers. It has been published in [HV].

Surprisingly, in a pure algebraic geometric manner and with a finite number of steps, we are able to read off almost from this weighted projective plane the

Fourier series

$$\begin{aligned}\mathbf{Heeg}_{\mathbf{C}}(\tau) &= \sum_{N=0}^{\infty}\left((\frac{3N}{2}-\frac{1}{8})a_2(N)+3\sum_{m=1}^{N}\sigma(m)a_2(N-m)\right)q^N \in \mathcal{M}_3(4,\chi),\\ &= -\frac{1}{8}\cdot\vartheta^6-\frac{17}{2}\cdot\vartheta^2\theta,\\ q &= \exp(2\pi i\tau),\ \tau\in\mathbb{H}\ \text{(complex upper halfplane)},\end{aligned} \tag{2}$$

with Jacobi's modular form ϑ and Hecke's modular form θ described in the appendix. This is an elliptic modular form of certain level, weight and Nebentypus χ. The N-th coefficient counts (with intersection multiplicities) the arithmetic curves of norm N on the Picard–Apollonius orbiplane. Thereby $\mathbf{C}$ is the orbital arithmetic curve (with above weights) sitting in the Apollonius cycle, where $\sigma(m)$ denotes the sum of divisors of m, and $a_2(k)$ is the number of $\mathbb{Z}$-solutions of $x^2+y^2=k$, and $\chi=\chi_8$ is the Dirichlet character on $\mathbb{Z}$ extending multiplicatively $\left(\frac{2}{p}\right)=(-1)^{(p^2-1)/8}$ for odd primes $p\in\mathbb{N}$ and 0 for even numbers. More precisely, let us extend the cycle and consider

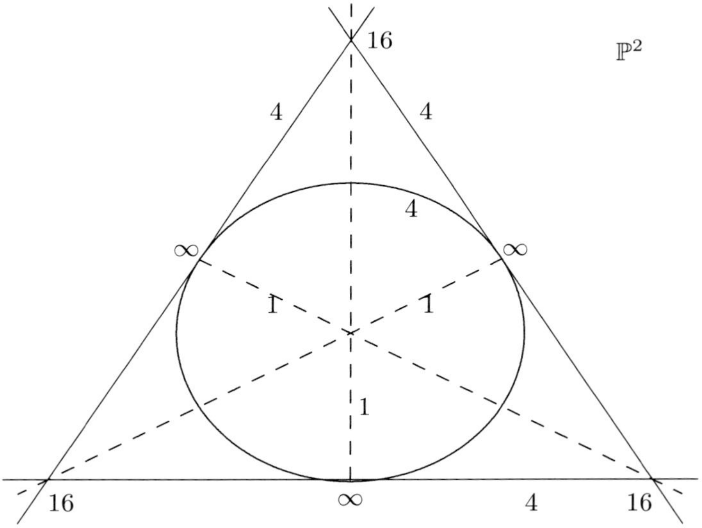

(3)

where all arithmetic curves of smallest norms 1 and 2 are drawn. One has the most difficulties with the algebraic geometric calculation of the constant coefficient of the Heegner series. For this purpose one has to consider rational orbital self-intersections on the so-called released Picard–Apollonius orbiplane. We draw the released Apollonius cycle:

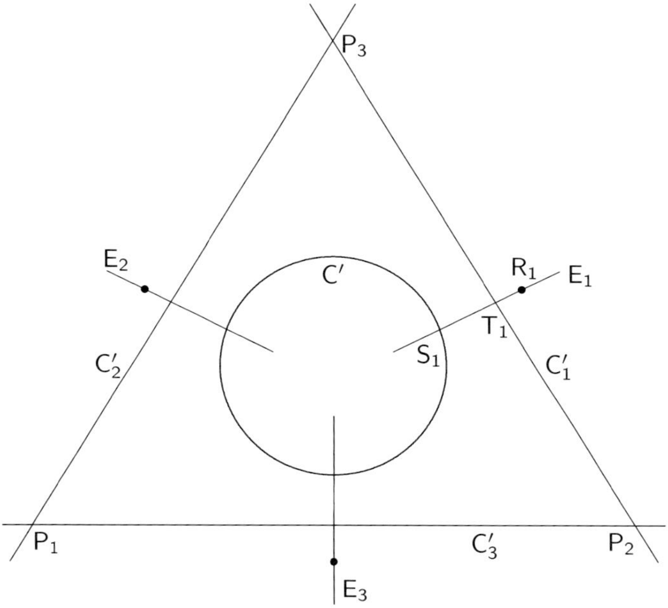

(4)

In the meantime we found another simple example. We call it the Hilbert–Cartesius orbiplane. The supporting **Cartesius configuration** lies also on $\mathbb{P}^2$. It consists of three quadrics touching each other (and nowhere crossing) together with four lines, each of them joining three of the six touch points. Take, for example, the projective closure of the affine curve described by the equation

$$(X^2 + Y^2 - 2)(XY - 1)(XY + 1)(X^2 - 1)(Y^2 - 1) = 0$$

consisting in the real affine plane of a circle, two hyperbolas and four lines parallel two the axis through pairs of the points $(\pm 1, \pm 1)$. Not visible are two further intersection points of same quality (two quadrics meet two lines) at infinity. It is easy to draw these seven curves.

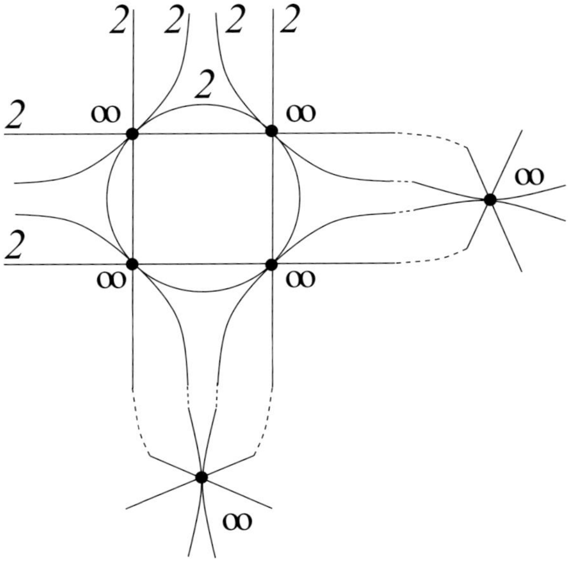

We endow the points of the plane with the following weights:

- ∞ for the six intersection points;
- all other points on the 7 curves get weight 2;
- the remaining points of the plane have "trivial weight" 1.

The picture represents an orbital Hilbert modular plane together with weighted branch locus of a bi-disc uniformization. A **Hilbert modular plane** is a Baily–Borel compactified Hilbert modular surface, which is a projective plane. The corresponding Hilbert modular group acting on the bi-product $\mathbb{H}^2$ of the upper half plane $\mathbb{H}$ can be found in [Hir1], [Hir2], [vdG]. It is commensurable with the full Hilbert modular group of the real quadratic number field $\mathbb{Q}(\sqrt{2})$. Also in this case it is possible in almost the same purely algebraic geometric manner in finite steps to read of the Fourier series

$$\mathbf{Heeg}_{\mathbf{C}}(\tau) = -1 + 2 \cdot \sum_{N=1}^{\infty} \left(\sum_{d|N} \chi(d)d \right) q^N \in \mathcal{M}_2(8, \chi), \tag{5}$$

connected with the plane quadric $C : X^2 + Y^2 - 2Z^2 = 0$. This is again an elliptic modular form of certain level, weight and Nebentypus χ. More precisely, this is an Eisenstein series. The coefficients again count (with degree multiplicities) the arithmetic curves of fixed norms and of Humbert type. We also need for the calculation of the rational constant coefficient of the series an orbital curve self-intersection on the released Hilbert–Cartesius orbiface, whose non-trivially weighted curves we draw in the following picture:

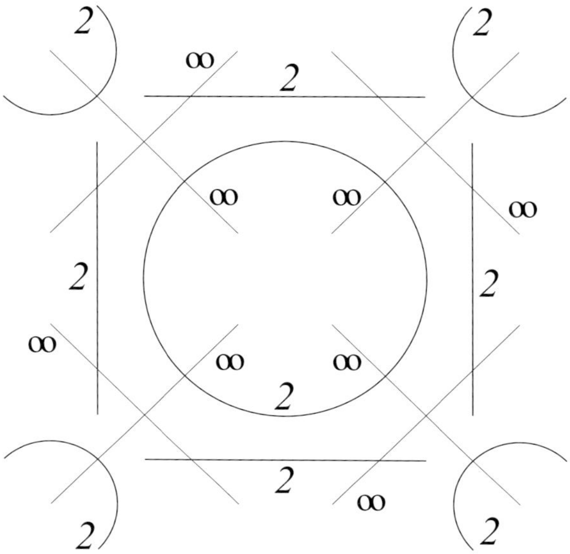

2. Introduction

We only consider normal surfaces, analytic or algebraic over the complex numbers. We need a precise geometric language. More precisely, we need objects, which are a little finer than varieties but not so abstract as Shimura varieties over number fields or stacks. The latter notions make much sense for fine number theoretic considerations. But here we are near only to the original geometric Shimura varieties over $\mathbb{C}$, especially to Picard and Hilbert modular surfaces, called Shimura surfaces in common, and modular or Shimura curves on them, which we call also arithmetic curves. For example, a Picard modular surfaces is a ball quotient $\Gamma\backslash\mathbb{B}$ by a lattice Γ, which is a discrete subgroup Γ of $\mathbb{PU}((2,1),\mathbb{C})$ acting properly discontinuously on the 2-dimensional complex unit ball $\mathbb{B}$. Let $\mathbb{D}$ be a linear disc on the ball $\mathbb{B}$ such that $\Gamma\backslash\mathbb{D}$ is an algebraic curve on $\Gamma\backslash\mathbb{B}$. Invariant metrics on the ball $\mathbb{B}$ (Bergmann) go down to metrics on the surfaces with degeneration cycles (join of irreducible curves and points). These curves and points are endowed with natural weights coming from ramification indices of the locally finite covering $\mathbb{B}\to\Gamma\backslash\mathbb{B}$. It can be proved that on canonical surface models (in Shimura's sense) these curves are defined over a ring the integers of a number field. At the same time these disc quotient curves are geodesics with respect to the degenerate metric. In the mean time it was proved (A.M. Uludag, [Ul]) that there are infinitely many Picard modular projective planes.

Especially, it would be very interesting to find plane equations with integral coefficients for the above arithmetic geodesics. At the moment we are only able to give a numerical characterization of such curves. Looking back to history we know

Plücker's formula for plane curves C with only cusps or transversal self-intersection points joining the numbers of them, the curve degree and the genus. Together with the idea of proof we remember Plücker's explicit relation

$$d(d-1) - 2\delta - 3\kappa = d^* = 2d + (2g-2) - \kappa, \tag{6}$$

where g denotes the genus, d the degree of C, d^* the number of tangents of C through a fixed general point of $\mathbb{P}^2$, δ the number of double points and κ the number of cusps of the curve. Forgetting d^* we get a relation between the degree d, the genus g, δ and κ.

We will define orbifaces, especially Picard and Hilbert modular orbifaces, and orbital curves $\mathbf{C}$ on the latter, all with help of branch weights. We will introduce orbital invariants for them, which are rational numbers explicitly expressed in terms of algebraic geometry. Of special interest are the two orbital invariants $\mathbf{E}$ and $\mathbf{S}$, which are singularity modifications of the Euler number and the selfintesection of the curve algebraically calculable on a special surface model, the so-called release model, which we must introduce. Restricting, for example to Picard orbifaces we get in analogy to (6) the relations

$$\mathbf{Eul}(\mathbf{C}) = \mathbf{vol}(\boldsymbol{\Gamma}_{\mathbb{D}}) = \mathbf{2} \cdot \mathbf{Self}(\mathbf{C}), \tag{7}$$

with supporting curve $C = \Gamma\backslash\mathbb{D}$. Forgetting the Euler–Bergman volume $vol(\Gamma_{\mathbb{D}})$ of a fundamental domain of the subgroup $\Gamma_{\mathbb{D}}$ of Γ of all elements operating on $\mathbb{D}$ one gets a characteristic relation between the orbital Euler invariant and the orbital self-intersection of $\mathbf{C}$:

Theorem 2.1. *With the above notation (and definitions of orbital invariants below) it holds that*

$$\mathbf{Eul}(\mathbf{C}) = \mathbf{2} \cdot \mathbf{Self}(\mathbf{C}).$$

The relation (7) generalizes the relative proportionality for modular curves and Shimura curves on neat ball quotient surfaces, see [H98]. For their importance we show that the relative orbital Euler invariants are the constant terms of Fourier series of modular quality joining infinitely many orbital invariants. Moreover, the (orbital invariant) coefficients count (with multiplicity depending on $\mathbf{C}$ embedded on $\mathbf{X}$) the number of arithmetic geodesics on X, see [H02].

It is not difficult to prove and write down the relative proportionality relation

$$\mathbf{Eul}(\mathbf{C}) = \mathbf{Self}(\mathbf{C}) \tag{8}$$

for orbital modular or Shimura curves on any Hilbert orbiface (including also symmetric Hilbert modular groups). It is also clear that the orbital invariant in (8) is the constant coefficient of a Hirzebruch–Zagier modular form [HZ]. The aim of the next section is to develop a common language for Hilbert and Picard modular surface with a common algebraic geometric understanding of the involved elliptic modular Hirzebruch–Zagier series [HZ] and Kudla–Cogdell series (ball case) [Cog]. To use this fine geometric language for common proofs and explicit calculations is the aim of the whole paper.

For a clear definition of orbital invariants we need a category $\mathcal{C}$ with a multiplicatively closed set of "finite coverings" $\mathbf{D}/\mathbf{C}$ and "degrees" $[\mathbf{D}:\mathbf{C}]$ of the latter, satisfying the degree formula

$$[\mathbf{E}:\mathbf{C}] = [\mathbf{E}:\mathbf{D}] \cdot [\mathbf{D}:\mathbf{C}]$$

for all double coverings $\mathbf{E}/\mathbf{D}/\mathbf{C}$. A **rational orbital invariant** on $\mathcal{C}$ is simply a non-constant map

$$\mathbf{h} : \mathbf{Ob}\,\mathcal{C} \longrightarrow \mathbb{Q}$$

from the objects of our "orbital category" satisfying the "orbital degree formula"

$$\mathbf{h}(\mathbf{D}) = [\mathbf{D}:\mathbf{C}] \cdot \mathbf{h}(\mathbf{C})$$

for all finite coverings $\mathbf{D}/\mathbf{C}$. Part of the work is to clarify, which orbital categories we can construct. In the paper [Ul] by A.M. Uludag I saw him working with weighted surfaces $\mathbf{X} = (X, w)$, $w : X \to \mathbb{N}_+$ a map from the surface X to the natural numbers. These object we will basically use for the construction of our orbital categories. After the definition procedure we are able to find explicitly infinitely many orbital invariants. We combine the rational intersection theory and Heegner cycles. But all these invariants are "modular dependent", which means, that they are connected by modular forms of known and fixed type with each other. It follows that it suffices only to know finitely many of the orbital invariants to determine the others together with the corresponding modular form. The interpretation of counting arithmetic curves on Picard or Hilbert modular surfaces is then general. Until now the series were only known in neat cases. So it was e.g. until now not possible to count arithmetic curves on modular planes. The extension from neat to general cases is the progress described in this paper. In contrast, in the neat Picard case the corresponding quotient surfaces are never rational. Explicit calculations there seem to be very difficult. But in the plane case the situation is much better, especially with a view to coding theory on explicit arithmetic curves.

Remark 2.2. Volumes of fundamental domains of Picard or Hilbert modular groups with respect to fixed volume forms on the the uniformizing domains are obviously orbital invariants. If we take the Euler volume form, then these volumes coincide with the orbital Euler invariants of the corresponding orbital surfaces. This is a theorem (Holzapfel, [H98], in the ball case, to be written down in the Hilbert case). The same is true for the signature volume form, which leads to the corresponding orbital signature invariants. The Euler and the signature forms are distinguished by a factor. This leads to a proportionality relation between the orbital invariants with different factors in the Picard and Hilbert surface cases.

For explicit calculations it is important to know that the orbital invariants of the modular surfaces of the full lattices can be expressed by special values of Zeta-functions (or L-series) of corresponding number fields. (Maass in the Hilbert case, Holzapfel [H98] in the Picard case).

3. The Language of Orbifaces

3.1. Galois weights

A **weighted (algebraic) surface** $\mathbf{X} = (X, w)$ is defined by:

- an irreducible normal complex algebraic surface X
- and a **weight map** $w: X \longrightarrow \mathbb{N}_+ \cup \{\infty\}$

with conditions

- almost everywhere (i.e. up to a proper closed algebraic subvariety of X) one has **absolute trivial weight** 1:
- w is almost constant on each closed irreducible curve C of X (that means up to a finite set of points $P_1, \ldots, P_r$ on C);

The corresponding constant $w(C) := w(C \setminus \{P_1, \ldots, P_r\})$ is called the **weight** of C on $\mathbf{X}$.

- If $P \in C$, C an irreducible curve on X, then $w(C)$ divides $w(C)$.

A point P on X is **relatively non-trivial weighted** if $w(P) > w(C)$ for all irreducible curves C though P. Otherwise it is called **relatively trivial weighted**. The formal (finite) double sum

$$B = B(X, w) := \sum w(C)C + \sum w(P)P = B_1 + B_0$$

over all irreducible curves C with non-trivial weight (> 1) respectively all relatively non-trivial weighted points P on X is called the **weight cycle** of $\mathbf{X}$. If the double sum is restricted to (all) finite non-trivial weights, then we call it the **finite weight cycle** of $\mathbf{X}$, denoted by $B^{fin} = B^{fin}(X, w)$. As above we have two partial sums, one over the curves, the other over the points of finite non-trivial weights:

$$B^{fin} = B_1^{fin} + B_0^{fin}.$$

Complementarily, we define the **infinite cycle** of $\mathbf{X}$ by

$$B^{\infty} = B^{\infty}(X, w) := B - B^{fin} = B_1^{\infty} + B_0^{\infty}$$

together with its 1, 0-dimensional decomposition in obvious manner.

As usual we call the union of component sets of a cycle D on X the **support** of D and denote it by $suppD$. The weighted surface $\overset{o}{\mathbf{X}} = (\overset{o}{X}, \overset{o}{w})$ with $\overset{o}{X} = X \setminus suppB^{\infty}$ and $\overset{o}{w} = w|_{\overset{o}{X}}$ is called **open finite part** of $\mathbf{X}$. We get the first examples of open subsurfaces $\overset{o}{\mathbf{X}} \subseteq \mathbf{X}$ and open embeddings $\overset{o}{\mathbf{X}} \hookrightarrow \mathbf{X}$ of weighted surfaces in this way. Generally, we take open subsurfaces U of X instead of $\overset{o}{X}$ and define in analogous manner **open weighted subvarieties** $\mathbf{U} \subseteq \mathbf{X}$ and **open embeddings** $\mathbf{U} \hookrightarrow \mathbf{X}$ using restrictions of the weight map w.

Let C be an irreducible curve on X. The system of open neighbourhoods U of C defines the surface germ of X along C. We imagine it as a small open neighbourhood of C or, more precisely, as refinement (equivalence) class of such neighbourhoods. Working additionally with weight restrictions we define the **weighted surface germ $\mathbf{C}$ of $\mathbf{X}$ along** C as refinement class of all $\mathbf{U} = (U, w|_U)$ with open neighbourhoods U of C. We write $\mathbf{C} \leftharpoonup \mathbf{X}$ in this situation and consider it as **closed embedding** of weighted objects (supported by the closed embedding of C into X in the usual sense).

The same can be done with open neighbourhoods of a point $P \in X$. Working with weight restrictions again we define the **weighted surface germ P** of **X at** P as refinement class of all $\mathbf{U} = (U, w|_U)$ with open neighbourhoods U of P. We write $\mathbf{P} \,\varepsilon\, \mathbf{X}$ in this situation. If P is a point on C, then we write $\mathbf{P} \,\varepsilon\, \mathbf{C}$ instead of the pair $(\mathbf{C}, \mathbf{P})$ and consider it as (closed) embedding of weighted surface germs.

Remark 3.1. We left it to the reader to work with complex or with Zariski topology. In the later definitions of orbital invariants there will be no difference. But there will be numerical differences between the embedded germs $\mathbf{P} \,\varepsilon\, \mathbf{X}$, $\mathbf{P} \,\varepsilon\, \mathbf{C}$, $\mathbf{P} \,\varepsilon\, \mathbf{D}$, where D is another irreducible curve through P. This is not surprising because already for unweighted surfaces it can happen that P is a singularity of one or two of the objects X, C or D, but may be a regular point of the other one(s).

An **isomorphism** of weighted surfaces $\mathbf{f} : \mathbf{X} \xrightarrow{\sim} \mathbf{Y}$, $\mathbf{Y} = (Y, v)$, $\mathbf{X} = (X, w)$ as above, is nothing else but a surface isomorphism $f : X \xrightarrow{\sim} Y$, which is weight compatible, that means $v \circ f = w$. If the isomorphism sends the irreducible curve C to D and the point P to $Q \in Y$, then it induces weighted surface germ **isomorphisms** $\mathbf{C} \xrightarrow{\sim} \mathbf{D}$ and $\mathbf{P} \xrightarrow{\sim} \mathbf{Q}$. They are not globally depending on $\mathbf{X}$ or $\mathbf{Y}$ but only on small open neighbourhoods U respectively V of the curves or points and on the isomorphic compatible weights around. For $P \in C$ we have $Q \in D$ and **isomorphisms** of embedded objects $(\mathbf{C}, \mathbf{P}) \xrightarrow{\sim} (\mathbf{D}, \mathbf{Q})$. As in scheme theory we visualize the situation by a commutative diagram:

$$\begin{array}{ccccccc} \mathbf{P} & \varepsilon & \mathbf{C} & \leftharpoonup & \mathbf{U} & \hookrightarrow & \mathbf{X} \\ \downarrow\wr & & \downarrow\wr & & \downarrow\wr & & \downarrow\wr \\ \mathbf{Q} & \varepsilon & \mathbf{D} & \leftharpoonup & \mathbf{V} & \hookrightarrow & \mathbf{Y} \end{array}$$

Trivially weighted surfaces $(Y, \mathbf{1})$, $\mathbf{1}$ the constant weight map (with weight 1 for each point), are identified with the surfaces Y themselves. So we will write Y again instead of $(Y, \mathbf{1})$. A smooth surface Y together with a finite covering $p : Y \to X$ is called a **finite uniformization** of $\mathbf{X} = (X, w)$, if and only if $p = p_G$ is the Galois covering with Galois group $G \subseteq Aut\, Y$, $X \cong Y/G$, $w(P) = \#G_Q$ (number of elements of the stabilizer subgroup of G of all elements fixing Q) for any p-preimage point $Q \in Y$ of P. We write also $\mathbf{p} : Y \to \mathbf{Y/G}$ in this situation and consider $\mathbf{p}$ as a **uniformization** morphism in the category of weighted surfaces. The weighted surface $\mathbf{X}$ is called (finitely) **uniformizable** if and only if a (finite) uniformization $Y \to \mathbf{X}$ exists. The weights $w(P)$ are called **Galois weights**. The weight of an irreducible curve on X coincides with the corresponding rami-fication index. It is also called **branch weight**. The branch weight of such a curve is non-trivial if and only if the curve belongs to the branch locus of p_G.

Point surface germs $\mathbf{P}$ on uniformizable weighted surfaces are called **orbital quotient points**. This definition is extended to isomorphic objects. With the following local-global diagram we introduce further notations.

$$\begin{array}{ccccccc} & \mathbf{Q} & \varepsilon & V & \hookrightarrow & Y & \\ & \downarrow{\scriptstyle G_Q} & & \downarrow{\scriptstyle G_Q} & & \downarrow{\scriptstyle G} & \\ \mathbf{Q/G_Q} = & \mathbf{P} & \varepsilon & \mathbf{U} & \hookrightarrow & \mathbf{X} & = \mathbf{Y/G} \end{array} \tag{9}$$

using small open neighbourhoods and quotient maps by G_Q or G in the columns (local and global uniformizations). The left column (together with the middle) is called a **uniformization of P**. The orbital point $\mathbf{P}$ is called **smooth** if and only if the supporting point P is a smooth surface point of X (or U). Orbital quotient points realized as above by abelian groups G_Q are called **abelian orbital points**. The others are called **non-abelian**.

A finitely weighted surface $\mathbf{X}$ is called **orbiface** (or finitely weighted orbiface), if each of its point surface germs $\mathbf{P}$ is an orbital quotient point. If thereby X is not compact, we call it an **open orbiface**. Let $Y \overset{H}{\longrightarrow} \mathbf{Z}$ be a uniformization of the orbiface $\mathbf{Z}$ with subgroup H of $G \subseteq Aut\, Y$. Then p_G factors through p_H defining a finite covering $f : Z \to X$ with $p_G = f \circ p_H$. We get a commutative orbital diagram

$$\begin{array}{ccc} Y & & \\ \downarrow{\scriptstyle H} & \searrow{\scriptstyle G} & \\ \mathbf{Z} & \xrightarrow{\mathbf{f}} & \mathbf{X} \end{array} \tag{10}$$

defining a **uniformizable orbital covering** or **uniformizable finite orbital morphism** $\mathbf{f} = \mathbf{f}_{G:H}$ on this way on the bottom. Working with Galois weight maps $w_H : Z \to \mathbb{N}_+$ and $w_G : X \to \mathbb{N}_+$ we define $f^*w_G := w_G \circ f$. This **lifted weight map** is obviously pointwise divisible by w_H, which means that $w_H(z)$ divides $f^*w_G(z) = w_G(f(z))$ for all $z \in Z$. We write $w_H \mid f^*w_G$ and define the **quotient weight map** $w_{G:H} := \frac{f^*w_G}{w_H} : Z \to \mathbb{N}_+$ pointwise. The curve weights on Z coincide with the corresponding ramification indices along f. If $w_{G:H}$ is constant on the fibres of f, then we can push it down to the weight map $w_f : X \to \mathbb{N}_+$. This happens, if H is a normal subgroup of G. Then we have $w_f = w_{G/H}$ on X and a new (reduced) kind of **finite orbital covering** $Z \to (X, w_f)$. We call it the **reduction** of $\mathbf{f} : (Z, w_H) \to (X, w_G)$ and w_f the **reduction of** w_G **through** Z. A geometric Galois problem for a given finite covering $f : Z \to X$ looks for a common uniformization Y as described in diagram (10). For this purpose it is worth to notice that

$$w_H = \frac{f^*w_G}{f^*w_f} \left(= \frac{f^*w_G}{w_{G:H}}\right).$$

So, if one knows w_G and f (hence w_f) one recognizes already the branch weights and curves of the possible uniformization p_H.

Weighted surface germs $\mathbf{C} \leftharpoonup \mathbf{X}$ along irreducible curves C on X are called **(open) orbital curves** if $\mathbf{X}$ is open orbital. The definition does not depend on X but, more precisely, on open neighbourhoods U of C on X. It may happen that $\mathbf{X}$ is not orbital but $\mathbf{U} \hookrightarrow \mathbf{X}$ is. Then $\mathbf{C}$ is orbital.

The notions of uniformization and of finite morphisms restrict to orbital curves and points. From scheme theory it is well-known that finite morphisms are surjective and open. So for orbital points, curves and orbifaces (open) we get restriction diagrams along finite coverings $\mathbf{f}: \mathbf{Y} \to \mathbf{X}$ via restrictions on germs

$$\begin{array}{ccccccc} \mathbf{Q} & \varepsilon & \mathbf{D} & \leftharpoonup & \mathbf{V} & \hookrightarrow & \mathbf{Y} \\ {\scriptstyle \mathbf{f}_{D,Q}}\downarrow & & \downarrow{\scriptstyle \mathbf{f}_D} & & \downarrow{\scriptstyle \mathbf{f}|_V} & & \downarrow{\scriptstyle \mathbf{f}} \\ \mathbf{P} & \varepsilon & \mathbf{C} & \leftharpoonup & \mathbf{U} & \hookrightarrow & \mathbf{X} \end{array} \qquad (11)$$

with vertical finite orbital coverings. These are orbitalizations of well-known diagrams in scheme theory or complex algebraic geometry. Here we used pairs (V_Q, V) of small open neighbourhoods of Q or D, respectively, to define **finite coverings $\mathbf{f}_{D,Q}$ of orbital points on orbital curves**. Working only with small open neighbourhoods V of Q we define **finite orbital point coverings $\mathbf{f}_Q$** algebraically visualized in the diagram

$$\begin{array}{ccccc} \mathbf{Q} & \varepsilon & \mathbf{V} & \hookrightarrow & \mathbf{Y} \\ {\scriptstyle \mathbf{f}_Q}\downarrow & & \downarrow{\scriptstyle \mathbf{f}|_V} & & \downarrow{\scriptstyle \mathbf{f}} \\ \mathbf{P} & \varepsilon & \mathbf{U} & \hookrightarrow & \mathbf{X} \end{array} \qquad (12)$$

In the special case of uniformizations we dispose on Diagram (9) for orbital points. Working with Galois group G again, the **normalizer group** (decomposition group)

$$N_G(D) := \{g \in G;\ g(D) = D\}$$

of D and $N_G(D)$-invariant small open neighbourhoods V of D. Orbital curves are said to be **smooth** if and only if the supporting curve is smooth and the supporting surface is smooth around the curve. Let D be a smooth curve on a trivially weighted surface Y with G-action such that the orbital curve $\mathbf{D} \leftharpoonup Y$ is smooth. Then we define an **orbital curve uniformization $\mathbf{f}_D$** of $\mathbf{C}$ by the vertical arrow on the left-hand side of the diagram

$$\begin{array}{ccccc} \mathbf{D} & \leftharpoonup & V & \hookrightarrow & Y \\ \downarrow{\scriptstyle N_G(D)} & & \downarrow{\scriptstyle N_G(D)} & & \downarrow{\scriptstyle G} \\ \mathbf{C} & \leftharpoonup & \mathbf{U} & \hookrightarrow & \mathbf{X} \end{array} \qquad (13)$$

The weight of C is equal to the order of the cyclic **centralizer group** (inertia group)

$$Z_G(D) := \{g \in G;\ g|_D = id|_D\}.$$

The curve C is isomorphic to the quotient curve D/G_D, where G_D is the effectively on D acting group defined by the exact group sequence

$$1 \longrightarrow Z_G(D) \longrightarrow N_G(D) \longrightarrow G_D = N_G(D)/Z_G(D) \longrightarrow 1. \tag{14}$$

3.2. Orbital releases

We want to introduce special birational morphisms for orbital points, curves and orbifaces. Changing special curve singularities by numerically manageable surface singularities. These will be **abelian singularities**, which are defined as supporting surface singularities of orbital abelian points $\mathbf{P}$. The latter are well-understood by linear algebra. Let G be a finite abelian subgroup of $\mathbb{G}l_2(\mathbb{C})$. It acts effectively on the complex affine plane $\mathbb{C}^2$ and around the origin $O = (0,0)$ of $\mathbb{C}^2$, hence on the trivially weighted smooth orbital point $\mathbf{O} \,\varepsilon\, \mathbb{C}^2$. Working in the analytic category, that means with small open analytic neighbourhoods of points, it is true that for each orbital quotient point $\mathbf{P}$ there is uniformization $\mathbf{O} \to \mathbf{P} \cong \mathbf{O}/\mathbf{G}$ for a suitable finite subgroup G of $\mathbb{G}l_2(\mathbb{C})$ (H. Cartan). Let us call it a **linear uniformization** of $\mathbf{P}$. If G is not abelian, then there are precisely three G-orbits of eigenlines in $\mathbb{C}^2$ of non-trivial elements of G. Going down to $\mathbf{P} = \mathbf{O}/\mathbf{G}$ they define precisely three orbital curve germs through $\mathbf{P}$ called **eigen germ triple** at $\mathbf{P}$. There are precisely two of them, called **eigen germ pair** at $\mathbf{P}$ if and only if G is an abelian group not belonging to the center of $\mathbb{G}l_2(\mathbb{C})$. If G is central, then we declare the germs of the images of any two different lines through O as eigen germ pair at $\mathbf{P}$. Different choices are isomorphic.

By the way, **orbital curve germs** on orbifaces $\mathbf{X}$ at a point P are defined in the same manner as orbital curves but working with orbital curve germs at P and small analytic open neighbourhoods of them instead of global curves and their open neighbourhoods. More precisely, it is the weighted analytic surface germ around a curve germ on X through P. Now let $\mathbf{P}$ be an orbital point on the orbital curve $\mathbf{C}$ on $\mathbf{X}$. If C is smooth at P, then $\mathbf{C}$ defines a unique orbital curve germ $\mathbf{C}_P$ at $\mathbf{P}$. Now let $\mathbf{P}$ be an abelian orbital point. We say that two orbital curve(germ)s $\mathbf{C}_1$ and $\mathbf{C}_2$ on $\mathbf{X}$ **cross (each other)** at $\mathbf{P} \,\varepsilon\, \mathbf{X}$ if and only if they form an eigen germ pair there. Necessarily C_1 and C_2 have to be smooth at P.

A curve C on a surface X is called **releasable** at $P \in C$ if and only if there is a birational morphism $\varphi_P : X' \to X$ such that the exceptional curve E_P of φ_P is smooth, irreducible, $\varphi_P(E_P) = P$ and the proper transform C' of C on X' crosses E_P at any common point. Observe that C' must be smooth at these intersection points. So $X' \to X$ resolves the (releasable) curve singularity P. If P is thereby a curve singularity, then we call φ_P the (honest) **release** of C at P. If P is a smooth curve and surface point, then one could take the σ-process at P, but this is not a honest release. Honest releases are only applied to curve singularities. Using the uniqueness of minimal singularity resolution for surfaces (here applied to the abelian surface singularities on $E_P \subset X'$) it is easy to see that this local release φ_P is uniquely determined by $P \in C \subset X$. The curve $C \subset X$ is called **releasable** if and only if it is releasable at each of its points. There are only finitely many honestly releasable points on each fixed curve. Therefore, if C is releasable, there exists a unique birational morphim $\varphi = \varphi_C : X' \to X$ releasing all singular points of C. This morphism is called the **release** of X **along** C.

Remark 3.2. The surface singularities on X of released curve points P are of special type. They are contractions of one curve E_P supporting finitely many abelian singularities (of X'). So P has a surface singularity resolution consisting of a (central) irreducible curve (the proper transform of E_P) crossed by some disjoint linear trees of lines (that means isomorphic to $\mathbb{P}^1$) with negative self-intersections smaller than -1. The linear trees are minimal resolutions of abelian surface singularities. Such a singularity resolutions of P is called **released**. It can happen that it is bigger than the minimal singularity resolution of P; for instance, if we are forced to release a smooth surface point, an abelian singularity or, more generally, a quotient singularity P.

Example 3.3. Let P be an **ordinary singularity** of a curve C on a surface X smooth at P. By definition, the branches of C at P cross each other there. Then the curve singularity P is released by the σ-process at P. The curve branches appear as (transversal) intersection points of the proper transform of C with the exceptional line over P.

Example 3.4. Hypercusp singularities of curves at smooth surface points are defined by local equations $y^n = x^m$, $m, n > 0$. They are releasable by a line (smooth rational curve) supporting at most two abelian surface points. This releasing line cuts the proper transform of the curve in precisely $gcd(m, n)$ smooth points. These intersections are transversal.

Idea of the proof. Stepwise resolution of the curve singularity by σ-processes. At the end one gets a tree of lines with negative self-intersections. One discovers that the proper transform of the curve crosses only one component of the tree. The two (or less) partial trees meeting this component contract to an abelian point. The resolution steps reduce the exponent pairs (m, n) following the euclidean algorithm. It stops by arriving equal exponents in the singularity equation, $y^d = x^d$, $d = gcd(m, n)$. This is the local equation of an ordinary curve singularity with d branches. □

Definition 3.5. *An orbital curve* **C** *is* **releasable** *if and only if the supporting surface embedded curve* C *is.*

The definition does not depend on the choice of orbiface **X** defining **C** by restriction.

Examples 3.6.

- Release of a curve cusp at smooth surface point.
  Let $[-3, -1, -2]$ represent a linear tree of three smooth rational curves on a smooth surface with the indicated self-intersections. It is contractable (stepwise) to the regular surface point P. But let us contract the first and last line

to cyclic singularities P_1, P_2 of types $< 3, 1 >$ respectively $< 2, 1 >$ on the middle line L. Now consider a curve C' intersecting L transversally at one point $P' \neq P_1, P_2$. Contracting L, the image point P is a curve cusp on the image curve C of C'. Altogether $L \to P$ is a release of (C, P) with exactly one branch point (C', P'), and $P' \in C'$ is totally smooth.

- A more complicated release of smooth point.
  Let P be a smooth surface point again. There is a release $L \to P$ with two honest cyclic singularities

$$P' : < 93, 76 > \quad \leftarrow \quad [-2, -2, -2, -2, -4, -2, -2, -2, -2, -2, -2, -2],$$

$$P'' : < 106, 17 > \quad \leftarrow \quad [-7, -2, -2, -2, -5],$$

  on L numerically resolved by continued fractions (Hirzebruch–Jung singularities). As in the previous example one has only to consider the composed linear resolution tree connected by a (-1)-line and its stepwise blowing down to a smooth point:

$$\begin{aligned} \tilde{E}' : & [-2, -2, -2, -2, -4, -2, -2, -2, -2, -2, -2, -2, -1, -7, -2, -2, -2, -5] \\ \to \quad & [-2, -2, -2, -2, -4, -1, -2, -2, -2, -5] \\ \to \quad & [-2, -2, -2, -2, -1, -5] \quad \to \quad [-1] \quad \to \quad P, \end{aligned} \tag{15}$$

Remark 3.7. It is easy to see now that each abelian point has infinitely many different releases.

- Hilbert cusps.
  An **irreducible neat Hilbert cusp curve** is a contractible rational curve H on a surface with a double point P as one and only curve singularity. Moreover, P has to be a cyclic surface point (including smooth ones), and the two branches of H cross each other at P. By the above remark each irreducible Hilbert cusp curve has infinitely many releases which — by abuse of language — are also called called **releases** of neat Hilbert cusp points. The irreducible Hilbert cusp curves are also called **simple releases** of neat Hilbert cusp points. There are also infinitely many simple releases of one and the same cusp point. One has only to consider the minimal resolution of such cusp point with smooth transversally intersecting components. It consists of a cycle of smooth rational curves. **Orbital Hilbert cusp points** in general are finite quotients of neat Hilbert cusp points. A release of one of them is nothing else but the quotient of a neat Hilbert cusp point release.

We say that the birational morphism $Y' \to Y$ is a **smooth release** of the curve $D \subset Y$, if it is a release of D and Y' is a smooth surface. Thereby we allow also non-honest releases at some points. Let G be a finite group acting effectively on Y and assume that the action lifts to Y' permuting the released points. The (smooth) proper transform of D on Y' is denoted by D'. Let $C =: D/G$ and $C' =: D'/G$ be the image curve of D on $X := Y/G$ or of D' on $X' := Y'/G$, respectively. We say that the release $Y' \to Y$ is G-**stable** if and only if additionally the induced morphism $X' \to X$ is a release of C (with proper transform C'). We endow X' and X with Galois weights by means of orders of stabilizer groups at points. Then we get an orbiface $\mathbf{X}'$ and its contraction $\mathbf{X}$. Such contractions will shortly also be called orbifaces. The induced orbital curve $\mathbf{C}'$ contracts to the weighted weighted surface germ $\mathbf{C}$, which we will also call orbital curve. Altogether we get a commutative diagram of orbital curves

$$\begin{array}{ccc} \mathbf{D}' & \longrightarrow & \mathbf{D} \\ \Big\downarrow{\scriptstyle N_G(D)} & & \Big\downarrow{\scriptstyle N_G(D)} \\ \mathbf{C}' & \longrightarrow & \mathbf{C} \end{array} \tag{16}$$

with trivially weighted release on the top, an orbital curve release on the bottom and an orbital curve uniformization on the left-hand side.

Definition 3.8. *The orbital curve* $\mathbf{C}$ *is called* **uniform releasable** *if and only if there exist a commutative diagram (16). The corresponding quotient release* $\mathbf{C}' \to \mathbf{C}$ *is called an* **orbital release** *of* $\mathbf{C}$. *The ambient map* $\mathbf{X}' \to \mathbf{X}$ *is called the* **orbital release** *of* $\mathbf{X}$ *along* C. *The morphisms* $\mathbf{D}' \to \mathbf{D}$, $Y' \to Y$ *are called* **release uniformizations** *of* $\mathbf{C}' \to \mathbf{C}$ *or of* $\mathbf{X}' \to \mathbf{X}$, *respectively.*

Notice that a release uniformization of $\mathbf{C}$ endows automatically the surface around C' with (Galois) weights. Therefore we get an orbital curve in this case. Starting from a smooth surface Y with G-action it is interesting to ask, which curves $D \subset Y$ have a releasable quotient curve $C = D/G \subset X = Y/G$? Keep in mind the Example 3.10 below, because it will play a central role.

Definition 3.9. *The action of* G *on* Y *as above is* **ordinary at** D *if and only if the curve*

$$GD := \bigcup_{g \in G} g(D)$$

has at most ordinary singularities. The action is **smooth at** D *if and only if* GD *is smooth. The action is* **separating at** D *if and only if for all* $g \in G$ *the curve* $g(D)$ *is either equal to* D *or has no common point with* D.

Obviously, a smooth action at D must be separating at D. For smooth curves D both notions coincide.

Example 3.10. Let Y be a smooth surface with G-action and D a smooth curve on X. If G acts ordinarily at D, then the orbital curve $\mathbf{C} = \mathbf{D}/\mathbf{G}$ is uniform releasable.

Proof. We release simultaneously the ordinary singularities of GD by σ-processes. Let $Y' \to Y$ be this simultaneous releasing morphism, and denote the proper transform of D on Y' by D'. Then G acts on Y' and thereby smoothly at D'. With $C' = D'/G = D'/N_G(D) = D'/G_{D'}$ we get a uniform release diagram (16). $\square$

Example 3.11. Let Y be a smooth surface with G-action separating at D, where D is a curve on X with at most hypercusp singularities. Then the orbital curve $\mathbf{C} = \mathbf{D}/\mathbf{G}$ is uniform releasable.

Proof. Because of the separating property we can assume that $G = N_G(D)$. Then G acts on the set of singularities of D. In a G-equivariant manner we resolve stepwise the curve singularities as described in Example 3.4, each by a linear tree of lines such that the smooth proper transform D' of D on the arising surface Y' intersects precisely one (central) line of the tree. All intersections of tree lines and such lines with D' are empty or transversal. The G-action on Y transfers to a G-action on Y'. Since $N_G(D') = G$ the quotient curve $C' = D'/G$ is smooth, or in other words, G acts smoothly at D'. Let $Q \in D$ be a curve singularity and $E_Q \subset Y'$ the resolving linear tree over Q. Then the Q-stabilizing group G_Q acts on E_Q and especially on the D'-crossing central line L_Q of E_Q. Now it is clear that G acts (via G_Q) separately, hence smooth at L_Q. This property refers also to the other line components of E_Q. The isotropy groups $G_{Q'}$ at points $Q' \in E_Q$ must be abelian because of transversal intersections of the components of $E_Q \cup D'$. Therefore the image of E_Q on Y'/G is again a linear tree of lines intersecting at abelian quotient singularities. If we take the minimal singularity resolutions of them, then we get again a linear tree of lines crossing each other. Now blow down the two partial linear trees outside of the proper transform of the central line $L_Q/G = L_Q/G_Q$ crossing $C' = D'/G$ in at most abelian singularities. Altogether one gets a uniform release diagram (16) for the orbital curve $\mathbf{C} = \mathbf{D}/\mathbf{G}$. The upper release $\mathbf{D}' \to \mathbf{D}$ is that of hypercusps of curves locally described in Example 3.4. □

3.3. Homogeneous points

Definition 3.12. *A* **simple surface singularity** *is a singularity, whose minimal resolution curve consists of one (smooth) irreducible curve only. A* **simple surface point** *is a simple singularity or a regular surface point.*

In the latter case we consider the exceptional line of the σ-process as resolution curve.

Now let G be a finite group acting on a surface Y with only simple points, and let $Q \in Y$ be one of them. The group action extends to the simultaneous minimal resolution Y' of all points of the orbit GQ, and the stabilizer group G_Q acts on the resolving curve $E_Q \subset Y'$. Moreover, G_Q acts on the normal bundle over E_Q respecting fibres. Therefore the stationary subgroups of G_Q at points on E_Q must be abelian (fibres and E_Q are diagonalizing). Take a G_Q-invariant open neighbourhood $V \subset Y$ of Q, smooth outside of Q. Locally the situation is described by the following commutative **local coniform release diagram**:

$$\begin{array}{ccc} V' & \longrightarrow & V \\ \downarrow & & \downarrow \\ U' = V'/G_Q & \longrightarrow & U = V/G_Q \end{array} \tag{17}$$

with vertical quotient morphisms and upper horizontal resolution. Let $P \in U$ be the image point of Q. We endow U', $U \setminus \{P\}$ with Galois weights coming from the finite G_Q-uniformizations $V' \to U'$, $V \setminus \{Q\} \to U \setminus \{P\}$. Finally, we set $w(P) := w(E_Q)$. The corresponding orbifaces are denoted by $\mathbf{U}'$ or $\mathbf{U}$, respectively. Our diagram can be written as

$$\begin{array}{ccc} V' & \longrightarrow & V \\ \downarrow & & \downarrow \\ \mathbf{U}' = \mathbf{V}'/\mathbf{G_Q} & \longrightarrow & \mathbf{U} = \mathbf{V}/\mathbf{G_Q} \end{array} \tag{18}$$

Again, we have a refinement equivalence class of weighted open neighbourhoods of P denoted by $\mathbf{P} = \mathbf{Q}/\mathbf{G_Q} \in \mathbf{U}$.

Definition 3.13. *Weighted surface points* $\mathbf{P}$ *constructed on this way are called* **(orbital) homogeneous points**.

It is clear that orbital quotient points are homogeneous. The morphism $Q \to \mathbf{P}$ of orbital points, defined as refinement class of $V \to \mathbf{U}$, is called a **coniformization** of $\mathbf{P}$. It is a **uniformization**, if the preimage point Q is regular on V. Homogeneous points which are coniformable but not uniformizable are called **honest homogeneous** points.

Remark 3.14. The supporting surface point P of a homogeneous point $\mathbf{P}$ is in any case a so-called **quasihomogeneous singularity**. The resolutions of these quasihomogeneous singularities are precisely known. We refer to [P], [D]. For graphical descriptions with weights see [H98].

Remark 3.15. Simple singularities are "cone-like". Namely, up to isomorphy (look at normal bundle of E_Q), they are contractions of a section C of a line bundle over C with negative self-intersection contractible to a singularity Q. In our imagination the contracting surface looks like a cone around the singularity Q. Therefore we introduced the notion "coniformization".

Remark 3.16. Observe that the $\mathbf{U}'$ supports finitely many abelian points sitting on the smooth quotient curve E_Q/G. These simpler orbital points "release" the homogeneous point $\mathbf{P}$, which explains our calling. Notice also that an abelian point $\mathbf{P}$ can loose its original weight after a coniformization. The old one is "released" by the new "coniform weight".

A global **coniform release diagram** looks like

$$\begin{array}{ccc} Y' & \longrightarrow & Y \\ \Big\downarrow & & \Big\downarrow \\ \mathbf{X}' = \mathbf{Y}'/\mathbf{G} & \longrightarrow & \mathbf{X} = \mathbf{Y}/\mathbf{G} \end{array} \tag{19}$$

Thereby Y is a surface with only simple singularities, G a finite group acting effectively on Y, $Y' \to Y$ resolves minimally all simple singularities of Y and, possibly, G-orbits of finitely many regular points by σ-processes. The weights of $\mathbf{X}'$ are Galois weights. We push forward the weights of the quotients of the exceptional curves of $Y' \to Y$ to get a birational morphism of weighted surfaces in the bottom of the diagram.

Definitions 3.17. *$Y \longrightarrow \mathbf{X}$ supported by a a quotient morphism $Y \longrightarrow X = Y/G$ is called a (global)* **coniformization**, *if and only if one can extend it to a coniform release diagram (19). Morphisms in the bottom of (19) are called* **coniform releases**, *and $Y' \longrightarrow Y$ is a* **coniformization** *of $\mathbf{X}' \longrightarrow \mathbf{X}$.*

An **(finitely weighted) orbiface** is a weighted surface supporting (finitely weighted) homogeneous points only.

At the end of this section we want to describe orbital cusp points. A **neat hyperbolic cusp point** is a simple elliptic surface point $Q \in V$ endowed with weight ∞. Its resolution curve $C \subset V$ is, by definition, elliptic. We use the notations of the diagrams (17) and (18) to get the homogeneous quotient point $\mathbf{P}$ of the elliptic point Q. We change the finite Galois weight of P by ∞ to get a **hyperbolic orbital cusp point** $\mathbf{P}_\infty$. Notice that the weights of $\mathbf{U}$ outside of P will not be changed. Also the weights of all points of the preimage curves of Q and P in diagram (17) will be changed to ∞. Then we get local coniform release diagrams (18) for hyperbolic orbital cusp points $\mathbf{P}_\infty$.

A **neat Hilbert cusp singularity** Q is a normal surface singularity which has a cycle of transversally intersecting smooth rational curves as resolution curve. The minimal resolution curve E_Q is of the same type or a rational curve with only one curve singularity which is an ordinary self-intersection of this curve. Locally, we have diagrams of type (17) again with a finite group action around Q. Endowing Q with weight ∞ and all other points on V outside Q with trivial weight 1, we get a **neat Hilbert cusp point** Q_∞. As in the hyperbolic cusp case we endow the quotient point P, the points of E_Q and of its quotient curve also with weight ∞ and all other points with the usual finite Galois weights to get a diagram (18) in the category of weighted surfaces called **local Hilbert cusp release diagrams**. The corresponding orbital point $\mathbf{P}_\infty$ itself is called a **Hilbert orbital cusp point**. After pulling back the weight ∞ to the points on the preimage curve of P we get, with obvious notations, representatives $\mathbf{U}'_\infty \to \mathbf{U}_\infty$ called **orbital cusp releases** of $\mathbf{P}_\infty$.

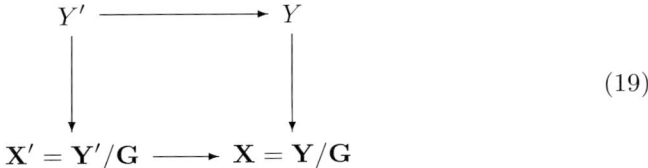

Altogether, in the **category of orbifaces** $\mathbf{Orb}^2$ we dispose on **orbital releases**, **orbital release diagrams**, **coniform releases**, **finite orbital morphisms**, **orbital open embeddings**, **birational orbital morphisms** by composition of orbital releases, **orbital morphisms** by composition of birational and finite orbital morphisms, and on **orbital correspondence classes**, which consist of orbital objects connected by finite orbital coverings.

The notions restrict in obvious manner to orbital curves on orbifaces via representative neighbourhoods. So we dispose on **category of orbital curves** $\mathbf{Orb}^{2,1}$ (on orbifaces) with all the types of orbital morphisms above.

Remark 3.18. Restricting to algebraic objects and morphisms one can work with Zariski-open sets only to define refinement classes and corresponding weighted (orbital) points. There will be no difference for the later definitions of orbital invariants.

3.4. Picard and Hilbert orbifaces

Let $\mathbb{B} \subset \mathbb{C}^2$ be a bounded domain and Γ a group of analytic automorphisms of $\mathbb{B}$ acting properly discontinuously. Then the quotient $\Gamma\backslash\mathbb{B}$ together with Galois weights is a finitely weighted orbiface, which we denote by $\mathbf{\Gamma}\backslash\mathbb{B}$.

There are two symmetric subdomains of $\mathbb{C}^2$: The irreducible **complex unit ball**

$$\mathbb{B}\ |z_1|^2 + |z_2|^2 < 1$$

and the product $\mathbb{D}^2$ of two unit discs. The latter is biholomorphic equivalent to the product

$$\mathbb{H}^2 = \mathbb{H} \times \mathbb{H}, \quad \mathbb{H}: \ Im\, z > 0,$$

of two upper half planes $\mathbb{H}$ of $\mathbb{C}$.

The automorphism groups are the projectivizations of the **unitary group** $\mathbb{U}((2,1),\mathbb{C}) \subset \mathbb{G}l_3(\mathbb{C})$ or of the symmetric extension (by transposition of coordinates) $\mathbb{G}S_2^+(\mathbb{R})$ of $\mathbb{G}l_2^+(\mathbb{R}) \times \mathbb{G}l_2^+(\mathbb{R})$, respectively. Both groups act transitively on the corresponding domain.

Now let $K = \mathbb{Q}(\sqrt{D})$ be a quadratic number field with discriminant $D = D_{K/\mathbb{Q}} \in \mathbb{Z}$ and ring of integers $\mathcal{O}_K$. In the ball case we let K be an imaginary quadratic field and in the splitting case a real quadratic field. The arithmetic groups acting (non-effectively) on $\mathbb{B}$ or $\mathbb{H}^2$,

$$\Gamma_K = \begin{cases} \mathbb{SU}((2,1),\mathcal{O}_K), & \text{if } K \text{ imaginary quadratic,} \\ \mathbb{S}l_2(\mathcal{O}_K), & \text{if } K \text{ real quadratic,} \end{cases}$$

are called **full Picard modular** or **full Hilbert modular group** of the field K, respectively. To be precise, we use, if nothing else is said, in the Picard case the hermitian metric on $\mathbb{C}^3$ of signature $(2,1)$ represented by the diagonal matrix $\begin{pmatrix} 1 & 0 & 0 \\ 0 & 1 & 0 \\ 0 & 0 & -1 \end{pmatrix}$. The action on $\mathbb{B}$ restricts the $\mathbb{G}l_3$-action on $\mathbb{P}^2$ in obvious manner. In the Hilbert case we have to restrict the action on $\mathbb{P}^1 \times \mathbb{P}^1$ of

$$\mathbb{G}l_2^+(K) \ni g: \ (z,w) \mapsto (g(z), g'(w)),$$

where $'$ denotes the non-trivial field automorphism of K applied to each coefficient of g.

Definitions 3.19. *A* **Picard modular group** *(of the imaginary quadratic field K) is a subgroup of $\mathbb{G}l_3(\mathbb{C})$ commensurable with Γ_K.*
A **Hilbert modular group** *(of the real quadratic field K) is a subgroup of $\mathbb{G}S_2^+(\mathbb{R})$ commensurable with Γ_K.*

Definitions 3.20. *The finitely weighted orbital quotient surfaces*

$$\overset{o}{\mathbf{X}}_\Gamma = \begin{cases} \mathbf{\Gamma}\backslash\mathbb{B} = \mathbb{P}\mathbf{\Gamma}\backslash\mathbb{B} \\ \mathbf{\Gamma}\backslash\mathbb{H}^2 = \mathbb{P}\mathbf{\Gamma}\backslash\mathbb{H}^2 \end{cases}$$

are called the **open Picard orbiface** *of Γ, if Γ is a Picard modular group, respectively the* **open Hilbert orbiface** *of Γ, if Γ is a Hilbert modular group.*

Forgetting Galois weights, the surfaces $\overset{o}{X}_\Gamma = \Gamma\backslash\mathbb{B}$ or $\Gamma\backslash\mathbb{H}^2$ are called **open Picard modular** or **open Hilbert modular surfaces**, respectively. Each of them has a unique analytic **Baily-Borel compactification** $\hat{X}_\Gamma := \widehat{\Gamma\backslash\mathbb{B}}$ adding finitely many hyperbolic respectively Hilbert cusp "singularities" (which may be regular). These are projective normal surfaces. We extend the Galois weight map of $\mathbf{X}_\Gamma$ to $\hat{X}_\Gamma$ endowing the cusps with weight ∞ to get the **orbital Baily-Borel model** $\hat{\mathbf{X}}_\Gamma$ of $\overset{o}{X}_\Gamma$ or $\overset{\mathbf{o}}{\mathbf{X}}_\Gamma$. Releasing all cusp points we get the **cusp released models** X_Γ with orbital versions $\mathbf{X}_\Gamma$.

Each arithmetic group Γ has a **neat normal subgroup** Δ of finite index. By definition, the eigenvalues of each element of a neat arithmetic linear group generate a free abelian subgroup of $\mathbb{C}^*$ (which may be trivial). Especially, because of absence of unit roots, a neat Picard and Hilbert modular group acts fixed point free on $\mathbb{B}$ or $\mathbb{H}^2$, respectively. Moreover, the cusp points of the corresponding modular surfaces are neat. With the groups Δ and $G := \Gamma/\Delta$ we get the global cusp release diagrams

$$\begin{array}{ccc} X_\Delta & \longrightarrow & \hat{X}_\Delta \\ {\scriptstyle /G}\downarrow & & \downarrow{\scriptstyle /G} \\ \mathbf{X}_\Gamma & \longrightarrow & \hat{\mathbf{X}}_\Gamma \end{array} \tag{20}$$

Notice that X_Δ is a smooth projective surface. Moreover, the objects and the orbital release morphism on the bottom of the diagram do not depend on the choice of Δ. In the Picard case we refer to [H98] for the complete classification of orbital hyperbolic cusps and their local releases working on $\hat{\mathbf{X}}_\Gamma$ only. The analogous work for the Hilbert case has not been done until now, but seems to be not difficult. Since X_Δ is smooth, all finitely weighted orbital points on $\hat{\mathbf{X}}_\Gamma$ are quotient points, also well-classified in [H98]. The non-abelian ones have also unique releases. Locally, they come from σ-processes at their preimage points on X_Δ, not depending on the choice of Δ again. The complete release of non-abelian orbital quotient points of $\mathbf{X}_\Gamma$ is denoted by $\mathbf{X}'_\Gamma$. We get global commutative orbital release diagrams

$$\begin{array}{ccc} X'_\Delta & \longrightarrow & X_\Delta \\ \big\downarrow{\scriptstyle /G} & & \big\downarrow{\scriptstyle /G} \\ \mathbf{X}'_\Gamma & \longrightarrow & \mathbf{X}_\Gamma \end{array} \tag{21}$$

Altogether we get orbital release diagrams

$$\begin{array}{ccccc} X'_\Delta & \longrightarrow & X_\Delta & \longrightarrow & \hat{X}_\Delta \\ \big\downarrow & & \big\downarrow{\scriptstyle /G} & & \big\downarrow \\ \mathbf{X}'_\Gamma & \longrightarrow & \mathbf{X}_\Gamma & \longrightarrow & \hat{\mathbf{X}}_\Gamma \end{array} \tag{22}$$

We can shorten them to one diagram

$$\begin{array}{ccc} X'_\Delta & \longrightarrow & \hat{X}_\Delta \\ \big\downarrow & & \big\downarrow \\ \mathbf{X}'_\Gamma & \longrightarrow & \hat{\mathbf{X}}_\Gamma \end{array} \tag{23}$$

with the releases of precisely all non-abelian orbital points, because all hyperbolic orbital cusp points are homogeneous, hence releasable.

Visualization:

1) Released Picard modular Apollonius plane (4);
2) Released Hilbert modular Cartesius plane.

From $\mathbf{Orb}^2$ we single out the correspondence classes

- $\overset{o}{\mathbf{Pic}}{}^2_K$ of **open Picard orbifaces** of the field K,
  with objects $\overset{\mathbf{o}}{\mathbf{X}}_{\boldsymbol{\Gamma}}$, Γ a Picard modular group of the field K;
- $\widehat{\mathbf{Pic}}{}^2_K$ of **Picard orbifaces** of the field K,
  with objects $\hat{\mathbf{X}}_\Gamma$;
- $\mathbf{Pic}^{2'}_K$ of **released Picard orbifaces** of the field K,
  with objects $\mathbf{X}'_\Gamma$.

As surviving orbital morphisms in each of these correspondence classes we take only the finite ones coming from pairs $\Delta \subset \Gamma$ of Picard modular groups of the same field K.

In the same manner we dispose in $\mathbf{Orb}^2$ on correspondence classes

- $\overset{\circ}{\mathbf{Hilb}}{}^2_K$ of **open Hilbert orbifaces** of the field K, with objects $\overset{\circ}{\mathbf{X}}_{\Gamma}$, Γ a Hilbert modular group of the field K;
- $\widehat{\mathbf{Hilb}}{}^2_K$ of **Hilbert orbifaces** of the field K, with objects $\hat{\mathbf{X}}_\Gamma$;
- $\mathbf{Hilb}^{2'}_K$ of **released Hilbert orbifaces** of the field K, with objects $\mathbf{X}'_\Gamma$.

In difference to the Picard case, the neat objects of $\mathbf{Hilb}_K$ come from minimal singularity resolution of the corresponding neat objects of $\widehat{\mathbf{Hilb}}_K$. One has to plug in at the Baily–Borel cusps a cycle of rational curves with negative self-intersection. In general we must plug in finite quotients of such cycles. For Picard and Hilbert cases we call it cusp released in common.

We denote by $\mathbf{Pic}^2$ or $\mathbf{Hilb}^2$ the complete subcategories of $\mathbf{Orb}^2$ with the above three types of Picard- or Hilbert objects, respectively. Both kinds of objects together form the complete subcategory $\mathbf{Shim}^2$ of irreducible Shimura orbifaces.

3.5. Orbital arithmetic curves

Let Γ be a Picard or Hilbert modular group of a quadratic number field K. We say that $\mathbb{D} \subset \mathbb{B}$ or $\mathbb{H}^2$ is a **K-arithmetic disc**, if and only if there is a holomorphic embedding of of the unit disc $\mathbb{D}^1 \hookrightarrow \mathbb{B}$ with image $\mathbb{D}$ such that $\mathbb{D}$ is closed in $\mathbb{B}$ and the **$\mathbb{D}$-normalizing subgroup** (decomposition group) of Γ

$$N_\Gamma(\mathbb{D}) := \{\gamma \in \Gamma;\ \gamma(\mathbb{D}) = \mathbb{D}\}$$

is a $\mathbb{D}$-lattice, that means $N_\Gamma(\mathbb{D})\backslash\mathbb{D} = \Gamma_{\mathbb{D}}\backslash\mathbb{D}$ is a quasiprojective algebraic curve, where

$$\Gamma_{\mathbb{D}} = N_\Gamma(\mathbb{D})/Z_\Gamma(\mathbb{D}), \text{ the effective decomposition group of } \mathbb{D},$$
$$Z_\Gamma(\mathbb{D}) = \{\gamma \in \Gamma;\ \gamma|_{\mathbb{D}} = id_{\mathbb{D}}\},\ \mathbb{D} - \text{centralizing (or inertia) group}.$$

To be more precise, we have commutative diagrams with algebraic groups defined over $\mathbb{Q}$ in the upper two rows

$$\begin{array}{ccccccc}
N_\Gamma(\mathbb{D}) & \longrightarrow & \mathbf{N} & \longrightarrow & \mathbf{G} & \longleftarrow & \Gamma \\
\uparrow & & \uparrow & & \uparrow & & \uparrow \\
\mathbb{S}N_\Gamma(\mathbb{D}) & \longrightarrow & \mathbb{S}\mathbf{N} & \longrightarrow & \mathbb{S}\mathbf{G} & \longleftarrow & \mathbb{S}\Gamma \\
 & & \downarrow & & \downarrow & & \\
 & & \mathbb{D} & \longrightarrow & \mathbb{B} & &
\end{array} \tag{24}$$

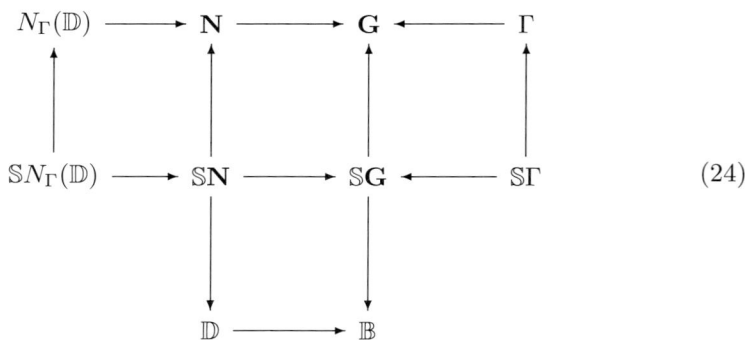

with algebraic Lie groups $\mathbf{N}$, $\mathbf{G}$ acting on the symmetric domains below. The algebraic groups in the middle are special: $\mathbb{S}N(\mathbb{R})$ isomorphic to $\mathbb{S}l_2(\mathbb{R})$ or $\mathbb{SU}((1,1),\mathbb{C})$. The $\mathbb{Q}$-algebraic group embeddings are lifted from the bottom row.

We denote the image curve of $\mathbb{D}$ on $\overset{o}{X}_\Gamma = \Gamma\backslash\mathbb{B}$ by $\Gamma\backslash\mathbb{D}$ or $\overset{o}{D}_\Gamma$. Its closure on X_Γ or $\hat{X}_\Gamma$ is denoted by D_Γ or $\hat{D}_\Gamma$, respectively, and their proper transform on X'_Γ by D'_Γ. We have birational curve morhisms

$$\widehat{\Gamma_{\mathbb{D}}\backslash\mathbb{D}} \longrightarrow D'_\Gamma \longrightarrow D_\Gamma \longrightarrow \hat{D}_\Gamma.$$

with (unique) smooth compact model on the left.

The corresponding weighted orbital curves on $\overset{\mathbf{o}}{\mathbf{X}}_\Gamma, \mathbf{X}'_\Gamma, \mathbf{X}_\Gamma$ or $\hat{\mathbf{X}}_\Gamma$ are denoted by $\overset{\mathbf{o}}{\mathbf{D}}_\Gamma, \mathbf{D}'_\Gamma, \mathbf{D}_\Gamma$ or $\hat{\mathbf{D}}_\Gamma$, respectively. We have birational orbital curve morphisms

$$\mathbf{D}'_\Gamma \longrightarrow \mathbf{D}_\Gamma \longrightarrow \hat{\mathbf{D}}_\Gamma, \quad \overset{\mathbf{o}}{\mathbf{D}}_\Gamma \hookrightarrow \mathbf{D}'_\Gamma, \quad \overset{\mathbf{o}}{\mathbf{D}}_\Gamma \hookrightarrow \mathbf{D}_\Gamma.$$

If $\overset{o}{D}_\Gamma$ is non-compact, then $\overset{\mathbf{o}}{\mathbf{D}}_\Gamma, \mathbf{D}'_\Gamma, \mathbf{D}_\Gamma, \hat{\mathbf{D}}_\Gamma$, are (weighted) **orbital modular curves**. If it is compact, then $\overset{\mathbf{o}}{\mathbf{D}}_\Gamma, \mathbf{D}'_\Gamma, \mathbf{D}_\Gamma, \hat{\mathbf{D}}_\Gamma = \overset{\mathbf{o}}{\mathbf{D}}_\Gamma$ are **orbital Shimura curves**. Altogether they are called **orbital arithmetic curves**. Their supporting curves are defined over an algebraic number field because these are one-dimensional Shimura varieties.

Theorem 3.21. *Each open orbital arithmetic curve* $\overset{\mathbf{o}}{\mathbf{D}}_\Gamma$ *on a Picard or Hilbert modular surface has a smooth model in its correspondence class. More precisely, one can find a finite uniformization of* $\overset{\mathbf{o}}{\mathbf{D}}_\Gamma$.

In the Picard case also $\mathbf{D}_\Gamma$ has a finitely covering smooth model in its correspondence class.

Theorem 3.22. *All orbital arithmetic curves* $\mathbf{D}_\Gamma$ *are orbital releasable.*

Proof steps for the last two theorems (e.g., for $\mathbb{B}$):

Definition 3.23. *A* $\mathbb{B}$*-lattice* Γ *is called* neat *if and only if each stationary group* Γ_P*,* $P \in \mathbb{B}$*, is torsion free.*

It is well known by a theorem of Borel, that each lattice Γ contains a normal sublattice Γ_0 (of finite index), which is neat. Especially, Γ_0 is torsion free.

Definition 3.24. *Let* $\mathbb{D}$ *be a* K*-disc on* $\mathbb{B}$*. A neat* $\mathbb{B}$*-lattice* Δ *is called* $\mathbb{D}$**-neat***, if and only if the implication*

$$\gamma(\mathbb{D}) \cap \mathbb{D} \neq \emptyset \Longrightarrow \gamma(\mathbb{D}) = \mathbb{D}$$

holds for all $\gamma \in \Delta$.

Since Δ is assumed to be neat there are no honest Galois weights, hence $\overset{o}{X}_\Delta = \overset{o}{\mathbf{X}}_\Delta$. Then our $\mathbb{D}$-neat condition is equivalent to the regularity of the image curve $\overset{o}{D}_\Delta = \Delta\backslash\mathbb{D}$ on $\overset{o}{X}_\Delta$. Thus Theorem 3.21 follows from the following

Proposition 3.25. Γ *has a normal* $\mathbb{D}$*-neat sublattice.*

It is well known that the arithmetic group Γ has a neat subgroup of finite index. Therefore we can assume Γ to be neat, hence $\Gamma_{\mathbb{D}} = N_{\Gamma}(\mathbb{D})$. Without loss of generality we can also assume that all elements of Γ are special, hence

$$\Gamma_{\mathbb{D}} \subset \mathbb{S}\mathbf{N}(\mathbb{Z}) \subset \mathbb{S}\mathbf{G}(\mathbb{Z}) \supset \mathbf{\Gamma} \tag{25}$$

with the notations of diagram (24).

Proof. We call a point $Q \in \mathbb{D}$ a **Γ-singular point** on $\mathbb{D}$, if and only if it belongs to $\mathbb{D} \cap \gamma(\mathbb{D})$ for a $\gamma \in \Gamma$ not belonging to $\Gamma_{\mathbb{D}}$. The Γ-singular points on $\mathbb{D}$ are precisely the preimages on $\mathbb{D}$ of the singular points of $\Gamma\backslash\mathbb{D}$. Therefore there are only finitely many $\Gamma_{\mathbb{D}}$-equivalence classes of Γ-singular points on $\mathbb{D}$. The branch set at Q of the orbit curve $\Gamma\mathbb{D} \subset \mathbb{B}$ corresponds bijectively to the set of branches of $\Gamma\backslash\mathbb{D}$ at the image point P. Let $Q_1, \ldots, Q_s$ be a complete set of $\Gamma_{\mathbb{D}}$- representatives of Γ-singular points on $\mathbb{D}$ and $\mathbb{D}_{ij} = \gamma_{ij}(\mathbb{D})$, $\gamma_{ij} \in \Gamma_{\mathbb{D}}$, $j = 1, \ldots, k_i$, all different $\Gamma\mathbb{D}$-branches at Q_i excluding $\mathbb{D}$.

It is a fact of algebraic group theory, see e.g. [B], II.5.1, that $\mathbf{N}$ is the normalizing group of a line L in a faithful linear representation space E of $\mathbf{G}$, all defined over $\mathbb{Q}$. Each $X \in L(\mathbb{Q})$ defines, over $\mathbb{Q}$ again, a weight character $\rho = \rho_X = \rho_L : \mathbf{N} \to \mathbb{G}\mathbf{l_1}$, and

$$\alpha : G \longrightarrow \mathbb{G}l(E), \quad \alpha(g)X = \rho(g)X,\ g \in \mathbf{N}(\mathbb{Q}),$$

Since $\mathbb{S}\mathbf{N}$ is simple, ρ restricts to the trivial character on $\mathbb{S}\mathbf{N}$, hence $\rho(\Gamma_{\mathbb{D}}) = \{1\}$. With the above chosen elements we define

$$E \ni X_{ij} := \gamma_{ij} X \neq X.$$

The latter non-incidence holds because each γ_{ij} does not normalize $\mathbb{D}$, hence does not belong to N, so it cannot normalize L. We can assume (by choice of $\mathbb{Q}$-base) that all X, X_{ij} belong to $E(\mathbb{Z})$. We find a natural number a such that

$$X \not\equiv X_{ij} \mod (a), \text{ for all } i = 1, \ldots, s,\ j = 1, \ldots, k_i.$$

Now we show that $\Gamma(a)$ is $\mathbb{D}$-neat:
Assume the existence of $\gamma' \in \Gamma(a)$ such that $\gamma'\mathbb{D}$ intersects $\mathbb{D}$ properly at P. The intersecting pair $(\mathbb{D}, \gamma'\mathbb{D})$ has to be $\Gamma_{\mathbb{D}}$-equivalent to one of the representative intersection pairs $(\mathbb{D}, \gamma_{ij}\mathbb{D})$, say

$$\delta\gamma'\mathbb{D} = \gamma_{ij}\mathbb{D},\ \delta \in \Gamma_{\mathbb{D}}.$$

Therefore $\gamma_{ij}^{-1}\delta\gamma' \in N$ because it normalizes $\mathbb{D}$.

$$\gamma_{ij}^{-1}\delta\gamma' X = X,\ \delta\gamma' X = X_{ij},\ \gamma_1 X = X_{ij},$$

with $\gamma_1 := \delta\gamma'\delta^{-1} \in \Gamma(a)$. Since $\gamma_1 \equiv id \mod (a)$, we get the contradiction

$$X = id(X) \equiv \gamma_1(X) = X_{ij} \mod (a).$$

□

4. Neat Proportionality

We want to define norms of orbital arithmetic curves $\overset{o}{\mathbf{D}}_{\mathbf{\Gamma}}$ and their birational companions on Picard or Hilbert modular surfaces. They sit in the specification of the discs $\mathbb{D} = \mathbb{D}_V \hookrightarrow \mathbb{B}$ or $\mathbb{H}^2$ defining the supporting curve as $\overset{o}{D}_\Gamma = \Gamma\backslash\mathbb{D}$, namely

$$\mathbb{D}_V = \mathbb{P}(V^\perp) \cap \mathbb{B} \text{ or } \mathbb{P} \times \mathbb{P}(V^\perp) \cap \mathbb{H}^2,$$

where in the Picard case : $V \in K^3$, $< V, V >$ positive, $\perp$ with respect to unitary $(2,1)$-metric on $\mathbb{C}^3$.

In the Hilbert case the situation is more complicated: Let $V \in \mathbb{G}l_2^+(K)$ be scew-hermitian with respect to the non-trivial $K/\mathbb{Q}$-isomorphism $'$, that means ${}^tV' = -V$, explicitly

$$V = \begin{pmatrix} a\sqrt{D} & \lambda \\ -\lambda' & b\sqrt{D} \end{pmatrix}, \tag{26}$$

$a, b \in \mathbb{Q}$, $\lambda \in K$. By abuse of language we define "orthogonality" with elements of $\mathbb{C}^2 \times \mathbb{C}^2$ in the following manner:

$$\mathbb{C}^2 \times \mathbb{C}^2 \ni (z_1, z_0; w_1, w_0) \perp V : \ (z_1, z_0) V \begin{pmatrix} w_1 \\ w_0 \end{pmatrix} = 0$$

and the "bi-projectivization" of $\mathbb{C}^2 \times \mathbb{C}^2$ by

$$\mathbb{P} \times \mathbb{P}(z_1, z_0; w_1, w_0) := (z_1 : z_0) \times (w_1 : w_0) \in \mathbb{P}^1 \times \mathbb{P}^1 \supset \mathbb{H} \times \mathbb{H}.$$

$\mathbb{D}_V \subset \mathbb{P}(V^\perp)$ or $\mathbb{P} \times \mathbb{P}(V^\perp)$ is complex 1-dimensional and analytically isomorphic to the upper half plane $\mathbb{H}$.

For later use of Heegner divisors we define norms of subdiscs and their quotient curves on this place.

Definition 4.1. *of* **norms**

Picard case: $N(V) \ := < V, V > \in \mathbb{N}_+$, $V \in \mathfrak{O}_K^{3,+}$,
Hilbert case: $N(V) \ := \det V \in \mathbb{N}_+$, $V \in Scew_2^+(\mathfrak{O}_K)$.

and of **norm sets** of arithmetic curves:

$$\mathcal{N}(\widehat{\Gamma\backslash\mathbb{D}}) = \mathcal{N}(\mathbb{D}) \ := \{N(V);\ \mathbb{D} = \mathbb{D}_V, V \text{ integral}\} \subset \mathbb{N}_+$$

Definition 4.2. *For $N \in \mathbb{N}_+$ the Weil divisor*

$$H_N = H_N(\Gamma) \ := \sum_{\substack{\mathbb{D} \\ \mathcal{N}(\mathbb{D}) \ni N}} \widehat{\Gamma\backslash\mathbb{D}}$$

is called the N-th **Heegner divisor** *on $\hat{X}_\Gamma$.*

The scew-symmetric elements (26) with fixed K form a quadratic vector space $\mathfrak{V}_\mathbb{Q} \ := (Scew_2(K), \det)$ with signature $(2,2)$ as well as $\mathfrak{V}_\mathbb{R} \cong \mathbb{R}^{2,2}$. The group $\mathbb{S}l_2(K)$ acts on $Scew_2(K)$ and on the positive part $Scew_2^+(K)$:

$$\mathbb{S}l_2(K) \ni g : \ V \mapsto {}^tg'Vg.$$

It defines an embedding

$$\begin{aligned} \mathbb{H}^2 &\longrightarrow Grass^+(2,\mathfrak{V}_\mathbb{R}) \subset Grass(2,\mathfrak{V}_\mathbb{R}) \\ (z,w) \mapsto (z,w)^\perp &:= \{V \in \mathfrak{V}_\mathbb{R};\ (z_1,z_0;w_1,w_0)\perp V\} \end{aligned}$$

and a homeomorphism

$$\mathbb{H}^2 \longleftrightarrow \mathbb{SO}_e(2,2)/(\mathbb{SO}(2)\times\mathbb{SO}(2)),$$

where the lower index $_e$ denotes the unit component. The normalizer of $\mathbb{D}_V$ in $\Gamma \subseteq \mathbb{S}l_2(\mathfrak{O}_K)$ is

$$N_\Gamma(\mathbb{D}_V) = \Gamma_V = \{g \in \Gamma;\ {}^t g' V g = \pm V\}.$$

In the special case $V = \left(\begin{smallmatrix} 0 & \lambda \\ -\lambda' & 0 \end{smallmatrix}\right)$, $\lambda \in \mathfrak{O}_K$, primitive, $\lambda\cdot\lambda' = N$, $\Gamma = \mathbb{S}l_2(\mathfrak{O}_K)$ the action on $\mathbb{D}_V$ is (conjugation) equivalent to the action of

$$\left(\begin{smallmatrix} \lambda & 0 \\ 0 & 1 \end{smallmatrix}\right)\mathbb{S}l_2(\mathfrak{O})\left(\begin{smallmatrix} \lambda^{-1} & 0 \\ 0 & 1 \end{smallmatrix}\right)$$

on the diagonal of $\mathbb{H}^2$. Then for $K = \mathbb{Q}(\sqrt{d})$

$$N_\Gamma(\mathbb{D}_V) \cong \begin{cases} \mathbb{S}l_2(\mathbb{Z})(N)_0 := \{\left(\begin{smallmatrix} a & b \\ c & d \end{smallmatrix}\right);\ a,b,c,d \in \mathbb{Z},\ N \mid c\}, & \text{if } \sqrt{d} \nmid \lambda \\ \text{index 2 extension of } \mathbb{S}l_2(\mathbb{Z})(N)_0, & \text{if } \sqrt{d} \mid \lambda \text{ in } \mathfrak{O}_K. \end{cases}$$

Any hermitian symmetric domain $\mathbb{B}$ is embedded in its dual symmetric space $\check{\mathbb{B}}$, which is compact, hermitian and of same dimension as $\mathbb{B}$. For $\mathbb{B}$, $\mathbb{D} \cong \mathbb{H}$ or $\mathbb{H}^2$ the duals are simply:

$$\check{\mathbb{B}} = \mathbb{P}^2,\ \check{\mathbb{D}} = \check{\mathbb{H}} = \mathbb{P}^1\ \check{\mathbb{H}}^2 = \mathbb{P}^1 \times \mathbb{P}^1.$$

The Lie algebra of the Lie group $G_\mathbb{C}$ of the dual symmetric space $\check{\mathbb{B}}$ is the complexification of the Lie algebra corresponding to the Lie group G of $\mathbb{B}$. For the splitting case $\mathbb{H}^2$ we use the isomorphy of Lie algebras $\mathfrak{so}(2,2) \cong \mathfrak{sl}_2(\mathbb{R}) \times \mathfrak{sl}_2(\mathbb{R})$ to get the pairs

$$\begin{aligned} G &\subset G_\mathbb{C} \\ \mathbb{SU}((2,1),\mathbb{C}) &\subset \mathbb{S}l_3(\mathbb{C}), \\ \mathbb{SO}_e(2,2) &\subset \mathbb{S}l_2(\mathbb{C}) \times \mathbb{S}l_2(\mathbb{C}) \end{aligned}$$

The conjugation classes of commutators of normalizing Lie group pairs $N' = [N,N]$ with complexifications $N'_\mathbb{C}$ corresponding to $\mathbb{D} \subset \mathbb{B}$ or $\mathbb{D} \subset \mathbb{H}^2$ are represented by:

$$\begin{aligned} G \supset\quad N' &\subset N'_\mathbb{C} \subset G_\mathbb{C} \\ \mathbb{SU}((1,1),\mathbb{C}) &\subset \mathbb{S}l_2(\mathbb{C}), \\ \mathbb{S}l_2(\mathbb{R}) &\subset \mathbb{S}l_2(\mathbb{C}) \text{ diagonal in } \mathbb{S}l_2(\mathbb{C})^2. \end{aligned}$$

We have the following corresponding commutative embedding diagrams:

$$\begin{array}{ccc} \check{\mathbb{B}} & \hookleftarrow & \mathbb{B} \\ \uparrow & & \uparrow \\ \check{\mathbb{D}} & \hookleftarrow & \mathbb{D} \end{array} \qquad\qquad \begin{array}{ccc} G_\mathbb{C} & \hookleftarrow & G \\ \uparrow & & \uparrow \\ N'_\mathbb{C} & \hookleftarrow & N' \end{array}$$

To be more explicit we consider the point-curve-surface flags

$$0 \times 0 = O \in 0 \times \mathbb{D} \subset \mathbb{B} \quad , \quad O \in \Delta \subset \mathbb{H} \times \mathbb{H} \text{ (diagonal)};$$
$$O \in 0 \times \mathbb{P}^1 \subset \mathbb{P}^2 \quad , \quad O \in \Delta' \subset \mathbb{P}^1 \times \mathbb{P}^1 \text{ (diagonal)}.$$

with embeddings

$$N'_{\mathbb{C}} \hookrightarrow G_{\mathbb{C}} : \begin{cases} \mathbb{S}l_2(\mathbb{C}) \ni h \mapsto \left(\begin{smallmatrix} 1 & o \\ o & h \end{smallmatrix}\right) & , \text{ Picard case} \\ \mathbb{S}l_2(\mathbb{C}) \ni g \mapsto (g, g) & , \text{ Hilbert case.} \end{cases}$$

The (special) compact stabilizer groups with complexifications are

Picard case:

$$K = Stab_O(G) = \mathbb{S}(\mathbb{U}(2) \times \mathbb{U}(1)) \ , \ K_{\mathbb{C}} = \mathbb{S}(\mathbb{S}l_2(\mathbb{C}) \times \mathbb{G}l_1(\mathbb{C}));$$
$$k = Stab_O(N') = \mathbb{S}(\mathbb{U}(1) \times \mathbb{U}(1)) \ , \ k_{\mathbb{C}} = \mathbb{S}(\mathbb{G}l_1(\mathbb{C}) \times \mathbb{G}l_1(\mathbb{C})).$$

Hilbert case:

$$K = Stab_O(G) = \mathbb{SO}(2) \times \mathbb{SO}(2) \ , \ K_{\mathbb{C}} = \mathbb{G}l_1(\mathbb{C})) \times \mathbb{G}l_1(\mathbb{C}) \subset Stab_O \mathbb{G}_{\mathbb{C}};$$
$$k = Stab_O(N') = \mathbb{SO}(2) \ , \ k_{\mathbb{C}} = \mathbb{S}(\mathbb{G}l_1(\mathbb{C}) \times \mathbb{G}l_1(\mathbb{C})) \subset Stab_O N'_{\mathbb{C}}.$$

Lie group diagram for $O \in \mathbb{D} \subset \mathbb{B}$:

$$\begin{array}{ccc} K = G_O & \longrightarrow & G \\ \uparrow & & \uparrow \\ k = N'_O & \longrightarrow & N' \end{array}$$

Complexification diagram for $O \in \check{\mathbb{D}} = \mathbb{P}^1 \in \check{\mathbb{B}} = \mathbb{P}^2$ or $\mathbb{P}^1 \times \mathbb{P}^1$:

$$\begin{array}{ccccc} K_{\mathbb{C}} & \longrightarrow & P_+ \cdot K_{\mathbb{C}} = G_{\mathbb{C},O} & \longrightarrow & G_{\mathbb{C}} \\ \uparrow & & \uparrow & & \uparrow \\ k_{\mathbb{C}} & \longrightarrow & p_+ \cdot k_{\mathbb{C}} = N'_{\mathbb{C},O} & \longrightarrow & N'_{\mathbb{C}} \end{array}$$

with suitable stabilizer splitting complex parabolic groups $P_+ \subset \mathbb{G}_{\mathbb{C}}$ or $p_+ \subset N'_{\mathbb{C}}$, respectively. These are the unipotent radicals of the corresponding stabilizer groups.

Example 4.3. 2-ball case:

$$P_+ = \{\left(\begin{smallmatrix} 1 & 0 & 0 \\ 0 & 1 & 0 \\ a & b & 1 \end{smallmatrix}\right) ; \ a, b \in \mathbb{C}\} \ , \ p_+ = \{\left(\begin{smallmatrix} 1 & 0 \\ c & 1 \end{smallmatrix}\right) ; \ c \in \mathbb{C}\}$$

Now take a G-vector bundle $\overset{o}{E}$ on $\mathbb{B}$. The stabilizer $K = G_O$ acts on the fibre E_O of E, and $\overset{o}{E}$ together with the G-action is received by extension of the K-action along the G/K-transport: $\overset{o}{E} = E_O \times_K G$.

We extend the (really represented, bi-unitary or bi-orthogonal) K-action on $\overset{o}{E}$ by complexification to a $K_{\mathbb{C}}$-action. Putting together with the trivially defined action of P_+ on E_O we get the $G_{\mathbb{C}}$-bundle

$$\check{E} := E_O \times_{P_+K} G_{\mathbb{C}} \text{ on } \check{\mathbb{B}}.$$

Now let Γ be a neat arithmetic subgroup of G, and $\overline{\Gamma\backslash\mathbb{B}} \longrightarrow \widehat{\Gamma\backslash\mathbb{B}}$ a singularity resolution of the Baily–Borel compactification of $\Gamma\backslash\mathbb{B}$ (already smooth) with (componentwise) normal crossing compactification divisor. The G-bundle $\overset{o}{E}$ goes down to the quotient bundle $E := \Gamma\backslash \overset{o}{E}$.

We endow $\overset{o}{E}$ with a G-equivariant hermitian metric $\overset{o}{h}$. It extends along the above $\mathbb{G}_{\mathbb{C}}/P_+K_{\mathbb{C}}$ transport from E_O to a hermitian $\mathbb{G}_{\mathbb{C}}$-equivariant metric $\check{h}$ on $\check{E}$. On the other hand it goes down along the quotient map $\mathbb{B} \to \Gamma\backslash\mathbb{B}$ to a hermitian metric h on E.

Theorem 4.4 (Mumford). *Up to isomorphy there is a unique hermitian vector bundle $\bar{E}$ extending E on $\overline{\Gamma\backslash\mathbb{B}}$, such that h is "logarithmically restricted" around the (smooth) compactification divisor X_Γ^∞.*

This means: Using coordinates z_i on a small polycylindric neighbourhood $\bar{U} = \mathbb{D}^{a+b}$ around $Q \in X_\Gamma^\infty$ with finite part $U = (\mathbb{D}\setminus 0)^a \times \mathbb{D}^b$, and a basis $\mathfrak{e}_j$ of $\bar{E}$ over $\bar{U}$, then

$$|h(\mathfrak{e}_j, \mathfrak{e}_k)|, |\det(h(\mathfrak{e}_j, \mathfrak{e}_k))| \le C \cdot \left(\sum_{i=1}^{a} \log|z_i|\right)^{2N}$$

with constants $C, N > 0$. Altogether we get bundle diagrams

$$\begin{array}{ccccccc}
\check{E} & \longleftarrow & \overset{o}{E} & \longrightarrow & E & \longrightarrow & \bar{E} \\
\downarrow & & \downarrow & & \downarrow & & \downarrow \\
\check{\mathbb{B}} & \longleftarrow & \mathbb{B} & \longrightarrow & \Gamma\backslash\mathbb{B} & \longrightarrow & \overline{\Gamma\backslash\mathbb{B}}
\end{array} \tag{27}$$

Let

$$1 + c_1(F) + \ldots + c_r(F) \in H^{even}(V, \mathbb{R})$$

be the total Chern class of a holomorphic vector bundle of rank r on the compact smooth complex algebraic variety V of dimension n, say. By Hodge theory we interpret $c_j(F)$ (uniquely up to exact forms) as a differential form $\gamma_j = \gamma_j(F)$ of degree $2j$ on X. We have the differential forms

$$\gamma_{\mathfrak{j}}(F) := \gamma_{j_1} \wedge \ldots \wedge \gamma_{j_k},\ \mathfrak{j} = (j_1, \ldots, j_k),\ \sum_i j_i \le n;$$

in particular the **Chern forms**, if $\sum_i j_i = n$. In the latter cases the **Chern numbers** are defined as

$$c_{\mathfrak{j}}(F) := \int_V \gamma_{\mathfrak{j}}(F).$$

Especially, the Chern number $c_n(V) = c_n(\mathcal{T}_V)$, where $\mathcal{T}_V$ is the tangent bundle on V, is the **Euler number** of V.

Coming back to our neat arithmetic quotient variety $\Gamma\backslash\mathbb{B}$ and vector bundle quadruples described in diagram (27), we come to the important

Theorem 4.5 (Mumford's Proportionality Theorem). *The Chern numbers of $\check{E}$ and $\bar{E}$ are related as follows:*

$$c_{\mathfrak{j}}(\bar{E}) \cdot c_n(\check{\mathbb{B}}) = c_{\mathfrak{j}}(\check{E}) \cdot c_n(\Gamma\backslash\mathbb{B}),$$

where $c_n(\Gamma\backslash\mathbb{B})$ denotes the Euler volume (with respect to Euler–Chern volume form of the Bergmann metric on $\mathbb{B}$) of a Γ-fundamental domain on $\mathbb{B}$.

Now take a symmetric subdomain $\mathbb{D}$ of $\mathbb{B}$ such that $\Gamma_{\mathbb{D}}$ is a neat arithmetic $\mathbb{D}$-lattice. We have an extended commutative diagram with vertical analytic embeddings

$$\begin{array}{ccccccc} \check{\mathbb{B}} & \longleftarrow & \mathbb{B} & \longrightarrow & \Gamma\backslash\mathbb{B} & \longrightarrow & \overline{\Gamma\backslash\mathbb{B}} \\ \uparrow & & \uparrow & & \uparrow & & \uparrow \\ \check{\mathbb{D}} & \longleftarrow & \mathbb{D} & \longrightarrow & \Gamma_{\mathbb{D}}\backslash\mathbb{D} & \longrightarrow & \overline{\Gamma_{\mathbb{D}}\backslash\mathbb{D}} \end{array} \tag{28}$$

satisfying the

Absolute and relative normal crossing conditions:

- All varieties in the diagram are smooth;
- the compactification divisor $X_\Gamma^\infty = \overline{\Gamma\backslash\mathbb{B}} \setminus (\Gamma\backslash\mathbb{B})$ is normal crossing;
- the compactification divisor $X_{\Gamma_{\mathbb{D}}}^\infty = \overline{\Gamma_{\mathbb{D}}\backslash\mathbb{D}} \setminus (\Gamma_{\mathbb{D}}\backslash\mathbb{D})$ is normal crossing,
- at each common point, the (small) compactification divisor $X_{\Gamma_{\mathbb{D}}}^\infty$ crosses transversally the big one X_Γ^∞.

As described in the beginning of this section we work with the (special) Lie group $N' \subset G$ acting on $\mathbb{D}$ and containing $\Gamma_{\mathbb{D}}$. Starting with a N'-equivariant hermitian vector bundle $\overset{o}{e}$ on $\mathbb{D}$ we get a commutative diagram as (27)

$$\begin{array}{ccccccc} \check{e} & \longleftarrow & \overset{o}{e} & \longrightarrow & e & \longrightarrow & \bar{e} \\ \downarrow & & \downarrow & & \downarrow & & \downarrow \\ \check{\mathbb{D}} & \longleftarrow & \mathbb{D} & \longrightarrow & \Gamma\backslash\mathbb{D} & \longrightarrow & \overline{\Gamma\backslash\mathbb{D}} \end{array} \tag{29}$$

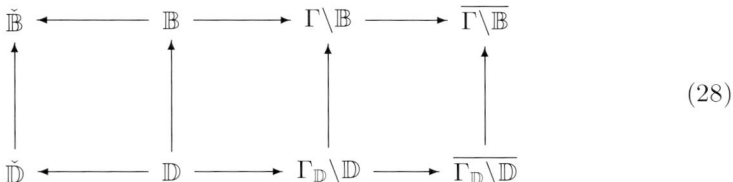

Mumford's Proportionality Theorem yields the Chern number relations

$$c_{\mathfrak{j}}(\bar{e}) \cdot c_m(\check{\mathbb{D}}) = c_{\mathfrak{j}}(\check{e}) \cdot c_m(\Gamma\backslash\mathbb{D}) \tag{30}$$

with $m = \dim \mathbb{D}$, $\mathfrak{j} = (j_1, \dots, j_k)$, $\sum_i j_i = m$.

If we start with a G-bundle $\overset{o}{E}$ on $\mathbb{B}$ and restrict it to the N'-bundle $\overset{o}{e} = \overset{o}{E}|_{\mathbb{D}}$ on $\mathbb{D}$, then we get two quadruples of bundles described in the diagrams (27) and (29). By construction, it is easy to see that the bundles $\check{e}$, e, $\bar{e}$ are restrictions of the bundles $\check{E}$, E or $\bar{E}$, respectively. The relations (30) specialize to the

Theorem 4.6 (Relative Proportionality Theorem). *For $\overset{o}{e} = \overset{o}{E}|_{\mathbb{D}}$, with the above notations, it holds that*

$$c_{\mathfrak{j}}(\bar{E}\,|_{\overline{\Gamma\backslash\mathbb{D}}}) \cdot c_m(\check{\mathbb{D}}) = c_{\mathfrak{j}}(\check{E}\,|_{\check{\mathbb{D}}}) \cdot c_m(\Gamma\backslash\mathbb{D}). \tag{31}$$

Corollary 4.7. *In particular, for* $\dim \mathbb{B} = 2$ *and* $\dim \mathbb{D} = 1$ *we get with the conditions of the theorem the relations*

$$c_1(\bar{E}\,|_{\overline{\Gamma\backslash\mathbb{D}}}) \cdot c_1(\check{\mathbb{D}}) = c_1(\check{E}\,|_{\check{\mathbb{D}}}) \cdot c_1(\Gamma\backslash\mathbb{D}). \tag{32}$$

Knowing $\check{\mathbb{D}} = \mathbb{P}^1$ and its Euler number $c_1(\mathbb{P}^1) = 2$ we get

$$2 \cdot c_1(\bar{E}\,|_{\overline{\Gamma\backslash\mathbb{D}}}) = c_1(\check{E}\,|_{\check{\mathbb{D}}}) \cdot vol_{EP}(\Gamma_{\mathbb{D}}), \tag{33}$$

where vol_{EP} denotes the Euler–Poincaré volume of a fundamental domain of a $\mathbb{D}$-lattice.

Now we work with canonical bundles $\mathcal{K} = \mathcal{T}^* \wedge \mathcal{T}^*$, $\mathcal{T}$ the tangent bundle on a smooth analytic variety with dual cotangent bundle $\mathcal{T}^*$. The above construction yields

$$\overset{o}{E} = \mathcal{K}_{\mathbb{B}}, \; \check{E} = \mathcal{K}_{\check{\mathbb{B}}}, \; E = \mathcal{K}_{\Gamma\backslash\mathbb{B}}.$$

For the restriction of $\check{E}$ to $\check{\mathbb{D}}$ we get

$$c_1(\check{E}\,|_{\check{\mathbb{D}}}) = (K_{\check{\mathbb{B}}} \cdot \check{\mathbb{D}}) = \begin{cases} (K_{\mathbb{P}^2} \cdot L) = -3(L^2) = -3, & \text{Picard case} \\ (K_{\mathbb{P}^1 \times \mathbb{P}^1} \cdot \Delta') = ((-2H - 2V) \cdot \Delta') = -4, & \text{Hilbert case} \end{cases}$$

Thereby, $H = \mathbb{P}^1 \times 0$ (horizontal line), $V = 0 \times \mathbb{P}^1$ (vertical line), Δ' the diagonal on $\mathbb{P}^1 \times \mathbb{P}^1$ and L is an arbitrary projective line on $\mathbb{P}^2$. Substituting in (33) we receive

$$c_1(\bar{E}\,|_{\overline{\Gamma\backslash\mathbb{D}}}) = \begin{cases} -\frac{3}{2} \cdot vol_{EP}(\Gamma_{\mathbb{D}}), & \text{Picard case} \\ -2 \cdot vol_{EP}(\Gamma_{\mathbb{D}}), & \text{Hilbert case.} \end{cases} \tag{34}$$

One knows that for cotangent bundles that $\bar{\mathcal{T}}^*$ is the extension of $\mathcal{T}^*_{\Gamma\backslash\mathbb{B}}$ by logarithmic forms along the compactification divisor X_Γ^∞ (allowing simple poles there). Wedging them we see that the Mumford-extended canonical bundle is nothing else but the logarithmic canonical bundle

$$\bar{E} = \Omega^2_{\overline{\Gamma\backslash\mathbb{B}}}(\log X_\Gamma^\infty)$$

corresponding to the **logarithmic canonical divisor** $K_{\overline{\Gamma\backslash\mathbb{B}}} + X_\Gamma^\infty$, where the first summand is a canonical divisor of $\overline{\Gamma\backslash\mathbb{B}}$ and the second summand is the compactification divisor (reduced divisor with the compactification set as support). Therefore we get

$$c_1(\bar{E}\mid_{\overline{\Gamma\backslash\mathbb{D}}}) = ((K_{\overline{\Gamma\backslash\mathbb{B}}} + X_\Gamma^\infty)\cdot\overline{\Gamma\backslash\mathbb{D}}).$$

Together with (34) we get

$$(K_{\overline{\Gamma\backslash\mathbb{B}}}\cdot\overline{\Gamma\backslash\mathbb{D}}) + (\overline{\Gamma\backslash\mathbb{D}}\cdot X_\Gamma^\infty) = -\begin{cases}\frac{3}{2}\cdot vol_{EP}(\Gamma_{\mathbb{D}}), & \text{Picard case}\\ 2\cdot vol_{EP}(\Gamma_{\mathbb{D}}), & \text{Hilbert case.}\end{cases} \tag{35}$$

The adjunction formula for curves on surfaces relates the Euler number of $\overline{\Gamma\backslash\mathbb{D}}$ with intersection numbers as follows:

$$-eul(\overline{\Gamma\backslash\mathbb{D}}) = (\overline{\Gamma\backslash\mathbb{D}}^2) + (K_{\overline{\Gamma\backslash\mathbb{B}}}\cdot\overline{\Gamma\backslash\mathbb{D}}).$$

On the other hand, by a very classical formula, the Euler number can be read off from the volume of a fundamental domain and the number of compactification points:

$$eul(\overline{\Gamma\backslash\mathbb{D}}) = vol_{EP}(\Gamma_{\mathbb{D}}) + (\overline{\Gamma\backslash\mathbb{D}}\cdot X_\Gamma^\infty) = eul(\Gamma_{\mathbb{D}}\backslash\mathbb{D}) + (\overline{\Gamma\backslash\mathbb{D}}\cdot X_\Gamma^\infty). \tag{36}$$

Adding the last two relations we get

$$(K_{\overline{\Gamma\backslash\mathbb{B}}}\cdot\overline{\Gamma\backslash\mathbb{D}}) + (\overline{\Gamma\backslash\mathbb{D}}\cdot X_\Gamma^\infty) = -(\overline{\Gamma\backslash\mathbb{D}}^2) - vol_{EP}(\Gamma_{\mathbb{D}}). \tag{37}$$

We define and obtain by substitution in (35)

$$Self(\overline{\Gamma\backslash\mathbb{D}}) := (\overline{\Gamma\backslash\mathbb{D}}^2) = \begin{cases}\frac{1}{2}\cdot vol_{EP}(\Gamma_{\mathbb{D}}), & \text{Picard case}\\ 1\cdot vol_{EP}(\Gamma_{\mathbb{D}}), & \text{Hilbert case.}\end{cases} \tag{38}$$

We call $Self(\overline{\Gamma\backslash\mathbb{D}})$ the **orbital self-intersection** and define also the **orbital Euler number**

$$Eul(\overline{\Gamma\backslash\mathbb{D}}) := eul(\Gamma\backslash\mathbb{D}) = vol_{EP}(\Gamma_{\mathbb{D}})$$

of $\overline{\Gamma\backslash\mathbb{D}} \hookleftarrow \overline{\Gamma\backslash\mathbb{B}}$ in the in the case of $\mathbb{D}$-neat lattices Γ. Then, together with (36), one gets

$$vol_{EP}(\Gamma_{\mathbb{D}}) = eul(\Gamma\backslash\mathbb{D}) = Eul(\overline{\Gamma\backslash\mathbb{D}}) = \begin{cases}2\cdot(\overline{\Gamma\backslash\mathbb{D}}^2) = 2\cdot Self(\overline{\Gamma\backslash\mathbb{D}}),\\ 1\cdot(\overline{\Gamma\backslash\mathbb{D}}^2) = 1\cdot Self(\overline{\Gamma\backslash\mathbb{D}}),\end{cases} \tag{39}$$

in the Picard or Hilbert case, respectively.

5. The General Proportionality Relation

5.1. Ten rules for the construction of orbital heights and invariants

We introduce relative orbital objects, and the corresponding notation:
These are finite orbital coverings: **Y**/**X** (surfaces), **D**/**C** (curves) **Q**/**P** (points) and also birational ones (relative releases, mainly) $\mathbf{X}'\to\mathbf{X}$, $\mathbf{C}'\to\mathbf{C}$, $\mathbf{P}'\to\mathbf{P}$.

The relative orbital morphisms joining these relative objects are commutative diagrams. We use the notation:

$$\mathbf{Y}'/\mathbf{X}'{\to}\mathbf{Y}/\mathbf{X},\quad \mathbf{D}'/\mathbf{C}'{\to}\mathbf{D}/\mathbf{C},\quad \mathbf{Q}'/\mathbf{P}'{\to}\mathbf{Q}/\mathbf{P}.$$

Definition 5.1. *A rational* **orbital invariant** *on* $\mathbf{Orb}^{2,1}$ *is a non-constant map* $\mathbf{h}$ *corresponding each orbital curve a rational number*

$$O \neq \mathbf{h}:\ \mathbf{Orb}^2 \longrightarrow \mathbb{Q} \ \textit{with}$$

R.1 deg(1/1)
$[\mathbf{D}:\mathbf{C}]\mathbf{h}(\mathbf{C})\mathbf{h}(\mathbf{D}) = [\mathbf{D}:\mathbf{C}]\cdot\mathbf{h}(\mathbf{C})$
for all finite orbital curve coverings $\mathbf{D}/\mathbf{C}$*. Thereby* $[\mathbf{D}:\mathbf{C}] := \frac{w_D}{w_C}[D:C]$ *is the* **orbital degree** *of the covering with usual covering degree* $[D:C]$.

If $0 \neq h:\ \mathbf{Orb}^2 \longrightarrow \mathbb{Q}$ *satisfies*

R.1 deg(1/1)
$h(\mathbf{D}) = [D:C]\cdot h(\mathbf{C})$,
for all finite orbital curve coverings $\mathbf{D}/\mathbf{C}$*, then we call it an* **orbital height**.

Remark 5.2. It is easy to see that we get immediately from an orbital height h an orbital invariant $\mathbf{h}$ setting $\mathbf{h}(\mathbf{C}) := w_{\mathbf{C}}\cdot h(\mathbf{C})$.

Convention. In this paragraph we restrict ourselves to compact curves or to open orbital curves $\overset{o}{\mathbf{C}}$ with cusp point compactification $\hat{\mathbf{C}}$. For simplicity we will write $\mathbf{C}$ instead of $\hat{\mathbf{C}}$ and set $h(\overset{o}{\mathbf{C}}) := h(\hat{\mathbf{C}}) = h(\mathbf{C})$ (same for $\mathbf{h}$).

For constructions we need:

- $h(\mathbf{P})$ local orbital invariants,
- relative orbital heights such that:
  - for orbital curves:

$$h(\mathbf{C}'{\to}\mathbf{C}) := h(\mathbf{C}') - h(\mathbf{C})\,,\ h(\mathbf{D}/\mathbf{C}) := h(\mathbf{D}) - [D:C]\cdot\mathbf{h}(\mathbf{C})$$

  - for orbital curve points:

$$h(\mathbf{P}'{\to}\mathbf{P}) := h(\mathbf{P}') - h(\mathbf{P})\,,\ h(\mathbf{Q}/\mathbf{P}) := h(\mathbf{Q}) - [Q:P]\cdot h(\mathbf{P}),$$

  where $[Q:P]$ has to be defined later.

and the **Decomposition laws** (absolute and relative):

R.2 Dec(1,0)
$$h(\mathbf{C}) = h_1(\mathbf{C}) + h_0(\mathbf{C})\,,\ h_0(\mathbf{C}) = \sum_{\mathbf{P}\in\mathbf{C}} h(\mathbf{P})$$

R.3 Dec(11,00)
$$h(\mathbf{C}'{\to}\mathbf{C}) = h_1(\mathbf{C}'{\to}\mathbf{C}) + h_0(\mathbf{C}'{\to}\mathbf{C})$$
$$h_0(\mathbf{C}'{\to}\mathbf{C}) = \sum_{\mathbf{P}'{\to}\mathbf{P}\in\mathbf{C}'{\to}\mathbf{C}} h(\mathbf{P}'{\to}\mathbf{P})$$

with finite sums h_0 and local incidence diagrams

$$\begin{array}{ccc} \mathbf{P}' & \longrightarrow & \mathbf{C}' \\ \downarrow & & \downarrow \\ \mathbf{P} & \longrightarrow & \mathbf{C} \end{array}$$

Relative orbital rules:

R.4 deg(11/11)

$$h(\mathbf{D}' \to \mathbf{D}) = [D : C] \cdot h(\mathbf{C}' \to \mathbf{C})$$

R.5 deg(00/00)

$$h(\mathbf{Q}' \to \mathbf{Q}) = [Q : P] \cdot h(\mathbf{P}' \to \mathbf{P})$$

Initial relations:

R.6 $\deg(1/1)_{sm}$
deg(00/00) holds for totally smooth **D**, **C**
(no curve and no surface singularities)

R.7 $\deg(00/00)_{sm}$
deg(00/00) holds for smooth releases of abelian point uniformizations
(background: stepwise resolution of singularities)

Shift techniques along releases:

R.8 $(\text{Shift})_{ab}$
shifts deg(1/1) along locally abelian releases

R.9 $(\text{Shift})_{*}^{ab}$
shifts deg(1/1) to all finitely weighted releasable orbital curves

R.10 $(\text{Shift})_{\infty}^{*}$
shifts deg(1/1) to all releasable orbital curves including infinite weights.

Implications (Geometric Local-Global Principle):

mp 1) Dec(11,00), deg(00/00) $\Longrightarrow$ deg(11/11) $\Longrightarrow$ $(\text{Shift})_{ab}$;
mp 2) $\deg(1/1)_{sm}$, $(\text{Shift})_{ab}^{sm}$ $\Longrightarrow$ $\deg(1/1)_{ab}$;

Later (via definitions)

np 3) $(\text{Shift})_{*}^{ab}$, $\deg(1/1)_{ab}$ $\Longrightarrow$ $\deg(1/1)_{*}$;
np 4) $(\text{Shift})_{\infty}^{*}$, $\deg(1/1)_{*}$ $\Longrightarrow$ deg(1/1).

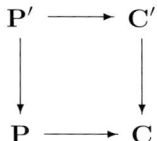

5.2. Rational and integral self-intersections

We need rational intersections of curves on (compact algebraic) normal surfaces. Let $\nu : Y \longrightarrow X$ be a birational morphism of normal surfaces. We denote by $Div\, X$ the space of Weil divisors on X with coefficients in $\mathbb{Q}$. There is a rational intersection theory for these divisor groups together with canonically defined orthogonal embeddings $\nu^{\#} : Div\, X \longrightarrow Div\, Y$ extending the integral intersection theory on smooth surfaces and the inverse image functor. Via resolution of singularities the intersection matrices are uniquely determined by the postulate of preserving intersections for all $\nu^{\#}$-preimages. Namely, let $E = E(\nu)$ the exceptional (reduced) divisor on Y, $Div_E Y$ the $\mathbb{Q}$-subspace of $Div\, Y$ generated by the irreducible components of E. For any Weil divisor C on X the generalized inverse image $\nu^{\#}C$ satisfies the conditions

$$Div_E Y \perp \nu^{\#}C = \nu' C + \nu_E^{\#} C\ ,\ \nu_E^{\#} C \in Div_E Y,$$

where $\nu' C$ is the proper transform of C on Y. Before proofs one has to define the rational intersections on normal surfaces. Let ν be a singularity resolution, $(\ \cdot\ )$ the usual intersection product on smooth surfaces, here on Y, and $E = E_1 + \cdots + E_r$ the decomposition into irreducible components. There is only one divisor $\nu' C + \sum c_i E_i \in Div\, Y$ orthogonal to $Div_E Y$ because the system of equations

$$\sum c_i (E_1 \cdot E_i) = -(E_1 \cdot \nu' C),$$

$$\cdots\cdots$$

$$\sum c_i (E_r \cdot E_i) = -(E_r \cdot \nu' C),$$

has a regular coefficient matrix (negative definite by a theorem of Mumford). The unique $\mathbb{Q}$-solution determines $\nu^{\#}C$. Then the rational intersection product on X is well defined by

$$< C \cdot D > := (\nu^{\#}C \cdot \nu^{\#}D)\ ,\ C, D \in Div\, X.$$

Using orthogonality we get for self-intersections the relations

$$\begin{aligned} < C^2 > &= (\nu^{\#}C \cdot \nu^{\#}C) = ((\nu' C + \nu_E^{\#}C) \cdot \nu^{\#}C)) = (\nu' C \cdot \nu^{\#}C) \\ &= (\nu' C \cdot (\nu' C + \nu_E^{\#}C)) = (\nu' C \cdot \nu' C) + (\nu' C \cdot \nu_E^{\#}C), \end{aligned}$$

hence

$$(\nu' C)^2 := (\nu' C \cdot \nu' C) = < C^2 > -(\nu' C \cdot \nu_E^{\#}C). \tag{40}$$

For minimal singularity resolutions μ we write briefly $(C^2) := (\mu' C)^2$ and notice

$$(C)^2 = < C^2 > -(\mu' C \cdot \mu_E^{\#}C). \tag{41}$$

This number is called the **minimal self-intersection** of C. The difference $(C)^2 - < C^2 >$ splits into a finite sum of point contributions $(\mu' C \cdot \mu_E^{\#}C)_Q$ at intersection points Q of $\mu' C$ and E.

Now we compare self-intersections of locally abelian orbital curves and their releases. The release at $\mathbf{P}$ is supported by a birational surface morphism, also denoted by ρ, precisely $\rho : (X', C') \longrightarrow (X, C)$. The intersection point of the proper transform C' of C and the exceptional line L is denoted by P'. With the weights $w(\mathbf{C}') = w(\mathbf{C})$ and of $\mathbf{L}$ it supports a well-defined abelian point $\mathbf{P}'$ of $\mathbf{C}'$. Let $\mu : \tilde{X} \to X$, $\mu_1 : X'' \to X'$ be the minimal resolutions of P or P', respectively and $\mu_2 : \tilde{X}' \to X'$ that of X'. (We have to resolve in general two cyclic singularities lying on L). For further notation we refer to the commutative diagram

$$\begin{array}{ccc} \tilde{X}' & \xrightarrow{\tilde{\rho}} & \tilde{X} \\ {\scriptstyle \mu_2}\downarrow & \searrow^{\nu} & \downarrow{\scriptstyle \mu} \\ X' & \xrightarrow[\rho]{} & X \end{array}$$

It restricts to morphisms along exceptional divisors

$$\begin{array}{ccc} \tilde{E}' & \xrightarrow{\tilde{\rho}} & \tilde{E} \\ {\scriptstyle \mu_2}\downarrow & \searrow^{\nu} & \downarrow{\scriptstyle \mu} \\ L & \xrightarrow[\rho]{} & P \end{array}$$

Since $< C^2 >$ is a birational constant we get from (40), (41) and by definition of minimal self-intersections the relation

$$(C^{'2}) - (C^2) = ((C^{'2}) - (C^2))_P := (\nu' C)^2 - (C)^2 = -(\nu' C \cdot \nu_E^{\#} C) + (\mu' C \cdot \mu_E^{\#} C)$$

with obvious notations. (The index E stands for the corresponding exceptional divisor).

Lemma 5.3. *The* **relative self-intersection**

$$s(P' \to P) := ((C^{'2}) - (C^2))_P \leq 0$$

does only depend on the local release $P' \to P$, not on the choice of C, C' crossing P or P', respectively. More precisely, it only depends on the exceptional resolution curves $E_P(\nu)$, $E_P(\mu)$, even only on the intersection graphs of $E(\nu) = E_P(\nu)$ and $E(\mu) = E_P(\mu)$.

Proof. Both linear trees resolve P, the latter minimally. Therefore $E(\nu) \to E(\mu)$ splits into a sequence of σ-processes. The stepwise contraction of a (-1)-line ascends $(\tilde{C}'^2)$ by 1 if and only if this exceptional line crosses $\tilde{C}'$. If not, then $(\tilde{C}'^2)$ is not changed. By stepwise blowing down (-1)-lines we see that $s(P' \to P)$ is equal to the number of steps contracting a line to a point on the image of $\tilde{C}'$. The curve $\tilde{C}'$ crossing the first line of $E(\nu)$ can be chosen arbitrarily. □

Definitions 5.4. *A release* $\mathbf{P}' \to \mathbf{P}$ *of an abelian orbital curve point is called* **smooth** *if and only if P' is a smooth surface point. A release* $\mathbf{C}' \to \mathbf{C}$ *of a locally abelian orbital curve* $\mathbf{C}$ *is called* **smooth** *if and only if it is smooth at each released abelian point on* $\mathbf{C}$*. A* **smooth release** *of the finite covering* $\mathbf{Q}/\mathbf{P}$ *is a relative finite covering* $\mathbf{Q}' \to \mathbf{Q}/\mathbf{P}' \to \mathbf{P}$ *with smooth releases* $\mathbf{Q}' \to \mathbf{Q}$ *and* $\mathbf{P}' \to \mathbf{P}$*. A* **smooth release** *of the* *finite covering* $\mathbf{D}/\mathbf{C}$ *is a relative finite covering* $\mathbf{D}'\to\mathbf{D}/\mathbf{C}'\to\mathbf{C}$ *with only smooth local releases* $\mathbf{Q}'\to\mathbf{Q}/\mathbf{P}'\to\mathbf{P}$.

Example 5.5. (Stepwise resolution of cyclic singularities) . Let (D,Q) be a uniformization of the abelian curve point (C,P), the latter of cyclic type $< d,e >$. It is realized by cyclic group action of $Z_{d,e} := \left\langle \begin{pmatrix} \zeta_d & 0 \\ 0 & \zeta_e \end{pmatrix} \right\rangle \subset \mathbb{G}l_2(\mathbb{C})$, where ζ_g denotes a g-th primitive unit root. Blow the point (D,Q) e times up, each time at intersection point of the exceptional line with the proper transform of C'. Blow down the arising $e-1$ exceptional (-2)-lines. Then we get a smooth release $(D',Q')\to(D,Q)$. The exceptional line E supports a cyclic surface singularity Q'' of type $< e,e-1 >$. Factorizing by $Z_{d.e}$ one gets a smooth release $P'\to P$ with singularity $P'' = Q''/Z_{d,e}$ of type $< e,d' >$ on the exceptional line L over P, where $d' \equiv -d mode$. Altogether we get a smooth release $Q'\to Q/P'\to P$ of the uniformization Q/P of P.

Proof. We enlarge $Z_{d,e}$ by the reflection group S generated by $< 1,\zeta_e >$, $< \zeta_e, 1 >$ to an abelian group A acting around $O \in \mathbb{C}^2$. We consider the σ-release $O'\to O$ by blowing up O to the (-1)-line N. The directions of x- or y-axis through O correspond to points O' and O'' on N, respectively. Easy coordinate calculations show that the double release $O'\to O\leftarrow O''$ goes down via factorization by S to the double release $\mathbf{Q}'\to\mathbf{Q}\leftarrow\mathbf{Q}''$. Thereby $\mathbf{Q}'\to\mathbf{Q}$ is a smooth release of the smooth abelian point $\mathbf{Q}$, and Q'' is of type $< e,e-1 >$. Furthermore, factorizing by A yields the double release $\mathbf{P}'\to\mathbf{P}\leftarrow\mathbf{P}''$ with smooth release $\mathbf{P}'\to\mathbf{P}$ and P'' of type $< e,d' > = < e,-d >$. Now forget the weights and the upper σ-double release to get $P'\to P\leftarrow P'' = Q'\to Q\leftarrow Q''/ < d,e >$. □

Proposition 5.6. *Let* $\mathbf{D}/\mathbf{C}$ *be a finite cover of orbital curves with smooth* $\mathbf{D}$. *There exists a smooth relative release* $\mathbf{D}'\to\mathbf{D}/\mathbf{C}'\to\mathbf{C}$.

Proof. Essentially, we have only to find local smooth relative releases $\mathbf{Q}'\to\mathbf{Q}/\mathbf{P}'\to\mathbf{P}$ over $\mathbf{Q}/\mathbf{P} \in \mathbf{D}/\mathbf{C}$, if $\mathbf{P}$ is not smooth. If the weights around $\mathbf{Q}$ and $\mathbf{P}$ are trivial (equal to 1), then we refer to the above proof. Otherwise the weights $w(\mathbf{C})$ and $w(\mathbf{D})$ are Galois weights coming from a common local uniformization O of $\mathbf{Q}$, $\mathbf{P}$. Since $\mathbf{Q}$ is smooth it is a quotient point of O by an abelian reflection group Σ. We lift $Q'\to Q$ to a release $O'\to O$ by normalization along the Σ-quotient map around the exceptional line supporting Q'. Then we get the coverings $O'\to O/\mathbf{Q}'\to\mathbf{Q}/\mathbf{P}'\to\mathbf{P}$ with the original weights we need. □

5.3. The decomposition laws

Definition 5.7. *Let* $\mathbf{P}'\to\mathbf{P}$ *be a release of the abelian cross point* $\mathbf{P}$ *of the abelian curve* $\mathbf{C}$ *of weight* w, *and* $< d',e' >$ *respectively* $< d,e >$ *the cyclic types of* P' *or* P. *The number*

$$h(\mathbf{P}'\to\mathbf{P}) = h_1(\mathbf{P}'\to\mathbf{P}) + h_0(\mathbf{P}'\to\mathbf{P}) := \frac{s(P'/P)}{w} + \left(\frac{e'}{wd'} - \frac{e}{wd}\right) \qquad (42)$$

is called the **(relative local) orbital self-intersection** *of the (local) release.*

For global releases $\mathbf{C}'\to\mathbf{C}$ of locally abelian orbital curves $\mathbf{C}$ we take sums over the (unique) local branches $(\mathbf{C}', \mathbf{P}')$ pulled back from $(\mathbf{C}, \mathbf{P})$ along local releases:

$$h_0(\mathbf{C}'\to\mathbf{C}) := \sum_{\mathbf{P}'\to\mathbf{P}} h_0(\mathbf{P}'\to\mathbf{P})$$

$$h_1(\mathbf{C}'\to\mathbf{C}) := \sum_{\mathbf{P}'\to\mathbf{P}} h_1(\mathbf{P}'\to\mathbf{P}).$$

5.8. *Relative Decomposition Law* $\mathrm{Dec}(11/00)_{ab}$.

$$h(\mathbf{C}'\to\mathbf{C}) := h_1(\mathbf{C}'\to\mathbf{C}) + h_0(\mathbf{C}'\to\mathbf{C}) = \sum_{\mathbf{P}'\to\mathbf{P}} h(\mathbf{P}'\to\mathbf{P}).$$

Definition 5.9. *The rational number* $h(\mathbf{C}'\to\mathbf{C})$ *is called the* **(relative) self-intersection height** *of the release* $\mathbf{C}'\to\mathbf{C}$.

On this way we presented us the relative Decomposition Law $\mathrm{Dec}(11/00)_{ab}$ for releases of locally abelian orbital curves by definition. Now we are well-motivated for the next absolute Decomposition Law given by definition again:

Definition 5.10. *Decomposition Law Dec(1,0)*$_{ab}$.
Let $\mathbf{C}$ *be a locally abelian orbital curve of weight* $w = w(\mathbf{C})$. *The* **signature height** *of* $\mathbf{C}$ *is*

$$h(\mathbf{C}) = h_1(\mathbf{C}) + h_0(\mathbf{C}) = h_1(\mathbf{C}) + \sum_{\mathbf{P}\in\mathbf{C}} h(\mathbf{P}) := \frac{1}{w}(C^2) + \sum_{\mathbf{P}\in\mathbf{C}} \frac{e_P}{wd_P}.$$

For a further motivation we refer to the rather immediately resulting relative degree formula (43) below for local releases.

We shift this definition now to general (finitely weighted locally releasable) orbital curves $\mathbf{C}$. By definition, there exists a (geometrically unique minimal) locally abelian release $\mathbf{C}'\to\mathbf{C}$. Splitting $\mathbf{C}$ at each blown up point $\mathbf{P}$ into finitely many orbital branch points $(\mathbf{C}', \mathbf{P}')$ we are able to generalize the above definition to the

Definition 5.11. *Decomposition Law Dec(1,0)*$_*$ *for releasable orbital curves.*
Set

$$h_1(\mathbf{C}) := h_1(\mathbf{C}') = \frac{1}{w}\cdot(C'^2),$$

$$h_0(\mathbf{C}) := \sum_{\mathbf{P}\in\mathbf{C}} h(\mathbf{P}) \quad \textit{with} \quad h(\mathbf{P}) := \sum_{\mathbf{P}'\to\mathbf{P}} (h(\mathbf{P}') + \delta_{P'}^{rls}),$$

$$h(\mathbf{C}) = h_1(\mathbf{C}) + \sum h(\mathbf{P}) = \frac{1}{w}\left((C'^2) + \sum_{\mathbf{P}'\to\mathbf{P}} (\delta_{P'}^{rls} + \frac{e_{P'}}{d_{P'}})\right)$$

with the **local release branch symbol**

$$\delta_{P'}^{rls} = \delta_{P'}^{rls}(\mathbf{C}') := \begin{cases} 1, & \textit{if } P' \in E_P \cap C',\ E_P \textit{ exceptional release curve over } P), \\ 0, & \textit{else.} \end{cases}$$

We call $\mathbf{h}(\mathbf{C})$ *the* **orbital self-intersection** *of* $\mathbf{C}$.

Remark 5.12. If $w = w(\mathbf{C}) > 1$, then each abelian point on $\mathbf{C}$ is automatically an abelian cross point of $\mathbf{C}$. In this case do not release $\mathbf{C}$ or consider the identical map as trivial release. So $\delta_{P'}^{rls} \neq 0$ appears only in the trivial weight case $w = 1$.

5.4. Relative local degree formula for smooth releases

Proof of deg(00/00)$_{sm}$. We start with a uniformization of a cyclic singularity P of type $< d, e >$ unramified outside P, see Example 5.5. From the covering exceptional lines

$$(E; Q', Q'') \to Q :< 1, 0 > \text{ over } (L; P', P'') \to P :< d, e >$$

supporting singular points $Q'' :< e, e-1 >$ respectively $P'' :< e, -d >$ we read off:

$$h(Q'/Q) = s(Q'/Q) + \frac{0}{1} - \frac{0}{1} = -e + 0 - 0 = -e,$$
$$h(P'/P) = s(Q'/Q) + \frac{0}{1} - \frac{e}{d} = 0 + 0 - \frac{e}{d} = -\frac{e}{d},$$

hence

$$h(Q'/Q) = d \cdot h(P'/P) = [Q : P] \cdot h(P'/P). \tag{43}$$

Now we allow $\mathbf{P} \in \mathbf{C}$ to come with honest weight $w = w(\mathbf{C}) > 1$. As demonstrated in the proof of Proposition 5.6 the situation is the same as above with additional weight w at $\mathbf{C}, \mathbf{C}', \mathbf{D}, \mathbf{D}'$. So we have only to divide the above identities by w to get

$$h(\mathbf{Q}'/\mathbf{Q}) = \frac{1}{w} \cdot d \cdot h(P'/P) = \frac{1}{w} \cdot [Q : P] \cdot h(P'/P) = [Q : P] \cdot h(\mathbf{P}'/\mathbf{P}). \qquad \square$$

5.5. Degree formula for smooth coverings

Proof of deg(1/1)$_{sm}$. Let $\mathbf{D}/\mathbf{C}$ be a finite covering of totally smooth orbital curves. By definition of orbital finite coverings and multiplicativity of covering degrees it suffices to assume that $\mathbf{D} =: D$ is trivially weighted and $\mathbf{C} = \mathbf{D}/\mathbf{G}$ with Galois group G. The supporting surfaces Y, X of D or C, respectively, are assumed to be smooth along D or C, and the Galois covering $D \to \mathbf{C}$ is the restriction of a global Galois covering $p : Y \to \mathbf{X} = \mathbf{Y}/\mathbf{G}$. We can assume that $G = N_G(D)$ because the self-intersection of a smooth curve on a smooth surface is locally defined as degree of its normal bundle restricted to the curve. Looking at the normal bundle surfaces we can also assume that D is the only preimage of C along p. The ramification index is equal to $w = w(\mathbf{C})$, hence $p^*C = w \cdot D$. Now we apply the well-known degree formula for inverse images of curves on smooth surface coverings to our situation:

$$w^2 \cdot (D^2) = (p^*C)^2 = [Y : X] \cdot (C^2) = \#N_G(D) \cdot (C^2) = [D : C] \cdot w \cdot (C^2).$$

Division by w^2, together with the definition of the orbital self-intersection invariants and absence of singularities, yields finally

$$h(D) = [D : C] \cdot \frac{1}{w} \cdot (C^2) = (\frac{1}{w} \cdot [D : C]) \cdot (w \cdot h(\mathbf{C})) = [\mathbf{D} : \mathbf{C}] \cdot \mathbf{h}(\mathbf{C}). \qquad \square$$

5.6. The shift implications and orbital self-intersection

Proof of Implication (Imp 1), *first part.* Suppose Dec(11,00), deg(00/00) to be satisfied for locally abelian orbital curves. Consider relative releases $\mathbf{D}'\to\mathbf{D}/\mathbf{C}'\to\mathbf{C}$, locally $\mathbf{Q}'\to\mathbf{Q}/\mathbf{P}'\to\mathbf{P}$, supported by relative finite orbiface coverings $\mathbf{Y}'\to\mathbf{Y}/\mathbf{X}'\to\mathbf{X}$. For degree formulas it is sufficient to consider uniformizing orbital Galois coverings $D'\to D$ of $\mathbf{C}'\to\mathbf{C}$ with trivially weighted objects D', D (omitting fat symbols) and Galois weights on $\mathbf{C}', \mathbf{C}$.

$$\mathbf{X} = \mathbf{Y}/\mathbf{G}\,,\ \mathbf{X}' = \mathbf{Y}'/\mathbf{G}\,,\ \mathbf{P} = \mathbf{Q}/\mathbf{G_Q}\,,\ \mathbf{P}' = \mathbf{Q}'/\mathbf{G_{Q'}}.$$

We use the notation of the following orbiface and orbital curve diagrams around orbital points.

$$\begin{array}{ccc} Y' & \longrightarrow & Y \\ \downarrow & & \downarrow \\ \mathbf{X}' & \longrightarrow & \mathbf{X} \end{array} \quad \text{restricting to} \quad \begin{array}{ccc} D' & \longrightarrow & D \\ \downarrow & & \downarrow \\ \mathbf{C}' & \longrightarrow & \mathbf{C} \end{array} \quad \text{restricting to} \quad \begin{array}{ccc} Q' & \longrightarrow & Q \\ \downarrow & & \downarrow \\ \mathbf{P}' & \longrightarrow & \mathbf{P} \end{array}$$

with vertical quotient morphisms and horizontal releases. The joining incidence diagram on the released side can be understood as locally abelian Galois diagram:

$$\begin{array}{ccc} Q' & \longrightarrow & D' \\ \downarrow & & \downarrow \\ \mathbf{P}' & \longrightarrow & \mathbf{C}' \end{array} \quad \cong \quad \begin{array}{ccc} Q' & \longrightarrow & D' \\ \downarrow & & \downarrow \\ \mathbf{Q}'/\mathbf{A}(\mathbf{Q}') & \longrightarrow & \mathbf{D}'/\mathbf{N_G}(\mathbf{D}) \end{array}$$

Especially we restrict to work along D, D' with

$$G = N_G(D),\ \text{abelian}\ A := G_Q = G_{Q'},$$
$$[Q:P] := \frac{\#A}{\#Z_A(D)} = \frac{\#A}{w(\mathbf{C})},$$

the number of preimage points of P on D around Q w.r.t. the local A-covering $(D,Q)\to(C,P)$.
For fixed $\mathbf{P}$ it holds that

$$\sum_{D\ni Q/\mathbf{P}} [Q:P] = [G:G_Q]\cdot\frac{\#G_Q}{w} = \frac{\#G}{w}. \tag{44}$$

Applying Dec(11,00), deg(00/00) and (44) we get

$$\begin{aligned} h(D'\to D) &= \sum_{Q\in D} h(Q'\to Q) = \sum_{\mathbf{P}}\sum_{Q/\mathbf{P}} h(Q'\to Q) \\ &= \sum_{\mathbf{P}}\sum_{Q/\mathbf{P}} [Q:P]\cdot h(\mathbf{P}'\to\mathbf{P}) = \frac{\#G}{w}\sum_{\mathbf{P}} h(\mathbf{P}'\to\mathbf{P}) \qquad (45) \\ &= [\mathbf{D}:\mathbf{C}]\cdot h(\mathbf{C}'\to\mathbf{C}). \end{aligned}$$

□

Proof of Implication (Imp 1), *second part.* Together with the definitions of h for relative objects one gets

$$\begin{aligned} h(\mathbf{D}'/\mathbf{C}') &= h(\mathbf{D}') - [D' : C'] \cdot h(\mathbf{C}') = h(\mathbf{D}') - [D : C] \cdot h(\mathbf{C}') \\ &= (h(\mathbf{D}) + h(\mathbf{D}' \to \mathbf{D})) - [D : C] \cdot (h(\mathbf{C}) + h(\mathbf{C}' \to \mathbf{C})) \\ &= (h(\mathbf{D}) + h(\mathbf{D}' \to \mathbf{D}) - [D : C] \cdot h(\mathbf{C}) - h(\mathbf{D}' \to \mathbf{D}) \\ &= h(\mathbf{D}) - [D : C] \cdot h(\mathbf{C}) = h(\mathbf{D}/\mathbf{C}). \end{aligned}$$

The degree formula deg(1/1) translates to the vanishing of relative degrees, by definition. This vanishing condition is shifted by the above identity. □

Remark 5.13. Via stepwise resolutions and contractions it is not difficult to extend the relations $\deg(11/11)_{sm}$, $\deg(00/00)_{sm}$ for smooth releases to $\deg(11/11)_{ab}$, $\deg(00/00)_{ab}$ for all abelian releases. Since we do not need it for the proof of $\deg(1/1)_{ab}$, the proof is left to the reader.

Proof of $\deg(1/1)_{ab}$ *via implication* (Imp 2). Especially we dispose on the shifting principle $(\text{Shift})_{ab}^{sm}$ for smooth releases $D'/\mathbf{C}' \to D/\mathbf{C}$ with locally abelian objects $D, \mathbf{C}$. The implication (Imp 2) is a simple application of $(\text{Shift})_{ab}^{sm}$. Since we proved already $\deg(1/1)_{sm}$, this degree formula shifts now to the covering $D/\mathbf{C}$. □

Proof of deg(1/1) via implications (Imp 3), (Imp 4) *in the coniform case.*
We want to shift the main orbital property $\deg(1/1)_{ab}$ to orbital curves supporting honest $*$-singularities for given coniform releasable orbital curve $\hat{\mathbf{C}} \subset \hat{\mathbf{X}}$. More precisely, we have globally the following situation:

$$\begin{array}{ccccc} Y' & \longrightarrow & Y & \longrightarrow & \hat{Y} \\ \downarrow & & \downarrow & & \downarrow \\ \mathbf{X}' & \longrightarrow & \mathbf{X} & \longrightarrow & \hat{\mathbf{X}} \end{array} \qquad (46)$$

with horizontal releases, vertical quotient maps by a Galois group G, $Y \to \hat{Y}$ releases all honest cone singularities, such that Y is smooth. Let $\hat{D}$ be a component of the preimage of $\hat{C}$ on $\hat{Y}$ and D its proper transform assumed to be smooth. The next release $Y' \to Y$ takes care for a smooth action of G along the proper transform D' of D by equivariant blowing up of some points of Y. Take the minimal set of such σ-processes. Locally along the orbital curves we get the following commutative diagram:

$$\begin{array}{ccccc} D' & \longrightarrow & D & \longrightarrow & \hat{D} \\ \downarrow & & \downarrow & & \downarrow \\ \mathbf{C}' & \longrightarrow & \mathbf{C} & \longrightarrow & \hat{\mathbf{C}} \end{array} \qquad (47)$$

Already C is smooth at (honest) $*$-points (contraction points of $X \to \hat{X}$), and C' is smooth everywhere. The surface singularities of X and X' are cyclic. All of them around C' are abelian cross points of $\mathbf{C}$ with released exceptional curves as opposite cross germ. Remember that we already defined $h(\mathbf{C})$ in 5.11 via releases $\mathbf{C}' \to \mathbf{C}$ which are unique up to weights of released exceptional curves. These weights play no role in 5.11.

We verify the degree property

$$h(D) = [D : C] \cdot h(\mathbf{C}),$$

which is sufficient for our coniform category (defining orbital curve coverings via smooth releasable coniformizations). In the case of $w > 1$ it is easy to see that the minimal release $\mathbf{C}' \to \mathbf{C}$ is the identity because all points on $\mathbf{C}$ are already abelian cross points of $\mathbf{C}$, see Remark 5.12. The degree formula is already proved. So we can assume that $w = 1$, hence

$$G_D = N_G(D) = N_G(D'), \ \#N_G(D) = \#N_G(D') = \#G_D = [D : C].$$

Using the same counting procedure as in (45) we can also assume that $\mathbf{C}' \to \mathbf{C}$ releases only one point $\mathbf{P}$. Choosing a preimage Q of P on D we have

$$h(D') = (D'^2) = (D^2) - \#G \cdot Q, \text{ hence } h(D) = h(D') + \#G \cdot Q.$$

On the other hand, from (45) we get

$$h(C) = h(C') + b_P^{rls}(C' \to C), \quad b_P^{rls}(C' \to C) = \#\{(\text{released}) \text{ branches of } C \text{ at } P\}.$$

This number of branches multiplied with $\#N_G(D)$ coincides with $\#G \cdot Q$:

$$b_P^{rls} \cdot \#G_D = \#G \cdot Q, \text{ hence } h(D) = h(D') + b_P^{rls} \cdot [\mathbf{D} : \mathbf{C}].$$

Now divide the latter identity by $[\mathbf{D} : \mathbf{C}]$ to get

$$\frac{h(D)}{[D : C]} = \frac{h(D')}{[D : C]} + b_P^{rls} = h(C') + b_P^{rls} = h(C),$$

which proves the degree formula $\deg(1/1)_*$.

The last shift to $\deg(1/1)$ including infinitely weighted points is simply done by definitions. Observe that for the definition of orbital self-intersections of orbital curves we never needed weights of points and of released exceptional curves. For points only the singularity types (of curves and points) were important.

Definitions 5.14. *If the orbital point $\mathbf{R} \in \hat{\mathbf{C}} \subset \hat{\mathbf{X}}$ is not a quotient point, then we set $w(\mathbf{R}) = \infty$. The same will be done for any exceptional curve $\mathbf{E}$ releasing $\mathbf{R}$: $w(\mathbf{E}) := \infty$.*

We break the releases $\mathbf{X} \to \hat{\mathbf{X}}$ and $\mathbf{C} \to \hat{\mathbf{C}}$ — and of its coniform Galois coverings — in the diagrams (46), (47) into two releases starting with releases $\mathbf{X}^* \to \mathbf{X}$ at infinitely weighted points. Altogether we get commutative orbital diagrams

$$\begin{array}{ccccccc} Y' & \longrightarrow & Y & \longrightarrow & Y^* & \longrightarrow & \hat{Y} \\ \downarrow & & \downarrow & & \downarrow & & \downarrow \\ \mathbf{X}' & \longrightarrow & \mathbf{X} & \longrightarrow & \mathbf{X}^* & \longrightarrow & \hat{\mathbf{X}} \end{array} \tag{48}$$

$$\begin{array}{ccccccc} D' & \longrightarrow & D & \longrightarrow & D^* & \longrightarrow & \hat{D} \\ \downarrow & & \downarrow & & \downarrow & & \downarrow \\ \mathbf{C}' & \longrightarrow & \mathbf{C} & \longrightarrow & \mathbf{C}^* & \longrightarrow & \hat{\mathbf{C}} \end{array} \tag{49}$$

Definition 5.15. *With the above notation, the* **signature height** *of* $\hat{\mathbf{C}}$ *is defined to be*

$$h(\hat{\mathbf{C}}) := h(\mathbf{C}^*) = \frac{1}{w}(C'^2) + \sum_{\mathbf{P}\in\mathbf{C}} h(\mathbf{P}), \quad w = w(\mathbf{C}),$$

$$h(\mathbf{P}) = \sum_{\mathbf{P}'\to\mathbf{P}} (h(\mathbf{P}') + \delta_{P'}^{rls}) = b_P^{rls} + \sum_{\mathbf{P}'\to\mathbf{P}} h(\mathbf{P}'), \quad h(\mathbf{P}') = \frac{e_{P'}}{wd_{P'}},$$

b_P^{rls} *the* **number of (released) curve branches** *of* $\mathbf{C}$ *at* P.

For the general degree formula (deg(1/1) there is nothing new to prove. We can restrict ourselves to Galois coverings as described in the above diagrams. Then we get

$$h(\hat{D}) = h(D^*) = [D : C] \cdot h(\mathbf{C}^*) = [D : C] \cdot h(\hat{\mathbf{C}})$$

by definition. □

For the signature height alone it makes not much sense to introduce infinite weights because it works only with the internal curve weights $w(\mathbf{C})$. But in the next section we will introduce orbital Euler invariants working with external weights around $\mathbf{C}$. Then infinite weights will become useful.

5.7. Orbital Euler heights for curves

Let first $\mathbf{C}'$ be an orbital curve with weight w having only abelian cross points. It follows that the supporting C' is a smooth curve. We follow the proof line of the ten rules. In detail it is then not difficult to follow the proof of degree formula of the signature height for orbital curves on orbifaces in the last subsection. Notice that we distinguish in this subsection h and $\hat{h}$ for local reasons.

Definition 5.16.

$$h(\mathbf{C}') := h_1(\mathbf{C}') - h_0(\mathbf{C}'),$$

$$h_1(\mathbf{C}') := eul(C') \text{ (Euler number)}, \; h_0(\mathbf{C}') := \sum_{\mathbf{P}'\in\mathbf{C}'} h(\mathbf{P}'),$$

$$h(\mathbf{P}') := 1 - \frac{1}{d_{P'}v_{P'}},$$

where $< d_{P'}, e_{P'} >$ is the type of the cyclic singularity P' and $v_{P'}$ is the weight of curve germ at P' opposite to $\mathbf{C}'$. The proof is given in [H98] by the same procedure as for orbital self-intersections through the first eight commandments. Basically, Hurwitz genus formula for the change of Euler numbers along finite curve coverings has to be applied.

Now we shift the definition as above along $\mathbf{C}' \to \mathbf{C}$ along a coniform orbital release as described in diagrams (48), (49), to the finitely weighted orbital curve $\mathbf{C}$ setting

$$h(\mathbf{C}) := h(\mathbf{C}') = h_1(\mathbf{C}) - h_0(\mathbf{C}),$$

$$h_1(\mathbf{C}) := h_1(\mathbf{C}') = eul(C'), \quad h_0(\mathbf{C}) := \sum_{\mathbf{P}\in\mathbf{C}} h(\mathbf{P}),$$

$$h(\mathbf{P}) := \sum_{\mathbf{C}'\ni\mathbf{P}'\to\mathbf{P}} h(\mathbf{P}') = b_P^{rls} - \frac{1}{v_P}\sum \frac{1}{d_{P'}}, \tag{50}$$

where v_P is the released weight of P defined as weight of the exceptional release curve $\mathbf{E}_P$ over P and b_P^{rls} is the number of exceptional curve branches of $\mathbf{C}$ at P.

$(\text{Shift})_*^{ab}$: $\deg(1/1)_s m = \deg(1/1)_a b \Rightarrow \deg(1/1)_*$

Let $\mathbf{D}'/\mathbf{C}' \to \mathbf{D}/\mathbf{C}$ a locally abelian (coniform) release. Then

$$h(\mathbf{D}) = h(\mathbf{D}') = [D' : C'] \cdot h(\mathbf{C}') = [D : C] \cdot h(\mathbf{C}).$$

$(\text{Shift})_\infty^*$: $\deg(1/1)_* \Rightarrow \deg(1/1) =: \deg(1/1)_\infty$.

We have only to check what happens at points R with new weight ∞. Changing to ∞ at some points we write $\hat{\mathbf{C}}$ instead of $\mathbf{C}$ and define $h(\hat{\mathbf{C}})$ as in (50) substituting the new weights ∞. So we get with obvious notations

$$\begin{aligned} \hat{h}(\hat{\mathbf{C}}) &= eul(C') - \sum_{\mathbf{P}\in\hat{\mathbf{C}}_{fin}} \hat{h}(\mathbf{P}) - \sum_{\mathbf{R}\in\hat{\mathbf{C}}_\infty} \hat{h}(\mathbf{R}) \\ &= eul(C') - \sum_{\mathbf{P}} \left(b_P^{rls} - \frac{1}{v_P} \sum_{\mathbf{P}'\to\mathbf{P}} \frac{1}{d_{P'}} \right) - \sum_{\mathbf{R}} b_R^{rls} \end{aligned} \tag{51}$$

defining $\hat{h}$ for orbital points and curves. In order to prove that $\hat{h}$ is orbital we have only to check the realtive local degree formula the following $\deg(00/00)_\infty^*$ over infinitely weighted points $\mathbf{R}$ for coniform coverings $D/\mathbf{C}$.

$$\sum_{D\ni S/R} (\hat{h}(S) - h(S)) = \sum_{D\ni S/R} b_S$$

$$\hat{h}(\hat{\mathbf{R}}) - h(\mathbf{R}) = \frac{1}{v_R} \sum_{\mathbf{R}'\to\mathbf{R}} \frac{1}{d_{R'}}.$$

The weight $\#Z_G(D)$ of $\mathbf{C}$ doesn't play any role. So we can assume that

$$G = N_G(D) = N_G(D') = G_D = G_{D'} = [D' : C'] = [D : C]$$

is the Galois group acting smoothly on D', where we find all the curve branches of D at points S over R we need. With the above notation we get

$$\sum_{D\ni S/R}(\hat{h}(S)-h(S))=[G:G_S]\cdot b_S=[D:C]\cdot\frac{b_S}{\#G_S}$$

$$\hat{h}(\hat{\mathbf{R}})-h(\mathbf{R})=\sum_{\mathbf{R}'\to\mathbf{R}}\frac{1}{\#G_{S'}}=\sum_{i=1}^{b_P^{rls}}\frac{1}{\#G_{S'_i}},$$

where S' is a (D-branch) point on the release curve L_S of S over $R'\in E_R=L_S/G_S$ and S'_i over R'_i after numeration. Since

$$b_S=\sum_{i=1}^{b_P^{rls}}|G_S\cdot S'_i|=\sum_{i=1}^{b_P^{rls}}\frac{\#G_S}{\#G_{S'_i}}=\#G_S\cdot\sum_{i=1}^{b_P^{rls}}\frac{1}{\#G_{S'_i}},$$

the relative local orbital property

$$(\hat{h}(S)-h(S))_R\ =\ \sum_{D\ni S/R}(\hat{h}(S)-h(S))=[D:C]\cdot\hat{h}(\hat{\mathbf{R}})-h(\mathbf{R})$$

follows immediately, and also the global one after summation over all infinitely weighted $\mathbf{R}\in\hat{\mathbf{C}}$:

$$\begin{aligned}\hat{h}(D)=h(D)+(\hat{h}(D)-h(D))&=[D:C]\cdot\Big(h(\mathbf{C})+((\hat{h}(\mathbf{C})-h(\mathbf{C}))\Big)\\&=[D:C]\cdot\hat{h}(\mathbf{C}).\end{aligned}$$

We have to distinguish abelian points $\mathbf{P}\in\hat{\mathbf{X}}$, which will not be released along $\mathbf{X}'\to\hat{\mathbf{X}}$ and those $\mathbf{P}'$, which arise from releasing. The former appear in (50) by identifying $\mathbf{P}'=\mathbf{P}$.

Convention 5.17. *For an abelian point $\mathbf{Q}=(\mathbf{C},Q,\mathbf{D})$, C,D crossing curve germs at Q with maximal weight product $w(\mathbf{C})\cdot w(\mathbf{D})$ around, we set in any case*

$$w(\mathbf{Q})\ :=d_Q\cdot w(\mathbf{C})\cdot w(\mathbf{D}),$$

where $<d_Q,e_Q>$ is the cyclic singularity type of Q. If $\mathbf{Q}$ is, more distinguished, understood as abelian cross point on $\mathbf{C}$, then we set

$$w_Q\ :=w(\mathbf{C}),\ v_Q\ :=w(\mathbf{D}),\ \textit{hence}\ w(\mathbf{Q})\ :=d_Q\cdot w_Q\cdot v_Q \tag{52}$$

and call v_Q the **opposite weight** *to w_Q (or to $w(\mathbf{C})$) at $\mathbf{Q}$.*

Definition 5.18. *We call the abelian point $\mathbf{Q}$ on $\hat{\mathbf{C}}$ a* **general point** *of $\hat{\mathbf{C}}$ if and only if $w(\mathbf{Q})=w(\hat{\mathbf{C}})$. The other orbital points on $\hat{\mathbf{C}}$ are called* **special**. *We use the notation $\hat{\mathbf{C}}^{gen}$ for the open orbital curve of general points and $\hat{\mathbf{C}}^{sp}$ for the complementary (orbital) cycle (or set) of special orbital points.*

Each abelian cross point $\mathbf{P}$ on $\hat{\mathbf{C}}$ yields the contribution $1-\frac{1}{d_Q\cdot v_Q}$ in the middle sum of (51), and the general points of $\hat{\mathbf{C}}$ are precisely those with contribution

0. The summands 1 in the point contributions disappear, if we change to the open curve $\hat{C}^{gen} \cong C^{gen} \cong C'^{gen}$ and its Euler number:

$$\begin{aligned}\hat{h}(\hat{\mathbf{C}}) &= eul(\hat{C}^{gen}) + \sum_{\mathbf{P}\in\hat{\mathbf{C}}^{sp}} \sum_{\mathbf{P}'\to\mathbf{P}} \frac{1}{d_P \cdot v_P} \\ &= eul(\hat{C}^{gen}) + \sum_{\mathbf{P}\in\mathbf{C}^{sp}_{fin}} \sum_{\mathbf{P}'\to\mathbf{P}} \frac{1}{d_P \cdot v_P},\end{aligned} \tag{53}$$

the latter because $v_P = \infty$ outside of the set $\mathbf{C}^{sp}_{fin}$ of finitely weighted special points.

We will write h_e for the orbital Euler height $\hat{h}$ and $\hat{\mathbf{h}}_\mathbf{e}$ for the corresponding orbital Euler invariant.

5.8. Released weights

Denote by $v = w(\mathbf{E}_P)$ the released weight of $\mathbf{P}$ defined as weight of the exceptional release curve $\mathbf{E}_P$ over P and b_P^{rls} is the number of exceptional curve branches of $\mathbf{C}$ at P, as above. The weight $w(\mathbf{E}_P)$ is uniquely determined by the coniform release. This follows from the self-intersection and Euler degree formulas applied to $L = L_Q \to \mathbf{E} = \mathbf{E}_P = L_Q/G_Q$, L_Q the releasing resolution curve of the cone singularity $Q \in D$ over P. Namely, the orbital degree formulas yield

$$0 > (L^2) = [L:E] \cdot h_\tau(L) = \frac{\#G_Q}{v} \cdot \frac{1}{v} \left((E^2) + \sum_i \frac{e_i}{d_i} \right)$$

$$2 - 2g(L) = eul(L) = [L:E] \cdot h_e(L) = \frac{\#G_Q}{v} \cdot \left(eul(E) - \sum_i (1 - \frac{1}{v_i d_i}) \right)$$

where the sum runs through the branches $\mathbf{P}'_i \in \mathbf{C}'$ of $(\mathbf{C}, \mathbf{P})$. It follows that

$$\frac{eul(L)}{(L^2)} = v \cdot \frac{eul(E) - \sum_i (1 - \frac{1}{v_i d_i})}{(E^2) + \sum_i \frac{e_i}{d_i}},$$

from where one gets v uniquely, if the numerators on both sides do not vanish. In the opposite case of an elliptic curve we work with the cusp weight $v = \infty$.

For uniform releases we have $L \cong \mathbb{P}^1$, $(L^2) = -1$, hence

$$-2 = v \cdot \frac{2 - \sum_i (1 - \frac{1}{v_i d_i})}{-1 + \sum_i \frac{e_i}{d_i}},$$

$$w(\mathbf{E}_P) = \begin{cases} 2(1 - \frac{e_1}{d_1} - \frac{e_2}{d_2}) / (\frac{1}{v_1 d_1} + \frac{1}{v_2 d_2}), & \text{if } \mathbf{P} \text{ is abelian,} \\ 2(1 - \frac{e_1}{d_1} - \frac{e_2}{d_2} - \frac{e_3}{d_3}) / (-1 + \frac{1}{v_1 d_1} + \frac{1}{v_2 d_2} + \frac{1}{v_3 d_3}), & \text{if } \mathbf{P} \text{ is non-abelian.} \end{cases}$$

6. Relative proportionality relations, explicit and general

Now we change notation to connect these numbers with the algebraically defined orbital invariants. We write D_Γ for the compactification of $\Gamma\backslash\mathbb{D}$ on the minimal surface singularity resolution X_Γ of the Baily–Borel compactification $\hat{X}_\Gamma$ of $\Gamma\backslash\mathbb{B}$. Since Γ is $\mathbb{D}$-neat we have $\Gamma\backslash\mathbb{D} = \Gamma_{\mathbb{D}}\backslash\mathbb{D}$ (smooth) and we have only to resolve the cusp singularities. In the Picard case the curve D_Γ is already smooth, but in the Hilbert case we have to release in general curve hypercusps at infinity. In any case we have a release diagram

$$\begin{array}{ccc} \overset{o}{X}_\Gamma & & \\ \downarrow & \searrow & \\ X'_\Gamma & \longrightarrow & \hat{X}_\Gamma \\ \uparrow & & \uparrow \\ D'_\Gamma & \longrightarrow & \hat{D}_\Gamma \\ \uparrow & \nearrow & \\ \overset{o}{D}_\Gamma & & \end{array}$$

with horizontal birational morphisms (releases) and vertical embeddings, closed in the middle part and open in the top and bottom parts. The only non-trivial weights are at infinity, especially D_Γ has weight 1. Therefore we get the algebraic orbital Euler height and orbital signature as volumes:

$$\begin{aligned} h_e(\mathbf{D}_\Gamma) &= eul(Reg\,\mathbf{D}_\Gamma) + \sum_{P\in D_\Gamma^\infty}\sum_{P'\to P}\frac{w(D_\Gamma)}{w(P')} \\ &= eul(\Gamma\backslash\mathbb{D}) + \sum_{P\in D_\Gamma^\infty}\sum_{P'\to P}\frac{1}{\infty} \\ &= eul(\Gamma\backslash\mathbb{D}) = vol_{EP}(\Gamma_{\mathbb{D}}). \end{aligned} \tag{54}$$

$$\begin{aligned} h_\tau(\mathbf{D}_\Gamma) &= \#Sing^1\overset{\mathbf{o}}{\mathbf{D}} + (D_\Gamma'^2) + \sum_{P\in D_\Gamma^\infty}\sum_{P'\to P}\frac{e(P')}{d(P')} \\ &= (D_\Gamma'^2) + \sum_{P\in D_\Gamma^\infty}\sum_{P'\to P}\frac{0}{1} \\ &= (D_\Gamma'^2) = (\overline{\Gamma\backslash\mathbb{D}}^2) = \begin{cases} \frac{1}{2}\cdot vol_{EP}(\Gamma_{\mathbb{D}}), & \text{Picard case} \\ 1\cdot vol_{EP}(\Gamma_{\mathbb{D}}), & \text{Hilbert case.} \end{cases} \end{aligned} \tag{55}$$

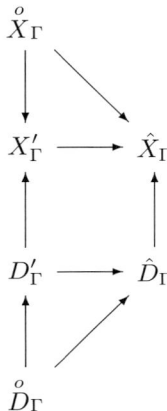

Comparing the identities (54) and (55) we come to

Theorem 6.1. *If* Γ *is a* $\mathbb{D}$*-neat arithmetic group of Picard or Hilbert modular type acting on* $\mathbb{B} = \mathbb{B}$ *or* $\mathbb{H}^2$*, respectively, then the orbital Euler and signature heights of* $\mathbf{D}_\Gamma$ *are in the following* **relative proportionality relation***:*

$$h_e(\mathbf{D}_\Gamma) = \begin{cases} 2 \cdot h_\tau(\mathbf{D}_\Gamma) & \textit{Picard case} \\ 1 \cdot h_\tau(\mathbf{D}_\Gamma) & \textit{Hilbert case.} \end{cases}$$

In the last paragraph we extended the heights to arbitrary Picard and Hilbert orbifaces satisfying the defining height rule $R.1$. With Remark 5.2 we orbitalize the heights of arithmetic curves $\mathbf{C} = \hat{\mathbf{C}}_\mathbf{\Gamma}$ or $\mathbf{C}'_\mathbf{\Gamma}$ to get the orbital Euler and self-intersection invariants setting

$$\mathbf{Eul}(\mathbf{C}) := \mathbf{h_e}(\mathbf{C}) = \frac{1}{w_\mathbf{C}} h_e(\mathbf{C}) \quad , \quad \mathbf{Self}(\mathbf{C})) := \mathbf{h}_\tau(\mathbf{C}) = \frac{1}{w_\mathbf{C}} h_\tau(\mathbf{C}).$$

They satisfy the defining orbital height rule **R.1**, also called orbital degree formula (see also subsection 5.1). The Main Theorem of the article is the

Relative Orbital Proportionality Theorem 6.2. *If* Γ *is an arbitrary arithmetic group of Picard or Hilbert modular type acting on* $\mathbb{B}$ *or* $\mathbb{H}^2$*, respectively, then the orbital Euler and self-intersection of the orbital arithmetic curve* $\mathbf{D}_\Gamma$ *satisfy the following* **relative proportionality relation***:*

$$\mathbf{Eul}(\mathbf{D}_\Gamma) = \begin{cases} 2 \cdot \mathbf{Self}(\mathbf{D}_\Gamma) & \textit{Picard case} \\ 1 \cdot \mathbf{Self}(\mathbf{D}_\Gamma) & \textit{Hilbert case.} \end{cases}$$

7. Orbital Heegner Invariants and Their Modular Dependence

We denote by $\mathbf{Pic^2}$ and $\mathbf{Hilb^2}$ the categories of all Picard respectively Hilbert orbifaces, including releases, finite orbital coverings and open embeddings corresponding to ball lattices commensurable with a group Γ_K, K a quadratic number field. If we restrict ourselves to Baily–Borel compactifications and orbital finite coverings only, then we write $\widehat{\mathbf{Pic}}^{\mathbf{2}}$. By restrictions we get the categories $\widehat{\mathbf{Pic}}^{\mathbf{2,1}}$, $\widehat{\mathbf{Hilb}}^{\mathbf{2,1}}$, of orbital arithmetic curves on the corresponding surfaces. Disjointly joint we denote the arising category by $\mathbf{Shim}^{2,1}$ because the objects are supported by surface embedded Shimura varieties of (co)dimension 1. We admit as finite coverings only those, which come from a restriction $\mathbf{X_\Delta} \to \mathbf{X_\Gamma}$ with objects from $\mathbf{Shim}^2$, where Δ is a sublattice of Γ. The notations for the subcategories $\widehat{\mathbf{Shim}}^{2,1}$, $\overset{\mathbf{o}}{\mathbf{Shim}}^{2,1}$, $\mathbf{Shim}^{2,1,\prime}$ of $\mathbf{Shim}^{2,1}$, should be clear, also for the correspondence classes $\widehat{\mathbf{Shim}}_K^{2,1}$, $\overset{\mathbf{o}}{\mathbf{Shim}}_K^{2,1}$, $\mathbf{Shim}_K^{2,1,\prime}$ in $\mathbf{Shim}_K^{2,1}$, K a quadratic number field.

We look for further orbital invariants for orbital arithmetic curves.

$$0 \neq \mathbf{h}: \ \mathbf{Shim^{2,1}} \longrightarrow \mathbb{Q}$$

satisfying, by definition, the orbital degree formula

$$\mathbf{h}(\hat{\mathbf{D}}) = [\hat{\mathbf{D}} : \hat{\mathbf{C}}] \cdot \mathbf{h}(\hat{\mathbf{C}})$$

with orbital degree

$$[\hat{\mathbf{D}} : \hat{\mathbf{C}}] := \frac{w(\hat{\mathbf{D}})}{w(\hat{\mathbf{C}})} \cdot [\hat{C} : \hat{D}]$$

for orbital finite coverings $\hat{\mathbf{D}}/\hat{\mathbf{C}}$ of orbital arithmetic curves (in $\widehat{\mathbf{Shim}}^{2,1}$). For each level group Γ we dispose on the $\mathbb{Q}$-vector space $\mathbf{Div}^{\mathbf{ar}}\hat{\mathbf{X}}_{\mathbf{\Gamma}}$ of orbital divisors generated by the (irreducible) arithmetic ones. The rational intersection product extends to the **orbital intersection product**

$$< \cdot >: \mathbf{Div}^{\mathbf{ar}}\hat{\mathbf{X}} \times \mathbf{Div}^{\mathbf{ar}}\hat{\mathbf{X}} \longrightarrow \mathbb{Q}$$

defined by

$$< \hat{\mathbf{C}} \cdot \hat{\mathbf{D}} > := \frac{< \hat{C} \cdot \hat{D} >}{w(\hat{\mathbf{C}})w(\hat{\mathbf{D}})}$$

for (irreducible) arithmetic curves and $\mathbb{Q}$-linear extension.

For finite orbital coverings $\mathbf{f} : \hat{\mathbf{Y}} \to \hat{\mathbf{X}}$ in $\widehat{\mathbf{Shim}}^{2}$ we dispose also on $\mathbb{Q}$-linear orbital direct and orbital inverse image homomorphisms

$$\mathbf{f}_{\#} : \mathbf{Div}^{\mathbf{ar}}\hat{\mathbf{Y}} \longrightarrow \mathbf{Div}^{\mathbf{ar}}\hat{\mathbf{X}}, \quad \mathbf{f}^{\#} : \mathbf{Div}^{\mathbf{ar}}\hat{\mathbf{X}} \longrightarrow \mathbf{Div}^{\mathbf{ar}}\hat{\mathbf{Y}}.$$

Restricting to coverings of arithmetic orbital curves $\hat{\mathbf{D}}/\hat{\mathbf{C}}$, the former is basically defined by

$$\mathbf{f}_{\#}\hat{\mathbf{D}} := [\hat{\mathbf{D}} : \hat{\mathbf{C}}] \cdot \hat{\mathbf{C}}, \quad (\hat{C} = f(\hat{D})).$$

The orbital inverse image of $\hat{\mathbf{C}}$ is nothing else but the reduced preimage divisor $f^{-1}C$ endowed componentwise with the weights on $\hat{\mathbf{Y}}$. In the orbital style of writing we set

$$\mathbf{f}^{\#}\hat{\mathbf{C}} := \mathbf{f}^{-1}\hat{\mathbf{C}}.$$

In [H02] we proved the projection formula in the Picard case. The proof transfers without difficulties to the Hilbert case, because it needed only the general orbital language. So

$$< \mathbf{f}_{\#}\mathbf{B} \cdot \mathbf{A} > = < \mathbf{B} \cdot \mathbf{f}^{\#}\mathbf{A} >$$

holds for all arithmetic orbital divisors $\mathbf{B}$ on $\hat{\mathbf{Y}}$ or $\mathbf{B}$ on $\hat{\mathbf{X}}$, respectively. It follows by $\mathbb{Q}$-linear extension after proving it for arithmetic orbital curves.

Definition 7.1. *The N-th* **Heegner divisor** *H_N on $\hat{X} = \hat{X}(\Gamma)$ is the reduced (Weil-) divisor with irreducible components $\widehat{\Gamma\backslash\mathbb{D}}$, $\mathbb{D}$ a K-disc on $\mathbb{B}$ of norm $N \in \mathbb{N}_+$ with respect to a maximal hermitian $\mathfrak{O}_K$-lattice in K^3. The N-th* **orbital Heegner divisor** *$\mathbf{H}_N = \mathbf{H}_N(\Gamma) \in \mathbf{Div}^{\mathbf{ar}}\hat{\mathbf{X}}$ is the sum of the orbitalized components $\widehat{\Gamma\backslash\mathbb{D}} \subset \widehat{\Gamma\backslash\mathbb{B}}$ of H_N.*

For finite coverings $\mathbf{f}:\hat{\mathbf{Y}}\to\hat{\mathbf{X}}$ corresponding to Picard lattices $\Gamma'\subset\Gamma$ it holds that

$$\mathbf{f}^{\#}\mathbf{H}_N(\Gamma)=\mathbf{H}_N(\Gamma'),$$

This property is called the **orbital preimage invariance** of Heegner divisors along finite coverings.

Theorem 7.2. *The correspondences*

$$\mathbf{h}_N:\widehat{\mathbf{Shim}}^{2,1}\longrightarrow\mathbb{Q},\quad \hat{\mathbf{C}}\mapsto<\hat{\mathbf{C}}\cdot\mathbf{H}_N>,$$

where $\hat{\mathbf{C}}\subset\hat{\mathbf{X}}(\Gamma)$ and $\mathbf{H}_N=\mathbf{H}_N(\Gamma)$ are taken on the same level Γ, are orbital invariants.

We use the neutral notation $\mathbf{h}$. We should denote it by $\hat{\mathbf{h}}$, and introduce $\overset{o}{\mathbf{h}}$ and $\mathbf{h}'$ by same values on corresponding open or released orbital surfaces. The reader should keep it in mind.

Proof. Let $\mathbf{f}:\hat{\mathbf{D}}\to\hat{\mathbf{C}}$ be a finite covering in $\widehat{\mathbf{Shim}}_K^{2,1}$ corresponding to $\Gamma'\subset\Gamma$, then

$$\begin{aligned}\mathbf{h}_N(\hat{\mathbf{D}})&=<\hat{\mathbf{D}}\cdot\mathbf{H}_N(\Gamma')>=<\hat{\mathbf{D}}\cdot\mathbf{f}^{\#}\mathbf{H}_N(\Gamma)>=<\mathbf{f}_{\#}\hat{\mathbf{D}}\cdot\mathbf{H}_N(\Gamma)>\\&=[\hat{\mathbf{D}}:\hat{\mathbf{C}}]\cdot<\hat{\mathbf{C}}\cdot\mathbf{H}_N(\Gamma)>=[\hat{\mathbf{D}}:\hat{\mathbf{C}}]\cdot\mathbf{h}_N(\hat{\mathbf{C}}).\end{aligned}$$ □

We look for a normalization of the three equal orbital invariants in the Proportionality Theorem 2.1 of Part 5 and a synchronization with the orbital Heegner invariants for establishing orbital power series.

Definition 7.3. *We call $\mathbf{h}_0:\widehat{\mathbf{Shim}}_{\mathbf{K}}^{\mathbf{2,1}}\to\mathbb{Q}$ with*

$$\begin{aligned}\mathbf{h}_0(\hat{\mathbf{C}})&:=\mathbf{Eul}(\hat{\mathbf{C}})=(1-\mathit{sign}\,D_{K/\mathbb{Q}})/2)\cdot\mathbf{Self}(\hat{\mathbf{C}})=\mathbf{vol_{EP}}(\Gamma_{\mathbb{D}})\\&:=\frac{1}{w(\hat{\mathbf{C}})}vol_{EP}(\Gamma_{\mathbb{D}})\end{aligned}$$

for all orbital arithmetic curves $\hat{\mathbf{C}}=\widehat{\mathbf{\Gamma}\backslash\mathbb{D}}$, the **0-th orbital Heegner invariant**.

We define the **Heegner Series** of $\hat{\mathbf{C}}$ by

$$\mathbf{Heeg}_{\hat{\mathbf{C}}}(\tau):=\sum_{N=0}^{\infty}\mathbf{h}_N(\hat{\mathbf{C}})\cdot q^N,\ q=\exp(2\pi iN\tau),\ Im\,\tau>0.$$

Theorem 7.4. *The Heegner series are elliptic modular forms belonging to $\mathcal{M}_3(D_{K/\mathbb{Q}},\chi_K)$ in the Picard case or $\mathcal{M}_2(D_{K/\mathbb{Q}},\chi_K)$ in the Hilbert case of the corresponding quadratic number field K with discriminant $D_{K/IQ}$.*

A detailed explanation of the vector spaces $\mathcal{M}_k(m,\chi_K)$ of elliptic modular forms of weight k, level m and Nebentypus χ_K can be found in the appendix.

Proof. We can refer to [H02] again. We used simply the orbital degree formula working simultaneously for each coefficient. We proved, that we find a $\mathbb{D}$-neat covering in any case. But then we get a Hirzebruch–Zagier series in the Hilbert case, or a Kudla–Cogdell series in the Picard case. These are elliptic modular forms of described type. The Heegner series we started with distinguish from the latter by the orbital property (orbital degree formula) for the coefficients only by a constant factor. □

Definition 7.5. *An infinite series* $\{\mathbf{h}_N\}_{N=0}^{\infty}$ *of orbital invariants on an orbital category (or correspondence class only) is called* **modular dependent**, *if the corresponding series* $\sum_{N=0}^{\infty}\mathbf{h}_N(\hat{\mathbf{C}})\cdot q^N$ *are elliptic modular forms of same type (weight, level, Nebentypus character) for all objects* $\hat{\mathbf{C}}$ *of the category. A countable set of orbital invariants is called* **modular dependent**, *if and only if there is a numeration such that the corresponding series is.*

Since the spaces of modular forms of same type are of finite dimension, it suffices to know the first coefficients of the series to know them completely, if the space is explicitly known. We proved

Theorem 7.6. *On each correspondence class* $\mathbf{Pic}_K^{2,1}$ *or* $\mathbf{Hilb}_K^{2,1}$ *are the corresponding orbital Heegner invariants* $\mathbf{h_N}$ *modular dependent. For each quadratic number field there is up to a constant factor only one Heegner series. The coefficients are rational numbers.*

8. Appendix: Relevant Elliptic Modular Forms of Nebentypus

We consider the congruence subgroups

$$\Gamma_0(m) := \{\left(\begin{smallmatrix} a & b \\ c & d \end{smallmatrix}\right) \in \mathbb{S}l_2(\mathbb{Z});\ c \equiv 0 \ mod\ m\}$$

of the modular group $\mathbb{S}l_2(\mathbb{Z})$ acting on the upper half plane $\mathbb{H} \subset \mathbb{C}$. We also need characters $\chi = \chi_K : \mathbb{Z} \to \{\pm 1\}$ of quadratic number fields K. They factorize through residue class rings of the corresponding discriminants. A holomorphic function $f = f(\tau)$, $\tau \in \mathbb{H}$, is called (elliptic) **modular form** of weight k, (scew) level m and Nebentypus χ, if and only if it satisfies the following functional equations:

$$f(\frac{a\tau + b}{c\tau + d}) = (c\tau + d)^k \chi(d)^k f(\tau) \quad \forall \left(\begin{smallmatrix} a & b \\ c & d \end{smallmatrix}\right) \in \Gamma_0(m),$$

and it must be regular at cusps. The space of these modular forms is denoted by $\mathcal{M}_k(m, \chi)$. This is a finite dimensional $\mathbb{C}$-vector space, which is O for $k < 0$. In [H02] we explained how to get

Example 8.1. . Take weight $k = 3$, level $m = 4 = |D_{K/\mathbb{Q}}|$ and the Dirichlet character $\chi = \chi_K$ of the Gauß number field $K = \mathbb{Q}(i)$.

$$\mathcal{M}_3(4, \chi) = \mathbb{C}\vartheta^6 + \mathbb{C}\vartheta^2\theta$$

with

$$\vartheta := \sum_{n\in\mathbb{Z}} q^{n^2} = 1 + 2\sum_{n>0} q^{n^2} \qquad \text{(Jacobi)}\ ,$$

$$\theta := \sum_{0<u\ odd} \sigma(u)q^u = q\cdot\prod_{m=1}^{\infty}(1-q^{4m})^4\prod_{n=1}^{\infty}(1+2q^n)^4 \qquad \text{(Hecke)}.$$

In [H02] we explained how to get the "Heegner-Apollonius modular form" (2) in the Introduction from the extended orbital Apollonius cycle on the projective plane, visualized in Figure 3.

Examples 8.2. Let D be the discriminant of a real quadratic number field K. Hecke [Heck] defined Eisenstein series in $\mathcal{M}_2(D,\chi_K)$ for prime discriminants. From [vdG], V, Appendix, we take more generally:

$$\frac{1}{2}L(-1,\chi_K) + \sum_{N=1}^{\infty}\left(\sum_{0<d|N}\chi_K(d)d\right)q^N,$$

$$= \sum_{N=1}^{\infty}\left(\sum_{0<d|N}\chi_{D_1}(d)\chi_{D_2}(N/d)\cdot d\right)q^N,$$

with honest decompositions $D = D_1\cdot D_2$ in two smaller discriminants.

Knowing dimensions of $\mathcal{M}_2(D,\chi_K)$ (see tables at the end of [vdG] and the first coefficient of the Heegner series for the Hilbert–Cartesius orbiplane ($K = \mathbb{Q}(\sqrt(2))$ of the introduction we come in the same manner as in the Picard–Apollonius orbiplane to the Heegner series (5).

On the orbiplanes we have a simple intersection theory. The intersection of an arbitrary plane curve with a quadric is nothing else but the double degree of the curve. In general for orbiplanes we left it as exercise for the reader to define the orbital degree **degree C** of arithmetic curves there, such that the following result holds.

Theorem 8.3. *For each orbital arithmetic curve* **C** *on an Picard or Hilbert orbiplane of the quadratic number field* K*, say, the Heegner series*

$$\mathbf{Heeg_C}(\tau) = \mathbf{Eul(C)} + \mathbf{degree\ C}\cdot\sum_{N=1}^{\infty}(\mathbf{degree\ H_N})q^N \qquad (56)$$

is an elliptic modular form belonging to $\mathcal{M}_2(D_{K/\mathbb{Q}},\chi_K)$ *or* $\mathcal{M}_2(D_{K/\mathbb{Q}},\chi_K)$*, respectively.*

Comparing the coefficients in (56) with those of the explicit arithmetic elliptic modular forms of Picard–Apollonius (2) and Hilbert–Cartesius (5) in the introduction we get a convenient counting of arithmetic curves sitting all in Heegner divisors with orbital degree multiplicities.

References

[B] A. Borel, *Introduction aux groupes arithmétiques*, Herman Paris, 1969.

[Cog] J.W. Cogdell, Arithmetic cycles on Picard modular surfaces and modular forms of Nebentypus, *Journ. f. Math.* **357** (1984), 115–137.

[D] I. Dolgachev, Automorphic forms and quasi-homogeneous singularities, *Funkt. Analysis y Priloz.* **9** (1975), 67–68.

[Heck] E. Hecke, Analytische Arithmetik der positiven quadratischen Formen, *Kgl. Danske Vid. Selskab. Math.-fys. Med.* **XII, 12** (1940), Werke, 789–918.

[Hir1] F. Hirzebruch, Überlagerungen der projektiven Ebene und Hilbertsche Modulflächen, *L'Einseigment mathématique* **XXIV**, fasc. 1–2 (1978), 63–78.

[Hir2] F. Hirzebruch, The ring of Hilbert modular forms for real quadratic fields of small discriminant, Proc. Intern. Summer School on modular functions, Bonn 1976, 287–323.

[HZ] F. Hirzebruch, D. Zagier, Intersection numbers of curves on Hilbert modular surfaces and modular forms of Nebentypus, *Invent. Math.* **36** (1976), 57–113.

[H98] R.-P. Holzapfel, Ball and surface arithmetics, Aspects of Mathematics **E 29**, Vieweg, Braunschweig -Wiesbaden, 1998.

[HPV] R.-P. Holzapfel, Picard-Einstein metrics and class fields connected with Apollonius cycle, HU-Preprint, **98-15** (with appendices by A. Piñeiro, N. Vladov).

[HV] R.-P. Holzapfel, N. Vladov, Quadric-line configurations degenerating plane Picard-Einstein metrics I–II, *Sitzungsber. d. Berliner Math. Ges.* 1997–2000, Berlin 2001, 79–142.

[H02] R.-P. Holzapfel, Enumerative Geometry for complex geodesics on quasihyperbolic 4-spaces with cusps, Proc. Conf. Varna 2002, In: Geometry, Integrability and Quantization, ed. by M. Mladenov, L. Naber, Sofia 2003, 42–87.

[P] H. Pinkham, Normal surface singularities with $\mathcal{C}^*$-action, *Math. Ann.* **227** (1977), 183–193.

[Ul] A.M. Uludag, Covering relations between ball-quotient orbifolds, *Math. Ann.* **328** no. 3 (2004), 503–523.

[vdG] G. van der Geer, Hilbert modular surfaces, Erg. Math. u. Grenzgeb., 3. Folge, **16**, Springer-Verlag, Berlin, Heidelberg, New York, London, Paris, Tokyo, 1987.

[Y-S] M- Yoshida, K. Sakurai, Fuchsian systems associated with $\mathbb{P}^2(\mathbb{F}_2)$-arrangement, *Siam J. Math. Anal.* **20** no. 6 (1989), 1490–1499.

Rolf-Peter Holzapfel
Institut für Mathematik
Humboldt-Universität
Unter den Linden 6
D-10099 Berlin
Germany
e-mail: `holzapfl@math.hu-berlin.de`

Progress in Mathematics, Vol. 260, 163–187

Hypergeometric Functions and Carlitz Differential Equations over Function Fields

Anatoly N. Kochubei

Abstract. The paper is a survey of recent results in analysis of additive functions over function fields motivated by applications to various classes of special functions including Thakur's hypergeometric function. We consider basic notions and results of calculus, analytic theory of differential equations with Carlitz derivatives (including a counterpart of regular singularity), umbral calculus, holonomic modules over the Weyl–Carlitz ring.

Mathematics Subject Classification (2000). Primary 11S80, 12H25, 33E50; Secondary 05A40, 11G09, 16S32, 32C38.

Keywords. Function fields, Thakur's hypergeometric function, differential equations with Carlitz derivative, umbral calculus, holonomic modules.

1. Introduction

Let K be the field of formal Laurent series $t = \sum_{j=N}^{\infty} \xi_j x^j$ with coefficients ξ_j from the Galois field $\mathbb{F}_q$, $\xi_N \neq 0$ if $t \neq 0$, $q = p^\upsilon$, $\upsilon \in \mathbf{Z}_+$, where p is a prime number. It is well known that any non-discrete locally compact field of characteristic p is isomorphic to such K. The absolute value on K is given by $|t| = q^{-N}$, $|0| = 0$. The ring of integers $O = \{t \in K : \ |t| \leq 1\}$ is compact in the topology corresponding to the metric $\operatorname{dist}(t,s) = |t-s|$. The absolute value $|\cdot|$ can be extended in a unique way onto the completion $\overline{K}_c$ of an algebraic closure of K.

Analysis over K and $\overline{K}_c$ initiated by Carlitz [5] and developed subsequently by Wagner, Goss, Thakur, the author, and others (see the bibliography in [14, 39]) is very different from the classical calculus. The new features begin with an appropriate version of the factorial invented by Carlitz — since the usual factorial $i!$,

This work was supported in part by CRDF (Grant UM1-2567-OD-03), DFG (Grant 436 UKR 113/72), and the Ukrainian Foundation for Fundamental Research (Grant 01.07/027).

seen as an element of K, vanishes for $i \geq p$, Carlitz introduced the new one as

$$D_i = [i][i-1]^q \dots [1]^{q^{i-1}}, \quad [i] = x^{q^i} - x \ (i \geq 1), \ D_0 = 1. \tag{1.1}$$

An important feature is the availability of many non-trivial $\mathbb{F}_q$-linear functions, that is such functions f defined on $\mathbb{F}_q$-subspaces $K_0 \subset K$ that

$$f(t_1 + t_2) = f(t_1) + f(t_2), \quad f(\alpha t) = \alpha f(t),$$

for any $t, t_1, t_2 \in K_0$, $\alpha \in \mathbb{F}_q$. Such are, for example, polynomials and power series of the form $\sum a_k t^{q^k}$, in particular, the Carlitz exponential

$$e_C(t) = \sum_{n=0}^{\infty} \frac{t^{q^n}}{D_n}, \quad |t| < 1, \tag{1.2}$$

and its composition inverse, the Carlitz logarithm

$$\log_C(t) = \sum_{n=0}^{\infty} (-1)^n \frac{t^{q^n}}{L_n}, \quad |t| < 1, \tag{1.3}$$

where $L_n = [n][n-1] \cdots [1]$ $(n \geq 1)$, $L_0 = 1$. The notion of the Carlitz exponential obtained a wide generalization in the theory of Drinfeld modules (see [14, 39]). On the other hand, in various problems going beyond the class of $\mathbb{F}_q$-linear functions, an extended version of the Carlitz factorial (and its Gamma function interpolations) is used, so that D_n can be seen as "an $\mathbb{F}_q$-linear part" of the full factorial; see [14, 39] and references therein for the details.

Among other special classes of $\mathbb{F}_q$-linear functions there are various polynomial systems (see below), an analog of the Bessel functions [6, 36], and Thakur's hypergeometric function [37, 38, 39]. The latter is defined as follows.

For $n \in \mathbb{Z}_+$, $a \in \mathbb{Z}$, denote

$$(a)_n = \begin{cases} D_{n+a-1}^{q^{-(a-1)}}, & \text{if } a \geq 1; \\ L_{-a-n}^{-q^n}, & \text{if } a \leq 0, n \leq -a; \\ 0, & \text{if } a \leq 0, n > -a. \end{cases} \tag{1.4}$$

Then, for $a_i, b_i \in \mathbb{Z}$, such that the series below makes sense, we set

$${}_rF_s(a_1, \dots, a_r; b_1, \dots, b_s; z) = \sum_{n=0}^{\infty} \frac{(a_1)_n \cdots (a_r)_n}{(b_1)_n \cdots (b_s)_n D_n} z^{q^n}. \tag{1.5}$$

Thakur [37, 38, 39] has carried out a thorough investigation of the functions (1.5) and obtained analogs of many properties known for the classical situation. In particular, he found an analog of the hypergeometric differential equation. Its main ingredients are the difference operator

$$(\Delta u)(t) = u(xt) - xu(t)$$

(an inner derivation of composition rings of $\mathbb{F}_q$-linear polynomials or more general $\mathbb{F}_q$-linear functions) introduced by Carlitz [5], the nonlinear ($\mathbb{F}_q$-linear) operator

$d = \sqrt[q]{\ } \circ \Delta$, and the $\mathbb{F}_q$-linear Frobenius operator $\tau u = u^q$. For example, the function $y = {}_2F_1(a, b; c; z)$ is a solution of the equation

$$(\Delta - [-a])(\Delta - [-b])y = d(\Delta - [1 - c])y. \tag{1.6}$$

Here we touch only a part of Thakur's results (he considered also hypergeometric functions corresponding to other places of $\mathbb{F}_q(x)$, a version of (1.5) with parameters from K and its extensions etc).

The Carlitz exponential e_C satisfies a much simpler equation of the same kind:

$$de_C = e_C, \tag{1.7}$$

so that the operator d may be seen as an analog of the derivative. The operator τ is an analog of the multiplication by t, so that Δ is the counterpart of $t\frac{d}{dt}$.

The same operators appear in the positive characteristic analogs of the canonical commutation relations of quantum mechanics [17, 18]. In the analog of the Schrödinger representation we consider, on the Banach space $C_0(O, \overline{K}_c)$ of continuous $\mathbb{F}_q$-linear functions on O, with values from $\overline{K}_c$ (with the supremum norm), the "creation and annihilation operators"

$$a^+ = \tau - I, \quad a^- = d$$

(I is the identity operator). Then

$$a^- a^+ - a^+ a^- = [1]^{1/q} I, \tag{1.8}$$

the operator a^+a^- possesses the orthonormal (in the non-Archimedean sense [34]) eigenbasis $\{f_i\}$,

$$(a^+a^-)f_i = [i]f_i, \quad i = 0, 1, 2, \ldots; \tag{1.9}$$

a^+ and a^- act upon the basis as follows:

$$a^+ f_{i-1} = [i]f_i, \;\; a^- f_i = f_{i-1}, \;\; i \geq 1; \;\; a^- f_0 = 0. \tag{1.10}$$

Here $\{f_i\}$ is the sequence of normalized Carlitz polynomials

$$f_i(s) = D_i^{-1} \prod_{\substack{m \in \mathbb{F}_q[x] \\ \deg m < i}} (s - m) \quad (i \geq 1), \quad f_0(s) = s, \tag{1.11}$$

which forms an orthonormal basis in $C_0(O, \overline{K}_c)$. The spectrum of the "number operator" a^+a^- is the set of elements $[i]$, so that even this notation (proposed by Carlitz in 1935) becomes parallel to the usual quantum mechanical situation.

An analog of the Bargmann–Fock representation is obtained if we consider the operators of almost the same form,

$$\tilde{a}^+ = \tau, \quad \tilde{a}^- = d,$$

but on the Banach space H of power series $u(t) = \sum\limits_{n=0}^{\infty} a_n \frac{t^{q^n}}{D_n}$ with $a_n \in \overline{K}_c$, $a_n \to 0$ as $n \to \infty$. These new operators satisfy the same relations (1.8)-(1.10), but this time instead of the Carlitz polynomials f_n we get the eigenfunctions $\tilde{f}_n = \frac{t^{q^n}}{D_n}$.

The above results motivated the author to begin to develop analysis and theory of differential equations for $\mathbb{F}_q$-linear functions over K and $\overline{K}_c$, that is for the case which can be seen as a concentrated expression of features specific for the analysis in positive characteristic. This paper is a brief survey of some achievements in this direction. In particular, we consider the counterparts of the basic notions of calculus, analytic theory of differential equations (in the regular case and the case of regular singularity), their applications to some special functions, like the power function, logarithm and polylogarithms, Thakur's hypergeometric function etc. An umbral calculus and a theory of holonomic modules are initiated for this case. Like in the classical situation (see [7]), it is shown that some basic objects of the function field arithmetic generate holonomic modules.

Note that some of the results can be easily extended to the case where the base field is a completion of $\mathbb{F}_q(x)$ with respect to a finite place determined by an irreducible polynomial $\pi \in \mathbb{F}_q[x]$ (the field K corresponds to $\pi(x) = x$); for some details see [23]. The situation is different for the "infinite" place widely used in function field arithmetic (see [39]). In this case some of the basic objects behave in a quite different way – absolute values of the Carlitz factorials D_n grow, as $n \to \infty$, the Carlitz exponential is an entire function, the Carlitz polynomials do not form an orthonormal basis etc. A thorough investigation of properties of the Carlitz differential equations for this situation has not been carried out so far.

2. Calculus

2.1. Higher Carlitz operators $\Delta^{(n)}$ are introduced recursively,

$$\left(\Delta^{(n)}u\right)(t) = \Delta^{(n-1)}u(xt) - x^{q^{n-1}}\Delta^{(n-1)}u(t), \quad n \geq 2. \tag{2.1}$$

For $n = 1$, the formula (2.1) coincides with the definition of $\Delta = \Delta^{(1)}$, if we set $\Delta^{(0)} = I$.

The first application of these operators is the reconstruction formula [18] for the coefficients a_n of a power series $u \in H$. Note that the classical formula does not make sense here because it contains the expression $u^{(n)}(t)/n!$ where both the numerator and denominator vanish.

Theorem 2.1. *If $u \in H$, then*

$$a_n = \lim_{t \to 0} \frac{\Delta^{(n)}u(t)}{t^{q^n}}, \quad n = 0, 1, 2, \ldots.$$

For a continuous non-holomorphic $\mathbb{F}_q$-linear function u the behaviour of the functions

$$\mathfrak{D}^k u(t) = t^{-q^k}\Delta^{(k)}u(t), \quad t \in O \setminus \{0\},$$

near the origin measures the smoothness of u. We say that $u \in C_0^{k+1}(O, \overline{K}_c)$ if $\mathfrak{D}^k u$ can be extended to a continuous function on O. This includes the case ($k = 0$) of differentiable functions.

The next theorem proved in [18] gives a characterization of the above smoothness in terms of coefficients of the Fourier–Carlitz expansion. It includes, as a particular case ($k = 0$), the characterization of differentiable $\mathbb{F}_q$-linear functions obtained by Wagner [43].

Theorem 2.2. *A function* $u = \sum_{n=0}^{\infty} c_n f_n \in C_0(O, \overline{K}_c)$ *belongs to* $C_0^{k+1}(O, \overline{K}_c)$ *if and only if*

$$q^{nq^k} |c_n| \to 0 \quad \text{for } n \to \infty.$$

In this case

$$\sup_{t\in O} |\mathfrak{D}^k u(t)| = \sup_{n\geq k} q^{(n-k)q^k} |c_n|.$$

For a generalization to some classes of not necessarily $\mathbb{F}_q$-linear functions see [45].

Similarly [18], a function u is analytic on the ball O (that is, $u(t) = \sum a_i t^{q^i}$, $a_i \to 0$) if and only if $q^{\frac{q^n}{q-1}} |c_n| \to 0$, as $n \to \infty$. A more refined result by Yang [44], useful in many applications, which makes it possible to find an exact domain of analyticity, is as follows (again we consider only $\mathbb{F}_q$-linear functions while in [44] a more general class is studied).

Theorem 2.3 (Yang). *A function* $u = \sum_{n=0}^{\infty} c_n f_n \in C_0(O, \overline{K}_c)$ *is locally analytic if and only if*

$$\gamma = \liminf_{n\to\infty} \left\{ -q^{-n} \log_q |c_n| \right\} > 0, \tag{2.2}$$

and if (2.2) *holds, then* u *is analytic on any ball of the radius* q^{-l},

$$l = \max(0, [-(\log(q-1) + \log\gamma)/\log q] + 1).$$

2.2. Viewing d as a kind of a derivative, it is natural to introduce an antiderivative S setting $Sf = u$ where u is a solution of the equation $du = f$, with the normalization $u(1) = 0$. It is easy to find Sf explicitly if f is given by its Fourier-Carlitz expansion (see [18]).

Next, we introduce a Volkenborn-type integral of a function $f \in C_0^1(O, \overline{K}_c)$ (see [34] for a similar integration theory over $\mathbb{Z}_p$) setting

$$\int_O f(t)\,dt \stackrel{\text{def}}{=} \lim_{n\to\infty} \frac{Sf(x^n)}{x^n} = (Sf)'(0).$$

The integral is a $\mathbb{F}_q$-linear continuous functional on $C_0^1(O, \overline{K}_c)$,

$$\int_O cf(t)\,dt = c^q \int_O f(t)\,dt, \quad c \in \overline{K}_c,$$

possessing the following "invariance" property (related, in contrast to the case of $\mathbf{Z}_p$, to the multiplicative structure):

$$\int_O f(xt)\,dt = x\int_O f(t)\,dt - f^q(1).$$

Calculating the integrals of some important functions we obtain new relations between them. In addition to the Carlitz exponential e_C and the Carlitz polynomials f_n (see (1.2) and (1.11)), we mention the Carlitz module function

$$C_s(z) = \sum_{i=0}^{\infty} f_i(s) z^{q^i}, \quad s \in O, |z| < 1. \tag{2.3}$$

Note that if $s \in \mathbb{F}_q[x]$, then only the terms with $i \le \deg s$ are different from zero in (2.3).

We have

$$\int_O t^{q^n}\,dt = -\frac{1}{[n+1]}, \quad n = 0,1,2,\dots;$$

$$\int_O f_n(t)\,dt = \frac{(-1)^{n+1}}{L_{n+1}}, \quad n = 0,1,2,\dots;$$

$$\int_O C_s(z)\,ds = \log_C(z) - z, \quad z \in K,\ |z| < 1;$$

$$\int_O e_C(st)\,ds = t - e_C(t), \quad t \in K, |t| < 1.$$

For the proofs see [18].

3. Differential Equations for $\mathbb{F}_q$-Linear Functions

3.1. Let us consider function field analogs of linear differential equations with holomorphic or polynomial coefficients. Note that in our situation the meaning of a polynomial coefficient is not a usual multiplication by a polynomial, but the action of a polynomial in the operator τ.

We begin with the regular case and consider an equation (actually, a system)

$$dy(t) = P(\tau)y(t) + f(t) \tag{3.1}$$

where for each $z \in \left(\overline{K}_c\right)^m$, $t \in K$,

$$P(\tau)z = \sum_{k=0}^{\infty} \pi_k z^{q^k}, \quad f(t) = \sum_{j=0}^{\infty} \varphi_j \frac{t^{q^j}}{D_j}, \tag{3.2}$$

π_k are $m\times m$ matrices with elements from $\overline{K}_c$, $\varphi_j \in \left(\overline{K}_c\right)^m$, and it is assumed that the series (3.2) have positive radii of convergence. The action of the operator τ

upon a vector or a matrix is defined component-wise, so that $z^{q^k} = \left(z_1^{q^k}, \dots, z_m^{q^k}\right)$ for $z = (z_1, \dots, z_m)$.

We seek a $\mathbb{F}_q$-linear solution of (3.1) on some neighbourhood of the origin, of the form

$$y(t) = \sum_{i=0}^{\infty} y_i \frac{t^{q^i}}{D_i}, \quad y_i \in \left(\overline{K}_c\right)^m, \tag{3.3}$$

where y_0 is a given element, so that the "initial" condition for our situation is

$$\lim_{t\to 0} t^{-1} y(t) = y_0. \tag{3.4}$$

The next theorem, proved in [19], is the function field analog of the Cauchy theorem from the classical analytic theory of differential equations.

Theorem 3.1. *For any $y_0 \in \left(\overline{K}_c\right)^m$ the equation* (3.1) *has a unique local solution of the form* (3.3), *which satisfies* (3.4), *with the series having a positive radius of convergence.*

Thus, regular equations with Carlitz derivatives behave more or less as their classical counterparts. The situation is different for singular equations. Let us consider scalar equations of arbitrary order

$$\sum_{j=0}^{m} A_j(\tau) d^j u = f \tag{3.5}$$

where $f(t) = \sum\limits_{n=0}^{\infty} \varphi_n \frac{t^{q^n}}{D_n}$, $A_j(\tau)$ are power series having (as well as the one for f) positive radii of convergence.

We investigate formal solutions of (3.5), of the form

$$u(t) = \sum_{n=0}^{\infty} u_n \frac{t^{q^n}}{D_n}, \quad u_n \in \overline{K}_c. \tag{3.6}$$

One can apply an operator series $A(\tau) = \sum\limits_{k=0}^{\infty} \alpha_k \tau^k$ (even without assuming its convergence) to a formal series (3.6), setting

$$\tau^k u(t) = \sum_{n=0}^{\infty} u_n^{q^k} [n+1]^{q^{k-1}} \dots [n+k] \frac{t^{q^{n+k}}}{D_{n+k}}, \quad k \geq 1,$$

and

$$A(\tau)u(t) = \sum_{l=0}^{\infty} \frac{t^{q^l}}{D_l} \sum_{n+k=l} \alpha_k u_n^{q^k} [n+1]^{q^{k-1}} \dots [n+k]$$

where the factor $[n+1]^{q^{k-1}} \dots [n+k]$ is omitted for $k=0$. These formal manipulations are based on the identity

$$\tau \left(\frac{t^{q^{i-1}}}{D_{i-1}} \right) = [i] \frac{t^{q^i}}{D_i}.$$

Using also the relation

$$d\left(\frac{t^{q^i}}{D_i}\right)=\frac{t^{q^{i-1}}}{D_{i-1}},$$

now we can give a meaning to the notion of a formal solution of the equation (3.5).

Theorem 3.2. *Let $u(t)$ be a formal solution* (3.6) *of the equation* (3.5), *where the series for $A_j(\tau)z$, $z\in\overline{K}_c$, and $f(t)$, have positive radii of convergence. Then the series* (3.6) *has a positive radius of convergence.*

This result (proved in [19]) is in a strong contrast to the classical theory. Note that in the p-adic case a similar phenomenon takes place for equations satisfying certain strong conditions upon zeros of indicial polynomials [1, 8, 27, 35]. In our case such a behavior is proved for any equation, which resembles the (much simpler) case [27] of differential equations over a field of characteristics zero, whose residue field also has characteristic zero.

3.2. The equations (3.1) and (3.5) behave like linear equations, though they are actually only $\mathbb{F}_q$-linear. Theorem 3.1 can be extended [22] to the case of strongly nonlinear equations (containing self-compositions $y\circ y\circ\cdots\circ y$).

On the other hand, it is natural to consider some equations of this kind in wider classes of $\mathbb{F}_q$-linear functions resembling meromorphic functions of a complex variables. The set $\mathcal{R}_K$ of locally convergent $\mathbb{F}_q$-linear holomorphic functions forms a non-commutative ring with respect to the composition operation (the pointwise multiplication violates the $\mathbb{F}_q$-linearity). The non-commutativity of $\mathcal{R}_K$ makes the algebraic structures related to Carlitz differential equations much more complicated compared to their classical counterparts. So far their understanding is only at its initial stage. It is known, however, that $\mathcal{R}_K$ can be imbedded into a skew field of $\mathbb{F}_q$-linear "meromorphic" series containing terms like $t^{q^{-k}}$ (see [22]). A deep investigation of bi-infinite series of this kind convergent on the whole of $\overline{K}_c$ has been carried out by Poonen [26].

A specific class of equations with solutions meromorphic in the above sense is the class of scalar Riccati-type equations

$$dy(t)=\lambda(y\circ y)(t)+(P(\tau)y)(t)+R(t) \tag{3.7}$$

where $\lambda\in\overline{K}_c$,

$$(P(\tau)y)(t)=\sum_{k=1}^{\infty}p_k y^{q^k}(t),\quad R(t)=\sum_{k=0}^{\infty}r_k t^{q^k},$$

$p_k,r_k\in\overline{K}_c$ (note that the right-hand side of (3.7) does not contain the linear term). The following theorem is proved in [22].

Theorem 3.3. *If $0<|\lambda|\le q^{-1/q^2}$, $|p_k|\le q^{-1/q^2}$, $|r_k|\le q^{-1/q^2}$ for all k, then the equation* (3.7) *possesses solutions of the form*

$$y(t)=ct^{1/q}+\sum_{n=0}^{\infty}a_n t^{q^n},\quad c,a_n\in\overline{K}_c,\ c\neq 0,$$

where the series converges on the open unit disk $|t| < 1$.

4. Regular Singularity

4.1. In analysis over $\mathbb{C}$, a typical class of systems with regular singularity at the origin $\zeta = 0$ over $\mathbb{C}$ consists of systems of the form

$$\zeta y'(\zeta) = \left(B + \sum_{k=1}^{\infty} A_k \zeta^k \right) y(\zeta) \tag{4.1}$$

where B, A_j are constant matrices, and the series converges on a neighbourhood of the origin. Such a system possesses a fundamental matrix solution of the form $W(\zeta)\zeta^C$ where $W(\zeta)$ is holomorphic on a neighbourhood of zero, C is a constant matrix, $\zeta^C = \exp(C \log \zeta)$ is defined by the obvious power series. Under some additional assumptions regarding the eigenvalues of the matrix B, one can take $C = B$. For similar results over $\mathbb{C}_p$ see [11].

In order to investigate such a class of equations in the framework of $\mathbb{F}_q$-linear analysis over K, one has to go beyond the class of locally analytic functions. Instead of power series expansions we can use the expansions in Carlitz polynomials on the compact ring $O \subset K$. The property of local analyticity, if it takes place, can be recovered with the use of Theorem 2.3. Note that our approach would fail if we consider equations over $\overline{K}_c$ instead of K (our solutions may take their values from $\overline{K}_c$, but they are defined over subsets of K). In this sense our techniques are different from the ones developed for both the characteristic zero cases.

We begin with the simplest model scalar equation

$$\tau du = \lambda u, \quad \lambda \in \overline{K}_c, \tag{4.2}$$

whose solution may be seen as a function field counterpart of the power function $t \mapsto t^\lambda$.

We look for a continuous $\mathbb{F}_q$-linear solution $u(t, \lambda)$ of the equation (4.2), with the "initial condition" $u(1, \lambda) = 1$, in the form

$$u(t) = \sum_{i=0}^{\infty} c_i f_i(t), \quad t \in O, \tag{4.3}$$

where $c_0 = 1$.

It is easy to see that the equation (4.2) has no continuous solutions if $|\lambda| \geq 1$. If $|\lambda| < 1$, then the solution $u(t, \lambda)$ is unique, continuous on O, and the coefficients from (4.3) have the form

$$c_n = \prod_{j=0}^{n-1} (\lambda - [j]).$$

The function $u(t, \lambda)$ is analytic on O if and only if $\lambda = [j]$ for some $j \geq 0$; in this case $u(t, \lambda) = u(t, [j]) = t^{q^j}$. If $\lambda \neq [j]$ for any integer $j \geq 0$, then $u(t, \lambda)$

is locally analytic on O if and only if $\lambda = -x$, and in that case $u(t,-x) = 0$ for $|t| \le q^{-1}$. The relation

$$u(t^{q^m},\lambda) = u(t,\lambda^{q^m} + [m]), \quad t \in O,$$

holds for all λ, $|\lambda| < 1$, and for all $m = 0,1,2,\ldots$. For the proofs see [20].

Similarly, if in (4.1) $\lambda = (\lambda_{ij})$ is is a $m \times m$ matrix with elements from $\overline{K}_c$, and we look for a matrix-valued solution of (4.1), then such a solution is given by the series (4.3) with the matrix coefficients

$$c_i = \left\{ \prod_{j=0}^{i-1} (\lambda - [j] I_m) \right\} c_0, \quad i \ge 1$$

(I_m is the unit matrix), if $|\lambda| \stackrel{\text{def}}{=} \max |\lambda_{ij}| < 1$.

4.2. The analog, for our situation, of the system (4.1) is the system

$$\tau du - P(\tau) u = 0 \tag{4.4}$$

where $P(\tau)$ is a matrix-valued analytic function, so that $P(\tau)z = \sum\limits_{k=0}^{\infty} \pi_k z^{q^k}$. We assume that $|\pi_k| \le \gamma$, $\gamma > 0$, for all k, $|\pi_0| < 1$. Denote by $g(t)$ a solution of the equation $\tau dg = \pi_0 g$. Let $\lambda_1, \ldots, \lambda_m \in \overline{K}_c$ be the eigenvalues of the matrix π_0.

Theorem 4.1. *If*

$$\lambda_i - \lambda_j^{q^k} \ne [k], \quad i,j = 1,\ldots,m; \ k = 1,2,\ldots, \tag{4.5}$$

then the system (4.4) *has a matrix solution*

$$u(t) = W(g(t)), \quad W(s) = \sum_{k=0}^{\infty} w_k s^{q^k}, \quad w_0 = I_m,$$

where the series for W *has a positive radius of convergence.*

The paper [20] contains, apart from the proof of Theorem 4.1, a discussion of some situations (the Euler type equations) where its conditions are violated, as well as of the meaning of the conditions (4.5). Here we only mention that in the scalar case $m = 1$ the condition (4.5) is equivalent to the assumption $\pi_0 \ne -x$, so that it excludes the case where solutions of the equation $\tau dg = \pi_0 g$ has pathological properties.

4.3. For the above equation, continuous solutions were found as Fourier–Carlitz expansions

$$u(t) = \sum_{n=0}^{\infty} c_n f_n(t), \tag{4.6}$$

and we had to impose certain conditions upon coefficients of the equation, in order to guarantee the uniform convergence of the series on O (which is equivalent to the fact that $c_n \to 0$). However formally we could write the series (4.6) for the solutions without those conditions. Thus, it is natural to ask whether the corresponding series (4.6) converge at some points $t \in O$. Note that (4.6) always makes sense for $t \in \mathbb{F}_q[x]$ (for each such t only a finite number of terms is different from zero). The question is whether the series converges on a wider set; if the answer is negative, such a formal solution is called *strongly singular.*

The available results regarding strong singularity of solutions of some equations are based on the following general fact [20].

Theorem 4.2. *If* $|c_i| \geq \rho > 0$ *for all* $i \geq i_0$ *(where* i_0 *is some natural number), then the function* (4.6) *is strongly singular.*

It follows from Theorem 4.2 that non-trivial formal solutions of the equation (4.2) with $|\lambda| \geq 1$ are strongly singular. A more complicated example is provided by the equation

$$(\Delta - [-a])(\Delta - [-b])u = d\Delta u, \quad a, b \in \mathbb{Z}, \tag{4.7}$$

for Thakur's hypergeometric function ${}_2F_1(a, b; 1; t)$.

A holomorphic solution of (4.7) is given by an appropriate specialization of (1.5). Classically (over $\mathbb{C}$), there exists the second solution with a logarithmic singularity. Here the situation is different. Looking for a solution of the form (4.6) we obtain a recursive relation

$$\begin{aligned}\left(c_{i+2}^{1/q} - c_{i+2}\right) + c_{i+1}^{1/q}[i+1]^{1/q} - c_{i+1}([i] + [i+1] - [-a] - [-b]) \\ - c_i([i] - [-a])([i] - [-b]) = 0, \quad i = 0, 1, 2, \dots. \end{aligned} \tag{4.8}$$

Taking arbitrary initial coefficients $c_0, c_1 \in \overline{K}_c$ we obtain a solution u defined on $\mathbb{F}_q[x]$. On each step we have to solve the equation

$$z^{1/q} - z = v. \tag{4.9}$$

If $|c_i| \leq 1$ and $|c_{i+1}| \leq 1$, then in the equation for c_{i+2} we have $|v| < 1$.

It can be shown [20] that the equation (4.9) has a unique solution $z_0 \in \overline{K}_c$, for which $|z_0| \leq |v|$, and $q - 1$ other solutions z, $|z| = 1$. It is natural to call a solution *generic* if, starting from a certain step of finding the coefficients c_n, we always take the most frequent option corresponding to a solution of (4.9) with $|z| = 1$. Now Theorem 4.2 implies the following fact.

Theorem 4.3. *A generic solution of the equation* (4.7) *is strongly singular.*

Of course, in some special cases the recursion (4.8) can lead to more regular solutions, in particular, to the holomorphic solutions found by Thakur.

5. Polylogarithms and a Zeta Function

5.1. The Carlitz differential equations can be used for defining new special functions with interesting properties. Some examples are given in this section.

An analog of the function $-\log(1-t)$ is defined via the equation

$$(1-\tau)du(t) = t, \quad t \in K_\pi, \tag{5.1}$$

a counterpart of the classical equation $(1-t)u'(t) = 1$. The next results are taken from [23] where the equation (5.1) is considered for an arbitrary finite place of $\mathbb{F}_q(x)$.

Let $l_1(t)$ be a $\mathbb{F}_q$-linear holomorphic solution of (5.1) with the zero initial condition (in the sense of (3.4)). Then it is easy to show that

$$l_1(t) = \sum_{n=1}^{\infty} \frac{t^{q^n}}{[n]}, \tag{5.2}$$

and the series in (5.2) converges for $|t| \le q^{-1}$.

Note that $l_1(t)$ is different from the well-known Carlitz logarithm $\log_C$ (see (1.3)). Analogies motivating the introduction of special functions are not so unambiguous, and, for instance, from the composition ring viewpoint, $\log_C$ is an analog of e^{-t}, though in other respects it is a valuable analog of the logarithm. By the way, another possible analog of the logarithm is a continuous function $u(t)$, $|t|_\pi \le 1$, satisfying the equation $\Delta u(t) = t$ (an analog of $tu'(t) = 1$) and the condition $u(1) = 0$. In fact, $u = \mathcal{D}_1$, the first hyperdifferential operator (the definition of $\mathcal{D}_1$ is given in Sect. 5.2 below); see [16].

Now we consider continuous non-holomorphic extensions of l_1.

Theorem 5.1. *The equation* (5.1) *has exactly q continuous solutions on O coinciding with* (5.2) *as $|t| \le q^{-1}$. These solutions have the expansions in the Carlitz polynomials $u = \sum_{i=0}^{\infty} c_i f_i$ where c_1 is an arbitrary solution of the equation $c_1^q - c_1 + 1 = 0$, higher coefficients are found from the relation*

$$c_n = \sum_{j=0}^{\infty} (c_{n-1}[n-1])^{q^{j+1}}, \quad n \ge 2,$$

and the coefficient c_0 is determined by the relation

$$c_0 = \sum_{i=1}^{\infty} (-1)^{i+1} \frac{c_i}{L_i}.$$

Below we denote by l_1 an arbitrary fixed "branch" of extensions of (5.2).

The polylogarithms $l_n(t)$ are defined recursively by the equations

$$\Delta l_n = l_{n-1}, \quad n \ge 2, \tag{5.3}$$

which agree with the classical ones $tl_n'(t) = l_{n-1}(t)$. Analytic $\mathbb{F}_q$-linear solutions of (5.3), such that $t^{-1}l_n(t) \to 0$ as $t \to 0$, are found easily by induction:

$$l_n(t) = \sum_{j=1}^{\infty} \frac{t^{q^j}}{[j]^n}, \quad |t| \le q^{-1}. \tag{5.4}$$

Theorem 5.2. *For each $n \geq 2$, there exists a unique continuous $\mathbb{F}_q$-linear solution of the equation (5.3) coinciding for $|t| \leq q^{-1}$ with the polylogarithm (5.4). The solution is given by the Carlitz expansion $l_n = \sum_{i=0}^{\infty} c_i^{(n)} f_i$ with*

$$\left| c_i^{(n)} \right| \leq C_n q^{-q^{i-1}}, \quad C_n > 0, \ i \geq 1,$$

$$c_0^{(n)} = \sum_{i=1}^{\infty} (-1)^{i+1} \frac{c_i^{(n)}}{L_i}.$$

5.2. Now that the above polylogarithms have been extended onto the disk $\{|t| \leq 1\}$, we can interpret their values at $t = 1$ as "special values" of a kind of a zeta function. In order to define the latter, we introduce the operator $\Delta^{(\alpha)}$, $\alpha \in O$, a function field analog of the Hadamard fractional derivative $\left(t\frac{d}{dt}\right)^{\alpha}$ from real analysis (see [33]).

Denote by $\mathcal{D}_k(t)$, $k \geq 0$, $t \in O$, the sequence of hyperdifferentiations defined initially on monomials by the relations $\mathcal{D}_0(x^n) = x^n$, $\mathcal{D}_k(1) = 0$, $k \geq 1$,

$$\mathcal{D}_k(x^n) = \binom{n}{k} x^{n-k},$$

where it is assumed that $\binom{n}{k} = 0$ for $k > n$. $\mathcal{D}_k$ is extended onto $\mathbb{F}_q[x]$ by $\mathbb{F}_q$-linearity, and then onto O by continuity [42]. The sequence $\{\mathcal{D}_k\}$ is an orthonormal basis of the space of continuous $\mathbb{F}_q$-linear functions on O [16, 9].

Let $\alpha \in O$, $\alpha = \sum_{n=0}^{\infty} \alpha_n x^n$, $\alpha_n \in \mathbb{F}_q$. Denote $\widehat{\alpha} = \sum_{n=0}^{\infty} (-1)^n \alpha_n x^n$. For an arbitrary continuous $\mathbb{F}_q$-linear function u on O we define its "fractional derivative" $\Delta^{(\alpha)} u$ at a point $t \in O$ by the formula

$$\left(\Delta^{(\alpha)} u\right)(t) = \sum_{k=0}^{\infty} (-1)^k \mathcal{D}_k(\widehat{\alpha}) u(x^k t).$$

The function $\alpha \mapsto \left(\Delta^{(\alpha)} u\right)(t)$ is continuous and $\mathbb{F}_q$-linear. As a function of t, $\Delta^{(\alpha)} u$ is continuous if, for example, u is Hölder continuous.

Our understanding of $\Delta^{(\alpha)}$ as a kind of a fractional derivative is justified by the following properties:

$$\Delta^{(x^n)} = \Delta^n, \quad n = 1, 2, \dots;$$

$$\Delta^{(\alpha)} \left(\Delta^{(\beta)} u\right)(t) = \left(\Delta^{(\alpha\beta)} u\right)(t),$$

for any $\alpha, \beta \in O$.

5.3. We define $\zeta(t)$, $t \in K$, setting $\zeta(0) = 0$,

$$\zeta(x^{-n}) = l_n(1), \quad n = 1, 2, \dots,$$

and

$$\zeta(t) = \left(\Delta^{(\theta_0 + \theta_1 x + \cdots)} l_n\right)(1), \quad n = 1, 2, \dots,$$

if $t = x^{-n}(\theta_0+\theta_1 x+\cdots)$, $\theta_j \in \mathbb{F}_q$. The function ζ is a continuous $\mathbb{F}_q$-linear function on K_x with values in $\overline{K}_c$.

In particular, we have

$$\zeta(x^m) = \left(\Delta^{m+1} l_1\right)(1), \quad m = 0, 1, 2, \ldots.$$

The above definition is of course inspired by the classical polylogarithm relation

$$\left(z\frac{d}{dz}\right)\sum_{n=1}^{\infty}\frac{z^n}{n^s} = \sum_{n=1}^{\infty}\frac{z^n}{n^{s-1}}.$$

In contrast to Goss's zeta function defined on natural numbers and interpolated onto $\mathbb{Z}_p$ (see [14, 39]), the above ζ is purely an object of the characteristic p arithmetic.

Let us write some relations for special values of our ζ; for the details see [23]. As we saw,

$$\sum_{j=1}^{\infty}\frac{t^{q^j}}{[j]^n} = \sum_{i=0}^{\infty}\zeta(x^{-n+i})\mathcal{D}_i(t), \quad |t| \le q^{-1}.$$

Next, let us consider the double sequence $A_{n,r} \in K$, $A_{n,1} = (-1)^{n-1}L_{n-1}$,

$$A_{n,r} = (-1)^{n+r}L_{n-1}\sum_{0<i_1<\ldots<i_{r-1}<n}\frac{1}{[i_1][i_2]\ldots[i_{r-1}]}, \quad r \ge 2.$$

These elements appear as the coefficients of the expansion [42] of a hyperdifferentiation $\mathcal{D}_r$ in the normalized Carlitz polynomials, as well as in the expression [15] of the operators $\Delta^{(n)}$ from (2.1) via the iterations Δ^r. Here we have the identity

$$\zeta(x^{-n}) = \sum_{i=1}^{\infty}(-1)^{i+1}L_i^{-1}\sum_{r=1}^{i}A_{i,r}\zeta(x^{r-n})$$

which may be seen as a distant relative of Riemann's functional equation for the classical zeta.

Finally, consider the coefficients c_i of the Carlitz expansion of l_1 (see Theorem 5.1). They are expressed via zeta values:

$$c_i = \sum_{r=1}^{i}A_{i,r}\zeta(x^{r-1}).$$

By Theorem 5.1, for $i \ge 2$ we have

$$c_i = \sum_{j=0}^{\infty}(z_i)^{q^j}, \quad z_i = c_{i-1}^q[i-1]^q. \tag{5.5}$$

The series in (5.5) may be seen as an analog of $\sum_j j^{-z}$. This analogy becomes clearer if, for a fixed $z \in \overline{K}_c$, $|z| < 1$, we consider the set S of all convergent power series $\sum_{n=1}^{\infty} z^{q^{j_n}}$ corresponding to sequences $\{j_n\} \subset \mathbb{N}$. Let us introduce the multiplication $\odot$ in S setting $z^{q^i} \odot z^{q^j} = z^{q^{ij}}$ and extending the operation distributively (for a similar construction in the framework of q-analysis in characteristic 0 see [25]). Denoting by $\prod_p^{\odot}$ the product in S of elements indexed by prime numbers we obtain in a standard way the identity

$$c_i = \prod_p{}^{\odot} \sum_{n=0}^{\infty} (z_i)^{q^{p^n}}$$

(the infinite product is understood as a limit of the partial products in the topology of $\overline{K}_c$), an analog of the Euler product formula. It would be interesting to study the algebraic structure of S in detail.

6. Umbral Calculus

6.1. Classical umbral calculus [31, 30] is a set of algebraic tools for obtaining, in a unified way, a rich variety of results regarding structure and properties of various polynomial sequences. There exists a lot of generalizations extending umbral methods to other classes of functions. However there is a restriction common to the whole literature on umbral calculus – the underlying field must be of zero characteristic. An attempt to mimic the characteristic zero procedures in the positive characteristic case [12] revealed a number of pathological properties of the resulting structures. More importantly, these structures were not connected with the existing analysis in positive characteristic based on a completely different algebraic foundation.

A version of umbral calculus inmplementing such a connection was developed by the author [21], and we summarize it in this section. Its basic notion is motivated by the following identity for the non-normalized Carlitz polynomials $e_i = D_i f_i$:

$$e_i(st) = \sum_{n=0}^{i} \binom{i}{n}_K e_n(t) \{e_{i-n}(s)\}^{q^n} \tag{6.1}$$

where the "K-binomial coefficients" $\binom{i}{n}_K$ are defined as

$$\binom{i}{n}_K = \frac{D_i}{D_n D_{i-n}^{q^n}}.$$

Computing the absolute values of the Carlitz factorials directly from their definition (1.1), it is easy to show that

$$\left| \binom{i}{n}_K \right| = 1, \quad 0 \le n \le i.$$

In fact, $\binom{i}{n}_K \in \mathbb{F}_q(x)$, and we can consider also other places of $\mathbb{F}_q(x)$, that is other non-equivalent absolute values. It can be proved [24] that $\binom{i}{n}_K$ belongs to the ring of integers for any finite place of $\mathbb{F}_q(x)$.

We see the relation (6.1) as a function field counterpart of the classical binomial identity [31, 30] satisfied by many classical polynomials. Now, considering a sequence u_i of $\mathbb{F}_q$-linear polynomials with coefficients from $\overline{K}_c$, we call it *a sequence of K-binomial type* if $\deg u_i = q^i$ and for all $i = 0, 1, 2, \dots$

$$u_i(st) = \sum_{n=0}^{i} \binom{i}{n}_K u_n(t) \{u_{i-n}(s)\}^{q^n}, \quad s, t \in K. \tag{6.2}$$

As in the conventional umbral calculus, the dual notion is that of a delta operator. However, in contrast to the classical situation, here the delta operators are only $\mathbb{F}_q$-linear, not linear.

Denote by ρ_λ the operator of multiplicative shift, $(\rho_\lambda u)(t) = u(\lambda t)$. We call a linear operator T, on the $\overline{K}_c$-vector space $\overline{K}_c\{t\}$ of all $\mathbb{F}_q$-linear polynomials, *invariant* if it commutes with ρ_λ for each $\lambda \in K$.

A $\mathbb{F}_q$-linear operator $\delta = \tau^{-1}\delta_0$, where δ_0 is a linear invariant operator on $\overline{K}_c\{t\}$, is called *a delta operator* if $\delta_0(t) = 0$ and $\delta_0(f) \neq 0$ for $\deg f > 1$. A sequence $\{P_n\}_0^\infty$ of $\mathbb{F}_q$-linear polynomials is called *a basic sequence* corresponding to a delta operator $\delta = \tau^{-1}\delta_0$, if $\deg P_n = q^n$, $P_0(1) = 1$, $P_n(1) = 0$ for $n \geq 1$,

$$\delta P_0 = 0, \quad \delta P_n = [n]^{1/q} P_{n-1}, \ n \geq 1, \tag{6.3}$$

or, equivalently,

$$\delta_0 P_0 = 0, \quad \delta_0 P_n = [n] P_{n-1}^q, \ n \geq 1. \tag{6.4}$$

It is clear that $d = \tau^{-1}\Delta$ is a delta operator. It follows from well-known identities for the Carlitz polynomials e_i [13] (see also (1.10)) that the sequence $\{e_i\}$ is basic with respect to the operator d.

Theorem 6.1. *For any delta operator $\delta = \tau^{-1}\delta_0$, there exists a unique basic sequence $\{P_n\}$, which is a sequence of K-binomial type. Conversely, given a sequence $\{P_n\}$ of K-binomial type, define the action of δ_0 on P_n by the relations* (6.4), *extend it onto $\overline{K}_c\{t\}$ by linearity and set $\delta = \tau^{-1}\delta_0$. Then δ is a delta operator, and $\{P_n\}$ is the corresponding basic sequence.*

The analogs of the higher Carlitz difference operators (2.1) in the present general context are the operators $\delta_0^{(l)} = \tau^l \delta^l$. The identity

$$\delta_0^{(l)} P_j = \frac{D_j}{D_{j-l}^{q^l}} P_{j-l}^{q^l} \tag{6.5}$$

holds for any $l \le j$. If f is a $\mathbb{F}_q$-linear polynomial, $\deg f \le q^n$, then a generalized Taylor formula

$$f(st) = \sum_{l=0}^{n} \frac{\left(\delta_0^{(l)} f\right)(s)}{D_l} P_l(t) \tag{6.6}$$

holds for any $s, t \in K$. For the Carlitz polynomials e_i, the formulas (6.5) and (6.6) are well known [13]. It is important that, in contrast to the classical umbral calculus, the linear operators involved in (6.6) are not powers of a single linear operator.

Any linear invariant operator T on $\overline{K}_c\{t\}$ admits a representation

$$T = \sum_{l=0}^{\infty} \sigma_l \delta_0^{(l)}, \quad \sigma_l = \frac{(TP_l)(1)}{D_l}. \tag{6.7}$$

The infinite series in (6.7) becomes actually a finite sum if both sides of (6.7) are applied to any $\mathbb{F}_q$-linear polynomial. Conversely, any such series defines a linear invariant operator on $\overline{K}_c\{t\}$.

Let us consider the case where $\delta = d$, so that $\delta_0^{(l)} = \Delta^{(l)}$. The next result leads to new delta operators and basic sequences.

Theorem 6.2. *The operator* $\theta = \tau^{-1}\theta_0$, *where*

$$\theta_0 = \sum_{l=1}^{\infty} \sigma_l \Delta^{(l)},$$

is a delta operator if and only if

$$S_n \stackrel{\text{def}}{=} \sum_{l=1}^{n} \frac{\sigma_l}{D_{n-l}^{q^l}} \neq 0 \quad \textit{for all} \quad n = 1, 2, \ldots. \tag{6.8}$$

Example 1. Let $\sigma_l = 1$ for all $l \ge 1$, that is

$$\theta_0 = \sum_{l=1}^{\infty} \Delta^{(l)}. \tag{6.9}$$

Estimates of $|D_n|$ which follow directly from (1.1) show that $|S_n| = q^{\frac{q^n - q}{q-1}}$, so that (6.8) is satisfied. Comparing (6.9) with a classical formula from [31] we may see the polynomials P_n for this case as analogs of the Laguerre polynomials.

Example 2. Let $\sigma_l = \dfrac{(-1)^{l+1}}{L_l}$. For this case it can be shown [21] that $S_n = D_n^{-1}$, $n = 1, 2, \ldots$; $\theta_0(t^{q^j}) = t^{q^j}$ for all $j \ge 1$ (of course, $\theta_0(t) = 0$), and $P_0(t) = t$, $P_n(t) = D_n\left(t^{q^n} - t^{q^{n-1}}\right)$ for $n \ge 1$.

6.2. As in the p-adic case [40, 41, 29], the umbral calculus can be used for constructing new orthonormal bases in $C_0(O, \overline{K}_c)$.

Let $\{P_n\}$ be the basic sequence corresponding to a delta operator $\delta = \tau^{-1}\delta_0$,

$$\delta_0 = \sum_{l=1}^{\infty} \sigma_l \Delta^{(l)}. \tag{6.10}$$

The sequence $Q_n = \dfrac{P_n}{D_n}$, $n = 0, 1, 2, \ldots$, called the normalized basic sequence, satisfies the identity

$$Q_i(st) = \sum_{n=0}^{i} Q_n(t) \{Q_{i-n}(s)\}^{q^n},$$

another form of the K-binomial property. Though it resembles its classical counterpart, the presence of the Frobenius powers is a feature specific for the case of a positive characteristic.

Theorem 6.3. *If $|\sigma_1| = 1$, $|\sigma_l| \le 1$ for $l \ge 2$, then the sequence $\{Q_n\}_0^\infty$ is an orthonormal basis of the space $C_0(O, \overline{K}_c)$ — for any $f \in C_0(O, \overline{K}_c)$ there is a uniformly convergent expansion*

$$f(t) = \sum_{n=0}^{\infty} \psi_n Q_n(t), \quad t \in O,$$

where $\psi_n = \left(\delta_0^{(n)} f\right)(1)$, $|\psi_n| \to 0$ as $n \to \infty$,

$$\|f\| = \sup_{n \ge 0} |\psi_n|.$$

By Theorem 6.3, the Laguerre-type polynomial sequence from Example 1 is an orthonormal basis of $C_0(O, \overline{K}_c)$. The sequence from Example 2 does not satisfy the conditions of Theorem 6.3.

Note that the conditions of Theorem 6.3 imply that $S_n \neq 0$ for all n, so that the series (6.10) considered in Theorem 6.3 always correspond to delta operators.

In [21] recursive formulas and generating functions for normalized basic sequences are also given.

7. The Weyl–Carlitz Ring and Holonomic Modules

7.1. The theory of holonomic modules over the Weyl algebra and more general algebras of differential or q-difference operators is becoming increasingly important, both as a crucial part of the general theory of D-modules and in view of various applications (see, for example, [4, 7, 32]). Usually, the holonomic property of the module corresponding to a system of differential equations is a sign of its "regular" behavior. Most of the classical special functions are associated (see [7]) with holonomic modules, which helps to investigate their properties.

It is clear from the above results that in the positive characteristic case a natural counterpart of the Weyl algebra is, for the case of a single variable, the ring $\mathfrak{A}_1$ generated by τ, d, and scalars from $\overline{K}_c$, with the relations

$$d\tau - \tau d = [1]^{1/q}, \quad \tau\lambda = \lambda^q\tau, \quad d\lambda = \lambda^{1/q}d \ (\lambda \in \overline{K}_c). \tag{7.1}$$

The ring consists of finite sums

$$a = \sum_{i,j} \lambda_{ij}\tau^i d^j, \quad \lambda_{ij} \in \overline{K}_c, \tag{7.2}$$

and the representation of an element in the form (7.2) is unique.

Basic algebraic properties of $\mathfrak{A}_1$ [19, 3] are similar to those of the Weyl algebra in characteristic 0 and quite different from the case of the algebra of usual differential operators over a field of positive characteristic [28].

The ring $\mathfrak{A}_1$ is left and right Noetherian, without zero divisors. $\mathfrak{A}_1$ possesses no non-trivial two-sided ideals stable with respect to the mapping

$$\sum_{i,j} \lambda_{ij}\tau^i d^j \mapsto \sum_{i,j} \lambda_{ij}^q \tau^i d^j.$$

The centre of $\mathfrak{A}_1$ is described explicitly in [3]; it contains countably many elements (this corrects an erroneous statement from [19]). In fact, $\mathfrak{A}_1$ belongs to the class of generalized Weyl algebras [2]. A well-developed theory available for them enabled Bavula [3] to classify ideals in $\mathfrak{A}_1$, as well as all simple modules over $\mathfrak{A}_1$.

A generalization of $\mathfrak{A}_1$ to the case of several variables is not straightforward because the Carlitz derivatives d_s and d_t do not commute on a monomial $f(s,t) = s^{q^m}t^{q^n}$, if $m \neq n$. Moreover, if $m > n$, then $d_s^m f$ is not a polynomial, nor even a holomorphic function in t (since the action of d is not linear and involves taking the q-th root).

A reasonable generalization is inspired by Zeilberger's idea (see [7]) to study holonomic properties of sequences of functions making a transform with respect to the discrete variables, which reduces the continuous-discrete case to the purely continuous one (simultaneously in all the variables). In our situation, if $\{P_k(s)\}$ is a sequence of $\mathbb{F}_q$-linear polynomials with $\deg P_k \leq q^k$, we set

$$f(s,t) = \sum_{k=0}^{\infty} P_k(s)t^{q^k}, \tag{7.3}$$

and d_s is well-defined. In the variable t, we consider not d_t but the linear operator Δ_t. The latter does not commute with d_s either, but satisfies the commutation relations

$$d_s\Delta_t - \Delta_t d_s = [1]^{1/q}d_s, \quad \Delta_t\tau - \tau\Delta_t = [1]\tau,$$

so that the resulting ring $\mathfrak{A}_2$ resembles a universal enveloping algebra of a solvable Lie algebra.

More generally, denote by $\mathcal{F}_{n+1}$ the set of all germs of functions of the form

$$f(s,t_1,\dots,t_n)=\sum_{k_1=0}^{\infty}\dots\sum_{k_n=0}^{\infty}\sum_{m=0}^{\min(k_1,\dots,k_n)} a_{m,k_1,\dots,k_n}s^{q^m}t_1^{q^{k_1}}\dots t_n^{q^{k_n}} \tag{7.4}$$

where $a_{m,k_1,\dots,k_n}\in\overline{K}_c$ are such that all the series are convergent on some neighbourhoods of the origin. We do not exclude the case $n=0$ where $\mathcal{F}_1$ will mean the set of all $\mathbb{F}_q$-linear power series $\sum\limits_m a_m s^{q^m}$ convergent on a neighbourhood of the origin. $\widehat{\mathcal{F}}_{n+1}$ will denote the set of all polynomials from $\mathcal{F}_{n+1}$, that is the series (7.4) in which only a finite number of coefficients is different from zero.

The ring $\mathfrak{A}_{n+1}$ is generated by the operators $\tau, d_s, \Delta_{t_1},\dots\Delta_{t_n}$ on $\mathcal{F}_{n+1}$, and the operators of multiplication by scalars from $\overline{K}_c$. To simplify the notation, we write Δ_j instead of Δ_{t_j} and identify a scalar $\lambda\in\overline{K}_c$ with the operator of multiplication by λ. The operators Δ_j are $\overline{K}_c$-linear, so that

$$\Delta_j\lambda=\lambda\Delta_j,\quad \lambda\in\overline{K}_c, \tag{7.5}$$

while the operators τ, d_s satisfy the commutation relations (7.1). In the action of each operator d_s,Δ_j (acting in a single variable), other variables are treated as scalars. The operator τ acts simultaneously on all the variables and coefficients. We have the relations involving Δ_j:

$$\Delta_j\tau-\tau\Delta_j=[1]\tau,\quad d_s\Delta_j-\Delta_j d_s=[1]^{1/q}d_s,\quad j=1,\dots,n. \tag{7.6}$$

Using the commutation relations (7.1), (7.5), and (7.6), we can write any $a\in\mathfrak{A}_{n+1}$, in a unique way, as a finite sum

$$a=\sum c_{l,\mu,i_1,\dots,i_n}\tau^l d_s^{\mu}\Delta_1^{i_1}\dots\Delta_n^{i_n}. \tag{7.7}$$

Let us introduce a filtration in $\mathfrak{A}_{n+1}$ (an analog of the Bernstein filtration) denoting by Γ_ν, $\nu\in\mathbb{Z}_+$, the $\overline{K}_c$-vector space of operators (7.7) with $\max\{l+\mu+i_1+\dots+i_n\}\le\nu$ where the maximum is taken over all the terms of (7.7). Then $\mathfrak{A}_{n+1}$ is a left and right Noetherian filtered ring.

In a standard way (see [10]) we define filtered left modules over $\mathfrak{A}_{n+1}$. All the basic notions regarding a filtered module M (like those of the graded module $\operatorname{gr}(M)$, dimension $d(M)$, multiplicity $m(M)$, good filtration etc) are introduced just as their counterparts in the theory of modules over the Weyl algebra.

If we consider $\mathfrak{A}_{n+1}$ as a left module over itself, then

$$d(\mathfrak{A}_{n+1})=n+2,\quad m(\mathfrak{A}_{n+1})=1. \tag{7.8}$$

For any finitely generated left $\mathfrak{A}_{n+1}$-module M, we have $d(M)\le n+2$. By (7.8), this bound cannot be improved in general. However, if I is a non-zero left ideal in $\mathfrak{A}_{n+1}$, then

$$d(\mathfrak{A}_{n+1}/I)\le n+1. \tag{7.9}$$

For the module $\widehat{\mathcal{F}}_{n+1}$ of $\mathbb{F}_q$-linear polynomials (7.4), we have

$$d\left(\widehat{\mathcal{F}}_{n+1}\right) = n+1, \quad m\left(\widehat{\mathcal{F}}_{n+1}\right) = n!$$

The proofs of all these results, as well as the ones given in this section below, can be found in [24].

It is natural to call an $\mathfrak{A}_{n+1}$-module M *holonomic* if $d(M) = n+1$. Thus, $\widehat{\mathcal{F}}_{n+1}$ is an example of a holonomic module.

The next theorem demonstrates, already for the case of $\mathfrak{A}_1$-modules, a sharp difference from the case of modules over the Weyl algebras. In particular, we see that an analog of the Bernstein inequality (see [10]) does not hold here without some additional assumptions.

Theorem 7.1. (i) *For any $k = 1, 2, \dots$, there exists such a nontrivial $\mathfrak{A}_1$-module M that* $\dim M = k$ *(*$\dim$ *means the dimension over $\overline{K}_c$), that is $d(M) = 0$.*

(ii) *Let M be a finitely generated $\mathfrak{A}_1$-module with a good filtration. Suppose that there exists a "vacuum vector" $v \in M$, such that $d_s v = 0$ and $\tau^m(v) \neq 0$ for all $m = 0, 1, 2, \dots$. Then $d(M) \geq 1$.*

7.2. Let us consider the case of holonomic submodules of the $\mathfrak{A}_{n+1}$-module $\mathcal{F}_{n+1}$, consisting of $\mathbb{F}_q$-linear functions (7.4) polynomial in s and holomorphic near the origin in $t_1, \dots, t_n$.

Let $0 \neq f \in \mathcal{F}_{n+1}$,

$$I_f = \{\varphi \in \mathfrak{A}_{n+1} : \varphi(f) = 0\}.$$

I_f is a left ideal in $\mathfrak{A}_{n+1}$. The left $\mathfrak{A}_{n+1}$-module $M_f = \mathfrak{A}_{n+1}/I_f$ is isomorphic to the submodule $\mathfrak{A}_{n+1} f \subset \mathcal{F}_{n+1}$ – an element $\varphi(f) \in \mathfrak{A}_{n+1} f$ corresponds to the class of $\varphi \in \mathfrak{A}_{n+1}$ in M_f. A natural good filtration in M_f is induced from that in $\mathfrak{A}_{n+1}$.

As we know (see (7.9)), if $I_f \neq \{0\}$, then $d(M_f) \leq n+1$. We call a function f *holonomic* if the module M_f is holonomic, that is $d(M_f) = n+1$. The condition $I_f \neq \{0\}$ means that f is a solution of a non-trivial "differential equation" $\varphi(f) = 0$, $\varphi \in \mathfrak{A}_{n+1}$. The case $n = 0$ is quite simple.

Theorem 7.2. *If a non-zero function $f \in \mathcal{F}_1$ satisfies an equation $\varphi(f) = 0$, $0 \neq \varphi \in \mathfrak{A}_1$, then f is holonomic.*

In particular, any $\mathbb{F}_q$-linear polynomial of s is holonomic, since it is annihilated by d_s^m, with a sufficiently large m.

If $n > 0$, the situation is more complicated. We call the module M_f (and the corresponding function f) *degenerate* if $D(M_f) < n+1$ (by the Bernstein inequality, there is no degeneracy phenomena for modules over the complex Weyl algebra). The simplest example of a degenerate function (for $n = 1$) is $f(s, t_1) = g(st_1) \in \mathcal{F}_2$ where the function g belongs to $\mathcal{F}_1$ and satisfies an equation $\varphi(g) = 0$, $\varphi \in \mathfrak{A}_1$. It can be shown that $d(M_f) = 1$.

In order to exclude the degenerate case, we introduce the notion of a non-sparse function.

A function $f \in \mathcal{F}_{n+1}$ of the form (7.4) is called *non-sparse* if there exists such a sequence $m_l \to \infty$ that, for any l, there exist sequences $k_1^{(i)}, k_2^{(i)}, \dots, k_n^{(i)} \geq m_l$ (depending on l), such that $k_\nu^{(i)} \to \infty$ as $i \to \infty$ $(\nu = 1, \dots, n)$, and $a_{m,k_1^{(i)},\dots,k_n^{(i)}} \neq 0$.

Theorem 7.3. *If a function f is non-sparse, then $d(M_f) \geq n+1$. If, in addition, f satisfies an equation $\varphi(f) = 0$, $0 \neq \varphi \in \mathfrak{A}_{n+1}$, then f is holonomic.*

7.3. We use Theorem 7.3 to prove that the functions (7.4) obtained via the sequence-to-function transform (7.3) or its multi-index generalizations, from some well-known sequences of polynomials over K are holonomic. In all the cases below the non-sparseness is evident, and we have only to prove that the corresponding function satisfies a non-trivial Carlitz differential equation.

a) *The Carlitz polynomials.* The transform (7.3) of the sequence $\{f_k\}$ is the Carlitz module function $C_s(t)$; see (2.3). It is easy to check that $d_s C_s(t) = C_s(t)$. Therefore the Carlitz module function is holonomic, jointly in both its variables.

b) *Thakur's hypergeometric polynomials.* We consider the polynomial case of Thakur's hypergeometric function (1.5), that is

$$ {}_lF_\lambda(-a_1, \dots, -a_l; -b_1, \dots, -b_\lambda; z) = \sum_m \frac{(-a_1)_m \dots (-a_l)_m}{(-b_1)_m \dots (-b_\lambda)_m D_m} z^{q^m} \tag{7.10} $$

where $a_1, \dots, a_l, b_1, \dots, b_\lambda \in \mathbb{Z}_+$. It is seen from (1.4) that the terms in (7.10), which make sense and do not vanish, are those with $m \leq \min(a_1, \dots, a_l, b_1, \dots, b_\lambda)$. Let the function $f \in \mathcal{F}_{l+\lambda+1}$ be given by

$$ \begin{aligned} & f(s, t_1, \dots, t_l, u_1, \dots, u_\lambda) \\ & \quad = \sum_{k_1=0}^\infty \dots \sum_{k_l=0}^\infty \sum_{\nu_1=0}^\infty \dots \sum_{\nu_\lambda=0}^\infty {}_lF_\lambda(-k_1, \dots, -k_l; -\nu_1, \dots, -\nu_\lambda; s) \\ & \qquad \times t_1^{q^{k_1}} \dots t_l^{q^{k_l}} u_1^{q^{\nu_1}} \dots u_\lambda^{q^{\nu_\lambda}}. \end{aligned} $$

It is known ([39], Sect. 6.5) that

$$ \begin{aligned} & d_s {}_lF_\lambda(-k_1, \dots, -k_l; -\nu_1, \dots, -\nu_\lambda; s) \\ & \quad = {}_lF_\lambda(-k_1+1, \dots, -k_l+1; -\nu_1+1, \dots, -\nu_\lambda+1; s) \end{aligned} \tag{7.11} $$

if all the parameters $k_1, \dots, k_l, \nu_1, \dots, \nu_\lambda$ are different from zero. If at least one of them is equal to zero, then the left-hand side of (7.11) equals zero. This property implies the identity $d_s f = f$, the same as that for the Carlitz module function. Thus, f is holonomic.

c) *K-binomial coefficients.* It can be shown [24] that the K-binomial coefficients $\binom{k}{m}_K$ (see Sect. 6) satisfy the Pascal-type identity

$$\binom{k}{m}_K = \binom{k-1}{m-1}_K^q + \binom{k-1}{m}_K^q D_m^{q-1} \tag{7.12}$$

where $0 \le m \le k$ and it is assumed that $\binom{k}{-1}_K = \binom{k-1}{k}_K = 0$.

Consider a function $f \in \mathcal{F}_2$ associated with the K-binomial coefficients, that is

$$f(s,t) = \sum_{k=0}^{\infty} \sum_{m=0}^{k} \binom{k}{m}_K s^{q^m} t^{q^k}. \tag{7.13}$$

The identity (7.12) implies the equation

$$d_s f(s,t) = \Delta_t f(s,t) + [1]^{1/q} f(s,t)$$

for the function (7.13). Therefore f is holonomic.

References

[1] F. Baldassarri, Differential modules and singular points of p-adic differential equations, *Adv. Math.* **44** (1982), 155–179.

[2] V. Bavula, Generalized Weyl algebras and their representations, *St. Petersburg Math. J.* **4**, No. 1 (1993), 71–92.

[3] V. Bavula, The Carlitz algebras, math.RA/0505397.

[4] Yu. Berest and A. Kasman, $\mathcal{D}$-modules and Darboux transformations, *Lett. Math. Phys.* **43** (1998), 279–294.

[5] L. Carlitz, On certain functions connected with polynomials in a Galois field, *Duke Math. J.* **1** (1935), 137–168.

[6] L. Carlitz, Some special functions over $GF(q,x)$, *Duke Math. J.* **27** (1960), 139–158.

[7] P. Cartier, Démonstration "automatique" d'identités et fonctions hypergéometriques (d'après D. Zeilberger), *Astérisque* **206** (1992), 41–91.

[8] D. N. Clark, A note on the p-adic convergence of solutions of linear differential equations, *Proc. Amer. Math. Soc.* **17** (1966), 262–269.

[9] K. Conrad, The digit principle, *J. Number Theory* **84** (2000), 230–237.

[10] S. C. Coutinho, *A Primer of Algebraic D-modules*, Cambridge University Press, 1995.

[11] B. Dwork, G. Gerotto, and F. J. Sullivan, *An Introduction to G-Functions*, Princeton University Press, 1994.

[12] L. Ferrari, An umbral calculus over infinite coefficient fields of positive characteristic, *Comp. Math. Appl.* **41** (2001), 1099–1108.

[13] D. Goss, Fourier series, measures, and divided power series in the theory of function fields, *K-Theory* **1** (1989), 533–555.

[14] D. Goss, *Basic Structures of Function Field Arithmetic*, Springer, Berlin, 1996.

[15] S. Jeong, Continuous linear endomorphisms and difference equations over the completions of $\mathbb{F}_q[T]$, *J. Number Theory* **84** (2000), 276–291.

[16] S. Jeong, Hyperdifferential operators and continuous functions on function fields, *J. Number Theory* **89** (2001), 165–178.

[17] A. N. Kochubei, Harmonic oscillator in characteristic p, *Lett. Math. Phys.* **45** (1998), 11–20.

[18] A. N. Kochubei, $\mathbb{F}_q$-linear calculus over function fields, *J. Number Theory* **76** (1999), 281–300.

[19] A. N. Kochubei, Differential equations for $\mathbb{F}_q$-linear functions, *J. Number Theory* **83** (2000), 137–154.

[20] A. N. Kochubei, Differential equations for $\mathbb{F}_q$-linear functions II: Regular singularity, *Finite Fields Appl.* **9** (2003), 250–266.

[21] A. N. Kochubei, Umbral calculus in positive characteristic, *Adv. Appl. Math.* **34** (2005), 175–191.

[22] A. N. Kochubei, Strongly nonlinear differential equations with Carlitz derivatives over a function field, *Ukrainian Math. J.* **57**, No. 5 (2005); math.NT/ 0405542.

[23] A. N. Kochubei, Polylogarithms and a zeta function for finite places of a function field, *Contemporary Math.* **384** (2005); math.NT/0405544.

[24] A. N. Kochubei, Holonomic modules in positive characteristic, math.RA/0503398.

[25] M. B. Nathanson, Additive number theory and the ring of quantum integers, math.NT/0204006.

[26] B. Poonen, Fractional power series and pairings on Drinfeld modules, *J. Amer. Math. Soc.* **9** (1996), 783–812.

[27] M. van der Put, Meromorphic differential equations over valued fields, *Indag. Math.* **42** (1980), 327–332.

[28] M. van der Put, Differential equations in characteristic p, *Compositio Math.* **97** (1995), 227–251.

[29] A. M. Robert, *A Course in p-Adic Analysis*, Springer, New York, 2000.

[30] S. Roman, *The Umbral Calculus*, Academic Press, London, 1984.

[31] G.-C. Rota, D. Kahaner and A. Odlyzko, On the foundations of combinatorial theory. VIII. Finite operator calculus, *J. Math. Anal. Appl.* **42** (1973), 684–760.

[32] C. Sabbah, Systèmes holonomes d'équations aux q-differences. In: *D-modules and Microlocal Geometry* (M. Kashiwara et al., eds.), Walter de Gruyter, Berlin, 1993, pp. 125–147.

[33] S. G. Samko, A. A. Kilbas, and O. I. Marichev, *Fractional Integrals and Derivatives: Theory and Applications*, Gordon and Breach, New York, 1993.

[34] W. Schikhof, *Ultrametric Calculus*, Cambridge University Press, 1984.

[35] M. Setoyanagi, Note on Clark's theorem for p-adic convergence, *Proc. Amer. Math. Soc.* **125** (1997), 717–721.

[36] D. Sinnou and D. Laurent, Indépendence algebrique sur les T-modules, *Compositio Math.* **122** (2000), 1–22.

[37] D. S. Thakur, Hypergeometric functions for function fields, *Finite Fields and Their Appl.* **1** (1995), 219–231.

[38] D. S. Thakur, Hypergeometric functions for function fields II, *J. Ramanujan Math. Soc.* **15** (2000), 43–52.

[39] D. S. Thakur, *Function Field Arithmetic*, World Scientific, Singapore, 2004.

[40] L. Van Hamme, Continuous operators, which commute with translations, on the space of continuous functions on $\mathbb{Z}_p$. In: *p-Adic Functional Analysis* (J. M. Bayod et al., eds.), Lect. Notes Pure Appl. Math. 137, Marcel Dekker, New York, 1992, pp. 75–88.

[41] A. Verdoodt, Umbral calculus in non-Archimedean analysis. In: *p-Adic Functional Analysis* (A. K. Katsaras et al., eds.), Lect. Notes Pure Appl. Math. 222, Marcel Dekker, New York, 2001, pp. 309–322.

[42] J. F. Voloch, Differential operators and interpolation series in power series fields, *J. Number Theory* **71** (1998), 106–108.

[43] C. G. Wagner, Linear operators in local fields of prime characteristic, *J. Reine Angew. Math.* **251** (1971), 153–160.

[44] Z. Yang, Locally analytic functions over completions of $\mathbf{F}_r(U)$, *J. Number Theory* **73** (1998), 451–458.

[45] Z. Yang, C^n-functions over completions of $\mathbb{F}_r[T]$ at finite places of $\mathbb{F}_r(T)$, *J. Number Theory* **108** (2004), 346–374.

Anatoly N. Kochubei
Institute of Mathematics,
National Academy of Sciences of Ukraine,
Tereshchenkivska 3
Kiev, 01601
Ukraine
e-mail: kochubei@i.com.ua

Progress in Mathematics, Vol. 260, 189–206

The Moduli Space of 5 Points on $\mathbb{P}^1$ and K3 Surfaces

Shigeyuki Kondō

Dedicated to Professor Yukihiko Namikawa on his 60th birthday

Abstract. We show that the moduli space of 5 ordered points on $\mathbb{P}^1$ is isomorphic to an arithmetic quotient of a complex ball by using the theory of periods of $K3$ surfaces. We also discuss a relation between our uniformization and the one given by Shimura [S], Terada [Te], Deligne–Mostow [DM].

Mathematics Subject Classification (2000). Primary 14J10; Secondary 14J28, 11F23, 33C80.

Keywords. Moduli, $K3$ surfaces, quartic del Pezzo surfaces, complex ball uniformization.

1. Introduction

The purpose of this note is to show that the moduli space of 5 ordered points on $\mathbb{P}^1$ is isomorphic to an arithmetic quotient of a 2-dimensional complex ball by using the theory of periods of $K3$ surfaces (Theorem 6.5). This was announced in [K2], Remark 6. The main idea is to associate a $K3$ surface with an automorphism of order 5 to a set of 5 ordered points on $\mathbb{P}^1$ (see §3). The period domain of such $K3$ surfaces is a 10-dimensional bounded symmetric domain of type IV. We remark that a non-zero holomorphic 2-form on the $K3$ surface is an eigen-vector of the automorphism, which implies that the period domain of the pairs of these $K3$ surfaces and the automorphism of order 5 is a 2-dimensional complex ball associated to a hermitian form of the signature $(1, 2)$ defined over $\mathbb{Z}[\zeta]$ where ζ is a primitive 5-th root of unity (see 6.3). Here we use several fundamental results of Nikulin [N1], [N2], [N3] on automorphisms of $K3$ surfaces and lattice theory. Note that this moduli space is isomorphic to the moduli space of nodal del Pezzo

Research of the author is partially supported by Grant-in-Aid for Scientific Research A-14204001, Japan.

surfaces of degree 4. For the moduli space of del Pezzo surfaces of degree 1, 2 or 3, the similar description holds. See [K2], Remark 5, [K1], [DGK], respectively.

On the other hand, Shimura [S], Terada [Te], Deligne–Mostow [DM] gave a complex ball uniformization by using the periods of the curve C which is the 5-fold cyclic covering of $\mathbb{P}^1$ branched along 5 points. We shall discuss a relation between their uniformization and ours in §7. In fact, the above $K3$ surface has an isotrivial pencil whose general member is the unique smooth curve D of genus 2 admitting an automorphism of order 5 (see Lemma 3.1). We show that the above $K3$ surface is birational to the quotient of $C \times D$ by a diagonal action of $\mathbb{Z}/5\mathbb{Z}$ in §7.

In this paper, a *lattice* means a $\mathbb{Z}$-valued non-degenerate symmetric bilinear form on a free $\mathbb{Z}$-module of finite rank. We denote by U or V the even lattice defined by the matrix $\begin{pmatrix} 0 & 1 \\ 1 & 0 \end{pmatrix}$, $\begin{pmatrix} 2 & 1 \\ 1 & -2 \end{pmatrix}$, respectively and by A_m, D_n or E_l the even negative definite lattice defined by the Dynkin matrix of type A_m, D_n or E_l respectively. If L is a lattice and m is an integer, we denote by $L(m)$ the lattice over the same $\mathbb{Z}$-module with the symmetric bilinear form multiplied by m. We also denote by $L^{\oplus m}$ the orthogonal direct sum of m copies of L, by L^* the dual of L and by A_L the finite abelian group L^*/L.

2. Quartic del Pezzo surfaces

2.1. Five points on $\mathbb{P}^1$

Consider the diagonal action of PGL(2) on $(\mathbb{P}^1)^5$. In this case, the semi-stable points and stable points in the sense of [Mu] coincide and the geometric quotient P_1^5 is smooth and compact. The stable points are $\{p_1, \dots, p_5\}$ no three of which coincide. It is known that P_1^5 is isomorphic to the quintic del Pezzo surface $\mathcal{D}_5$, that is, a smooth surface obtained by blowing up four points $\{q_1, \dots, q_4\}$ in general position on $\mathbb{P}^2$ (e.g. Dolgachev [D], Example 11.5). The quintic del Pezzo surface $\mathcal{D}_5$ contains 10 lines corresponding to the 4 exceptional curves over $q_1, \dots, q_4$ and the proper transforms of 6 lines through two points from $\{q_1, \dots, q_4\}$. These ten lines correspond to the locus consisting of $\{p_1, \dots, p_5\}$ with $p_i = p_j$ for some i, j. The group of automorphisms of $\mathcal{D}_5$ is isomorphic to the Weyl group $W(A_4) \simeq S_5$ which is induced from the natural action of S_5 on $(\mathbb{P}^1)^5$.

2.2. Quartic del Pezzo surfaces

Let S be a smooth quartic del Pezzo surface. It is known that S is a complete intersection of two quadrics in $\mathbb{P}^4$. Consider the pencil of quadrics whose base locus is S. Its discriminant is a union of distinct five points of $\mathbb{P}^1$. Conversely any distinct five points $(1 : \lambda_i)$ on $\mathbb{P}^1$, the intersection of quadrics

$$\sum_{i=1}^{5} z_i^2 = \sum_{i=1}^{5} \lambda_i z_i^2 = 0 \tag{2.1}$$

is a smooth quartic del Pezzo surface. Thus the moduli space of smooth quartic del Pezzo surfaces is isomprphic to $(((\mathbb{P}^1)^5 \setminus \Delta)/\mathrm{PGL}(2))/S_5$ where Δ is the locus consisting of points $(x_1, \ldots, x_5)$ with $x_i = x_j$ for some i, j $(i \neq j)$. If five points are not distinct, but stable, the equation (2.1) defines a quartic del Pezzo surface with a node. Thus P_1^5 is the coarse moduli space of nodal quartic del Pezzo surfaces.

3. K3 surfaces associated to five points on $\mathbb{P}^1$

3.1. A plane quintic curve

Let $\{p_1, \ldots, p_5\}$ be an ordered stable point in $(\mathbb{P}^1)^5$. It defines a homogenious polynomial $f_5(x_1, x_2)$ of degree 5. Let C be the plane quintic curve defined by

$$x_0^5 = f_5(x_1, x_2) = \prod_{i=1}^{5}(x_1 - \lambda_i x_2). \tag{3.1}$$

The projective transformation

$$g : (x_0 : x_1 : x_2) \longrightarrow (\zeta x_0 : x_1 : x_2) \tag{3.2}$$

acts on C as an automorphism of C of order 5 where ζ is a primitive 5-th root of unity. Let E_0, L_i $(1 \leq i \leq 5)$ be lines defined by

$$E_0 : x_0 = 0,$$

$$L_i : x_1 = \lambda_i x_2.$$

Note that all L_i are members of the pencil of lines through $(1 : 0 : 0)$ and L_i meets C at $(0 : \lambda_i : 1)$ with multiplicity 5.

3.2. K3 surfaces

Let X be the minimal resolution of the double cover of $\mathbb{P}^2$ branched along the sextic curve $E_0 + C$. Then X is a $K3$ surface. We denote by τ the covering transformation. The projective transformation g in (3.2) induces an automorphism σ of X of order 5. We denote by the same symbol E_0 the inverse image of E_0.

Case (i) Assume that the equation $f_5 = 0$ has no multiple roots. In this case there are 5 (-2)-curves, denoted by E_i $(1 \leq i \leq 5)$, obtained as exceptional curves of the minimal resolution of singularities of type A_1 corresponding to the intersection of C and E_0. The inverse image of L_i is the union of two smooth rational curves F_i, G_i such that F_i is tangent to G_i at one point. Let p, q be the inverse image of $(1 : 0 : 0)$. We may assume that all F_i (resp. G_i) are through p (resp. q). Obviously σ preserves each curve E_i, F_j, G_j $(0 \leq i \leq 5, 1 \leq j \leq 5)$ and τ preserves each E_i and $\tau(F_i) = G_i$.

Case (ii) If $f_5 = 0$ has a multiple root, then the double cover has a rational double point of type D_7. Hence X contains 7 smooth rational curves E'_j, $(1 \leq j \leq 7)$ whose dual graph is of type D_7. We assume that E'_1 meets E_0 and $\langle E'_1, E'_2 \rangle = \langle E'_2, E'_3 \rangle = \langle E'_3, E'_4 \rangle = \langle E'_4, E'_5 \rangle = \langle E'_5, E'_6 \rangle = \langle E'_5, E'_7 \rangle = 1$. If λ_i is a multiple root, then F_i and G_i are disjoint and each of them meets one componet of D_7, for example, F_i meets E'_6 and G_i meets E'_7.

3.3. A pencil of curves of genus two

The pencil of lines on $\mathbb{P}^2$ through $(1 : 0 : 0)$ gives a pencil of curves of genus two on X. Each member of this pencil is invariant under the action of the automorphism σ of order 5. Hence a general member is a smooth curve of genus two with an automorphism of order five. Such a curve is unique up to isomorphism and is given by

$$y^2 = x(x^5 + 1) \tag{3.3}$$

(see Bolza [Bol]). If λ_i is a simple root of the equation $f_5 = 0$, then the line L_i defines a singular member of this pencil consisting of three smooth rational curves $E_i + F_i + G_i$. We call this singular member a *singular member of type* I. If λ_i is a multiple root of $f_5 = 0$, then the line L_i defines a singular member consisting of nine smooth rational curves $E'_1, \ldots, E'_7, F_i, G_i$. We call this a *singular member of type* II. The two points p, q are the base points of the pencil. After blowing up at p, q, we have a base point free pencil of curves of genus two. The singular fibers of such pencils are completly classified by Namikawa and Ueno [NU]. The type I (resp. type II) corresponds to [IX-2] (resp. [IX-4]) in [NU]. We now conclude:

Lemma 3.1. *The pencil of lines on $\mathbb{P}^2$ through $(1 : 0 : 0)$ gives a pencil of curves of genus two on X. A general member is a smooth curve of genus two with an automorphism of order five. In case that $f_5 = 0$ has no multiple roots, it has five singular members of type I. In case that $f_5 = 0$ has a multiple root (resp. two multiple roots), it has three singular members of type I and one singular member of type II (resp. one of type I and two of type II).*

3.4. A 5-fold cyclic cover of $\mathbb{P}^1 \times \mathbb{P}^1$

The above $K3$ surface has an automorphism of order 5 by construction. This implies that X is obtained as a 5-fold cyclic cover of a rational surface. In this subsection we shall explain such a construction of X due to I. Dolgachev. We use the same notation as in 3.2, Case (i).

First blow up X at p, q and 5 points $F_i \cap G_i$ $(i = 1, 2, \ldots, 5)$, and then blow up at infinitely near points of $F_i \cap G_i$. Then we have a surface $\tilde{X}$ which contains the following curves: we have ten (-5)-curves $\tilde{F}_i, \tilde{G}_i$ which are the strictly transforms of F_i, G_i. Also we have five (-3)-curves $\tilde{E}_i$ which are the strictly transformation of E_i $(i = 1, \ldots, 5)$. We denote by H_i and H'_i the (-2)- and (-1)-exceptional curves over $F_i \cap G_i$ respectively. We also denote by H_p, H_q the exceptional curve over p, q respectively. Then the strict transform $\tilde{C}$ of C has the self intersection number 0.

We can construct $\tilde{X}$ as a 5-cyclic cover of the smooth quadric surface as follows. Let D be a divisor of $\mathbb{P}^1 \times \mathbb{P}^1$ defined by

$$D = 4(l_1 + \cdots + l_5) + m_1 + m_2 + 3m_3 \tag{3.4}$$

where $l_1, \ldots, l_5$ are the fibers of the first projection from $\mathbb{P}^1 \times \mathbb{P}^1$ over the five points determined by the polynomial $f_5(x_1, x_2)$ in (3.1), and m_1, m_2, m_3 are three fibers of the second projection which are unique up to projective transformations. Take the 5-cyclic cover of $\mathbb{P}^1 \times \mathbb{P}^1$ branched along D. Then taking the normalization and resolving the singularities we have $\tilde{X}$. Locally the singularities over the intersection points of l_i and m_1, m_2 are given by $z^5 = x^4 y$ and those over the intersection points of l_i and m_3 are given by $z^5 = x^4 y^3$. The exceptional curves over $l_i \cap m_1$, $l_i \cap m_2$ correspond to ten (-5)-curves $\tilde{F}_i, \tilde{G}_i$ and those over $l_i \cap m_3$ correspond to the sum of $(-3)-$ and (-2)-curves $\tilde{E}_i + H_i$. Note that the ruling of the first projection from $\mathbb{P}^1 \times \mathbb{P}^1$ gives a pencil of curves of genus 2 on $\tilde{X}$. l_i corresponds to H_i' and m_1, m_2 or m_3 corresponds to H_p, H_q or E_0 respectively. On the other hand, consider the involution of $\mathbb{P}^1 \times \mathbb{P}^1$ which changes m_1 and m_2, and fixes m_3. Let m_4 be the another fixed fiber of this involution. Then this involution induces an involution of $\tilde{X}$ which fixes the inverse image $\tilde{C}$ of m_4.

We can also write down the map from X to $\mathbb{P}^1 \times \mathbb{P}^1$ as follows (due to the referee). Consider an affine equation $y^2 = x_0(x_0^5 + f_5(x_1, 1))$ of X ($x_2 = 1$). Then the automorphism σ of order 5 acts on X as

$$\sigma(y, x_0, x_1) = (\zeta^3 y, \zeta x_0, x_1)$$

and the rational map from X to $\mathbb{P}^1 \times \mathbb{P}^1$ is given by

$$(y, x_0, x_1) \to ((x_0^3 : y), (x_1 : 1)).$$

4. Picard and transcendental lattices

In this section we shall study the Picard lattice and the transcendental lattice of $K3$ surfaces X given in 3.2. We denote by S_X the Picard lattice of X and by T_X the transcendental lattice of X.

4.1. The Picard lattice

Lemma 4.1. *Assume that $f_5 = 0$ has no multiple roots. Let S be the sublattice generated by E_0 and components of the singular members of the pencil in Lemma 3.1. Then rank$(S) = 10$ and det$(S) = 5^3$. Moreover if X is generic in the sense of moduli, then the Picard lattice $S_X = S$.*

Proof. First note that the dimension of P_1^5 is 2. On the other hand, X has an automorphism σ of order 5 induced from g given in (3.2) which acts non trivially on $H^0(X, \Omega^2)$. Nowhere vanishing holomorphic 2-forms are eigenvectors of σ^*. We can see that the dimension of the period domain is $(22 - \text{rank}(S_X))/(5-1)$ ([N2], Theorem 3.1. Also see the following section 6). Hence the local Torelli theorem implies that $\text{rank}(S_X) = 10$ for generic X. Let S_0 be the sublattice of S_X generated by $E_i, F_i, (1 \leq i \leq 5)$. Then a direct calculation shows that $\text{rank}(S_0) = 10$ and $\det(S_0) = \pm 5^5$. The first assertion now follows from the relations:

$$5E_0 = \sum_{i=1}^{5} (F_i - 2E_i),$$

$$G_i + F_i = 2E_0 + \sum_{j \neq 0,i} E_j.$$

Note that $S^*/S \simeq (\mathbb{Z}/5\mathbb{Z})^3$. Now assume that $\mathrm{rank}(S_X) = 10$. If $S_X \neq S$, then $S \subset S_X \subset S^*$ and hence there exists an algebraic cycle C not contained in S and satisfying

$$5C = \sum_{i=0}^{5} a_i E_i + \sum_{i=1}^{4} b_i F_i, \ a_i, b_i \in \mathbb{Z}.$$

By using the relations

$$\langle 5C, E_i \rangle \equiv 0 \pmod 5, \quad \langle 5C, F_i \rangle \equiv 0 \pmod 5,$$

we can easily show that

$$a_i \equiv 0 \pmod 5, \quad b_i \equiv 0 \pmod 5.$$

This is a contradiction. □

4.2. Discriminant quadratic forms

Let L be an even lattice. We denote by L^* the dual of L and put $A_L = L^*/L$. Let

$$q_L : A_L \to \mathbb{Q}/2\mathbb{Z}$$

be the discriminant quadratic form defined by

$$q_L(x \bmod L) = \langle x, x \rangle \bmod 2\mathbb{Z}$$

and

$$b_L : A_L \times A_L \to \mathbb{Q}/\mathbb{Z}$$

the discriminant bilinear form defined by

$$b_L(x \bmod L, y \bmod L) = \langle x, y \rangle \bmod \mathbb{Z}.$$

Let S be as in Lemma 4.1. Then A_S is isomorphic to $(\mathbb{Z}/5\mathbb{Z})^3$ generated by

$$\alpha = (E_1 + 2F_1 + 3F_2 + 4E_2)/5, \ \beta = (E_1 + 2F_1 + 3F_3 + 4E_3)/5,$$
$$\gamma = (E_1 + 2F_1 + 3F_4 + 4E_4)/5$$

with $q_S(\alpha) = q_S(\beta) = q_S(\gamma) = -4/5$, and $b_S(\alpha, \beta) = b_S(\beta, \gamma) = b_S(\gamma, \alpha) = 3/5$.

4.3. The transcendental lattice

Let T be the orthogonal complement of S in $H^2(X, \mathbb{Z})$. For generic X, T is isomorphic to the transcendental lattice T_X of X which consists of transcendental cycles, that is, cycles not perpendicular to holomorphic 2-forms on X.

Lemma 4.2. *Assume that f_5 has no multiple roots. Then*

$$S \simeq V \oplus A_4 \oplus A_4, \quad T \simeq U \oplus V \oplus A_4 \oplus A_4$$

where V or U is the lattice defined by the matrix $\begin{pmatrix} 2 & 1 \\ 1 & -2 \end{pmatrix}$, $\begin{pmatrix} 0 & 1 \\ 1 & 0 \end{pmatrix}$, *respectively.*

Proof. We can see that q_S and the discriminant quadratic form of $V \oplus A_4 \oplus A_4$ coincide. Also note that $q_T = -q_S$ (Nikulin [N1], Corollary 1.6.2). Now the assertion follows from Nikulin [N1], Theorem 1.14.2. □

Lemma 4.3. *Let S_i be the sublattice generated by E_0 and components of the singular members of the pencil in Lemma* 3.1 *where $i = 1$ or 2 is the number of multiple roots of $f_5 = 0$. Let T_i be the orthogonal complement of S_i in $H^2(X, \mathbb{Z})$. Then*

$$S_1 \simeq V \oplus E_8 \oplus A_4, \quad T_1 \simeq U \oplus V \oplus A_4,$$

$$S_2 \simeq V \oplus E_8 \oplus E_8, \quad T_2 \simeq U \oplus V.$$

Proof. The proof is similar to those of Lemmas 4.1, 4.2. □

4.4. The Kähler cone

Let S_X be the Picard lattice of X. Denote by $P(X)^+$ the connected component of the set $\{x \in S_X \otimes \mathbb{R} : \langle x, x \rangle > 0\}$ which contains an ample class. Let $\Delta(X)$ be the set of effective classes r with $r^2 = -2$. Let

$$C(X) = \{x \in P(X)^+ : \langle x, r \rangle > 0, r \in \Delta(X)\}$$

which is called the Kähler cone of X. It is known that $C(X) \cap S_X$ consists of ample classes. Let $W(X)$ be the subgroup of $O(S_X)$ generated by reflections defined by

$$s_r : x \rightarrow x + \langle x, r \rangle r, \quad r \in \Delta(X).$$

Note that the action of $W(X)$ on S_X can be extended to $H^2(X, \mathbb{Z})$ acting trivially on T_X because $r \in S_X = T_X^{\perp}$. The Kähler cone $C(X)$ is a fundamental domain of the action of $W(X)$ on $P(X)^+$.

5. Automorphisms

We use the same notation as in §3, 4. In this section we study the covering involution τ of X over $\mathbb{P}^2$ and the automorphism σ of X of order 5.

5.1. The automorphism of order 2

Lemma 5.1. *Let $\iota = \tau^*$. Then the invariant sublattice $M = H^2(X, \mathbb{Z})^{<\iota>}$ is generated by E_i $(0 \leq i \leq 5)$.*

Proof. Note that M is a 2-elementary lattice, that is, its discriminant group $A_M = M^*/M$ is a finite 2-elementary abelian group. Let r be the rank of M and let l be the number of minimal generator of $A_M \cong (\mathbb{Z}/2\mathbb{Z})^l$. The set of fixed points of τ is the union of C and E_0. It follows from Nikulin [N3], Theorem 4.2.2 that $(22 - r - l)/2 = g(C) = 6$ and the number of components of fixed points set of τ other than C is $(r - l)/2 = 1$. Hence $r = 6, l = 4$. On the other hand we can easily see that $\{E_i : 0 \leq i \leq 5\}$ generates a sublattice of M with rank 6 and discriminant 2^4. Now the assertion follows. □

Lemma 5.2. *Let N be the orthogonal complement of $M = H^2(X, \mathbb{Z})^{<\iota>}$ in S. Then N is generated by the classes of $F_i - G_i$ $(1 \leq i \leq 5)$ and contains no (-2)-vectors.*

Proof. Since $\tau(F_i) = G_i$, the classes of $F_i - G_i$ are contained in N. A direct calculation shows that their intersection matrix $(\langle F_i - G_i, F_j - G_j\rangle)_{1\leq i,j\leq 4}$ is

$$\begin{pmatrix} -8 & 2 & 2 & 2 \\ 2 & -8 & 2 & 2 \\ 2 & 2 & -8 & 2 \\ 2 & 2 & 2 & -8 \end{pmatrix}$$

whose discriminant is $\pm 2^4 \cdot 5^3$. On the other hand, N is the orthogonal complement of M in S, and M (resp . S) has the discriminant $\pm 2^4$ (resp. $\pm 5^3$). Hence the discriminant of N is $\pm 2^4 \cdot 5^3$. Therefore the first assertion follows. It follows from the above intersection matrix that N contains no (-2)-vectors. □

Lemma 5.3. *Let r be a (-2)-vector in $H^2(X,\mathbb{Z})$. Assume that $r \in M^\perp$ in $H^2(X,\mathbb{Z})$. Then $\langle r, \omega_X\rangle \neq 0$.*

Proof. Assume that $\langle r, \omega_X\rangle = 0$. Then r is represented by a divisor. By Riemann-Roch theorem, we may assume that r is effective. By assumption $\iota(r) = -r$. On the other hand the automorphism preserves effective divisors, which is a contradiction. □

Lemma 5.4. *Let $P(M)^+$ be the connected component of the set*

$$\{x \in M \otimes \mathbb{R} : \langle x, x\rangle > 0\}$$

which contains the class of C where C is the fixed curve of τ of genus 6. Put

$$C(M) = \{x \in P(M)^+ : \langle x, E_i\rangle > 0, i = 0, 1, \ldots, 5\}.$$

Let $W(M)$ be the subgroup generated by reflections associated with (-2)-vectors in M. Then $C(M)$ is a fundamental domain of the action of $W(M)$ on $P(M)^+$ and $O(M)/\{\pm 1\} \cdot W(M) \cong S_5$ where S_5 is the symmetry group of degree 5 which is the automorphism group of $C(M)$.

Proof. First consider the dual graph of $E_0, \ldots, E_5$. Note that any maximal extended Dynkin diagram in this dual graph is $\tilde{D}_4$ which has the maximal rank 4 $(= \mathrm{rank}(M) - 2)$. It follows from Vinberg [V], Theorem 2.6 that the group $W(M)$ is of finite index in $O(M)$ the orthogonal group of M. The assertion now follows from Vinberg [V], Lemma 2.4. □

Lemma 5.5. *Let $\tilde{W}(M)$ be the subgroup of $O(M)$ generated by all reflections associated with negative norm vectors in M. Then $C(M)$ is a fundamental domain of the action of $\tilde{W}(M)$ on $P(M)^+$. Moreover $\tilde{W}(M) = W(M) \cdot S_5$.*

Proof. First note that $W(M) \subset \tilde{W}(M) \subset O(M)$. Let $r = E_i - E_j$ $(1 \leq i < j \leq 5)$ which is a (-4)-vector in M. Since $\langle E_i - E_j, M\rangle \subset 2\mathbb{Z}$, the reflection defined by

$$s_r : x \to x + \langle x, r\rangle r/2$$

is contained in $O(M)$. These reflections generate S_5 acting on $C(M)$ as the automorphism group of $C(M)$. Now Lemma 5.4 implies that $O(M) = \{\pm 1\} \cdot \tilde{W}(M)$. □

Lemma 5.6. *Let $C(X)$ be the Kähler cone of X. Then*

$$C(M) = C(X) \cap P(M)^+.$$

Proof. Since the class of C is contained in the closure of $C(X)$, $C(X) \cap P(M)^+ \subset C(M)$, and hence it suffices to see that any face of $C(X)$ does not cut $C(M)$ along proper interior points of $C(M)$. Let r be the class of an effective cycle with $r^2 = -2$. If $r \in M$, Lemma 5.4 implies the assertion. Now assume $\iota(r) \neq r$. Then $r = (r + \iota(r))/2 + (r - \iota(r))/2$. By Hodge index theorem, $(r - \iota(r))^2 < 0$. Since $r^2 = -2$, this implies that $((r + \iota(r))/2)^2 \geq 0$ or -1. If $((r + \iota(r))/2)^2 \geq 0$, again by Hodge index theorem, $\langle x, r + \iota(r) \rangle > 0$ for any $x \in C(M)$. Since ι acts trivially on M, we have $\langle x, r \rangle > 0$. If $((r + \iota(r))/2)^2 = -1$, then $\langle r, \iota(r) \rangle = 0$. Note that for any $x \in M$, $\langle x, r + \iota(r) \rangle = 2\langle x, r \rangle \in 2\mathbb{Z}$. Hence the (-4)-vector $r + \iota(r)$ defines a reflection in $\tilde{W}(M)$. It follows from Lemma 5.5 that $\langle r + \iota(r), x \rangle > 0$ for any $x \in C(M)$. Since ι acts trivially on M, $\langle r, x \rangle > 0$ for any $x \in C(M)$. Thus we have proved the assertion. □

5.2. An isometry of order five

Let σ be the automorphism of X of order 5 induced by the automorphism given in 3.2. In the following Lemma 5.7 we shall show that $\sigma^* \mid T$ is conjugate to the isometry ρ defined as follows:

Let e, f be a basis of $U = \begin{pmatrix} 0 & 1 \\ 1 & 0 \end{pmatrix}$ satisfying $e^2 = f^2 = 0, \langle e, f \rangle = 1$. Let x, y be a basis of $V = \begin{pmatrix} 2 & 1 \\ 1 & -2 \end{pmatrix}$ satisfying $x^2 = -y^2 = 2, \langle x, y \rangle = 1$, and let e_1, e_2, e_3, e_4 be a basis of A_4 so that $e_i^2 = -2, \langle e_i, e_{i+1} \rangle = 1$ and other e_i and e_j are orthogonal.

Let ρ_0 be an isometry of $U \oplus V$ defined by

$$\rho_0(e) = -f, \quad \rho_0(f) = -e - f - y, \quad \rho_0(x) = f - x, \quad \rho_0(y) = 3f - x + y. \quad (5.1)$$

Also let ρ_4 be an isometry of A_4 defined by

$$\rho_4(e_1) = e_2, \quad \rho_4(e_2) = e_3, \quad \rho_4(e_3) = e_4, \quad \rho_4(e_4) = -(e_1 + e_2 + e_3 + e_4). \quad (5.2)$$

Combining ρ_0 and ρ_4, we define an isometry ρ of $T = U \oplus V \oplus A_4 \oplus A_4$. By definition, ρ is of order 5 and has no non-zero fixed vectors in T. Moreover the action of ρ on the discriminant group T^*/T is trivial. Hence ρ can be extended to an isometry ρ (we use the same symbol) of $H^2(X, \mathbb{Z})$ acting trivially on S (Nikulin [N1], Corollary 1.5.2).

Lemma 5.7. *The isometry σ^* is conjugate to ρ.*

Proof. By the surjectivity of the period map of $K3$ surfaces, there exists a $K3$ surface X' whose transcendental lattice $T_{X'}$ is isomorphic to T. Moreover we may assume that $\omega_{X'}$ is an eigenvector of ρ under the isomorphism $T_{X'} \cong T$. Since ρ acts trivially on $S_{X'}$, there exists an automorphism σ' of X' with $(\sigma')^* = \rho$ ([PS]).

Since $S_{X'} \cong S$, there exist 16 (-2)-classes in $S_{X'}$ whose dual graph coincides with that of E_i, $(0 \leq i \leq 5)$, F_j, G_k $(1 \leq j, k \leq 5)$ on X in 3.2. We denote by E'_i, F'_j, G'_k these divisors corresponding to E_i, F_j, G_k. We shall show that if necessary by changing them by $w(E'_i), w(F'_j), w(G'_k)$ for a suitable $w \in W(X')$, all E'_i, F'_j, G'_k are smooth rational curves. Consider the divisor $D = 2E'_0 + E'_1 + E'_2 + E'_3 + E'_4$. Obviously $D^2 = 0$. If necessary, by replacing D by $w(D)$ where $w \in W(X')$, we may assume that D defines an elliptic fibration. Then D is a singular fiber of type I_0^* and E'_i $(0 \leq i \leq 4)$ are components of singular fibers. Thus we may assume that E'_i $(0 \leq i \leq 4)$ are smooth rational curves. Next consider the divisor $D' = 2E'_0 + E'_1 + E'_2 + E'_3 + E'_5$. By replacing D' by $w(D')$, $w \in W(X')$ with $w(E'_i) = E'_i, 0 \leq i \leq 4$, D' defines an elliptic fibration. Thus we may assume that all E'_i are smooth rational curves. Since $\mid F'_i + G'_i \mid = \mid 2E'_0 + E'_1 + E'_2 + E'_3 + E'_4 + E'_5 - E'_i \mid$, all F'_i, G'_i are also smooth rational curves.

Next we shall show that the incidence relation of E'_i, F'_j, G'_k is the same as that of E_i, F_j, G_k. Obviously E'_0 is pointwisely fixed by σ'. Recall that σ' acts on $H^0(X', \Omega_{X'})$ non trivially. By considering the action of σ' on the tangent space of $E'_i \cap E'_0$, σ' acts on E'_i non trivially. Now consider the elliptic fibration defined by the linear system $\mid 2E'_0 + E'_1 + E'_2 + E'_3 + E'_4 + E'_5 - E'_i \mid$ with sections F'_j, G'_j $(j \neq i)$. Since no elliptic curves have an automorphism of order 5, σ' acts on the sections F'_j and G'_j non trivially. Note that σ' has exactly two fixed points on each of E'_i, F'_j, G'_k, $(1 \leq i, j, k \leq 5)$. Hence F'_i and G'_i meets at one point with multiplicity 2. Now we can easily see that not only the dual graph, but also the incident relation of E'_i, F'_j, G'_k coincides with that of E_i, F_j, G_k.

Finally define the isometry ι' of order 2 of S_Y by $\iota'(F'_i) = G'_i$ $(1 \leq i \leq 5)$ and $\iota'(E'_i) = E'_i$. Then ι' can be extended to an isometry of $H^2(X', \mathbb{Z})$ acting on $T_{X'}$ as $-1_{T_{X'}}$ as ι. By definition of ι', it preserves $C(M)$, and hence preserves the Kähler cone $C(X)$ (Lemma 5.5). By the Torelli theorem, there exists an automorphism τ' with $(\tau')^* = \iota'$. It follows from Nikulin [N3], Theorem 4.2.2 that the set of fixed points of τ' is the disjoint union of E_0 and a smooth curve of genus 6. By taking the quotient of X' by τ', we have the same configuration as in 3.2. Thus X' can be deformed to X smoothly and hence σ^* is conjugate to ρ. □

Lemma 5.8. *Let $e \in T$ with $e^2 = 0$. Let K be the sublattice generated by $\rho^i(e)$ $(0 \leq i \leq 4)$. Then K contains a vector with positive norm.*

Proof. First note that e, $\rho(e)$, $\rho^2(e)$ are linearly independent isotoropic vectors. Since the signature of T is $(2, 8)$, we may assume that $\langle e, \pm\rho(e)\rangle > 0$. Then $e \pm \rho(e)$ is a desired one. □

Lemma 5.9. *Let $r \in T$ with $r^2 = -2$. Let R be the lattice generated by $\rho^i(r)$ $(0 \leq i \leq 4)$. Assume that R is negative definite. Then R is isometric to the root lattice A_4.*

Proof. Put $m_i = \langle r, \rho^i(r)\rangle$, $1 \leq i \leq 4$. Then by assumption $\mid m_i \mid \leq 1$. Also obviously $m_1 = m_4, m_2 = m_3$ and $\sum_{i=0}^{4} \rho^i(r) = 0$. Then

$$-2 = r^2 = \langle r, -\sum_{i=1}^{4} \rho^i(r)\rangle = -2m_1 - 2m_2.$$

Hence $(m_1, m_2) = (1, 0)$ or $(0, 1)$. Therefore $\{\rho^i(r) : 0 \leq i \leq 3\}$ is a basis of the root lattice A_4. □

Lemma 5.10. *Let $R \cong A_4$ be a sublattice of T. Assume that R is invariant under the action of ρ. Then the orthogonal complement $R^\perp$ of R in T is isomorphic to $U \oplus V \oplus A_4$.*

Proof. Let T' be the orthogonal complement of R in T. Then $T = R \oplus T'$ or T contains $R \oplus T'$ as a sublattice of index 5. We shall show that the second case does not occur. Assume that $[T : R \oplus T'] = 5$. Then $A_{T'} = (T')^*/T' \cong (\mathbb{F}_5)^{\oplus 4}$ because $| A_T | \cdot [T : R \oplus T']^2 = | A_R | \cdot | A_{T'} |$. Let ρ' be an isometry of L so that $\rho' \mid T' = \rho \mid T'$ and $\rho' \mid (T')^\perp = 1$. The existence of such ρ' follows from [N1], Corollary 1.5.2. It follows from the surjectivity of the period map of $K3$ surfaces that there exists a $K3$ surface Y whose transcendental lattice isomorphic to T' and whose period is an eigen-vector of ρ' under a suitable marking. Since ρ' acts trivially on the Picard lattice $(T')^\perp$ of Y, ρ' is induced from an automorphism σ' of Y. It follows from Vorontsov's theorem [Vo] that the number of minimal generator of $A_{T'}$ is at most $\mathrm{rank}(T')/\varphi(5) = 2$ where φ is the Euler function. This contradicts the fact $A_{T'} \cong (\mathbb{F}_5)^{\oplus 4}$. Thus we have proved that $T = R \oplus T'$. Since $q_{T'} \cong q_{U \oplus V \oplus A_4}$, the assertion now follows from Nikulin [N1], Theorem 1.14.2. □

5.3. Discriminant locus

Let $r \in T$ with $r^2 = -2$. Let R be the sublattice generated by $\rho^i(r)$ $(0 \leq i \leq 4)$. Assume that R is negative definite. Then $R \cong A_4$ and the orthogonal complement of R in T is isomorphic to $T' = U \oplus V \oplus A_4$ (Lemmas 5.9, 5.10). Let ρ' be an isometry of L so that $\rho' \mid T' = \rho \mid T'$ and $\rho' \mid (T')^\perp = 1$. Then there exists an $K3$ surface Y and an automorphism σ' such that the transcendental lattice of Y is isomorphic to T', the period of Y is an eigen-vector of ρ' and σ' acts trivially on the Picard lattice of Y. By the same argument as in the proof of Lemma 5.7, we can see that Y is corresponding to the case that $f_5 = 0$ has a multiple root in 3.2.

6. A complex ball uniformization

6.1. Hermitian form

Let $\zeta = e^{4\pi\sqrt{-1}/5}$. We consider T as a free $\mathbb{Z}[\zeta]$-module Λ by

$$(a + b\zeta)x = ax + b\rho(x).$$

Let

$$h(x, y) = \frac{2}{5 + \sqrt{5}} \sum_{i=0}^{4} \zeta^i \langle x, \rho^i(y) \rangle.$$

Then $h(x, y)$ is a hermitian form on $\mathbb{Z}[\zeta]$-module Λ. With respect to a $\mathbb{Z}[\zeta]$-basis e_1 of A_4, the hermitian matrix of $h \mid A_4$ is given by -1. And with respect to a

$\mathbb{Z}[\zeta]$-basis e of $U \oplus V$, the hermitian matrix of $h \mid U \oplus V$ is given by $(\sqrt{5}-1)/2$. Thus h is given by

$$\begin{pmatrix} \frac{\sqrt{5}-1}{2} & 0 & 0 \\ 0 & -1 & 0 \\ 0 & 0 & -1 \end{pmatrix}. \tag{6.1}$$

Let

$$\varphi : \Lambda \to T^*$$

be a linear map defined by

$$\varphi(x) = \sum_{i=0}^{3} (i+1)\rho^i(x)/5.$$

Note that $\varphi((1-\zeta)x) = \varphi(x - \rho(x)) = -\rho^4(x) \in T$. Hence φ induces an isomorphism

$$\Lambda/(1-\zeta)\Lambda \simeq A_T = T^*/T. \tag{6.2}$$

6.2. Reflections

Let $a \in \Lambda$ with $h(a,a) = -1$. Then the map

$$R_a^{\pm} : v \to v - (-1 \pm \zeta)h(v,a)a$$

is an automorphism of Λ. This automorphism R_a^+ has order 5 and R_a^- has order 10 both of which fix the orthogonal complement of a. They are called reflections. Consider a decomposition

$$T = U \oplus V \oplus A_4 \oplus A_4.$$

If $a = e_1$ of the last component A_4 as in (5.2), we can easily see that

$$R_a^{\pm} = \pm s_{e_1} \circ s_{e_2} \circ s_{e_3} \circ s_{e_4}$$

where s_{e_i} is a reflection in $O(T)$ associated with (-2)-vector e_i defined by

$$s_{e_i} : x \to x + \langle x, e_i \rangle e_i.$$

In other words,

$$R_{e_1}^{\pm} = 1_U \oplus 1_V \oplus 1_{A_4} \oplus (\pm\rho_4).$$

Since s_{e_i} acts trivially on A_T, R_a^+ acts trivially on $A_T \simeq \Lambda/(1-\zeta)\Lambda$. On the other hand, R_a^- acts on A_T as a reflection associated with $\alpha = (e_1 + 2e_2 + 3e_3 + 4e_4)/5 \in A_T$.

6.3. The period domain and arithmetic subgroups

We use the same notation as in 5.2. Let

$$T \otimes \mathbb{C} = T_\zeta \oplus T_{\zeta^2} \oplus T_{\zeta^3} \oplus T_{\zeta^4}$$

be the decomposition of ρ-eigenspaces where ζ is a primitive 5-th root of unity (see Nikulin [N2], Theorem 3.1). An easy calculation shows that

$$\xi = e_1 + (\zeta^4 + 1)e_2 + (-\zeta - \zeta^2)e_3 - \zeta e_4$$

is an eigenvector of ρ_4 with the eigenvalue ζ and

$$\langle \xi, \bar{\xi} \rangle = -5.$$

On the other hand,

$$\mu = e - (\zeta^4 + 1)f + (-\zeta - \zeta^2)(e + f + y) - \zeta(-e + f - x)$$

is an eigenvector of ρ_0 with the eigenvalue ζ and

$$\langle \mu, \bar{\mu} \rangle = 5(\zeta^2 + \zeta^3).$$

Thus if $\zeta = e^{\pm 4\pi\sqrt{-1}/5}$, the hermitian form $\langle \omega, \bar{\omega} \rangle/5$ on T_ζ is of signature $(1, 2)$ and is given by

$$\begin{pmatrix} \frac{\sqrt{5}-1}{2} & 0 & 0 \\ 0 & -1 & 0 \\ 0 & 0 & -1 \end{pmatrix}. \tag{6.3}$$

For other ζ, the hermitian form is negative definite. Now we take $\zeta = e^{4\pi\sqrt{-1}/5}$ and define

$$\mathcal{B} = \{z \in \mathbb{P}(T_\zeta) : \langle z, \bar{z} \rangle > 0\}. \tag{6.4}$$

Then $\mathcal{B}$ is a 2-dimensional complex ball. For a (-2)-vector r in T, we define

$$\mathcal{H}_r = r^\perp \bigcap \mathcal{B}, \quad \mathcal{H} = \bigcup_r \mathcal{H}_r$$

where r runs over (-2)-vectors in T. Let

$$\Gamma = \{\phi \in O(T) : \phi \circ \rho = \rho \circ \phi\}, \quad \Gamma' = \{\phi \in \Gamma : \phi \mid A_T = 1\}. \tag{6.5}$$

Remark 6.1. The hermitian form h in (6.1) coincides with the one of Shimura [S], Yamazaki and Yoshida [YY]. This and the isomorphism (6.2) imply that our groups Γ, Γ' coincide with Γ, $\Gamma(1-\mu)$ in Yamazaki and Yoshida [YY].

Proposition 6.2. (1) Γ *is generated by reflections* R_a^- *with* $h(a,a) = -1$ *and* Γ' *is generated by* R_a^+ *with* $h(a,a) = -1$. *The quotient* Γ/Γ' *is isomorphic to* $O(3, \mathbb{F}_5) \simeq \mathbb{Z}/2\mathbb{Z} \times S_5$.

(2) $\mathcal{B}/\Gamma'$ *is isomorphic to the quintic del Pezzo surface and* $\mathcal{H}/\Gamma'$ *consists of* 10 *smooth rational curves corresponding to* 10 *lines on the quintic del Pezzo surface. The factor* $\mathbb{Z}/2\mathbb{Z}$ *in* Γ/Γ' *acts trivially on* $\mathcal{B}$ *and* S_5 *corresponds to the group of automorphisms of the quintic del Pezzo surface.*

Proof. The assertions follow from the above Remark 6.1 and Propositions 4.2, 4.3, 4.4 in Yamazaki and Yoshida [YY]. □

6.4. Discriminant quadratic forms and discriminant locus

Let

$$q_T : A_T \to \mathbb{Q}/2\mathbb{Z}$$

be the discriminant quadratic form of T. The discriminant group A_T consists of the following 125 vectors:

Type (00) : $\alpha = 0,\ \#\alpha = 1$;
Type (0) : $\alpha \neq 0,\ q_T(\alpha) = 0,\ \#\alpha = 24$;
Type (2/5) : $q_T(\alpha) = 2/5,\ \#\alpha = 30$;
Type (−2/5) : $q_T(\alpha) = -2/5,\ \#\alpha = 30$;
Type (4/5) : $q_T(\alpha) = 4/5,\ \#\alpha = 20$;
Type (−4/5) : $q_T(\alpha) = -4/5,\ \#\alpha = 20$.

Let A_4 be a component of T with a basis e_1, e_2, e_3, e_4 as in 5.2. Then $(e_1 + 2e_2 + 3e_3 + 4e_4)/5 = (e_1 2\rho(e_1) + 3\rho^2(e_1) + 4\rho^3(e_1))/4 \bmod T$ is a vector in A_T with norm $-4/5$. It follows from Proposition 6.2 that Γ/Γ' acts transitively on the set of $(-4/5)$-vectors in A_T. Hence for each $\alpha \in A_T$ with $q_T(\alpha) = -4/5$ there exists a vector $r \in T$ with $r^2 = -2$ satisfying $\alpha = (r+2\rho(r)+3\rho^2(r)+4\rho^3(r))/5$. Moreover $\pm\alpha$ defines

$$\mathcal{H}_\alpha = \bigcup_r H_r$$

where r moves over the set

$$\{r \in T : r^2 = -2, \alpha = (r + 2\rho(r) + 3\rho^2(r) + 4\rho^3(r))/5 \bmod T\}.$$

Thus the set

$$\{\alpha \in A_T : q_T(\alpha) = -4/5\}/\pm 1$$

bijectively corresponds to the set of components of $\mathcal{H}/\Gamma'$. Let

$$\tilde{\Gamma} = \{\tilde{\phi} \in O(L) : \tilde{\phi} \circ \rho = \rho \circ \tilde{\phi}\}.$$

Lemma 6.3. *The restriction map $\tilde{\Gamma} \to \Gamma$ is surjective.*

Proof. We use the same notation as in 3.2. The symmetry group S_5 of degree 5 naturally acts on the set $\{E_1, \ldots, E_5\}$ as permutations. This action can be extended to the one on S. Together with the action of ι, the natural map

$$O(S) \to O(q_S) \cong \{\pm 1\} \times S_5$$

is surjective. Let $g \in \Gamma$. Then the above implies that there exists an isometry g' in $O(S)$ whose action on $A_S \cong A_T$ coincides with the one of g on A_T. Then it follows from Nikulin [N1], Proposition 1.6.1 that the isometry (g', g) of $S \oplus T$ can be extended to an isometry in $\tilde{\Gamma}$ which is the desired one. □

6.5. Period map

We shall define an S_5-equivariant map

$$p : P_1^5 \to \mathcal{B}/\Gamma'$$

called the *period map*. Denote by $(P_1^5)^0$ the locus of five distinct ordered points on $\mathbb{P}^1$. Let $\{p_1, \ldots, p_5\} \in (P_1^5)^0$. Let X be the corresponding $K3$ surface with the automorphism σ of order 5 as in 3.2. The order of $\{p_1, \ldots, p_5\}$ defines an order of smooth rational curves

$$E_i, \quad (0 \leq i \leq 5) \quad F_j, G_j, \quad (1 \leq j \leq 5)$$

modulo the action of the covering involution ι. It follows from Lemma 5.7 that there exists an isometry

$$\alpha : L \to H^2(X, \mathbb{Z})$$

satisfying $\alpha \circ \rho = \sigma^* \circ \alpha$. Now we define

$$p(X, \alpha) = (\alpha \otimes \mathbb{C})^{-1}(\omega_X).$$

Lemma 6.4. $p(X, \alpha) \in \mathcal{B} \setminus \mathcal{H}$.

Proof. If not, there exists a vector $r \in T$ with $r^2 = -2$ which is represented by an effective divisor on X as in the proof of Lemma 5.3. Obviously $r + \sigma^*(r) + \cdots + (\sigma^*)^4(r) = 0$. On the other hand $r + \sigma^*(r) + \cdots + (\sigma^*)^4(r)$ is non-zero effective because σ is an automorphism. Thus we have a contradiction. □

Thue we have a holomorphic map

$$p : (P_1^5)^0 \to (\mathcal{B} \setminus \mathcal{H})/\Gamma'.$$

The group S_5 naturally acts on P_1^5 which induces an action on S as permutations of $E_1, \ldots, E_5$. On the other hand, $S_5 \cong \Gamma/\{\pm 1\} \cdot \Gamma'$ naturally acts on $\mathcal{B}/\Gamma'$. Under the natural isomorphism $O(q_S) \cong O(q_T) \cong \{\pm 1\} \cdot S_5$, p is equivariant under these actions of S_5.

It is known that the quotient $\mathcal{B}/\Gamma'$ is compact (see Shimura [S]). We remark that cusps of $\mathcal{B}$ correspond to totally isotropic sublattices of T invariant under ρ. Hence the compactness also follows from Lemma 5.8.

Theorem 6.5 (Main theorem). *The period map p can be extended to an S_5-equivariant isomorphism*

$$\tilde{p} : P_1^5 \to \mathcal{B}/\Gamma'.$$

Proof. Let $\mathcal{M}$ be the space of all 5 stable points on $\mathbb{P}^1$ and $\mathcal{M}_0$ the space of all distinct 5 points on $\mathbb{P}^1$. We can easily see that $\mathcal{M} \setminus \mathcal{M}_0$ is locally contained in a divisor with normal crossing. By construction, p is locally liftable to $\mathcal{B}$. It now follows from a theorem of Borel [Borel] that p can be extended to a holomorphic map from $\mathcal{M}$ to $\mathcal{B}/\Gamma'$ which induces a holomorphic map $\tilde{p}$ from P_1^5 to $\mathcal{B}/\Gamma'$. Next we shall show the injectivity of the period map over $(\mathcal{B} \setminus \mathcal{H})/\Gamma'$. Let C, C' be two plane quintic curves as in 3.1. Let (X, α) (resp. (X', α')) be the associated marked

$K3$ surfaces with automorphisms τ, σ (resp. τ', σ'). Assume that the periods of (X, α) and (X', α') coincide in $\mathcal{B}/\Gamma'$. Then there exists an isometry

$$\varphi : H^2(X', \mathbb{Z}) \to H^2(X, \mathbb{Z})$$

preserving the periods and satisfying $\varphi \circ (\tau')^* = \tau^* \circ \varphi$ and $\varphi \circ (\sigma')^* = \sigma^* \circ \varphi$ (Lemma 6.3). It follows from Lemma 5.4 that φ preserves the Kähler cones. The Torelli theorem for $K3$ surfaces implies that there exists an isomorphism $f : X \to X'$ with $f^* = \varphi$. Then f induces an isomorphism between the corresponding plane quintic curves C and C'. Thus we have proved the injectivity of the period map.

Since both P_1^5 and $\mathcal{B}/\Gamma'$ are compact, $\tilde{p}$ is surjective. Recall that both $P_1^5 \setminus (P_1^5)^0$ and $\mathcal{H}/\Gamma'$ consist of 10 smooth rational curves. The surjectivity of $\tilde{p}$ implies that no components of $P_1^5 \setminus (P_1^5)^0$ contract to a point. Now the Zariski main theorem implies that $\tilde{p}$ is isomorphic. By construction, $\tilde{p}$ is S_5-equivariant over the Zariski open set $(P_1^5)^0$. Hence $\tilde{p}$ is S_5-equivariant isomorphism between P_1^5 and $\mathcal{B}/\Gamma'$. □

7. Shimura–Terada–Deligne–Mostow's reflection groups

The plane quintic curve C defined by (3.1) appeared in the papers of Shimura [S], Terada [Te], Deligne–Mostow [DM], and the moduli space of these curves has a complex ball uniformization. As we remarked, the hermitian form (6.1) coincides with those of Shimura [S], Terada [Te], Deligne–Mostow [DM] (see Remark 6.1). This implies

Theorem 7.1. *The arithmetic subgroup* Γ *is the one appeared in Deligne–Mostow's list* [DM]:

$$\left(\tfrac{2}{5}, \ \tfrac{2}{5}, \ \tfrac{2}{5}, \ \tfrac{2}{5}, \ \tfrac{2}{5}\right).$$

A geometric meaning of this theorem is as follows. Recall that X has an *isotrivial* pencil of curves of genus two whose general member is the smooth curve D of genus two with an automorphism of order 5 given by the equation (3.3). The X is given by

$$s^2 = x_0(x_0^5 - f_5(x_1, x_2))$$

where x_1/x_2 is the parameter of this pencil. On the other hand, we consider C as a base change $C \to \mathbb{P}^1$ given by

$$(v, x_1, x_2) \to (x_1, x_2).$$

Then over C, $v^5 = f_5(x_1, x_2)$ and hence the pencil is given by

$$s^2 = x_0(x_0^5 + v^5)$$

which is nothing but the equation of the curve D. Thus the $K3$ surface X is birational to the quotient of $C \times D$ by an diagonal action of $\mathbb{Z}/5\mathbb{Z}$. This correspondence gives a relation between the Hodge structures of C and X.

7.1. Problem

Let μ_i be a positive rational number ($0 \leq i \leq d+1$ or $i = \infty$) satisfying $\sum_i \mu_i = 2$. Set

$$F_{gh}(x_2, \ldots, x_{d+1}) = \int_g^h u^{-\mu_0}(u-1)^{-\mu_1} \prod_{i=2}^{d+1} (u - x_i)^{-\mu_i} du$$

where $g, h \in \{\infty, 0, 1, x_2, \ldots, x_{d+1}\}$. Then F_{gh} is a multivalued function on

$$M = \{(x_i) \in (\mathbb{P}^1)^{d+3} \mid x_i \neq \infty, 0, 1,\ x_i \neq x_j\ (i \neq j)\}.$$

These functions generate a $(d+1)$-dimensional vector space which is invariant under monodromy. Let $\Gamma_{(\mu_i)}$ be the image of $\pi_1(M)$ in $\mathrm{PGL}(d+1, \mathbb{C})$ under the monodromy action. In Deligne–Mostow [DM] and Mostow [Mo], they gave a sufficient condition for which $\Gamma_{(\mu_i)}$ is a lattice in the projective unitary group $PU(d,1)$, that is, $\Gamma_{(\mu_i)}$ is discrete and of finite covolume, and gave a list of such (μ_i) (see [Th] for the correction of their list).

Denote $\mu_i = \mu_i'/D$ where D is the common denominator. As remarked in Theorem 7.1, in the case $D = 5$, $\Gamma_{(\mu_i)}$ is related to $K3$ surfaces. In case of $D = 3, 4$ or 6, $\Gamma_{(\mu_i)}$ is also related to $K3$ surfaces (see [K2], [K3], [DGK]). In these cases, the corresponding $K3$ surfaces have an isotrivial elliptic fibration whose general fiber is an elliptic curve with an automorphism of order 4 or 6.

For the remaining arithmetic subgroups $\Gamma_{(\mu_i)}$ with $D > 6$ in the Deligne–Mostow's list, are they related to $K3$ surfaces ?

References

[Bol] O. Bolza, *On binary sextics with linear transformations into themselves*, Amer. J. Math., **10** (1888), 47–70.

[Borel] A. Borel, *Some metric properties of arithmetic quotients of symmetric spaces and an extension theorem*, J. Diff. Geometry, **6** (1972), 543–560.

[DM] P. Deligne, G. W. Mostow, *Monodromy of hypergeometric functions and non-lattice integral monodromy*, Publ. Math. IHES, **63** (1986), 5–89.

[D] I. Dolgachev, *Lectures on Invariant Theory*, London Math. Soc. Lecture Note Ser., **296** , Cambridge 2003.

[DGK] I. Dolgachev, B. van Geemen, S. Kondō, *A complex ball uniformaization of the moduli space of cubic surfaces via periods of $K3$ surfaces*, math.AG/0310342, J. reine angew. Math. (to appear).

[K1] S. Kondō, *A complex hyperbolic structure of the moduli space of curves of genus three*, J. reine angew. Math., **525** (2000), 219–232.

[K2] S. Kondō, *The moduli space of curves of genus 4 and Deligne-Mostow's complex reflection groups*, Adv. Studies Pure Math., **36** (2002), Algebraic Geometry 2000, Azumino, 383–400.

[K3] S. Kondō, *The moduli space of 8 points on $\mathbb{P}^1$ and automorphic forms*, to appear in the Proceedings of the Conference "Algebraic Geometry in the honor of Igor Dolgachev".

[Mo] G. W. Mostow, *Generalized Picard lattices arising from half-integral conditions*, Publ. Math. IHES, **63** (1986), 91–106.

[Mu] D. Mumford, K. Suominen, *Introduction to the theory of moduli*, Algebraic Geometry, Oslo 1970, F. Oort, ed., Wolters-Noordholff 1971.

[Na] Y. Namikawa, *Periods of Enriques surfaces*, Math. Ann., **270** (1985), 201–222.

[NU] Y. Namikawa, K. Ueno, *The complete classification of fibres in pencils of curves of genus two*, Manuscripta math., **9** (1973), 143–186.

[N1] V. V. Nikulin, *Integral symmetric bilinear forms and its applications*, Math. USSR Izv., **14** (1980), 103–167.

[N2] V. V. Nikulin, *Finite automorphism groups of Kähler $K3$ surfaces*, Trans. Moscow Math. Soc., **38** (1980), 71–135.

[N3] V. V. Nikulin, *Factor groups of groups of automorphisms of hyperbolic forms with respect to subgroups generated by 2-reflections*, J. Soviet Math., **22** (1983), 1401–1475.

[PS] I. Piatetski-Shapiro, I. R. Shafarevich, *A Torelli theorem for algebraic surfaces of type $K3$*, Math. USSR Izv., **5** (1971), 547–587.

[S] G. Shimura, *On purely transcendental fields of automorphic functions of several variables*, Osaka J. Math., **1** (1964), 1–14.

[Te] T. Terada, *Fonction hypergéométriques F_1 et fonctions automorphes*, J. Math. Soc. Japan **35** (1983), 451–475; II, ibid., **37** (1985), 173–185.

[Th] W. P. Thurston, *Shape of polyhedra and triangulations of the sphere*, Geometry & Topology Monograph, **1** (1998), 511–549.

[V] E. B. Vinberg, *Some arithmetic discrete groups in Lobachevskii spaces*, in "Discrete subgroups of Lie groups and applications to moduli", Tata-Oxford (1975), 323–348.

[Vo] S. P. Vorontsov, *Automorphisms of even lattices that arise in connection with automorphisms of algebraic $K3$ surfaces*, Vestnik Mosk. Univ. Mathematika, **38** (1983), 19–21.

[YY] T. Yamazaki, M. Yoshida, *On Hirzebruch's examples of surfaces with $c_1^2 = 3c_2$*, Math. Ann., **266** (1984), 421–431.

Acknowledgment

The author thanks to Igor Dolgachev for stimulating discussions and useful suggestions. In particular the result in 3.4 is due to him. The author thanks to the referee for suggesting an improvement of 3.4.

Shigeyuki Kondō
Graduate School of Mathematics
Nagoya University
Nagoya 464-8602
Japan
e-mail: kondo@math.nagoya-u.ac.jp

Progress in Mathematics, Vol. 260, 207–244

Uniformization by Lauricella Functions — An Overview of the Theory of Deligne–Mostow

Eduard Looijenga

Abstract. This is a survey of the Deligne–Mostow theory of Lauricella functions, or what almost amounts to the same, of the period map for cyclic coverings of the Riemann sphere.

Mathematics Subject Classification (2000). Primary: 33C65, 22E40; Secondary: 32G20.

Keywords. Lauricella function, ball quotient.

Introduction

These notes are about a chapter in the theory of hypergeometric functions in several variables. The functions in question generalize the Gauß hypergeometric function and are obtained as integrals of a multivalued differential of the form

$$\eta_z := (z_0 - \zeta)^{-\mu_0} \cdots (z_n - \zeta)^{-\mu_n} d\zeta.$$

Here $z_0, \dots, z_n$ are pairwise distinct complex numbers that are allowed to vary and the exponents μ_k are taken in the open unit interval $(0, 1)$ and are kept fixed. If we choose a simple arc in $\mathbb{C}$ connecting two zeroes of η_z but avoiding any zero on its relative interior and if a branch of η_z along that arc is chosen, then η_z can be integrated along that arc (the integral will indeed converge). The value of this integral will depend holomorphically on $z = (z_0, \dots, z_n)$, for if we vary z a little, then we can let the arc and the branch of η_z follow this variation in a continuous manner. This type of (multivalued) function of z is called a *Lauricella* function. Another choice of branch of η_z will change it by phase factor, but its dependence on the arc is of course more substantial. The Lauricella functions have a beautiful and fascinating associated geometry, and it is this geometry that this paper is mostly concerned with. Let us briefly explain how this geometry enters.

One readily finds that it is better not to focus on one such integral, but to consider all of them simultaneously, or rather, to consider for every $z = (z_0, \dots, z_n)$

as above (and given exponents), the space L_z of power series expansions in $n+1$ complex variables at z that are linear combinations of the Lauricella functions associated to η_z. It turns out that this subspace $L_z \subset \mathbb{C}\{z_0, \dots, z_n\}$ has dimension n and that the 'tautological' map-germ $(\mathbb{C}^{n+1}, z) \to L_z^*$ never takes the value $0 \in L_z^*$ and has the following regularity property: if $\mathcal{M}_{0,n+2}$ stands for the configuration space of $(n+1)$-tuples in $\mathbb{C}$ modulo affine-linear equivalence (which is also the configuration space of $(n+2)$-tuples on the Riemann sphere modulo projective-linear equivalence), then this map-germ descends to a local isomorphism $(\mathcal{M}_{0,n+2}, [z]) \to \mathbb{P}(L_z^*)$. By analytic continuation we have an identification of L_z with $L_{z'}$ for nearby z' and the multivalued nature of the hypergeometric functions is reflected by the fact that if we let z traverse a loop in the space of pairwise distinct $(n+1)$-tuples and the elements of L_z follow that loop by analytic continuation, then there results a linear (monodromy) transformation of L_z which need not be the identity. The transformations of L_z^* thus obtained form a subgroup Γ of $\mathrm{GL}(L_z^*)$, called the monodromy group of the system. The main questions addressed here are:

1. When does Γ leave invariant a Hermitian form on L_z^* which is positive definite, semidefinite or of hyperbolic signature? If such a form exists, then we would actually ask for a bit more: The set of vectors in L_z^* with positive self-inner product defines a domain $\mathbb{D} \subset \mathbb{P}(L_z^*)$ that is left invariant by Γ. This domain in fact is a complex symmetric manifold of constant holomorphic curvature on which Γ acts by isometries: we get respectively all of $\mathbb{P}(L_z^*)$ with its Fubini-Study metric, an affine space in $\mathbb{P}(L_z^*)$ with a translation invariant Hermitian metric or an open ball with its complex hyperbolic metric. We would like the multi-valued map defined on $\mathcal{M}_{0,n+2}$ to take its values in this domain, so that $\mathcal{M}_{0,n+2}$ inherits a Kähler metric of constant holomorphic curvature.

2. When is Γ discrete as a subgroup of $\mathrm{GL}(L_z^*)$? If Question 1 has been answered positively, then this is essentially equivalent to: when acts Γ properly on $\mathbb{D}$? A positive answer triggers the next question, namely: when is Γ arithmetic (in a naturally defined $\mathbb{Q}$-algebraic group that contains Γ)?

The answer to the first question in its strong form is short enough to give here: when $\mu_0 + \cdots + \mu_n$ is < 1, $= 1$ or in the interval $(1, 2)$ respectively (although we are not claiming the converse). Question 2 is harder to deal with. If Γ is discrete as well, then the exponents μ_k must be rational numbers and one of the main results of the theory states that $\mathcal{M}_{0,n+2}$ has then finite invariant volume and that its natural metric completion is an algebraic variety (we get a projective space in the elliptic and parabolic cases and in the hyperbolic case it is obtained by adding the stable orbits in a setting of geometric invariant theory). Deligne and Mostow gave sufficient conditions for discreteness, which were later weakened by Mostow and Sauter to make them necessary as well.

If the μ_k's are all rational, then there is a connection with the theory of period maps (even if Γ fails to be discrete): if m is their smallest common denominator and if we write $\mu_k = d_k/m$, then the hypergeometric functions become periods of the cyclic cover of $\mathbb{C}$ defined by $w^m = (z_0 - \zeta)^{d_0} \cdots (z_n - \zeta)^{d_n}$. For η_z then lifts to a regular single-valued differential on this affine curve (regular resp. with simple poles at infinity when $\sum_k \mu_k$ is greater than resp. equal to 1) and γ is covered by a cycle such that the hypergeometric integral is the period of the lift over this cycle.

As the reader will have gathered, this is mostly an account of work of Mostow (and his student Sauter) and of Deligne–Mostow. It is basically self-contained in the sense that we have included proofs (except for a technical lemma needed for an arithmeticity criterion). Occasionally our treatment somewhat differs from theirs. For instance, our discussion of invariant Hermitian forms does not use the approach in [8] inspired by Hodge theory, but rather follows the more pedestrian path in [6]. We also found it natural to use the language of orbifolds throughout. For some of the history of the material expounded here, we refer to the first and the last section of [8] as well as to the review [5]. In Section 5 we — albeit very sketchily — mention some recent developments.

Acknowledgements. This paper is based on a series of talks I gave at the CIMPA summer school (2005) in Istanbul. I thank my hosts, in particular Professor Uludag, for their hospitality and for making this summer school such a pleasant and fruitful experience.

I have been so fortunate to have this paper checked by a very careful referee. The reader has also good reason to be grateful, for the paper benefited greatly from his or her comments. But all errors and inaccuracies that remained are my responsibility, of course.

Some notation. Throughout this paper we denote by $(\mathbb{C}^{n+1})^\circ$ the set of $(z_0, \dots, z_n) \in \mathbb{C}^{n+1}$ whose components are pairwise distinct. So this is simply the configuration space of $n+1$ distinct numbered points in $\mathbb{C}$. We shall also use

$$V_n := \mathbb{C}^{n+1}/\text{main diagonal} \quad \text{and} \quad V_n^\circ := (\mathbb{C}^{n+1})^\circ/\text{main diagonal}. \tag{0.1}$$

The latter can be interpreted as the configuration space of $n+1$ distinct numbered points in $\mathbb{C}$, given up to a (common) translation. It is also clear that the orbit space of $(\mathbb{C}^{n+1})^\circ$ with respect to the whole affine-linear group (assuming now $n \geq 2$), is equal to $\mathbb{P}(V_n^\circ) \subset \mathbb{P}(V_n)$. As the affine group of $\mathbb{C}$ is the stabilizer of ∞ in $\mathrm{Aut}(\mathbb{P}^1)$, we thus have an identification of $\mathbb{P}(V_n^\circ)$ with the moduli space $\mathcal{M}_{0,n+1}$ of $(n+1)$-pointed smooth rational curves. But this description breaks the natural $\mathcal{S}_{n+1}$-symmetry of $\mathcal{M}_{0,n+1}$ as only the action of the subgroup $\mathcal{S}_n$ is immediately visible in $\mathbb{P}(V_n^\circ)$.

If $\mathbb{C}^\times$ acts on a variety X, then we often write $\mathbb{P}(X)$ for the orbit space of the subspace of X where $\mathbb{C}^\times$ acts with finite isotropy groups. This convention is also used if we are given a covering $\tilde{X}$ of X to which the $\mathbb{C}^\times$ action on X lifts to one of a covering of $\mathbb{C}^\times$: we may then denote the space of orbits in $\tilde{X}$ with finite isotropy groups by $\mathbb{P}(\tilde{X})$.

Contents

1. The Lauricella differential

1.1. Definition and first properties

Assume given real numbers $\mu_0, \dots, \mu_n$ in the interval $(0,1)$, where $n > 0$. We shall refer to the $(n+1)$-tuple $\mu = (\mu_0, \dots, \mu_n)$ as a *weight system* and we call its sum $|\mu| := \sum_{i=0}^{n} \mu_i$ the *total weight* of μ. The *Lauricella differential* of weight μ is

$$\eta_z := (z_0 - \zeta)^{-\mu_0} \cdots (z_n - \zeta)^{-\mu_n} d\zeta, \quad \text{with } z = (z_0, \dots, z_n) \in (\mathbb{C}^{n+1})^\circ.$$

(We recall that $(\mathbb{C}^{n+1})^\circ$ stands for the set of $(z_0, \dots, z_n) \in \mathbb{C}^{n+1}$ whose components are pairwise distinct.) If we wish to view this as a multivalued over $(\mathbb{C}^{n+1})^\circ$ we will simply write η. Although this differential is multivalued, it has a natural branch on a left half plane (where it is like $(-\zeta)^{-|\mu|} d\zeta$) by taking there the value of $(-\zeta)^{-|\mu|}$ whose argument lies in $(-\pi/|\mu|, \pi/|\mu|)$. We further note that η_z is locally integrable as a multivalued function: near z_k, η_z is of the form $(\zeta - z_k)^{-\mu_k} \exp(holom) d\zeta$; this is the differential of a function of the form $const + (\zeta - z_k)^{1-\mu_k} \exp(holom)$ and since $1 - \mu_k > 0$, that function takes a well-defined value in z_k. This implies that η_z can be integrated along every *relative arc* of $(\mathbb{C}, \{z_0, \dots, z_n\})$; by the latter we mean an oriented piecewise differentiable arc in $\mathbb{C}$ whose end points lie in $\{z_0, \dots, z_n\}$, but which does not meet this set elsewhere.

The behavior of a differential at infinity is studied (as usual) by means of the substitution $\zeta = \omega^{-1}$ and examining the result at $\omega = 0$; here we get

$$\eta_z = -(\omega z_0 - 1)^{-\mu_0} \cdots (\omega z_n - 1)^{-\mu_n} \omega^{|\mu|-2} d\omega,$$

which suggests to put $z_{n+1} := \infty$ and $\mu_{n+1} := 2 - |\mu|$. In case $\mu_{n+1} < 1$ (equivalently, $|\mu| > 1$), η_z is also (multivalued) integrable at z_{n+1}.

Remark 1.1. Following Thurston [16], we may think of η_z as a way of putting a flat Euclidean structure on $\mathbb{P}^1$ with singularities at $z_0, \dots, z_{n+1}$: a local primitive of η_z defines a metric chart with values in $\mathbb{C}$, but now regarded as the Euclidean plane (so the associated metric is simply $|\eta_z|^2$). At z_k, $k \leq n$, the metric space is isometric to a Euclidean cone with total angle $2\pi(1 - \mu_k)$; this is also true for $k = n+1$ in case $\mu_{n+1} < 1$, or equivalently, $|\mu| > 1$; if $|\mu| = 1$ (resp. $|\mu| < 1$), then a punctured neighborhood of ∞ is isometric to a flat cylinder (resp. the complement of a compact subset of a Euclidean cone with total angle $2\pi(1 - |\mu|)$).

Let a relative arc γ_z of $(\mathbb{C}, \{z_0, \dots, z_n\})$ be given and a branch of η_z on γ_z so that $\int_{\gamma_z} \eta_z$ is defined. Choose open disks D_k about z_k in $\mathbb{C}$ such that the $D_0, \dots, D_n$ are pairwise disjoint. Then we can find for every $z' \in D_0 \times \cdots \times D_n$, a relative arc $\gamma_{z'}$ of $(\mathbb{C}, \{z'_0, \dots, z'_n\})$ and a branch of $\eta_{z'}$ on $\mathrm{supp}(\gamma_{z'})$ such that both depend continuously on z' and yield the prescribed value for $z = z'$. Any primitive of η near (z, z_k) with respect to its second variable is (as a function of (z', ζ)) of the form $g(z') + (\zeta - z'_k)^{1-\mu_k} h(\zeta, z')$, with g and h holomorphic and so the function

$$z' \in D_0 \times \cdots \times D_n \mapsto \int_{\gamma_{z'}} \eta_{z'} \in \mathbb{C}$$

is holomorphic. We call such a function (or some analytic extension of it) a *Lauricella function*. The Lauricella functions (with given weight system μ) define a local system of $\mathbb{C}$-vector spaces: its stalk L_z at z is the space of germs of holomorphic functions at $z \in (\mathbb{C}^{n+1})^\circ$ that are in fact germs of Lauricella functions and it is clear that for $z' \in D_0 \times \cdots \times D_n$, we can naturally identify $L_{z'}$ with L_z.

Here are some elementary properties of Lauricella functions (the proofs are left to the reader, who should be duely careful with exchanging differentiation and integration in the proof of (c)).

Proposition 1.2. *Any* $f \in L_z$

(a) *is translation invariant:* $f(z_0 + a, \dots, z_n + a) = f(z_0, \dots, z_n)$ *for small* $a \in \mathbb{C}$,

(b) *is homogeneous of degree* $1 - |\mu|$*:* $f(e^t z_0, \dots, e^t z_n) = e^{(1-|\mu|)t} f(z_0, \dots, z_n)$ *for small* $t \in \mathbb{C}$ *and*

(c) *obeys the system of differential equations*

$$\frac{\partial^2 f}{\partial z_k \partial z_l} = \frac{1}{z_k - z_l}\left(\mu_l \frac{\partial f}{\partial z_k} - \mu_k \frac{\partial f}{\partial z_l}\right), \quad 0 \le k < l \le n. \tag{1.1}$$

The translation invariance of the Lauricella functions shows that they are in fact defined (as multivalued functions) on V_n°. In other words, the local system is a pull-back of a local system on V_n°. The homogeneity implies that when $|\mu| = 1$, these functions are also constant on the $\mathbb{C}^\times$-orbits and hence define a local system on $\mathbb{P}(V_n^\circ)$; for reasons which will become clear later, we call this the *parabolic* case.

An important consequence of part (c) of the preceding proposition is

Corollary 1.3. *The map which assigns to $f \in L_z$ its 1-jet at z is injective.*

Proof. If $f \in L_z$, then its partial derivatives $f_k := \frac{\partial f}{\partial z_k}$ satisfy the system of ordinary differential equations

$$\frac{\partial f_k}{\partial z_l} = \frac{1}{z_k - z_l}\left(\mu_l f_k - \mu_k f_l\right), \quad k \neq l. \tag{1.2}$$

We can complete this system in order to get such equations also for $\frac{\partial f_k}{\partial z_k}$, by using the fact $\sum_k f_k = 0$ (which follows from the translation invariance). The elementary theory of such systems of ODE's says that there is precisely one solution for it, once the initial conditions $f_k(z)$ have been prescribed. To such a solution corresponds at most one element of L_z up to a constant. □

1.2. L-slits

Definition 1.4. Given $(z_0, \dots, z_n) \in \mathbb{C}^{n+1}$, we define an *$L$-slit* to be an oriented arc in the Riemann sphere $\mathbb{P}^1 = \mathbb{C} \cup \{\infty\}$ from z_0 to $z_{n+1} = \infty$ which passes successively through $z_1, \dots z_n$ and near ∞ follows a line parallel to the real axis in the positive direction. If δ is such an L-slit, then we denote the piece connecting z_{k-1} with z_k by δ_k and we often let δ also stand for the system of arcs $(\delta_1, \dots, \delta_{n+1})$.

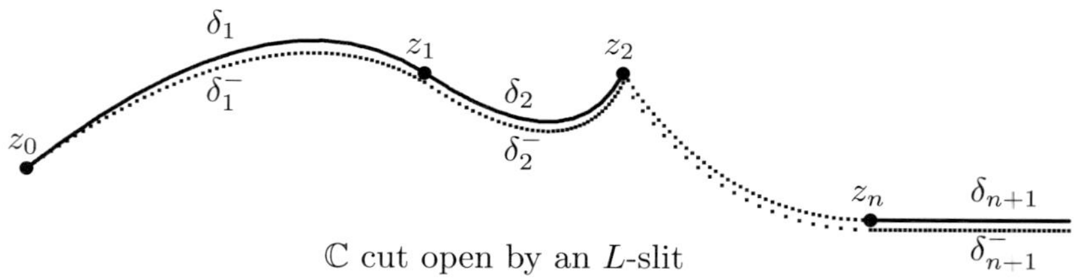

$\mathbb{C}$ cut open by an L-slit

The complement of the support of δ is simply connected and so we have a well-defined branch of η_z on this complement which extends the one we already have on a left half plane. We also extend η_z to the support of δ itself by insisting that η_z be continuous from the left along δ when we traverse that arc from z_0 to ∞. This makes the branch of η_z discontinuous along δ. With these conventions, $\int_{\delta_k} \eta_z$

has for $k = 1, \dots, n$ a well-defined meaning (and also makes sense for $k = n+1$ in case $\mu_{n+1} < 1$). If we let z vary in a small neighborhood, we get an element of L_z that we simply denote by $\int_{\delta_k} \eta$. We denote by δ_k^- the arc connecting z_{k-1} with z_k that is 'infinitesimally' to the right of δ_k. By this we really mean that η_z is given on δ_k^- the value it gets as a limit from the right. Notice that

$$\eta_z|\delta_k^- = \exp(-2\pi\sqrt{-1}(\mu_0 + \cdots + \mu_{k-1}))\eta_z|\delta_k.$$

Theorem 1.5. *The functions $\int_{\delta_1} \eta, \dots, \int_{\delta_n} \eta$ define a basis for L_z. Moreover, L_z contains the constant functions if and only if we are in the parabolic case: $|\mu| = 1$.*

Proof. Suppose γ is a closed piecewise differentiable path in $\mathbb{C}$ that is the boundary of an embedded disk $B \subset \mathbb{C}$ whose interior contains no z_k. Then we can choose a branch of η_z over B and subsequently a primitive of η_z on B so that we have $\int_\gamma \eta_z = 0$ on this branch. Since $\int_\gamma \eta_z$ is a sum of Lauricella functions associated to simple relative arcs, this yields a relation of linear dependence among such functions.

This proves that L_z is generated by any set of Lauricella functions whose underlying relative arcs generate the relative homology group $H_1(\mathbb{C}, \{z_0, \dots, z_n\})$. Clearly, the set of arcs $\delta_1, \dots, \delta_n$ has that last property.

If $|\mu| = 1$, then near ∞, η_z is equal to $-\zeta^{-1}d\zeta$. So then for a loop γ which encircles $z_0, \dots, z_n$ in the clockwise direction, we have

$$\int_\gamma \eta_z = \int_\gamma -\zeta^{-1}d\zeta = 2\pi\sqrt{-1},$$

which proves that L_z contains the constant $2\pi\sqrt{-1}$.

It remains to show that if $a_1, \dots, a_k, c \in \mathbb{C}$ are such that $\sum_{k=1}^n a_k \int_{\delta_k} \eta = c$, then $c \neq 0$ implies $|\mu| = 1$ and $c = 0$ implies that all a_i vanish as well. We prove this with induction on n. To this end, we consider a curve $z(s)$ in $(\mathbb{C}^{n+1})^\circ$ of the form $(z_0, \dots, z_{n-2}, 0, s)$, with $s > 0$ and an L-slit $\delta(s)$ for $z(s)$ with $\delta_1, \dots, \delta_{n-1}$ fixed and $\delta_n = [0, s]$. By analytic continuation we may assume that $\sum_{k=1}^{n-1} a_k \int_{\delta_k} \eta_{z(s)} + a_n \int_0^s \eta_{z(s)} = c$. We multiply this identity by s^{μ_n} and investigate what happens for $s \to \infty$. For $k < n$,

$$s^{\mu_n} \int_{\delta_k} \eta_{z(s)} = \int_{\delta_k} (z_0 - \zeta)^{-\mu_0} \cdots (z_{n-2} - \zeta)^{-\mu_{n-2}} (-\zeta)^{-\mu_{n-1}} (1 - s^{-1}\zeta)^{-\mu_n} d\zeta,$$

which for $s \to \infty$ tends to $\int_{\delta_k} \eta_{z'}$, where $z' = (z_0, \dots, z_{n-2}, 0)$. On the other hand,

$$\begin{aligned}\int_{\delta_k} \eta_{z(s)} &= \int_0^s (z_0 - \zeta)^{-\mu_0} \cdots (-t)^{-\mu_{n-1}} (s - t)^{-\mu_n} d\zeta \\ &= s(-s)^{-|\mu|} \int_0^1 (-s^{-1}z_0 + t)^{-\mu_0} \cdots (t)^{-\mu_{n-1}} (-1 + t)^{-\mu_n} dt \\ &= s(-s)^{-|\mu|} + o(|s|^{1-|\mu|}), \quad s \to \infty.\end{aligned}$$

So we find that

$$s^{\mu_n}\left(c + a_n\left((-s)^{1-|\mu|} + o(|s|^{1-|\mu|})\right)\right) = \sum_{k=1}^{n-1} a_k \int_{\delta_k} \eta_{z'}, \qquad s \to \infty.$$

This shows that $c \neq 0$ implies $|\mu| = 1$ (and $a_n = (-1)^{-|\mu|}c$). Suppose now $c = 0$. If $\mu_n < |\mu| - 1$, then the left-hand side tends to zero as $s \to \infty$ and so the right-hand side must be zero. Since in z' can represent any element of V'_{n-1}, our induction hypothesis applied to η', then implies that $a_1 = \cdots = a_{n-1} = 0$ and from this we see that $a_n = 0$, too. If $\mu_n > |\mu| - 1$, then we clearly must have $a_n = 0$ and the induction hypothesis implies that $a_1 = \cdots = a_{n-1} = 0$, also. □

Remark 1.6. The space of solutions of the (completed) system of differential equations (1.1) contains the constants and is of dimension $\leq n+1$. It follows that in the non-parabolic case, this solution space is equal to $L_z \oplus \mathbb{C}$. (In fact, the dimension of the solution space is always $n+1$, so that in the parabolic case, it also contains L_z as a hyperplane.)

1.3. The rank of the Schwarz map

We find it convenient to modify our basis of Lauricella functions by a scalar factor by putting

$$\begin{aligned} F_k(z,\delta) &:= \int_{\delta_k} (\zeta - z_0)^{-\mu_0} \cdots (\zeta - z_{k-1})^{-\mu_{k-1}} (z_k - \zeta)^{-\mu_k} \cdots (z_n - \zeta)^{-\mu_n} d\zeta \\ &= \bar{w}_k \int_{\delta_k} \eta_z, \quad \text{where } w_k := e^{\sqrt{-1}\pi(\mu_0 + \cdots + \mu_{k-1})}. \end{aligned} \tag{1.3}$$

The notation now also displays the fact that the value of the integral depends on the whole L-slit (which was needed to make η_z single-valued) and not just on δ_k. Notice that if $z = x$ is real and $x_0 < x_1 < \cdots < x_n$ and δ consists of real intervals, then the integrand is real valued and positive and hence so is F_k. Let us also observe that

$$\int_{\delta_k} \eta_z = w_k F_k(z,\delta) \text{ and } \int_{\delta_k^-} \eta_z = \bar{w}_k F_k(z,\delta),$$

where the second identity follows from the fact that $\eta_z|\delta_k^- = \bar{w}_k^2 \eta_z|\delta_k$. So if we are in the parabolic case, then the integral of η_z along a clockwise loop which encloses $\{z_0,\dots,z_n\}$ yields the identity $\sum_{k=1}^n (w_k - \bar{w}_k)F_k(z,\delta) = 2\pi\sqrt{-1}$, or equivalently,

$$\sum_{k=1}^{n} \mathrm{Im}(w_k) F_k(z,\delta) = \pi. \tag{1.4}$$

In other words, $F = (F_1,\dots,F_n)$ then maps to the affine hyperplane $\mathbb{A}^{n-1}$ in $\mathbb{C}^n$ defined by this equation.

Corollary 1.7. *If we are not in the parabolic case, then $F = (F_1, \dots, F_n)$, viewed as a multivalued map from V_n° to $\mathbb{C}^n$, is a local isomorphism taking values in $\mathbb{C}^n - \{0\}$. In the parabolic case, $F = (F_1, \dots, F_n)$ factors through a local (multivalued) isomorphism from $\mathbb{P}(V_n^\circ)$ to the affine hyperplane $\mathbb{A}^{n-1}$ in $\mathbb{C}^n$ defined by $\sum_{k=1}^n \mathrm{Im}(w_k) F_k = \pi$.*

Proof. Given (z, δ), consider the n covectors $dF_1(z, \delta), \dots, dF_n(z, \delta)$ in the cotangent space of z. According to Corollary 1.3, a linear relation among them must arise from a linear relation among the function germs $F_1, \dots, F_n \in L_z$ and the constant function 1. According to Theorem 1.5, such a relation exists if and only if $|\mu| = 1$. The corollary easily follows, except perhaps the claim that F is nowhere zero. But if $F_k(z, \delta) = 0$ for all k, then we must have $|\mu| \neq 1$; since F is identically then zero on the $\mathbb{C}^\times$-orbit through z, this contradicts the fact that F is a local isomorphism. □

Definition 1.8. We call the multivalued map F from V_n° to $\mathbb{C}^n$ the *Lauricella map* and its projectivization $\mathbb{P}F$ from $\mathbb{P}(V_n^\circ)$ to $\mathbb{P}^{n-1}$ the *Schwarz map* for the weight system μ.

The above corollary tells us that the Schwarz map always is a local isomorphism (which in the parabolic case takes values in the affine open $\mathbb{A}^{n-1} \subset \mathbb{P}^{n-1}$).

1.4. When points coalesce

We investigate the limiting behavior of F when some of the z_k's come together. To be specific, fix $0 < r < n$, let us identify V_n with $\mathbb{C}^n$ by taking $z_0 = 0$, put

$$z_\varepsilon = (0, \varepsilon z_1, \dots \varepsilon z_r, z_{r+1}, \dots z_n) \in V_n \tag{1.5}$$

and let us see what happens when $\varepsilon \in (0, 1] \to 0$. We assume here that $0 = z_0, \dots, z_r$ lie inside the unit disk, whereas the others are outside that disk and choose δ accordingly: $\delta_1, \dots, \delta_r$ (resp. $\delta_{r+2}, \dots, \delta_{n+1}$) lie inside (resp. outside) the unit disk. We let $\delta(\varepsilon)$ be an L-slit for z_ε, that depends continuously on ε, with $\delta(\varepsilon)_k = \delta_k$ for $k > r+1$ and $\delta(\varepsilon)_k = \varepsilon^{-1}\delta_k$ for $k \leq r$.

If we let $\mu' := (\mu_0, \dots, \mu_r)$, then

$$\begin{aligned} F_k(z_\varepsilon, \delta) &= \bar{w}_k \int_{\delta(\varepsilon)_k} (\varepsilon z_0 - \zeta)^{-\mu_0} \cdots (\varepsilon z_r - \zeta)^{-\mu_r} (z_{r+1} - \zeta)^{-\mu_{r+1}} \cdots (z_n - \zeta)^{-\mu_n} d\zeta \\ &= \varepsilon^{1-|\mu'|} \bar{w}_k \int_{\varepsilon\delta(\varepsilon)_k} (z_0 - \zeta)^{-\mu_0} \cdots (z_r - \zeta)^{-\mu_r} (z_{r+1} - \varepsilon\zeta)^{-\mu_{r+1}} \cdots (z_n - \varepsilon\zeta)^{-\mu_n} d\zeta. \end{aligned}$$

So if $z' = (z_0, \dots, z_r)$, then for $k \leq r$,

$$\varepsilon^{|\mu'|-1} z_{r+1}^{\mu_{r+1}} \cdots z_n^{\mu_n} F_k(z_\varepsilon, \delta) = (1 + O(\varepsilon)) F'_k(z', \delta'), \tag{1.6}$$

where F'_k is a component of the Lauricella map with weight system μ' and δ' equals δ, but with fewer bus stops:

$$F'_k(z', \delta') = \bar{w}_k \int_{\delta_k} (z_0 - \zeta)^{-\mu_0} \cdots (z_k - \zeta)^{-\mu_k} d\zeta.$$

When $k > r+1$ or $k = r+1$ and $|\mu'| < 1$, we find

$$F_k(z_\varepsilon, \delta(\varepsilon)) = (1+O(\varepsilon))\bar{w}_k \int_{\delta_k} (-\zeta)^{-|\mu'|}(z_{r+1}-\zeta)^{-\mu_{r+1}} \cdots (z_n-\zeta)^{-\mu_n} d\zeta. \quad (1.7)$$

Elliptic clustering. Assume now $|\mu'| < 1$. Then these estimates suggest to replace in $F = (F_1, \dots, F_n)$, for $k \le r$, $F_k(z)$ by $\varepsilon^{|\mu'|-1} z_{r+1}^{\mu_{r+1}} \cdots z_n^{\mu_n} F_k(z, \delta)$. In geometric terms, this amounts to enlarging the domain and range of F: now view it as a multivalued map defined on an open subset of the blowup $\mathrm{Bl}_{D^r} V_n$ of the diagonal D^r defined $z_0 = \cdots = z_r$ and as mapping to the blowup $\mathrm{Bl}_{0\times\mathbb{C}^{n-r}} \mathbb{C}^n$ of the subspace of $\mathbb{C}^n$ defined by $u_1 = \cdots = u_r = 0$. To be concrete, $\mathrm{Bl}_{D^r} V_n = \mathrm{Bl}_{0\times\mathbb{C}^{n-r}} \mathbb{C}^n$ is the set of $([Z_1 : \cdots : Z_r], 0, z_1, \dots, z_n) \in \mathbb{P}^{r-1} \times V_n$ satisfying $Z_i z_j = W_j z_i$ for $1 \le i < j \le r$. Likewise, $\mathrm{Bl}_{0\times\mathbb{C}^{n-r}} \mathbb{C}^n$ is defined as a subset of $\mathbb{P}^{r-1} \times \mathbb{C}^n$. The lift of F is then on $V_n - D^r$ of course given by

$$([z_1 : \cdots : z_r], 0, z_1, \dots, z_n) \mapsto ([F_1(z) : \cdots : F_r(z)], F(z)).$$

(We omitted δ here in the argument of F_k.) But on the exceptional divisor $\mathbb{P}^{r-1} \times D^r$ it is

$$\begin{aligned}&([Z_1 : \cdots : Z_r], (0, \dots, 0, z_r, \dots z_n)) \mapsto \\ &\qquad ([F_0'(0, Z_1, \dots, Z_r) : \cdots : F_r'(0, Z_1, \dots, Z_r)], 0, \dots, 0, F_{r+1}(z), \dots, F_n(z)),\end{aligned}$$

and hence takes its values in the exceptional divisor $\mathbb{P}^{r-1}\times 0\times\mathbb{C}^{n-r}$ of $\mathrm{Bl}_{0\times\mathbb{C}^{n-r}} \mathbb{C}^n$. So if we identify D^r with V_{1+n-r} (via $(0, \dots 0, z_{r+1}, \dots, z_n) \mapsto (0, z_r, \dots, z_n)$), then we see that the first component of this restriction is the Schwarz map $\mathbb{P}F'$ for the weight system μ' and the second component is $\bar{w}_r$ times the Lauricella map for the weight system $(|\mu'|, \mu_{r+1}, \dots, \mu_n)$.

If several such clusters are forming, then we have essentially a product situation.

Parabolic clustering. We shall also need to understand what happens when $|\mu'| = 1$. Then taking the limit for $\varepsilon \to 0$ presents a problem for F_{r+1} only (the other components have well-defined limits). This is related to the fact that η_z is single-valued on the unit circle S^1; by the theory of residues we then have

$$\int_{S^1} \eta_z = \int_{S^1} (z_0-\zeta)^{-\mu_0} \cdots (z_n-\zeta)^{-\mu_n} d\zeta = 2\pi\sqrt{-1} z_{r+1}^{-\mu_{r+1}} \cdots z_n^{-\mu_n}.$$

We therefore replace η_z by $\hat{\eta}_z := z_{r+1}^{\mu_{r+1}} \cdots z_n^{\mu_n} \eta_z$ and F by $\hat{F} := z_{r+1}^{\mu_{r+1}} \cdots z_n^{\mu_n} F$. This does not change the Schwarz map, of course. Notice however, that now $\int_{S^1} \hat{\eta}_z = 2\pi\sqrt{-1}$.

Lemma 1.9. *Assume that μ' is of parabolic type: $|\mu'| = 1$. Define Lauricella data $\mu'' := (\mu_{r+1}, \dots, \mu_{n+1})$, $z'' := (z_{r+1}^{-1}, \dots, z_n^{-1}, 0)$ and let $\delta'' = (\delta_1'', \dots, \delta_{n-r}'')$ be the image of $(\delta_{r+2}, \dots, \delta_{n+1})$ under the map $z \mapsto z^{-1}$. Then we have*

$$\hat{F}_k(z_\varepsilon, \delta) = \begin{cases} (1+O(\varepsilon))F_k'(z', \varepsilon\delta') & \textit{when } 1 \le k \le r, \\ (1+O(\varepsilon))F_{k-r-1}''(z'', \delta'') & \textit{when } r+2 \le k \le n, \end{cases}$$

whereas $\lim_{\varepsilon\to 0}\hat F_{r+1}(z_\varepsilon,\delta)/\log\varepsilon$ *is a nonzero constant.*

Moreover, we have $\sum_{k=1}^r \operatorname{Im}(w_k)\hat F_k(z,\delta)=\pi$.

Proof. The assertion for $k\le r$ is immediate from our previous calculation. For $1\le i\le n-r-1$ we find

$$\begin{aligned}
&\hat F_{r+1+i}(z_\varepsilon,\delta)\\
&=\bar w_{r+1+i}\int_{\delta_{r+1+i}}(\varepsilon z_0-\zeta)^{-\mu_0}\cdots(\varepsilon z_r-\zeta)^{-\mu_r}(1-\frac{\zeta}{z_{r+1}})^{-\mu_{r+1}}\cdots(1-\frac{\zeta}{z_n})^{-\mu_n}d\zeta\\
&=-\bar w_i''\int_{\delta_i''}(\varepsilon z_0-\zeta^{-1})^{-\mu_0}\cdots(\varepsilon z_r-\zeta^{-1})^{-\mu_r}(1-\frac{1}{\zeta z_{r+1}})^{-\mu_{r+1}}\cdots(1-\frac{1}{\zeta z_n})^{-\mu_n}\frac{d\zeta}{-\zeta^2}\\
&=\bar w_i''\int_{\delta_i''}(1-\varepsilon z_0\zeta)^{-\mu_0}\cdots(1-\varepsilon z_r\zeta)^{-\mu_r}(\frac{1}{z_{r+1}}-\zeta)^{-\mu_{r+1}}\cdots(\frac{1}{z_n}-\zeta)^{-\mu_n}(-\zeta)^{-\mu_{n+1}}d\zeta\\
&=(1+O(\varepsilon))F_i''(z'',\delta'').
\end{aligned}$$

As to the limiting behavior of $\hat F_{r+1}$, observe that

$$\begin{aligned}
&\hat F_{r+1}(z_\varepsilon,\delta(\varepsilon))\\
&=-\int_{\delta(\varepsilon)_{r+1}}(\varepsilon z_0-\zeta)^{-\mu_0}\cdots(\varepsilon z_r-\zeta)^{-\mu_r}(1-\frac{\zeta}{z_{r+1}})^{-\mu_{r+1}}\cdots(1-\frac{\zeta}{z_n})^{-\mu_n}d\zeta\\
&=\int_{\delta_{r+1}(\varepsilon)}(\zeta-\varepsilon z_0)^{-\mu_0}\cdots(\zeta-\varepsilon z_r)^{-\mu_r}\left[(1-\frac{\zeta}{z_{r+1}})^{-\mu_{r+1}}\cdots(1-\frac{\zeta}{z_n})^{-\mu_n}\right]d\zeta.
\end{aligned}$$

In order to understand the behaviour of this integral for $\varepsilon\to 0$, we may ignore the bracketed factor and continue with

$$\int_{\varepsilon z_r}^1(\zeta-\varepsilon z_0)^{-\mu_0}\cdots(\zeta-\varepsilon z_r)^{-\mu_r}d\zeta=\int_{z_r}^{1/\varepsilon}(\zeta-\varepsilon z_0)^{-\mu_0}\cdots(\zeta-\varepsilon z_r)^{-\mu_r}d\zeta,$$

where we indicated not the path of integration but only its end points. The last integral is as a function of ε for large values of ε well approximated by $\int_{z_r}^{1/\varepsilon}\zeta^{-1}d\zeta$ and hence by $-\log\varepsilon$. The claimed limiting behaviour follows from this.

The last assertion follows from the fact that $\int_{S^1}\hat\eta_z=2\pi\sqrt{-1}$ (see the derivation of Equation (1.4)). □

We can interpret this calculation geometrically as well: it tells us that the n-tuple $(\hat F_1,\dots,\hat F_r,\exp(-\hat F_{r+1}),\hat F_{r+2},\dots,\hat F_n)$ defines a (multi-valued) map from an open subset of $\mathrm{Bl}_{D^r}V_n$ (which contains an open-dense subset of the exceptional divisor) to $\mathbb{A}^{r-1}\times\mathbb{C}\times\mathbb{C}^{n-r-1}$, where $\mathbb{A}^{r-1}\subset\mathbb{C}^r$ is the affine space defined by $\sum_{k=1}^r\operatorname{Im}(w_k)u_k=0$. On the exceptional divisor we have almost separation of variables, as it is there given by

$$\begin{aligned}
&([Z_1:\cdots:Z_r],0,\dots,0,z_{r+1},\dots,z_n)\mapsto\\
&\qquad(F'(Z),0,F''(z_{r+1}^{-1},\dots,z_n^{-1}))\in\mathbb{A}^{r-1}\times\mathbb{C}\times\mathbb{C}^{n-r-1}.\quad(1.8)
\end{aligned}$$

Our reason for replacing $\hat{F}_{r+1}$ by $\exp(-\hat{F}_{r+1})$ is that we thus make it single-valued and holomorphic at the exceptional divisor: the monodromy around this divisor is obtained by letting by letting $z_0 = 0, z_1, \dots, z_r$ rotate simultaneously and counter clockwise around the 0 and although this does not affect the integrand, it does affect the path of integration that defines $\hat{F}_{r+1}$ and causes $\hat{F}_{r+1}$ to change by the constant $2\pi\sqrt{-1}$.

The map defined by $(\hat{F}_1, \dots, \hat{F}_r, \exp(-\hat{F}_{r+1}), \hat{F}_{r+2}, \dots, \hat{F}_n)$ is constant on $\mathbb{C}^\times$-orbits and hence factors through the blowup of $\mathbb{P}(D_r)$ in $\mathbb{P}(V_n)$. The induced map on the latter is a local isomorphism.

1.5. Monodromy group and monodromy cover

We begin with making a few remarks about the fundamental group of $(\mathbb{C}^{n+1})^\circ$. We take $[n] = (0, 1, 2, \dots, n)$ as a base point for $(\mathbb{C}^{n+1})^\circ$ and use the same symbol for its image in V_n° (see Equation (0.1)). The projection $(\mathbb{C}^{n+1})^\circ \to V_n^\circ$ induces an isomorphism on fundamental groups: $\pi_1((\mathbb{C}^{n+1})^\circ, [n]) \cong \pi_1(V_n^\circ, [n])$. This group is known as the *pure* (also called *colored*) *braid group with* $n+1$ *strands*; we denote it by PBr_{n+1}. Another characterization of PBr_{n+1} is that as the group of connected components of the group of diffeomorphisms $\mathbb{C} \to \mathbb{C}$ that are the identity outside a compact subset of $\mathbb{C}$ and fix each z_k.

If α is a path in $(\mathbb{C}^{n+1})^\circ$ from z to z', and if we are given an L-slit δ for z, then we can carry that system continuously along when we follow α; we end up with an L-slit δ' for z' and this L-slit will be unique up to isotopy. In this way PBr_{n+1} acts on the set of isotopy classes of L-slits. It is not hard to see that this action is simply transitive: for every ordered pair of isotopy classes of L-slits, there is a unique element of PBr_{n+1} which carries the first one onto the second one.

The group PBr_{n+1} has a set of distinguished elements, called *Dehn twists*, defined as follows. The basic Dehn twist is a diffeomorphism of the annulus $D_{1,2} \subset \mathbb{C} : 1 \le |z| \le 2$; it is defined by $re^{\sqrt{-1}\theta} \mapsto re^{\sqrt{-1}(\theta+\phi(r))}$, where ϕ is a differentiable function which is zero (resp. 2π) on a neighborhood of 1 (resp. 2) (all such diffeomorphisms of $D_{1,2}$ are isotopic relative to the boundary $\partial D_{1,2}$). If S is an oriented surface, and we are given an orientation preserving diffeomorphism $h : D_{1,2} \to S$, then the Dehn twist on the image and the identity map on its complement define a diffeomorphism of S, which is also called a Dehn twist. Its isotopy class only depends on the isotopy class of the image of the counter clockwise oriented unit circle (as an oriented submanifold of S). These embedded circles occur here as the isotopy classes of embedded circles in $\mathbb{C} - \{z_1, \dots, z_n\}$. A particular case of interest is when such a circle encloses precisely two points of $\{z_1, \dots, z_n\}$, say z_k and z_l. The isotopy class of such a circle defines (and is defined by) the isotopy class of an unoriented path in $\mathbb{C} - \{z_1, \dots, z_n\}$ that connects z_k and z_l (the boundary of a regular neighborhood of such a path gives an embedded circle). The element of the pure braid group associated to this is called *simple*; if we choose for every pair $0 \le k < l \le n$ a simple element, then the resulting collection of simple elements is known to generate PBr_{n+1}.

There is a standard way to obtain a covering of V_n° on which F is defined as a single-valued map. Let us recall this in the present case. First notice that if α is a path in $(\mathbb{C}^{n+1})^\circ$ from z to z', then analytic continuation along this path gives rise to an isomorphism of vector spaces $\rho_\mu(\alpha) : L_z \to L_{z'}$. This is compatible with composition: if β is a path in $(\mathbb{C}^{n+1})^\circ$ from z' to z'', then $\rho_\mu(\beta)\rho_\mu(\alpha) = \rho_\mu(\beta\alpha)$ (we use the functorial convention for composition of paths: $\beta\alpha$ means α followed by β). A loop in $(\mathbb{C}^{n+1})^\circ$ based at $[n]$ defines an element $\rho_\mu(\alpha) \in \mathrm{GL}(L_{[n]})$ and we thus get a representation ρ_μ of PBr_{n+1} in $L_{[n]}$. The image of this *monodromy representation* is called the *monodromy group* (of the Lauricella system with weight system μ); we shall denote that group by Γ_μ, or simply by Γ. The monodromy representation defines a Γ-covering $\widetilde{V_n^\circ}$ of V_n° on which the F_k's are single-valued. It is the covering whose fundamental group is the kernel of the monodromy representation: a point of $\widetilde{V_n^\circ}$ can be represented as a pair (z, α), where α is a path in $\mathbb{C}^{n+1}$ from $[n]$ to z, with the understanding that (z', α') represents the same point if and only if $z - z'$ lies on the main diagonal (so that $L_{z'} = L_z$) and $\rho_\mu(\alpha) = \rho_\mu(\alpha')$. The action of Γ on $\widetilde{V_n^\circ}$ is then given as follows: if $g \in \Gamma$ is represented by the loop α_g in $\mathbb{C}^{n+1}$ from $[n]$, then $g.[(z,\alpha)] = [(z, \alpha\alpha_g^{-1})]$. But it is often more useful to represent a point of $\widetilde{V_n^\circ}$ as a pair (z, δ), where δ is an L-slit for z, with the understanding that (z', δ') represents the same point if and only if $z - z'$ lies on the main diagonal and $F_k(z, \delta) = F_k(z', \delta')$ for all $k = 1, \dots, n$. For this description we see right away that the basic Lauricella functions define a single-valued holomorphic map

$$F = (F_1, \dots F_n) : \widetilde{V_n^\circ} \to \mathbb{C}^n.$$

Since $[(z, \delta)]$ only depends on the isotopy class of δ, the action of Γ is also easily explicated in terms of the last description. The germ of F at the base point defines an isomorphism $L_{[n]}^* \cong \mathbb{C}^n$: $c = (c_1, \dots, c_n) \in \mathbb{C}^n$ defines the linear form on L_z which sends F_k to c_k. If we let Γ act on $\mathbb{C}^n$ accordingly (i.e., as the dual of $L_{[n]}$), then F becomes Γ-equivariant.

The $\mathbb{C}^\times$-action on V_n° given by scalar multiplication will lift not necessarily to a $\mathbb{C}^\times$-action on $\widetilde{V_n^\circ}$, but to one of a (possibly) infinite covering $\widetilde{\mathbb{C}^\times}$. For this action, F is homogeneous of degree $1 - |\mu|$. Let us denote by $\mathbb{P}(\widetilde{V_n^\circ})$ the $\widetilde{\mathbb{C}^\times}$-orbit space of $\widetilde{V_n^\circ}$.

1.6. Invariant Hermitian forms

Our goal is to prove the following theorem.

Theorem 1.10. *If $|\mu| < 1$, then the monodromy group Γ leaves invariant a positive definite Hermitian form H on $\mathbb{C}^n$.*

If $|\mu| = 1$ (the parabolic case), then Γ leaves invariant a positive definite Hermitian form H on the (linear) translation hyperplane of the affine hyperplane $\mathbb{A}^{n-1}$ in $\mathbb{C}^n$, defined by $\sum_{k=1}^n \mathrm{Im}(w_k) F_k = 0$.

If $1 < |\mu| < 2$, then the monodromy group Γ leaves invariant a hyperbolic Hermitian form H on $\mathbb{C}^n$ (i.e., of signature $(n-1,1)$) with the property that $H(F(\tilde{z}), F(\tilde{z})) < 0$ for all $\tilde{z} \in \widetilde{V_n^\circ}$.

Before we begin the proof, let us make the following observation. If W is a finite dimensional complex vector space, then by definition a point u of $\mathbb{P}(W)$ is given by a one-dimensional subspace $L_p \subset W$. An exercise shows that the complex tangent space $T_p\mathbb{P}(W)$ of $\mathbb{P}(W)$ at p is naturally isomorphic to $\mathrm{Hom}(L_p, W/L_p)$. If we are also given a Hermitian form H on W which is nonzero on L_p, then it determines a Hermitian form H_p on $T_p\mathbb{P}(W) \cong \mathrm{Hom}(L_p, W/L_p)$ as follows: since H is nonzero on L_p, the H-orthogonal complement $L_p^\perp$ maps isomorphically onto W/L_p. If we choose a generator $u \in L_p$ and think of a tangent vector as a linear map $\phi : L_p \to L_p^\perp$, then we put $H_p(\phi, \phi') := |H(u,u)|^{-1} H(\phi(u), \phi'(u))$. This is clearly independent of the generator u. It is also clear that H_p only depends on the conformal equivalence class of H: it does not change if we multiply H by a positive scalar.

If H is positive definite, then so is H_p for every $p \in \mathbb{P}(W)$. In this way $\mathbb{P}(W)$ acquires a Hermitian metric, known as the *Fubini–Study* metric. It is in fact a Kähler manifold of constant holomorphic curvature 1 on which the unitary group of (W, H) acts transitively.

There is another case of interest, namely when H has hyperbolic signature: if we restrict ourselves to the set $\mathbb{B}(W)$ of $p \in \mathbb{P}(W)$ for which H is negative on L_p, then H_p is positive definite as well. This defines a metric on $\mathbb{B}(W)$ which is invariant under the unitary group of (W, H). If we choose a basis of linear forms $u_0, \dots, u_m$ on W such that H takes the standard form $H(u,u) = -|u_0|^2 + |u_1|^2 + \cdots + |u_m|^2$, then we see that $\mathbb{B}(W)$ is defined in $\mathbb{P}(W)$ by the inequality $|u_1/u_0|^2 + \cdots + |u_m/u_0|^2 < 1$, which is simply the open unit ball in complex m-space. We call $\mathbb{B}(W)$ a *complex-hyperbolic space* and the metric defined above, the *complex-hyperbolic metric*. As in the Fubini–Study case, this metric makes $\mathbb{B}(W)$ into a Kähler manifold of constant holomorphic curvature (here equal to -1) on which the unitary group of (W, H) acts transitively. For $m = 1$ we recover the complex unit disk with its Poincaré metric.

Returning to the situation of Theorem 1.10, we see that in all three cases $\mathbb{P}F$ is a local isomorphism mapping to a homogeneous Kähler manifold: when $|\mu| < 1$, the range is a Fubini–Study space $\mathbb{P}_{n-1}$ (this notation is a private one: the subscript is supposed to distinguish it from the metricless projective space $\mathbb{P}^{n-1}$), for $|\mu| = 1$ we get a complex affine space with a translation invariant metric (indeed, denoted here by $\mathbb{A}_{n-1}$) and when $|\mu| > 1$ we get a complex ball $\mathbb{B}_{n-1}$ with its complex-hyperbolic metric. Since these structures are Γ-invariant, we can state this more poignantly: the weight system μ endows $\mathbb{P}(V_n^\circ)$ with a natural Kähler metric locally isometric with a Fubini–Study metric, a flat metric or a complex-hyperbolic metric. We will therefore use the corresponding terminology for the cases $|\mu| < 1$ and $1 < |\mu| < 2$ and call them the *elliptic* and *hyperbolic* case, respectively. Recall that $|mu| = 1$ defined the parabolic case, so that we are in the elliptic, parabolic or, hyperbolic case, according to whether μ_{n+1} is greater, equal or smaller than 1.

Theorem 1.10 follows from a more specific result that takes some of preparation to formulate. We shall associate to the weight system μ a Hermitian form H on $\mathbb{C}^n$ or on the hyperplane $A^{n-1} \subset \mathbb{C}^n$ defined by $\sum_{k=1}^n \operatorname{Im}(w_k)F_k = 0$ in $\mathbb{C}^n$ (recall that $w_k := e^{\sqrt{-1}\pi(\mu_0+\cdots+\mu_{k-1})}$), depending on whether $|\mu|$ is integral. We do this somewhat indirectly. Let $\tilde{H}$ be the Hermitian form on $\mathbb{C}^{n+1}$ defined by

$$\tilde{H}(F,G) = \sum_{1\leq j<k\leq n+1} \operatorname{Im}(w_j\bar{w}_k)F_k\bar{G}_j.$$

The $\tilde{H}$-orthogonal complement in $\mathbb{C}^{n+1}$ of the last basis vector e_{n+1} is the hyperplane $A^n \subset \mathbb{C}^{n+1}$ defined by $\sum_{k=1}^n \operatorname{Im}(w_k)F_k = 0$. Consider the composite map

$$pr : A^n \subset \mathbb{C}^{n+1} = \mathbb{C}^n \times \mathbb{C} \to \mathbb{C}^n,$$

where the second map is a projection. When $|\mu| \notin \mathbb{Z}$, we have $\operatorname{Im}(w_{n+1}) \neq 0$ (because $w_{n+1} = e^{\pi\sqrt{-1}|\mu|}$) and so pr is an isomorphism; we then let H then be the restriction of $\tilde{H}$ to A^n transfered to $\mathbb{C}^n$ via this isomorphism.

If $|\mu| \in \mathbb{Z}$, then $\operatorname{Im}(w_{n+1}) = 0$ and hence $\ker(pr) = \mathbb{C}e_{n+1}$ and $\operatorname{Im}(pr) = A^{n-1} \subset \mathbb{C}^n$. Since e_{n+1} is $\tilde{H}$-isotropic, we thus obtain an induced a Hermitian form on A^{n-1}. The following proposition implies Theorem 1.10.

Proposition 1.11. *For all weight systems μ, the form H is Γ-invariant. For $0 < |\mu| \leq 1$, the form H is positive definite. For $1 < |\mu| < 2$, H is of hyperbolic signature and we have $H(F(z,\delta), F(z,\delta)) = N(z)$, where*

$$N(z) = -\frac{\sqrt{-1}}{2}\int_{\mathbb{C}} \eta_z \wedge \bar{\eta}_z = -\int_{\mathbb{C}} |z_0-\zeta|^{-2\mu_0}\cdots|z_n-\zeta|^{-2\mu_n}\, d(\text{area}).$$

Proof. The assertions about the signature of H involve a linear algebra calculation that we leave to the reader (who may consult [7]). We treat the hyperbolic case first, so assume $1 < |\mu| < 2$. First notice that the integral defining $N(z)$ converges (here we use that $|\mu| > 1$) and takes is real and negative. We claim that

$$N(z) = \sum_{1\leq j<k\leq n+1} w_j\bar{w}_k\bar{F}_j(z,\delta)F_k(z,\delta). \tag{1.9}$$

To see this, let us integrate η_z, using the branch defined by δ: $\Phi_z(\zeta) := \int_{z_0}^{\zeta}\eta_z$, where the path of integration is not allowed to cross $\operatorname{supp}(\delta)$. We have $d\Phi_z = \eta_z$ outside $\operatorname{supp}(\delta)$ and by Stokes theorem

$$N(z) = -\frac{\sqrt{-1}}{2}\int_{\mathbb{C}}\eta_z\wedge\bar{\eta}_z = \frac{\sqrt{-1}}{2}\int_{\mathbb{C}} d(\bar{\Phi}_z\eta_z) = \frac{\sqrt{-1}}{2}\sum_{k=1}^{n+1}\left(\int_{\delta_k}\bar{\Phi}_z\eta_z - \int_{\delta_k^-}\bar{\Phi}_z\eta_z\right).$$

As to the last terms, we observe that on δ_k we have $\Phi_z(\zeta) = \sum_{j<k} w_jF_j + \int_{z_{k-1}}^{\zeta}\eta_z$ (we abbreviate $F_j(z,\delta)$ by F_j), where the last integral is taken over a subarc of

δ_k. Likewise, on δ_k^-: $(\Phi_z|\delta_k^-)(\zeta) = \sum_{j<k} \bar{w}_j F_j + \int_{z_{k-1}}^{\zeta} \bar{w}_k^2 \eta_z$. Hence on δ_k we have

$$\begin{aligned}\bar{\Phi}_z\eta_z - (\bar{\Phi}_z\eta_z|\delta_k^-) &= \sum_{j<k}\left(\bar{w}_j\bar{F}_j + \int_{z_{k-1}}^{\zeta}\bar{\eta}_z\right)\eta_z - \sum_{j<k}\left(w_j\bar{F}_j + \int_{z_{k-1}}^{\zeta} w_k^2\bar{\eta}_z\right)\bar{w}_k^2\eta_z \\ &= \sum_{j<k}\left(\bar{w}_j - w_j\bar{w}_k^2\right)\bar{F}_j\eta_z = \sum_{j<k}\left(\bar{w}_j w_k - w_j\bar{w}_k\right)\bar{F}_j\bar{w}_k\eta_z,\end{aligned}$$

which after integration over δ_k yields

$$\int_{\delta_k}\bar{\Phi}_z\eta_z - \int_{\delta_k^-}\bar{\Phi}_z\eta_z = \sum_{j<k}\left(\bar{w}_j w_k - w_j\bar{w}_k\right)\bar{F}_j F_k = \frac{2}{\sqrt{-1}}\sum_{j<k}\operatorname{Im}(w_j\bar{w}_k)\bar{F}_j F_k.$$

Our claim follows if we substitute this identity in the formula for N above.

We continue the proof. The claim implies that $H(F(z,\delta), F(z,\delta)) = N(z)$. The function N is obviously Γ-invariant (it does not involve δ). Since N determines H, so is H. So this settles the hyperbolic case.

For the elliptic and parabolic cases we may verify by hand that it is invariant under a generating set of monodromy transformations, but a computation free argument, based on analytic continuation as in [7], is perhaps more satisfying. It runs as follows: if we choose a finite set of generators $\alpha_1, \dots, \alpha_N$ of PBr_{n+1}, then for every weight system μ we have a projective linear transformation $\mathbb{P}\rho_\mu(\alpha_i)$ of $\mathbb{P}^{n-1}$ that depends in a real-analytic manner on μ. The Hermitian forms h_μ defined on an open subset of the tangent bundle of $\mathbb{P}^{n-1}$ also depend real-analytically on μ; so if h_μ is preserved by the $\mathbb{P}\rho_\mu(\alpha_i)$'s for a nonempty open subset of μ's, then it is preserved for all weight systems for which this makes sense. Hence $\mathbb{P}\rho_\mu(\alpha_i)$ multiplies H by a scalar. For $1 < |\mu| < 2$ this scalar is constant 1. Another analytic continuation argument implies that it is 1 for all μ. □

1.7. Cohomological interpretation via local systems of rank one

We sketch a setting in terms of which the Hermitian form H is best understood. It will not play a role in what follows (hence may be skipped), although it will reappear in a more conventional context (and formally independent of this discussion) in Section 4. The reader should consult § 2 of [8] for a more thorough treatment.

Fix complex numbers $\alpha_0, \dots, \alpha_n$ in $\mathbb{C}^\times$. Let $\mathbb{L}$ be a local system of rank one on $U := \mathbb{C} - \{z_0, \dots, z_n\} = \mathbb{P}^1 - \{z_0, \dots, z_{n+1}\}$ such that the (counterclockwise) monodromy around z_k is multiplication by α_k. It is unique up to isomorphism. We fix a nonzero multivalued section e of $\mathbb{L}$ by choosing a nonzero section of $\mathbb{L}$ on some left half plane and then extend that section to the universal cover of U (defined by that left half plane). Denote by $\mathcal{L} := \mathcal{O}_U \otimes_{\mathbb{C}} \mathbb{L}$ the underlying holomorphic line bundle. So if $\mu_k \in \mathbb{C}$ is such that $\exp(2\pi\mu_k\sqrt{-1}) = \alpha_k$, then $s(\zeta) := \prod_{k=1}^n (z_k - \zeta)^{-\mu_k} \otimes e$ can be understood as a generating section of $\mathcal{L}$. Likewise, $s d\zeta$ is a generating section of $\Omega(\mathcal{L}) = \Omega_U \otimes_{\mathbb{C}} \mathbb{L}$. Notice that $\mathcal{L}$ comes with a connection $\nabla : \mathcal{L} \to \Omega(\mathcal{L})$ characterized by

$$\nabla(s) = \left(\sum_{k=0}^{n} \frac{\mu_k}{z_k - \zeta} \right) s d\zeta$$

and that $\mathbb{L}$ is recovered from the pair $(\mathcal{L}, \nabla)$ as the kernel of ∇.

The topological Euler characteristic of a rank one local system on a space homotopy equivalent to a finite cell complex is independent of that local system and hence equal to the topological Euler characteristic of that space. So the topological Euler characteristic of $\mathbb{L}$ is $-n$. Now assume that $\alpha_k \neq 1$ for all k. This ensures that $\mathbb{L}$ has no nonzero section. As there is no cohomology in degrees $\neq 0, 1$, this implies that $\dim H^1(\mathbb{L}) = n$. Moreover, if $j : U \subset \mathbb{P}^1$ is the inclusion, then the stalk of $j_*\mathbb{L}$ in z_k is represented by the sections of $\mathbb{L}$ on a punctured neighborhood of z_k, hence is zero unless $k = n+1$ and $\alpha_0 \cdots \alpha_n = 1$: then it is nonzero. So the map of complexes $u : j_!\mathbb{L} \to j_*\mathbb{L}$ has cokernel a one-dimensional skyscraper sheaf at ∞ or is an isomorphism. This implies that for the natural map

$$H^1(u) : H^1_c(\mathbb{L}) \to H^1(\mathbb{L})$$

$\dim \mathrm{Ker}(H^1(u)) = \dim \mathrm{Coker}(H^1(u))$ is 1 or 0, depending on whether or not $\alpha_0 \cdots \alpha_n = 1$. It is customary to denote the image of $H^1(u)$ by $IH^1(\mathbb{L})$.

A relative arc α plus a section of $\mathbb{L}^\vee$ over its relative interior defines a relative cycle of $(\mathbb{P}^1, \{z_0, \ldots, z_{n+1}\})$ with values in $\mathbb{L}^\vee$ and hence an element $[\alpha]$ of the relative homology space $H_1(\mathbb{P}^1, \{z_0, \ldots, z_{n+1}\}; \mathbb{L}^\vee)$. Alexander duality identifies the latter cohomology space with the dual of $H^1(\mathbb{L})$. To make the connection with the preceding section, let us identify η with $s d\zeta$ (we need not assume here that $\mu_k \in (0,1)$), so that we have a De Rham class $[\eta] \in H^1(\mathbb{L})$. If we are given an L-slit δ and choose the determination of e on δ_k prescribed by the slit, then $\{\bar{w}_k[\delta_k]\}_{k=1}^n$ is a basis of $H_1(\mathbb{P}^1, \{z_0, \ldots, z_{n+1}\}; \mathbb{L}^\vee)$ and the value of $[\eta]$ on $\bar{w}_k[\delta_k]$ is just $F_k(z, \delta)$.

We have a perfect (Poincaré) duality $H^1_c(\mathbb{L}) \times H^1(\mathbb{L}^\vee) \to \mathbb{C}$, which, if cohomology is represented by means of forms, is given by integration over U of the cup product. Suppose now in addition that $|\alpha_k| = 1$ for all k. This implies that $\mathbb{L}$ carries a flat metric; we choose this metric to be the one for which e has unit length. The metric may be viewed as a $\mathbb{C}$-linear isomorphism of sheaves $\overline{\mathbb{L}} \to \mathbb{L}^\vee$ (here $\overline{\mathbb{L}}$ stands for the local system $\mathbb{L}$ with its complex conjugate complex structure) so that our perfect duality becomes a bilinear map $H^1_c(\mathbb{L}) \times \overline{H^1(\mathbb{L})} \to \mathbb{C}$. We multiply that map by $\frac{1}{2}\sqrt{-1}$ and denote the resulting sesquilinear map $h : H^1_c(\mathbb{L}) \times H^1(\mathbb{L}) \to \mathbb{C}$. Then h is Hermitian in the sense that if $\alpha, \beta \in H^1_c(\mathbb{L})$, then $h(\alpha, i_*\beta) = \overline{h(\beta, i_*\alpha)}$, in particular, it induces a nondegenerate Hermitian form on $IH^1(\mathbb{L})$. This is just minus the form we defined in Subsection 1.6. If we take $\mu_k \in (0,1)$ for $k = 0, \ldots, n$ and assume $1 < |\mu| < 2$ (so that $\mu_{n+1} \in (0,1)$ also and u is an isomorphism), then $h([\eta], [\eta])$ equals $\frac{1}{2}\sqrt{-1} \int_{\mathbb{C}} \eta \wedge \bar{\eta}$ indeed and hence equals $-N(z) = -H(F(z), F(z))$.

2. Discreteness of monodromy and orbifolds

2.1. Monodromy defined by a simple Dehn twist

Fix a simple arc γ_0 in $(\mathbb{C}, \{z_0, \dots, z_n\})$ which connects z_k with z_l, $k \neq l$. This defines a Dehn twist $D(\gamma_0)$ and hence an element T of PBr_{n+1}. We determine the action of T on $\mathbb{C}^n$. For this we need to make η_z single-valued. For this purpose it is convenient to have also given a (closed) disk-like neighborhood B of γ_0 in $\mathbb{C}$ which does not contain any of the z_i , $i \neq k, l$ and a straight piece of arc γ_1 connecting z_l with a boundary point $p_1 \in B$ such that $\gamma_0\gamma_1$ is a simple arc whose complement in B is simply connected. In other words, $\gamma_0\gamma_1$ defines a slit for η_z in B. Choose any branch for η_z on this complement. Let us abbreviate $u_k := e^{2\pi\sqrt{-1}\mu_k}$ and $u_l := e^{2\pi\sqrt{-1}\mu_l}$. Then the value of η_z on γ_0 (resp. γ_1) when approached from the right is $\bar{u}_k$ (resp. $\bar{u}_k\bar{u}_l$) times the value that we get when we approach it from the left. We may assume that the Dehn twist has support in the interior of B.

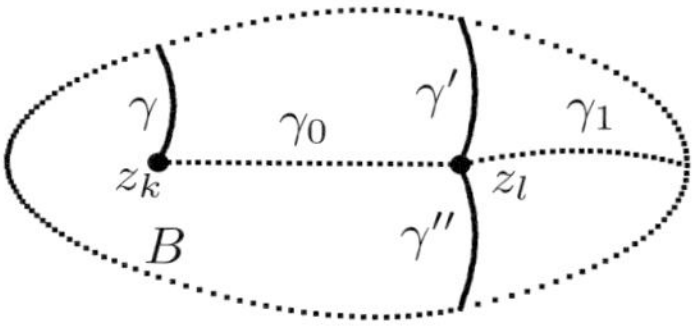

B and its slit

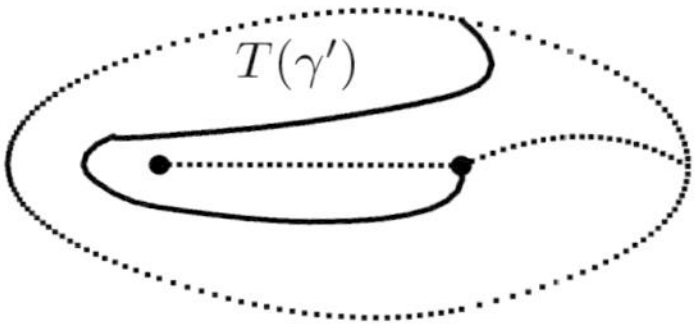

Effect of the Dehn twist T on γ'

There is essentially one way in which an arc can enter B and end up in z_k while avoiding the slit before hitting its destination: the final stretch is just a simple arc γ from a point of ∂B to z_k. For an arc ending in z_l, there are two such: one stays at the left and the other stays at the right of the slit. Let γ' and γ'' be such arcs. Then one sees that

$$\begin{aligned}\int_{T(\gamma)} \eta_z &= \int_{\gamma} \eta_z + (\bar{u}_k - \bar{u}_k\bar{u}_l) \int_{\gamma_0} \eta_z, \\ \int_{T(\gamma')} \eta_z &= \int_{\gamma'} \eta_z + (-1 + \bar{u}_k) \int_{\gamma_0} \eta_z, \\ \int_{T(\gamma'')} \eta_z &= \int_{\gamma''} \eta_z + (-u_k u_l + u_l) \int_{\gamma_0} \eta_z.\end{aligned} \tag{2.1}$$

These formulae cover the general case by taking linear combinations of composites of such arcs. For instance, γ_0 itself is isotopic to a difference of an arc of the second type and an arc of the first type and therefore

$$\int_{T(\gamma_0)} \eta_z = \int_{\gamma_0} \eta_z + ((-1 + \bar{u}_k) - (\bar{u}_k - \bar{u}_k\bar{u}_l)) \int_{\gamma_0} \eta_z = \bar{u}_k\bar{u}_l \int_{\gamma_0} \eta_z.$$

Another example is an arc $\tilde{\gamma}$ which simply crosses γ_0. This arc is isotopic to the difference of an arc of the third type and an arc of the second type and so

$$\int_{T(\tilde{\gamma})} \eta_z = \int_{\tilde{\gamma}} \eta_z + (-(-u_k u_l + u_l) - (-1 + \bar{u}_k)) \int_{\gamma_0} \eta_z$$
$$= \int_{\tilde{\gamma}} \eta_z + (u_k u_l - 1)(\bar{u}_k - 1) \int_{\gamma_0} \eta_z .$$

Recall that $\mu_k, \mu_l \in (0,1)$, so that $u_k \neq 1 \neq u_l$ and $\mu_k + \mu_l \in (0,2)$. By a *complex reflection* we mean here a semisimple transformation which fixes a hyperplane pointwise. One now easily deduces from the monodromy formulae (2.1):

Corollary 2.1. *If $\mu_k + \mu_l \neq 1$, then T acts in $\mathbb{C}^n$ semisimply as a complex reflection over an angle $e^{2\pi\sqrt{-1}(\mu_k+\mu_l)}$. If $\mu_k + \mu_l = 1$, then T acts in $\mathbb{C}^n$ as a nontrivial unipotent transformation. In particular, T acts with finite order if and only if $\mu_k + \mu_l$ is a rational number $\neq 1$.*

In the elliptic and hyperbolic cases, T will be an orthogonal reflection with respect the Hermitian form H; in the parabolic case, it will be restrict to $\mathbb{A}_{n-1}$ as an orthogonal affine reflection.

The Dehn twist T has a natural square root $\sqrt{T}$ which has the effect of interchanging z_k and z_l. It preserves η_z up to a scalar factor if $\mu_k = \mu_l$, and in that case a similar discussion shows:

$$\int_{\sqrt{T}(\gamma)} \eta_z = \int_{\gamma} \eta_z + \bar{u}_k \int_{\gamma_0} \eta_z ,$$
$$\int_{\sqrt{T}(\gamma')} \eta_z = \int_{\gamma'} \eta_z - \int_{\gamma_0} \eta_z , \quad \int_{\sqrt{T}(\gamma'')} \eta_z = \int_{\gamma''} \eta_z - u_k^2 \int_{\gamma_0} \eta_z . \tag{2.2}$$

Corollary 2.2. *If $\mu_k = \mu_l \neq \frac{1}{2}$, then $\sqrt{T}$ acts in $\mathbb{C}^n$ semisimply as a complex reflection over an angle $e^{2\pi\sqrt{-1}\mu_k}$. If $\mu_k = \mu_l = \frac{1}{2}$, then $\sqrt{T}$ acts in $\mathbb{C}^n$ as a nontrivial unipotent transformation. In particular, $\sqrt{T}$ acts with finite order if and only if $\mu_k = \mu_l$ is a rational number $\neq \frac{1}{2}$.*

2.2. Extension of the evaluation map

The Γ-covering $\widetilde{V_n^\circ} \to V_n^\circ$ can sometimes be extended as a ramified Γ-covering over a bigger open subset $V_n^f \supset V_n^\circ$ of V_n (the superscript f stands for $f$inite ramification; we may write $V_n^{f(\mu)}$ instead of V_n^f if such precision is required). This means that we find a normal analytic variety $\widetilde{V_n^f}$ which contains $\widetilde{V_n^\circ}$ as an open-dense subset and to which the Γ-action extends such that the Γ-orbit space can be identified with V_n^f. This involves a standard tool in analytic geometry that presumably goes back to Riemann and now falls under the heading of *normalization.* It goes like this. If $v \in V_n$ has a connected neighborhood U_v in V_n such that one (hence every) connected component of its preimage in $\widetilde{V_n^\circ}$ is finite over $U_v \cap V_n^\circ$, then the Γ-covering over $U_v \cap V_n^\circ$ extends to a ramified Γ-covering over U_v. The property imposed on U_v is equivalent to having finite monodromy over

$U_v \cap V_n^\circ$. The extension is unique and so if V_n^f denotes the set of $v \in V_n$ with this property, then a ramified Γ-covering $\widetilde{V_n^f} \to V_n^f$ exists as asserted. The composite map $\widetilde{V_n^f} \to V_n^f \to V_n$ is often called the *normalization of* V_n *in* $\widetilde{V_n^\circ}$, because this is the geometric counterpart of taking integral closure of a ring in finite extension of that ring, at least when the maps involved have finite degree. The naturality of the construction also ensures that the $\tilde{\mathbb{C}}^\times$-action on $\widetilde{V_n^\circ}$ (which covers the $\mathbb{C}^\times$-action on V_n°) extends to $\widetilde{V_n^f}$.

The space V_n receives a natural stratification from the stratification of $\mathbb{C}^{n+1}$ by its diagonals and since the topology of V_n^f along strata does not change, V_n^f is a union of strata that is open in V_n. The codimension one strata are of the form $D_{k,l}$, $0 \le k < l \le n$, parameterizing the z for which $z_k = z_l$, but no other equality among its components holds.

Lemma 2.3. *The stratum $D_{k,l}$ lies in V_n^f if and only if $\mu_k + \mu_l$ is a rational number $\neq 1$. The Schwarz map extends over the preimage of $\mathbb{P}(D_{k,l}))$ holomorphically if and only if $\mu_k + \mu_l < 1$ and it does so as a local isomorphism if and only if $1 - \mu_k - \mu_l$ is the reciprocal of a positive integer. If $|\mu| \neq 1$, then the corresponding assertions also hold for the Lauricella map.*

Proof. In order that $D_{k,l} \subset V_n^f$, it is necessary and sufficient that we have finite monodromy along a simple loop around $D_{k,l}$. This monodromy is the image of a Dehn twist along a circle separating z_k and z_l from the other elements of $\{z_0, \dots, z_n\}$. So the first assertion follows from Corollary 2.1.

If γ_0 connects z_k with z_l within the circle specified above, then $\int_{\gamma_0} \eta_z = (z_k - z_l)^{1-\mu_k-\mu_l} \exp(\text{holom})$. This is essentially a consequence of the identity

$$\int_0^\varepsilon t^{-\mu_k}(t-\varepsilon)^{-\mu_l}\,dt = \varepsilon^{1-\mu_k-\mu_l} \int_0^1 t^{-\mu_k}(t-1)^{-\mu_l}\,dt.$$

Suppose now that $\mu_k + \mu_l \in \mathbb{Q} - \{1\}$ and write $1 - \mu_k - \mu_l = p/q$ with p, q relatively prime integers with $q > 0$. So the order of the monodromy is q and over the preimage of a point of $D_{k,l}$, we have a coordinate $\tilde{z}_{k,l}$ with the property that $z_k - z_l$ pulls back to $\tilde{z}_{k,l}^q$. Hence $\int_{\gamma_0} \eta_z$ pulls back to $\tilde{z}_{k,l}^p$. In order that the Schwarz map extends over the preimage of $D_{k,l}$ holomorphically (resp. as a local isomorphism), a necessary condition is that the Lauricella function $\int_{\gamma_0} \eta_z$ (which after all may be taken as part of a basis of Lauricella functions) is holomorphic (resp. has a nonzero derivative everywhere). This means that $p > 0$ (resp. $p = 1$). It is not hard to verify that this is also sufficient. □

2.3. The elliptic and parabolic cases

Here the main result is:

Theorem 2.4 (Elliptic case). *Suppose that $|\mu| < 1$ and that for all $0 \le k < l \le n$, $1 - \mu_k - \mu_l$ is the reciprocal of an integer. Then Γ is a finite complex reflection group* *in* $\mathrm{GL}(n,\mathbb{C})$ *(so that in particular* $V_n^f = V_n$*) and* $F : \widetilde{V_n} \to \mathbb{C}^n$ *is a* Γ*-equivariant isomorphism which descends to an isomorphism* $V_n \to \Gamma\backslash\mathbb{C}^n$*.*

So $\mathbb{P}(V_n)$ acquires in these cases the structure of an orbifold modeled on Fubini–Study space. We remark that the hypotheses of this theorem are quite strong: only very few weight functions μ satisfy its hypotheses. At the same time we prove a proposition that will be also useful later. Observe that a stratum of V_n is given by a partition of $\{0,\dots,n\}$: for z in this stratum we have $z_k = z_l$ if and only if k and l belong to the same part. Let us say that this stratum is *stable relative to* μ if its associated partition has the property that every part has μ-weight < 1. We denote by $V_n^{\mathrm{st}} \subset V_n$ (or $V_n^{\mathrm{st}(\mu)} \subset V_n$) the union of stable strata.

Proposition 2.5. *Suppose that whenever* $0 \le k < l \le n$ *are such that* $\mu_k + \mu_l < 1$*, then* $1-\mu_k-\mu_l$ *is the reciprocal of an integer. Then* $V_n^{\mathrm{st}} \subset V_n^f$ *and* $\widetilde{V_n^{\mathrm{st}}}$ *is a complex manifold. The Lauricella admits a holomorphic extension over this manifold which has the same regularity properties as the map it extends: it is a local isomorphism when we are not in the parabolic case, whereas in the parabolic case, the Schwarz map defines a local isomorphism to* $\mathbb{A}_{n-1}$*.*

We shall need:

Lemma 2.6. *Let* $f : X \to Y$ *be a local diffeomorphism from a manifold to a connected Riemannian manifold. Assume that* X *is complete for the induced metric. Then* f *is a covering map.*

Proof. We use the theorem of Hopf–Rinow which says that completeness is equivalent to the property that every geodesic extends indefinitely as a geodesic. Let $y \in Y$. Choose $\varepsilon > 0$ such that the closed ε-ball $\bar{B}(y,\varepsilon)$ is geodesically convex (i.e., any points of $\bar{B}(y,\varepsilon)$ are joined by a unique geodesic in $\bar{B}(y,\varepsilon)$) and is the diffeomorphic image of the closed ε-ball in T_yY under the exponential map. It is enough to show that for every $x \in f^{-1}\bar{B}(y,\varepsilon)$, the connected component C_x of x in $f^{-1}\bar{B}(y,\varepsilon)$ is mapped by f diffeomorphically onto $\bar{B}(y,\varepsilon)$. Since X is complete, so is C_x. The geodesic convexity of $\bar{B}(y,\varepsilon)$ is then easily shown to imply the same property for C_x. It follows that C_x maps diffeomorphically onto $\bar{B}(y,\varepsilon)$. □

We now begin the proofs of Theorem 2.4 and Proposition 2.5. Let us write A_k for the assertion of Theorem 2.4 for $k+1$ points and B_k for the assertion of Proposition 2.5 for elliptic strata of codimension $\le k$. Let us observe that B_1 holds: an elliptic stratum of codimension one is a stratum of the form $D_{k,l}$ satisfying the hypotheses of Lemma 2.3. We now continue with induction following the scheme below.

Proof that A_k *implies* B_k*.* Consider a stratum of codimension k. Let us first assume that it is irreducible in the sense that it is given by a single part. Without loss of generality we may then assume that it is open-dense in the locus $z_0 = \cdots = z_k$. This is the situation we studied in Subsection 1.4 (mainly for this reason, as we can now confess). We found that F extends as a multivalued map defined on an open subset of the blowup $\mathrm{Bl}_{D_k} V_n$ going to the blowup $\mathrm{Bl}_{0\times\mathbb{C}^{n-k}}\mathbb{C}^n$. On the exceptional divisor, F is the product of the Schwarz map for $\mu'=(\mu_0,\dots,\mu_k)$ and the Lauricella map for $(|\mu'|,\mu_{k+1},\dots,\mu_n)$. Our hypothesis A_k then implies that the projectivized monodromy near a point of the stratum is finite. Equation (1.6) shows that in the transversal direction (the ε coordinate) the multivaluedness is like that of $(\varepsilon)^{1-|\mu'|}$. Since $\mu_i+\mu_j\in\mathbb{Q}$ for all $0\le i<j\le k$ and the sum of these numbers is $k|\mu'|$, it follows that $|\mu'|\in\mathbb{Q}$. So we also have finite order monodromy along the exceptional divisor. This implies that we have finite local monodromy at a point of the stratum: the stratum lies in V^f. With this in mind, we now see from Equation (1.6) that we proved slightly more, namely that this local monodromy group is the one associated to the Lauricella system of type μ'. So we may invoke A_k to conclude that $\widetilde{V_n^{\mathrm{st}}}$ is in fact smooth over this stratum.

In the general case, with a stratum corresponding to several clusters forming, we have topologically a product situation: the local monodromy group near a point of that stratum decomposes as a product with each factor corresponding to a cluster being formed. It is clear that if each cluster is of elliptic type, then so is the stratum. Its preimage in $\widetilde{V_n^{\mathrm{st}}}$ will be smooth.

The asserted regularity properties of this extension of the Lauricella map hold on codimension one strata by Lemma 2.3. But then they hold everywhere, because the locus where a holomorphic map between complex manifolds of the same dimension fails to be a local isomorphism is of codimension ≤ 1. □

Proof that B_{k-1} implies A_k. Since B_{k-1} holds, it follows that V_k^f contains $V_k-\{0\}$. Thus $\mathbb{P}F:\mathbb{P}(\widetilde{V_k})\to\mathbb{P}_{k-1}$ is defined. The latter is a Γ-equivariant local isomorphism with Γ acting on $\mathbb{P}(\widetilde{V_k})$ with compact fundamental domain (for its orbit space is the compact $\mathbb{P}(V_k)$) and on the range as a group of isometries. This implies that $\mathbb{P}(\widetilde{V_k})$ is complete. According to Lemma 2.6, $\mathbb{P}F$ is then a covering projection. Hence so is $F:\widetilde{V_k-\{0\}}\to\mathbb{C}^k-\{0\}$. Since the domain of the latter is connected and the range is simply connected, this map is an isomorphism. In particular, $\mathbb{P}(\widetilde{V_k})$ is compact, so that the covering $\mathbb{P}(\widetilde{V_k})\to\mathbb{P}(V_k)$ is finite. This means that the projectivization of Γ is finite. On the other hand, the $\mathbb{C}^\times$-action on $V_k-\{0\}$ needs a finite cover (of degree equal to the denominator of $1-|\mu|$) to lift to $\widetilde{V_k-\{0\}}$. This implies that Γ is finite, so that $V_k^f=V_k$. It is now clear that $F:\widetilde{V_k}\to\mathbb{C}^k$ is an isomorphism. It is Γ-equivariant and descends to an isomorphism $V_k\to\Gamma\backslash\mathbb{C}^k$ of affine varieties. □

In the parabolic case $\mathbb{P}(V_n)$ acquires the structure of an orbifold modeled on flat space:

Proposition 2.7 (Parabolic case). *Suppose that $|\mu|=1$ and that for all $0\le k<l\le n$, $1-\mu_k-\mu_l$ is the reciprocal of an integer. Then Γ acts as a complex Bieberbach group in $\mathbb{A}_{n-1}$, $V_n^f=V_n-\{0\}$ and $\mathbb{P}F:\mathbb{P}(\widetilde{V_n})\to\mathbb{A}_{n-1}$ is a Γ-equivariant isomorphism which descends to an isomorphism $\mathbb{P}(V_n)\to\Gamma\backslash\mathbb{A}_{n-1}$.*

Proof. It follows from Proposition 2.5 that V_n^f contains $V_n - \{0\}$ so that $\mathbb{P}F : \mathbb{P}(\widetilde{V_n}) \to \mathbb{A}_{n-1}$ is defined. The latter is a Γ-equivariant local isomorphism with Γ acting on the $\mathbb{P}(\widetilde{V_n})$ with compact fundamental domain and on the range as a group of isometries. Hence $\mathbb{P}(\widetilde{V_n})$ is complete. It follows from Lemma 2.6 that $\mathbb{P}F$ is a Γ-equivariant isomorphism. It also follows that Γ acts on $\mathbb{A}_{n-1}$ discretely with compact fundamental domain. This group is generated by complex reflections, in particular it is a complex Bieberbach group. Clearly, $\mathbb{P}F$ induces an isomorphism $\mathbb{P}(V_n) \cong \Gamma\backslash\mathbb{A}_{n-1}$. □

We have also partial converses of Theorem 2.4 and Proposition 2.7. They will be consequences of

Lemma 2.8. *The Lauricella map extends holomorphically over any stable stratum contained in* V_n^f.

Proof. Let $S \subset \{0,\dots,n\}$ be such that $\sum_{k\in S}\mu_k < 1$ and assume that the stratum D_S that is open dense in the diagonal defined by S is contained in V_n^f. For $0 \le k < l \le n$ in S, we have $\mu_k + \mu_l \le |\mu| < 1$ and so the associated monodromy transformation T is according to Corollary 2.1 a reflection over an angle $2\pi(\mu_k+\mu_l)$. Since $D_S \subset V_n^f$, we must have $\mu_k+\mu_l \in \mathbb{Q}$. Lemma 2.3 tells us that F then extends holomorphically over the preimage of $D_{k,l}$. The usual codimension argument then shows that this is also the case over the preimage of D_S. □

Proposition 2.9. *If* $|\mu| < 1$ *and* Γ *is finite, then the Lauricella map descends to a finite map* $V_n \to \Gamma\backslash\mathbb{C}^n$.

If $|\mu| = 1$, $n > 1$ *and* Γ *acts on the complex Euclidean space* $\mathbb{A}_{n-1}$ *as a complex Bieberbach group, then* $V_n^f = V_n$ *and the Schwarz map descends to a finite map* $\mathbb{P}(V_n) \to \Gamma\backslash\mathbb{A}_{n-1}$.

Proof. In the elliptic case, it follows from Lemma 2.8 that the map F descends to a map $V_n \to \Gamma\backslash\mathbb{C}^n$ which exists in the complex-analytic category. The map in question is homogeneous (relative to the natural $\mathbb{C}^\times$-actions) and the preimage of 0 is 0. Hence it must be a finite morphism. In the parabolic case, the lemma implies that the Schwarz map determines a map $\mathbb{P}(V_n) \to \Gamma\backslash\mathbb{A}_{n-1}$ which lives in the complex-analytic category. This map will be finite, because its fibers are discrete and its domain is compact. □

3. The hyperbolic case

Throughout this section we always suppose that $1 < |\mu| < 2$.

3.1. A projective set-up

An important difference with the elliptic and the parabolic cases is that $z_{n+1} = \infty$ is now of the same nature as the finite singular points, since we have $\mu_{n+1} = 2 - |\mu| \in (0,1)$. This tells us that we should treat all the points $z_0,\dots,z_{n+1}$ on the same footing. In more precise terms, instead of taking $z_{n+1} = \infty$ and study the transformation behavior of the Lauricella integrals under the affine group $\mathbb{C}^\times \ltimes \mathbb{C}$ of $\mathbb{C}$, we should let $z_0, \dots, z_{n+1}$ be distinct, but otherwise arbitrary points of $\mathbb{P}^1$ and let $\mathrm{PGL}(2,\mathbb{C})$ play the role of the affine group. This means in practice that we will sometimes allow some finite z_k to coalesce with z_{n+1} (that is, to fly off to infinity). For this we proceed as follows. Let $Z_0, \dots, Z_{n+1}$ be nonzero linear forms on $\mathbb{C}^2$ defining distinct points $z_0, \dots, z_{n+1}$ of $\mathbb{P}^1$. Consider the multivalued 2-form on $\mathbb{C}^2$ defined by

$$Z_0(\zeta)^{-\mu_0} \cdots Z_{n+1}(\zeta)^{-\mu_{n+1}} d\zeta_0 \wedge d\zeta_1.$$

Let us see how this transforms under the group $\mathrm{GL}(2,\mathbb{C})$. The subgroup $\mathrm{SL}(2,\mathbb{C})$ leaves $d\zeta_0 \wedge d\zeta_1$ invariant, and so the above form transforms under $\mathrm{SL}(2,\mathbb{C})$ via the latter's diagonal action on the $(\mathbb{C}^2)^{n+2}$ (the space that contains $Z = (Z_0, \dots, Z_{n+1})$). The subgroup of scalars, $\mathbb{C}^\times \subset \mathrm{GL}(2,\mathbb{C})$ leaves the 2-form invariant, as $\sum_{k=0}^{n+1} = 2$. So the form has a pole of order one along the projective line $\mathbb{P}^1$ at infinity. We denote the residue of that form on $\mathbb{P}^1$ by η_Z. It is now clear, that a Lauricella function $\int_\gamma \eta_Z$ will be $\mathrm{GL}(2,\mathbb{C})$-invariant. Since the 2-form (and hence η_Z) is homogeneous of degree $-\mu_k$ in Z_k, it follows that the quotient of two Lauricella functions will only depend on the $\mathrm{GL}(2,\mathbb{C})$-orbit of $(z_0, \dots, z_{n+1})$.

Let $\mathcal{Q}_\mu^\circ$ denote the $\mathrm{SL}(2,\mathbb{C})$-orbit space of the subset of $(\mathbb{P}^1)^{n+2}$ parameterizing distinct $(n+2)$-tuples in $\mathbb{P}^1$. This is in a natural way a smooth algebraic variety which can be identified with $\mathbb{P}(V_n^\circ)$ (every orbit is represented by an $(n+2)$-tuple of which the last point is ∞). So we have a Γ-covering $\widetilde{\mathcal{Q}}_\mu^\circ \to \mathcal{Q}_\mu^\circ$ and a local isomorphism $\mathbb{P}F : \widetilde{\mathcal{Q}}_\mu^\circ \to \mathbb{B}_{n-1}$. Thus far our treatment of z_{n+1} as one of the other z_i's has not accomplished anything, but it will matter when we seek to extend it as a ramified covering.

We say that $z = (z_0, \dots, z_{n+1}) \in (\mathbb{P}^1)^{n+2}$ is *μ-stable* (resp. *μ-semistable*) if the $\mathbb{R}$-divisor $\mathrm{Div}(z) := \sum_{k=0}^{n+1} \mu_k(z_k)$ has no point of weight ≥ 1 (resp. > 1). Let us denote the corresponding (Zariski open) subsets of $(\mathbb{P}^1)^{n+2}$ by U_μ^{st} resp. U_μ^{sst}. Notice that when z is μ-stable, the support of $\sum_{k=0}^{n+1} \mu_k(z_k)$ has at least three points. This implies that the $\mathrm{SL}(2,\mathbb{C})$-orbit space (denoted $\mathcal{Q}_\mu^{\mathrm{st}}$) of U_μ^{st} is in a natural manner a nonsingular algebraic variety: given a μ-stable point z, we can always pick three pairwise distinct components for use as an affine coordinate for $\mathbb{P}^1$. By means of this coordinate we get a nonempty Zariski-open subset in $(\mathbb{P}^1)^{n-1}$ which maps bijectively to an open subset of $\mathcal{Q}_\mu^{\mathrm{st}}$. These bijections define an atlas for the claimed structure. In the strictly semistable case, we can choose a coordinate for $\mathbb{P}^1$ such that ∞ has weight 1 and the weighted sum of the remaining points is 0; using the $\mathbb{C}^\times$-action on $\mathbb{P}^1$ with 0 and ∞ as fixed points we see that this divisor has in its closure the divisor $(0) + (\infty)$. Geometric Invariant Theory tells us that the stable orbits and the minimal strictly semistable orbits are the points of a projective variety (at least when the μ_k's are all rational): $\mathcal{Q}_\mu^{\mathrm{st}}$ is compactified to a normal projective variety by adding just finitely many points: one for each orbit containing a point whose associated divisor is $(0) + (\infty)$ or equivalently, one for each splitting of $\{0, \dots, n+1\}$ into two subsets, each of which of total μ-weight

1. (So if no such splitting exists, then $\mathcal{Q}_\mu^{\mathrm{st}}$ is already a projective variety.) Let us denote that projective compactification by $\mathcal{Q}_\mu^{\mathrm{sst}}$.

Theorem 3.1. *Assume that for every pair $0 \leq k < l \leq n+1$ for which $\mu_k + \mu_l < 1$, $1 - \mu_k - \mu_l$ is the reciprocal of an integer. Then the monodromy covering $\widetilde{\mathcal{Q}}_\mu^\circ \to \mathcal{Q}_\mu^\circ$ extends to a ramified covering $\widetilde{\mathcal{Q}}_\mu^{\mathrm{st}} \to \mathcal{Q}_\mu^{\mathrm{st}}$ and F extends to a Γ-equivariant isomorphism $\widetilde{\mathcal{Q}}_\mu^{\mathrm{st}} \to \mathbb{B}_{n-1}$. Moreover Γ acts in $\mathbb{B}_m$ discretely and with finite covolume; this action is with compact fundamental domain if and only if no subsequence of μ has weight 1.*

Remarks 3.2. Our hypotheses imply that the μ_k's are all rational so that the GIT compactification $\mathcal{Q}_\mu^{\mathrm{sst}}$ makes sense. The compactification of $\Gamma\backslash\mathbb{B}_{n-1}$ that results by $\Gamma\backslash\mathbb{B}_{n-1} \cong \mathcal{Q}_\mu^{\mathrm{st}} \subset \mathcal{Q}_\mu^{\mathrm{sst}}$ coincides with the *Baily–Borel compactification* of $\Gamma\backslash\mathbb{B}_{n-1}$.

The cohomology and intersection homology of the variety $\mathcal{Q}_\mu^{\mathrm{sst}}$ has been investigated by Kirwan–Lee–Weintraub [11].

Before we begin the proof of Theorem 3.1 we need to know a bit about the behavior of the complex hyperbolic metric on a complex ball near a cusp. Let W be a finite dimensional complex vector space equipped with a nondegenerate Hermitian form H of hyperbolic signature so that $H(w,w) > 0$ defines a complex ball $\mathbb{B}(W) \subset \mathbb{P}(W)$. Let $e \in W$ be a nonzero isotropic vector. Since its orthogonal complement is negative semidefinite, every positive definite line will meet the affine hyperplane in W defined by $H(w,e) = -1$. In this way we find an open subset Ω in this hyperplane which maps isomorphically onto $\mathbb{B}(W)$. This is what is called a realization of $\mathbb{B}(W)$ as a Siegel domain of the second kind.

Lemma 3.3. *This subset Ω of the affine space $H(w,e) = -1$ is invariant under the half group of translations over τe with $\mathrm{Re}(\tau) \geq 0$. If $K \subset \Omega$ is compact and measurable, then $K + \sqrt{-1}\mathbb{R}_{\geq 0} e$ is as a subset of Ω complete and of finite volume. More precisely, if K_o is a compact ball in a hyperplane section of Ω that is not parallel to e and $S(R) \subset \mathbb{C}$ is the set of $\tau \in \mathbb{C}$ with $\mathrm{Re}(\tau) \geq 0$, $|\mathrm{Im}(\tau)| \leq R$, then*

$$(w,\tau) \in K_o \times S(R) \mapsto w + \tau e_0 \in \Omega$$

is an embedding and the pull-back of the complex hyperbolic metric of Ω is for $\mathrm{Re}(\tau) \geq 1$ comparable to the 'warped' metric $(\mathrm{Re}\,\tau)^{-1}(g_\Omega|_{K_o}) + (\mathrm{Re}\,\tau)^{-2}|d\tau|^2$. In particular, the pull-back of the volume element of Ω is there comparable to $\mathrm{Re}(\tau)^{-\dim\Omega} dvol_{K_o} dvol_{\mathbb{C}}$.

Proof. This well-known result is straightforward to verify. We first check the invariance of Ω under the half group. Let $e_1 \in W$ be another isotropic vector such that $H(e,e_1) = 1$, and denote by U the orthogonal complement of the span of e, e_1. So if we write $\omega = we + w_1 e_1 + u$ with $u \in U$, then Ω is defined by $w_1 = -1$ and $\mathrm{Re}(w) > \frac{1}{2}H(w',w')$. This shows in particular that Ω is invariant under translation by τe, when $\mathrm{Re}(\tau) \geq 0$. The other assertions follow easily from an explicit description of the metric on Ω in terms of the coordinates (w,z): a calculation shows that it is on Ω at (w_o, u_o) given by

$$\frac{|dw - H(du, u_o)|^2}{(\operatorname{Re} w_o - \frac{1}{2}H(u_o, u_o))^2} + c\frac{H(du, du)}{\operatorname{Re} w_o - \frac{1}{2}H(u_o, u_o)}$$

for some positive constant c (which we did not bother to work out). □

It follows from Proposition 2.5 that $\mathcal{Q}_\mu^{\mathrm{st}} \subset \mathcal{Q}_\mu^f$ and that the Schwarz map $\mathbb{P}F : \widetilde{\mathcal{Q}}_\mu^{\mathrm{st}} \to \mathbb{B}_{n-1}$ is a local isomorphism. So $\mathcal{Q}_\mu^{\mathrm{st}}$ inherits a metric from $\mathbb{B}_{n-1}$. We need to show that $\mathcal{Q}_\mu^{\mathrm{st}}$ is complete and has finite volume. The crucial step toward this is:

Lemma 3.4. *Let $0 < r < n$ be such that $\mu_0 + \cdots + \mu_r = 1$. Denote by D° the set of $(z_0, \dots, z_n) \in \mathbb{C}^{n+1}$ satisfying $z_0 = 0 < |z_1| < \cdots < |z_r| \leq \frac{1}{2}$, $z_{r+1} = 1 < |z_{r+2}| < \cdots < |z_n|$. Then D° embeds in $\mathcal{Q}_\mu^\circ$ and its closure D in $\mathcal{Q}_\mu^{\mathrm{st}}$ is complete and of finite volume.*

Proof. First notice that D° is contained in the affine space defined by $z_0 = 0$, $z_{r+1} = 1$. This affine space embeds in $\mathbb{P}(V_n)$ with image the complement of the hyperplane defined by $Z_0 = Z_{r+1}$. It is clear that under this embedding, D° lands in $\mathcal{Q}_\mu^\circ$. Let $D' \subset D^\circ$ be the open-dense subset of $z \in D^\circ$ for which none of z_k, $k = 1, \dots, n$ is real ≤ 0. This subset of D° is simply connected and there is a natural isotopy class of L-slits δ for every $z \in D'$ characterized by the property that δ_k never crosses the negative real axis and $|\delta_k|$ is monotonous. This defines a lift $\tilde{D}'$ of D' to $\widetilde{\mathcal{Q}}_\mu^{\mathrm{st}}$. We denote its closure by $\tilde{D}$ so that is defined $F : \tilde{D} \to \mathbb{C}^n$. Notice that the image of $\tilde{D}$ in $\mathcal{Q}_\mu^{\mathrm{st}}$ equals D.

Let D'_o the part of D' for which $z_r = \frac{1}{2}$. It is easy to see that D'_o has compact closure in $\mathcal{Q}_\mu^{\mathrm{st}}$ and so the closure of its preimage in $\tilde{D}$ (which we shall denote by $\tilde{D}_o$) is compact as well. We parametrize $\tilde{D}$ by $\tilde{D}_o \times S(\pi)$ with the help of the map introduced in Equation (1.5):

$$\Phi : (\tilde{z}, \tau) \in \tilde{D}_o \times S(\pi) \mapsto (e^{-\tau}\tilde{z}_0, \dots, e^{-\tau}\tilde{z}_r, \tilde{z}_{r+1}, \dots, \tilde{z}_n),$$

where we recall that $S(\pi)$ is the set of $\tau \in \mathbb{C}$ with $\operatorname{Re} \tau \geq 0$ and $|\operatorname{Im} \tau| \leq \pi$. Since $\mu_0 + \cdots + \mu_r = 1$, Lemma 1.9 applies here. As in that lemma, we put $\hat{F} := z_{r+1}^{-\mu_{r+1}} \cdots z_n^{-\mu_n} F$. According to that lemma we have $\sum_{k=1}^r \operatorname{Im}(w_k)\hat{F}_k(z) = 0$. This amounts to saying that $H(\hat{F}, e_{r+1}) = -\pi$, where e_{r+1} denotes the $(r+1)$-st basis vector of $\mathbb{C}^n$. (For $H(F, G) = \sum_{1 \leq j < k \leq n+1} \operatorname{Im}(w_j \bar{w}_k)\bar{G}_j F_k$ and so $H(e_{r+1}, G) = \sum_{1 \leq j \leq r} \operatorname{Im}(w_j)\bar{G}_j$.) We also notice that $H(e_{r+1}, e_{r+1}) = 0$. So $\hat{F}$ maps to the Siegel domain Ω defined in Lemma 3.4 if we take $e := \pi^{-1}e_{r+1}$. From Lemma 1.9 we see that the coordinates $\hat{F}_k\Phi$ stay bounded for all $k \neq r+1$. Since $\hat{F}_k$ is holomorphic, the same is true for its derivatives. On the other hand, $\tau^{-1}\hat{F}_{r+1}\Phi(z, \tau)$ has a limit for $\operatorname{Re}(\tau) \to \infty$. From Lemma 3.4 we then learn that the image of $\hat{F}\Phi$ is complete and that the volume element of Ω pulled back by $\Phi\hat{F}$ is comparable with $\operatorname{Re}(\tau)^{1-n}$ times the volume element of a Euclidean product volume element on $\tilde{D}_o \times S(\pi)$. These properties imply that D is complete and has finite volume. □

Proof of Theorem 3.1. The GIT compactification $\mathcal{Q}_\mu^{\rm sst}$ of $\mathcal{Q}_\mu^{\rm st}$ adds a point for every permutation σ of $\{0,\dots,n\}$ for which $\mu_{\sigma(0)}+\cdots+\mu_{\sigma(r)}=1$ for some $0<r<n$. If σ is such a permutation, then we have defined an open subset $D_\sigma\subset\mathcal{Q}_\mu^\circ$ as in Lemma 3.4 and according to that Lemma, the closure of D_σ in $\mathcal{Q}_\mu^{\rm st}$ is complete and of finite volume. The complement in $\mathcal{Q}_\mu^{\rm st}$ of the union of these closures is easily seen to be compact. Hence $\mathcal{Q}_\mu^{\rm st}$ is complete and of finite volume. The theorem now follows from Lemma 2.6 (bearing in mind that $\mathcal{Q}_\mu^{\rm sst}=\mathcal{Q}_\mu^{\rm st}$ if and only if no subsequence of μ has weight 1). □

3.2. Extending the range of applicability

We begin with stating a partial converse to Theorem 3.1, the hyperbolic counterpart of Proposition 2.9:

Proposition 3.5. *Suppose that* $1<|\mu|<2$, $n>1$ *and* Γ *acts on* $\mathbb{B}_{n-1}$ *as a discrete group. Then* Γ *has finite covolume and the Schwarz map descends to a finite morphism* $\mathcal{Q}_\mu^{\rm st}\to\Gamma\backslash\mathbb{B}_{n-1}$.

Proof. It follows from Lemma 2.8 that the Schwarz map is defined over $\mathcal{Q}_\mu^{\rm st}$ and hence descends to a map $\mathcal{Q}_\mu^{\rm st}\to\Gamma\backslash\mathbb{B}_{n-1}$. It follows from Lemma 3.4 (by arguing as in the proof of Theorem 3.1) that $\mathcal{Q}_\mu^{\rm st}$ is complete as a metric orbifold and of finite volume. This implies that $\mathcal{Q}_\mu^{\rm st}\to\Gamma\backslash\mathbb{B}_{n-1}$ is a finite morphism. □

This immediately raises the question which weight systems μ satisfy the hypotheses of Proposition 3.5. The first step toward the answer was taken by Mostow himself [12], who observed that if some of the weights μ_k coincide, then the conditions of Theorem 2.4, Proposition 2.7 and Theorem 3.1 may be relaxed, while still ensuring that Γ is a discrete subgroup of the relevant Lie group. The idea is this: if $\mathcal{S}_\mu$ denotes the group of permutations of $\{0,\dots,n+1\}$ which preserve the weights, then we should regard the Lauricella map F as being multivalued on $\mathcal{S}_\mu\backslash V_n^\circ$, rather than on V_n°. This can make a difference, for the monodromy cover of $\mathcal{S}_\mu\backslash V_n^\circ$ need not factor through V_n°. We get the following variant of Lemma 2.3

Lemma 3.6. *Suppose that in Lemma* 2.3 *we have* $\mu_k=\mu_l\in\mathbb{Q}-\{\frac12\}$. *Then the Lauricella map (the Schwarz map if* $|\mu|=1$*) extends over the image in* $D_{k,l}$ *in* $\mathcal{S}_\mu\backslash V_n^\circ$ *as a local isomorphism if and only if* $\frac12-\mu_k$ *is the reciprocal of a positive integer.*

Definition 3.7. We say that μ satisfies the *half integrality conditions* if whenever for $0\le k<l\le n+1$ we have $\mu_k+\mu_l<1$, then $(1-\mu_k-\mu_l)^{-1}$ is an integer or in case $\mu_k=\mu_l$, just half an integer.

This notion is a priori weaker than Mostow's ΣINT condition, but in the end it leads to the same set of weight systems. Now Proposition 2.5 takes the following form.

Proposition 3.8. *If* μ *satisfies the half integrality conditions, then* $V_n^{\rm st}\subset V_n^f$, $\widetilde{\mathcal{S}_\mu\backslash V_n^{\rm st}}$ *is nonsingular, and the Lauricella map extends holomorphically to* $\widetilde{\mathcal{S}_\mu\backslash V_n^{\rm st}}$.

This extension has the same regularity properties as the map it extends: it is a local isomorphism when we are not in the parabolic case, whereas in the parabolic case, the Schwarz map defines a local isomorphism to $\mathbb{A}_{n-1}$.

This leads to the theorem stated below (see [12] and for the present version, [7]).

Theorem 3.9. *Suppose that μ satisfies the half integrality conditions.*

ell: If $|\mu| < 1$, then Γ is a finite complex reflection group in $\mathrm{GL}(n,\mathbb{C})$ and $F : \widetilde{\mathcal{S}_\mu\backslash V_n} \to \mathbb{C}^n$ is a Γ-equivariant isomorphism which descends to an isomorphism $\mathcal{S}_\mu\backslash V_n \to \Gamma\backslash\mathbb{C}^n$.

par: If $|\mu| = 1$, then Γ acts as a complex Bieberbach group in $\mathbb{A}_{n-1}$, $V_n^f = V_n - \{0\}$ and $\mathbb{P}F : \mathbb{P}(\widetilde{\mathcal{S}_\mu\backslash V_n}) \to \mathbb{A}_{n-1}$ is a Γ-equivariant isomorphism which descends to an isomorphism $\mathbb{P}(\mathcal{S}_\mu\backslash V_n) \to \Gamma\backslash\mathbb{A}_{n-1}$.

hyp: If $1 < |\mu| < 2$, then the monodromy covering $\widetilde{\mathcal{S}_\mu\backslash \mathcal{Q}_\mu^\circ} \to \mathcal{S}_\mu\backslash \mathcal{Q}_\mu^\circ$ extends to a ramified covering $\widetilde{\mathcal{S}_\mu\backslash \mathcal{Q}_\mu^{\mathrm{st}}} \to \mathcal{S}_\mu\backslash \mathcal{Q}_\mu^{\mathrm{st}}$ and F extends to a Γ-equivariant isomorphism $\widetilde{\mathcal{S}_\mu\backslash \mathcal{Q}_\mu^{\mathrm{st}}} \to \mathbb{B}_{n-1}$. Moreover Γ acts discretely in $\mathbb{B}^m$ and with finite covolume.

Example. Let us take $n \leq 10$ and $\mu_k = \frac{1}{6}$ for $k = 0, \dots, n$. So we have $\mu_{n+1} = \frac{11-n}{6}$. The half integrality conditions are fulfilled for all $n \leq 10$ with $1 \leq n \leq 4$, $n = 5$, $6 \leq n \leq 11$ yielding an elliptic, parabolic and hyperbolic case, respectively and $\mathcal{S}_\mu$ is the permutation group of $\{0, \dots, n\}$ for $n \leq 9$ and the one of $\{0, \dots, 11\}$ for $n = 10$.

Mostow subsequently showed that in the hyperbolic range with $n \geq 3$ we thus find all but ten of the discrete monodromy groups of finite covolume: one is missed for $n = 4$ (namely $(\frac{1}{12}, \frac{3}{12}, \frac{5}{12}, \frac{5}{12}, \frac{5}{12}, \frac{5}{12})$) and nine for $n = 3$ (see [13], (5.1)). He conjectured that in these nine cases Γ is always commensurable with a group obtained from Theorem 3.9. This was proved by his student Sauter [14]. It is perhaps no surprise that things are a bit different when $n = 2$ (so that we are dealing with discrete groups of automorphism of the unit disk): indeed, the exceptions then make up a number of infinite series ([13], Theorem 3.8). It turns out that for $n > 10$ the monodromy group is never discrete and that for $n = 10$ this happens only when $\mu_k = \frac{1}{6}$ for $k = 0, \dots, 10$.

Other examples of discrete complex reflection groups of finite covolume have been found (among others) by Barthel–Hirzebruch–Höfer [3], Allcock [1], [2] and Couwenberg–Heckman–Looijenga [7]. A particular interesting example (acting on the complex ball of dimension 13) is described by Allcock in [2]. No higher dimensional example seems to be known. A piece of the family tree of such ball quotients is given by Doran in his thesis [10].

4. Modular interpretation

We assume here that we are in the $\mathbb{Q}$-hyperbolic case: $\mu_k \in (0,1)$ and rational for $k = 0, \dots, n+1$ (where we recall that $\mu_{n+1} = 2 - \sum_{k=0}^{n} \mu_k$).

4.1. Cyclic covers of $\mathbb{P}^1$

We will show that the Schwarz map can be interpreted as a 'fractional period' map. This comes about by passing to a cyclic cover of $\mathbb{P}^1$ on which the Lauricella integrand becomes a regular differential. Concretely, write $\mu_k = d_k/m$ with d_k, m positive integers such that the d_k's have no common divisor, and write m_k for the denominator of the reduced fraction μ_k. Consider the cyclic cover $C \to \mathbb{P}^1$ of order m which has ramification over z_k of order m_k. In affine coordinates, C is given as the normalization of the curve defined by

$$w^m = \prod_{k=0}^{n} (z_k - \zeta)^{d_k}.$$

This is a cyclic covering which has the group G_m of mth roots of unity as its Galois group: $g^*(w, z) = (\chi(g)w, z)$, where $\chi : G_m \subset \mathbb{C}^\times$ stands for the tautological character. The Lauricella integrand pulls back to a single-valued differential $\tilde{\eta}$ on C, represented by $w^{-1}d\zeta$ so that $g^*(\tilde{\eta}) = \bar{\chi}(g)\tilde{\eta}$. Hence, if we let G_m act on forms in the usual manner ($g \in G_m$ acts as $(g^{-1})^*$), then $\tilde{\eta}$ is an eigenvector with character χ. It is easily checked that $\tilde{\eta}$ is regular everywhere.

In order to put this in a period setting, we recall some generalities concerning the Hodge decomposition of C: its space of holomorphic differentials, $\Omega(C)$, has dimension equal to the genus g of C and $H^1(C;\mathbb{C})$ is canonically represented on the form level by the direct sum $\Omega(C) \oplus \overline{\Omega}(C)$ (complex conjugation on forms corresponds to complex conjugation in $H^1(C;\mathbb{C})$ with respect to $H^1(C;\mathbb{R})$). The intersection product on $H^1(C;\mathbb{Z})$ defined by $(\alpha, \beta) \mapsto (\alpha \cup \beta)[C]$ (where the fundamental class $[C] \in H_2(C, \mathbb{Z})$ is specified by the complex orientation of C), is on the level of forms given by $\int_C \alpha \wedge \beta$. The associated Hermitian form on $H^1(C;\mathbb{C})$ defined by $h(\alpha, \beta) := \frac{\sqrt{-1}}{2}(\alpha \cup \bar{\beta})[C] = \frac{\sqrt{-1}}{2}\int_C \alpha \wedge \bar{\beta}$ has signature (g, g). The Hodge decomposition $H^1(C;\mathbb{R}) = \Omega(C) \oplus \overline{\Omega}(C)$ is h-orthogonal with the first summand positive definite and the second negative definite. The Hodge decomposition, the intersection product and (hence) the Hermitian form h are all left invariant by the action of G_m.

Proposition 4.1. *The eigenspace $\Omega(C)^\chi$ is of dimension one and spanned by $\tilde{\eta}$ and the eigenspace $\overline{\Omega}(C)^\chi$ is of dimension $n-1$. The eigenspace $H^1(C, \mathbb{C})^\chi$ has signature $(1, n-1)$ and we have $h(\tilde{\eta}, \tilde{\eta}) = -mN(z)$.*

Lemma 4.2. *Let $r \in \{0, 1, \dots, m-1\}$. Then the eigenspace $\Omega(C)^{\chi^r}$ is spanned by the forms $w^{-r}f(\zeta)d\zeta$ where f runs over the polynomials of degree $< -1 + r\sum_{k=0}^{n} \mu_k$ that have a zero of order $\geq \lfloor r\mu_k \rfloor$ at z_k, $k = 0, \dots, n$. In particular, if r is relatively prime to m, then* $\dim \Omega(C)^{\chi^r} = -1 + \sum_{k=0}^{n+1} \{r\mu_k\}$ *(recall that $\{a\} := a - \lfloor a \rfloor$).*

Proof. Any meromorphic differential on C which transforms according to the character χ^r, $r = 0, 1, \dots, m-1$, is of the form $w^{-r}f(\zeta)d\zeta$ with f meromorphic. A local computation shows that in order that such a differential be regular, it is necessary and sufficient that f be a polynomial of degree $< -1 + r\sum_{k=0}^n \mu_k$ which has a zero of order $> -1 + r\mu_k$ at z_k, that is, of order $\geq \lfloor r\mu_k \rfloor$. Hence $\dim \Omega(C)^{\chi^r}$ is the largest integer smaller than $\sum_{k=0}^n \{r\mu_k\}$. Suppose now that r is relatively prime to m. Then $r\mu_k \notin \mathbb{Z}$ for every k. Since $\sum_{k=0}^{n+1} r\mu_k = 2r$ it follows that the largest integer smaller than $\sum_{k=0}^n \{r\mu_k\}$ is $-1 + \sum_{k=0}^{n+1}\{r\mu_k\}$. □

Proof of Proposition 4.1. If we apply Lemma 4.2 to the case $r = 1$, then we find that f must have degree $< -1 + \sum_{k=0}^n \mu_k = 1 - \mu_{n+1}$ and as $\mu_{n+1} \in (0,1)$, this means that f is constant. So $\tilde\eta$ spans $\Omega(C)^\chi$.

For $r = m-1$, we find that $\dim \Omega(C)^{\bar\chi} = -1 + \sum_{k=0}^{n+1}\{(m-1)\mu_k\} = -1 + \sum_{k=0}^{n+1}(1-\mu_k) = n+1-\sum_{k=0}^{n+1}\mu_k = n-1$. Since $\overline{\Omega}(C)^\chi$ is the complex conjugate of $\Omega(C)^{\bar\chi}$, it follows that this space has dimension $n-1$ also. The fact that $H^1(C,\mathbb{C})^\chi$ has signature $(1, n-1)$ is now a consequence of its orthogonal decomposition into $\Omega(C)^\chi$ and $\overline{\Omega}(C)^\chi$. Finally,

$$h(\tilde\eta,\tilde\eta) = \frac{\sqrt{-1}}{2}\int_C \tilde\eta \wedge \bar{\tilde\eta} = \frac{m\sqrt{-1}}{2}\int_{\mathbb{C}} \eta\wedge\bar\eta = -mN(z)(>0). \qquad \square$$

Thus the Schwarz map $\mathbb{P}F : \widetilde{\mathcal{Q}}_\mu^{\mathrm{st}} \to \mathbb{B}_{n-1}$ can now be understood as associating to the curve C with its G_m-action the Hodge decomposition of $H^1(C;\mathbb{C})^\chi$.

4.2. Arithmeticity

The above computation leads to an arithmeticity criterion for Γ:

Theorem 4.3. *The monodromy group Γ is arithmetic if and only if for every $r \in (\mathbb{Z}/m)^\times - \{\pm 1\}$ we have $\sum_{k=0}^{n+1}\{r\mu_k\} \in \{1, n+1\}$.*

We need the following density lemma.

Lemma 4.4. *The Zariski closure of Γ in $\mathrm{GL}(H^1(C,\mathbb{C})^\chi \oplus H^1(C,\mathbb{C})^{\bar\chi})$ is defined over $\mathbb{R}$ and the image of its group of real points in the general linear group of $H^1(C,\mathbb{C})^\chi$ contains the special unitary group of $H^1(C,\mathbb{C})^\chi$.*

The proof amounts to exhibiting sufficiently many complex reflections in Γ. It is somewhat technical and we therefore omit it.

Proof of Theorem 4.3. Let us abbreviate $H^1(C,\mathbb{C})^{\chi^r}$ by H_r. The smallest subspace of $H^1(C,\mathbb{C})$ which contains H_1 and is defined over $\mathbb{Q}$ is the sum of the eigenspaces $H := \oplus_{r\in(\mathbb{Z}/m)^\times} H_r$. We may identify H with the quotient of $H^1(C,\mathbb{C})$ by the span of the images of the maps $H^1(G_k\backslash C,\mathbb{C}) \to H^1(C,\mathbb{C})$, where k runs over the divisors $\neq 1$ of m. In particular, $H(\mathbb{Z}) := H^1(C,\mathbb{Z})\cap H$ spans H. The monodromy group Γ may be regarded as a subgroup of $\mathrm{GL}(H_{\mathbb{Z}})$. On the other hand, Γ preserves each summand H_r. So if we denote by $\mathcal{G}$ the $\mathbb{Q}$-Zariski closure of Γ in $\mathrm{GL}(H)$, then $\Gamma \subset$

$\mathcal{G}(\mathbb{Z})$ and $\mathcal{G}(\mathbb{C})$ decomposes as $\mathcal{G}(\mathbb{C}) = \prod_{r\in(\mathbb{Z}/m)^\times} \mathcal{G}_r(\mathbb{C})$ with $\mathcal{G}_r(\mathbb{C}) \subset \mathrm{GL}(H_r)$. To say that Γ is arithmetic is to say that Γ is of finite index in $\mathcal{G}(\mathbb{Z})$.

Since $H_r \oplus H_{-r}$ is defined over $\mathbb{R}$, so is $\mathcal{G}_{r,-r} := \mathcal{G}_r \times \mathcal{G}_{-r}$. According to Lemma 4.4, the image of $\mathcal{G}_{1,-1}(\mathbb{R})$ in $\mathcal{G}_r(\mathbb{C})$ contains the special unitary group of H_1. For $r \in (\mathbb{Z}/m)^\times$, the summand H_r with its Hermitian form is a Galois conjugate of H_1 and so the image of $\mathcal{G}_{r,-r}(\mathbb{R})$ in $\mathcal{G}_r(\mathbb{C})$ then contains the special unitary group of H_r.

Suppose now that Γ is arithmetic. The projection $\mathcal{G}(\mathbb{R}) \to \mathcal{G}_{1,-1}(\mathbb{R})$ is injective on Γ and so the kernel of this projection must be anisotropic: $\mathcal{G}_{r,-r}(\mathbb{R})$ is compact for $r \neq \pm 1$. This means that the Hermitian form on H_r is definite for $r \neq \pm 1$. Since $H_r = \Omega(C)^{\chi^r} \oplus \overline{\Omega(C)^{\chi^{-r}}}$ with the first summand positive and the second summand negative, this means that for every $r \in (\mathbb{Z}/m)^\times - \{\pm 1\}$ (at least) one of the two summands must be trivial. Following Lemma 4.2 this amounts to $\sum_{k=0}^{n+1}\{r\mu_k\} = 1$ or $\sum_{k=0}^{n+1}\{-r\mu_k\} = 1$. The last identity is equivalent to $\sum_{k=0}^{n+1}\{r\mu_k\} = n+1$.

Suppose conversely, that for all $r \in (\mathbb{Z}/m)^\times - \{\pm 1\}$ we have $\sum_{k=0}^{n}\{r\mu_k\} < 1$ or $\sum_{k=0}^{n}\{-r\mu_k\} < 1$. As we have just seen, this amounts to $\mathcal{G}_{r,-r}(\mathbb{R})$ being compact for all $r \in (\mathbb{Z}/m)^\times - \{\pm 1\}$. In other words, the projection, $\mathcal{G}(\mathbb{R}) \to \mathcal{G}_{1,-1}(\mathbb{R})$ has compact kernel. Since $\mathcal{G}(\mathbb{Z})$ is discrete in $\mathcal{G}(\mathbb{R})$, it follows that its image in $\mathcal{G}_{1,-1}(\mathbb{R})$ is discrete as well. In particular, Γ is discrete in $\mathrm{GL}(H_1)$. Following Proposition 3.5 this implies that Γ has finite covolume in $\mathcal{G}_{1,-1}(\mathbb{R})$. Hence it also has finite covolume in $\mathcal{G}(\mathbb{R})$. This implies that Γ has finite index in $\mathcal{G}(\mathbb{Z})$. □

Example. The case for which $n = 3$, $(\mu_0, \mu_1, \mu_2, \mu_3) = (\frac{3}{12}, \frac{3}{12}, \frac{3}{12}, \frac{7}{12})$ (so that $\mu_4 = \frac{8}{12}$) satisfies the hypotheses of Theorem 3.1, hence yields a monodromy group which operates on $\mathbb{B}_2$ discretely with compact fundamental domain. But the group is not arithmetic since $\sum_{k=0}^{4}\{5\mu_k\} = 2 \notin \{1, 4\}$.

4.3. Working over a ring of cyclotomic integers

If we are given an L-slit δ, then $C \to \mathbb{P}^1$ comes with a section (continuous outside δ) in much the same way we found a branch of η_z: for ζ in a left half plane, $\prod_{k=0}^{n}(z_k - \zeta)^{d_k}$ has argument $< \pi/2$ in absolute value and so it has there a natural mth root (with argument $< \pi/2m$ in absolute value); the resulting section we find there is then extended in the obvious way. We identify δ_k with its image in C under the section and thus regard it as a chain on C. For $k = 1, \dots, n$, we introduce a $\mathbb{Z}[\zeta_m]$-valued 1-chain on C:

$$\varepsilon_k := \bar{w}_k \sum_{g\in G_m} \chi(g) g_* \delta_k.$$

Notice that the coefficient $\bar{w}_k$ is an mth root of unity and so a unit of $\mathbb{Z}[\zeta_m]$. We put it in, in order to maintain the connection with the Lauricella map. It will also have the effect of keeping some of the formulae simple.

Lemma 4.5. *The element ε_k is a 1-cycle on C with values in $\mathbb{Z}[\zeta_m]$ and has the property that $g_*\varepsilon_k = \bar{\chi}(g)\varepsilon_k$ (and hence defines an element of $H_1(C, \mathbb{Z}[\zeta_m])^{\bar{\chi}}$).*

We have $\int_{\varepsilon_k} \tilde{\eta} = mF_k(z,\delta)$*. Moreover,* $H_1(C,\mathbb{Z}[\zeta_m])^{\bar\chi}$ *is as a* $\mathbb{Z}[\zeta_m]$*-module freely generated by* $\varepsilon_1,\dots,\varepsilon_n$*.*

Proof. The identity involving integrals is verified by

$$\begin{aligned}\int_{\varepsilon_k} \tilde{\eta} = \bar{w}_k \sum_{g\in G_m} \chi(g) \int_{g_*\delta_k} \tilde{\eta} = \bar{w}_k \sum_{g\in G_m} \chi(g) \int_{\delta_k} g^*\tilde{\eta} \\ = \bar{w}_k \sum_{g\in G_m} \chi(g) \int_{\delta_k} \bar{\chi}(g)\eta = m\bar{w}_k \int_{\delta_k} \eta = mF_k(z,\delta).\end{aligned}$$

Give $\mathbb{P}^1$ the structure of a finite cell complex by taking the singletons $\{z_0,\dots,z_n\}$ as 0-cells, the intervals $\delta_1,\dots,\delta_n$ minus their end points as 1-cells and $\mathbb{P}^1 - \bigcup_{i=k}^n \delta_k$ as 2-cell. The connected components of the preimages of cells in C give the latter the structure of a finite cell complex as well (over the 2-cell we have one point of ramification, namely ∞, and so connected components of its preimage are indeed 2-cells). The resulting cellular chain complex of C,

$$0 \to C_2 \to C_1 \to C_0 \to 0,$$

comes with a G_m-action. Notice that C_1 is the free $\mathbb{Z}[G_m]$-module generated by $\delta_1,\dots,\delta_n$. On the other hand, $C_0 \cong \bigoplus_{k=0}^n \mathbb{Z}[G_m/G_{m_k}]$ (see Subsection 4.1 for the definition of m_k) and $C_2 \cong \mathbb{Z}[G_m/G_{m_{n+1}}]$, so that $(C_0)^{\bar\chi} = (C_2)^{\bar\chi} = 0$. The remaining assertions of the lemma follow from this. □

We describe the Hermitian form on the free $\mathbb{Z}[\zeta_m]$-module $H_1(C,\mathbb{Z}[\zeta_m])^{\bar\chi}$:

Proposition 4.6. *The Hermitian form* $H = -\frac{1}{m}h$ *is given in the basis* $(\varepsilon_1,\dots,\varepsilon_n)$ *as follows: for* $1 \le l \le k \le n$ *we have*

$$H(\varepsilon_k,\varepsilon_l) = \begin{cases} 0 & \text{if } l < k-1, \\ -\frac{1}{4}\sin(\pi/m)^{-1} & \text{if } l = k-1, \\ \frac{1}{4}(\cot(\pi/m_{k-1}) + \cot(\pi/m_k)) & \text{if } l = k. \end{cases}$$

It is perhaps noteworthy that this proposition shows that the matrix of H on $\varepsilon_1,\dots,\varepsilon_n$ only involves the denominators of the weigths $\mu_0,\dots,\mu_n$. The proof relies on a local computation of intersection multiplicities with values in $\mathbb{Z}[\zeta_m]$. The basic situation is the following. Consider the G_m-covering X over the complex unit disk Δ defined by $w^m = z^d$, where $d \in \{1,\dots,m-1\}$ and $g \in G_m$ acts as $g^*w = \chi(g)w$. The normalization $\tilde{X}$ of X consists of $e := \gcd(d,m)$ copies Δ, $\{\Delta_k\}_{k\in\mathbb{Z}/e}$, as follows: if we write $m = e\bar{m}$ and $d = e\bar{d}$ and t_k is the coordinate of Δ_k, then $\Delta_k \to X$ is given by $z = t_k^{\bar m}$ and $w = \zeta_m^k t_k^{\bar d}$, so that on Δ_k, $w^{\bar m} = \zeta_m^{k\bar m} t_k^{\bar d \bar m} = \zeta_e^k z^{\bar d}$. If $g_1 \in G_m$ is such that $\chi(g_1) = \zeta_m$, then $g_1^*(t_{k+1}) = t_k$, $k = 0,1\dots,e-1$ and $g_1^* t_0 = \zeta_m t_{e-1}$ (because $w|\Delta_{k+1} = \zeta_m^{k+1} t_{k+1}^{\bar d}$ and $(g_1^* w)|\Delta_k = \zeta_m w|\Delta_k = \zeta_m^{k+1} t_k^{\bar d}$).

Choose $\theta \in (0,2\pi)$ and let δ resp. δ' be the ray on Δ_0 defined by $t_0 = r$ (resp. $t_0 = r\exp(\sqrt{-1}\theta/\bar m)$ with $0 \le r < 1$). We regard both as chains with closed support. Notice that z maps δ (resp. δ') onto $[0,1)$ (resp.onto a ray $\neq [0,1)$). Consider the $\mathbb{Z}[\zeta_m]$-valued chains with closed support

$$\tilde{\delta} := \sum_{g\in G_m} \chi(g) g_*\delta, \quad \tilde{\delta}' := \sum_{g\in G_m} \chi(g) g_*\delta'.$$

These are in fact 1-cycles with closed support which only meet in the preimage of the origin (a finite set). So they have a well-defined intersection number.

Lemma 4.7. *We have* $\tilde{\delta}\cdot\overline{\tilde{\delta}'} = m\zeta_m(\zeta_m-1)^{-1} = \frac{1}{2}m(1-\sqrt{-1}\cot(\pi/\bar{m}))$.

Proof. This intersection product gets a contribution from each connected component Δ_k. Because of the G_m-equivariance these contributions are the same and so it is enough to show that the contribution coming from one of them is $(m/2e)(1+\sqrt{-1}\cot(\pi/2\bar{m})) = \frac{1}{2}\bar{m}(1+\sqrt{-1}\cot(\pi/2\bar{m}))$. This means that there is no loss in generality in assuming that d and m are relative prime. Assuming that this is the case, then we can compute the intersection product if we write $\tilde{\delta}$ and $\tilde{\delta}'$ as a sum of closed 1-cycles with coefficients in $\mathbb{Z}[\zeta_m]$. This is accomplished by

$$\begin{aligned}\tilde{\delta} &= \sum_{g\in G_m} \chi(g) g_*\delta \\ &= \sum_{k=1}^{m}(1+\zeta_m+\cdots+\zeta_m^{k-1})(g_{1*}^{k-1}\delta - g_{1*}^{k}\delta) = \sum_{k=1}^{m}\frac{1-\zeta_m^k}{1-\zeta_m}(g_{1*}^{k-1}\delta - g_{1*}^{k}\delta),\end{aligned}$$

(notice that $g_{1*}^{k-1}\delta - g_{1*}^{k}\delta$ is closed, indeed) and likewise for $\tilde{\delta}'$. We thus reduce our task to computing the intersection numbers $(g_{1*}^{k-1}\delta - g_{1*}^{k}\delta)\cdot(g_{1*}^{l-1}\delta' - g_{1*}^{l}\delta')$. This is easy: we find that this equals 1 if $l=k$, -1 if $l=k-1$ and 0 otherwise. Thus

$$\tilde{\delta}\cdot\overline{\tilde{\delta}'} = \sum_{k=1}^{m}\frac{1-\zeta_m^k}{1-\zeta_m}\bar{\zeta}_m^{k-1} = \frac{m\zeta_m}{\zeta_m-1} = \frac{m\zeta_{2m}}{\zeta_{2m}-\bar{\zeta}_{2m}} = \tfrac{1}{2}m(1-\sqrt{-1}\cot(\pi/m)). \qquad \square$$

Proof of 4.6. We may of course assume that each z_k is real: $z_k = x_k \in \mathbb{R}$ with with $x_0 < x_1 < \cdots < x_n$ and that $\delta_k = [x_{k-1}, x_k]$. Let us put $\tilde{\delta}_k := w_k\varepsilon_k = \sum_{g\in G_m}\chi(g)g_*\delta_k$ and compute $\tilde{\delta}_k\cdot\overline{\tilde{\delta}_l}$ for $1\le l\le k\le n$. It is clear that this is zero in case $l<k-1$. For $l=k$, we let δ_k' go along a straight line from x_{k-1} to a point in the upper half plane (with real part $\frac{1}{2}x_{k-1}+\frac{1}{2}x_k$, say) and then straight to x_k. We have a naturally defined $\mathbb{Z}[\zeta_m]$-valued 1-chain $\tilde{\delta}_k'$ on C homologous to $\tilde{\delta}_k$ and with support lying over δ_k. So $\tilde{\delta}_k\cdot\overline{\tilde{\delta}_k} = \tilde{\delta}_k\cdot\overline{\tilde{\delta}_k'}$. The latter is computed with the help of Lemma 4.7: the contribution over x_{k-1} is $\frac{1}{2}m(1-\sqrt{-1}\cot(\pi/m_{k-1}))$ and over x_k it is $-\frac{1}{2}m(1-\sqrt{-1}\cot(\pi/m_k))$ and so $\varepsilon_k\cdot\varepsilon_k = \tilde{\delta}_k\cdot\overline{\tilde{\delta}_k'} = -\frac{1}{2}m\sqrt{-1}\cot(\pi/m_{k-1}))+\frac{1}{2}m\sqrt{-1}\cot(\pi/m_k)$. We now do the case $l=k-1$. The 1-chains on C given by δ_{k-1} and δ_k make an angle over x_{k-1} of $\pi\mu_{k-1} = \pi d_{k-1}/m$. In terms of the local picture of Lemma 4.7 this means that the pair (δ_k, δ_{k-1}) corresponds to $(\delta, -\bar{\zeta}_{2m}^{d_{k-1}-1}\delta')$.

It follows that

$$\begin{aligned}\tilde{\delta}_k \cdot \bar{\tilde{\delta}}_{k-1} &= \tilde{\delta} \cdot \overline{-\zeta_{2m}^{d_{k-1}-1}\tilde{\delta}'} = -\zeta_{2m}^{d_{k-1}-1}\tilde{\delta} \cdot \bar{\tilde{\delta}}' \\ &= -\zeta_{2m}^{d_{k-1}-1} m\zeta_m(\zeta_m - 1)^{-1} = -m(\zeta_{2m} - \bar{\zeta}_{2m})^{-1} e^{\sqrt{-1}\pi\mu_{k-1}}.\end{aligned}$$

Hence $\varepsilon_k \cdot \bar{\varepsilon}_{k-1} = -m(\zeta_{2m} - \bar{\zeta}_{2m})^{-1}$ and so $H(\varepsilon_k, \varepsilon_{k-1}) = -\frac{1}{2m\sqrt{-1}}\varepsilon_k \cdot \bar{\varepsilon}_{k-1} = (2\sqrt{-1}(\zeta_{2m} - \bar{\zeta}_{2m}))^{-1} = -\frac{1}{4}(\sin(\pi/m))^{-1}$ is as asserted. □

5. Generalizations and other view points

5.1. Higher dimensional integrals

This refers to the situation where $\mathbb{P}^1$ and the subset $\{z_0, \dots, z_{n+1}\}$ are replaced by a projective arrangement; such generalizations were considered by Deligne, Varchenko [17] and others. To be specific, fix an integer $N \geq 1$, a finite set K with at least $N+2$ elements and a *weight function* $\mu : k \in K \mapsto \mu_k \in (0,1)$. Given an injective map $z : k \in K \mapsto z_k \in \check{\mathbb{P}}^N$, choose for every $k \in K$ a linear form $Z_k : \mathbb{C}^{N+1} \to \mathbb{C}$ whose zero set is the hyperplane H_{z_k} defined by z_k and put

$$\eta_z = \mathrm{Res}_{\mathbb{P}^N}\left(\prod_{k \in K} Z_k(\zeta)^{-\mu_k}\right) d\zeta_0 \wedge \cdots \wedge d\zeta_N.$$

This is a multivalued holomorphic N-form on $U_z := \mathbb{P}^N - \bigcup_{k\in K} H_{z_k}$. If σ is a sufficiently regular relative N-chain of the pair $(\mathbb{P}^N, \mathbb{P}^N - U_z)$ and we are given a branch of η over σ, then η is integrable over σ so that $\int_\sigma \eta$ is defined. Here it pays however to take the more cohomological approach that we briefly described in Subsection 1.7. So we let $\mathbb{L}_z$ be the rank one local system on U_z such that its monodromy around H_{z_k} is multiplication by $\exp(2\pi\mu_k\sqrt{-1})$ and endow it with a flat Hermitian metric. Then after the choice of a multivalued section of $\mathbb{L}_z$ of unit norm, η_z can be interpreted as a section of $\Omega^N_{U_z} \otimes_{\mathbb{C}} \mathbb{L}_z$. It thus determines an element $[\eta_z] \in H^N(\mathbb{L}_z)$. Similarly, σ plus the branch of η_z over σ defines an element $[\sigma] \in H_N(\mathbb{P}^N, \mathbb{P}^N - U_z; \mathbb{L}_z^\vee)$. The latter space is dual to $H^N(\mathbb{L}_z)$ by Alexander duality in such a manner that $\int_\sigma \eta_z$ is the value of the Alexander pairing on $([\eta_z], [\sigma])$. In order for η_z to be square integrable it is necessary and sufficient that for every nonempty intersection L of hyperplanes H_{z_k} we have $\sum_{\{k \mid H_{z_k} \supset L\}} \mu_k < \mathrm{codim}(L)$. Assume that this is the case. Then η_z defines in a class in the intersection homology space $IH^m(\mathbb{P}^N, \mathbb{L}_z)$. This space comes with a natural hermitian form h for which $h(\eta_z, \eta_z) > 0$. (It is clear that the line spanned by η_z only depends z; Hodge theory tells us that the image of that line is $F^N IH^N(\mathbb{P}^N, \mathbb{L})$.) In order for the situation to be like the one we studied, we would want the orthogonal complement of η_z in $IH^N(\mathbb{P}^N, \mathbb{L}_z)$ to be negative. Unfortunately this seems rarely to be the case when $N > 1$. When that is so, then we might vary z over the connected constructible set S of injective maps $K \to \check{\mathbb{P}}^N$ for which the topological type of the arrangement it defines stays constant. Then over S we have a local system $\mathbb{H}_S$ whose stalk at

$z \in S$ is $IH^N(\mathbb{P}^N, \mathbb{L}_z)$ and the Schwarz map which assigns to z the line in $\mathbb{H}_z$ defined by η_z will take values in a ball. The first order of business should be to determine the cases for which the associated monodromy group is discrete, but we do not know whether that has been done yet.

5.2. Geometric structures on arrangement complements

A generalization of the Deligne–Mostow theory based on a different point of view (and going in a different direction) was developed by Couwenberg, Heckman and the author in [7]. The point of departure is here a finite dimensional complex inner product space V, a finite collection $\mathcal{H}$ of linear hyperplanes in V and a map κ which assigns to every $H \in \mathcal{H}$ a positive real number κ_H. These data define a connection ∇^κ on the tangent bundle of the arrangement complement $V^\circ := V - \cup_{h\in\mathcal{H}} H$ as follows. For $H \in \mathcal{H}$ denote by $\pi_H \in \mathrm{End}(V)$ the orthogonal projection with kernel H and by ω_H the logarithmic differential on V defined by $\phi_H^{-1} d\phi_H$, where ϕ_H is a linear form on V with kernel H. Form $\Omega^\kappa := \sum_{H\in\mathcal{H}} \kappa_H \pi_H \otimes \omega_H$ and regard it as a differential on V° which takes values in the tangent bundle of V°, or rather, as a connection form on this tangent bundle: a connection is defined by

$$\nabla^\kappa := \nabla^0 - \Omega^\kappa,$$

where ∇^0 stands for the usual affine connection on V restricted to V°. This connection is easily verified to be torsion free. It is well-known that such a connection defines an affine structure (that is, it defines an atlas of charts whose transition maps are affine-linear) precisely when the connection is flat; the sheaf of affine-linear functions is then the sheaf of holomorphic functions whose differential is flat for the connection (conversely, an affine structure is always given by a flat torsion free connection on the tangent bundle). There is a simple criterion for the flatness of ∇^κ in terms of linear algebra. Let $\mathcal{L}(\mathcal{H})$ denote the collection of subspaces of V that are intersections of members of $\mathcal{H}$ and, for $L \in \mathcal{L}(\mathcal{H})$, let $\mathcal{H}_L$ be the set of $H \in \mathcal{H}$ containing L. Then the following properties are equivalent:

(i) ∇ is flat,

(ii) $\Omega \wedge \Omega = 0$,

(iii) for every pair $L, M \in \mathcal{L}(\mathcal{H})$ with $L \subset M$, the endomorphisms $\sum_{H\in\mathcal{H}_L} \kappa_H \pi_H$ and $\sum_{H\in\mathcal{H}_M} \kappa_H \pi_H$ commute,

(iv) for every $L \in \mathcal{L}(\mathcal{H})$ of codimension 2, the sum $\sum_{H\in\mathcal{H}_L} \kappa_H \pi_H$ commutes with each of its terms.

If these mutually equivalent conditions are satisfied we call the triple $(V, \mathcal{H}, \kappa)$ a *Dunkl system.*

Suppose that $(V, \mathcal{H}, \kappa)$ is such a system so that V° comes with an affine structure. If $L \in \mathcal{L}(\mathcal{H})$ is irreducible (in the sense that there is no nontrivial decomposition of $\mathcal{H}_L$ such that the corresponding intersections are perpendicular), then the fact that $\sum_{H\in\mathcal{H}_L} \kappa_H \pi_H$ commutes with each of its terms implies that this sum must be proportional to the orthogonal projection π_L with kernel L. A trace computation shows that the scalar factor must be $\kappa_L := \mathrm{codim}(L)^{-1} \sum_{H\in\mathcal{H}_L} \kappa_H$. Let us now assume that the whole system is irreducible in the sense that the intersection of all members of $\mathcal{H}$ is reduced to the origin and that this intersection is irreducible. We then have defined $\kappa_0 = \dim(V)^{-1}\sum_{H\in\mathcal{H}}\kappa_H$. The connection is invariant under scalar multiplication by $e^t \in \mathbb{C}^\times$ and one verifies that for t close to 0, the corresponding affine-linear transformation is like scalar multiplication by $e^{(1-\kappa_0)t}$ if $\kappa_0 \neq 1$ and by a translation if $\kappa_0 = 1$. This means that if $\kappa_0 \neq 1$, the affine structure on V° is in fact a linear structure and that this determines a (new) projective structure on $\mathbb{P}(V^\circ)$, whereas when $\kappa_0 = 1$ (the *parabolic* case), $\mathbb{P}(V^\circ)$ inherits an affine structure which makes the projection $V^\circ \to \mathbb{P}(V^\circ)$ affine-linear. Notice that $(V, \mathcal{H}, t\kappa)$ will be a Dunkl system for every $t > 0$. The behavior of that system (such as its monodromy) may change dramatically if we vary t.

Before we proceed, let us show how a weight system μ that gives rise to the Lauricella differential also gives rise to such an irreducible Dunkl system: we take $V = V_n = \mathbb{C}^{n+1}/\text{main diagonal}$, $\mathcal{H}$ to be the collection of diagonal hyperplanes $H_{k,l} := (z_k = z_l)$, $0 \le k < l \le n$, and $\kappa(H_{k,l}) = \mu_k + \mu_l$. The inner product on V_n comes from the inner product on $\mathbb{C}^{n+1}$ for which $\langle e_k, e_l\rangle = \mu_k\delta_{k,l}$ and is the one which makes the projection $\mathbb{C}^{n+1} \to V_n$ self-adjoint. It is an amusing exercise to verify that the connection is flat indeed and that the space of affine-linear functions at $z \in V_n^\circ$ is precisely the space of solutions of the system of differential equations we encountered in part (c) of Proposition 1.2. So the Schwarz map is now understood as a multivalued chart (in standard terminology, a developing map) for the new projective structure on $\mathbb{P}(V_n^\circ)$. We also find that $\kappa_0 = |\mu|$; more generally, an irreducible member $L \in L(\mathcal{H})$ is given by a subset $I \subset \{0, \dots, n\}$ with at least two elements (so that $L = L(I)$ is the locus where all z_k, $k \in I$ coincide) and $\kappa_{L(I)} = \sum_{k\in I}\mu_k$.

Another interesting class of examples is provided by the finite complex reflection groups: let G be a finite complex reflection group operating irreducibly and unitarily in a complex inner product space V, $\mathcal{H}$ the collection of complex hyperplanes of G and $H \in \mathcal{H} \mapsto \kappa_H$ constant on the G-orbits. Then $(V, \mathcal{H}, \kappa)$ is a Dunkl system.

It turns out that in many cases of interest (including the examples mentioned above), one can show that there exists a ∇^κ-flat Hermitian form h on V° with the following properties

- **ell.** if $0 < \kappa_0 < 1$, then h is positive definite,
- **par.** if $\kappa_0 = 1$, then h is positive semidefinite with kernel the tangent spaces to the $\mathbb{C}^\times$-orbits,
- **hyp.** if $1 < \kappa_0 < m_{\text{hyp}}$ for some $m_{\text{hyp}} > 1$, then h is nondegenerate hyperbolic and such that the tangent spaces to the $\mathbb{C}^\times$-orbits are negative.

This implies that $\mathbb{P}(V^\circ)$ acquires a geometric structure which is respectively modeled on Fubini–Study space, flat complex Euclidean space and complex hyperbolic space. A suitable combination of rationality and symmetry conditions which generalizes the half integrality condition 3.7 (and is called the *Schwarz condition*), yields a generalization of Theorem 3.9. We thus obtain new examples of groups operating discretely and with finite covolume on a complex ball (see the tables at the end of [7]). For the real finite reflection arrangements, all groups thus obtained are arithmetic.

References

[1] D. Allcock, *New complex- and quaternion-hyperbolic reflection groups*, Duke Math. J. **103** (2000), 303–333.

[2] D. Allcock, *The Leech lattice and complex hyperbolic reflections*, Invent. Math. **140** (2000), 283–301.

[3] G. Barthel, F. Hirzebruch, T. Höfer, *Geradenkonfigurationen und algebraische Flächen*, Aspects of Mathematics, Vieweg, Braunschweig–Wiesbaden (1987).

[4] W. Casselman, *Families of curves and automorphic forms*, Thesis, Princeton University, 1966 (unpublished).

[5] P.B. Cohen, F. Hirzebruch, *Review of* Commensurabilities among lattices in $\mathrm{PU}(1,n)$ *by Deligne and Mostow*, Bull. Amer. Math. Soc. **32** (1995), 88–105.

[6] W. Couwenberg, *Complex Reflection Groups and Hypergeometric Functions*, Thesis (123 p.), University of Nijmegen, 1994, also available at `http://members.chello.nl/~w.couwenberg/`.

[7] W. Couwenberg, G. Heckman, E. Looijenga, *Geometric structures on the complement of a projective arrangement*, Publ. Math. IHES 101 (2005), pp. 69–161, also available at `arXiv math.AG/0311404`.

[8] P. Deligne, G.D. Mostow, *Monodromy of hypergeometric functions and non-lattice integral monodromy*, Publ. Math. IHES **63** (1986), 1–89.

[9] P. Deligne, G.D. Mostow, *Commensurabilities among lattices in* $\mathrm{PU}(1,n)$, Ann. of Math. Studies **132**, Princeton U.P., Princeton, 1993.

[10] B.R. Doran, *Intersection Homology, Hypergeometric Functions, and Moduli Spaces as Ball Quotients*, Thesis, Princeton University (93 p.), 2003.

[11] F.C. Kirwan, R. Lee, S.H. Weintraub, *Quotients of the complex ball by discrete groups*, Pacific J. of Math. **130** (1987), 115–141.

[12] G.D. Mostow, *Generalized Picard lattices arising from half-integral conditions*, Inst. Hautes Études Sci. Publ. Math. **63** (1986), 91–106.

[13] G.D. Mostow, *On discontinuous action of monodromy groups on the complex* n*-ball*, J. Amer. Math. Soc. **1** (1988), 555–586.

[14] J.K. Sauter, Jr., *Isomorphisms among monodromy groups and applications to lattices in* $\mathrm{PU}(1,2)$, Pacific J. Math. **146** (1990), 331–384.

[15] H.A. Schwarz, *Über diejenigen Fälle in welchen die Gaussische hypergeometrische Reihe eine algebraische Funktion ihres vierten Elementes darstellt*, J. f. d. reine u. angew. Math. **75** (1873), 292–335.

[16] W.P. Thurston, *Shapes of polyhedra and triangulations of the sphere*, Geometry & Topology Monographs **1** (1998), 511–549.

[17] A.N. Varchenko: *Hodge filtration of hypergeometric integrals associated with an affine configuration of general position and a local Torelli theorem*, in: I.M. Gelfand Seminar, Adv. Soviet Math. **16**; Part 2, 167–177, Amer. Math. Soc., Providence, RI, 1993.

Eduard Looijenga
Betafaculteit Universiteit Utrecht
Departement Wiskunde
Postbus 80.010
NL-3508 TA Utrecht
Nederland
e-mail: looijeng@math.uu.nl

Progress in Mathematics, Vol. 260, 245–271

Invariant Functions with Respect to the Whitehead-Link

Keiji Matsumoto

Abstract. We survey our construction of invariant functions on the real 3-dimensional hyperbolic space $\mathbb{H}^3$ for the Whitehead-link-complement group $W \subset GL_2(\mathbb{Z}[i])$ and for a few groups commensurable with W. We make use of theta functions on the bounded symmetric domain $\mathbb{D}$ of type $I_{2,2}$ and an embedding $\imath : \mathbb{H}^3 \to \mathbb{D}$. The quotient spaces of $\mathbb{H}^3$ by these groups are realized by these invariant functions. We review classical results on the λ-function, the j-function and theta constants on the upper half space; our construction is based on them.

Mathematics Subject Classification (2000). Primary 11F55; Secondary 14P05, 57M25.

Keywords. Whitehead link, hyperbolic structure, automorphic forms, theta functions.

Contents

1. Introduction

The ratio of solutions of the Gauss hypergeometric differential equation

$$E(\alpha,\beta,\gamma): \quad x(1-x)\frac{d^2f}{dx^2}+\{\gamma-(\alpha+\beta+1)x\}\frac{df}{dx}-\alpha\beta f=0$$

for $(\alpha,\beta,\gamma)=(1/2,1/2,1)$ induces an isomorphism

$$\psi:\mathbb{C}-\{0,1\}\to\mathbb{H}/M$$

where $\mathbb{H}$ is the upper half space $\{\tau\in\mathbb{C}\mid \mathrm{Im}(\tau)>0\}$ and M is the monodromy group of $E(1/2,1/2,1)$. Note that M is the level 2 principal congruence subgroup of $SL_2(\mathbb{Z})$, which can be identified with the fundamental group $\pi_1(\mathbb{C}-\{0,1\})$. The inverse of ψ is the λ-function, which is a modular function on $\mathbb{H}$ with respect to M. In particular, the real and imaginary parts of the λ-function are real analytic on the real 2-dimensional hyperbolic space $\mathbb{H}$ and invariant under the action of M.

In this lecture note, I explain the construction of real analytic functions on the real 3-dimensional hyperbolic space $\mathbb{H}^3=\{(z,t)\in\mathbb{C}\times\mathbb{R}\mid t>0\}$ in [MNY] and [MY], which are invariant under the action of some discrete subgroups of the isometry group $GL_2^T(\mathbb{C})$ of $\mathbb{H}^3$. We use theta functions $\Theta\binom{a}{b}$ on the symmetric domain $\mathbb{D}$ of type $I_{2,2}$ over the ring $\mathbb{Z}[i]$ discussed in [F] and [M1], and an embedding $\imath:\mathbb{H}^3\to\mathbb{D}$. We are interested in the Whitehead link L, see Figure 1; its complement S^3-L is known to admit a hyperbolic structure: there is a finite

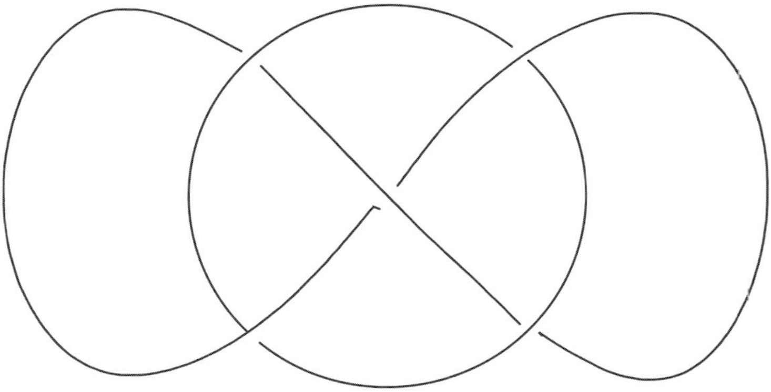

FIGURE 1. Whitehead link

index subgroup $W\subset GL_2(\mathbb{Z}[i])$ isomorphic to $\pi_1(S^3-L)$ and a homeomorphism

$$\varphi:\mathbb{H}^3/W\xrightarrow{\cong}S^3-L.$$

Note that the situation is quite similar to the isomorphism

$$\lambda:\mathbb{H}/M\xrightarrow{\cong}\mathbb{C}-\{0,1\}.$$

We explain our construction of the homeomorphism φ in [MNY]; it is given in terms of invariant functions on $\mathbb{H}^3$ with respect to W, and its image is explicitly presented as part of a real algebraic set (we need some inequalities) in $\mathbb{R}^{13}$.

Since our construction of invariant functions is based on classical results on the λ-function, the j-function and theta constants $\vartheta\binom{a}{b}$ on $\mathbb{H}$, we review them in Sections 2 and 3. In Section 4, fundamental properties of $\Theta\binom{a}{b}$ on $\mathbb{D}$ are given. In these sections, some important facts are proved. In the rest sections, an outline of [MNY] is given; proofs of theorems, and details are omitted.

It is known that the Borromean-rings-complement $S^3 - R$ also admits a hyperbolic structure: there is a discrete group B in $GL_2(\mathbb{C})$ isomorphic to the fundamental group of $S^3 - R$ such that the quotient $\mathbb{H}^3/B$ is homeomorphic to $S^3 - R$. This homeomorphism is explicitly given in [M3] by invariant functions for B.

2. The λ-function and the j-function

In this section, I explain the λ-function and the j-function, which help us understand results in Sections 10 and 11.

The group $SL_2(\mathbb{Z})$ acts on the upper half space $\mathbb{H}$ as linear fractional transformations:

$$g\cdot\tau = \frac{a\tau+b}{c\tau+d},\quad \tau\in\mathbb{H},\ g=\begin{pmatrix} a & b\\ c & d\end{pmatrix}\in SL_2(\mathbb{Z}).$$

Fundamental domains of $SL_2(\mathbb{Z})$ and the level 2 principal congruence subgroup

$$M=\left\{\begin{pmatrix} a & b\\ c & d\end{pmatrix}\in SL_2(\mathbb{Z}) \mid a-1,b,c,d-1\equiv 0 \bmod 2\right\}$$

are given in Figure 2. The elements patching their boundaries give generators of these groups

$$\begin{array}{lll} SL_2(\mathbb{Z}) & : & \begin{pmatrix} 1 & 1\\ 0 & 1\end{pmatrix}: \ell_1\to\ell_2,\quad \begin{pmatrix} 0 & -1\\ 1 & 0\end{pmatrix}: \ell_3\to\ell_4,\\ M & : & \begin{pmatrix} 1 & 2\\ 0 & 1\end{pmatrix}: m_1\to m_2,\quad \begin{pmatrix} 1 & 0\\ 2 & 1\end{pmatrix}: m_3\to m_4,\end{array}$$

where ℓ_j and m_j are drawn in Figure 2. The sequence

$$1\longrightarrow M\longrightarrow SL_2(\mathbb{Z})\longrightarrow SL_2(\mathbb{F}_2)\longrightarrow 1$$

is exact and $SL_2(\mathbb{F}_2)$ is isomorphic to the symmetric group S_3. Thus the quotient group $SL_2(\mathbb{Z})/M$ is isomorphic to S_3, and the quotient space $\mathbb{H}/M$ is an S_3 covering of $\mathbb{H}/SL_2(\mathbb{Z})$ with two branching points i and $\omega=\frac{-1+\sqrt{3}i}{2}$.

On the other hand, the monodromy representation of the hypergeometric differential equation $E(\alpha,\beta,\gamma)$ is well known.

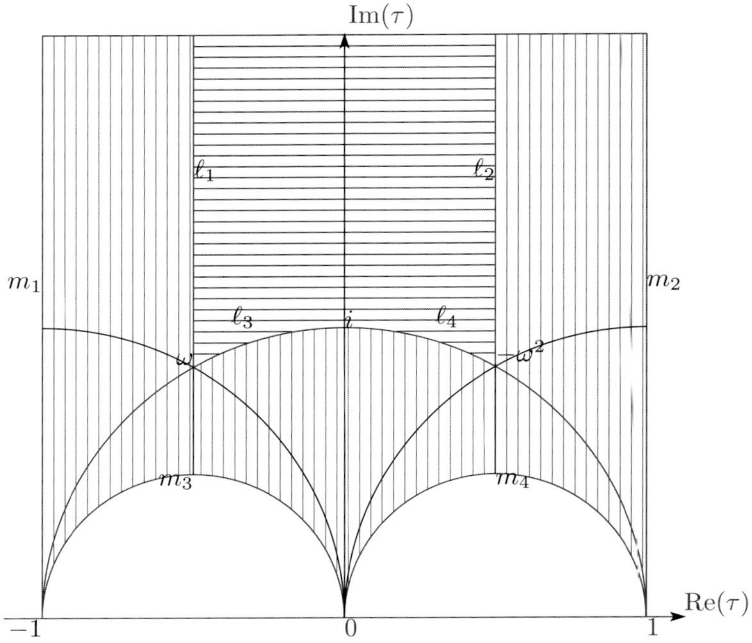

FIGURE 2. Fundamental domains of $SL_2(\mathbb{Z})$ and M

Fact 1 (Theorem 6.1 in [K]). If none of $\alpha, \beta, \gamma-\alpha$ and $\gamma-\beta$ is an integer, then there exists a fundamental system of solutions of $E(\alpha,\beta,\gamma)$ such that the monodromy group with respect to this system is generated by

$$\rho_1 = \begin{pmatrix} 1 & 0 \\ -1+\exp(-2\pi i\beta) & \exp(-2\pi i\gamma) \end{pmatrix}, \quad \rho_2 = \begin{pmatrix} 1 & 1-\exp(-2\pi i\alpha) \\ 0 & \exp(-2\pi i(\alpha+\beta-\gamma)) \end{pmatrix}.$$

Remark 2.1. When one of $\alpha, \beta, \gamma-\alpha$ and $\gamma-\beta$ is an integer, the fundamental system of $E(\alpha,\beta,\gamma)$ in Fact 1 degenerates. The monodromy groups of $E(\alpha,\beta,\gamma)$ for such parameters are studied in Section 2.4 in [IKSY].

We have

$$\rho_1 = \begin{pmatrix} 1 & 0 \\ -2 & 1 \end{pmatrix}, \quad \rho_2 = \begin{pmatrix} 1 & 2 \\ 0 & 1 \end{pmatrix}$$

for the parameter $(\alpha,\beta,\gamma) = (1/2, 1/2, 1)$, and

$$P^{-1}\rho_1 P = \begin{pmatrix} 1 & 1 \\ 0 & 1 \end{pmatrix}, \quad P^{-1}\rho_2 P = i\begin{pmatrix} 0 & 1 \\ -1 & 0 \end{pmatrix}$$

for the parameter $(\alpha,\beta,\gamma) = (1/12, 5/12, 1)$, where

$$P = \begin{pmatrix} 0 & (\sqrt{3}-1)i\omega \\ 1+i & 1-i \end{pmatrix}.$$

The group M is isomorphic to the fundamental group $\pi_1(\mathbb{C}-\{0,1\})$, which is the free group generated by two elements.

The ratio of the fundamental system of solutions $E(1/2,1/2,1)$ induces an isomorphism from $\mathbb{C}-\{0,1\}$ to $\mathbb{H}/M$, and that of $E(1/12,5/12,1)$ induces an isomorphism from $\mathbb{C}-\{0,1\}$ to $\mathbb{H}/SL_2(\mathbb{Z})$. Their inverses are the λ-function and the j-function. The j-function can be expressed by the λ-function:

$$j(\tau)=\frac{4}{27}\frac{(\lambda(\tau)^2-\lambda(\tau)+1)^3}{\lambda(\tau)^2(1-\lambda(\tau))^2}.$$

We can regard this expression as the 6-fold branched covering of $\mathbb{P}^1$. The branching information of this covering is given as

$$\begin{array}{ccc} \lambda\text{-space} & & j\text{-space} \\ 0,1,\infty & \mapsto & \infty \\ -1,2,\frac{1}{2} & \mapsto & 1 \\ -\omega,-\omega^2 & \mapsto & 0, \end{array}$$

and the covering transformation group consists of the projective transformations keeping the set $\{0,1,\infty\}$ invariant, which is isomorphic to S_3. We have the diagram:

$$\begin{array}{ccc} \mathbb{H}/M & \xrightarrow{\lambda} & \mathbb{C}-\{0,1\} \\ S_3\downarrow & \searrow^{j} & \downarrow S_3 \\ \mathbb{H}/SL_2(\mathbb{Z}) & \xrightarrow{j} & \mathbb{C}-\{0,1\}. \end{array}$$

In the next section, we give an expression of the λ-function in terms of theta constants, which shows the compatibility of the S_3-actions in this diagram.

3. Theta constants

The theta constant with characteristics $a,b\in\mathbb{Q}$ is defined as

$$\vartheta\begin{pmatrix}a\\b\end{pmatrix}(\tau)=\sum_{n\in\mathbb{Z}}\mathbf{e}[(n+a)^2\tau+2(n+a)b],$$

where $\tau\in\mathbb{H}$ and $\mathbf{e}[x]=\exp(\pi i x)$. Since this series converges absolutely and uniformly on any compact set in $\mathbb{H}$, $\vartheta\binom{a}{b}(\tau)$ is holomorphic on $\mathbb{H}$. We denote $\vartheta\binom{a}{b}(\tau)$ by $\vartheta\begin{bmatrix}2a\\2b\end{bmatrix}(\tau)$ for $a,b\in\frac{1}{2}\mathbb{Z}$. We give some fundamental properties of $\vartheta\binom{a}{b}(\tau)$.

1. $\vartheta\binom{-a}{-b}(\tau)=\vartheta\binom{a}{b}(\tau)$, and if $a,b\in\frac{1}{2}\mathbb{Z}$, then $\vartheta\binom{-a}{b}(\tau)=\vartheta\binom{a}{b}(\tau)$;
2. $\vartheta\binom{a+m}{b+n}(\tau)=\mathbf{e}[2an]\vartheta\binom{a}{b}(\tau)$ for $m,n\in\mathbb{Z}$;
3. if $2ab\notin\mathbb{Z}$ for $a,b\in\frac{1}{2}\mathbb{Z}$, then $\vartheta\binom{a}{b}(\tau)=0$;
4. $\vartheta\binom{a}{b}(\tau+1)=\mathbf{e}[-a^2-a]\vartheta\binom{a}{b+a+1/2}(\tau)$;
5. $\vartheta\binom{a}{b}(-\tau^{-1})=\mathbf{e}[-2ab]\sqrt{\frac{\tau}{i}}\vartheta\binom{-b}{a}(\tau)$.

The identities 1–4 can be seen easily by the definition of the theta constants. In order to show the last one, we use Poisson's summation formula

$$\sum_{n\in\mathbb{Z}} f(n) = \sum_{n\in\mathbb{Z}} \tilde{f}(n),$$

where f goes to zero fast enough at $\pm\infty$, and $\tilde{f}$ is the Fourier transform of f

$$\tilde{f}(n) = \int_{-\infty}^{\infty} f(x)e^{-2\pi i x n}dx.$$

For $\tau = it$ $(t > 0)$ and $f(n) = \exp[-\pi(n+a)^2 t + 2\pi i(n+a)b]$,

$$\int_{-\infty}^{\infty} f(x)e^{-2\pi i x n}dx = \frac{\mathbf{e}[2ab]}{\sqrt{t}}\mathbf{e}[-\frac{1}{it}(n-b)^2 + 2(n-b)a]$$

which implies the identity 5.

A holomorphic function f on $\mathbb{H}$ is called a modular form of weight k for a subgroup $G \subset SL_2(\mathbb{Z})$ if f satisfies

$$f((a\tau+b)/(c\tau+d)) = (c\tau+d)^k f(\tau)$$

and some boundedness conditions around the cusps.

Fact 2. The functions $\vartheta\left[\begin{smallmatrix}0\\0\end{smallmatrix}\right](\tau)^4$, $\vartheta\left[\begin{smallmatrix}0\\1\end{smallmatrix}\right](\tau)^4$, $\vartheta\left[\begin{smallmatrix}1\\0\end{smallmatrix}\right](\tau)^4$ are modular forms of weight 2 for M.

There is a linear relation among them:

Fact 3 (Jacobi's identity).

$$\vartheta\begin{bmatrix}0\\0\end{bmatrix}(\tau)^4 = \vartheta\begin{bmatrix}0\\1\end{bmatrix}(\tau)^4 + \vartheta\begin{bmatrix}1\\0\end{bmatrix}(\tau)^4.$$

Proof. We give a proof which can be generalized to that of quadratic relations among theta functions on the symmetric domain $\mathbb{D}$ of type $I_{2,2}$ in Section 4. Let $L_1 = \mathbb{Z}^4$, $L_2 = \mathbb{Z}^4 A$, and L be the lattice $\langle L_1, L_2\rangle$ generated by L_1 and L_2, where

$$A = \frac{1}{2}\begin{pmatrix}1 & 1 & 1 & 1\\ 1 & 1 & -1 & -1\\ 1 & -1 & 1 & -1\\ 1 & -1 & -1 & 1\end{pmatrix}.$$

Note that ${}^tA = A$, $A^2 = I_4$ and that the lattices L and L_2 are expressed as

$$\begin{aligned} L &= \{(n_1,\dots,n_4) \in \tfrac{1}{2}\mathbb{Z}^4 \mid n_j - n_k \in \mathbb{Z}\},\\ L_2 &= \{(n_1,\dots,n_4) \in L \mid n_1+n_2+n_3+n_4 \in 2\mathbb{Z}\}.\end{aligned}$$

By definition,

$$\begin{aligned}
\vartheta\begin{bmatrix}0\\0\end{bmatrix}(\tau)^4 &= \sum_{n_1,\dots,n_4\in\mathbb{Z}} \mathbf{e}[(n_1^2+\cdots+n_4^2)\tau],\\
\vartheta\begin{bmatrix}0\\1\end{bmatrix}(\tau)^4 &= \sum_{n_1,\dots,n_4\in\mathbb{Z}} \mathbf{e}[(n_1^2+\cdots+n_4^2)\tau]\mathbf{e}[n_1+\cdots+n_4],\\
\vartheta\begin{bmatrix}1\\0\end{bmatrix}(\tau)^4 &= \sum_{n_1,\dots,n_4\in\mathbb{Z}} \mathbf{e}[(m_1^2+\cdots+m_4^2)\tau],\\
\vartheta\begin{bmatrix}1\\1\end{bmatrix}(\tau)^4 &= \sum_{n_1,\dots,n_4\in\mathbb{Z}} \mathbf{e}[(m_1^2+\cdots+m_4^2)\tau]\mathbf{e}[m_1+\cdots+m_4],
\end{aligned}$$

where $m_j = n_j + \frac{1}{2}$. Since we have $[L, L_1] = [L_1, L_1 \cap L_2] = 2$ for $L = \langle L_1, L_2\rangle$, $\mathbf{e}[n_1 + \cdots + n_4]$ is the non-trivial character on L/L_2. The summation of the four right-hand sides of the above can be regarded as the summation over L and

$$\mathbf{e}[n_1+\cdots+n_4] = \begin{cases} 1 & \text{if} \quad (n_1,\dots,n_4)\in L_2,\\ -1 & \text{if} \quad (n_1,\dots,n_4)\notin L_2.\end{cases}$$

Thus it becomes

$$\begin{aligned}
& 2\sum_{(m_1,\dots,m_4)\in L_2} \mathbf{e}[(m_1^2+\cdots+m_4^2)\tau]\\
=\ & 2\sum_{(n_1,\dots,n_4)\in L_1} \mathbf{e}[(n_1,\dots,n_4)A\ {}^t((n_1,\dots,n_4)A)\tau]\\
=\ & 2\sum_{(n_1,\dots,n_4)\in L_1} \mathbf{e}[(n_1,\dots,n_4)\ {}^t(n_1,\dots,n_4)\tau] = 2\vartheta\begin{bmatrix}0\\0\end{bmatrix}(\tau)^4.
\end{aligned}$$

Since $\vartheta\left[\begin{smallmatrix}1\\1\end{smallmatrix}\right](\tau) = 0$, we have Jacobi's identity. □

The λ-function can be expressed as

$$\lambda(\tau) = \frac{\vartheta\left[\begin{smallmatrix}0\\1\end{smallmatrix}\right](\tau)^4}{\vartheta\left[\begin{smallmatrix}0\\0\end{smallmatrix}\right](\tau)^4};$$

we can easily see that this functions is invariant under the action of M by Fact 2:

$$\lambda(\frac{a\tau+b}{c\tau+d}) = \frac{\vartheta\left[\begin{smallmatrix}0\\1\end{smallmatrix}\right](\frac{a\tau+b}{c\tau+d})^4}{\vartheta\left[\begin{smallmatrix}0\\0\end{smallmatrix}\right](\frac{a\tau+b}{c\tau+d})^4} = \frac{(c\tau+d)^2\vartheta\left[\begin{smallmatrix}0\\1\end{smallmatrix}\right](\tau)^4}{(c\tau+d)^2\vartheta\left[\begin{smallmatrix}0\\0\end{smallmatrix}\right](\tau)^4} = \lambda(\tau)$$

for any $\begin{pmatrix} a & b\\ c & d\end{pmatrix} \in M$.

The group $SL_2(\mathbb{Z})$ acts on the vector space of modular forms f of weight k for M by

$$f^\gamma(\tau) = \frac{1}{(c\tau+d)^k} f((a\tau+b)/(c\tau+d)),$$

where $\gamma = \begin{pmatrix} a & b \\ c & d \end{pmatrix} \in SL_2(\mathbb{Z})$. By the actions of the generators $\gamma_1 = \begin{pmatrix} 1 & 1 \\ 0 & 1 \end{pmatrix}$ and $\gamma_2 = \begin{pmatrix} 0 & -1 \\ 1 & 0 \end{pmatrix}$ of $SL_2(\mathbb{Z})$, the vector

$$(\vartheta \begin{bmatrix} 0 \\ 0 \end{bmatrix} (\tau)^4, \vartheta \begin{bmatrix} 0 \\ 1 \end{bmatrix} (\tau)^4, \vartheta \begin{bmatrix} 1 \\ 0 \end{bmatrix} (\tau)^4)$$

is multiplied the matrix $\tilde{\gamma}_1$ and $\tilde{\gamma}_2$ from the right, respectively, where

$$\tilde{\gamma}_1 = \begin{pmatrix} & 1 & \\ 1 & & \\ & & -1 \end{pmatrix}, \quad \tilde{\gamma}_2 = \begin{pmatrix} -1 & & \\ & & -1 \\ & -1 & \end{pmatrix}.$$

The group generated by $\tilde{\gamma}_1$ and $\tilde{\gamma}_2$ is isomorphic to S_3, since they satisfy

$$\tilde{\gamma}_1^2 = \tilde{\gamma}_2^2 = I_3, \quad \tilde{\gamma}_1\tilde{\gamma}_2\tilde{\gamma}_1 = \tilde{\gamma}_2\tilde{\gamma}_1\tilde{\gamma}_2.$$

By Jacobi's identity, the lambda function $\lambda(\tau)$ changes into

$$\frac{1}{\lambda(\tau)}, \quad 1 - \lambda(\tau),$$

by the actions of γ_1 and γ_2, respectively.

Remark 3.1. In order to see the S_3-action on the λ-function, it is convenient to regard it as the map

$$\mathbb{H}/M \ni \tau \mapsto \left[\vartheta \begin{bmatrix} 0 \\ 0 \end{bmatrix} (\tau)^4, \vartheta \begin{bmatrix} 0 \\ 1 \end{bmatrix} (\tau)^4, \vartheta \begin{bmatrix} 1 \\ 0 \end{bmatrix} (\tau)^4\right] \in Y,$$

where $Y = \{[t_0, t_1, t_2] \in \mathbb{P}^2 \mid t_0 - t_1 - t_2 = 0, \quad t_0t_1t_2 \neq 0\}$ is isomorphic to $\mathbb{C} - \{0, 1\}$.

Since the group $SL_2(\mathbb{Z})$ acts on the set $\{\vartheta \left[\begin{smallmatrix} 0 \\ 0 \end{smallmatrix}\right] (\tau)^8, \vartheta \left[\begin{smallmatrix} 0 \\ 1 \end{smallmatrix}\right] (\tau)^8, \vartheta \left[\begin{smallmatrix} 1 \\ 0 \end{smallmatrix}\right] (\tau)^8\}$ as permutations, their fundamental symmetric polynomials

$$\vartheta \begin{bmatrix} 0 \\ 0 \end{bmatrix} (\tau)^8 + \vartheta \begin{bmatrix} 0 \\ 1 \end{bmatrix} (\tau)^8 + \vartheta \begin{bmatrix} 1 \\ 0 \end{bmatrix} (\tau)^8,$$

$$\vartheta \begin{bmatrix} 0 \\ 1 \end{bmatrix} (\tau)^8 \vartheta \begin{bmatrix} 1 \\ 0 \end{bmatrix} (\tau)^8 + \vartheta \begin{bmatrix} 0 \\ 0 \end{bmatrix} (\tau)^8 \vartheta \begin{bmatrix} 1 \\ 0 \end{bmatrix} (\tau)^8 + \vartheta \begin{bmatrix} 0 \\ 0 \end{bmatrix} (\tau)^8 \vartheta \begin{bmatrix} 0 \\ 1 \end{bmatrix} (\tau)^8,$$

$$\vartheta \begin{bmatrix} 0 \\ 0 \end{bmatrix} (\tau)^8 \vartheta \begin{bmatrix} 0 \\ 1 \end{bmatrix} (\tau)^8 \vartheta \begin{bmatrix} 1 \\ 0 \end{bmatrix} (\tau)^8,$$

are invariants under the action of $SL_2(\mathbb{Z})$. By Jacobi's identity, the j-function can be expressed as the following ratio of the symmetric polynomials of $\vartheta \left[\begin{smallmatrix} 0 \\ 0 \end{smallmatrix}\right] (\tau)^8$, $\vartheta \left[\begin{smallmatrix} 0 \\ 1 \end{smallmatrix}\right] (\tau)^8$ and $\vartheta \left[\begin{smallmatrix} 1 \\ 0 \end{smallmatrix}\right] (\tau)^8$:

$$j(\tau) = \frac{4}{27} \frac{(\vartheta \left[\begin{smallmatrix} 0 \\ 0 \end{smallmatrix}\right] (\tau)^8 + \vartheta \left[\begin{smallmatrix} 0 \\ 1 \end{smallmatrix}\right] (\tau)^8 + \vartheta \left[\begin{smallmatrix} 1 \\ 0 \end{smallmatrix}\right] (\tau)^8)^3}{\vartheta \left[\begin{smallmatrix} 0 \\ 0 \end{smallmatrix}\right] (\tau)^8 \vartheta \left[\begin{smallmatrix} 0 \\ 1 \end{smallmatrix}\right] (\tau)^8 \vartheta \left[\begin{smallmatrix} 1 \\ 0 \end{smallmatrix}\right] (\tau)^8}.$$

It is clear that $j(\tau)$ is invariant under the action of $SL_2(\mathbb{Z})$.

Remark 3.2. By Jacobi's identity, every symmetric polynomial of $\vartheta\left[\begin{smallmatrix}0\\0\end{smallmatrix}\right](\tau)^8$, $\vartheta\left[\begin{smallmatrix}0\\1\end{smallmatrix}\right](\tau)^8$ and $\vartheta\left[\begin{smallmatrix}1\\0\end{smallmatrix}\right](\tau)^8$ of 2nd order is a constant multiple of $(\vartheta\left[\begin{smallmatrix}0\\0\end{smallmatrix}\right](\tau)^8 + \vartheta\left[\begin{smallmatrix}0\\1\end{smallmatrix}\right](\tau)^8 + \vartheta\left[\begin{smallmatrix}1\\0\end{smallmatrix}\right](\tau)^8)^2$.

4. Theta functions on $\mathbb{D}$

The symmetric domain $\mathbb{D}$ of type $I_{2,2}$ is defined as

$$\mathbb{D} = \left\{\tau \in M_{2,2}(\mathbb{C}) \mid \frac{\tau - \tau^*}{2i} \text{ is positive definite}\right\},$$

where $\tau^* = {}^t\bar{\tau}$. The group

$$U_{2,2}(\mathbb{C}) = \left\{h \in GL_4(\mathbb{C}) \mid gJg^* = J = \begin{pmatrix} O & -I_2 \\ I_2 & O \end{pmatrix}\right\}$$

and an involution T act on $\mathbb{D}$ as

$$h \cdot \tau = (h_{11}\tau + h_{12})(h_{21}\tau + h_{22})^{-1}, \quad T \cdot \tau = {}^t\tau,$$

where $h = \begin{pmatrix} h_{11} & h_{12} \\ h_{21} & h_{22} \end{pmatrix} \in U_{2,2}(\mathbb{C})$, and h_{jk} are 2×2 matrices. We define some discrete subgroups of $U_{2,2}(\mathbb{C})$:

$$\begin{aligned} U_{2,2}(\mathbb{Z}[i]) &= U_{2,2}(\mathbb{C}) \cap GL_4(\mathbb{Z}[i]), \\ U_{2,2}(1+i) &= \{h \in U_{2,2}(\mathbb{Z}[i]) \mid h \equiv I_4 \bmod (1+i)\}. \end{aligned}$$

The theta function with characteristics a, b is defined as

$$\Theta\begin{pmatrix} a \\ b \end{pmatrix}(\tau) = \sum_{n \in \mathbb{Z}[i]^2} \mathbf{e}[(n+a)\tau(n+a)^* + 2\mathrm{Re}(nb^*)],$$

where $\tau \in \mathbb{D}$, $a, b \in \mathbb{Q}[i]^2$, and n, a, b are represented by row vectors. Since this series converges absolutely and uniformly on any compact set in $\mathbb{D}$, $\Theta\binom{a}{b}(\tau)$ is holomorphic on $\mathbb{D}$. By the definition, we have the following fundamental properties.

Fact 4. 1. If $b \in \frac{1}{1+i}\mathbb{Z}[i]^2$, then $\Theta\binom{a}{ib}(\tau) = \Theta\binom{a}{b}(\tau)$.
If $b \in \frac{1}{2}\mathbb{Z}[i]^2$, then $\Theta\binom{a}{-b}(\tau) = \Theta\binom{a}{b}(\tau)$.

2. For $k \in \mathbb{Z}$ and $m, n \in \mathbb{Z}[i]^2$, we have

$$\begin{aligned} \Theta\begin{pmatrix} i^k a \\ i^k b \end{pmatrix}(\tau) &= \Theta\begin{pmatrix} a \\ b \end{pmatrix}(\tau), \\ \Theta\begin{pmatrix} a+m \\ b+n \end{pmatrix}(\tau) &= \mathbf{e}[-2\mathrm{Re}(mb^*)]\Theta\begin{pmatrix} a \\ b \end{pmatrix}(\tau). \end{aligned}$$

3. If $(1+i)ab^* \notin \mathbb{Z}[i]$ for $a, b \in \frac{1}{1+i}\mathbb{Z}[i]^2$, then $\Theta\binom{a}{b}(\tau) = 0$.

It is known that any element of $U_{2,2}(\mathbb{Z}[i])$ on $\tau \in \mathbb{D}$ is a composition of the following transformations:

1. $\tau \mapsto \tau + s$, where $s = (s_{jk})$ is a 2×2 matrix over $\mathbb{Z}[i]$ satisfying $s^* = s$;

2. $\tau \mapsto g\tau g^*$, where $g \in GL_2(\mathbb{Z}[i])$;
3. $\tau \mapsto -\tau^{-1}$.

Fact 5. By T and these actions, $\Theta\binom{a}{b}(\tau)$ changes into as follows:

$$\begin{aligned}
\Theta\begin{pmatrix} a \\ b \end{pmatrix}(T\cdot\tau) &= \Theta\begin{pmatrix} \bar{a} \\ \bar{b} \end{pmatrix}(\tau), \\
\Theta\begin{pmatrix} a \\ b \end{pmatrix}(\tau+s) &= \mathbf{e}[asa^*]\Theta\begin{pmatrix} a \\ b+as+\frac{1+i}{2}(s_{11},s_{22}) \end{pmatrix}(\tau), \\
\Theta\begin{pmatrix} a \\ b \end{pmatrix}(g\tau g^*) &= \Theta\begin{pmatrix} ag \\ b(g^*)^{-1} \end{pmatrix}(\tau) \quad \text{for } g \in GL_2(\mathbb{Z}[i]), \\
\Theta\begin{pmatrix} a \\ b \end{pmatrix}(-\tau^{-1}) &= -\det(\tau)\mathbf{e}[2\mathrm{Re}(ab^*)]\Theta\begin{pmatrix} -b \\ a \end{pmatrix}(\tau).
\end{aligned}$$

Proof. We can show the first and second equalities by the definition. We can show the last equality by the multi-variable version of Poisson's summation formula. We here show the 3rd equality, which will be used. We have

$$\begin{aligned}
&\Theta\begin{pmatrix} a \\ b \end{pmatrix}(g\tau g^*) \\
= &\sum_{n\in\mathbb{Z}[i]^2} \mathbf{e}[(n+a)(g\tau g^*)(n+a)^* + 2\mathrm{Re}(n(gg^{-1})b^*)] \\
= &\sum_{n\in\mathbb{Z}[i]^2} \mathbf{e}[(ng+ag)\tau(ng+ag)^* + 2\mathrm{Re}(ng(b(g^*)^{-1})^*)] \\
= &\Theta\begin{pmatrix} ag \\ b(g^*)^{-1} \end{pmatrix}(\tau),
\end{aligned}$$

since $m = ng$ runs over $\mathbb{Z}[i]^2$ for any $g \in GL_2(\mathbb{Z}[i])$. □

These transformation formulas imply the following.

Proposition 4.1. *If $a, b \in \frac{1}{1+i}\mathbb{Z}[i]^2$, then $\Theta^2\binom{a}{b}(\tau)$ is a modular from of weight* 2 *with character* det *for $U_{2,2}(1+i)$, i.e., it is holomorphic on $\mathbb{D}$ and it satisfies*

$$\Theta^2\begin{pmatrix} a \\ b \end{pmatrix}(T\cdot\tau) = \Theta^2\begin{pmatrix} a \\ b \end{pmatrix}(\tau),$$

$$\Theta^2\begin{pmatrix} a \\ b \end{pmatrix}(h\cdot\tau) = \det(h)\det(h_{21}\tau+h_{22})^2\Theta^2\begin{pmatrix} a \\ b \end{pmatrix}(\tau),$$

for any $h = (h_{jk}) \in U_{2,2}(1+i)$.

Remark 4.2. We assumed some boundedness conditions around the cusps in the definition of a modular form on the upper half space $\mathbb{H}$. We do not need this kind of hypothesis in the definition of a modular form on $\mathbb{D}$ by the Koecher principle.

By following the proof of Jacobi' identity for lattices $L_1 = M_{2,2}(\mathbb{Z}[i])$, $L_2 = L_1 A$, $L = \langle L_1, L_2 \rangle$, where

$$A = \frac{1+i}{2}\begin{pmatrix} 1 & 1 \\ 1 & -1 \end{pmatrix}, \quad AA^* = I_2, \; A^2 = iI_2,$$

we have quadratic relations among theta functions $\Theta\binom{a}{b}(\tau)$.

Theorem 4.3 (Theorem 1 in [M2]**).** *We have quadratic relations among theta functions for any $a, b \in \mathbb{Q}[i]^2$:*

$$\begin{aligned} &4\Theta\begin{pmatrix} a \\ b \end{pmatrix}(\tau)^2 \\ = &\sum_{e,f \in \frac{1+i}{2}\mathbb{Z}[i]^2/\mathbb{Z}[i]^2} \mathbf{e}[2\mathrm{Re}((1+i)be^*)]\Theta\begin{pmatrix} e + (1+i)a \\ f + (1+i)b \end{pmatrix}(\tau)\Theta\begin{pmatrix} e \\ f \end{pmatrix}(\tau). \end{aligned}$$

For $a, b \in (\frac{\mathbb{Z}[i]}{1+i}/\mathbb{Z}[i])^2$, there are ten $\Theta\binom{a}{b}(\tau)$ which do not vanish identically.

Corollary 4.4. *The ten $\Theta\binom{a}{b}(\tau)^2$ satisfy the same linear relations as the Plücker relations for the $(3,6)$-Grassmann manifold, which is the linear relations among the ten products $D_{ijk}(X)D_{lmn}(X)$ of the Plücker coordinates, where*

$$X = \begin{pmatrix} x_{11} & \dots & x_{16} \\ x_{21} & \dots & x_{26} \\ x_{31} & \dots & x_{36} \end{pmatrix}, \quad D_{ijk}(X) = \det\begin{pmatrix} x_{1i} & x_{1j} & x_{1k} \\ x_{2i} & x_{2j} & x_{2k} \\ x_{3i} & x_{3j} & x_{3k} \end{pmatrix}$$

and $\{i, j, k, l, m, n\} = \{1, \dots, 6\}$. There are five linearly independent $\Theta\binom{a}{b}(\tau)^2$.

Remark 4.5. The element $\tau \in \mathbb{D}$ can be regarded as the periods of the $K3$-surface coming from the double cover of $\mathbb{P}^2$ branching along 6 lines given by the 6 columns of X, refer to [MSY].

5. A hyperbolic structure on the complement of the Whitehead link

Let $\mathbb{H}^3$ be the upper half space model

$$\mathbb{H}^3 = \{(z,t) \in \mathbb{C} \times \mathbb{R} \mid t > 0\}$$

of the 3-dimensional real hyperbolic space. The group $GL_2(\mathbb{C})$ and an involution T act on $\mathbb{H}^3$ as

$$g \cdot (z,t) = \left(\frac{g_{11}\bar{g}_{21}t^2 + (g_{11}z + g_{12})\overline{(g_{21}z + g_{22})}}{|g_{21}|^2t^2 + (g_{21}z + g_{22})\overline{(g_{21}z + g_{22})}}, \frac{|\det(g)|t}{|g_{21}|^2t^2 + (g_{21}z + g_{22})\overline{(g_{21}z + g_{22})}} \right),$$

$$T \cdot (z,t) = (\bar{z}, t),$$

where

$$g = \begin{pmatrix} g_{11} & g_{12} \\ g_{21} & g_{22} \end{pmatrix} \in GL_2(\mathbb{C}).$$

Let $GL_2^T(\mathbb{C})$ be the group generated by $GL_2(\mathbb{C})$ and T with relations

$$T \cdot g = \bar{g} \cdot T$$

for any $g \in GL_2(\mathbb{C})$.

The Whitehead-link-complement $S^3 - L$ admits a hyperbolic structure. We have a homeomorphism

$$\varphi : \mathbb{H}^3/W \overset{\cong}{\longrightarrow} S^3 - L,$$

where

$$W := \langle g_1, g_2 \rangle, \quad g_1 = \begin{pmatrix} 1 & i \\ 0 & 1 \end{pmatrix}, \; g_2 = \begin{pmatrix} 1 & 0 \\ 1+i & 1 \end{pmatrix}.$$

We call W the *Whitehead-link-complement group.* A fundamental domain FD for W in $\mathbb{H}^3$ is the union of two pyramids given in Figure 3 (cf. [W]). Put the two pictures very close to your eyes and gradually move them away. At a certain distance you will see the third picture, which should be 3-dimensional. The group W has

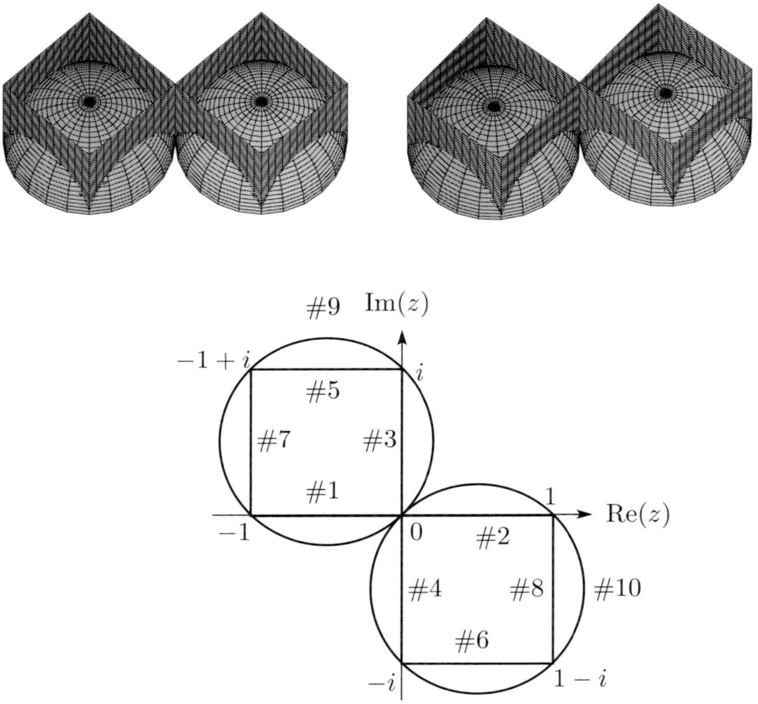

FIGURE 3. Stereographic figures of fundamental domain FD of W in $\mathbb{H}^3$

two cusps. They are represented by the vertices of the pyramids:

$$(z,t) = (*, +\infty), \quad (0,0) \sim (\pm i, 0) \sim (\pm 1, 0) \sim (\mp 1 \pm i, 0).$$

Remark 5.1. The monodromy groups of $E(\alpha,\beta,\gamma)$ for parameters satisfying

$$\cos(2\pi\alpha) = \frac{1+i}{2}, \quad \beta = -\alpha, \quad \gamma \in \mathbb{Z}$$

are conjugate to W. In fact, under this condition for α,β,γ, we have

$$\begin{aligned} P^{-1}\rho_1 P &= \begin{pmatrix} 1 & 0 \\ -i(-1+\exp(-2\pi i\beta))(1-\exp(-2\pi i\alpha)) & \exp(-2\pi i\gamma) \end{pmatrix} = g_2, \\ P^{-1}\rho_2 P &= \begin{pmatrix} 1 & i \\ 0 & \exp(-2\pi i(\alpha+\beta-\gamma)) \end{pmatrix} = g_1, \end{aligned}$$

for

$$P = \begin{pmatrix} 1 & 0 \\ 0 & i(1-\exp(-2\pi i\alpha))^{-1} \end{pmatrix}$$

by Fact 1.

6. Discrete subgroups of $GL_2(\mathbb{C})$, in particular Λ

We define some discrete subgroups of $GL_2(\mathbb{C})$:

$$\begin{aligned} \Gamma &= GL_2(\mathbb{Z}[i]), \\ S\Gamma_0(1+i) &= \{g=(g_{jk})\in\Gamma \mid \det(g)=\pm1,\ g_{21}\in(1+i)\mathbb{Z}[i]\}, \\ S\Gamma(1+i) &= \{g\in S\Gamma_0(1+i) \mid g_{12}\in(1+i)\mathbb{Z}[i]\}, \\ \Gamma(2) &= \{g\in\Gamma \mid g_{11}-g_{22}, g_{12}, g_{21}\in 2\mathbb{Z}[i]\}, \\ \overline{W} &= TWT = \{\bar{g} \mid g\in W\}, \\ \hat{W} &= W\cap\overline{W}, \\ \breve{W} &= \langle W,\overline{W}\rangle. \end{aligned}$$

Convention. 1. We regard these groups as subgroups of the projective group $PGL_2(\mathbb{C})$; in other words, every element of the groups represented by a scalar matrix is regarded as the identity. Note that each center of the groups Γ, $S\Gamma_0(1+i)$, $S\Gamma(1+i)$ and $\Gamma(2)$ is $\{\pm I_2, \pm iI_2\}$. For an element g of these groups with $\det(g)=-1$, we have $\det(ig)=1$.
2. For any subgroup G of Γ, we denote G^T the group $\langle G,T\rangle$ generated by G and T in $GL_2^T(\mathbb{C})$.

The group $\Gamma^T(2)$ is a Coxeter group generated by the eight reflections, whose mirrors form an octahedron in $\mathbb{H}^3$, see Figure 4.

Let Λ be the group $\langle\Gamma^T(2), W\rangle$ generated by $\Gamma^T(2)$ and W. So far we defined many subgroups of $\Gamma^T = GL_2^T(\mathbb{Z}[i])$; their inclusion relation can be depicted as

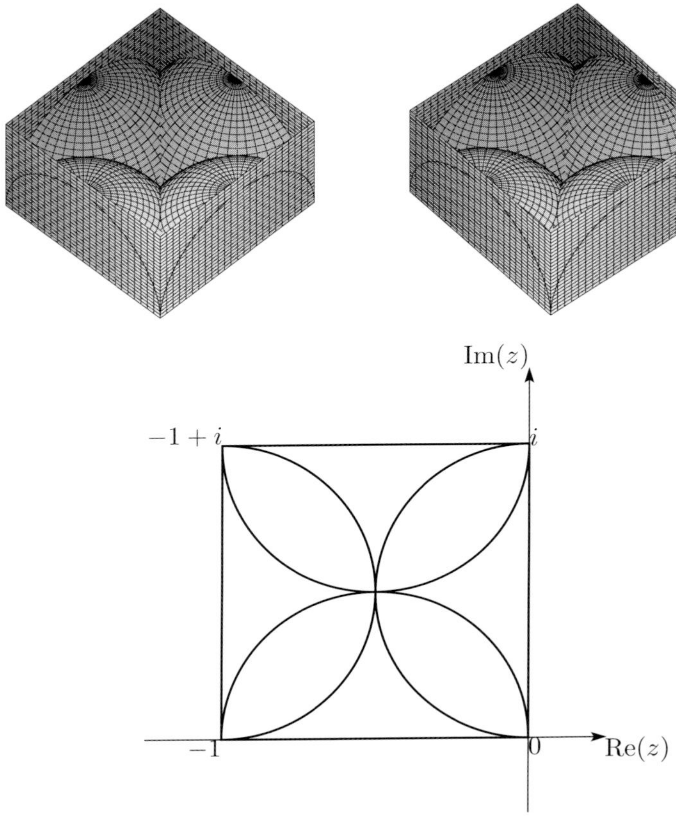

FIGURE 4. Stereographic figures of a Weyl chamber of $\Gamma^T(2)$

follows

$$
\begin{array}{ccccc}
 & & & \Lambda = S\Gamma_0^T(1+i) & \\
 & / & & | & \\
S\Gamma^T(1+i) & & & S\Gamma_0(1+i) & \\
| & & & | & \\
* & & & \breve{W} = \langle W, \overline{W} \rangle & \\
| & & / & & \backslash \\
\Gamma^T(2) & & W & & \overline{W} \\
| & & \backslash & & / \\
\Gamma(2) & & & \hat{W} = W \cap \overline{W} &
\end{array}
$$

When two groups are connected by a segment, the one below is a subgroup of the one above of index 2.

Lemma 6.1. 1. *The group* $\Gamma^T(2)$ *is normal in* Λ*;*
the quotient $\Lambda/\Gamma^T(2)$ *is isomorphic to the dihedral group* D_8 *of order 8.*

2. *We have* $[\Lambda, W] = 8$, W *is not normal in* Λ: $TWT = \overline{W}$.
3. *The domain bounded by the four walls*

$$a : \mathrm{Im}(z) = 0, \quad b : \mathrm{Re}(z) = 0,$$

$$c : \mathrm{Im}(z) = \frac{1}{2}, \quad d : \mathrm{Re}(z) = -\frac{1}{2},$$

and by the hemisphere

$$\#9 : |z - \frac{-1+i}{2}| = \frac{1}{\sqrt{2}}.$$

is a fundamental domain of Λ, *see Figure* 5.

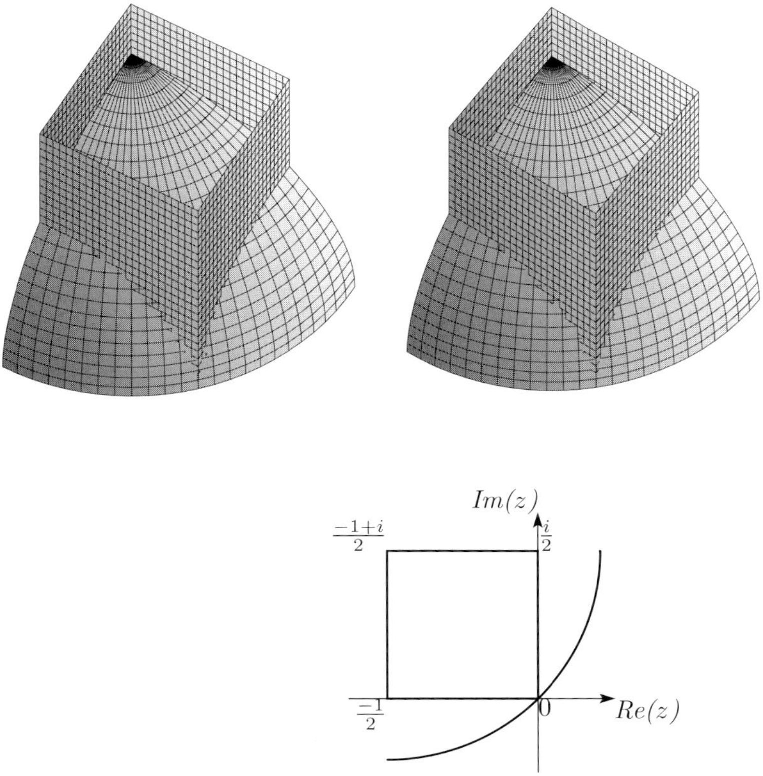

FIGURE 5. Stereographic figures of fundamental domain of Λ

4. *The group* Λ *coincides with* $S\Gamma_0^T(1+i)$, *and* $[S\Gamma_0(1+i), W] = 4$.
(We will see $S\Gamma_0(1+i)/W = (\mathbb{Z}/2\mathbb{Z})^2$.*)*

Our strategy for the rest of this paper is as follows. At first, we realize the quotient space $\mathbb{H}^3/\Gamma^T(2)$ by using theta functions $\Theta\binom{a}{b}(\tau)$ on $\mathbb{D}$ in Section 10. Next we construct D_8-invariant functions which realize $\mathbb{H}^3/\Lambda$ in Section 11. This is a generalization of the construction of the j-function from the λ-function. Finally, we construct the 3 double covers in the right line step by step. In order to know the branch locus of each of double covers, we investigate the symmetry of the Whitehead link.

$\mathbb{H}^3/\Gamma^T(2)$ $\qquad\qquad\qquad$ $\mathbb{H}^3/W$

$\setminus$ $\qquad\qquad\qquad$ $|_{\mathbb{Z}/(2\mathbb{Z})}$

$\mathbb{H}^3/\langle W, \overline{W}\rangle$

$\setminus_{D_8}$ $\qquad\qquad\qquad$ $|_{\mathbb{Z}/(2\mathbb{Z})}$

$\mathbb{H}^3/S\Gamma_0(1+i)$

$\setminus$ $\qquad\qquad\qquad$ $/_{\mathbb{Z}/(2\mathbb{Z})}$

$\mathbb{H}^3/\Lambda$

7. Symmetry of the Whitehead link

The π-rotations with axes F_1, F_2 and F_3 in Figure 6 are orientation preserving homeomorphisms of S^3 keeping L fixed; they form a group $(\mathbb{Z}/2\mathbb{Z})^2$. These rotations can be represent as elements of Λ. We give the axes in the fundamental domain of $\mathbb{H}^3/W$.

Proposition 7.1. *The three π-rotations with axes F_1, F_2 and F_3 are represented by the transformations*

$$\gamma_1 : \begin{pmatrix} -1 & 1 \\ 0 & 1 \end{pmatrix}, \quad \gamma_2 : \begin{pmatrix} 1 & 1 \\ 0 & 1 \end{pmatrix}, \quad \gamma_3 : \begin{pmatrix} -1 & 0 \\ 0 & 1 \end{pmatrix},$$

respectively, of $\mathbb{H}^3$ modulo W. The fixed loci in FD, as well as in $\mathbb{H}^3/W$, of the rotations γ_1, γ_2 and γ_3 are also called the axes F_1, F_2 and F_3. The axis F_1 consists of the geodesics in FD given by the inverse images of $z = \pm\frac{1}{2}$ and $z = \pm\frac{1-i}{2}$ under the projection $p : \mathbb{H}^3 \ni (z,t) \mapsto z \in \mathbb{C}$, F_2 consists of the geodesic in FD joining points $(0,0)$ and $(i,0)$ in $\partial\mathbb{H}^3$ and the geodesic joining $(-1,0), (i,0) \in \partial\mathbb{H}^3$, F_3 consists of the geodesics in FD given by the inverse images of $z = 0, -1, \frac{i}{2}$, and $-1 + \frac{i}{2}$ under the projection p and the geodesic joining $(0,0), (-1+i,0) \in \partial\mathbb{H}^3$. They are depicted in FD as in Figure 7, where $\square = (\frac{-1+i}{2}, \frac{1}{\sqrt{2}})$ $\bigcirc = (\frac{i}{2}, \frac{1}{2})$ are points of $\mathbb{H}^3$, a bullet $\bullet$ stands for a vertical line: the inverse image of the point under the projection p, and a thick segment stands for a geodesic curve in $\mathbb{H}^3$

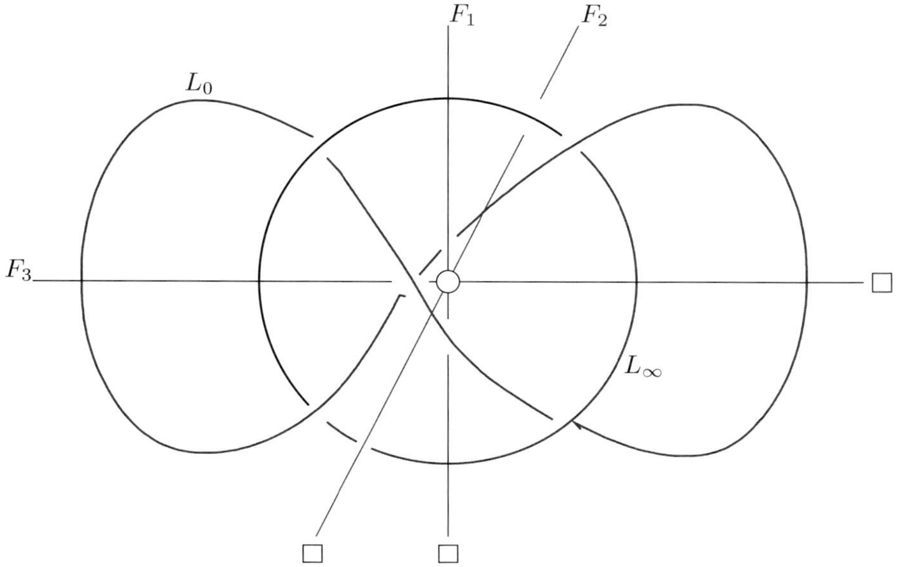

FIGURE 6. The Whitehead link with its symmetry axes

joining its terminal points in $\partial\mathbb{H}^3$, which is the vertical semicircle with the given segment as its diameter; its image under p is the given segment.

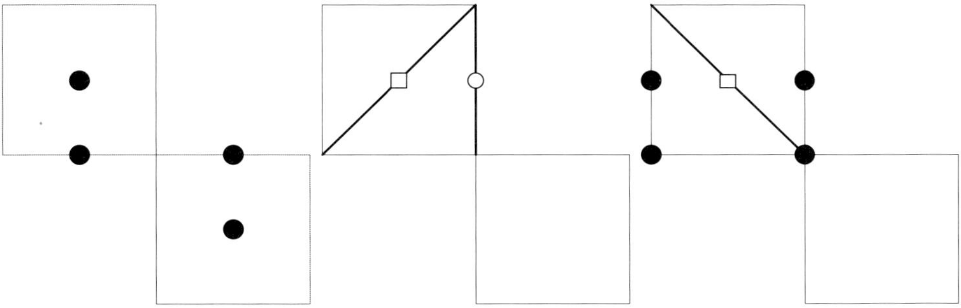

FIGURE 7. The fixed loci of γ_1, γ_2, γ_3

8. Orbit spaces under $\breve{W}$, $S\Gamma_0(1+i)$ and Λ

We give fundamental domains for $\breve{W}$, $S\Gamma_0(1+i)$ and Λ, and the orbifolds $\mathbb{H}^3/\breve{W}$, $\mathbb{H}^3/S\Gamma_0(1+i)$ and $\mathbb{H}^3/\Lambda$ in Figures 8, 9, 11 and 10.

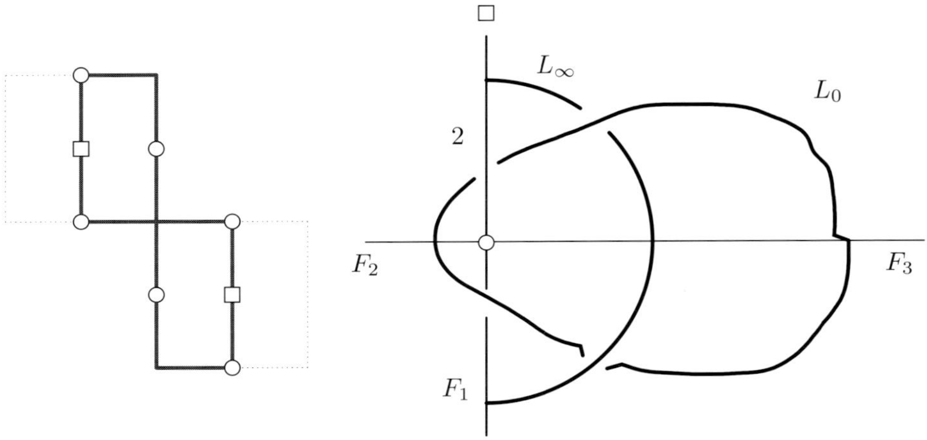

FIGURE 8. A fundamental domain for $\breve{W}$ and the orbifold $\mathbb{H}^3/\breve{W}$

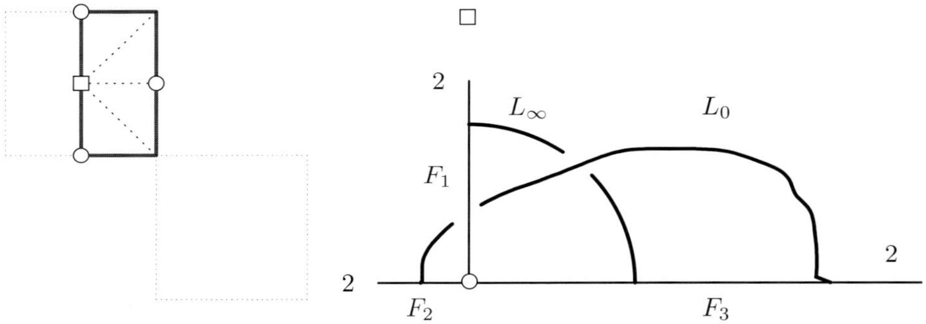

FIGURE 9. A fundamental domain for $S\Gamma_0(1+i)$ and the orbifold $\mathbb{H}^3/S\Gamma_0(1+i)$

Proposition 8.1. 1. *The branch locus of the double cover* $\mathbb{H}^3/S\Gamma_0(1+i)$ *of* $\mathbb{H}^3/\Lambda$ *is the union of the walls* a, b, c, d *given in Lemma* 6.1.

2. *That of the double cover* $\mathbb{H}^3/\breve{W}$ *of* $\mathbb{H}^3/S\Gamma_0(1+i)$ *is the union of the axes* F_2 *and* F_3 *(the axes* F_2 *and* F_3 *are equivalent in the space* $\mathbb{H}^3/\breve{W}$*).*
3. *That of the double cover* $\mathbb{H}^3/W$ *of* $\mathbb{H}^3/\breve{W}$ *is the axis* F_1.

9. Embedding of $\mathbb{H}^3$ into $\mathbb{D}$

We embed $\mathbb{H}^3$ into $\mathbb{D}$ by

$$\imath : \mathbb{H}^3 \ni (z,t) \mapsto \frac{i}{t}\begin{pmatrix} t^2+|z|^2 & z \\ \bar{z} & 1 \end{pmatrix} \in \mathbb{D};$$

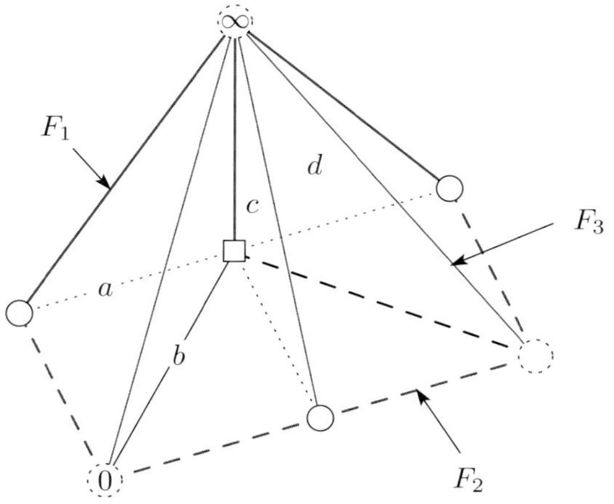

FIGURE 10. A better picture of the fundamental domain for $S\Gamma_0(1+i)$ corresponding to the left figure in Figure 9

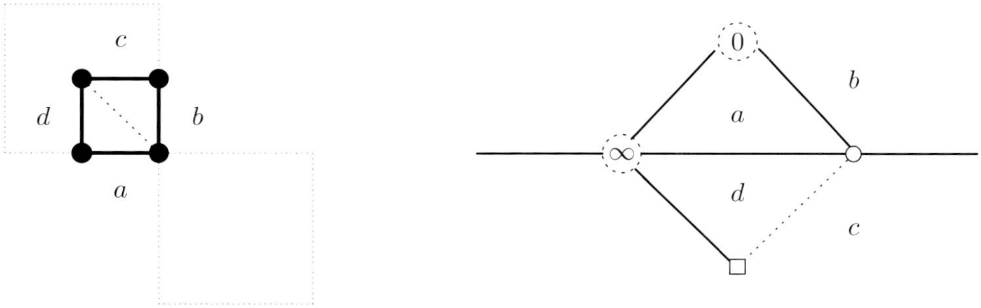

FIGURE 11. A fundamental domain for Λ and the boundary of $\mathbb{H}^3/\Lambda$

we define a homomorphism

$$\jmath : GL_2(\mathbb{C}) \ni g \mapsto \begin{pmatrix} g/\sqrt{|\det(g)|} & O \\ O & (g^*/\sqrt{|\det(g)|})^{-1} \end{pmatrix} \in U_{2,2}(\mathbb{C}).$$

They satisfy

$$\begin{aligned} \imath(g\cdot(z,t)) &= \jmath(g)\cdot\imath(z,t) \quad \text{for any } g \in GL_2(\mathbb{C}), \\ \imath(T\cdot(z,t)) &= T\cdot\imath(z,t), \\ -(\imath(z,t))^{-1} &= \left(\jmath\begin{pmatrix} 0 & -1 \\ 1 & 0 \end{pmatrix}\cdot T\right)\cdot\imath(z,t). \end{aligned}$$

We denote the pull back of $\Theta\binom{a}{b}(\tau)$ under the embedding $\imath : \mathbb{H}^3 \to \mathbb{D}$ by $\Theta\binom{a}{b}(z,t)$. By definition, we have the following.

Fact 6. 1. For $a, b \in \frac{1}{2}\mathbb{Z}[i]^2$, each $\Theta\binom{a}{b}(z,t)$ is real valued. If $2\mathrm{Re}(ab^*)+2\mathrm{Im}(ab^*) \notin \mathbb{Z}$, then $\Theta\binom{a}{b}(z,t) \equiv 0$.
2. For $a, b \in \frac{1}{1+i}\mathbb{Z}[i]^2$, each $\Theta\binom{a}{b}(z,t)$ is invariant under the action of $\Gamma^T(2)$.
3. The function $\Theta = \Theta\binom{00}{00}(z,t)$ is positive and invariant under the action of Γ^T.

10. Invariant functions for $\Gamma^T(2)$ and an embedding of $\mathbb{H}^3/\Gamma^T(2)$

For theta functions with characteristics $\frac{p}{2}, \frac{q}{2} \in \frac{1}{2}\mathbb{Z}[i]^2$ on $\mathbb{H}^3$, we set

$$\Theta\begin{bmatrix} p \\ q \end{bmatrix} = \Theta\begin{bmatrix} p \\ q \end{bmatrix}(z,t) = \Theta\begin{pmatrix} \frac{p}{2} \\ \frac{q}{2} \end{pmatrix}(z,t), \quad p, q \in \mathbb{Z}[i]^2$$

and

$$x_0 = \Theta\begin{bmatrix} 0,0 \\ 0,0 \end{bmatrix}, \; x_1 = \Theta\begin{bmatrix} 1+i, 1+i \\ 1+i, 1+i \end{bmatrix}, \; x_2 = \Theta\begin{bmatrix} 1+i, 0 \\ 0, 1+i \end{bmatrix}, \; x_3 = \Theta\begin{bmatrix} 0, 1+i \\ 1+i, 0 \end{bmatrix}.$$

Theorem 10.1 (Corollary 3.1 in [MY]). *The map*

$$\mathbb{H}^3 \ni (z,t) \mapsto \frac{1}{x_0}(x_1, x_2, x_3) \in \mathbb{R}^3$$

induces an isomorphism between $\mathbb{H}^3/\Gamma^T(2)$ and the octahedron

$$Oct = \{(t_1, t_2, t_3) \in \mathbb{R}^3 \mid |t_1| + |t_2| + |t_3| \leq 1\}$$

minus the six vertices $(\pm 1, 0, 0), (0, \pm 1, 0), (0, 0, \pm 1)$. The restriction of this isomorphism to the complement of the union of the mirrors of reflections in $\Gamma^T(2)$ is a real analytic diffeomorphism to the interior of Oct.

We can regard the map in this theorem as a generalization of the λ-function.

11. Invariant functions for Λ and an embedding of $\mathbb{H}^3/\Lambda$

Once an embedding of $\mathbb{H}^3/\Gamma^T(2)$ is obtained, in terms of x_j, for a supergroup Λ of $\Gamma^T(2)$, an embedding of $\mathbb{H}^3/\Lambda$ can be obtained by polynomials of the x_j's invariant under the finite group $\Lambda/\Gamma^T(2)$ isomorphic to the dihedral group D_8 of order 8.

Proposition 11.1. *By the actions of the generators g_1 and g_2 of W, the functions x_1, x_2, x_3 are transformed into as follows:*

$$(x_1, x_2, x_3) \cdot g_1 = (x_1, x_2, x_3) \begin{pmatrix} & -1 & \\ -1 & & \\ & & 1 \end{pmatrix},$$

$$(x_1,x_2,x_3)\cdot g_2=(x_1,x_2,x_3)\begin{pmatrix}-1&&\\&1&\\&&-1\end{pmatrix}.$$

The group generated by these matrices is isomorphic to the dihedral group D_8.

Theorem 11.2. *The functions* $x_1^2+x_2^2,\ x_1^2x_2^2,\ x_3^2,\ x_1x_2x_3$ *are* Λ*-invariant. The map*

$$\lambda:\mathbb{H}^3\ni(z,t)\longmapsto(\lambda_1,\lambda_2,\lambda_3,\lambda_4)=(\xi_1^2+\xi_2^2,\ \xi_1^2\xi_2^2,\ \xi_3^2,\ \xi_1\xi_2\xi_3)\in\mathbb{R}^4$$

induces an embedding of $\mathbb{H}^3/\Lambda$ *into the subdomain of the variety* $\lambda_2\lambda_3=\lambda_4^2$, *where* $\xi_j=x_j/x_0$.

We can regard the map in this theorem as a generalization of the j-function expressed in terms of symmetric polynomials of three theta constants.

12. Invariant functions for W

In this section, we construct invariant functions Φ_1,Φ_2 and Φ_3 for W by utilizing theta functions with characteristics $a,b\in\frac{1}{2}\mathbb{Z}[i]^2$. Set

$$y_1=\Theta\begin{bmatrix}0,1\\1+i,0\end{bmatrix},\quad y_2=\Theta\begin{bmatrix}1+i,1\\1+i,0\end{bmatrix},\quad z_1=\Theta\begin{bmatrix}0,1\\1,0\end{bmatrix},\quad z_2=\Theta\begin{bmatrix}1+i,1\\1,1+i\end{bmatrix}.$$

We define functions as

$$\begin{aligned}\Phi_1&=x_3z_1z_2,\\ \Phi_2&=(x_2-x_1)y_1+(x_2+x_1)y_2,\\ \Phi_3&=(x_1^2-x_2^2)y_1y_2.\end{aligned}$$

Theorem 12.1. *The functions* Φ_1,Φ_2 *and* Φ_3 *are* W*-invariant. By the actions* $g=I_2+2\begin{pmatrix}p&q\\r&s\end{pmatrix}\in\Gamma(2)$ *and* T, *we have*

$$\begin{aligned}\Phi_1\cdot g&=\mathbf{e}[\mathrm{Re}((1+i)p+(1-i)s)]\Phi_1,\\ \Phi_2\cdot g&=\mathbf{e}[\mathrm{Re}(r(1-i))]\Phi_2,\\ \Phi_3\cdot g&=\Phi_3.\end{aligned}$$

$$\Phi_1\cdot T=\Phi_1,\quad \Phi_3\cdot T=-\Phi_3.$$

Remark 12.2. We have $\Phi_2\cdot T=(x_2-x_1)y_1-(x_2+x_1)y_2$. This is not invariant under W but invariant under $\overline{W}$.

Let Iso_j be the isotropy subgroup of $\Lambda=S\Gamma_0^T(1+i)$ for Φ_j.

Theorem 12.3. *We have*

$$S\Gamma_0(1+i)=\mathrm{Iso}_3,\quad \breve{W}=\mathrm{Iso}_1\cap\mathrm{Iso}_3,\quad W=\mathrm{Iso}_1\cap\mathrm{Iso}_2\cap\mathrm{Iso}_3.$$

This theorem implies the isomorphism $S\Gamma_0(1+i)/W\simeq(\mathbb{Z}/2\mathbb{Z})^2$ and the following arithmetical characterizations for $\breve{W}$, W and $\hat{W}=W\cap\bar{W}$.

Theorem 12.4. *An element* $g = \begin{pmatrix} p & q \\ r & s \end{pmatrix} \in S\Gamma_0(1+i)$ *satisfying* $\mathrm{Re}(s) \equiv 1 \bmod 2$ *belongs to* $\breve{W}$ *if and only if*

$$\begin{aligned} &\frac{\mathrm{Re}(p) + \mathrm{Im}(s) - (-1)^{\mathrm{Re}(q)+\mathrm{Im}(q)}(\mathrm{Im}(p) + \mathrm{Re}(s))}{2} \\ \equiv\ &\frac{((-1)^{\mathrm{Re}(r)} + 1)\mathrm{Im}(q) + (\mathrm{Re}(q) + \mathrm{Im}(q))(\mathrm{Re}(r) + \mathrm{Im}(r))}{2} \bmod 2. \end{aligned}$$

The element $g \in \breve{W}$ *belongs to* W *if and only if*

$$\mathrm{Re}(p+q) + \frac{\mathrm{Re}(r) - (-1)^{\mathrm{Re}(q)+\mathrm{Im}(q)}\mathrm{Im}(r)}{2} \equiv 1 \bmod 2.$$

The element $g \in W$ *belongs to* $\hat{W}$ *if and only if* $r \in 2\mathbb{Z}[i]$.

13. Embeddings of the quotient spaces

In order to embed the quotient spaces $\mathbb{H}^3/S\Gamma_0(1+i)$, $\mathbb{H}^3/\breve{W}$ and $\mathbb{H}^3/W$, we construct, for each $j = 1, 2, 3$, invariant functions $f_{j1}, f_{j2}, \dots$ for W such that their common zero is $F_k \cup F_l$, where $\{j, k, l\} = \{1, 2, 3\}$. We use W-invariant functions as follows:

$$\begin{aligned}
f_{00} &= (x_2^2 - x_1^2)y_1y_2 = \Phi_3, \\
f_{01} &= (x_2^2 - x_1^2)z_1z_2z_3z_4, \\
f_{11} &= x_3z_1z_2 = \Phi_1, \\
f_{12} &= x_1x_2z_1z_2, \\
f_{13} &= x_3(x_2^2 - x_1^2)z_3z_4, \\
f_{14} &= x_1x_2(x_2^2 - x_1^2)z_3z_4, \\
f_{20} &= (x_2 - x_1)z_2z_3 + (x_2 + x_1)z_1z_4, \\
f_{21} &= z_1z_2\{(x_2 - x_1)z_1z_3 + (x_2 + x_1)z_2z_4\}, \\
f_{22} &= (x_2^2 - x_1^2)\{(x_2 - x_1)z_1z_4 + (x_2 + x_1)z_2z_3\}, \\
f_{30} &= (x_2 - x_1)y_1 + (x_2 + x_1)y_2 = \Phi_2, \\
f_{31} &= (x_2 - x_1)z_1z_3 - (x_2 + x_1)z_2z_4, \\
f_{32} &= z_3z_4\{-(x_2 - x_1)z_1z_4 + (x_2 + x_1)z_2z_3\},
\end{aligned}$$

where

$$z_1 = \Theta\begin{bmatrix} 0, 1 \\ 1, 0 \end{bmatrix}, \quad z_2 = \Theta\begin{bmatrix} 1+i, 1 \\ 1, 1+i \end{bmatrix}, \quad z_3 = \Theta\begin{bmatrix} 0, i \\ 1, 0 \end{bmatrix}, \quad z_4 = \Theta\begin{bmatrix} 1+i, i \\ 1, 1+i \end{bmatrix}.$$

Proposition 13.1. *We have*

$$\begin{aligned} 4z_1^2 &= (x_0+x_1+x_2+x_3)(x_0-x_1-x_2+x_3),\\ 4z_2^2 &= (x_0+x_1-x_2-x_3)(x_0-x_1+x_2-x_3),\\ 4z_3^2 &= (x_0+x_1-x_2+x_3)(x_0-x_1+x_2+x_3),\\ 4z_4^2 &= (x_0+x_1+x_2-x_3)(x_0-x_1-x_2-x_3). \end{aligned}$$

Proposition 13.2. *The functions f_{jp} are W-invariant. These change the signs by the actions of γ_1, γ_2 and γ_3 as in the table*

	γ_1	γ_2	γ_3
f_{0j}	$+$	$+$	$+$
f_{1j}	$+$	$-$	$-$
f_{2j}	$-$	$+$	$-$
f_{3j}	$-$	$-$	$+$

Theorem 13.3. *The analytic sets V_1, V_2, V_3 of the ideals*

$$I_1 = \langle f_{11}, f_{12}, f_{13}, f_{14}\rangle, \quad I_2 = \langle f_{21}, f_{22}\rangle, \quad I_3 = \langle f_{31}, f_{32}\rangle$$

are $F_2 \cup F_3$, $F_1 \cup F_3$, $F_1 \cup F_2$.

Corollary 13.4. *The analytic set V_{jk} of the ideals $\langle I_j, I_k\rangle$ is F_l for $\{j,k,l\} = \{1,2,3\}$.*

Theorem 13.5. *The map*

$$\varphi_0 : \mathbb{H}^3/S\Gamma_0(1+i) \ni (z,t) \mapsto (\lambda_1, \ldots, \lambda_4, \eta_{01}) \in \mathbb{R}^5$$

is injective, where $\eta_{01} = f_{01}/x_0^6$. Its image Image(φ_0) *is determined by the image* Image(λ) *under $\lambda : \mathbb{H}^3 \ni (z,t) \mapsto (\lambda_1, \ldots, \lambda_4)$ and the relation*

$$\begin{aligned} &256 f_{01}^2\\ = \ &(\lambda_1^2 - 4\lambda_2) \prod_{\varepsilon_3=\pm 1} (\lambda_3^2 - 2(x_0^2+\lambda_1)\lambda_3 + \varepsilon_3 8 x_0 \lambda_4 + x_0^4 - 2x_0^2\lambda_1 + \lambda_1^2 - 4\lambda_2), \end{aligned}$$

as a double cover of Image(λ) *branching along its boundary.*

The axes F_1, F_2 and F_3 can be illustrated as in Figure 12. Each of the two cusps $\bar{\infty}$ and $\bar{0}$ is shown as a hole. These holes can be deformed into sausages as in Figure 13.

Theorem 13.6. *The map*

$$\varphi_1 : \mathbb{H}^3/\breve{W} \ni (z,t) \mapsto (\varphi_0, \eta_{11}, \ldots, \eta_{14}) \in \mathbb{R}^9$$

is injective, where $\eta_{1j} = f_{1j}/x_0^{\deg(f_{1j})}$. The products $f_{1p}f_{1q}$ $(1 \le p \le q \le 4)$ can be expressed as polynomials of x_0, $\lambda_1, \ldots, \lambda_4$ and f_{01}. The image Image(φ_0) *together with these relations determines the image* Image(φ_1) *under the map φ_1.*

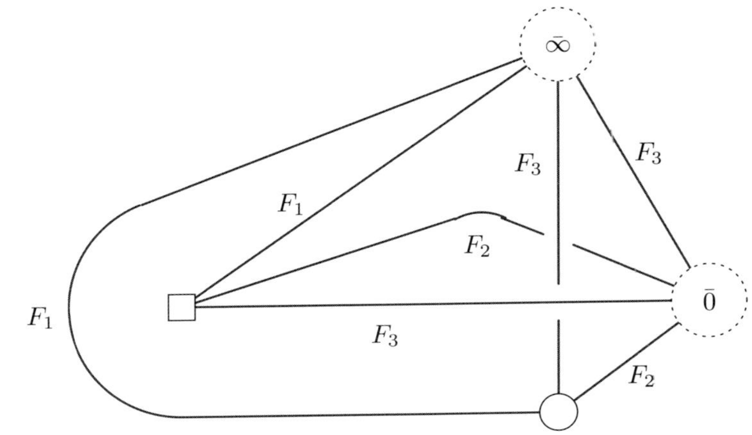

FIGURE 12. Orbifold singularities in Image(φ_0) and the cusps $\bar{\infty}$ and $\bar{0}$

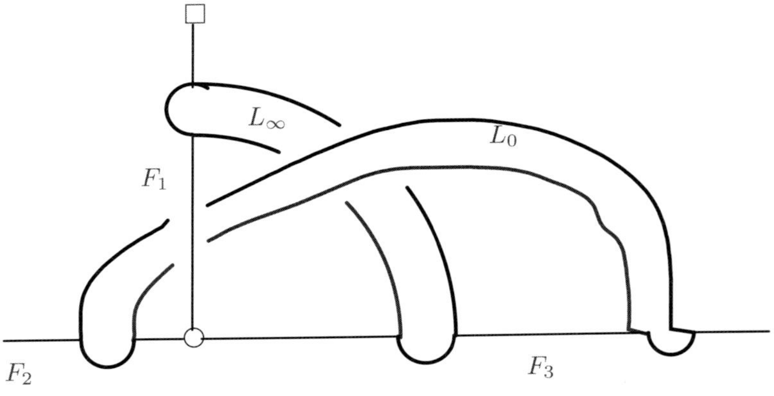

FIGURE 13. The cusp-holes are deformed into two sausages

The boundary of a small neighborhood of the cusp $\varphi_1(0)$ is a torus, which is the double cover of that of the cusp $\varphi_0(0)$; note that two F_2-curves and two F_3-curves stick into $\varphi_0(0)$. The boundary of a small neighborhood of the cusp $\varphi_1(\infty)$ remains to be a 2-sphere; note that two F_1-curves and two F_3-curves stick into $\varphi_0(\infty)$, and that four F_1-curves stick into $\varphi_1(\infty)$, see Figures 13, 14. Note also that the sausage and the doughnut in Figure 14 are obtained by the two sausages in Figure 13 and their copies by the π-rotation with axis F_3.

Theorem 13.7. *The map*

$$\varphi : \mathbb{H}^3/W \ni (z,t) \mapsto (\varphi_1, \eta_{21}, \eta_{22}, \eta_{31}, \eta_{32}) \in \mathbb{R}^{13}$$

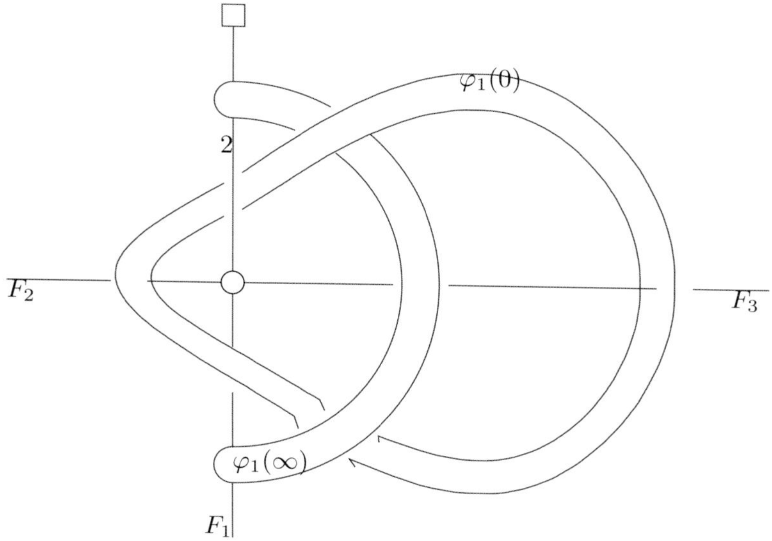

FIGURE 14. The double covers of the cusp holes in Figure 13

is injective, where $\eta_{ij} = f_{ij}/x_0^{\deg(f_{ij})}$. *The products* $f_{2q}f_{2r}$ $f_{3q}f_{3r}$ *and* $f_{1p}f_{2q}f_{3r}$ $(p = 1, \ldots, 4,\ q, r = 1, 2)$ *can be expressed as polynomials of* x_0, $\lambda_1, \ldots, \lambda_4$ *and* f_{01}. *The image* Image(φ_1) *together with these relations determines the image* Image(φ) *under the map* φ.

The boundary of a small neighborhood of the cusp $\varphi(\infty)$ is a torus, which is the double cover of that of the cusp $\varphi_1(\infty)$; recall that four F_1-curves stick into $\varphi_1(\infty)$. The boundary of a small neighborhood of the cusp $\varphi(0)$ is a torus, which is the unbranched double cover of that of the cusp $\varphi_1(0)$, a torus. Note that the boundary of a small neighborhood of the cusp $\varphi(\infty)$ and that of the cusp $\varphi(0)$ are obtained by the sausage and the doughnut in Figure 14 and their copies by the π-rotation with axis F_1.

Ultimately, the sausage and the doughnut in Figure 14 are covered by two linked doughnuts, tubular neighborhoods of the curves L_0 and L_∞ of the Whitehead link.

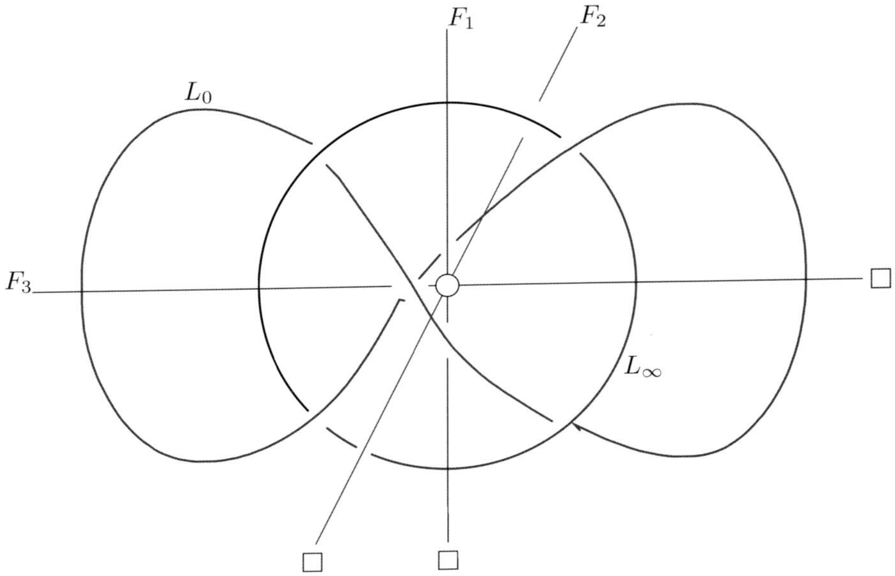

Acknowledgement. The author would like to thank to a referee for valuable comments on this paper.

References

[F] E. Freitag, Modulformen zweiten Grades zum rationalen und Gaußschen Zahlkörper, *Sitzungsber. Heidelb. Akad. Wiss.*, **1** (1967), 1–49.

[IKSY] K. Iwasaki, H. Kimura, S. Shimomura and M. Yoshida, *From Gauss to Painlevé — A modern theory of special functions*, Aspects of Mathematics, E16, Friedrich Vieweg & Sohn, Braunschweig, 1997.

[I] J. Igusa, *Theta Functions*, Springer-Verlag, Berlin, Heidelberg, New York, 1972.

[K] T. Kimura, *Hypergeometric Functions of Two Variables*, Seminar Note in Math. Univ. of Tokyo, 1973.

[M1] K. Matsumoto, Theta functions on the bounded symmetric domain of type $I_{2,2}$ and the period map of 4-parameter family of K3 surfaces, *Math. Ann.*, **295** (1993), 383–408.

[M2] K. Matsumoto, Algebraic relations among some theta functions on the bounded symmetric domain of type $I_{r,r}$, to appear in *Kyushu J. Math.*.

[M3] K. Matsumoto, Automorphic functions for the Borromean-rings-complement group, preprint, 2005.

[MNY] K. Matsumoto, H. Nishi and M. Yoshida, Automorphic functions for the Whitehead-link-complement group, preprint, 2005.

[MSY] K. Matsumoto, T. Sasaki and M. Yoshida, The monodromy of the period map of a 4-parameter family of K3 surfaces and the Aomoto-Gel'fand hypergeometric function of type (3,6), *Internat. J. of Math.*, **3** (1992), 1–164.

[MY] K. Matsumoto and M. Yoshida, Invariants for some real hyperbolic groups, *Internat. J. of Math.*, **13** (2002), 415–443.

[T] W. Thurston, *Geometry and Topology of 3-manifolds*, Lecture Notes, Princeton Univ., 1977/78.

[W] N. Wielenberg, The structure of certain subgroups of the Picard group. *Math. Proc. Cambridge Philos. Soc.*, **84** (1978), no. 3, 427–436.

[Y] M. Yoshida, *Hypergeometric Functions, My Love*, Aspects of Mathematics, E32, Friedrich Vieweg & Sohn, Braunschweig, 1997.

Keiji Matsumoto
e-mail: matsu@math.sci.hokudai.ac.jp
Department of Mathematics
Hokkaido University
Sapporo 060-0810
Japan

Progress in Mathematics, Vol. 260, 273–285

On the Construction of Class Fields by Picard Modular Forms

Thorsten Riedel

Abstract. The goal of this article is to construct modular functions living on the complex ball of dimension two such that the values in special points — similar to the elliptic modular function — generate class fields. For this purpose we are well prepared by the papers [5] and [6]. The first one classifies the moduli space of abelian 3-folds with a multiplication by $\mathbb{Q}(i)$ of type $(2,1)$ as projective surface. Via Jacobians we connect this Shimura surface with the moduli space of a family of curves of Shimura equation type. Thus we are able to continue the construction of the inverse period map of the family by theta constants given in [6]. Knowing the action of the modular group we reach a modular function j by modular forms with respect to the congruence subgroup of level $(1+i)$ of the full Picard modular group of Gauß numbers. If τ is the period of a (Jacobian of a) curve with complex multiplication the corresponding moduli field is generated over the rational numbers by $j(\tau)$. Hence the values in CM points of this function generate abelian extensions of the associated reflex field.

Mathematics Subject Classification (2000). 11G15, 14D05, 14H10, 14H42.

Keywords. Please provide some keywords.

1. Introduction

Every abelian extension of the rational numbers is contained in a cyclotomic field. This famous result is the statement of the Theorem of Kronecker and Weber. Cyclotomic fields are generated by roots of the unity, i.e., by special values of the exponential function. The corresponding assertion for imaginary quadratic number fields, known as Kronecker's Jugendtraum and proved by Takagi, says that in the imaginary quadratic case Hilbert class fields are obtained by adjoining values of the elliptic modular function in moduli of elliptic curves with complex multiplication. The idea of generating abelian extensions of number fields by special values of

I am grateful to Professor Rolf-Peter Holzapfel for his support and advice.

analytic functions Hilbert took up in his twelth problem: Find those functions that play the role of the exponential function and the j-function whatever is laid down as field of rationality.

If we want to generalize the imaginary quadratic case we have to replace elliptic curves by abelian varieties of higher dimension. The moduli τ of an abelian g-fold A_τ with fixed (principally) polarization belongs to a quotient of the Siegel upper halfspace $\mathbb{H}_g$ of degree g by a certain discrete subgroup $\Gamma \subset \mathbb{S}\mathrm{p}(2g,\mathbb{R})$ commensurable with $\mathbb{S}\mathrm{p}(2g,\mathbb{Z})$. We have to find a Γ-invariant map j from $\mathbb{H}_g$ into a projective space such that the field of moduli of A_τ is generated over $\mathbb{Q}$ by the coordinates of $j(\tau)$. If A_τ is a variety with multiplication by a CM field k of absolute degree $2g$ then $k^*(j(\tau))$ is a class field of the corresponding reflex field k^*.

For the family of Picard curves $C(\zeta) : w^3 = z(z-\zeta_0)(z-\zeta_1)(z-\zeta_2)$ this was done by R.-P. Holzapfel [2] and H. Shiga [7]. Here the two-dimensional complex ball serves as space of periods and the modular group acting on the ball is the Picard modular group of Eisenstein numbers.

In this article we will give a similar result for a second family F of smooth projective curves $\widetilde{C}(x,y)$ of genus three with model

$$C(x,y) : w^4 = z^2(z-1)^2(z-x)(z-y)$$

and parameters $(x,y) \in \Lambda := \mathbb{C}^2\backslash\{(x,y) \mid xy(x-1)(y-1)(x-y) = 0\}$.

Fortunately, the main problems leading to the announced modular function are solved by K. Matsumoto [6] and R.-P. Holzapfel [5]. In [6] we find the inverse period map explicitely given by products of theta constants. These products are modular forms with respect to the monodromy group of the family which is an index-2 subgroup of the congruence subgroup of level $(1+i)$ of the Picard modular group $\Gamma = \mathbb{SU}(H,\mathbb{Z}[i])$, $H = \begin{pmatrix} 0 & -i & 0 \\ i & 0 & 0 \\ 0 & 0 & 1 \end{pmatrix}$, of Gauß numbers. The latter group is not only the modular group of our family but the modular group of principally polarized abelian 3-folds with multiplication by $\mathbb{Q}(i)$ of type $(2,1)$ too, and the Jacobians $\mathrm{Jac}(\widetilde{C})$, $\widetilde{C} \in F$, are of this type. The (Baily–Borel compactified) quotients of the ball $\mathbb{B}_H : {}^t z H \bar{z} < 0$ by the congruence subgroup $\Gamma(1+i)$ and by the Picard modular group Γ, are $\mathbb{P}^2$ respectively $\mathbb{P}^2/S_3$ with concrete given ramification data of the uniformizing morphism, see [5]. Using these properties, we are able to continue Matsumoto's construction to get a Picard modular function $j = (j_1 : j_2 : j_3)$ from the ball into $\mathbb{P}^2$ by $\Gamma(1+i)$-modular forms. The resulting map induces a birational transformation of the moduli space $\Gamma\backslash\mathbb{B}_H$ and $\mathbb{P}^2/S_3$ such that the moduli field of an abelian variety is generated by the coordinates of the modular function (outside a certain set). So we apply Shimura–Taniyama's theory to deduce that the j-values of CM-points generate abelian extensions of the associated reflex field.

First let us give an informal definition of the kind of modular forms we want to construct in the sequel.

Definition 1.1. Let F be a family of curves with moduli space M, modular group Γ, and let Γ' denote a congruence subgroup of Γ.

(i) By *Schottky forms* with respect to Γ' we understand Γ'-modular forms leading for a given period to an equation for a curve with this period;
(ii) *Taniyama forms* with respect to Γ' are defined as Γ'-modular forms which serve for the construction of class fields;
(iii) *Schottky–Taniyama forms* of the family are Γ'-modular forms that are both, Schottky forms and Taniyama forms;
(iv) *Schottky, Taniyama* or *Schottky–Taniyama rings* are defined as graduate rings generated by Schottky, Taniyama respectively Schottky–Taniyama forms.

This definition is motivated by the Eisenstein series in the elliptic case. So we take a look at some results about the arithmetic of elliptic curves.

Example 1.2. The family of elliptic curves defined over $\mathbb{C}$ consists of all smooth algebraic curves defined over the complex numbers of genus one. The periods are parametrized by the symmetric space $\mathbb{H} = \{\tau \in \mathbb{C} \mid \operatorname{Im}(\tau) > 0\}$, and the compactified space of moduli is the modular curve $X(1) := \widehat{\Gamma\backslash\mathbb{H}} \cong \mathbb{P}^1$, where the arithmetic group $\Gamma = \mathbb{PSl}_2(\mathbb{Z})$ denotes the (full) elliptic modular group. The *Eisenstein series* $g_2, g_3 : \mathbb{H} \longrightarrow \mathbb{C}$,

$$g_2(\tau) := 60 \sum_{0\neq\lambda\in L_\tau} \frac{1}{\lambda^4},\ g_3(\tau) := 140 \sum_{0\neq\lambda\in L_\tau} \frac{1}{\lambda^6},\ L_\tau = \mathbb{Z} + \tau\mathbb{Z},$$

are known to be Γ-modular forms which have the following properties:

(i) g_2, g_3 are Schottky forms; each isomorphism class of curves is represented by a curve C_τ with period point τ defined by a Weierstraß equation

$$C_\tau : y^2z = 4x^3 - g_2(\tau)xz^2 - g_3(\tau)z^3,\ \Delta(\tau) = g_2^3(\tau) - 27g_3^2(\tau) \neq 0;$$

(ii) the ring of modular forms $R[\Gamma] = \mathbb{C}[g_2, g_3]$ is generated by g_2, g_3;
(iii) the elliptic modular function $j := 12^3g_2^3/\Delta : \mathbb{H} \longrightarrow \mathbb{P}^1$ induces an isomorphism $X(1) \cong \mathbb{P}^1$. It holds that

$$C_\tau \cong C_\sigma \iff \sigma \in \Gamma\tau \iff j(\tau) = j(\sigma);$$

(iv) g_2, g_3 are Taniyama forms; let τ be a CM-point, $M(C_\tau)$ the moduli field. $M(C_\tau) = \mathbb{Q}(j(\tau))$ is an abelian extension (*class field*) of the imaginary quadratic number field $\mathbb{Q}(\tau)$;
(v) the field $Mer(X(1))$ of meromorphic functions on $X(1)$ is $\mathbb{C}(j)$; each modular function is a quotient of polynomials in g_2^3, g_3^2 homogeneous of same degree.

In the next section we mainly summarize known results about the family F most of them obtained by K. Matsumoto, cf. [6]. The culminating point is the arithmetic description of the monodromy group that forms the bridge to arithmetic properties of (Jacobians of) our curves and enables us to go further in the construction of Schottky-Taniyama forms. That will be done in the last section.

2. Monodromy and arithmetic groups

The aim of this section is the arithmetic description of the monodromy group of the family F (Proposition 2.3). Before we are able to do that, we have to collect some results most of them obtained by K. Matsumoto. Hence we reduce our explanations and refer to [6]. The arithmetic description is presented there too, but we will give a different proof using R.-P. Holzapfel's classification of ball quotients, see [5].

Let us consider the family F consisting of smooth hyperelliptic curves $\widetilde{C}(x,y)$ of genus three with model

$$C(x,y) : w^4 = z^2(z-1)^2(z-x)(z-y) \subset \mathbb{P}^1 \times \mathbb{P}^1$$

and parameters $(x,y) \in \Lambda := \mathbb{C}^2 \backslash \{(x,y) \mid xy(x-1)(y-1)(x-y) = 0\} \subset \mathbb{P}^1 \times \mathbb{P}^1$.

First we notice that the projective balls

$$\mathbb{P}^2 \supset \mathbb{B} : |z_1|^2 + |z_2|^2 - |z_0|^2 < 0 \text{ resp. } \mathbb{B}_H : \ {}^t z H \bar{z} < 0, \ H = \left(\begin{smallmatrix} 0 & -i & 0 \\ i & 0 & 0 \\ 0 & 0 & 1 \end{smallmatrix}\right),$$

serve as spaces of periods of the family F.

To see this, we consider the automorphism $\rho : (z,w) \mapsto (z, iw) \in \mathrm{Aut}(\widetilde{C}(x,y))$ and define generators of the space of holomorphic differential forms $\eta_1 = dz/w$, $\eta_2 = z(z-1)dz/w^3$, $\eta_3 = z\eta_2$. For fixed $(x_0, y_0) \in \Lambda$, we can choose cycles A_1, A_3, B_1 on $\widetilde{C}(x_0,y_0)$ such that $\mathfrak{b}_0 = \{A_1, \rho A_1, A_3, B_1, \rho B_1, \rho A_3\}$ forms a basis of the homology group $H_1(\widetilde{C}(x_0,y_0), \mathbb{Z})$ with intersection matrix in canonical form, see [6]. For a general curve $\widetilde{C}(x,y) \in F$, we take a path s in Λ starting in (x_0,y_0) and ending in (x,y). Define a basis $\mathfrak{b}(x,y) = \{A_1(x,y), \dots, \rho A_3(x,y)\}$ for $H_1(\widetilde{C}(x,y), \mathbb{Z})$ by continuation of $\mathfrak{b}_0$ along s. A straightforward calculation — using the relations between the cycles and Riemann period relations — shows that the resulting period matrix $\Omega(x,y) \in \mathbb{H}_3 = \{\Omega \in M_3(\mathbb{C}) \mid {}^t\Omega = \Omega, \mathrm{Im}(\Omega) > 0\}$ depends only on the chosen path s and on integrals $\int_{Z(x,y)} \eta_1$ with $Z(x,y) \in \{A_1(x,y), A_3(x,y), B_1(x,y)\}$.

On this way we get the multivalued map, cf. [6], Proposition 1.2,

$$\Psi : \Lambda \longrightarrow \mathbb{H}_3, \ (x,y) \mapsto \Omega(x,y) = \begin{pmatrix} u + \frac{i}{2}v^2 & -\frac{1}{2}v^2 & -iv \\ -\frac{1}{2}v^2 & u - \frac{i}{2}v^2 & v \\ -iv & v & i \end{pmatrix} \tag{2.1}$$

with $u := u(x,y) := \int_{B_1(x,y)} \eta_1 / \int_{A_1(x,y)} \eta_1$, $v := \int_{A_3(x,y)} \eta_1 / \int_{A_1(x,y)} \eta_1$.
We define an embedding $\mu : \mathbb{B}_H \hookrightarrow \mathbb{H}_3$ by

$$\mu(\zeta_0 : \zeta_1 : \zeta_2) := \begin{pmatrix} u + \frac{i}{2}v^2 & -\frac{1}{2}v^2 & -iv \\ -\frac{1}{2}v^2 & u - \frac{i}{2}v^2 & v \\ -iv & v & i \end{pmatrix}, \ u = \frac{\zeta_1}{\zeta_0}, \ v = \frac{\zeta_2}{\zeta_0}. \tag{2.2}$$

We have $\Psi(\Lambda) \subset \mu(\mathbb{B}_H)$, thus

$$\widetilde{\Psi} := \mu^{-1} \circ \Psi : \Lambda \longrightarrow \mathbb{B}_H, \ (x,y) \mapsto (\mu^{-1} \circ \Psi)(x,y)$$

defines a multivalued (period) map.

In a certain sense there is a $Z_2 \times S_3$-action on the parameter space preserving isomorphism classes of curves, where Z_2 is cyclic of order two and S_3 denotes the symmetric group of three letters. To be more precise, let K be a subgroup of $\mathrm{Aut}(\mathbb{P}^1 \times \mathbb{P}^1)$ generated by $k_1(x,y) = (y,x)$, $k_2(x,y) = (1-x, 1-y)$ and $k_3(x,y) = (\frac{1}{x}, \frac{1}{y})$. A. Piñeiro proved the following

Proposition 2.1. *(see [4], Appendix 1) The (compactified) space of moduli of the family F is given by $\Lambda/K \subset \mathbb{P}^1 \times \mathbb{P}^1/K$, $K \cong Z_2 \times S_3$.*

The next goal is to compare monodromy and arithmetic groups. For this purpose we recall from Deligne and Mostow [1] or Yoshida [11] that curves of a special equation type and parameter space Λ are connected with systems of hypergeometric differential equations.

Let $\mu = (\mu_0, \mu_1, \mu_2, \mu_3, \mu_4) \in \mathbb{Q}^5$, $0 < \mu_j < 1$ with $\sum_{j=0}^4 \mu_j = 2$ and let d denote the least common denominator of the μ_j. Consider projective smooth curves $\widetilde{C}_\mu(x,y)$ with affine model

$$C_\mu(x,y) : w^d = z^{d\mu_0}(z-1)^{d\mu_1}(z-x)^{d\mu_2}(z-y)^{d\mu_3}$$

and parameter $(x,y) \in \Lambda$. On each of these curves $\eta = dz/w = z^{-\mu_0}(z-1)^{-\mu_1}(z-x)^{-\mu_2}(z-y)^{-\mu_3}$ defines a holomorphic differential form. As function in x, y the period $\int_{Z(x,y)} \eta$ of $C_\mu(x,y)$ satisfies the system $E(a,b,b',c)$ of linear partial differential equations

$$x(1-x)\frac{\partial^2 u}{\partial x^2} + y(1-x)\frac{\partial^2 u}{\partial x \partial y} + (c-(a+b+1)x)\frac{\partial u}{\partial x} - by\frac{\partial u}{\partial y} - abu = 0,$$

$$y(1-y)\frac{\partial^2 u}{\partial y^2} + x(1-y)\frac{\partial^2 u}{\partial y \partial x} + (c-(a+b'+1)y)\frac{\partial u}{\partial y} - b'x\frac{\partial u}{\partial x} - ab'u = 0,$$

where $\mu_0 = c - b - b'$, $\mu_1 = 1 + a - c$, $\mu_2 = b$, $\mu_3 = b'$. This is the system for the Appell hypergeometric function $F_1(a,b,b',c;x,y)$. Given a fundamental solution for $(x,y) \in \Lambda$ — the system is non-singular and the space of solutions has dimension three on Λ — by analytic continuation along paths $\delta \in \pi_1(\Lambda,(x,y))$ we reach a representation of the fundamental group in $\mathbb{G}\mathrm{l}(V) \cong \mathbb{G}\mathrm{l}_3(\mathbb{C})$. The image of $\pi_1(\Lambda,(x,y))$ under this representation is called *monodromy group* of the system, the *projective monodromy group* is the image of the monodromy group under the projection $\mathbb{G}\mathrm{l}_3(\mathbb{C}) \to \mathbb{P}\mathbb{G}\mathrm{l}_3(\mathbb{C})$.

In our case the associated system has values $\mu_0 = \mu_1 = \mu_4 = 1/2$ and $\mu_2 = \mu_3 = 1/4$ and therefore $E(a,b,b',c) = E(\frac{1}{2}, \frac{1}{4}, \frac{1}{4}, 1)$. The μ_j are positive rational numbers less than one and sum equal two satisfying the condition

INT) for $i \neq j$: $(1 - \mu_i - \mu_j)^{-1} \in \mathbb{Z} \cup \{\infty\}$.

Taking into account that the period points $(\int_{A_1(x,y)} \eta_1 : \int_{B_1(x,y)} \eta_1 : \int_{A_3(x,y)} \eta_1)$ yield projective solutions of the system, we know from [1], Theorem 11.4 and List 14.4, see also [11], PTMD-Theorem, that the projective monodromy group $\mathbb{P}Mon =: \Gamma_2'$ is an arithmetic lattice in $\mathrm{Aut}(\mathbb{B}_H)$, and that $\widetilde{\Psi}$ induces an injective map from Λ to $\Gamma_2'\backslash\mathbb{B}_H$. Up to compactification, $\widetilde{\Psi}^{-1}$ (continued over the ramification and into the cusps) coincides with the quotient map $\mathbb{B}_H \longrightarrow \Gamma_2'\backslash\mathbb{B}_H$ and leads to a biholomorphic map between $\Gamma_2'\backslash\mathbb{B}_H$ and $\mathbb{P}^1\times\mathbb{P}^1\backslash\{(0,0),(1,1),(\infty,\infty)\}$. Let us identify the latter objects. The compactified branch locus S of the quotient map consists of seven lines $x=0,1,\infty$, $y=0,1,\infty$, $x=y$ with ramification numbers two over the diagonal and four over the remaining lines outside the triple points. If we define a weight map $b:\mathbb{P}^1\times\mathbb{P}^1\longrightarrow\mathbb{N}\cup\{\infty\}$ by $b(P)=\infty$ for all $P\in\{(0,0),(1,1),(\infty,\infty)\}=:S_\infty$ and $b(P)$ equal to the ramification number of $\widetilde{\Psi}^{-1}$ at Q, where $Q\in\mathbb{B}_H$ is a point in the preimage of P, $P\notin S_\infty$, then $(\mathbb{P}^1\times\mathbb{P}^1,S,b)$ is an orbifold uniformized by the ball $\mathbb{B}_H$ in Yoshida's sense [11].

To describe the monodromy explicitly Matsumoto fixes again a pair of parameters (x_0,y_0) and chooses a basis of $\pi_1(\Lambda,(x_0,y_0))$. Deformation of the curve $\widetilde{C}(x_0,y_0)$ along the loops leads to symplectic transformations of the Siegel domain and finally to automorphisms of the ball. This can be done — with a slight modification — for the map $\bar{\Psi}:\Lambda/K\longrightarrow\mathbb{B}_H$ too. Let us denote the resulting transformation groups by $Mon(\widetilde{\Psi})$ and $Mon(\bar{\Psi})$. The calculation leads to (see [6], (2.5) and (2.7))[1]
$Mon(\widetilde{\Psi}) =<\gamma_1,\dots,\gamma_5> \subset Mon(\bar{\Psi}) =<\gamma_1,\dots,\gamma_8>$, where

$$\gamma_1=\begin{pmatrix}1&0&0\\1+i&1&1-i\\-1-i&0&i\end{pmatrix},\qquad \gamma_2=\begin{pmatrix}2+i&-1-i&-1-i\\1+i&-i&-1-i\\1-i&-1+i&i\end{pmatrix},$$
$$\gamma_3=\begin{pmatrix}1&0&0\\0&1&0\\0&0&-1\end{pmatrix},\qquad \gamma_4=\begin{pmatrix}i&1-i&1-i\\0&i&0\\0&-1-i&-1\end{pmatrix},$$
$$\gamma_5=\begin{pmatrix}2+i&-1-i&1-1\\1+i&-i&1-i\\-1-i&1+i&i\end{pmatrix},\qquad \gamma_6=\begin{pmatrix}1&0&0\\0&1&0\\0&0&i\end{pmatrix},$$
$$\gamma_7=\begin{pmatrix}1&-1&0\\0&1&0\\0&0&1\end{pmatrix},\qquad \gamma_8=\begin{pmatrix}-1&0&0\\-1&-1&0\\0&0&-1\end{pmatrix}.$$

Let Z be the center of the unitary group $\mathbb{U}(H,\mathbb{Z}[i])=\{g\in\mathbb{G}\mathrm{l}_3(\mathbb{Z}[i]);\ {}^t\bar{g}Hg=H\}$. We identify the factor group $\mathbb{U}(H,\mathbb{Z}[i])/Z$ with the *Picard modular group of Gauß numbers* $\Gamma:=\mathbb{SU}(H,\mathbb{Z}[i])\cong\mathbb{PU}(H,\mathbb{Z}[i])$ acting effectively on $\mathbb{B}_H$. The congruence subgroup $\Gamma(1+i)=\mathbb{P}\{g\in\mathbb{U}(H,\mathbb{Z}[i]);g\equiv I_3 \mod (1+i)\}$ will play an important role in the following considerations.

We want to connect the (extended) projective monodromy groups

$$\Gamma_2':=\mathbb{P}Mon(\widetilde{\Psi})\subset\Gamma_1':=\mathbb{P}<\gamma_1,\dots,\gamma_6>\subset\Gamma':=\mathbb{P}Mon(\bar{\Psi})$$

[1] For a better comparison with Matsumoto's article we should mention that γ_5 given below differs from Matsumoto's $g(\delta_5)$. Looking at the corresponding symplectic matrix $N(\delta_5)$, see [6], (2.4), it is easy to check that $g(\delta_5)$ is provided with a typing error.

with the arithmetic groups

$$\Gamma_1 := \Gamma(1+i) \subset \Gamma = \mathbb{PU}(H, \mathbb{Z}[i]).$$

Given a ball lattice $G \subset \mathrm{Aut}(\mathbb{B}_H)$ the corresponding quotient surface is denoted by $X_G = G \backslash \mathbb{B}_H$. For the Baily–Borel compactification of X_G, that is a projective surface obtained by adding finitely many points to X_G, we write $\hat{X}_G$. In the sequel we will make extensive use of the following facts proved by R.-P. Holzapfel and N. Vladov.

Theorem 2.2. *(see [5], Chapter 8 and 9) The Baily–Borel compactification $\hat{X}_{\Gamma_1}$, $\Gamma_1 = \Gamma(1+i)$, of $\Gamma_1 \backslash \mathbb{B}_H$ is equal to $\mathbb{P}^2$. The compactified branch locus of the quotient map $p_{\Gamma_1} : \mathbb{B}_H \longrightarrow X_{\Gamma_1}$ consists of a plane* Apollonius configuration, *i.e., a quadric $\hat{C}_0$ and three tangents $\hat{C}_j$, $j = 1, 2, 3$. Γ_1 has three inequivalent cusps $\kappa_1, \kappa_2, \kappa_3 \in \partial\mathbb{B}_H$ which correspond to the intersection points $\hat{C}_0 \cap \hat{C}_j$, $j = 1, 2, 3$.*

Outside the images of the cusps the weight attached to each curve $\hat{C}_j$ is four, so we have Apoll-3 in the notation of [5]. Knowing these results we are able to verify the following

Proposition 2.3. *(cf. [6], Prop. 2.2) Arithmetic description of the monodromy.*

$$\text{(i) } \Gamma_1' = \Gamma_1, \quad \text{(ii) } \Gamma' = \Gamma, \quad \text{(iii) } [\Gamma_1 : \Gamma_2'] = 2.$$

Proof. (i) Looking at the generators it is easy to check that $\Gamma_1' \subset \Gamma_1$, hence we have a projection $\hat{\pi}_1 : \hat{X}_{\Gamma_1'} \longrightarrow \hat{X}_{\Gamma_1} = \mathbb{P}^2$. We will show that the covering is unramified outside of finitely many points. Since $\mathbb{P}^2$ remains simply connected if we remove a finite number of points, this implies that $\hat{\pi}_1$ is of degree one.

The Γ_1-inequivalent cusps are represented by $\kappa_1 = (1:0:0)$, $\kappa_2 = (0:1:0)$, $\kappa_3 = (1:1:0)$. The calculation of the isotropy groups of κ_2 yields $\Gamma_{1,\kappa_2} = \Gamma'_{1,\kappa_2}$, see [6], proof of Proposition 2.2. Hence we conclude from the properly discontinuous action of the transformation groups that $\hat{\pi}_1$ is locally unramified around this cusp, and it is unramified around all cusps because isotropy groups of cusps are conjugated.

Suppose $\hat{\pi}_1$ ramifies over an irreducible curve $\hat{C} \subset \hat{X}_{\Gamma_1}$. Then $\hat{C}$ has to be a curve of the branch locus of the projection $p_{\Gamma_1} : \mathbb{B}_H \to X_{\Gamma_1}$ too. In other words $\hat{C}$ belongs to the Apollonius configuration. But on each such component there is a cusp and therefore a non-singular point over which $\hat{\pi}_1$ is unramified. So we deduce that $\hat{\pi}_1$ is unramified over the curve outside of singularities.

We have obtained that only a subset of the singular locus $Sg(\hat{X}_{\Gamma_1})$ may remain as branch locus. But $Sg(\hat{X}_{\Gamma_1})$ consists of a finite number of points. So the restriction to $\hat{X}_{\Gamma_1'} \backslash \hat{\pi}_1^{-1}(Sg(\hat{X}_{\Gamma_1})) \longrightarrow \hat{X}_{\Gamma_1} \backslash Sg(\hat{X}_{\Gamma_1}) = \mathbb{P}^2 \backslash \{\text{finitely many points}\}$ is an unramified cover of a simply connected domain, therefore of degree one and that has to be the degree of $\hat{\pi}_1$ too.

(ii) Γ'/Γ_1 is a subgroup of Γ/Γ_1. The latter quotient is isomorphic to the orthogonal group $\mathbb{O}(3, \mathbb{F}_2)$ and therefore isomorphic to the symmetric group S_3.

Γ'/Γ_1 is generated by two non-commuting elements $\gamma_7\Gamma_1 \neq \gamma_8\Gamma_1$ of order two. It follows $\Gamma'/\Gamma_1 = \Gamma/\Gamma_1$.

(iii) $\gamma_6 \notin \Gamma_2'$, because $\hat{X}_{\Gamma_2'} = \mathbb{P}^1 \times \mathbb{P}^1 \longrightarrow \mathbb{P}^2 = \hat{X}_{\Gamma_1}$ cannot have degree one. But $\gamma_6^2 = \gamma_3 \in \Gamma_2'$, hence Γ_2' has index two in Γ_1. □

3. Schottky–Taniyama forms

3.1. Schottky forms

The construction of Schottky forms goes essentially back to the given connection between ball quotients and an expression of $\widetilde{\Psi}^{-1}$ by $\mathrm{Mon}(\widetilde{\Psi})$-modular forms. Via the composition of $\widetilde{\Psi}^{-1}$ with the Segre embedding and a suitable projection we will describe the uniformizing morphism $p_{\Gamma_1} : \mathbb{B}_H \longrightarrow \mathbb{P}^2$ by Γ_1-modular forms. The Schottky property follows immediately from the construction of the modular forms.

Let us recall the definition of theta constants. For p, q in $(\frac{1}{2}\mathbb{Z})^g$ *theta constants* with characteristic $\begin{bmatrix} {}^t p \\ {}^t q \end{bmatrix}$ are defined by

$$\Theta \begin{bmatrix} {}^t p \\ {}^t q \end{bmatrix} (\Omega) \quad := \textstyle\sum_{n \in \mathbb{Z}^g} \exp(\pi i \, {}^t(n+p)\Omega(n+p) + 2\pi i \, {}^t(n+p)q), \; \Omega \in \mathbb{H}_g.$$

Looking back at the embedding $\mu : \mathbb{B}_H \hookrightarrow \mathbb{H}_3$ (see 2.2) we define theta constants on the ball by

$$\Theta \begin{bmatrix} {}^t p \\ {}^t q \end{bmatrix} (\tau) := \Theta \begin{bmatrix} {}^t p \\ {}^t q \end{bmatrix} (\mu(\tau)).$$

By $(1 : u : v :) \mapsto (u, v)$ we may identify $\mathbb{B}_H$ with the unrestricted complex domain $\{(u, v) \in \mathbb{C}^2;\ 2\mathrm{Im}(u) - |v|^2 > 0\}$. The action of Γ on $\mathbb{B}_H$ is given by

$$g = (g_{ij}) : (u, v) \mapsto \left(\frac{g_{21} + g_{22}u + g_{23}v}{g_{11} + g_{12}u + g_{13}v}, \frac{g_{31} + g_{32}u + g_{33}v}{g_{11} + g_{12}u + g_{13}v} \right).$$

Let $G \subset \Gamma$ be an arithmetic subgroup. A holomorphic function ϕ on $\mathbb{B}_H$ is called modular form with respect to G or G-modular form of weight k if

$$\phi(g(u, v)) = (g_{11} + g_{12}u + g_{13}v)^k \phi(u, v) \qquad \forall g = (g_{ij}) \in G.$$

Matsumoto has shown the following

Theorem 3.1. *(see* [6], *Theorems* 3.4 *and* 4.1*)* $\widetilde{\Psi}^{-1} : \mathbb{B}_H \longrightarrow \mathbb{P}^1 \times \mathbb{P}^1$ *has a representation*

$$\widetilde{\Psi}^{-1}(u, v) = \left(\frac{\phi_1}{\phi_0}(u, v), \frac{\phi_3}{\phi_2}(u, v) \right) = (x, y)$$

with

$$\begin{aligned}
\phi_0 &:= \left(\Theta\begin{bmatrix} 0 & \frac12 & \frac12 \\ 0 & \frac12 & \frac12 \end{bmatrix}\Theta\begin{bmatrix} \frac12 & 0 & 0 \\ 0 & \frac12 & 0 \end{bmatrix}\Theta\begin{bmatrix} 0 & \frac12 & \frac12 \\ \frac12 & 0 & 0 \end{bmatrix}\Theta\begin{bmatrix} \frac12 & 0 & 0 \\ 0 & \frac12 & \frac12 \end{bmatrix}\right)^2, \\
\phi_1 &:= \left(\Theta\begin{bmatrix} 0 & \frac12 & \frac12 \\ \frac12 & \frac12 & \frac12 \end{bmatrix}\Theta\begin{bmatrix} \frac12 & 0 & 0 \\ 0 & 0 & 0 \end{bmatrix}\Theta\begin{bmatrix} \frac12 & 0 & 0 \\ 0 & 0 & \frac12 \end{bmatrix}\Theta\begin{bmatrix} 0 & \frac12 & \frac12 \\ 0 & 0 & 0 \end{bmatrix}\right)^2, \\
\phi_2 &:= \left(\Theta\begin{bmatrix} 0 & \frac12 & \frac12 \\ 0 & \frac12 & \frac12 \end{bmatrix}\Theta\begin{bmatrix} \frac12 & 0 & 0 \\ 0 & \frac12 & 0 \end{bmatrix}\Theta\begin{bmatrix} 0 & \frac12 & 0 \\ \frac12 & 0 & \frac12 \end{bmatrix}\Theta\begin{bmatrix} \frac12 & 0 & \frac12 \\ 0 & \frac12 & 0 \end{bmatrix}\right)^2, \\
\phi_3 &:= \left(\Theta\begin{bmatrix} 0 & \frac12 & \frac12 \\ \frac12 & \frac12 & \frac12 \end{bmatrix}\Theta\begin{bmatrix} \frac12 & 0 & 0 \\ 0 & 0 & 0 \end{bmatrix}\Theta\begin{bmatrix} 0 & \frac12 & 0 \\ 0 & 0 & \frac12 \end{bmatrix}\Theta\begin{bmatrix} \frac12 & 0 & \frac12 \\ 0 & 0 & 0 \end{bmatrix}\right)^2,
\end{aligned}$$

where $\Theta\begin{bmatrix} {}^tp \\ {}^tq \end{bmatrix} := \Theta\begin{bmatrix} {}^tp \\ {}^tq \end{bmatrix}(\mu(u,v))$.

Moreover, the functions ϕ_j, $j = 0,\dots,3$, *are* $Mon(\widetilde{\Psi})$*-modular forms of weight* 8.

Let $v : \mathbb{P}^1 \times \mathbb{P}^1 \to \mathbb{P}^3$, $(\frac{x_1}{x_0}, \frac{y_1}{y_0}) \mapsto (x_0y_0 : x_0y_1 : x_1y_0 : x_1y_1)$ be the Segre embedding. We denote the image (Segre quadric) by $\mathbb{P}^3 \supset V : Z_0Z_3 = Z_1Z_2$. Thus $v \circ \widetilde{\Psi}^{-1} : \mathbb{B}_H \longrightarrow V$ sends τ to $(\chi_0(\tau) : \chi_1(\tau) : \chi_2(\tau) : \chi_3(\tau))$ with

$$\chi_0 = \phi_0\phi_2,\ \chi_1 = \phi_0\phi_3,\ \chi_2 = \phi_1\phi_2,\ \chi_3 = \phi_1\phi_3.$$

We remember Proposition 2.3 and the connections between ball quotients

$$\begin{array}{ccccc}
\hat{X}_{\Gamma_2'} & = & \mathbb{P}^1 \times \mathbb{P}^1 & = & V \\
\downarrow & & \downarrow & & \downarrow \\
\hat{X}_{\Gamma_1} & = & (\mathbb{P}^1 \times \mathbb{P}^1)/{<}k_1{>} & = & \mathbb{P}^2 \\
\downarrow & & \downarrow & & \downarrow \\
\hat{X}_{\Gamma} & = & (\mathbb{P}^1 \times \mathbb{P}^1)/K & = & \mathbb{P}^2/S_3.
\end{array}$$

Next we want to describe the vertical maps. The map from the Segre quadric V into $\mathbb{P}^2$ corresponding to the projection $\mathbb{P}^1 \times \mathbb{P}^1 \longrightarrow \mathbb{P}^1 \times \mathbb{P}^1/ < k_1 >$ is given by $pr : (z_0 : z_1 : z_2 : z_3) \mapsto (z_0 : z_1 + z_2 : z_3)$. To realize this one has to find the group $2\Sigma_3 \subset \mathrm{Aut}(V)$ such that the operation on V corresponds to the K-action on $\mathbb{P}^1 \times \mathbb{P}^1$. This group has generators $S, R, T \in \mathbb{P}\mathbb{G}l_4(\mathbb{C})$ represented by

$$S = \begin{pmatrix} 1 & 0 & 0 & 0 \\ 0 & 0 & 1 & 0 \\ 0 & 1 & 0 & 0 \\ 0 & 0 & 0 & 1 \end{pmatrix}, \quad R = \begin{pmatrix} 1 & 0 & 0 & 0 \\ 1 & -1 & 0 & 0 \\ 1 & 0 & -1 & 0 \\ 1 & -1 & -1 & 1 \end{pmatrix}, \quad T = \begin{pmatrix} 0 & 0 & 0 & 1 \\ 0 & 0 & 1 & 0 \\ 0 & 1 & 0 & 0 \\ 1 & 0 & 0 & 0 \end{pmatrix},$$

since $v \circ k_1 = S \circ v,\ v \circ k_2 = R \circ v,\ v \circ k_3 = T \circ v$. So the projection by the cyclic group $< k_1 >$ leads to $V/{<}S{>}$ which is isomorphic to $\mathbb{P}^2$, and an isomorphism is induced by the map pr. That can be seen if one considers the extended action of S on $\mathbb{C}[\chi_0, \chi_1, \chi_2, \chi_3] =: \mathbb{C}[\chi]$ changing χ_1 and χ_2. We have $\ker(S - id) = (\chi_1 - \chi_2)$ and $\ker(S + id) = (\chi_0, \chi_1 + \chi_2, \chi_3)$. Setting $t_0 := \chi_0$, $t_1 := \chi_1 + \chi_2$, $t_2 := \chi_3$ we get $\mathbb{C}[\chi]^{<S>} = \mathbb{C}[t_0, t_1, t_2]$ and $\mathbb{C}[\chi] = \mathbb{C}[t_0, t_1, t_2] + \mathbb{C}[t_0, t_1, t_2](\chi_1 - \chi_2)$. Thus

$$pr \circ v \circ \widetilde{\Psi}^{-1} = (t_0 : t_1 : t_2) : \mathbb{B}_H \longrightarrow \mathbb{P}^2 = \mathrm{Proj}(\mathbb{C}[t_0, t_1, t_2]). \tag{3.1}$$

We know already — or check it in a minute — that the branch locus, i.e., the $pr \circ v$-image of the seven lines $x = 0, 1, \infty$, $y = 0, 1, \infty$, $x = y$, is a plane Apollonius configuration. But the configuration is not in S_3-*normalized* form yet, that means that the branch tangent lines at the branch quadric do not coincide with the coordinate axes, or in other words, the action of the modular group doesn't lead to the canonical S_3-permutation of projective coordinates. To achieve this we have to transform the projective coordinates in a suitable manner. If one takes a look at the equations defining the (non-normalized) configuration or alternatively on the Σ_3-action on $\mathbb{P}^2$ coming from the $< R, T >$-action on V one easily concludes that this is done by sending $(t_0 : t_1 : t_2)$ to $(t_0 : t_0 - t_1 + t_2 : t_2)$. Altogether we have

Proposition 3.2. *Set* $\mathrm{pr}_{\Gamma_1} = (g_0 : g_1 : g_2) : \mathbb{B}_H \longrightarrow \mathbb{P}^2$, $g_0 = t_0$, $g_1 = t_0 - t_1 + t_2$, $g_2 = t_2$ *and let* $\tau \in \mathbb{B}_H$ *be outside of the ramification locus of* pr_{Γ_1}. *It holds that*

(i) g_0, g_1, g_2 *are* $\Gamma(1+i) = \Gamma_1$*-modular forms of weight* 16;

(ii) *the curve* $\widetilde{C}_\tau$ *defined by the equation*

$$C_\tau : w^4 = z^2(z-1)^2 \left(z^2 - \frac{g_0(\tau) - g_1(\tau) + g_2(\tau)}{g_0(\tau)} z + \frac{g_2(\tau)}{g_0(\tau)} \right)$$

has period point τ, *i.e.,* g_0, g_1, g_2 *are* Γ_1*-Schottky forms;*

(iii) $\widetilde{C}_\tau \cong \widetilde{C}_{\tau'} \iff (g_0(\tau') : g_1(\tau') : g_2(\tau')) \sim_{S_3} (g_0(\tau) : g_1(\tau) : g_2(\tau))$, *where* S_3 *acts by permutation of the coordinates.*

Proof. To prove the Γ_1-modularity one has to write down the t_j in terms of theta constants and to apply Lemma 4.2 in [6]. Weight 16 of the t_j is trivial, because they are products of two forms of weight 8. Thus modularity and weight of the g_j are clear. The Schottky property and the last statement follow immediately from Theorem 3.1 and the construction of the g_j. □

For a group $G \subset \Gamma$ of finite index let $R[G]^k$ be the space of G-modular forms of weight k, $R[G]_0^k$ the space of cusp forms, i.e., modular forms which vanish in the cusps, of weight k. $R[G] = \oplus_{k \geq 0} R[G]^k$ denotes the graduate ring of G-modular forms. Consider the graduate subrings $\mathbb{C}[\chi] \subset R[\Gamma_2']$ and $\mathbb{C}[\mathfrak{t}] \subset R[\Gamma_1]$ generated by $\chi_0, \chi_1, \chi_2, \chi_3$ respectively t_0, t_1, t_2. Let $\mathbb{C}[\chi]^d := \mathbb{C}[\chi] \cap R[\Gamma_2']^{16d}$, $\mathbb{C}[\mathfrak{t}]^d := \mathbb{C}[\mathfrak{t}] \cap R[\Gamma_1]^{16d}$ be the space of modular forms of weight $16d$ in $\mathbb{C}[\chi]$ and $\mathbb{C}[\mathfrak{t}]$, $\mathbb{C}[\chi]_0^d$, $\mathbb{C}[\mathfrak{t}]_0^d$ the subspaces of cusp forms. $\mathbb{C}[\mathfrak{t}]^d$ consists of the homogeneous elements of degree d in $\mathbb{C}[\mathfrak{t}]$ (cf. 3.1) and therefore it has dimension $\binom{d+2}{2}$. Taking into account that the condition to vanish in the cusps yields three additional linearly independent conditions on a modular form, that we have a decomposition in S-eigenspaces $\mathbb{C}[\chi] = \mathbb{C}[\mathfrak{t}] + \mathbb{C}[\mathfrak{t}](\chi_1 - \chi_2)$ (thus $\dim \mathbb{C}[\chi]^d = \dim \mathbb{C}[\mathfrak{t}]^d + \dim \mathbb{C}[\mathfrak{t}]^{d-1}$) and that $\chi_1 - \chi_2$ is a cusp form, we conclude

$$\begin{aligned} \dim \mathbb{C}[\mathfrak{t}]^d &= d^2/2 + 3d/2 + 1, & \dim \mathbb{C}[\mathfrak{t}]_0^d &= d^2/2 + 3d/2 - 2, \\ \dim \mathbb{C}[\chi]^d &= d^2 + 2d + 1, & \dim \mathbb{C}[\chi]_0^d &= d^2 + 2d - 2. \end{aligned}$$

For the ring of $2\Sigma_3$-invariants we obtain

$$\mathbb{C}[\chi]^{2\Sigma_3} = \mathbb{C}[\mathfrak{t}]^{\Sigma_3} = \mathbb{C}[g_0, g_1, g_2]^{S_3} = \mathbb{C}[s_1, s_2, s_3],$$

where $s_1 = t_0 + t_2 + (t_0 - t_1 + t_2)$, $s_2 = (t_0 + t_2)(t_0 - t_1 + t_2)$, $s_2 = t_0 t_2 (t_0 - t_1 + t_2)$.

3.2. Taniyama forms

Via the quotient map $p_{\Gamma_1} = (g_0 : g_1 : g_2) : \mathbb{B}_H \longrightarrow \mathbb{P}^2$ given by Γ_1-Schottky forms we want to define a Picard modular function $j = (j_1 : j_2 : j_3) : \mathbb{B}_H \longrightarrow \mathbb{P}^2$ such that the values classify the curves up to isomorphism and such that the field of moduli (for τ outside a certain subset) of the corresponding Jacobian is $\mathbb{Q}\left(\frac{j_1}{j_3}(\tau), \frac{j_2}{j_3}(\tau)\right)$. Since the (principally polarized) Jacobians are abelian 3-folds with multiplication by $\mathbb{Q}(i)$ of type $(2, 1)$, and these varieties are parametrized by the Shimura surface $\hat{X}_\Gamma$, the values of j in periods of CM-curves can be used to generate class fields. We will follow the construction given in [4].

In order to clarify notations we recall some definitions and facts about complex multiplication of abelian varieties, cf. [10].

Let A be a principally polarized abelian variety, k a number field. We say that A has *multiplication by* k if there is a $\mathbb{Q}$-algebra embedding of k into the algebra of endomorphisms $\mathrm{End}_{\mathbb{Q}}(A) := \mathrm{End}(A) \otimes \mathbb{Q}$. If k is a *CM field*, i.e., a totally imaginary quadratic extension of a totally real number field, of absolute degree $[k : \mathbb{Q}] = 2 \cdot \dim A$, we call A a variety with *complex multiplication* or *CM-variety* for short. Curves C_τ and period points τ are called *CM-curves* respectively *CM-points* if the corresponding varieties $A_\tau = \mathrm{Jac}(C_\tau)$ are CM-varieties. An abelian variety with *decomposed complex multiplication* or *DCM* is defined as variety which splits up to isogeny into simple abelian varieties with complex multiplication.

Each endomorphism f of an abelian g-fold A lifts to a linear map l_f on the tangent space $T_0(A) \cong \mathbb{C}^g$ of A at O. Let $R_{\mathbb{C}} : \mathrm{End}_{\mathbb{Q}}(A) \longrightarrow M_g(\mathbb{C})$, $f \mapsto l_f$, be the complex representation, k a CM field of absolute degree $2g$ and A a CM-variety with multiplication by $k \cong \mathrm{End}_{\mathbb{Q}}(A)$. The diagonalized complex representation $\mathrm{diag}(\varphi_1, \ldots, \varphi_g)$ yields a *CM-type* $\Phi = \{\varphi_1, \ldots, \varphi_g\}$ for k, that means that the complete set of complex embeddings of k is given by the φ_j and their complex conjugated $\bar{\varphi}_j$, $j = 1, \ldots, g$. The *reflex field* k^* of k is defined as the field generated over $\mathbb{Q}$ by the Φ-traces $Tr_\Phi(f) = \varphi_1(f) + \ldots + \varphi_g(f)$, $f \in k$.

It is known that each abelian CM-variety has a model defined over a number field L. Let $G_{\mathbb{Q}} = Gal(\bar{\mathbb{Q}}/\mathbb{Q})$ denote the group of automorphisms of the algebraic closure $\bar{\mathbb{Q}}$ of $\mathbb{Q}$ fixing $\mathbb{Q}$. We consider a model A/L defined by finitely many polynomials $F_j(X) \in L[X]$, $X = (X_1, \ldots, X_n)$. For $\sigma \in G_{\mathbb{Q}}$ the σ-transformed variety A^σ is defined by equations $F_j^\sigma(X) = 0$, where σ acts on the coefficients of the polynomials. We set $Stab(A) := \{\sigma \in G_{\mathbb{Q}};\ A^\sigma \cong A\}$. The *moduli field* $M(A) := \bar{\mathbb{Q}}^{Stab(A)}$ of A is the subfield of $\bar{\mathbb{Q}}$ fixed by $Stab(A)$. For a CM-variety A with k-multiplication the k^*-extended moduli field $M(A)k^*$ is an abelian extension (Shimura class field) of k^*, see [10].

Let us return to our genus three curves. We saw already that the moduli space of the family F — and hence the moduli space of the principally polarized Jacobians by Torelli's Theorem — is equal to $\hat{X}_\Gamma = \mathbb{P}^2/S_3$. On the other hand this object is the moduli space of principally polarized abelian 3-folds with $\mathbb{Q}(i)$-multiplication of type (2,1), see [9]. Obviously, the Jacobians $\mathrm{Jac}(\widetilde{C})$ of curves

$\widetilde{C} \in F$ are abelian 3-folds. The automorphism of order four $\rho(z,w) = (z, iw) \in \mathrm{Aut}(\widetilde{C})$ induces an embedding $\mathbb{Q}(i) \hookrightarrow \mathrm{End}_{\mathbb{Q}}(\mathrm{Jac}(\widetilde{C}))$ coming from the (dual) basis of differential-1-forms $\{\eta_1 = dz/w, \eta_2 = z(z-1)dz/w^3, \eta_3 = z\eta_2\}$. Since the restriction of the complex representation of $\mathrm{End}_{\mathbb{Q}}(\mathrm{Jac}(\widetilde{C}))$ to $\mathbb{Q}(i)$ is $\mathrm{diag}(id, id, \bar{id})$ we have type $(2,1)$.

Applying a Theorem of H. Shiga and J. Wolfart, A_τ has decomposed complex multiplication if and only if τ and $p_{\Gamma_1}(\tau)$ have algebraic coordinates, see [8]. In our situation a DCM-variety A_τ is a CM-variety if and only if $K(\tau)$ has degree 3 over $K = \mathbb{Q}(i)$. Moreover it holds that p_{Γ_1}-images of CM-points do not belong to the Apollonius configuration. This follows from the fact that the ramification locus of p_{Γ_1} consists of four Γ_1-reflection discs which are $\mathbb{Q}(i)$-discs, i.e., intersections of $\mathbb{B}_H$ with a projective line on $\mathbb{P}^2$ through two different points $P_1, P_2 \in \mathbb{P}^2(\mathbb{Q}(i))$. But ball points belonging to $\mathbb{Q}(i)$-discs cannot correspond to simple abelian varieties, see [3] or [4].

Now we are prepared for the construction of the modular function. Remember that the branch locus of $p_{\Gamma_1} = (g_0 : g_1 : g_2) : \mathbb{B}_H \longrightarrow \mathbb{P}^2$, $\tau \mapsto (z_0 : z_1 : z_2)$ consists of a S_3-normalized Apollonius configuration. Let s_i be the elementary symmetric functions in z_0, z_1, z_2 of degree $i = 1,2,3$. We define

$$H_1 := s_1 s_2 - 3s_3,\ H_2 := s_1^3 - s_1 s_2 + 3s_3,\ H_3 := s_1 s_2 - s_3$$

and rational functions $H_i/s_3,\ i = 1,2,3$. The map

$$J := \left(\frac{H_1}{s_3}, \frac{H_2}{s_3}\right) : \mathbb{P}^2 \longrightarrow \mathbb{P}^2$$

is (well-)defined outside of three points $Q_1 = (0:1:-1)$, $Q_2 = (1:0:-1)$, $Q_3 = (1:-1:0)$ belonging to the Apollonius configuration, and J factors over $\mathbb{P}^2/S_3$. Hence

$$j := (j_1 : j_2 : j_3) := J \circ p_{\Gamma_1} : \mathbb{B}_H \longrightarrow \mathbb{P}^2,$$

$j_i := H_i \circ p_{\Gamma_1}$, $i = 1,2$, $j_3 := s_3 \circ p_{\Gamma_1}$, is well-defined outside the p_{Γ_1}-preimage of the three points. Especially, j is defined in all CM-points, since they do not belong to the preimage of the Apollonious configuration. We repeat the proof of

Proposition 3.3. *(see [4], Proposition 9.4) Let $p_{\Gamma_1}(\tau) = (z_0 : z_1 : z_2) =: \zeta$ be outside the Apollonius configuration, $\zeta \notin L : Z_0 + Z_1 + Z_2 = 0$ and $z = (z_0, z_1, z_2)$. It holds that $M(\widetilde{C}_\tau) := M(\mathrm{Jac}(\widetilde{C})) = \mathbb{Q}\left(\frac{j_1}{j_3}(\tau), \frac{j_2}{j_3}(\tau)\right) =: \mathbb{Q}(j(\tau))$.*

Proof. Consider $\sigma \in Gal(\bar{\mathbb{Q}}/\mathbb{Q})$. By Torelli's Theorem, it holds that $\widetilde{C}_\tau \cong \widetilde{C}_\tau^\sigma$ if and only if $\mathrm{Jac}(\widetilde{C}_\tau) \cong \mathrm{Jac}(\widetilde{C}_\tau^\sigma)$.

$$\begin{aligned} \widetilde{C}_\tau \cong \widetilde{C}_\tau^\sigma \quad &\Longleftrightarrow \sigma \text{ permutes } z_0,\ z_1,\ z_2 \\ &\Longleftrightarrow \sigma \text{ fixes } W^3 - \tfrac{H_1}{s_3}(z)W^2 + \tfrac{H_2}{s_3}(z)W - \tfrac{H_3}{s_3}(z) \in \bar{\mathbb{Q}}[W] \\ &\Longleftrightarrow \sigma|_{\mathbb{Q}(J(\zeta))} = id. \end{aligned}$$

□

Summarizing the above we have proved

Theorem 3.4. *Let* g_0, g_1, g_2 *be as in Proposition* 3.2, $j := J \circ p_{\Gamma_1}$ *defined as above,* $L : Z_0 + Z_1 + Z_2$, $k_\tau = \mathbb{Q}(i, \tau)$ *with reflex field* k_τ^*. *Then* g_0, g_1, g_2 *are Schottky-Taniyama forms for* $\Gamma_1 = \Gamma(1+i)$. *For CM-points* $\tau \notin p_{\Gamma_1}^{-1}(L)$ *the extended moduli field* $M(A_\tau)k_\tau^* = \mathbb{Q}(j(\tau))k_\tau^*$ *is a class field of the reflex field* k_τ^*. *Outside the preimage of the Apollonius configuration it holds that* $\widetilde{C}_\tau \cong \widetilde{C}_{\tau'}$ *if and only if* $j(\tau) = j(\tau')$. *The field of* Γ*-modular functions is generated by* j, $\mathbb{C}(j) = Mer(\hat{X}_\Gamma)$.

References

[1] P. Deligne, G. D. Mostow, Monodromy Of Hypergeometric Functions And Non-Lattice Integral Monodromy, *Publ. Math. IHES* **63** (1986), 5–88.

[2] R.-P. Holzapfel, *Geometry and Arithmetic around Euler partial differential equations*, Dt. Verlag d. Wiss./Reidel Publ. Comp., Berlin-Dordrecht, 1986.

[3] R.-P. Holzapfel, Hierarchies of endomorphism algebras of abelian varieties corresponding to Picard modular surfaces, *Schriftenreihe Komplexe Mannigfaltigkeiten* **190**, Univ. Erlangen, 1994.

[4] R.-P. Holzapfel, A. Piñeiro, N. Vladov, Picard–Einstein Metrics and Class Fields Connected with Apollonius Cycle, HU preprint Nr. 98-15, 1998.

[5] R.-P. Holzapfel, N. Vladov, Quadric-line configurations degenerating plane Picard–Einstein metrics I–II, *Sitzungsber. d. Berliner Math. Ges.*, Jahrgänge 1997–2000, Berlin 2001, 79–142.

[6] K. Matsumoto, On Modular Functions in 2 Variables Attached to a Family of Hyperelliptic Curves of Genus 3, *Sc. Norm. Sup. Pisa* **16** no. 4 (1989), 557–578.

[7] H. Shiga, On the representation of the Picard modular function by θ constants I–II, *Publ. R.I.M.S. Kyoto Univ.* **24** (1988), 311–360.

[8] H. Shiga, J. Wolfart, Criteria for complex multiplication and transcendence properties of automorphic functions, *J. reine angew. Math.* **463** (1995), 1–25.

[9] G. Shimura, On analytic families of polarized abelian varieties and automorphic functions, *Ann. Math.* **78** no. 1 (1963), 149–192.

[10] G. Shimura, Y. Taniyama, Complex Multiplication of Abelian Varieties and its Applications to Number Theory, *Publ. Math. Soc. Japan* **6**, Tokyo, 1961.

[11] M. Yoshida, *Fuchsian Differential Equations*, Vieweg, Braunschweig-Wiesbaden, 1987

Thorsten Riedel
Humboldt-University of Berlin
Department of Mathematics
Rudower Chaussee 25
D-10099 Berlin
Germany
e-mail: thorsten_riedel@gmx.de

Progress in Mathematics, Vol. 260, 287–312

Algebraic Values of Schwarz Triangle Functions

Hironori Shiga and Jürgen Wolfart

Abstract. We consider Schwarz maps for triangles whose angles are rather general rational multiples of π. Under which conditions can they have algebraic values at algebraic arguments? The answer is based mainly on considerations of complex multiplication of certain Prym varieties in Jacobians of hypergeometric curves. The paper can serve as an introduction to transcendence techniques for hypergeometric functions, but contains also new results and examples.

Mathematics Subject Classification (2000). Primary 14K22; Secondary 30C20, 33C05, 11J91, 11J95.

Keywords. Schwarz triangle functions, hypergeometric functions, algebraic values, transcendence, complex multiplication.

Hypergeometric functions have many interesting relations to arithmetics, for example to modular forms, diophantine approximation, continued fractions and so on. In the following contribution we will concentrate on transcendence questions and explain the relevant techniques in the framework of a question concerning the classical Schwarz triangle functions.

These triangle functions $D(\nu_0, \nu_1, \nu_\infty; z)$ are defined as quotients of two linearly independent solutions of Gauss' hypergeometric differential equations. If their *angular parameters* ν_0, ν_1, ν_∞ are real and have absolute value in the open interval $]0,1[$, they define biholomorphic mappings of the complex upper half plane $\mathcal{H}$ onto triangles in the Riemann sphere bounded by circular arcs. The singular points $0, 1, \infty$ of the differential equation are sent by D to the vertices of the triangle including there angles $\pi|\nu_0|, \pi|\nu_1|, \pi|\nu_\infty|$, respectively. Particularly interesting special cases are those where ν_0, ν_1, ν_∞ are the inverses of positive integers p, q, r because then D is the inverse function of an automorphic function for the triangle group with signature $\langle p, q, r\rangle$, isomorphic to the (projective) monodromy group of the hypergeometric differential equation.

The present paper considers the question if Schwarz triangle functions can have algebraic values at algebraic arguments. The problem has its origins in the natural general question if or under which conditions (suitably normalized) transcendental functions have transcendental values at algebraic arguments, and in this special context it is related to automorphic functions and periods of abelian varieties. For a general survey about algebraic and transcendental periods in number theory see Waldschmidt's recent article [20]. In the cases related to automorphic functions mentioned above the problem is treated already in our previous paper [16, Cor. 5]. It turned out that a positive answer is directly related to the condition if certain Prym varieties are of complex multiplication (CM) type, the Pryms being defined in a natural way via the integral representation of the associated hypergeometric functions. Now we generalize the setting and consider arbitrary rational angular parameters ν_0, ν_1, ν_∞, restricted only by some mild technical condition excluding logarithmic singularities and some other very special situations. The main results will show that we have still 'CM' as necessary condition for 'algebraic values at algebraic arguments', but that even under the CM condition this algebraicity is rather exceptional. However, we will give examples that such exceptions occur.

This more general type of triangle functions has still images of $\mathcal{H}$ bounded by parts of circles but they are in general not globally biholomorphic — the image domains may overlap with themselves. We treated in [15] an analogous problem admitting apparent singularities in the associate Fuchsian differential equations. In many cases, the triangle functions of the present paper may in fact be considered as limit cases of those of [15], and many techniques developed there are useful also for the problem treated in the present paper. Therefore we collect in Section 1 some known material mainly from [15], [16], [21]. Section 2 presents the necessary tools from transcendence theory, and in Section 3 we state and prove the main results. The methods rely in part on the classical theory of hypergeometric functions, in part on the consideration of families of abelian varieties, and in part on Wüstholz' transcendence techniques [24]. Sections 4 and 5 present instructive examples.

Notation: we will call *Propositions* the statements we took from the literature and *Theorems* the new results presented here even if they might be less important than the *Propositions.*

1. Families of Prym varieties and associate functions

1.1. Integral representation by the periods on curves

Throughout this paper we will suppose that the angular parameters satisfy

$$\nu_0,\ \nu_1,\ \nu_\infty \in \mathbb{Q}-\mathbb{Z}, \quad \nu_0 \pm \nu_1 \pm \nu_\infty \notin \mathbb{Z}\,. \tag{1.1}$$

We will use the integral representation of the Gauss hypergeometric function $F(a,b,c;z)$ — omitting the usual normalizing Beta factor and some algebraic nonzero factors, see Section 5 of [15] for a careful discussion — in the form

$$\int_\gamma u^{a-c}(u-1)^{c-b-1}(u-z)^{-a}du = \int_\gamma u^{-\mu_0}(u-1)^{-\mu_1}(u-z)^{-\mu_z}du = \int_\gamma \eta(z)$$

with the (rational) exponents

$$\begin{aligned}
\mu_0 &= \frac{1}{2}(1-\nu_0+\nu_1-\nu_\infty)\\
\mu_1 &= \frac{1}{2}(1+\nu_0-\nu_1-\nu_\infty)\\
\mu_z &= \frac{1}{2}(1-\nu_0-\nu_1+\nu_\infty)\\
\mu_\infty &= \frac{1}{2}(1+\nu_0+\nu_1+\nu_\infty)\\
\mu_0+\mu_1+\mu_z+\mu_\infty &= 2
\end{aligned}$$

for some Pochhammer cycle γ around two of the singularities $0, 1, z, \infty$. As already remarked by Klein [10, §19], analytic continuation of $F(a, b, c; z)$ means only to replace γ by another cycle of integration, and a basis of solutions of the corresponding hypergeometric differential equation will be obtained by taking two Pochhammer cycles around different pairs of singularities: remark that our hypothesis on the sums of the angular parameters guarantees that no exponent μ_j is an integer, whence all singularities are nontrivial. For fixed arguments $z \neq 0, 1, \infty$ this integral representation can be seen as a period integral on a nonsingular projective model $X(k, z)$ of the algebraic curve

$$y^k = u^{k\mu_0}(u-1)^{k\mu_1}(u-z)^{k\mu_z} \tag{1.2}$$

where k is the least common denominator of the μ_j, γ some homology cycle on $X(k, z)$, and η a differential given on the singular model as

$$\eta = \eta(z) = \frac{du}{y}.$$

It is a second kind differential what can be seen using appropriate local variables ([21]; N. Archinard [1] explains in more detail the desingularization procedure). Our Schwarz triangle map is a multivalued analytic function on $\mathbb{C} - \{0, 1\}$ defined by

$$D(\nu_0, \nu_1, \nu_\infty; z) = D(\eta; z) = D(z) = \frac{\int_{\gamma_1} \eta(z)}{\int_{\gamma_2} \eta(z)} \tag{1.3}$$

for some *independent* cycles γ_1, γ_2 on $X(k, z)$.

In the next subsection we will give a precise definition of *independence* for these cycles, for the moment we can assume that they come from Pochhammer cycles around different pairs of singularities and are locally independent of $z \neq 0, 1, \infty$. The triangle functions extend continuously to the arguments excluded here, and our normalization guarantees that $D(0), D(1), D(\infty)$ become algebraic or ∞, see [15, Section 3.1]. For later use recall the relation between angular and exponential parameters and a, b, c.

$$\begin{aligned}
\nu_0 &= \quad 1-c \quad = 1-\mu_0-\mu_z = \mu_1+\mu_\infty-1\\
\nu_1 &= c-a-b = 1-\mu_1-\mu_z = \mu_0+\mu_\infty-1\\
\nu_\infty &= \quad a-b \quad = \mu_z+\mu_\infty-1 = 1-\mu_0-\mu_1\,.
\end{aligned} \tag{1.4}$$

1.2. The family of Prym varieties

The family of Prym varieties in question can be described as follows. For all proper divisors d of k there is an obvious morphism of the curve $X(k,z)$ onto the curve $X(d,z)$ in whose definition (1.2) we keep fixed the exponential parameters $k\mu_i$ on the right hand side and replace k by d as exponent of y. These morphisms induce epimorphisms

$$\mathrm{Jac}\, X(k,z) \;\to\; \mathrm{Jac}\, X(d,z)\,.$$

Let $T(k,z)$ be the connected component of 0 in the intersection of all kernels of these epimorphisms. Then it is known by [21], [1] that $T(k,z)$ is an abelian variety of dimension $\varphi(k)$ where φ denotes Euler's function. $T(k,z)$ has a special endomorphism structure called *generalized complex multiplication* (complex multiplication in the narrow sense will be treated in Subsection 2.3) by the cyclotomic field

$$\mathbb{Q}(\zeta_k) \;\subseteq\; \mathrm{End}_0 T(k,z) \;:=\; \mathbb{Q}\otimes_{\mathbb{Z}} \mathrm{End}\, T(k,z)$$

induced by an automorphism of the curve $X(k,z)$ described on its singular model by

$$\sigma \;:\; (u,y) \;\mapsto\; (u,\zeta_k^{-1}y)\,, \qquad \zeta_k \;=\; e^{\frac{2\pi i}{k}}\,.$$

If $\langle s\rangle$ denotes the fractional part $s-[s]$ of $s\in\mathbb{Q}$, the CM type of $T(k,z)$ can be easily calculated in terms of the μ_j by

$$r_n \;=\; \dim W_n \;=\; -1+\sum_j \langle \mu_j n\rangle\,, \tag{1.5}$$

where W_n denotes the eigenspace for the eigenvalue ζ_k^n for the action of σ on the vector space $H^0(T(k,z),\Omega)$ of the first kind differentials, see, e.g., [16] (on p. 23 use formula (4) with $N=2$) or [3]. Note that r_n can take the values $0,1,2$ only and satisfies $r_n+r_{-n}=2$ for all n.

In the following we will consider the second kind differentials η always as differentials on $T(k,z)$ and the cycles γ_1,γ_2 as cycles of the homology in $T(k,z)$. This homology $H_1(T(k,z),\mathbb{Z})$ is a $\mathbb{Z}[\zeta_k]$-module of rank two, and *independence* of the cycles in the definition of the normalized Schwarz triangle function $D(z)=\int_{\gamma_1}\eta(z)/\int_{\gamma_2}\eta(z)$ means now $\mathbb{Q}(\zeta_k)$-linear independence in the $\mathbb{Q}(\zeta_k)$-module $H_1(T(k,z),\mathbb{Q}) = \mathbb{Q}\otimes_{\mathbb{Z}} H_1(T(k,z),\mathbb{Z})$. Note that for algebraic z the curve, its Jacobian, its Prym variety $T(k,z)$ and the differential $\eta(z)$ are all defined over number fields.

1.3. Associate functions

As common in the literature about hypergeometric functions, we call two hypergeometric functions $F(a,b,c;z)$, $F(a',b',c';z)$ *associate* if

$$a\equiv a'\,,\ b\equiv b'\,,\ c\equiv c' \bmod \mathbb{Z}$$

or equivalently, if the respective angular parameters satisfy

$$\nu_0\equiv\nu_0'\,,\ \nu_1\equiv\nu_1'\,,\ \nu_\infty\equiv\nu_\infty' \bmod \mathbb{Z} \qquad \text{and}$$
$$\nu_0+\nu_1+\nu_\infty\equiv\nu_0'+\nu_1'+\nu_\infty' \bmod 2\mathbb{Z}$$

or if the respective exponential parameters satisfy

$$\mu_j \equiv \mu'_j \bmod \mathbb{Z} \quad \text{for all} \quad j = 0, 1, z, \infty \quad \text{and} \quad \sum_j \mu_j = \sum_j \mu'_j = 2 \, .$$

All functions associate to $F(a, b, c; z)$ generate a vector space of dimension two over the field of rational functions $\mathbb{C}(z)$, and since our parameters are supposed to be rational, between any three associate functions there is a linear relation with coefficients in $\mathbb{Q}(z)$. These relations can explicitely produced by means of Gauss' relations between contiguous functions, see [8]. Any two associate hypergeometric functions generate the vector space over $\mathbb{C}(z)$ (obvious exceptions like $F(a, a{+}1, c; z)$, $F(a{+}1, a, c; z)$ are excluded by our assumptions about the angular parameters). The congruences for the exponential parameters imply that the differentials η, η' differ only by factors which are rational functions $R(u, z) \in \mathbb{Q}(u, z)$. As second kind differentials on the Prym variety $T(k, z)$ they belong therefore to the same $\mathbb{Q}(\zeta_k)$-eigenspace V_1 in its de Rham cohomology. In our normalization, the differentials of this eigenspace are characterized by $\eta \circ \sigma = \zeta_k \eta$. The intersection of V_1 with $H^0(T(k, z), \Omega)$ gives the eigenspace W_1 mentioned in the definition of the CM type. This observation extends to the other eigenspaces V_n, $n \in (\mathbb{Z}/k\mathbb{Z})^*$, and the fact that all associate hypergeometric functions generate a 2-dimensional vector space over $\mathbb{C}(z)$ has an obvious interpretation for the eigenspaces V_n in the de Rham cohomology:

Lemma 1.1. $\dim V_n = 2$ *for all* $n \in (\mathbb{Z}/k\mathbb{Z})^*$.

Dimension means here the dimension over $\mathbb{C}$, but for algebraic z we can give another useful interpretation: as already mentioned, $T(k, z)$ is then defined over $\overline{\mathbb{Q}}$, all differentials η in the integral representation are defined over $\overline{\mathbb{Q}}$ as well whence we consider the vector spaces

$$H^0(T(k, z), \Omega) \, , \; H^1_{DR}(T(k, z)) \, , \; V_n$$

of differentials of the first and second kind defined over $\overline{\mathbb{Q}}$ as vector spaces over $\overline{\mathbb{Q}}$. In this sense, the Lemma remains true as a statement about $\overline{\mathbb{Q}}$-dimensions. For $z \in \overline{\mathbb{Q}}$ we will follow this interpretation.
In the proof of Lemma 1.1 there is only one point which is not obvious: even if associate differentials $\eta(z)$ generate a 2-dimensional $\mathbb{C}(z)$-vector space modulo exact differentials, it could be possible that for some fixed value $z = \tau$ the $\mathbb{C}$-dimension would be smaller if, e.g., all differentials in question vanish for $z = \tau$. This breakdown of the dimension can be seen to be impossible for $\tau \neq 0, 1, \infty$ either by a careful analysis of the possible relations between contiguous functions or by the fact that the genus of $X(k, z)$ is the same for all $z \neq 0, 1, \infty$, hence also $\dim H_{DR}(X(k, z))$ is independent of z, see [9, Ch. 3.5].

1.4. Shimura varieties, monodromy groups, and modular groups

In general, our Prym varieties $T(k,z)$ are only special cases of principally polarized complex abelian varieties A of dimension $\varphi(k)$, with period lattice isomorphic to $\mathbb{Z}[\zeta_k]^2$, and with an action of $\mathbb{Q}(\zeta_k) \subseteq \mathrm{End}_0 A$ of *(generalized) CM type* $(r_n)_{n\in(\mathbb{Z}/k\mathbb{Z})^*}$, see (1.5). This CM type encodes the complex representation of $\mathbb{Q}(\zeta_k)$ on the space of holomorphic differentials $H^0(A,\Omega)$ such that r_n is the dimension of the eigenspace on which ζ_k acts via

$$\zeta_k \ : \ \omega \ \mapsto \zeta_k^n \cdot \omega \ .$$

If we denote the family of all these abelian varieties by $\mathcal{A}$, we know by work of Shimura [17] and Siegel [18]:

Proposition 1.2. *The family $\mathcal{A}$ is parametrized by the product $\mathcal{H}^r$ of upper half planes $\mathcal{H}$ with dimension*

$$r \ = \ \frac{1}{2} \sum_{n\in(\mathbb{Z}/k\mathbb{Z})^*} r_n r_{-n} \ .$$

Since $0 \le r_n \le 2$ and $r_n + r_{-n} = 2$ for all n, we may rephrase this statement by saying that the dimension r is half of the number of the one-dimensional $\mathbb{Q}(\zeta_k)$-eigenspaces in $H^0(A,\Omega)$. For the special case of the Prym varieties $A = T(k,z)$ in question, we may take generators ω_j, $j = 1,\dots,r$ of one-dimensional eigenspaces $W_n \subset H^0(T(k,z),\Omega)$, $n \in (\mathbb{Z}/k\mathbb{Z})^*/\{\pm 1\}$. Then — up to linear fractional transformations — the values of the triangle functions $D(\omega_j; z)$ defined by period quotients in (1.3) serve as coordinates of the point in $\mathcal{H}^r$ corresponding to $T(k,z)$.

Two points in $\mathcal{H}^r$ correspond to isomorphic abelian varieties in $\mathcal{A}$ if and only if they belong to one Γ-orbit where Γ denotes the (arithmetically defined) *modular group* acting discontinuously on $\mathcal{H}^r$. The quotient space $\Gamma\backslash\mathcal{H}^r$ is therefore a classifying space for $\mathcal{A}$, the *Shimura variety* of $\mathcal{A}$. In the case $r = 1$ we call it a *Shimura curve*, of course (we neglect many interesting questions about algebraic or arithmetic stucture of these spaces). One subgroup of Γ is well known in the context of hypergeometric functions:

Proposition 1.3. *Let $\omega_1(z)$ be a generator of a one-dimensional $\mathbb{Q}(\zeta_k)$-eigenspace of $H^0(T(k,z),\Omega)$ and let Δ be the (projective) monodromy group of the hypergeometric functions $\int_{\gamma_1} \omega_1(z)$, $\int_{\gamma_2} \omega_1(z)$ used in Definition (1.3). Then Δ has a natural embedding into the modular group Γ of $\mathcal{A}$.*

As already explained in Section 1.1, the monodromy group — defined by analytic continuation of the hypergeometric functions — acts on the homology of $X(k,z)$ without changing the curve, hence leaving fixed the isomorphism class of its Jacobian and of the Prym variety. By consequence they embed into the modular group of the family, acting by fractional linear transformations on the coordinates of $\mathcal{H}^r$. For a more detailed explanation and a much stronger version of this proposition see [4]; in fact, there is even a holomorphic *modular embedding* of $\mathcal{H}$ into $\mathcal{H}^r$ compatible with the actions of Δ and Γ.

One extreme case of Proposition 1.2 will be very useful in Proposition 2.8 below. It can happen that the dimension of the family $\mathcal{A}$ is $r = 0$. This is the case if and only if the modular group and a fortiori the monodromy group Δ is finite. By the classical reasoning of H.A. Schwarz [14], the fact that the hypergeometric functions and their triangle functions have only finitely many branches is equivalent to state that they are algebraic functions: just observe that the elementary symmetric functions of their branches are single-valued meromorphic, hence rational functions, and note that by the hypotheses (1.1) our hypergeometric differential equations are irreducible. For other arguments in that direction and their generalization to hypergeometric functions in several variables see [5].

The next interesting case is that of Shimura curves, i.e., the case $r = 1$. Then the modular group Γ and a fortiori the monodromy group Δ act as arithmetically defined Fuchsian groups. In Section 3 it will become clear why the arithmeticity of Δ is so important for our question, and Sections 4 and 5 will discuss in great detail one example, i.e., the family of hypergeometric curves (4.2) with angular parameters $\nu_0 = -\nu_1 = \nu_\infty = 1/5$ and

$$k = 5\,, \quad r_1 = 0\,,\ r_2 = r_3 = 1\,,\ r_4 = 2\,, \quad r = 1\,.$$

Caution. On the other hand, there are arithmetically defined monodromy groups for which $r > 1$. In these cases the $T(k,z)$ belong to some subfamily of *Hodge type* of $\mathcal{A}$, i.e., to a Shimura subvariety described by a special splitting behaviour of $T(k,z)$ or — equivalently — by the fact that the common endomorphism algebra of all $T(k,z)$ is strictly larger than $\mathbb{Q}(\zeta_k)$. As an example, take

$$\nu_0 = \frac{1}{2}\,,\ \nu_1 = \frac{1}{3}\,,\ \nu_\infty = \frac{1}{10}\,. \tag{1.6}$$

An obvious calculation leads to $k = 30$ and $r = 2$. We have $W_{\pm 1}, W_{\pm 11}$ as one-dimensional eigenspaces in $H^0(T(k,z)\Omega)$. The generators of W_1 and W_{11} lead with (1.3) to triangle functions

$$D(\frac{1}{2},\frac{1}{3},\frac{1}{10};z) \quad \text{and} \quad D(\frac{1}{2},-\frac{1}{3},\frac{1}{10};z)$$

which are constant multiples of each other, see [21, (16)]. Recall that these triangle functions give the coordinates of the point in $\mathcal{H}^2$ corresponding to $T(k,z)$. Therefore these Pryms are parametrized by an upper half plane $\mathcal{H}$ linearly embedded in $\mathcal{H}^2$, and a more detailed analysis shows that they split into two factors, both isogenous to the Pryms of the family (4.2); this is not surprising since the monodromy group for the example (1.6) is the triangle group of signature $\langle 2,3,10\rangle$, an index 6 extension of that one in (4.2) of signature $\langle 5,5,5\rangle$.

2. Tools from transcendence

2.1. The analytic subgroup theorem

The main instrument to obtain transcendence results for hypergeometric functions is Wüstholz' analytic subgroup theorem, see [23] and [24].

Proposition 2.1. *Let G be a connected commutative algebraic group defined over $\overline{\mathbb{Q}}$ of dimension $\dim G > r > 0$ and*

$$\varphi \,:\, \mathbb{C}^r \;\to\; G$$

an analytic homomorphism whose tangential map $d\varphi$ is an homomorphism of $\overline{\mathbb{Q}}$-vectorspaces. If the image contains a nontrivial algebraic point, i.e., if $\varphi(\mathbb{C}^r)(\overline{\mathbb{Q}}) \neq \{0\}$, there is an algebraic subgroup $H \subseteq \varphi(\mathbb{C}^r)$ defined over $\overline{\mathbb{Q}}$ with $\dim H > 0$.

The unexperienced reader may wonder why this is a theorem about transcendental numbers. Let us explain it first with a classical example: Let G be the product $\mathbb{C}^* \times \mathbb{C}$ of the multiplicative and the additive group of complex numbers and observe that $\{1\} \times \mathbb{C}$ and $\mathbb{C}^* \times \{0\}$ are the only nontrivial connected algebraic subgroups of G. As analytic homomorphism take the exponential map

$$\varphi \,:\, (z,w) \;\mapsto\; (e^z, w)\,,$$

restricted to the one-dimensional subspace $z = bw$ of the tangent space $\mathbb{C}^2$ of G. Now suppose there were an algebraic number $a \neq 0, 1$ with an algebraic logarithm $b = \log a$. Then $d\varphi$ and our one-dimensional subspace are defined over $\overline{\mathbb{Q}}$ and the φ-image contains and algebraic point $(a, 1)$. On the other hand, it does not contain any proper algebraic subgroup of G in contradiction to Proposition 2.1. So we obtain the Lindemann–Weierstrass theorem that e^b is transcendental for all algebraic $b \neq 0$. With $b = i\pi$ we get the transcendence of π as well.

2.2. Application to periods

The application of Proposition 2.1 needed for the values of the Schwarz maps is a powerful theorem about linear independence of periods over $\overline{\mathbb{Q}}$ first stated as Theorem 5 of [23]. The proof has been worked out by Paula Cohen in the appendix of [15].

Proposition 2.2. *Let A be an abelian variety isogenous over $\overline{\mathbb{Q}}$ to the direct product $A_1^{k_1} \times \ldots \times A_N^{k_N}$ of simple, pairwise non-isogenous abelian varieties A_ν defined over $\overline{\mathbb{Q}}$, with A_ν of dimension n_ν, $\nu = 1, \ldots, N$. Then the $\overline{\mathbb{Q}}$-vector space $\widehat{V}_A$ generated by $1, 2\pi i$ together with all periods of differentials, defined over $\overline{\mathbb{Q}}$, of the first and the second kind on A, has dimension*

$$\dim_{\overline{\mathbb{Q}}} \widehat{V}_A \;=\; 2 \,+\, 4 \sum_{\nu=1}^{N} \frac{n_\nu^2}{\dim_{\mathbb{Q}} End_0 A_\nu}\,.$$

We will not repeat the proof here. To give an impression how linear independence of periods follows from Wüstholz' analytic subgroup theorem, we will however state and prove a simpler and very special case, see also [22, Satz 1].

Proposition 2.3. *Let A be a simple abelian variety defined over $\overline{\mathbb{Q}}$, $\omega_1, \ldots, \omega_n \in H^0(A, \Omega)$ a basis of holomorphic differentials on A, also defined over $\overline{\mathbb{Q}}$, and let $\gamma \in H_1(A, \mathbb{Z})$ be a nonzero cycle on A. Then the periods*

$$\int_\gamma \omega_1 \,, \; \ldots \,, \; \int_\gamma \omega_n$$

are linearly independent over $\overline{\mathbb{Q}}$.

Assume the statement to be wrong. Then there is a linear relation

$$a_1 \int_\gamma \omega_1 + \ldots + a_n \int_\gamma \omega_n = 0$$

with algebraic coefficients a_i not all $= 0$. Consider the exponential map

$$\varphi : \mathbb{C}^n \to A \cong \mathbb{C}^n/\Lambda$$

where Λ denotes the period lattice $\{(\int_\delta \omega_1, \ldots, \int_\delta \omega_n) \,|\, \delta \in H^1(A, \mathbb{Z})\}$ and restrict φ to the $(n-1)$-dimensional subspace S given by

$$a_1 z_1 + \ldots + a_n z_n = 0 \,.$$

This subspace S and $d\varphi$ are defined over $\overline{\mathbb{Q}}$. By our assumption, the nonzero vector

$$v := (\int_\gamma \omega_1, \ldots, \int_\gamma \omega_n)$$

belongs to the kernel of φ and $\varphi(\mathbb{Q}v)$ consists of torsion points of A, hence belongs to $A(\overline{\mathbb{Q}})$. Therefore Proposition 2.1 applies, but A is simple and has no proper algebraic subgroup of positive dimension, contradiction.

2.3. Complex multiplication

Proposition 2.2 indicates that the splitting of $T(k, z)$ and the endomorphism algebra of its simple components will be very important for the understanding of linear dependence or independence of periods. An extreme case is the situation that the abelian variety A has *complex multiplication* or *CM* in short. This means that there is a number field $K \subseteq \mathrm{End}_0 A$ of the (maximal possible) degree $[K : \mathbb{Q}] = 2 \dim A$. For the convenience of the reader, we collect here some facts well known from the literature (see, e.g., [13]).

The field K is necessarily a *CM field*, that is a totally imaginary quadratic extension of some totally real number field F of degree $g = \dim A$. The space $H_{DR}(A)$ of all first and second kind differentials splits into $2 \dim A$ one-dimensional subspaces V_σ where σ runs over all embeddings $K \to \mathbb{C}$ and every $\alpha \in K$ acts on V_σ by multiplication with $\sigma(\alpha)$. The subspace $H^0(A, \Omega)$ of first kind differentials splits under the action of K into g one-dimensional eigenspaces $W_\sigma = V_\sigma$ among them for which σ runs over a system of representatives of all embeddings $K \to \mathbb{C}$ modulo complex conjugation. (In the case of a cyclotomic field $\mathbb{Q}(\zeta_k)$ we may caracterize the embeddings σ as usual by representatives of prime residue classes in $\mathbb{Z}/k\mathbb{Z}$ modulo ± 1.) The collection of these representatives σ are called the *CM type* of A and determine A uniquely up to isogeny. The abelian varieties with this endomorphism structure form a zero-dimensional Shimura variety, and A is defined over a (particularly interesting!) number field.

It can happen that A with CM is not simple: it may be isogeneous to some power B^m of a simple abelian variety B with CM by a subfield L of K of degree $[L:\mathbb{Q}] = \frac{1}{m}[K:\mathbb{Q}]$. The CM type of A arises from that of B by extending the embeddings of L to K. Therefore symmetries of the CM type of A show whether A is simple or not.

An abelian variety T is called *of CM type* if it is isogenous to a direct product of factors with complex multiplication. The corresponding points in a Shimura variety are called *CM points* or *special points*. In the easiest example where the upper half plane $\mathcal{H}$ parametrizes the family of all elliptic curves, the imaginary quadratic points give the CM points if we pass to the Shimura variety $\Gamma\backslash\mathcal{H}$, Γ denoting the elliptic modular group.

2.4. The splitting pattern of the Pryms

We come back to the Prym varieties defined in Section 1.2 and collect results of [21, Satz 4] and [2, Exemple 3, Thm. 1, Lemme 1].

Proposition 2.4. *Let C be the subalgebra of $\mathrm{End}_0 T(k,z)$ of elements commuting with $\mathbb{Q}(\zeta_k) \subseteq \mathrm{End}_0 T(k,z)$. This subalgebra belongs to one of the following three types.*

1. *$C = \mathbb{Q}(\zeta_k)$. Then $T(k,z)$ is isogenous to a power D^m of a simple abelian variety D whose endomorphism algebra S is a subfield $S \subseteq \mathbb{Q}(\zeta_k)$ with*
$$m = [\mathbb{Q}(\zeta_k) : S] \quad \textit{and} \quad \dim D = [S:\mathbb{Q}]\,.$$
In particular, no simple factor of $T(k,z)$ has complex multiplication.
2. *$C = K$ is a quadratic extension of $\mathbb{Q}(\zeta_k)$. The Prym variety has complex multiplication by K and is isogenous to a power B^m of a simple abelian variety with CM by a subfield $L \subseteq K$ with $m = [K:L]$.*
3. *C has zero divisors. Then $T(k,z)$ is isogenous to $A_1 \oplus A_2$ with two abelian varieties A_i of dimension $\frac{1}{2}\varphi(k)$ and with endomorphism algebra $\mathrm{End}_0 A_i \subseteq \mathbb{Q}(\zeta_k)$. Both A_i have complex multiplication by $\mathbb{Q}(\zeta_k)$.*

The proof can be sketched as follows. If C has a zero divisor, its image of $T(k,z)$ gives a proper $\mathbb{Q}(\zeta_k)$-invariant abelian subvariety A_1 and a $\mathbb{Q}(\zeta_k)$-invariant complement A_2. It is well known that for such abelian varieties $[\mathbb{Q}(\zeta_k):\mathbb{Q}]$ divides $2\dim A_i$, therefore we have equality, hence CM — the third case of the classification.

If C has no zero divisors, it is a (commutative) field by [2, Lemme 1] and by reasons of divisibility again, it is either $\mathbb{Q}(\zeta_k)$ or a quadratic extension of it. If $C = \mathbb{Q}(\zeta_k)$, [2, Exemple 3] applies to give the first case of our classification. The second case is now obvious by the information given in the last subsection.

2.5. Pryms not of CM type

Now we suppose $z = \tau \in \overline{\mathbb{Q}}$ and consider all eigenspaces V_n as vector spaces over $\overline{\mathbb{Q}}$. Proposition 2.2 implies in particular

Lemma 2.5. *Suppose $\tau \in \overline{\mathbb{Q}}, \neq 0, 1$, and suppose that $T(k, \tau)$ is an abelian variety not of CM type, see Proposition 2.4.1. Then all periods*

$$\int_\gamma \eta\,, \quad \gamma \in H_1(T(k,\tau),\mathbb{Z})$$

of a fixed nonzero $\eta \in V_n \subset H^1_{DR}(T(k,\tau))$ generate a $\overline{\mathbb{Q}}$-vector space Π_η of dimension 2.

The upper bound ≤ 2 for this dimension follows directly from the facts that $H_1(T(k,\tau),\mathbb{Z})$ is a $\mathbb{Z}[\zeta_k]$-module of rank 2 and that η is an eigendifferential. On the other hand, dimension $= 1$ would lead to a contradiction as follows. Recall that by Proposition 2.4.1, $T(k,z)$ has only one simple factor D of dimension $g = \varphi(k)/m$ and with $\mathrm{End}_0 D = S$, S a number field of degree g. Complete η to a basis of $H_{DR}(D)$ consisting of $2g$ eigendifferentials for the action of $\mathbb{Q}(\zeta_k)$. As η, all of them have their periods in an at most 2-dimensional $\overline{\mathbb{Q}}$-vector space. On the other hand, this upper bound is attained because Proposition 2.2 shows that all periods on $T(k,z)$ together with 1 and π generate a $\overline{\mathbb{Q}}$-vector space of dimension $2 + 4g$ [1].

2.6. Pryms of CM type

Next we consider case 2 of Proposition 2.4.

Lemma 2.6. *Suppose $\tau \in \overline{\mathbb{Q}}, \neq 0, 1$ and suppose that $T(k,\tau)$ has complex multiplication by a CM field K, $[K : \mathbb{Q}(\zeta_k)] = 2$. All periods*

$$\int_\gamma \eta\,, \quad \gamma \in H_1(T(k,\tau),\mathbb{Z})$$

of a nonzero second kind $\mathbb{Q}(\zeta_k)$-eigendifferential $\eta \in V_n \subset H^1_{DR}(T(k,\tau))$ generate a $\overline{\mathbb{Q}}$-vector space Π_η of dimension

- 1 *if η is a K-eigendifferential,*
- 2 *if not.*

The first case happens in precisely two onedimensional subspaces of V_n.

For the proof recall that $T(k,\tau)$ is isogenous to a power B^m of a simple abelian variety with complex multiplication by some subfield L of K, and V_n splits into two L-eigenspaces for factors B but for different eigenvalues. Then the result follows again from Proposition 2.2.

The last possibility is case 3 of Proposition 2.4.

[1] In [15, Prop. 4.1] we treated only the case that $T(k,z)$ is simple without CM, i.e., we overlooked the possibility that it can be isogenous to D^m with $m > 1$ as described in Proposition 2.4.1. However, the result remains true (see Lemma 2.5) also in the non-simple case, so all consequences drawn in [15] are correct.

Lemma 2.7. *Suppose* $\tau \in \overline{\mathbb{Q}}, \neq 0, 1$ *and that* $T(k,\tau)$ *is isogenous to* $A_1 \oplus A_2$ *for two abelian varieties of* A_i *dimension* $\frac{1}{2}\varphi(k)$ *and with complex multiplication by* $\mathbb{Q}(\zeta_k)$.

1. *If* A_1 *and* A_2 *have the same CM type, all periods*

$$\int_\gamma \eta\,, \quad \gamma \in H_1(T(k,\tau),\mathbb{Z})\ ,\ \eta \in V_n$$

of any eigenspace $V_n \subset H^1_{DR}(T(k,\tau))$ *generate a* $\overline{\mathbb{Q}}$*-vector space* Π_n *of dimension* 1 , *and* $\Pi_n = \Pi_\eta$ *for all nonzero* $\eta \in V_n$.

2. *If* A_1 *and* A_2 *have different CM types, we have* $\dim \Pi_n = 2$, *and the periods of every fixed* $0 \neq \eta \in V_n$ *generate a* 2*-dimensional vector space* Π_η *over* $\overline{\mathbb{Q}}$, *except in the case that* η *belongs to one of the factors in the decomposition*

$$H^1_{DR}(T(k,\tau)) \;=\; H^1_{DR}(k,A_1) \oplus H^1_{DR}(k,A_2)\ .$$

In this case (happening in precisely two onedimensional subspaces of V_n*)* Π_η *is of dimension* 1 .

In both cases the A_i are isogenous to pure powers $B_i^{m_i}$ of simple abelian varieties B_i with complex multiplication. In the first case, B_1 and B_2 are isogenous, and in the second case not. Then the result follows again by Wüstholz' analytic subgroup theorem in the version of Proposition 2.2, similar to Lemma 2.6. For more details the reader my consult also the proof of [15, Prop. 4.4]. Finally we give precise conditions under which the first case of Lemma 2.7 occurs. Note that these conditions do not depend on the algebraicity of z .

Proposition 2.8. *The following statements are equivalent.*

- *The Shimura family* $\mathcal{A}$ *in Proposition* 1.2 *has dimension* $r = 0$.
- *For one (hence for all)* $z \neq 0, 1,$ *the CM type of* $T(k,z)$ *satisfies*

$$r_n \;=\; 0 \quad or \quad 2 \qquad for\ all \quad n \in (\mathbb{Z}/k\mathbb{Z})^*\ .$$

- *For one (hence for all)* $z \neq 0, 1$, *the abelian variety* $T(k,z)$ *is isogenous to* $A_1 \oplus A_2$, *both* A_i *have dimension* $\frac{1}{2}\varphi(k)$ *and complex multiplication by* $\mathbb{Q}(\zeta_k)$ *with equal CM type.*
- *The monodromy group of the corresponding hypergeometric differential equation is finite.*
- *The corresponding triangle function* $D(\nu_0,\nu_1,\nu_\infty;z)$ *is an algebraic function of* z .

The equivalence between the first and the second point follows from Proposition 1.2. The equivalence between the second and the third is known by work of Shimura [17, Thm. 5, Prop. 14], the equivalence between the first and the last two points has been discussed already in Subsection 1.4. The equivalence between the last two points is classical, of course, see, e.g., [10, §57].

3. Special values of Schwarz triangle functions

3.1. The role of complex multiplication

We work still under the hypothesis $z = \tau \in \overline{\mathbb{Q}}$ and recall that the cycles γ_1, γ_2 in the definition $D(\nu_0, \nu_1, \nu_\infty; \tau) = D(\tau) = \int_{\gamma_1} \eta(\tau) / \int_{\gamma_2} \eta(\tau)$ are generators of the 2-dimensional $\mathbb{Q}(\zeta_k)$-module $H_1(T(k,\tau), \mathbb{Q}) = \mathbb{Q} \otimes_{\mathbb{Z}} H_1(T(k,\tau), \mathbb{Z})$. Numerator and denominator generate the period vector space Π_η discussed in the last section. We conclude from Lemmata 2.5, 2.6 and 2.7:

Theorem 3.1. *Suppose* $\tau \in \overline{\mathbb{Q}}, \neq 0, 1$.

$$D(\nu_0, \nu_1, \nu_\infty; \tau) \;=\; D(\tau) \;=\; \frac{\int_{\gamma_1} \eta(\tau)}{\int_{\gamma_2} \eta(\tau)}$$

is algebraic or ∞ *if and only if* $T(k,\tau)$ *is of CM type and* $\dim_{\overline{\mathbb{Q}}} \Pi_{\eta(\tau)} = 1$, *i.e., if* $\eta(\tau)$ *is a*

- K*-eigendifferential under the hypotheses of Proposition* 2.4.2, *or a*
- $\mathbb{Q}(\zeta_k)$*-eigendifferential on one of the factors* A_1, A_2 *under the hypotheses of Proposition* 2.4.3.

In two special situations we can give more explicit conditions. The first is obvious by Proposition 2.8 and Lemma 2.7.1

Theorem 3.2. *If the monodromy group* Δ *of the corresponding differential equation is finite, all values* $D(\tau)$ *of the triangle function at algebraic arguments* τ *are algebraic or* ∞.

In the following we will therefore restrict our attention to infinite monodromy groups Δ. In these cases, we know by Proposition 2.8 that at least one $r_n = 1$, in other words one $W_n = V_n \cap H^0(T(k,\tau), \Omega)$ contains a nonzero differential $\eta = \omega$ of the first kind, unique up to multiples. For periods of the first kind we can apply a sharper version of Wüstholz' theorem giving a period vector space Π_ω of dimension 1 if the abelian variety has CM type. Another way to prove $\dim \Pi_\omega = 1$ is a second look on Lemma 2.6 and Lemma 2.7: in Lemma 2.6, $H^0(T(k,\tau), \Omega)$ is K-invariant, therefore W_n is one of the onedimensional subspaces of K-eigendifferentials. In Lemma 2.7, W_n belongs to precisely one of the homology factors $H^0(A_i, \Omega)$ since only one of them contains eigendifferentials ω with $\omega \circ \sigma = \zeta_k^n \omega$, otherwise we would have $\dim W_n = 2$. Summing up we get (see also [16, Cor. 5] for a different argument)

Theorem 3.3. *Suppose* $\tau \in \overline{\mathbb{Q}}, \neq 0, 1$, *and that* $T(k,\tau)$ *is of CM type, let* W_n *be a one-dimensional* $\mathbb{Q}(\zeta_k)$*-eigenspace in* $H^0(T(k,\tau), \Omega)$. *If* $0 \neq \omega = \eta(\tau) \in W_n$, *the value of the corresponding triangle function* $D(\tau) = \int_{\gamma_1} \eta(\tau) / \int_{\gamma_2} \eta(\tau)$ *is algebraic.*

The first natural question is now: how to control that η is of first kind? For simplicity, take $n = 1$. There $\eta = du/y$ — see Section 1.1 — is of first kind if and only if the exponential parameters μ_j are all < 1. The second question is already much more difficult: for which $\tau \in \overline{\mathbb{Q}}$ is $T(k,\tau)$ of CM type? The answer depends on the nature of the monodromy group Δ and, unfortunately, does not give a general explicit criterion for the distinction between CM and non-CM cases.

1. If Δ is finite, $T(k,z)$ is of CM type for every z, see Proposition 2.8.
2. If Δ is an arithmetic group, there is an infinity of $T(k,\tau)$ of CM type and an infinity of $T(k,\tau)$ not of CM type. In these cases — classified by Takeuchi [19] — Δ is commensurable to the modular group for a complex onedimensional family of polarized abelian varieties with a certain endomorphism structure. Our $T(k,z)$, $z \neq 0,1$, form a dense subset of this family, and the Schwarz triangle function D is the inverse function of an arithmetic automorphic function for this modular group, possibly up to composition with an algebraic function. See also our remarks about Shimura curves in Section 1.4 and about CM points in Section 2.3.
3. If Δ is infinite and non-arithmetic, the $T(k,z)$ form a subfamily not of Hodge type in the Shimura variety of all polarized abelian varieties of their endomorphism structure. In this case, the André–Oort conjecture predicts that there are only finitely many $T(k,\tau)$ of CM type. This conjecture is proven by Edixhoven and Yafeev [7] for those CM types discussed in Proposition 2.4.3, but it is open in general. For more information and applications to other hypergeometric questions see [6].

3.2. Other algebraic values at algebraic arguments

The aim of this part is to show that Theorems 3.2 and 3.3 describe very exceptional situations, i.e., that in general for $\tau \in \overline{\mathbb{Q}} - \{0,1\}$

$$D(\nu;\tau) \;=\; D(\nu_0,\nu_1,\nu_\infty;\tau) \notin \overline{\mathbb{Q}}$$

even if the necessary condition given by Theorem 3.1 is satisfied that $T(k,\tau)$ is of CM type. We used here an abbreviated notation $\nu := (\nu_0,\nu_1,\nu_\infty)$ for the rational triplets of angular parameters (always under the restriction (1.1)). We call two such triplets ν,ν' *associate* if they belong to associate hypergeometric functions, see the conditions on their components given in Section 1.3. Observe that triangle functions with associate angular parameters belong to the same monodromy group.

Theorem 3.4. *Let P be a finite set of associate rational angular parameter triplets ν, belonging to an infinite monodromy group Δ. There is a finite set $E_P \subset \overline{\mathbb{Q}}$ of exceptional arguments such that for all other $\tau \in \overline{\mathbb{Q}} - E_P$ at most two of the values $D(\nu;\tau)$, $\nu \in P$, are algebraic or ∞.*

We may assume that P contains more than two elements and that $0,1 \in E_P$, and for the proof we may assume moreover that $T(k,\tau)$ is of CM type since we know by Theorem 3.1 that otherwise all values in question are transcendental. Theorem 3.4 uses Lemma 1.1 and classical facts about associate hypergeometric functions: denote the differentials in the integral representation of $D(\nu;z), \nu \in P$, by $\eta(\nu;z)$ and observe that all these $\eta(\nu;z), \nu \in P$, belong to one eigenspace V_n. By Gauss' relations among contiguous hypergeometric functions, any two of them generate V_n as a $\mathbb{C}(z)$-vector space. The only obstacle is mentioned already in Section 1.3 that for a fixed value $z = \tau$ they may fail to be a basis over $\mathbb{C}$ or $\overline{\mathbb{Q}}$. A closer look into Gauss' relations [8] shows that this can happen only at finitely many algebraic points since the relations always have coefficients in $\mathbb{Q}(z)$: For any three different fixed associate $\eta(\nu; z), \eta(\nu'; z), \eta(\nu''; z)$ we get a representation

$$\eta(\nu; z) \;=\; r'(z)\eta(\nu'; z) \;+\; r''(z)\eta(\nu''; z)$$

with nonvanishing rational functions $r', r'' \in \mathbb{Q}(z)$. We can use these relations in all special points $\tau \in \overline{\mathbb{Q}}$ as relations over $\overline{\mathbb{Q}}$, except for the (algebraic) poles of r', r''. For $\nu, \nu', \nu'' \in P$ we obtain finitely many such poles and also finitely many algebraic zeros of all such r, r'. If we include these finitely many exceptions in our exceptional set E_P, in all other points $\tau \in \overline{\mathbb{Q}}$ the $\eta(\nu, \tau)$, $\nu \in P$, generate pairwise different one-dimensional subspaces of V_n. Lemma 2.6 and Lemma 2.7.2 show that in only two such one-dimensional subspaces the period vector spaces Π_η are of dimension 1, and this is equivalent to the algebraicity of the period quotient $D(\nu; \tau)$.

It seems to be very likely that $E_P \supseteq \{0, 1\}$ is finite even for infinite sets P of associate parameter triplets because in exceptional points $\tau \neq 0, 1$ three quite different conditions have to be satisfied. First,

- two $\eta(\nu; \tau), \eta(\nu'; \tau)$, $\nu, \nu' \in P$, have to be multiples of each other.

As an example that this can happen take relation (28) on p. 103 of [8]

$$\begin{aligned}(c-a)F(a-1, b, c; z) \;+\; &(2a - c - az + bz)F(a, b, c; z) \\ &+ a(z-1)F(a+1, b, c; z) \;=\; 0\,.\end{aligned}$$

Translated to the language of differentials and angular parameters it says that in the special point $\tau = (2a - c)/(a - b) = (\nu_\infty - \nu_1)/\nu_\infty$ the associate differentials

$$\eta(\nu_0, \nu_1 + 1, \nu_\infty + 1; \tau)\,, \quad \eta(\nu_0, \nu_1 - 1, \nu_\infty - 1; \tau)$$

are $\overline{\mathbb{Q}}$-linearly dependent and give there the same period quotient $D(\tau)$. Whether or not this value is really algebraic depends of course on two further conditions, namely

- if $T(k, \tau)$ is of CM type and
- if $\eta(\nu; \tau)$ generates one of the two one-dimensional eigenspaces mentioned in Lemma 2.6 or Lemma 2.7.2 .

In general, the second and the third condition are difficult to verify, for examples see the next Sections. In the case treated in Theorem 3.3 we can better localize at least one of these one-dimensional eigenspaces: it is *the* subspace of differentials of the first kind but containing no other $\eta(\nu; \tau)$, if we have no coincidences coming from the degeneration of Gauss' relations discussed in the proof of Theorem 3.4. Therefore we get the following sharper result.

Theorem 3.5. *Let P be a finite set of associate rational angular parameter triplets ν, belonging to an infinite monodromy group Δ, and suppose further that there is precisely one first kind differential $\omega = \eta(\nu'; z)$ associate to these $\eta(\nu; z)$, $\nu \in P$, but with $\nu' \notin P$. Then there is a finite set $E_P \subset \overline{\mathbb{Q}}$ of exceptional arguments such* *that for all* $\tau \in \overline{\mathbb{Q}} - E_P$ *at most one of the values* $D(\nu;\tau)$, $\nu \in P$, *is algebraic or* ∞.

4. Examples of algebraic values: Pryms of CM type

As for the special values of the Schwarz map $D(z)$ for a differential $\eta(z)$ on a family of hypergeometric curves, we have established the general properties in preceding sections. Here we consider examples explaining the situation in question. They all arise from specializations of a family of curves studied in the framework of ball quotients and Appell–Lauricella hypergeometric functions in two variables.

4.1. Pentagonal curves and their degeneration

Let us consider a family of hypergeometric curves given by (1.2):

$$X(p,z) = X(z) :$$
$$y^p = x^{p\mu_0}(x-1)^{p\mu_1}(x-z)^{p\mu_z} \quad (z \in \mathbb{C} - \{0,1\}), \tag{4.1}$$

where we suppose p to be a prime and $\mu_0, \mu_1, \mu_z, \mu_\infty \notin \frac{1}{2}\mathbb{Z}$, equivalent to the non-integrality condition (1.1). We defined the Prym variety $T(p,z)$ for $X(z)$ induced from the Jacobi variety $\mathrm{Jac}(X(z))$. Since $k = p$ is prime, $T(p,z)$ coincides with $\mathrm{Jac}(X(z))$. So in our case the field $\mathbb{Q}(\zeta_p)$ acts on the space of holomorphic differentials $H^0(\mathrm{Jac}(X(z)),\Omega) \cong H^0(X(z),\Omega)$ with parameter z. We note also that the $\mathbb{Q}(\zeta_p)$-action on $X(z)$ induces a $\mathbb{Q}(\zeta_p)$ module structure on $H_1(X(z),\mathbb{Q})$ of rank two. Let γ_1, γ_2 be two 1-cycles on $X(z)$ independent over $\mathbb{Q}(\zeta_p)$ and let as in Subsection 1.1

$$\eta(z) = x^{-\mu_0}(x-1)^{-\mu_1}(x-z)^{-\mu_z}dx$$

be a differential of second kind on $X(z)$. Then the corresponding Schwarz map is defined by (1.3). Let $P(\lambda_1,\lambda_2)$ be a projective nonsingular model of the affine curve

$$y^5 = x(x-1)(x-\lambda_1)(x-\lambda_2), \quad (\lambda_1, \lambda_2, \lambda_1/\lambda_2, \in \mathbb{C} - \{0,1\}).$$

$P(\lambda_1,\lambda_2)$ is a curve of genus 6 and is called a pentagonal curve. There are many articles concerned with this family. We cite here just one by K. Koike [12]. We have a basis of $H^0(P(\lambda_1,\lambda_2),\Omega)$:

$$\varphi_1 = \frac{dx}{y^2},\ \varphi_2 = \frac{dx}{y^3},\ \varphi_3 = \frac{xdx}{y^3},\ \varphi_4 = \frac{dx}{y^4},\ \varphi_5 = \frac{xdx}{y^4},\ \varphi_6 = \frac{x^2dx}{y^4}.$$

Let $DegP(z)$ be the compact nonsingular model of

$$y^5 = x^2(x-1)(x-z) \quad (z \in \mathbb{C} - \{0,1\}). \tag{4.2}$$

It is a degenerate pentagonal curve of genus 4. There is a natural ζ_5-action

$$\sigma : (x,y) \mapsto (x, \zeta_5^{-1}y).$$

So we have

$$\mathbb{Q}(\zeta_5) \subseteq \mathrm{End}_0(\mathrm{Jac}\ (DegP(z))).$$

We have a basis of $H^0(DegP(z), \Omega)$:

$$\omega_1 = \frac{dx}{y^2},\ \omega_2 = \frac{xdx}{y^3},\ \omega_3 = \frac{xdx}{y^4},\ \omega_4 = \frac{x^2dx}{y^4} \tag{4.3}$$

consisting of eigendifferentials for the action of $\mathbb{Q}(\zeta_5)$.

Remark 4.1. *We note that ω_3 and ω_4 are mutually associate.*

In general we have a solution for the Gauss hypergeometric differential equation

$$E(a, b, c) \quad : \quad z(1-z)f'' + (c - (1+a+b)z)f' - abf = 0$$

given by the integrals

$$\begin{aligned} & e^{-\pi i(-c+b+1-a)} \int_1^\infty x^{a-c}(x-1)^{c-b-1}(x-z)^{-a}dx \\ = \ & e^{-\pi i(-c+b+1-a)} \int_0^1 u^{b-1}(1-u)^{c-b-1}(1-zu)^{-a}du \\ = \ & \int_1^\infty x^{a-c}(1-x)^{c-b-1}(z-x)^{-a}dx \ = \ F_{1\infty}(a, b, c; z) \end{aligned}$$

with

$$ux = 1\,,\ 1 - x = e^{-\pi i}(x-1)\,,\ z - x = e^{\pi i}(x - z)\,.$$

That solution is single valued holomorphic at $z = 0$ and

$$F(a, b, c; z) \ = \ e^{\pi i(1-c+b-a)} \frac{\Gamma(c)}{\Gamma(b)\Gamma(c-b)} F_{1\infty}(a, b, c; x)\,.$$

Then the integral

$$\int_1^\infty \omega_1(z)$$

is a holomorphic solution of $E(2/5, 3/5, 6/5)$ at $z = 0$. The absolute values of the angular parameters are given by

$$(|1-c|, |c-a-b|, |a-b|) \ = \ (\frac{1}{5}, \frac{1}{5}, \frac{1}{5})\,.$$

By putting $x = 1/x_1, y = z^{1/5}y_1/x_1, z = 1/z_1$ in (4.2) we obtain an isomorphic nonsingular curve given by

$$y_1^5 = x_1(x_1 - 1)(x_1 - z_1)\,,$$

the integral $\int dx_1/y_1^2$ gives a solution for $E(2/5, 1/5, 4/5)$ with the same angular parameters in absolute values. So the inverse of the Schwarz map becomes an automorphic function on the upper half plane with respect to a cocompact arithmetic triangle group $\Delta(5, 5, 5)$. An explicit expression of this automorphic function is given by Koike [12, Theorem 6.3].

4.2. First example

Theorem 4.2. *In the family of curves* (4.2), $T(5,-\zeta_3)$ *has complex multiplication by the field* $\mathbb{Q}(\zeta_{15})$. *For the differentials* (4.3), *the value of the Schwarz map* $D(\omega_3,-\zeta_3)=D(-\frac{2}{5},-\frac{3}{5},-\frac{2}{5};-\zeta_3)$ *is transcendental, but*

$$\begin{aligned} D(\omega_1,-\zeta_3) &= D(-\frac{1}{5},\frac{1}{5},-\frac{1}{5};-\zeta_3)\\ D(\omega_2,-\zeta_3) &= D(\frac{1}{5},-\frac{1}{5},\frac{1}{5};-\zeta_3)\\ D(\omega_4,-\zeta_3) &= D(\frac{3}{5},-\frac{3}{5},\frac{3}{5};-\zeta_3) \end{aligned}$$

are algebraic numbers.

Proof. Define

$$\Sigma \,:\, w^5 \;=\; t^2(t^3-1)\,.$$

It is a singular model of a curve of genus 4 and we have on this model a basis of the space of holomorphic differentials:

$$\varphi_1=\frac{dt}{w^2},\ \varphi_2=\frac{tdt}{w^3},\ \varphi_3=\frac{tdt}{w^4},\ \varphi_4=\frac{t^2dt}{w^4}\,.$$

There are actions of ζ_3 and ζ_5:

$$t'=\zeta_3 t\,,\ w'=\zeta_3 w\,,$$

$$t'=t\,,\ w'=\zeta_5 w$$

on Σ. They generate a cyclic group of automorphisms on Σ generated by a single action

$$t'=\zeta_{15}^5 t\,,\ w'=\zeta_{15}^2 w$$

and induces an action of $\mathbb{Q}(\zeta_{15})$ on the space of holomorphic differentials. Any φ_i $(i=1,2,3,4)$ is an eigendifferential for this action, and $\mathbb{Q}(\zeta_{15})$ acts faithfully on the space of holomorphic differentials. We have

$$[\mathbb{Q}(\zeta_{15}):\mathbb{Q}]=8=2\cdot\text{ genus of }\Sigma\,.$$

It means $\mathrm{End}_0\mathrm{Jac}(\Sigma)=\mathbb{Q}(\zeta_{15})$ and that $\mathrm{Jac}(\Sigma)$ is an abelian variety with complex multiplication by $\mathbb{Q}(\zeta_{15})$. A more detailed analysis of its CM type shows that it is simple and that the φ_i generate the eigenspaces in $H^0(\Sigma,\Omega)$.

Defining

$$T: t(x)=\frac{x}{\zeta_3\,(-1+\zeta_3+x)}\,,\quad w(x,y)=\frac{(-1)^{\frac{1}{10}}\,3^{\frac{3}{10}}\,y}{-1+\zeta_3+x}\,,$$

the CM curve Σ is transformed to the degenerated pentagonal hypergeometric curve

$$DegP(-\zeta_3)\,:\,y^5\;=\;x^2(x-1)(x+\zeta_3)$$

whose Prym variety $T(5, -\zeta_3)$ (here it is just the Jacobian) belongs therefore to those discussed in case 2 of Proposition 2.4.

The converse transformation T^{-1} is given by

$$x(t) = \frac{(1-\zeta_3)\, t}{-\zeta_3^2 + t}\ ,\ y(t,w) = \frac{3^{\frac{1}{5}}\, (\zeta_3)^{\frac{1}{5}}\, w}{-1+\zeta_3\, t}\ .$$

The pullback of the differentials under the transformation T is

$$T^*(\omega_1) = \frac{\left(-\left(\frac{1}{3}\right)\right)^{\frac{2}{5}} \left(-1+(-1)^{\frac{2}{3}}\right)}{w^2} dt$$

$$T^*(\omega_2) = \frac{(-1)^{\frac{23}{30}} \left(-1+(-1)^{\frac{2}{3}}\right) t}{3^{\frac{1}{10}}\, w^3} dt$$

$$T^*(\omega_3) = \frac{(-1)^{\frac{19}{30}} \left(-1+(-1)^{\frac{2}{3}}\right) t \left(-1+(-1)^{\frac{2}{3}}\, t\right)}{3^{\frac{3}{10}}\, w^4} dt$$

$$T^*(\omega_4) = -\left(\frac{(-1)^{\frac{2}{15}}\, 3^{\frac{1}{5}} \left(-1+(-1)^{\frac{2}{3}}\right) t^2}{w^4}\right) dt\ .$$

So via the transformation T, ω_1, ω_2 and ω_4 are equal to φ_1, φ_2 and φ_4 up to a constant factor, respectively. But ω_3 is a linear combination of φ_3 and φ_4 and it is not an eigendifferential for the action of the CM field $\mathbb{Q}(\zeta_{15})$.

If we consider the Schwarz map

$$D(\omega_j, z) = \frac{\int_{\gamma_1} \omega_j}{\int_{\gamma_2} \omega_j} \quad (j = 1,2,3,4)$$

for the family $\{DegP(z)\}$ with respect to the differentials ω_j, Theorem 4.2 follows directly from Theorem 3.1 and Lemma 2.6. □

4.3. Second example

Now we study the same family of curves at the point $\tau = -1$

$$DegP(-1) \ = \ \Sigma' \ : \ y^5 \ = \ x^2(x^2-1)$$

and show that its Jacobian belongs to those studied in case 3 of Proposition 2.4.

Theorem 4.3.

$$\mathrm{Jac}(\Sigma') = T(5,-1) \quad \textit{is isogenous to} \quad A_1 \oplus A_2$$

with $\mathrm{End}_0(A_i) = \mathbb{Q}(\zeta_5)$. *For all differentials in* (4.3) *the Schwarz maps have algebraic values* $D(\omega_i, -1)$ $(i = 1,2,3,4)$.

Consider

$$HypE \ : \ y^5 \ = \ u(u-1)\ .$$

We have a natural map

$$\Sigma' \to HypE$$

by $x \mapsto u = x^2$. It shows that $\mathrm{Jac}(\Sigma')$ is not simple and $A_1 = \mathrm{Jac}(HypE)$ is a component. The differentials

$$\omega_2 = \frac{xdx}{y^3}\,, \qquad \omega_3 = \frac{xdx}{y^4}$$

are the lifts from those on $HypE$. The action of ζ_5 is given by $\sigma : (x, y) \mapsto (x, \zeta_5^{-1} y)$. So

$$\sigma(\omega_2) = \sigma(\frac{xdx}{y^3}) = \zeta_5^3 \omega_2\,, \qquad \sigma(\omega_3) = \sigma(\frac{xdx}{y^4}) = \zeta_5^4 \omega_3\,.$$

Hence A_1 is an abelian variety of CM type with the field $\mathbb{Q}(\zeta_5)$ and simple CM type $(3, 4)$. As we see later the cofactor A_2 is of CM type $(4, 2)$. By the change of a primitive 5-th root of unity A_1 and A_2 are isogenous. We will see below by a period matrix calculation that we have even an isomorphism.

We consider the special values of the Schwarz maps $D(\omega_2, -1)$ and $D(\omega_3, -1)$. They are reduced to consider the periods

$$\int_1^\infty \frac{du}{y^3}\,, \quad \int_0^1 \frac{du}{y^3}$$

and

$$\int_1^\infty \frac{du}{y^4}\,, \quad \int_0^1 \frac{du}{y^4}$$

on the CM hyperelliptic curve $y^5 = u(u-1)$. The differentials du/y^3 and du/y^4 are eigendifferentials for the action of the corresponding CM field $\mathbb{Q}(\zeta_5)$ on the factor A_1. According to Theorems 3.1, 3.3 and Lemma 2.7.2 the values $D(\omega_2, -1)$ and $D(\omega_3, -1)$ are algebraic. Theorem 3.3 shows the algebraicity of $D(\omega_1, -1)$ as well. Only ω_4 cannot be seen directly to be a differential on A_2.

We have to consider the following question: are the two associate differentials ω_3, ω_4 (see Remark 4.1) just those two differentials of Theorem 3.4 generating the two one-dimensional eigenspaces in V_4 needed according to Lemma 2.7? The answer will be "yes" by explicit calculation of the period matrix of $\Sigma' : y^5 = x^2(x^2-1)$. Set

$$x = \frac{1}{x_1}\,, \qquad y = -\frac{y_1}{x_1}\,.$$

So we get an isomorphic curve $\Sigma_1 : y_1^5 = x_1(x_1^2 - 1)$. We have the expression of the basis $\{\omega_1, \omega_2, \omega_3, \omega_4\}$ on Σ_1 :

$$\omega_1 = -\frac{dx_1}{y_1^2},\ \omega_2 = \frac{dx_1}{y_1^3},\ \omega_3 = -\frac{x_1 dx_1}{y_1^4},\ \omega_4 = -\frac{dx_1}{y_1^4}\,.$$

Let r, r' be arcs on Σ_1 given by the oriented lines $[0,1], [-1,0]$ with real negative and real positive value y_1, respectively. Remember that σ denotes the change of sheets induced by $y_1 \mapsto \zeta_5^{-1} y_1$ and let $r^{(i)}$ be the arc $\sigma^{1-i} r$ $\quad (i = 1, 2, 3, 4)$. Set

$\alpha^{(i)} = r^{(i)} - r^{(i+1)}$ and $\beta^{(i)} = r'^{(i)} - r'^{(i+1)}$. Set

$$M_1 = \begin{pmatrix} 0 & -1 & 0 & 0 & 0 & 1 & 0 & 1 \\ 0 & -1 & 0 & 0 & 0 & 1 & 0 & 0 \\ 0 & -1 & 0 & 0 & 0 & 0 & 0 & 0 \\ 0 & -1 & 0 & 1 & 0 & 0 & 0 & 0 \\ 1 & -1 & 0 & 1 & 0 & 1 & 0 & 0 \\ 0 & 0 & 0 & 0 & 1 & 0 & 0 & -1 \\ 0 & -1 & 1 & 1 & 0 & 0 & 0 & 0 \\ 0 & 0 & 0 & 1 & 1 & 0 & 1 & -1 \end{pmatrix}.$$

Then

$$(A_2, A_3, A_5, A_6, B_2, B_3, B_5, B_6) = (\alpha^{(1)}, \dots, \alpha^{(4)}, \beta^{(1)}, \dots, \beta^{(4)}) M_1$$

is a homology basis of Σ_1 with the intersection matrix

$$\begin{pmatrix} 0 & 0 & 0 & 0 & 1 & 0 & 0 & 0 \\ 0 & 0 & 0 & 0 & 0 & 1 & 0 & 0 \\ 0 & 0 & 0 & 0 & 0 & 0 & 1 & 0 \\ 0 & 0 & 0 & 0 & 0 & 0 & 0 & 1 \\ -1 & 0 & 0 & 0 & 0 & 0 & 0 & 0 \\ 0 & -1 & 0 & 0 & 0 & 0 & 0 & 0 \\ 0 & 0 & -1 & 0 & 0 & 0 & 0 & 0 \\ 0 & 0 & 0 & -1 & 0 & 0 & 0 & 0 \end{pmatrix}.$$

These cycles $A_2, A_3, A_5, A_6, B_2, B_3, B_5, B_6$ are the same ones as those given by K. Koike in [12] on the general pentagonal curve going to the limit

$$\lim_{\lambda \to -\infty} y_1^5 = x_1(x_1 - 1)(x_1 + 1)(x_1 + \lambda) .$$

Put

$$p_i = \int_{\alpha^{(1)}} \omega_i \, , \quad q_i = \int_{\beta^{(1)}} \omega_i \quad (i = 1, 2, 3, 4) \, ,$$

then we have

$$q_1 = -p_1 \, , \; q_2 = p_2 \, , \; q_3 = p_3 \, , \; q_4 = -p_4 \, .$$

Setting $\omega_i' = \omega_i / p_1 \quad (i = 1, 2, 3, 4)$, we have the period matrix of ω_i' for the cycles $(\alpha^{(1)}, \dots, \alpha^{(4)}, \beta^{(1)}, \dots, \beta^{(4)})$.

The period matrix of $\Sigma_1 : y_1^5 = x_1(x_1^2 - 1)$ with respect to the basis $\{\omega_1' \dots, \omega_4'\}$ of $H^0(\Sigma_1, \Omega)$ and the basis

$$\{\alpha^{(1)}, \dots, \alpha^{(4)}, \beta^{(1)}, \dots, \beta^{(4)}\}$$

of $H^1(\Sigma_1, \mathbb{Z})$ is given by

$$\begin{pmatrix} 1 & \zeta_5^3 & \zeta_5^1 & \zeta_5^4 & -1 & -\zeta_5^3 & -\zeta_5^1 & -\zeta_5^4 \\ 1 & \zeta_5^2 & \zeta_5^4 & \zeta_5^1 & 1 & \zeta_5^2 & \zeta_5^4 & \zeta_5^1 \\ 1 & \zeta_5 & \zeta_5^2 & \zeta_5^3 & 1 & \zeta_5 & \zeta_5^2 & \zeta_5^3 \\ 1 & \zeta_5 & \zeta_5^2 & \zeta_5^3 & -1 & -\zeta_5 & -\zeta_5^2 & -\zeta_5^3 \end{pmatrix}.$$

By changing the $\mathbb{Q}$-homology basis to

$$\{\alpha^{(1)} + \beta^{(1)}, \dots, \alpha^{(4)} + \beta^{(4)}, \alpha^{(1)} - \beta^{(1)}, \dots, \alpha^{(4)} + \beta^{(4)}\}$$

we know that $\mathrm{Jac}(\Sigma_1) = T(5,-1)$ is isogenous to the direct sum

$$\mathbb{C}^2 / \left(\mathbb{Z}\begin{pmatrix}1\\1\end{pmatrix} + \mathbb{Z}\begin{pmatrix}\zeta_5^2\\ \zeta_5\end{pmatrix} + \mathbb{Z}\begin{pmatrix}\zeta_5^4\\ \zeta_5^2\end{pmatrix} + \mathbb{Z}\begin{pmatrix}\zeta_5\\ \zeta_5^3\end{pmatrix} \right)$$
$$+ \quad \mathbb{C}^2 / \left(\mathbb{Z}\begin{pmatrix}1\\1\end{pmatrix} + \mathbb{Z}\begin{pmatrix}\zeta_5^3\\ \zeta_5\end{pmatrix} + \mathbb{Z}\begin{pmatrix}\zeta_5\\ \zeta_5^2\end{pmatrix} + \mathbb{Z}\begin{pmatrix}\zeta_5^4\\ \zeta_5^3\end{pmatrix} \right).$$

That means $\mathrm{Jac}(\Sigma')$ is $\mathbb{Q}$-isogenous to a direct sum of two 2-dimensional abelian varieties of CM type with the CM field $\mathbb{Q}(\zeta_5)$ of type $(3,4)$ and of type $(4,2)$, and these types are the same under the isomorphism $\zeta_5 \mapsto \zeta_5^3$.

5. Examples of algebraic values: symmetry and degeneration

The results in Theorems 4.2 and 4.3 concerning $D(\omega_1;\tau)$ and $D(\omega_2;\tau)$ in the points $\tau = -\zeta_3, -1$ are not at all surprising since they are easily proved with Theorem 3.3 provided we know that $T(5,\tau)$ is of CM type. Even if we know that fact, the results of the preceding section concerning ω_3 and ω_4 needed much more effort since they do not generate a one-dimensional eigenspace of holomorphic differentials W_n (hypothesis of Theorem 3.3). For them it is quite remarkable that ω_4 was a K-eigendifferential for Lemma 2.6 in the case $\tau = -\zeta_3$ or that even both belonged to the two factors in Lemma 2.7.2 in the case $\tau = -1$. In this section, we will shed some further light on these phenomena, extend parts of the previous results and explain why both τ in question are exceptional arguments in the sense of Theorems 3.4 and 3.5.

Theorem 5.1. 1. *Suppose* $\rho = \nu_0 = \nu_1 = \nu_\infty \in \mathbb{Q} - \frac{1}{3}\mathbb{Z}$. *Then* $D(\rho,\rho,\rho;-\zeta_3)$ *is algebraic or* ∞.
2. *Suppose* ν_1 *and* $\rho = \nu_0 = \nu_\infty \in \mathbb{Q} - \mathbb{Z}$ *satisfy the non-integrality condition* (1.1). *Then* $D(\nu_0,\rho,\rho;-1)$ *is algebraic or* ∞.

5.1. Symmetry arguments

To prove the first statement, observe that in this case $\rho \in \mathbb{Q} - \frac{1}{3}\mathbb{Z}$ is a restatement of the non-integrality condition. The corresponding hypergeometric differential equation is invariant under the fractional linear transformation

$$z \mapsto \frac{z-1}{z}$$

inducing a cyclic permutation of the singularities $0, 1, \infty$; fixed points are ζ_6 and $-\zeta_3$. The image of the lower half plane has therefore a symmetry of order 3, hence its vertices $D(0), D(1), D(\infty)$ under the Schwarz map D have a *midpoint* $D(-\zeta_3)$, i.e., a fixed point of an order 3 $\mathrm{PSL}_2(\mathbb{C})$-transformation μ providing an automorphism of the D-image and a cyclic permutation of the vertices. (To see that it is really a fractional linear transformation, observe that μ extends to either a disc or the Riemann sphere if one considers all analytic continuations of D.) These vertices are algebraic or ∞ what can be seen either by direct calculation as in [21, (15)] or by the fact that in these points the Prym varieties $T(k,z)$ degenerate to abelian varieties of dimension $\frac{1}{2}\varphi(k)$ with complex multiplication by $\mathbb{Q}(\zeta_k)$. Therefore the midpoint also has to be algebraic or ∞. The same argument works also for ζ_6, but the analytic continuation of D to the upper half plane changes at least one of the three vertices.

The second statement can be proved similarly but with the anticonformal transformation

$$z \mapsto \overline{z^{-1}}$$

exchanging 0 and ∞ and fixing the unit circle, in particular the point -1 which can be considered as the midpoint of the border edge $D(]\infty, 0[)$. The triangle function D maps the unit circle to a symmetry axis of the D-images of upper and lower half plane and again the algebraicity of the vertices implies $D(-1) \in \overline{\mathbb{Q}} \cup \{\infty\}$.

Another version of these symmetry arguments has been indicated for the special case $\rho = \frac{1}{5}$ already in the end of Subsection 1.4: by nonlinear relations, D is related to other triangle functions for the parameter triplets $\frac{1}{2}, \frac{1}{3}, \frac{\rho}{2}$ (first part) or $\frac{1}{2}, \frac{\nu_0}{2}, \rho$ (second part). In both cases, the D-values in question belong to the (algebraic!) vertices of the new image triangles.

5.2. Corollaries and Remarks

1. By the "only if" part of Theorem 3.1, Theorem 5.1 implies that the respective Pryms $T(k, -\zeta_3), T(k, -1)$ are of CM type — but without the precise information given in the last section, of course.

2. The same kind of symmetry arguments as in the second part of Theorem 5.1 works for the argument $\tau = \frac{1}{2}$ if $\nu_0 = \nu_1$ and for $\tau = 2$ if $\nu_1 = \nu_\infty$.

3. Sign changes of the angular parameters change the triangle functions at most by fractional linear transformations defined over $\overline{\mathbb{Q}}$, see [21, (16)–(18)], so Theorem 5.1 covers the algebraicity results of the last section.

4. For Theorem 5.1 it does not matter whether the underlying differential is of first kind or only of second kind. For example, the parameter triplets

$$(\tfrac{2}{5}, \tfrac{3}{5}, \tfrac{2}{5}) \quad , \quad (\tfrac{2}{5}, -\tfrac{2}{5}, \tfrac{2}{5})$$

belong to generators of the eigenspace V_1 for the curve family (4.2), both of second kind but the algebraicity of their values are covered by Theorem 5.1.

5. As already explained in the end of subsection 3.1, we expect only finitely many $T(k,\tau)$ to be of CM type if the corresponding monodromy group is non-arithmetic. Apparently $\tau = -1, \frac{1}{2}, 2, \zeta_6, -\zeta_3$ lead to these cases if suitable symmetry conditions are satisfied, independently of the arithmeticity of the monodromy group.

5.3. Degeneration of contiguity

As we explained in Subsection 3.2, associate differentials $\eta(\nu; z)$ generate one-dimensional subspaces of V_n which are generically pairwise different. Therefore —

if the monodromy group is infinite — at most two of them give algebraic values $D(\nu;\tau)$, see Theorems 3.4 and 3.5, if the argument τ is not an "exceptional" one where several $\eta(\nu;\tau)$ are multiples of each other. Theorem 5.1 gives examples for such exceptional arguments because arbitrarily many associate angular parameters lead to algebraic values.

Theorem 5.2. 1. *Suppose* $\rho \in \mathbb{Q} - \frac{1}{3}\mathbb{Z}$ *and let* P *be a set of associate angular parameter triplets*

$$(\rho + 2k, \rho + 2k, \rho + 2k) , \quad k \in \mathbb{Z} .$$

Then $\tau = \zeta_6$ *and* $-\zeta_3$ *are exceptional arguments.*

2. *Suppose* ν_1 *and* $\rho = \nu_0 = \nu_\infty \in \mathbb{Q} - \mathbb{Z}$ *satisfy the non-integrality condition (1.1) and let* P *be a set of associate parameter triplets*

$$(\rho + k, \nu_1, \rho + k) , \quad k \in \mathbb{Z} .$$

Then $\tau = -1$ *is an exceptional argument.*

The truth of Theorem 5.2 follows from Theorems 3.4 and 5.1. As we explained in the proof of Theorem 3.4, the statement implies in particular that the corresponding differentials in these exceptional points are multiples of each other. This is in turn equivalent to a degeneration of contiguity relations: generically, any two different associate differentials generate their two-dimensional eigenspace because any other can be written as a $\mathbb{C}(z)$-linear combination of them. But for a fixed argument $z = \tau$ this may fail if the coefficient functions have poles. These relations can be produced explicitely using Gauss' relations between contiguous hypergeometric functions. We illustrate this degeneration phenomenon in the second case. From the contiguity relations in [8, (28)–(45)] one may deduce with Mathematica^{TR} the relation

$$\begin{aligned} & (1+c)\,(c + z(1-a+b))\, F(a, b+1, c+1; z) \\ = \;& c(1+c)F(a,b,c;z) + z(1+b)(1-a+c)F(a, b+2, c+2; z) . \end{aligned}$$

In the case $\nu_0 = \nu_\infty = \rho \notin \mathbb{Z}$ we have $1 - a + b = c \neq -1$, and the left-hand side coefficient vanishes precisely for $z = -1$. Passing to the angular parameters and to the differentials, it means that $\eta(\rho, \nu_1, \rho; -1)$ and $\eta(\rho - 2, \nu_1, \rho - 2; -1)$ are multiples of each other. By induction, we see that for $\tau = -1$ in this family of associate differentials all elements with even k are multiples of each other, and similarly all elements with k odd.

An analogous argument for the first case of Theorem 5.2 should be possible, but would need explicit relations between associate hypergeometric functions

$$F(a,b,c;z) , \; F(a+k, b+3k, c+2k; z) \quad \text{and} \quad F(a+2k, b+6k, c+4k; z) ,$$

an extremely difficult task. In several easier cases we expect to be able to perform the calculation with computer support.

5.4. Gamma values

Finally another access to Theorem 5.1 has to be mentioned. In the symmetric situations discussed here, special values of hypergeometric functions in the fixed points of the symmetries are known, see, e.g., [8, (46)–(56)]. One may use them — often together with Kummer's relations between different representations of hypergeometric functions — to produce explicit formulas describing $D(-1)$ or $D(\zeta_6)$ in terms of products of values of the Gamma function at rational arguments. For the normalization of D used in the present paper compare also [15, Thm. 5.3 and p. 649]. Take, e.g., ω_3 in (4.3); we know already by different reasons (Theorems 4.3 and 5.1) that the corresponding value of the triangle function in -1 is algebraic. Up to algebraic nonzero factors (indicated by "$\sim$") we can write it as

$$D(-\frac{2}{5}, -\frac{3}{5}, -\frac{2}{5}; -1) \sim \frac{\Gamma(\frac{4}{5})\Gamma(\frac{1}{10})}{\Gamma(\frac{1}{5})\Gamma(\frac{3}{10})}$$

and verify that this quotient is algebraic. This verification can be done either explicitely using functional equation, parity relation and Gauss–Legendre's distribution relations (see, e.g., [22, p. 6]; Serge Lang conjectures moreover that all algebraic relations between Gamma values at rational arguments follow from these classical relations), but this requires patience and luck. However there is an easy criterion due to Koblitz and Ogus ([11] or [22, Prop. 1]) to decide whether the algebraicity of such a product follows from classical Gamma relations. This criterion applies here and leads as well to the result; we leave it as an exercise for the reader.

References

[1] N. Archinard, *Hypergeometric Abelian Varieties,* Canad. J. Math. **55** (5) (2003), 897–932.

[2] D. Bertrand, *Endomorphismes de groupes algébriques; applications arithmétiques,* pp. 1–45 in *Approximations Diophantiennes et Nombres Transcendants*, Progr. Math. 31, Birkhäuser, 1983.

[3] Cl. Chevalley, A. Weil, *Über das Verhalten der Integrale 1. Gattung bei Automorphismen des Funktionenkörpers*, Abh. Hamburger Math. Sem. **10** (1934), 358–361.

[4] P. Cohen, J. Wolfart, *Modular embeddings for some non–arithmetic Fuchsian groups,* Acta Arithmetica **56** (1990), 93–110.

[5] P.B. Cohen, J. Wolfart, *Algebraic Appell–Lauricella Functions*, Analysis **12** (1992), 359–376.

[6] P.B. Cohen, G. Wüstholz, *Application of the André–Oort Conjecture to some Questions in Transcendence,* pp. 89–106 in *A Panorama in Number Theory or The View from Baker's Garden,* ed.: G. Wüstholz, Cambridge Univ. Press, 2002.

[7] B. Edixhoven, A. Yafeev, *Subvarieties of Shimura varieties,* Ann. Math. **157** (2003), 621–645.

[8] A. Erdélyi, W. Magnus, F. Oberhettinger, F.G. Tricomi, *Higher Transcendental Functions, Bateman Manuscript Project,* Vol. 1., McGraw–Hill, 1953.

[9] Ph. Griffiths, J. Harris, *Principles of Algebraic Geometry*, Wiley, 1978.

[10] F. Klein, *Vorlesungen über die hypergeometrische Funktion,* Springer, 1933.

[11] N. Koblitz, A. Ogus, *Algebraicity of some products of values of the* Γ *function*, Proc. Symp. Pure Math. **33** (1979), 343–346.

[12] K. Koike, *On the family of pentagonal curves of genus* 6 *and associated modular forms on the ball,* J. Math. Soc. Japan, **55** (2003), 165–196.

[13] S. Lang, *Complex Multiplication,* Springer, 1983.

[14] H.A. Schwarz, *Über diejenigen Fälle, in welchen die Gaußische hypergeometrische Reihe eine algebraische Funktion ihres vierten Elements darstellt*, J. Reine Angew. Math. **75** (1873), 292–335.

[15] H. Shiga, T. Tsutsui, J. Wolfart, *Fuchsian differential equations with apparent singularities,* Osaka J. Math. **41** (2004), 625–658.

[16] H. Shiga, J. Wolfart, *Criteria for complex multiplication and transcendence properties for automorphic functions,* J. reine angew. Math. **463** (1995), 1–25.

[17] G. Shimura, *On analytic families of polarized abelian varieties and automorphic functions,* Ann. Math. **78** (1963), 149–192.

[18] C.L. Siegel, *Lectures on Riemann Matrices*, Tata Inst., Bombay, 1963.

[19] K. Takeuchi, *Arithmetic triangle groups,* J. Math. Soc. Japan **29** (1977), 91–106.

[20] M. Waldschmidt, *Transcendance de périodes: état de connaissances,* to appear in the proceedings of a conference in Mahdia 2003.

[21] J. Wolfart, *Werte hypergeometrischer Funktionen,* Invent. math. **92** (1988), 187–216.

[22] J. Wolfart, G. Wüstholz, *Der Überlagerungsradius algebraischer Kurven und die Werte der Betafunktion an rationalen Stellen*, Math. Ann. **273** (1985), 1–15.

[23] G. Wüstholz, *Algebraic Groups, Hodge Theory, and Transcendence*, pp. 476–483 in *Proc. of the ICM Berkeley 1986* (ed.: A.M. Gleason), AMS, 1987.

[24] G. Wüstholz, *Algebraische Punkte auf analytischen Untergruppen algebraischer Gruppen,* Ann. of Math. **129** (1989), 501–517.

Hironori Shiga
Inst. of Math. and Physics
Chiba University
Yayoi–cho 1–33, Inage–ku
Chiba 263–8522
Japan
e-mail: shiga@math.s.chiba-u.ac.jp

Jürgen Wolfart
Math. Sem. der Univ.
Postfach 111932
D–60054 Frankfurt a.M.
Germany
e-mail: wolfart@math.uni-frankfurt.de

Progress in Mathematics, Vol. 260, 313–371

GKZ Hypergeometric Structures

Jan Stienstra

Abstract. This text is based on lectures by the author in the Summer School *Algebraic Geometry and Hypergeometric Functions* in Istanbul in June 2005. It gives a review of some of the basic aspects of the theory of hypergeometric structures of Gelfand, Kapranov and Zelevinsky, including Differential Equations, Integrals and Series, with emphasis on the latter. The Secondary Fan is constructed and subsequently used to describe the 'geography' of the domains of convergence of the Γ-series. A solution to certain Resonance Problems is presented and applied in the context of Mirror Symmetry. Many examples and some exercises are given throughout the paper.

Mathematics Subject Classification (2000). Primary 33C70, 14M25; Secondary 14N35.

Keywords. GKZ hypergeometric, Γ-series, secondary fan, resonant, mirror symmetry.

1. Introduction

GKZ stands for *Gelfand, Kapranov* and *Zelevinsky*, who discovered fascinating generalizations of the classical hypergeometric structures of Euler, Gauss, Appell, Lauricella, Horn [10, 12, 14]. The main ingredient for these new hypergeometric structures is a finite subset $\mathcal{A} \subset \mathbb{Z}^{k+1}$ which generates $\mathbb{Z}^{k+1}$ as an abelian group and for which there exists a group homomorphism $h : \mathbb{Z}^{k+1} \to \mathbb{Z}$ such that $h(\mathcal{A}) = \{1\}$. The latter condition means that $\mathcal{A}$ lies in a k-dimensional affine hyperplane in $\mathbb{Z}^{k+1}$. Figure 1 shows $\mathcal{A}$ (the black dots) sitting in this hyperplane for some classical hypergeometric structures. In [12, 14] these new structures were called *$\mathcal{A}$-hypergeometric systems.* Nowadays many authors call them *GKZ hypergeometric systems.* The original name indeed seems somewhat unfortunate, since *$\mathcal{A}$-hypergeometric* sounds negative, like 'αγεομετριτος μη 'εισιτω (a non-geometer should not enter), written over the entrance of Plato's academy and in the logo of the American Mathematical Society. Besides the set $\mathcal{A}$ the construction of GKZ hypergeometric structures requires a vector $\mathbf{c} \in \mathbb{C}^{k+1}$.

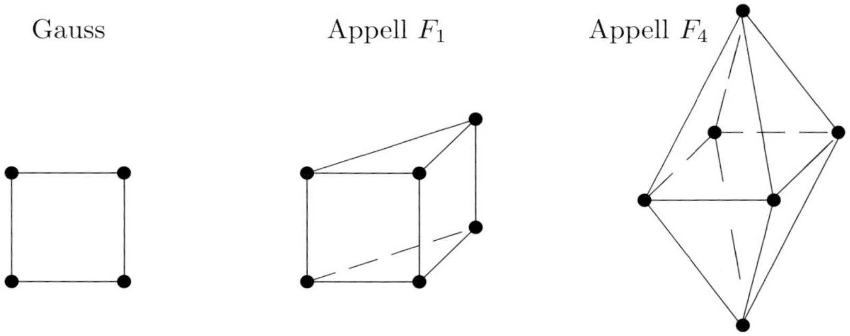

FIGURE 1.

In these notes we report on the basic theory of GKZ hypergeometric structures and show how the traditional aspects *differential equations, integrals, series* are attached to the data $\mathcal{A}$, $\mathbf{c}$. In Section 2 we introduce the GKZ differential equations and give examples of GKZ hypergeometric integrals. In Section 3 we discuss GKZ hypergeometric series (so-called Γ-series). We have put details of the GKZ theory for Lauricella's F_D together in Section 7, so that the reader can compare results and view-points on F_D for various lectures in this School (e.g., [19]).

The beautiful insight of Gelfand, Kapranov and Zelevinsky was that hypergeometric structures greatly simplify if one introduces extra variables and balances this with an appropriate torus action. More precisely the variables in GKZ theory are the natural coordinates on the space $\mathbb{C}^{\mathcal{A}} := \mathrm{Maps}(\mathcal{A}, \mathbb{C})$ of maps from $\mathcal{A}$ to $\mathbb{C}$. The torus $\mathbb{T}^{k+1} := \mathrm{Hom}(\mathbb{Z}^{k+1}, \mathbb{C}^*)$ of group homomorphisms from $\mathbb{Z}^{k+1}$ to $\mathbb{C}^*$, acts naturally on $\mathbb{C}^{\mathcal{A}}$ and on functions on $\mathbb{C}^{\mathcal{A}}$: *for* $\sigma \in \mathbb{T}^{k+1}$ *and* $\Phi : \mathbb{C}^{\mathcal{A}} \to \mathbb{C}$

$$(\sigma \cdot \mathbf{u})(\mathbf{a}) = \sigma(\mathbf{a})\mathbf{u}(\mathbf{a}), \qquad (\Phi \cdot \sigma)(\mathbf{u}) = \Phi(\sigma \cdot \mathbf{u}), \qquad \forall \mathbf{a} \in \mathcal{A},\ \forall \mathbf{u} \in \mathbb{C}^{\mathcal{A}}. \quad (1)$$

The GKZ hypergeometric functions associated with $\mathcal{A}$ and $\mathbf{c}$ are defined on open domains in $\mathbb{C}^{\mathcal{A}}$, but they are not invariant under the action of $\mathbb{T}^{k+1}$, unless $\mathbf{c} = 0$. Rather, for $\mathbf{c} \in \mathbb{Z}^{k+1}$ they transform according to the character of $\mathbb{T}^{k+1}$ given by $\mathbf{c}$. For $\mathbf{c} \notin \mathbb{Z}^{k+1}$ there is only an infinitesimal analogue of this transformation behavior, encoded in one part of the GKZ system of differential equations (see (20)). On the other hand, the quotient of any two GKZ hypergeometric functions with a common domain of definition associated with $\mathcal{A}$ and $\mathbf{c}$ is always $\mathbb{T}^{k+1}$-invariant (see (21)).

The important role of the $\mathbb{T}^{k+1}$-action in GKZ hypergeometric structures motivates a study of the orbit space. Without going into details, this can be described as follows. First take the complement of the coordinate hyperplanes $(\mathbb{C}^*)^{\mathcal{A}} := \mathrm{Maps}(\mathcal{A}, \mathbb{C}^*) = \{\mathbf{u} \in \mathbb{C}^{\mathcal{A}} \mid \mathbf{u}(\mathbf{a}) \neq 0,\ \forall \mathbf{a} \in \mathcal{A}\}$. The above action of $\mathbb{T}^{k+1}$ preserves this set. In fact $(\mathbb{C}^*)^{\mathcal{A}}$ is a complex torus and $\mathbb{T}^{k+1}$ can be identified with a subtorus, acting by left multiplication. The quotient is the torus

$$(\mathbb{C}^*)^{\mathcal{A}} / \mathbb{T}^{k+1} = \operatorname{Hom}(\mathbb{L}, \mathbb{C}^*), \tag{2}$$

where $\mathbb{L}$ is the lattice (= free abelian group) of linear relations in $\mathcal{A}$. It is often convenient to fix a numbering for the elements of $\mathcal{A}$, i.e., $\mathcal{A} = \{\mathbf{a}_1, \ldots, \mathbf{a}_N\}$. Then $\mathbb{L}$ can be described as

$$\mathbb{L} := \{(\ell_1, \ldots, \ell_N) \in \mathbb{Z}^N \mid \ell_1\mathbf{a}_1 + \ldots + \ell_N\mathbf{a}_N = 0\}. \tag{3}$$

The rank of $\mathbb{L}$ and the dimension of the torus in (2) are $d := N - k - 1$. In order to obtain the natural space on which the GKZ hypergeometric structures live one must compactify the complex torus in (2). For this purpose Gelfand, Kapranov and Zelevinsky developed the theory of the *Secondary Fan*. This is a complete fan of rational polyhedral cones in the real vector space $\mathbb{L}_{\mathbb{R}}^{\vee} := \operatorname{Hom}(\mathbb{L}, \mathbb{R})$. Sections 4 and 5 give full details about the Secondary Fan and the associated toric variety $\mathcal{V}_{\mathcal{A}}$. Since the Secondary Fan has interesting applications outside the theory of hypergeometric systems Sections 4 and 5 are written so that they can be read independently of other sections. The toric variety $\mathcal{V}_{\mathcal{A}}$ provides a very clear picture of the 'geography' for the domains of convergence of the various GKZ hypergeometric series, since these match exactly with discs about the special points of $\mathcal{V}_{\mathcal{A}}$ coming from the maximal cones in the secondary fan (see Proposition 7). For the examples in Figure 1 the toric varieties and special points associated with the maximal cones in the secondary fan are: for Gauss the projective line $\mathbb{P}^1$ with points $[1,0]$, $[0,1]$, for F_4 the projective plane $\mathbb{P}^2$ with points $[1,0,0]$, $[0,1,0]$, $[0,0,1]$, for F_1 the projective plane blown up in the three points $[1,0,0]$, $[0,1,0]$, $[0,0,1]$ equiped with the six points of intersection of the exceptional divisors and the proper transforms of the coordinate axes in $\mathbb{P}^2$.

For most $\mathcal{A}$ the dimension of local solution spaces for the GKZ differential equations equals the volume of the k-dimensional polytope $\Delta_{\mathcal{A}} :=$ convex hull of $\mathcal{A}$ (see Section 2.7); here the volume is normalized as $k!\times$ the Euclidean volume. Thus for the examples in Figure 1 the local solution spaces have dimension 2, 3, 4, respectively. For generic $\mathcal{A}$ and $\mathbf{c}$ the Γ-series provide bases of local solutions for the GKZ differential equations. However, in some exceptional, but very important, cases there are not enough Γ-series, due to a phenomenon called *resonance*. In Section 6 we discuss resonance and demonstrate how one *sometimes* can obtain enough solutions by considering infinitesimal deformations of Γ-series. *Sometimes* here means under the severe restrictions that $\mathbf{c} = 0$ and that one works in the neighborhood of a point on $\mathcal{V}_{\mathcal{A}}$ which corresponds to a unimodular triangulation of the polytope $\Delta_{\mathcal{A}}$. Very recently Borisov and Horja [5] found a way to obtain enough solutions for any $\mathbf{c} \in \mathbb{Z}^{k+1}$ and any triangulation. Their method is close in spirit to the method in Section 6 and [5] gives an up-to-date presentation of this aspect of GKZ hypergeometric structures.

In the 1980s, while Gelfand, Kapranov and Zelevinsky were working on new hypergeometric structures, physicists discovered fascinating new structures in string theory: the so-called *string dualities*. One of these string dualities, known as *Mirror Symmetry*, soon attracted the attention of mathematicians, because it claimed very striking consequences for enumerative geometry. Especially the paper [7] of Candelas, de la Ossa, Green and Parkes with a detailed study of the quintic in $\mathbb{P}^4$ played a pivotal role. Batyrev [1] pointed out that many examples of the Mirror Symmetry phenomenon dealt with pairs of families of Calabi–Yau hypersurfaces in toric varieties coming from dual polytopes. In [3] Batyrev and Borisov extended this kind of Mirror Symmetry to Calabi–Yau complete intersections in toric varieties. Batyrev ([2], Thm. 14.2) also noticed that the solutions to the differential equations which appeared in Mirror Symmetry, were solutions to GKZ hypergeometric systems constructed from the same data as the toric varieties. The converse is, however, not true: the GKZ system can have solutions which are not solutions to the system of differential equations in Mirror Symmetry. This means that the latter system contains extra differential equations in addition to those of the underlying GKZ system (see [16] §3.3). On the other hand, the solutions to the differential equations which one encounters in Mirror Symmetry, can all be obtained by a few differentiations from solutions to extremely resonant GKZ hypergeometric systems with $\mathbf{c} = \mathsf{0}$. Thus we do not need those extended GKZ systems. In Section 8 we discuss some examples of this intriguing application of GKZ hypergeometric structures to String Theory.

The quotient of two solutions of a GKZ system of differential equations associated with $\mathcal{A}$ and $\mathbf{c}$ is $\mathbb{T}^{k+1}$-invariant. So one can define (at least locally) a *Schwarz map* from the toric variety $\mathcal{V}_{\mathcal{A}}$ to the projectivization of the vector space of (local) solutions. For Gauss's system, and more generally for Lauricella's F_D, the toric variety $\mathcal{V}_{\mathcal{A}}$ and the projectivized solution space have the same dimension, equal to the rank of $\mathbb{L}$. The Schwarz map for Gauss's system and Lauricella's F_D is discussed extensively in other lectures in this school, e.g., [19]. Quite in contrast with F_D is the situation for GKZ systems associated with families of Calabi–Yau threefolds. For these the toric variety $\mathcal{V}_{\mathcal{A}}$ has dimension equal to $\operatorname{rank}\mathbb{L}$, but the projectivized local solution space has dimension $1+2\operatorname{rank}\mathbb{L}$. The discussion about the *canonical coordinates* and the *pre-potential* in Section 8.5 can be seen as a description of the image of the (local) Schwarz map. This is closely related to what in the (physics) literature is called *Special Kähler Geometry.*

All this basically concerns only local aspects of GKZ systems of differential equations. About singularities, global solutions or global monodromy of the system not much seems to be known, except for classically studied systems like Gauss's and Lauricella's F_D.

Since these notes are intended as an introduction to GKZ hypergeometric structures, we have included throughout the text many examples and a few exercises. On the other hand we had to omit many topics. One of these omissions concerns $\mathcal{A}$-*discriminants.* These come up when one identifies $\mathbb{C}^{\mathcal{A}}$ with the space of Laurent polynomials in $k+1$ variables with exponents in $\mathcal{A}$,

$$\mathbf{u} \in \mathbb{C}^{\mathcal{A}} \qquad \leftrightarrow \qquad \sum_{\mathbf{a}\in\mathcal{A}} \mathbf{u}_{\mathbf{a}}\mathbf{x}^{\mathbf{a}},$$

and then wonders about the Laurent polynomials with singularities, i.e., for which there is a point at which all partial derivatives vanish. For the $\mathcal{A}$-discriminant and its relation to the secondary fan we refer to [13]. Another omission concerns *symplectic geometry* in connection with the secondary fan. We recommend Guillemin's book [15] for further reading on this topic.

Acknowledgments. This text is an expanded version of the notes for my lectures in the summer school on *Arithmetic and Geometry around Hypergeometric Functions* at Galatasaray University in Istanbul, june 2005. I want to thank the organizers, in particular Prof. Uludag, for the hospitality and for the opportunity to lecture in this summer school. I am also much indebted to the authors of the many papers on hypergeometric structures, triangulations and mirror symmetry from which I myself learnt this subject.

Contents

2. GKZ systems via examples

2.1. Roots of polynomial equations

It is clear that in general the zeros of a polynomial

$$P_{\mathbf{u}}(x) := u_0 + u_1 x + u_2 x^2 + \ldots + u_n x^n \tag{4}$$

are functions of the coefficients $\mathbf{u} = (u_0, \ldots, u_n)$. One wonders: *What kind of functions?* For instance, it has been known since ancient times that the zeros of a quadratic polynomial $ax^2 + bx + c$ are $\frac{1}{2a}(-b \pm \sqrt{b^2 - 4ac})$. Similar formulas exist for polynomials of degrees 3 and 4, but, according to Galois theory, the zeros of a general polynomial of degree ≥ 5 can not be obtained from the polynomial's coefficients by a finite number of algebraic operations. Changing the point of view K. Mayr proved that the roots of polynomials are solutions of certain systems of differential equations:

Theorem 1. (Mayr [20]) *If all roots of the equation $P_{\mathbf{u}}(\xi) = 0$ are simple, then a root ξ satisfies the differential equations: for $i_1 + \ldots + i_r = j_1 + \ldots + j_r$:*

$$\frac{\partial^r \xi}{\partial u_{i_1} \ldots \partial u_{i_r}} = \frac{\partial^r \xi}{\partial u_{j_1} \ldots \partial u_{j_r}}.$$

Proof. By differentiating the equation $P_{\mathbf{u}}(\xi) = 0$ with respect to u_i we find $P'_{\mathbf{u}}(\xi)\frac{\partial \xi}{\partial u_i} + \xi^i = 0$. This implies $\frac{\partial \xi}{\partial u_i} = \xi^i \frac{\partial \xi}{\partial u_0} = \frac{1}{1+i}\frac{\partial \xi^{1+i}}{\partial u_0}$. Induction now gives

$$\frac{\partial^r \xi}{\partial u_{i_1} \ldots \partial u_{i_r}} = \frac{1}{1 + i_1 + \ldots + i_r} \frac{\partial^r \xi^{1+i_1+\ldots+i_r}}{\partial u_0^r}. \qquad \square$$

It obviously suffices to use only those Mayr's differential equations for which $\{i_1,\ldots,i_r\}\cap\{j_1,\ldots,j_r\}=\emptyset$. These can also be written as

$$\prod_{\ell_i<0}\left(\frac{\partial}{\partial u_i}\right)^{-\ell_i}\xi = \prod_{\ell_i>0}\left(\frac{\partial}{\partial u_i}\right)^{\ell_i}\xi \qquad \text{if} \qquad \sum_{i=0}^{n}\ell_i\begin{bmatrix}1\\ i\end{bmatrix}=\begin{bmatrix}0\\ 0\end{bmatrix}. \tag{5}$$

A second system of differential equations, satisfied by the roots of polynomials, follows from the easily checked fact that for all $s\in\mathbb{C}^*$:

$$\xi(su_0,su_1,\ldots,su_n) = \xi(u_0,\ldots,u_n)\,, \quad \xi(u_0,su_1,\ldots,s^nu_n) = s^{-1}\xi(u_0,\ldots,u_n)\,.$$

When we differentiate this with respect to s and set $s=1$, we find:

$$\begin{aligned} u_0\frac{\partial\xi}{\partial u_0}+u_1\frac{\partial\xi}{\partial u_1}+u_2\frac{\partial\xi}{\partial u_2}+\ldots+u_n\frac{\partial\xi}{\partial u_n} &= 0\,,\\ 0\,u_0\frac{\partial\xi}{\partial u_0}+1\,u_1\frac{\partial\xi}{\partial u_1}+2\,u_2\frac{\partial\xi}{\partial u_2}+\ldots+n\,u_n\frac{\partial\xi}{\partial u_n} &= -\xi\,,\end{aligned}$$

This can be written more transparently as:

$$\xi(tu_0,tsu_1,ts^2u_2\ldots,ts^nu_n) = s^{-1}\xi(u_0,\ldots,u_n) \quad \text{for} \quad (t,s)\in(\mathbb{C}^*)^2\,, \tag{6}$$

$$\sum_{i=0}^{n}\begin{bmatrix}1\\ i\end{bmatrix}u_i\frac{\partial\xi}{\partial u_i} = \begin{bmatrix}0\\ -1\end{bmatrix}\xi\,. \tag{7}$$

For more on zeros of 1-variable polynomials and hypergeometric functions see [21].

2.2. Integral with polynomial integrand

Consider the integral

$$I_\sigma^{(m)} = I_\sigma^{(m)}(u_0,\ldots,u_n) := \int_\sigma P_{\mathbf{u}}(x)^m\,\frac{dx}{x}$$

with $m\in\mathbb{Z}$, $P_{\mathbf{u}}(x)$ as in (4) and σ a circle in $\mathbb{C}$, with radius >0, centred at 0, independent of $u_0,\ldots,u_n$, not passing through any zero of $P_{\mathbf{u}}(x)$.

By differentiating under the integral sign we see

$$\frac{\partial I_\sigma^{(m)}}{\partial u_i} = m\int_\sigma x^iP_{\mathbf{u}}(x)^{m-1}\,\frac{dx}{x}$$

and hence, if $i_1+\ldots+i_r=j_1+\ldots+j_r$, then

$$\frac{\partial^r I_\sigma^{(m)}}{\partial u_{i_1}\ldots\partial u_{i_r}} = \frac{\partial^r I_\sigma^{(m)}}{\partial u_{j_1}\ldots\partial u_{j_r}}\,.$$

As before this can also be written as

$$\prod_{\ell_i<0}\left(\frac{\partial}{\partial u_i}\right)^{-\ell_i}I_\sigma^{(m)} = \prod_{\ell_i>0}\left(\frac{\partial}{\partial u_i}\right)^{\ell_i}I_\sigma^{(m)} \quad \text{if} \quad \sum_{i=0}^{n}\ell_i\begin{bmatrix}1\\ i\end{bmatrix}=\begin{bmatrix}0\\ 0\end{bmatrix}. \tag{8}$$

For $s \in \mathbb{C}^*$ close to 1 one checks: $I_\sigma^{(m)}(su_0, \ldots, su_n) = s^m I_\sigma^{(m)}(u_0, \ldots, u_n)$ and

$$I_\sigma^{(m)}(u_0, su_1, \ldots, s^n u_n) = \int_\sigma P_{\mathbf{u}}(sx)^m \frac{dx}{x} = \int_{s\sigma} P_{\mathbf{u}}(x)^m \frac{dx}{x} = I_\sigma^{(m)}(u_0, \ldots, u_n).$$

More transparently: for $(t, s) \in (\mathbb{C}^*)^2$ sufficiently close to $(1, 1)$

$$I_\sigma^{(m)}(tu_0, tsu_1, ts^2u_2 \ldots, ts^n u_n) = t^m I_\sigma^{(m)}(u_0, \ldots, u_n). \tag{9}$$

By differentating (9) with respect to t and s and setting $t = s = 1$ we find, similar to (7),

$$\sum_{i=0}^{n} \begin{bmatrix} 1 \\ i \end{bmatrix} u_i \frac{\partial I_\sigma^{(m)}}{\partial u_i} = \begin{bmatrix} m \\ 0 \end{bmatrix} I_\sigma^{(m)}. \tag{10}$$

Note the fundamental role of the set $\mathcal{A} = \left\{ \begin{bmatrix} 1 \\ i \end{bmatrix} \mid i = 0, 1, \ldots, n \right\}$ in (5)–(10). Notice also the torus action (1) on the left-hand sides of (6) and (9).

2.3. Integral with k-variable Laurent polynomial integrand

Let us take a Laurent polynomial in k variables

$$P_{\mathbf{u}}(x_1, x_2, \ldots, x_k) := \sum_{\mathsf{a} \in \mathsf{A}} u_{\mathsf{a}}\, x_1^{a_1} x_2^{a_2} \cdots x_k^{a_k} \tag{11}$$

where $\mathsf{a} = (a_1, a_2, \ldots, a_k)$ and $\mathsf{A} = \{\mathsf{a}_1, \ldots, \mathsf{a}_N\}$ is a finite subset of $\mathbb{Z}^k$. Consider the integral

$$I_\sigma^{(m)}(\mathbf{u}) := \int_\sigma P_{\mathbf{u}}(x_1, \ldots, x_k)^m \frac{dx_1}{x_1} \cdots \frac{dx_k}{x_k} \tag{12}$$

with $\mathbf{u} = (u_{\mathsf{a}})_{\mathsf{a} \in \mathsf{A}}$, $m \in \mathbb{Z}$ and with $\sigma = \sigma_1 \times \ldots \times \sigma_k$ a product of k circles $\sigma_1, \ldots, \sigma_k$ in $\mathbb{C}$, centred at 0, independent of $\mathbf{u}$, so that $P_{\mathbf{u}}(x_1, \ldots, x_k) \neq 0$ for all $(x_1, \ldots, x_k) \in \sigma_1 \times \ldots \times \sigma_k$.

By differentiating under the integral sign we see, for $\mathsf{a} = (a_1, \ldots, a_k)$,

$$\frac{\partial I_\sigma^{(m)}(\mathbf{u})}{\partial u_{\mathsf{a}}} = m \int_\sigma x_1^{a_1} \cdots x_k^{a_k} P_{\mathbf{u}}(x_1, \ldots, x_k)^{m-1} \frac{dx_1}{x_1} \cdots \frac{dx_k}{x_k}.$$

From this one derives that for every vector $(\ell_1, \ldots, \ell_N) \in \mathbb{Z}^N$ which satisfies

$$\ell_1 + \ldots + \ell_N = 0, \qquad \ell_1 \mathsf{a}_1 + \ldots + \ell_N \mathsf{a}_N = 0, \tag{13}$$

the following differential equation holds:

$$\prod_{\ell_i < 0} \left(\frac{\partial}{\partial u_i} \right)^{-\ell_i} I_\sigma^{(m)}(\mathbf{u}) = \prod_{\ell_i > 0} \left(\frac{\partial}{\partial u_i} \right)^{\ell_i} I_\sigma^{(m)}(\mathbf{u}); \tag{14}$$

for simplicity of notation we write here and henceforth u_i instead of u_{a_i}.

For $s \in \mathbb{C}^*$ sufficiently close to 1 and for $i = 1, \ldots, k$ one calculates:

$$\begin{aligned} I_\sigma^{(m)}(s^{a_{i1}} u_1, s^{a_{i2}} u_2, \ldots, s^{a_{iN}} u_N) &= \int_\sigma P_{\mathbf{u}}(x_1, \ldots, sx_i, \ldots, x_k)^m\, \omega \\ = \int_{\sigma_1 \times \ldots \times s\sigma_i \times \ldots \times \sigma_k} P_{\mathbf{u}}(x_1, \ldots, x_k)^m\, \omega &= \int_{\sigma_1 \times \ldots \times \sigma_i \times \ldots \times \sigma_k} P_{\mathbf{u}}(x_1, \ldots, x_k)^m\, \omega = I_\sigma^{(m)}(u_1, \ldots, u_N); \end{aligned}$$

here a_{ij} denotes the i-th coordinate of the vector a_j and $\omega = \frac{dx_1}{x_1} \cdots \frac{dx_k}{x_k}$. This together with $I_\sigma^{(m)}(su_1, \ldots, su_N) = s^m I_\sigma^{(m)}(u_1, \ldots, u_N)$ can also be written as:

$$I_\sigma^{(m)}(ts_1^{a_{11}} s_2^{a_{21}} \cdots s_k^{a_{k1}} u_1, \ldots, ts_1^{a_{1N}} s_2^{a_{2N}} \cdots s_k^{a_{kN}} u_N) = t^m I_\sigma^{(m)}(u_0, \ldots, u_n) \quad (15)$$

for $(t, s_1, \ldots, s_k) \in (\mathbb{C}^*)^{k+1}$ close to $(1, \ldots, 1)$. By differentiating with respect to $t, s_1, \ldots, s_k$ and setting $t = s_1 = \ldots = s_k = 1$ we find

$$\begin{bmatrix} 1 \\ \mathsf{a}_1 \end{bmatrix} u_1 \frac{\partial I_\sigma^{(m)}(\mathbf{u})}{\partial u_1} + \ldots + \begin{bmatrix} 1 \\ \mathsf{a}_N \end{bmatrix} u_N \frac{\partial I_\sigma^{(m)}(\mathbf{u})}{\partial u_N} = \begin{bmatrix} m \\ 0 \end{bmatrix} I_\sigma^{(m)}(\mathbf{u}) \,. \quad (16)$$

Note the appearance of the set $\mathcal{A} = \left\{ \begin{bmatrix} 1 \\ \mathsf{a} \end{bmatrix} \in \mathbb{Z}^{k+1} \mid \mathsf{a} \in \mathsf{A} \right\}$ in (13) and (16). Notice also the torus action (1) on the left hand side of (15).

Remark. For $m > 0$ one can evaluate $I_\sigma^{(m)}(\mathbf{u})$ using the multinomial and residue theorems. One finds that $I_\sigma^{(m)}(\mathbf{u})$ is actually a polynomial:

$$\frac{1}{(2\pi \mathsf{i})^k} I_\sigma^{(m)}(\mathbf{u}) = \sum_{(m_1, \ldots, m_N)} \frac{m!}{(m_1)! \cdots (m_N)!} u_1^{m_1} \cdots u_N^{m_N} \quad (17)$$

where the sum runs over all N-tuples of non-negative integers $(m_1, \ldots, m_N)$ satisfying $m_1 + \ldots + m_N = m$ and $m_1 \mathsf{a}_1 + \ldots + m_N \mathsf{a}_N = 0$.

In Section 8 one can find explicit examples of these integrals with $m = -1$.

2.4. Generalized Euler integrals

In [12, 14] Gelfand, Kapranov and Zelevinsky investigate integrals of the form

$$\int_\sigma \prod_i P_i(x_1, \ldots, x_k)^{\alpha_i} \, x_1^{\beta_1} \cdots x_k^{\beta_k} \, dx_1 \cdots dx_k \,, \quad (18)$$

which they call *generalized Euler integrals*. Here the P_i are Laurent polynomials, α_i and β_j are complex numbers and σ is a k-cycle. Since the integrand can be multivalued and can have singularities one must carefully give the precise meaning of Formula (18) (see [12] §2.2). Having dealt with the technicalities of the precise definition Gelfand, Kapranov and Zelevinsky view the integrals (18) as functions of the coefficients of the Laurent polynomials P_i. Using the same arguments as we used in Section 2.3 they then verify that these functions satisfy a system of differential equations (19)–(20) for the appropriate data $\mathcal{A}$ and $\mathbf{c}$. Examples can be found in Sections 7.2 and 8.3.

2.5. General GKZ systems of differential equations

The systems of differential equations (5)–(7), (8)–(10) and (14)–(16) found in the preceding examples are special cases of systems of differential equations discovered by Gelfand, Kapranov and Zelevinsky [10, 12, 14]. The general GKZ system for functions Φ of N variables $u_1, \ldots, u_N$ is constructed from a vector $\mathbf{c} \in \mathbb{C}^{k+1}$ and an N-element subset $\mathcal{A} = \{\mathbf{a}_1, \ldots, \mathbf{a}_N\} \subset \mathbb{Z}^{k+1}$ which generates $\mathbb{Z}^{k+1}$ as an abelian group and for which there exists a group homomorphism $h : \mathbb{Z}^{k+1} \to \mathbb{Z}$ such that $h(\mathbf{a}) = 1$ for all $\mathbf{a} \in \mathcal{A}$. Let $\mathbb{L} \subset \mathbb{Z}^N$ denote the lattice of relations in $\mathcal{A}$:

$$\mathbb{L} := \{(\ell_1, \dots, \ell_N) \in \mathbb{Z}^N \mid \ell_1 \mathbf{a}_1 + \dots + \ell_N \mathbf{a}_N = \mathsf{0}\}.$$

Note that the condition $h(\mathbf{a}) = 1$ for all $\mathbf{a} \in \mathcal{A}$, implies that $\ell_1 + \dots + \ell_N = 0$ for every $(\ell_1, \dots, \ell_N) \in \mathbb{L}$.

Definition 1. *The GKZ system associated with $\mathcal{A}$ and $\mathbf{c}$ consists of*

- *for every $(\ell_1, \dots, \ell_N) \in \mathbb{L}$ one differential equation*

$$\prod_{\ell_i<0} \left(\frac{\partial}{\partial u_i}\right)^{-\ell_i} \Phi = \prod_{\ell_i>0} \left(\frac{\partial}{\partial u_i}\right)^{\ell_i} \Phi, \tag{19}$$

- *the system of $k+1$ differential equations*

$$\mathbf{a}_1 u_1 \frac{\partial \Phi}{\partial u_1} + \dots + \mathbf{a}_N u_N \frac{\partial \Phi}{\partial u_N} = \mathbf{c}\Phi. \tag{20}$$

Remark. It is natural to view $u_1, \dots, u_N$ as coordinates on the space $\mathbb{C}^{\mathcal{A}} :=$ Maps$(\mathcal{A}, \mathbb{C})$. Then the left-hand side of the equation (20) is the infinitesimal version of the torus action (1). If Φ_1 and Φ_2 are two solutions of (20) on some open set $U \subset \mathbb{C}^{\mathcal{A}}$, their quotient satisfies

$$\mathbf{a}_1 u_1 \frac{\partial}{\partial u_1} \left(\frac{\Phi_1}{\Phi_2}\right) + \dots + \mathbf{a}_N u_N \frac{\partial}{\partial u_N} \left(\frac{\Phi_1}{\Phi_2}\right) = 0 \tag{21}$$

and is therefore constant on the intersections of U with the $\mathbb{T}^{k+1}$-orbits.
Thus a basis $\Phi_1, \dots, \Phi_r$ of the solution space of (19)–(20) induces map from the orbit space $\mathbb{T}^{k+1} \cdot U / \mathbb{T}^{k+1}$ into the projective space $\mathbb{P}^{r-1}$, like the Schwarz map *for Gauss's hypergeometric systems.*

Another simple, but nevertheless quite useful, consequence of the GKZ differential equations is:

Proposition 1. *If function Φ satisfies the differential equations* (19)–(20) *for $\mathcal{A}$ and $\mathbf{c}$, then $\frac{\partial \Phi}{\partial u_j}$ satisfies the differential equations* (19)–(20) *for $\mathcal{A}$ and $\mathbf{c} - \mathbf{a}_j$.*

Proof. The derivation $\frac{\partial}{\partial u_j}$ commutes with all derivations involved in (19). On the other hand, by applying $\frac{\partial}{\partial u_j}$ to both sides of (20) we get

$$\mathbf{a}_1 u_1 \frac{\partial}{\partial u_1} \left(\frac{\partial \Phi}{\partial u_j}\right) + \dots + \mathbf{a}_N u_N \frac{\partial}{\partial u_N} \left(\frac{\partial \Phi}{\partial u_j}\right) + \mathbf{a}_j \frac{\partial \Phi}{\partial u_j} = \mathbf{c} \frac{\partial \Phi}{\partial u_j}. \qquad \square$$

2.6. Gauss's hypergeometric differential equation as a GKZ system

The most classical hypergeometric differential equation, due to Euler and Gauss, is:

$$z(z-1)F'' + ((a+b+1)z - c)F' + abF = 0. \tag{22}$$

Here F is a function of one variable z, $' = \frac{d}{dz}$ and a, b, c are additional complex parameters. It is reproduced in the GKZ formalism by $\mathbf{c} = (1-c, -a, -b)$ and

$$\mathcal{A} = \left\{ \begin{bmatrix} 1 \\ 1 \\ 1 \end{bmatrix}, \begin{bmatrix} -1 \\ 0 \\ 0 \end{bmatrix}, \begin{bmatrix} 0 \\ 1 \\ 0 \end{bmatrix}, \begin{bmatrix} 0 \\ 0 \\ 1 \end{bmatrix} \right\} \subset \mathbb{Z}^3$$

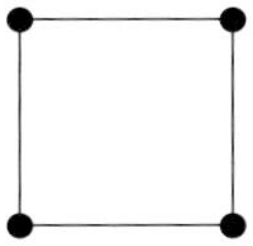

and, hence, $\mathbb{L} = \mathbb{Z}(1,1,-1,-1) \subset \mathbb{Z}^4$. Indeed, for these data the GKZ system boils down to the following four differential equations for a function Φ of four variables (u_1, u_2, u_3, u_4):

$$\begin{aligned} \frac{\partial^2 \Phi}{\partial u_1 \partial u_2} &= \frac{\partial^2 \Phi}{\partial u_3 \partial u_4} \\ u_1 \frac{\partial \Phi}{\partial u_1} - u_2 \frac{\partial \Phi}{\partial u_2} &= (1-c)\,\Phi \\ u_1 \frac{\partial \Phi}{\partial u_1} + u_3 \frac{\partial \Phi}{\partial u_3} &= -a\,\Phi \\ u_1 \frac{\partial \Phi}{\partial u_1} + u_4 \frac{\partial \Phi}{\partial u_4} &= -b\,\Phi. \end{aligned}$$

From the second equation we get

$$\frac{\partial^2 \Phi}{\partial u_1 \partial u_2} = u_2^{-1} \left(u_1 \frac{\partial^2 \Phi}{\partial u_1^2} + c \frac{\partial \Phi}{\partial u_1} \right).$$

From the third and fourth equations we get

$$\frac{\partial^2 \Phi}{\partial u_3 \partial u_4} = u_3^{-1} u_4^{-1} \left(-u_1 \frac{\partial}{\partial u_1} - a \right) \left(-u_1 \frac{\partial}{\partial u_1} - b \right) \Phi.$$

Together with the first equation this yields

$$u_3^{-1} u_4^{-1} \left(u_1^2 \frac{\partial^2 \Phi}{\partial u_1^2} + (1+a+b) u_1 \frac{\partial \Phi}{\partial u_1} + ab\Phi \right) = u_2^{-1} \left(u_1 \frac{\partial^2 \Phi}{\partial u_1^2} + c \frac{\partial \Phi}{\partial u_1} \right).$$

Setting $u_2 = u_3 = u_4 = 1$, $u_1 = z$ and $F(z) = \Phi(z,1,1,1)$ we find that F satisfies the differential equation (22).

2.7. Dimension of the solution space of a GKZ system

The spaces of (local) solutions of the GKZ differential equations (19)–(20) are complex vector spaces. Theorems 2 and 5 in [10] state that the dimension of the space of (local) solutions of (19)–(20) near a generic point is equal to the normalized volume of the k-dimensional polytope $\Delta_{\mathcal{A}} :=$ convex hull$(\mathcal{A})$; here 'normalized volume' means $k!$ times the usual Euclidean volume. In [11] it is pointed out that the proof in [10] requires an additional condition on $\mathcal{A}$. Corollary 8.9 and Proposition 13.15 in [25] show that this additional condition is satisfied if the polytope $\Delta_{\mathcal{A}}$ admits a unimodular triangulation. Triangulations of $\Delta_{\mathcal{A}}$ and their importance in GKZ hypergeometric structures are discussed in Section 4.2.2.

3. Γ-series

As before we consider a subset $\mathcal{A} = \{\mathbf{a}_1, \ldots, \mathbf{a}_N\} \subset \mathbb{Z}^{k+1}$ which generates $\mathbb{Z}^{k+1}$ as an abelian group and for which there exists a group homomorphism $h : \mathbb{Z}^{k+1} \to \mathbb{Z}$ such that $h(\mathbf{a}) = 1$ for all $\mathbf{a} \in \mathcal{A}$. And, still as before, we write:

$$\mathbb{L} := \{(\ell_1, \ldots, \ell_N) \in \mathbb{Z}^N \mid \ell_1\mathbf{a}_1 + \ldots + \ell_N\mathbf{a}_N = 0\}.$$

The condition $h(\mathbf{a}) = 1$ for all $\mathbf{a} \in \mathcal{A}$, implies that $\ell_1 + \ldots + \ell_N = 0$ for every $(\ell_1, \ldots, \ell_N) \in \mathbb{L}$. With $\mathbb{L}$ and a vector $\underline{\gamma} = (\gamma_1, \ldots, \gamma_N) \in \mathbb{C}^N$ Gelfand, Kapranov and Zelevinsky [10] associate what they call a Γ-*series*:

Definition 2. *The* Γ*-series associated with* $\mathbb{L}$ *and* $\underline{\gamma} = (\gamma_1, \ldots, \gamma_N) \in \mathbb{C}^N$ *is*

$$\Phi_{\mathbb{L},\underline{\gamma}}(u_1, \ldots, u_N) = \sum_{(\ell_1,\ldots,\ell_N)\in\mathbb{L}} \prod_{j=1}^{N} \frac{u_j^{\gamma_j+\ell_j}}{\Gamma(\gamma_j + \ell_j + 1)}. \tag{23}$$

Here Γ is the Γ-function; its definition and main properties are recalled in Section 3.1. In Section 3.2 we demonstrate how the classical hypergeometric series of Gauss, Appell and Lauricella appear in the Γ-series format. In Section 3.3 we give estimates for the growth of the coefficients in (23). Formula (23) requires for $\underline{\gamma} \notin \mathbb{Z}^{k+1}$ choices of logarithms for $u_1, \ldots, u_N$. By carefully manoeuvreing conditions on $\underline{\gamma}$ and substitutions setting some u_j equal to 1, we can avoid problems and show in Section 3.6 how a Γ-series can be viewed as a power series in $d = N - k - 1$ variables with positive radii of convergence. Nevertheless, a formula avoiding choices of logarithms is desirable. For that reason we introduce Fourier Γ-series in Section 3.5. In Section 3.6 we prove that $\Phi_{\mathbb{L},\underline{\gamma}}(u_1, \ldots, u_N)$ can be viewed as a function on some domain in $(u_1, \ldots, u_N)$-space and satisfies the GKZ differential equations.

3.1. The Γ-function

The Γ-function is defined for complex numbers s with $\Re s > 0$ by the integral

$$\Gamma(s) := \int_0^\infty t^{s-1}e^{-t}dt. \tag{24}$$

Using partial integration one immediately checks $\Gamma(s+1) = s\Gamma(s)$ and, hence, for $n \in \mathbb{Z}$, $n > 0$

$$\Gamma(s+n) = s(s+1)\ldots(s+n-1)\Gamma(s). \tag{25}$$

Formulas (24) and (25) imply in particular

$$\Gamma(1) = 1, \qquad \Gamma(n+1) = n! \qquad \text{for } n \in \mathbb{N}. \tag{26}$$

One can extend the Γ-function to a meromorphic function on all of $\mathbb{C}$ by setting

$$\Gamma(s) = \frac{\Gamma(s+n)}{s(s+1)\ldots(s+n-1)} \qquad \text{with} \quad n \in \mathbb{Z},\, n > -\Re s. \tag{27}$$

The functional equation (25) shows that this does not depend on the choice of n. Formula (24) shows $\Gamma(s) \neq 0$ if $\Re s > 1$ and hence Formula (27) shows that the extended Γ-function is holomorphic on $\mathbb{C} \setminus \mathbb{Z}_{\leq 0}$ and has at $s = -m \in \mathbb{Z}_{\leq 0}$ a first order pole with residue

$$\mathrm{Res}_{s=-m}\Gamma(s) \;=\; \frac{(-1)^m}{m!}\,. \tag{28}$$

The function $\frac{1}{\Gamma(s)}$ is holomorphic on the whole complex plane. Its zero set is $\mathbb{Z}_{\leq 0}$ and its Taylor series at $-m \in \mathbb{Z}_{\leq 0}$ starts like

$$\frac{1}{\Gamma(s-m)} \;=\; (-1)^m m!\, s \;+\; \ldots \tag{29}$$

The coefficients of (classical) hypergeometric series are usually expressed in terms of *Pochhammer symbols* $(s)_n$. These are defined by $(s)_n \;=\; s(s+1)\cdots(s+n-1)$ and can be rewritten as quotients of Γ-values:

$$(s)_n \;=\; s(s+1)\cdots(s+n-1) \;=\; \frac{\Gamma(s+n)}{\Gamma(s)} \;=\; (-1)^n \frac{\Gamma(1-s)}{\Gamma(1-n-s)}\,. \tag{30}$$

Note, however, that for integer values of s the Pochhammer symbol $(s)_n$ is perfectly well defined, while some of the individual Γ-values in (30) may become ∞.

3.2. Examples of Γ-series

3.2.1. Gauss's hypergeometric series. As in the example of Gauss's hypergeometric differential equation (Section 2.6) we take $\mathbb{L} \;=\; \mathbb{Z}(1, 1, -1, -1)$ in $\mathbb{Z}^4$ and $\underline{\gamma} = (0, c-1, -a, -b) \in \mathbb{C}^4$. If c is not an integer ≤ 0, then, by (23) and (30),

$$\begin{aligned}
\Phi_{\mathbb{L},\underline{\gamma}}(u_1, u_2, u_3, u_4) \;&=\; \sum_{n\in\mathbb{Z}} \frac{u_1^n u_2^{c-1+n} u_3^{-a-n} u_4^{-b-n}}{\Gamma(1+n)\Gamma(c+n)\Gamma(1-n-a)\Gamma(1-n-b)} \\
&=\; \frac{u_2^{c-1} u_3^{-a} u_4^{-b}}{\Gamma(c)\Gamma(1-a)\Gamma(1-b)} \sum_{n\geq 0} \frac{(a)_n (b)_n}{n!(c)_n} (u_1 u_2 u_3^{-1} u_4^{-1})^n
\end{aligned}$$

and, hence, $\Phi_{\mathbb{L},\underline{\gamma}}(z, 1, 1, 1) = \frac{1}{\Gamma(c)\Gamma(1-a)\Gamma(1-b)} F(a, b, c|z)$ with

$$F(a, b, c|z) := \sum_{n\geq 0} \frac{(a)_n (b)_n}{n!(c)_n} z^n\,.$$

The power series $F(a, b, c|z)$ is Gauss's hypergeometric series. Note that if a or b is a positive integer, the Γ-series is 0, but Gauss's hypergeometric series is not 0.

3.2.2. The hypergeometric series ${}_pF_{p-1}$. Quite old generalizations of Gauss's hypergeometric series are the series

$${}_pF_{p-1}\left(\begin{array}{ccc} a_1 & ,\ldots, & a_p \\ c_1 & ,\ldots, & c_{p-1} \end{array} \middle|\, z \right) \;:=\; \sum_{n\geq 0} \frac{(a_1)_n \cdots (a_p)_n}{n!(c_1)_n \cdots (c_{p-1})_n} z^n\,.$$

Like for Gauss's series one easily finds that the series ${}_pF_{p-1}$ match (up to a constant factor) the Γ-series for $\mathbb{L} = \mathbb{Z}(1, \ldots, 1, -1, \ldots, -1)$ with p 1's and p (-1)'s.

3.2.3. The case ${}_1F_0$. The simplest, yet not totally trivial, case of a Γ-series arises for $\mathbb{L} = \mathbb{Z}(1,-1) \subset \mathbb{Z}^2$. The Γ-series with $\underline{\gamma} = (0,a)$, $a \in \mathbb{C}$ is

$$\Phi_{\mathbb{Z}(1,-1),(0,a)}(u_1,u_2) = \sum_{n\in\mathbb{Z}} \frac{u_1^n u_2^{a-n}}{\Gamma(1+n)\Gamma(1+a-n)} = \frac{1}{\Gamma(1+a)}(u_1+u_2)^a\,;$$

here we use the generalized binomial theorem and (30):

$$\binom{a}{n} = \frac{a(a-1)\dots(a-n+1)}{n!} = \frac{\Gamma(1+a)}{\Gamma(1+n)\Gamma(1+a-n)}\,.$$

Remark. Note that $\mathbb{L} = \mathbb{Z}(1,-1)$ implies that the two elements of $\mathcal{A}$ are equal. The GKZ differential equations in this case imply that the hypergeometric functions are in fact just functions of the single variable $u_1 + u_2$. This illustrates a general fact: when setting up the theory of GKZ hypergeometric systems one could take for $\mathcal{A}$ a list of vectors in $\mathbb{Z}^{k+1}$ instead of just a subset (i.e., the elements may occur more than once). But any such apparently more general set up, arises from a case with $\mathcal{A}$ a genuine set by simply replacing a variable by a sum of new variables. So by allowing for $\mathcal{A}$ a list instead of a set one does not get a seriously more general theory. Therefore we ignore this option in these notes.

3.2.4. The Appell–Lauricella hypergeometric series. These are generalizations of Gauss's series to n variables defined by Appell for $n = 2$ and Lauricella for general n. With the notations $\mathbf{z}^{\mathbf{m}} := z_1^{m_1}\cdots z_n^{m_n}$, $(\mathbf{x})_{\mathbf{m}} := (x_1)_{m_1}\cdots(x_n)_{m_n}$, $\mathbf{m}! := m_1!\cdots m_n!$, $|\mathbf{m}| := m_1+\ldots+m_n$ for n-tuples of complex numbers $\mathbf{z} = (z_1,\ldots,z_n)$, $\mathbf{x} = (x_1,\ldots,x_n)$ and of non-negative integers $\mathbf{m} = (m_1,\ldots,m_n)$, the four Lauricella series are

$$\begin{aligned}
F_A(a,\mathbf{b},\mathbf{c}|\mathbf{z}) &:= \sum_{\mathbf{m}} \frac{(a)_{|\mathbf{m}|}(\mathbf{b})_{\mathbf{m}}}{(\mathbf{c})_{\mathbf{m}}\mathbf{m}!}\,\mathbf{z}^{\mathbf{m}}\\
F_B(\mathbf{a},\mathbf{b},c|\mathbf{z}) &:= \sum_{\mathbf{m}} \frac{(\mathbf{a})_{\mathbf{m}}(\mathbf{b})_{\mathbf{m}}}{(c)_{|\mathbf{m}|}\mathbf{m}!}\,\mathbf{z}^{\mathbf{m}}\\
F_C(a,b,\mathbf{c}|\mathbf{z}) &:= \sum_{\mathbf{m}} \frac{(a)_{|\mathbf{m}|}(b)_{|\mathbf{m}|}}{(\mathbf{c})_{\mathbf{m}}\mathbf{m}!}\,\mathbf{z}^{\mathbf{m}}\\
F_D(a,\mathbf{b},c|\mathbf{z}) &:= \sum_{\mathbf{m}} \frac{(a)_{|\mathbf{m}|}(\mathbf{b})_{\mathbf{m}}}{(c)_{|\mathbf{m}|}\mathbf{m}!}\,\mathbf{z}^{\mathbf{m}};
\end{aligned}$$

in the summations $\mathbf{m}$ runs over $\mathbb{Z}^n_{\geq 0}$ and the c-parameters are not integers ≤ 0. In Appell's notation (for $n = 2$) these series are called F_2, F_3, F_4, F_1 respectively.

One can use (30) to explore the relations between the Lauricella series and Γ-series. For the coefficients in F_D, for instance, we find

$$\frac{(a)_{|\mathbf{m}|}(\mathbf{b})_{\mathbf{m}}}{(c)_{|\mathbf{m}|}\mathbf{m}!} = \Gamma(1-a)\Gamma(c)\prod_{j=1}^{n}\Gamma(1-b_j)\cdot\prod_{j=1}^{N}\frac{1}{\Gamma(1+\gamma_j+\ell_j)}$$

with $N = 2n+2$, $\quad \underline{\gamma} = (\gamma_1, \ldots, \gamma_N) = (c-1, -b_1, \ldots, -b_n, -a, 0, \ldots, 0)$,

$$(\ell_1, \ldots, \ell_N) = (m_1, \ldots, m_n) \begin{pmatrix} 1 & -1 & 0 & \ldots & 0 & -1 & 1 & 0 & \ldots & 0 \\ 1 & 0 & -1 & \ddots & \vdots & -1 & 0 & 1 & \ddots & \vdots \\ \vdots & \vdots & \ddots & \ddots & 0 & \vdots & \vdots & \ddots & \ddots & 0 \\ 1 & 0 & \ldots & 0 & -1 & -1 & 0 & \ldots & 0 & 1 \end{pmatrix}.$$

So for $\mathbb{L}$ we take the lattice which is spanned by the rows of the above $n \times N$-matrix. Substituting $u_j = 1$ for $1 \leq j \leq n+2$ and $u_j = z_{j-n-2}$ for $n+3 \leq j \leq 2n+2$ turns the Γ-series into a power series:

$$\Phi_{\mathbb{L},\underline{\gamma}}(1, \ldots, 1, z_1, \ldots, z_n) = \left(\prod_{j=1}^{n+2} \Gamma(1+\gamma_j)^{-1} \right) F_D(a, \mathbf{b}, c|\mathbf{z}).$$

Exercise. Note that the matrix describing $\mathbb{L}$ for Lauricella's F_D is $(\mathbf{1}_n, -\mathbb{I}_n, -\mathbf{1}_n, \mathbb{I}_n)$ where $\mathbf{1}_n$ is the column vector with n components 1 and $\mathbb{I}_n$ is the $n \times n$-identity matrix. Now find the lattice $\mathbb{L}$ for the Lauricella functions F_A, F_B and F_C.

3.3. Growth of coefficients of Γ-series

Here is first a simple lemma about the growth behavior of the Γ-function.

Lemma 1. *For every $C \in \mathbb{C} \setminus \mathbb{Z}$ there are real constants P, $R, \kappa_1, \kappa_2 > 0$ (depending on C) such that for all $M \in \mathbb{Z}_{\geq 0}$:*

$$|\Gamma(C+M)| \geq \kappa_1 R^M M^M \qquad \textit{and} \qquad |\Gamma(C-M)| \geq \kappa_2 P^{-M} M^{-M}. \tag{31}$$

Proof. From (25) one derives

$$\begin{aligned} |\Gamma(C+M)| &= \prod_{j=1}^{M} |C-1+j| \cdot |\Gamma(C)| \geq Q^M M! |\Gamma(C)| \geq \kappa R^M M^M |\Gamma(C)|, \\ |\Gamma(C-M)| &\geq \prod_{j=1}^{M} (|C|+j)^{-1} \cdot |\Gamma(C)| \geq \frac{|\Gamma(C)|}{(|C|+M)^M} \geq P^{-M} M^{-M} |\Gamma(C)| \end{aligned}$$

with $Q := \min_{k \in \mathbb{N}} \frac{|C-1+k|}{k}$, $R := Qe^{-1}$, $P := 2(1+|C|)$ and some constant κ (from Stirling's formula). □

Now consider the coefficient $\prod_{j=1}^{N} \Gamma(\gamma_j + \ell_j + 1)^{-1}$ in the Γ-series (23). Set $\gamma_j' = \gamma_j$ if $\gamma_j \notin \mathbb{Z}$ and $\gamma_j' = \gamma_j - \frac{1}{2}$ if $\gamma_j \in \mathbb{Z}$. Note that $\Gamma(k-\frac{1}{2}) = (k-\frac{3}{2}) \cdots \frac{1}{2}\Gamma(\frac{1}{2}) \leq (k-1)!\Gamma(\frac{1}{2}) = \Gamma(k)\sqrt{\pi}$ for $k \in \mathbb{Z}_{\geq 1}$. Then, using the above lemma, one sees that there are real constants $K, S > 0$ such that

$$\left| \prod_{j=1}^{N} \frac{1}{\Gamma(\gamma_j + \ell_j + 1)} \right| \leq \left| \prod_{j=1}^{N} \frac{\sqrt{\pi}}{\Gamma(\gamma_j' + \ell_j + 1)} \right| \leq K S^D \prod_{j=1}^{N} |\ell_j|^{-\ell_j}$$

with $D := \frac{1}{2} \sum_{j=1}^{N} |\ell_j| = \sum_{\ell_j > 0} \ell_j = -\sum_{\ell_j < 0} \ell_j$. Since $\prod_{\ell_j < 0} |\ell_j|^{-\ell_j} \leq D^D$ and $\prod_{\ell_j > 0} |\ell_j|^{-\ell_j} \leq N^D D^{-D}$, our final estimate becomes:

Proposition 2. *There are real numbers $K, T > 0$, depending on $\underline{\gamma} = (\gamma_1, \dots, \gamma_N)$, but independent of $\underline{\ell} = (\ell_1, \dots, \ell_N)$, such that*

$$\left| \prod_{j=1}^{N} \frac{1}{\Gamma(\gamma_j + \ell_j + 1)} \right| \leq K\, T^{\sum_{j=1}^{N} |\ell_j|}. \tag{32}$$

□

3.4. Γ-series and power series

Let $J \subset \{1, \dots, N\}$ be a set with $k+1$ elements, such that the vectors $\mathbf{a}_j$ with $j \in J$ are linearly independent. Write $J' := \{1, \dots, N\} \setminus J$. Let $\underline{\gamma} = (\gamma_1, \dots, \gamma_N) \in \mathbb{C}^N$ be such that $\gamma_j \in \mathbb{Z}$ for $j \in J'$. Since $\frac{1}{\Gamma(s)} = 0$ if $s \in \mathbb{Z}_{\leq 0}$, the Γ-series (23) constructed with such a $\underline{\gamma}$ involves only terms from the set

$$\mathbb{L}_{J,\underline{\gamma}} := \{(\ell_1, \dots, \ell_N) \in \mathbb{L} \mid \gamma_j + \ell_j \geq 0 \quad \text{if} \quad j \in J'\}. \tag{33}$$

The substitution

$$u_j = z_j \quad \text{if} \quad j \in J', \qquad u_j = 1 \quad \text{if} \quad j \in J \tag{34}$$

therefore turns the Γ-series into the power series

$$\sum_{(\ell_1, \dots, \ell_N) \in \mathbb{L}_{J,\underline{\gamma}}} \left(\prod_{j=1}^{N} \frac{1}{\Gamma(\gamma_j + \ell_j + 1)} \right) \prod_{j \in J'} z_j^{\gamma_j + \ell_j}. \tag{35}$$

The following lemma is needed to convert (32) into estimates for the radii of convergence of this power series.

Lemma 2. *Let $J \subset \{1, \dots, N\}$ be a set with $k+1$ elements, such that the vectors $\mathbf{a}_j$ with $j \in J$ are linearly independent. Write $J' := \{1, \dots, N\} \setminus J$. Then there is a positive real constant β such that for every $(\ell_1, \dots, \ell_N) \in \mathbb{L}$*

$$|\ell_1| + \dots + |\ell_N| \leq \beta \sum_{j \in J'} |\ell_j|. \tag{36}$$

Proof. Take any $d \times N$-matrix B whose rows form a $\mathbb{Z}$-basis of $\mathbb{L}$. Let $\mathbf{b}_1, \dots, \mathbf{b}_N$ be its columns. Let $\mathsf{B}_{J'}$ denote the $d \times d$-matrix with columns $\mathbf{b}_j$ $(j \in J')$. Then the matrix $\mathsf{B}_{J'}$ is invertible over $\mathbb{Q}$; indeed, if it were not, its rows would be linearly dependent and there would be a vector $(\ell_1, \dots, \ell_N) \in \mathbb{L}$ such that $\ell_j = 0$ for $j \in J'$; the relation $\ell_1 \mathbf{a}_1 + \dots + \ell_N \mathbf{a}_N = \mathsf{0}$ would contradict the linear independence of the vectors $\mathbf{a}_j$ with $j \in J$. Now we have the equality of row vectors for every $(\ell_1, \dots, \ell_N) \in \mathbb{L}$

$$(\ell_1, \dots, \ell_N) = (\ell)_{J'} (\mathsf{B}_{J'})^{-1} \mathsf{B}$$

where $(\ell)_{J'}$ is the row vector with components ℓ_j $(j \in J')$. So for β in (36) one can take the maximum of the absolute values of the entries of the matrix $(\mathsf{B}_{J'})^{-1}\mathsf{B}$. □

Proposition 3. *Let $J \subset \{1, \ldots, N\}$ be a set with $k+1$ elements, such that the vectors $\mathbf{a}_j$ with $j \in J$ are linearly independent. Let $\underline{\gamma} = (\gamma_1, \ldots, \gamma_N) \in \mathbb{C}^N$ be such that $\gamma_j \in \mathbb{Z}$ for $j \in J' := \{1, \ldots, N\} \setminus J$. Then there is an $R \in \mathbb{R}_{>0}$ such that the power series (35) converges on the polydisc given by $|z_j| < R$ for $j = 1, \ldots, d$.*

Proof. This follows, with $R = T^{-\beta}$, from Proposition 2 and Lemma 2. □

3.5. Fourier Γ-series

The substitutions in (34) depend too rigidly on the choice of the set J and make it difficult to combine series constructed with different J's. In order to get a more flexible framework we make in the Γ-series (23) the substitution of variables $u_j = e^{2\pi i w_j}$ for $j = 1, \ldots, N$. We write $\mathbf{w} = (w_1, \ldots, w_N)$, $\underline{\gamma} = (\gamma_1, \ldots, \gamma_N)$ and $\underline{\ell} = (\ell_1, \ldots, \ell_N)$. We also use the dot-product:

$$\mathbf{w} \cdot \underline{\ell} = w_1\ell_1 + w_2\ell_2 + \ldots + w_N\ell_N \,.$$

With these new variables and notations the Γ-series (23) becomes

$$\Psi_{\mathbb{L},\underline{\gamma}}(\mathbf{w}) \;=\; \sum_{\underline{\ell} \in \mathbb{L}} \frac{e^{2\pi i \mathbf{w} \cdot (\underline{\gamma} + \underline{\ell})}}{\prod_{j=1}^{N} \Gamma(\gamma_j + \ell_j + 1)} \,. \tag{37}$$

As in Section 3.4 we take a set $J \subset \{1, \ldots, N\}$ with $k+1$ elements, such that the vectors $\mathbf{a}_j$ with $j \in J$ are linearly independent and let $\underline{\gamma} = (\gamma_1, \ldots, \gamma_N) \in \mathbb{C}^N$ be such that $\gamma_j \in \mathbb{Z}$ for $j \in J' := \{1, \ldots, N\} \setminus J$. The vector $\sum_{i \in J'} \mathbf{a}_i$ is a $\mathbb{Z}$-linear combination of the vectors $\mathbf{a}_j$ with $j \in J$. Such a relation is an element of $\mathbb{L}$. Thus one sees that $\mathbb{L}$ contains an element $\underline{\ell} = (\ell_1, \ldots, \ell_N)$ with $\ell_j = 1$ for all $j \in J'$. Since Γ-series do not change if one adds to $\underline{\gamma}$ an element of $\mathbb{L}$, we can assume without loss of generality that $\underline{\gamma} = (\gamma_1, \ldots, \gamma_N) \in \mathbb{C}^N$ is such that $\gamma_j \in \mathbb{Z}_{\leq 0}$ for $j \in J' := \{1, \ldots, N\} \setminus J$. Then the series $\Psi_{\mathbb{L},\underline{\gamma}}(\mathbf{w})$ in (23) involves only terms from the set

$$\mathbb{L}_J := \{(\ell_1, \ldots, \ell_N) \in \mathbb{L} \mid \ell_j \geq 0 \quad \text{if} \quad j \in J'\} \,. \tag{38}$$

Using the estimates (32) we see that the series $\Psi_{\mathbb{L},\underline{\gamma}}(\mathbf{w})$ converges if the imaginary part $\Im\mathbf{w}$ of $\mathbf{w}$ satisfies $\Im\mathbf{w} \cdot \underline{\ell} > \frac{\log T}{2\pi}$ for every non-zero $\underline{\ell} \in \mathbb{L}_J$. We return to this issue and put it an appropriate perspective in Section 5.2.

3.6. Γ-series and GKZ differential equations

As in the previous section we consider a $k+1$-element subset $J \subset \{1, \ldots, N\}$ such that the vectors $\mathbf{a}_j$ with $j \in J$ are linearly independent, and a vector $\underline{\gamma} = (\gamma_1, \ldots, \gamma_N) \in \mathbb{C}^N$ such that $\gamma_j \in \mathbb{Z}$ for $j \in J' := \{1, \ldots, N\} \setminus J$. The Γ-series constructed with such a $\underline{\gamma}$ involves only terms from the set $\mathbb{L}_{J,\underline{\gamma}}$ (see (23), (33)). For $M \in \mathbb{N}$ we define the M-th partial Γ-series $\Phi_{\mathbb{L},\underline{\gamma},M}(u_1, \ldots, u_N)$ to be the subseries of (23) consisting of the terms with $|\ell_1| + \ldots + |\ell_N| \leq M$. Then it follows, as in Proposition 3 from Proposition 2 and Lemma 2, that the sequence $\{\Phi_{\mathbb{L},\underline{\gamma},M}(u_1, \ldots, u_N)\}_{M \in \mathbb{N}}$ converges for $M \to \infty$ to $\Phi_{\mathbb{L},\underline{\gamma}}(u_1, \ldots, u_N)$ if $|u_j| \leq (2T)^{-\beta}$ for $j \in J'$ and $\frac{1}{2} \leq |u_j| \leq 2$ for $j \in J$. So on this domain the Γ-series

$\Phi_{\mathbb{L},\underline{\gamma}}(u_1,\ldots,u_N)$ becomes a function of $(u_1,\ldots,u_N)$ that can be differentiated term by term. This shows

- for $(\lambda_1,\ldots,\lambda_N)\in\mathbb{L}$

$$\prod_{\lambda_i<0}\left(\frac{\partial}{\partial u_i}\right)^{-\lambda_i}\Phi_{\mathbb{L},\underline{\gamma}} = \sum_{(\ell_1,\ldots,\ell_N)\in\mathbb{L}}\prod_{j=1}^{N}\frac{u_j^{\gamma_j+\ell_j+\min(0,\lambda_j)}}{\Gamma(\gamma_j+\ell_j+1+\min(0,\lambda_j))}$$

$$= \sum_{(\ell_1,\ldots,\ell_N)\in\mathbb{L}}\prod_{j=1}^{N}\frac{u_j^{\gamma_j+\ell_j+\lambda_j-\max(0,\lambda_j)}}{\Gamma(\gamma_j+\ell_j+\lambda_j+1-\max(0,\lambda_j))} = \prod_{\lambda_i>0}\left(\frac{\partial}{\partial u_i}\right)^{\lambda_i}\Phi_{\mathbb{L},\underline{\gamma}}.$$

- for $(a_1,\ldots,a_N)\in\mathbb{Z}^N$ such that $\sum_{j=1}^{N}a_j\ell_j=0$ for every $(\ell_1,\ldots,\ell_N)\in\mathbb{L}$:

$$\sum_{j=1}^{N}a_j\,u_j\frac{\partial\Phi_{\mathbb{L},\underline{\gamma}}}{\partial u_j} = \sum_{(\ell_1,\ldots,\ell_N)\in\mathbb{L}}\left(\sum_{j=1}^{N}a_j(\gamma_j+\ell_j)\right)\prod_{j=1}^{N}\frac{u_j^{\gamma_j+\ell_j}}{\Gamma(\gamma_j+\ell_j+1)}$$

$$= (\sum_{j=1}^{N}a_j\gamma_j)\Phi_{\mathbb{L},\underline{\gamma}}$$

The latter system of differential equations is equivalent with the system (20) with $\mathbf{c}=\sum_{j=1}^{N}\gamma_j\mathbf{a}_j$. This shows:

Proposition 4. *As a function on its domain of convergence $\Phi_{\mathbb{L},\underline{\gamma}}$ satisfies all differential equations of the GKZ system associated with $\mathcal{A}$ and $\mathbf{c}=\sum_{j=1}^{N}\gamma_j\mathbf{a}_j$.* □

Note that the Γ-series $\Phi_{\mathbb{L},\underline{\gamma}}$ does not change if one adds to $\underline{\gamma}$ an element of $\mathbb{L}$ whereas the differential equations (20) with $\mathbf{c}=\sum_{j=1}^{N}\gamma_j\mathbf{a}_j$ do not change if one adds to $\underline{\gamma}$ an element of $\mathbb{L}\otimes\mathbb{C}$.

4. The Secondary Fan

As before we consider a subset $\mathcal{A}=\{\mathbf{a}_1,\ldots,\mathbf{a}_N\}\subset\mathbb{Z}^{k+1}$ which generates $\mathbb{Z}^{k+1}$ as an abelian group and for which there exists a group homomorphism $h:\mathbb{Z}^{k+1}\to\mathbb{Z}$ such that $h(\mathbf{a})=1$ for all $\mathbf{a}\in\mathcal{A}$. Still as before, we write

$$\mathbb{L}:=\{(\ell_1,\ldots,\ell_N)\in\mathbb{Z}^N\mid \ell_1\mathbf{a}_1+\ldots+\ell_N\mathbf{a}_N=0\},$$

and note that $\ell_1+\ldots+\ell_N=0$ for every $(\ell_1,\ldots,\ell_N)\in\mathbb{L}$. In order to better keep track of the various spaces involved we write $\mathbb{M}$ instead of $\mathbb{Z}^{k+1}$. Thus the input data is a short exact sequence

$$0\longrightarrow\mathbb{L}\longrightarrow\mathbb{Z}^N\longrightarrow\mathbb{M}\longrightarrow 0. \tag{39}$$

The vectors $\mathbf{a}_1,\ldots,\mathbf{a}_N\in\mathbb{M}$ are the images of the standard basis vectors of $\mathbb{Z}^N$. We set $d:=\operatorname{rank}\mathbb{L}$ and $k+1:=\operatorname{rank}\mathbb{M}=N-d$.

Apart from the common input data this section is independent of the sections on GKZ-systems and Γ-series. It concentrates on geometric and combinatorial structures associated with $\mathcal{A}$ (or equivalently $\mathbb{L}$).

4.1. Construction of the secondary fan

We write $\mathbb{L}_{\mathbb{R}}^{\vee} := \mathrm{Hom}(\mathbb{L}, \mathbb{R})$, $\mathbb{M}_{\mathbb{R}}^{\vee} := \mathrm{Hom}(\mathbb{M}, \mathbb{R})$ and identify $\mathrm{Hom}(\mathbb{Z}^N, \mathbb{R})$ and $\mathbb{R}^N$ via the standard bases. The $\mathbb{R}$-dual of the exact sequence (39) is

$$0 \longrightarrow \mathbb{M}_{\mathbb{R}}^{\vee} \longrightarrow \mathbb{R}^N \xrightarrow{\pi} \mathbb{L}_{\mathbb{R}}^{\vee} \longrightarrow 0\,. \tag{40}$$

Let $\mathcal{P} := \{(x_1, \ldots, x_N) \in \mathbb{R}^N \mid x_i \geq 0\,,\ \forall i\}$ be the positive orthant in $\mathbb{R}^N$ and let

$$\widehat{\pi} : \mathcal{P} \longrightarrow \mathbb{L}_{\mathbb{R}}^{\vee} \tag{41}$$

denote the restriction of π. Since the vector $(1, 1, \ldots, 1)$ lies in $\ker \pi$ the map $\widehat{\pi}$ is also surjective.

Example. Take $\mathbb{L} = \mathbb{Z}(-2, 1, 1) \subset \mathbb{R}^3$. Then π can be identified with the map

$$\pi : \mathbb{R}^3 \longrightarrow \mathbb{R}\,, \qquad \pi(x_1, x_2, x_3) = -2x_1 + x_2 + x_3\,.$$

For $t \in \mathbb{R}$ the polytope $\widehat{\pi}^{-1}(t)$ is the intersection of the positive octant and the plane with equation $-2x_1 + x_2 + x_3 = t$. Figure 2 illustrates this for $t = 1$ and $t = -1$ (with the x_1-axis drawn vertically).

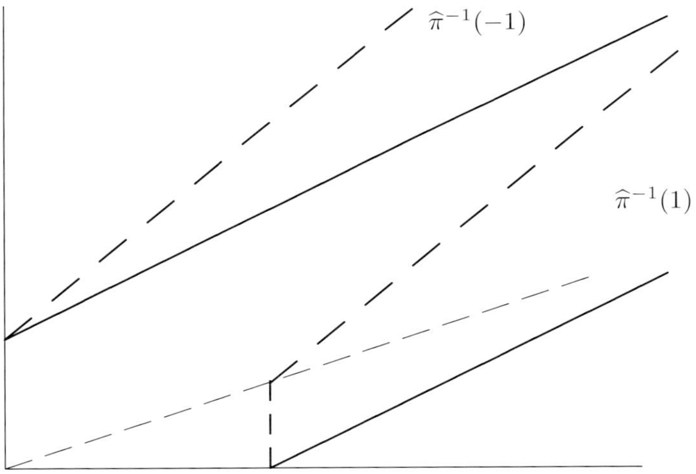

FIGURE 2. Fibres $\widehat{\pi}^{-1}(1)$ and $\widehat{\pi}^{-1}(-1)$ for $\mathbb{L} = \mathbb{Z}(-2, 1, 1) \subset \mathbb{R}^3$

Let $\mathbf{b}_1, \ldots, \mathbf{b}_N \in \mathbb{L}_{\mathbb{R}}^{\vee}$ be the images of the standard basis vectors of $\mathbb{R}^N$ under the map π. Then, for $\mathsf{t} \in \mathbb{L}_{\mathbb{R}}^{\vee}$,

$$(x_1, \ldots, x_N) \in \widehat{\pi}^{-1}(\mathsf{t}) \qquad \Longleftrightarrow \qquad \mathsf{t} = x_1\mathbf{b}_1 + \ldots + x_N\mathbf{b}_N \quad \text{and} \quad x_i \geq 0\,,\ \forall i\,.$$

We see that the fiber $\widehat{\pi}^{-1}(\mathsf{t})$ is a convex (unbounded) polyhedron.

Lemma 3. $\mathbf{v} = (v_1, \ldots, v_N) \in \mathcal{P}$ *is a vertex of* $\widehat{\pi}^{-1}(\mathsf{t})$ *if and only if* $\mathsf{t} = \sum_{j=1}^{N} v_j \mathbf{b}_j$ *and the vectors* $\mathbf{b}_j$ *with* $v_j \neq 0$ *are linearly independent over* $\mathbb{R}$.

Proof. Suppose $\mathsf{t} = \sum_{j=1}^{N} v_j \mathbf{b}_j$, all $v_j \geq 0$ and the vectors $\mathbf{b}_j$ with $v_j \neq 0$ are linearly *dependent* over $\mathbb{R}$. Then there is a non-trivial relation $\sum_{j=1}^{N} x_j \mathbf{b}_j = 0$ with $|x_j| \leq v_j$ for all j and the whole interval $\{\mathbf{v} + s(x_1, \ldots, x_N) \,|\, |s| \leq 1\}$ lies in $\widehat{\pi}^{-1}(\mathsf{t})$. Therefore $\mathbf{v}$, being the midpoint of this interval, can not be a vertex of $\widehat{\pi}^{-1}(\mathsf{t})$.

Suppose $\mathbf{v} = (v_1, \ldots, v_N) \in \widehat{\pi}^{-1}(\mathsf{t})$ is not a vertex of $\widehat{\pi}^{-1}(\mathsf{t})$. Then there is a non-zero vector $\mathbf{x} = (x_1, \ldots, x_N) \in \mathbb{R}^N$ such that the interval $\{\mathbf{v} + s\mathbf{x} \mid |s| \leq 1\}$ lies in $\widehat{\pi}^{-1}(\mathsf{t}) = \mathcal{P} \cap \pi^{-1}(\mathsf{t})$. This implies $|x_j| \leq v_j$ for all j and $\sum_{j=1}^{N} x_j \mathbf{b}_j = 0$. Consequently, the vectors $\mathbf{b}_j$ with $v_j \neq 0$ are linearly *dependent* over $\mathbb{R}$. □

For a vertex $\mathbf{v} = (v_1, \ldots, v_N)$ of $\widehat{\pi}^{-1}(\mathsf{t})$ we set

$$I_{\mathbf{v}} := \{i \mid v_i = 0\} \subset \{1, 2, \ldots, N\}. \tag{42}$$

In this way every $\mathsf{t} \in \mathbb{L}_{\mathbb{R}}^{\vee}$ yields a list of subsets of $\{1, 2, \ldots, N\}$:

$$T_{\mathsf{t}} := \{I_{\mathbf{v}} \mid \mathbf{v} \text{ vertex of } \widehat{\pi}^{-1}(\mathsf{t})\}. \tag{43}$$

Since $\pi^{-1}(\mathsf{t})$ has dimension $N - d$, the cardinality of each $I_{\mathbf{v}}$ must be at least $N - d$.

The above lemma provides an alternative description of the list T_{t}:

Corollary 1. *A subset $I \subset \{1, \ldots, N\}$ is on the list T_{t} if and only if the vectors $\mathbf{b}_j$ with $j \notin I$ are linearly independent over $\mathbb{R}$ and $\mathsf{t} = \sum_{j \notin I} \tau_j \mathbf{b}_j$ with all $\tau_j \in \mathbb{R}_{>0}$.* □

We now define an equivalence relation on $\mathbb{L}_{\mathbb{R}}^{\vee}$ by: $\mathsf{t} \sim \mathsf{t}' \iff T_{\mathsf{t}} = T_{\mathsf{t}'}$. From Corollary 1 one sees that the equivalence class containing t is

$$\mathcal{C} = \bigcap_{I \in T_{\mathsf{t}}} (\text{positive span of } \{\mathbf{b}_i\}_{i \notin I}). \tag{44}$$

So the equivalence classes are strongly convex polyhedral cones in $\mathbb{L}_{\mathbb{R}}^{\vee}$.

Definition 3. *This collection of cones is called the* secondary fan of $\mathcal{A}$ (or $\mathbb{L}$).

For an equivalence class $\mathcal{C}$ we set $T_{\mathcal{C}} := T_{\mathsf{t}}$ for any $\mathsf{t} \in \mathcal{C}$. It follows from (44) that an equivalence class $\mathcal{C}$ is an open cone of dimension d if and only if all sets on the list $T_{\mathcal{C}}$ have exactly $N - d$ elements.

Example. In the example of $\mathbb{L} = \mathbb{Z}(-2, 1, 1) \subset \mathbb{R}^3$ (see Figure 2) the vertices are given by the lists

$$T_t = \begin{cases} \{\{2,3\}\} & \text{if} \quad t < 0 \\ \{\{1,2,3\}\} & \text{if} \quad t = 0 \\ \{\{1,2\},\{1,3\}\} & \text{if} \quad t > 0. \end{cases}$$

Example. For Gauss's hypergeometric structures $\mathbb{L} = \mathbb{Z}(1, 1, -1, -1) \subset \mathbb{R}^4$ and, hence, $\mathbf{b}_1 = \mathbf{b}_2 = 1$, $\mathbf{b}_3 = \mathbf{b}_4 = -1$ in $\mathbb{R}$. Corollary 1 now yields the lists

$$T_t = \begin{cases} \{\{1,2,3\},\{1,2,4\}\} & \text{if} \quad t < 0 \\ \{\{1,2,3,4\}\} & \text{if} \quad t = 0 \\ \{\{2,3,4\},\{1,3,4\}\} & \text{if} \quad t > 0. \end{cases}$$

Example. For Appell's F_1 the lattice $\mathbb{L} \subset \mathbb{Z}^6$ has rank 2 and is generated by the two vectors $(1, -1, 0, -1, 1, 0)$ and $(1, 0, -1, -1, 0, 1)$ which express that the three vertical segments in Figure 1 are parallel. The vectors $\mathbf{b}_1, \dots, \mathbf{b}_6 \in \mathbb{Z}^2$ are therefore

$$\mathbf{b}_1 = \begin{bmatrix} 1 \\ 1 \end{bmatrix}, \mathbf{b}_2 = \begin{bmatrix} -1 \\ 0 \end{bmatrix}, \mathbf{b}_3 = \begin{bmatrix} 0 \\ -1 \end{bmatrix}, \mathbf{b}_4 = \begin{bmatrix} -1 \\ -1 \end{bmatrix}, \mathbf{b}_5 = \begin{bmatrix} 1 \\ 0 \end{bmatrix}, \mathbf{b}_6 = \begin{bmatrix} 0 \\ 1 \end{bmatrix}.$$

Figure 3 shows the secondary fan for F_1 and gives for each maximal cone $\mathcal{C}$ the corresponding list $T_{\mathcal{C}}$ according to Corollary 1.

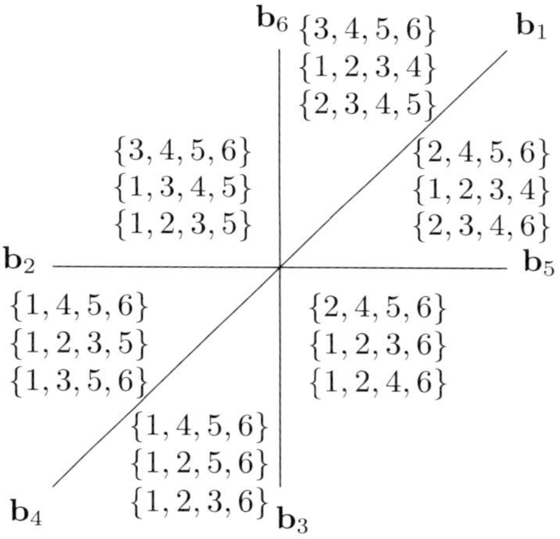

FIGURE 3. Secondary fan for F_1

Example. For Appell's F_4 the lattice $\mathbb{L} \subset \mathbb{Z}^6$ has rank 2 and is generated by the two vectors $(1, -1, 1, -1, 0, 0)$ and $(1, 0, 1, 0, -1, -1)$ which express that the three diagonals in Figure 1 intersect at the centre. The vectors $\mathbf{b}_1, \dots, \mathbf{b}_6 \in \mathbb{Z}^2$ are

$$\mathbf{b}_1 = \mathbf{b}_3 = \begin{bmatrix} 1 \\ 1 \end{bmatrix}, \mathbf{b}_2 = \mathbf{b}_4 = \begin{bmatrix} -1 \\ 0 \end{bmatrix}, \mathbf{b}_5 = \mathbf{b}_6 = \begin{bmatrix} 0 \\ -1 \end{bmatrix}.$$

Figure 4 shows the secondary fan for F_4 and gives for each maximal cone $\mathcal{C}$ the corresponding list $T_{\mathcal{C}}$ according to Corollary 1.

Example/Exercise. The reader is invited to apply the techniques demonstrated in the previous examples to the examples in Section 8.1, Table 1.

4.2. Alternative descriptions for secondary fan constructions

We are going to present geometrically appealing alternative descriptions for the polyhedra $\widehat{\pi}^{-1}(\mathfrak{t})$ and for the lists $T_{\mathcal{C}}$ associated with the maximal cones in the secondary fan. Whereas the constructions in Section 4.1 were completely presented in terms of $\mathbb{L}$, the alternative descriptions use $\mathcal{A}$ only.

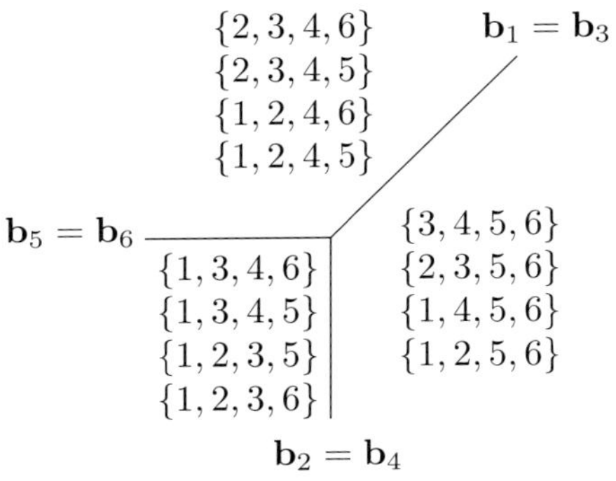

FIGURE 4. Secondary fan for F_4

4.2.1. Piecewise linear functions associated with $\mathcal{A}$. The vectors $\mathbf{a}_1, \ldots, \mathbf{a}_N \in \mathbb{M}$ are linear functions on the space $\mathbb{M}_{\mathbb{R}}^{\vee} := \mathrm{Hom}(\mathbb{M}, \mathbb{R})$. Let $\mathbb{M}_{\mathbb{R}} := \mathbb{M} \otimes \mathbb{R}$ and denote the pairing between $\mathbb{M}_{\mathbb{R}}$ and $\mathbb{M}_{\mathbb{R}}^{\vee}$ by $\langle , \rangle$. The inclusion $\mathbb{M}_{\mathbb{R}}^{\vee} \hookrightarrow \mathbb{R}^N$ is then given by

$$\mathbb{M}_{\mathbb{R}}^{\vee} \hookrightarrow \mathbb{R}^N, \qquad \mathbf{v} \mapsto (\langle \mathbf{a}_1, \mathbf{v} \rangle, \ldots, \langle \mathbf{a}_N, \mathbf{v} \rangle) . \tag{45}$$

For an N-tuple $\underline{\alpha} = (\alpha_1, \ldots, \alpha_N) \in \mathbb{R}^N$ one has the polyhedron

$$K_{\underline{\alpha}} := \{ \mathbf{v} \in \mathbb{M}_{\mathbb{R}}^{\vee} \mid \langle \mathbf{a}_j, \mathbf{v} \rangle \geq -\alpha_j, \ \forall j \} . \tag{46}$$

Recall that $\mathbf{b}_1, \ldots, \mathbf{b}_N \in \mathbb{L}_{\mathbb{R}}^{\vee}$ denote the images of the standard basis vectors of $\mathbb{R}^N$ under the map π.

Proposition 5. *For $\underline{\alpha} = (\alpha_1, \ldots, \alpha_N) \in \mathbb{R}^N$ set $\mathsf{t} = \alpha_1 \mathbf{b}_1 + \ldots + \alpha_N \mathbf{b}_N$. Then*

$$\widehat{\pi}^{-1}(\mathsf{t}) \;=\; \underline{\alpha} + K_{\underline{\alpha}} .$$

Proof. Since $\underline{\alpha}$ is in $\pi^{-1}(\mathsf{t})$ a point $\mathbf{x}$ is in $\pi^{-1}(\mathsf{t})$ if and only if $\mathbf{x} - \underline{\alpha}$ is in $\ker \pi = \mathbb{M}_{\mathbb{R}}^{\vee}$. By definition, a point $\mathbf{x} = (x_1, \ldots, x_N) \in \pi^{-1}(\mathsf{t})$ lies in $\widehat{\pi}^{-1}(\mathsf{t})$ if and only if $x_j \geq 0$ for all j. Thus, in view of (45),

$$\mathbf{x} \in \widehat{\pi}^{-1}(\mathsf{t}) \qquad \Longleftrightarrow \qquad \mathbf{v} := \mathbf{x} - \underline{\alpha} \text{ satisfies } \langle \mathbf{a}_j, \mathbf{v} \rangle + \alpha_j \geq 0, \ \forall j . \qquad \square$$

Recall that throughout these notes we assume the existence of a group homomorphism $h : \mathbb{Z}^{k+1} \to \mathbb{Z}$ such that $h(\mathbf{a}) = 1$ for all $\mathbf{a} \in \mathcal{A}$. In the present terminology this amounts to the existence of an element $\mathbf{h} \in \mathbb{M}_{\mathbb{R}}^{\vee}$ such that $\langle \mathbf{a}_j, \mathbf{h} \rangle = 1$ for $j = 1, \ldots, N$. Now fix a direct sum decomposition of real vector spaces

$$\mathbb{M}_{\mathbb{R}}^{\vee} \;=\; \mathbb{M}_{\mathbb{R}}^{\circ} \oplus \mathbb{R}\mathbf{h} \tag{47}$$

and consider the function

$$\mu_{\underline{\alpha}} : \mathbb{M}_{\mathbb{R}}^{\circ} \longrightarrow \mathbb{R}, \qquad \mu_{\underline{\alpha}}(\mathbf{u}) = \min_j \left(\langle \mathbf{a}_j, \mathbf{u} \rangle + \alpha_j \right) . \tag{48}$$

Proposition 6. *For every $\mathbf{u} \in \mathbb{M}_{\mathbb{R}}^{\circ}$ the vector $\mathbf{u} - \mu_{\underline{\alpha}}(\mathbf{u})\mathbf{h}$ lies in the boundary $\partial K_{\underline{\alpha}}$ of $K_{\underline{\alpha}}$. In other words $\partial K_{\underline{\alpha}}$ is the graph of the function $-\mu_{\underline{\alpha}}$ on $\mathbb{M}_{\mathbb{R}}^{\circ}$.*

Proof. Take $\mathbf{u} \in \mathbb{M}^{\circ}_{\mathbb{R}}$. Then one checks for every j

$$\langle \mathbf{a}_j, \mathbf{u} - \mu_{\underline{\alpha}}(\mathbf{u})\mathbf{h} \rangle = \langle \mathbf{a}_j, \mathbf{u} \rangle - \mu_{\underline{\alpha}}(\mathbf{u}) \geq \langle \mathbf{a}_j, \mathbf{u} \rangle - (\langle \mathbf{a}_j, \mathbf{u} \rangle + \alpha_j) = -\alpha_j .$$

So $\mathbf{u} - \mu_{\underline{\alpha}}(\mathbf{u})\mathbf{h}$ lies in $K_{\underline{\alpha}}$. If j is such that $\mu_{\underline{\alpha}}(\mathbf{u}) = \langle \mathbf{a}_j, \mathbf{u} \rangle + \alpha_j$, then the above computation shows $\langle \mathbf{a}_j, \mathbf{u} - \mu_{\underline{\alpha}}(\mathbf{u})\mathbf{h} \rangle = -\alpha_j$. Therefore $\mathbf{u} - \mu_{\underline{\alpha}}(\mathbf{u})\mathbf{h}$ lies in $\partial K_{\underline{\alpha}}$. □

If $\mu_{\underline{\alpha}}(\mathbf{u}) = \langle \mathbf{a}_j, \mathbf{u} \rangle + \alpha_j$, then the point $\mathbf{u} - \mu_{\underline{\alpha}}(\mathbf{u})\mathbf{h}$ lies in the affine hyperplane

$$\mathcal{H}^{\underline{\alpha}}_j := \{ \mathbf{v} \in \mathbb{M}^{\vee}_{\mathbb{R}} \mid \langle \mathbf{a}_j, \mathbf{v} \rangle = -\alpha_j \} , \qquad j = 1, \ldots, N . \tag{49}$$

For generic $\mathbf{u} \in \mathbb{M}^{\circ}_{\mathbb{R}}$ (i.e., outside some codimension 1 closed subset) the minimum in (48) is attained for exactly one j. Therefore each codimension 1 face of the polyhedron $K_{\underline{\alpha}}$ lies in some unique hyperplane $\mathcal{H}^{\underline{\alpha}}_j$.

Remark. $K_{\underline{\alpha}}$ can also be described as the closure of that connected component of $\mathbb{M}^{\vee}_{\mathbb{R}} \setminus \bigcup_{j=1}^{N} \mathcal{H}^{\underline{\alpha}}_j$ that contains the points $t\mathbf{h}$ for sufficiently large t.

Example. Figure 5 shows (a piece of) the polyhedron $K_{\underline{\alpha}}$ for

$$\mathcal{A} = \left\{ \begin{bmatrix} 1 \\ 0 \\ 1 \end{bmatrix}, \begin{bmatrix} 1 \\ 1 \\ 1 \end{bmatrix}, \begin{bmatrix} 1 \\ -1 \\ 0 \end{bmatrix}, \begin{bmatrix} 1 \\ 0 \\ 0 \end{bmatrix}, \begin{bmatrix} 1 \\ 1 \\ 0 \end{bmatrix}, \begin{bmatrix} 1 \\ 0 \\ -1 \end{bmatrix} \right\} \tag{50}$$

and $\underline{\alpha} = (21, 35, 35, 14, 21, 28)$. Matching the faces of $K_{\underline{\alpha}}$ with the vectors in $\mathcal{A}$ one checks that the list of vertices is $\{\{1,2,5\}, \{1,4,5\}, \{1,3,4\}, \{4,5,6\}, \{3,4,6\}\}$.

4.2.2. Regular triangulations. Assume $\underline{\alpha} = (\alpha_1, \ldots, \alpha_N) \in \mathbb{R}^N$ with all $\alpha_j > 0$. Then the dual of the polyhedron $K_{\underline{\alpha}}$ in (46) is, by definition

$$K^{\vee}_{\underline{\alpha}} := \{ \mathbf{w} \in \mathbb{M}_{\mathbb{R}} \mid \langle \mathbf{w}, \mathbf{v} \rangle > -1 , \ \forall \mathbf{v} \in K_{\underline{\alpha}} \} . \tag{51}$$

Lemma 4. $K^{\vee}_{\underline{\alpha}} = \text{convex hull} \, \{ 0, \frac{1}{\alpha_1}\mathbf{a}_1, \ldots, \frac{1}{\alpha_N}\mathbf{a}_N \}$.

Proof. The inclusion $\supset$ follows directly from the definition of $K_{\underline{\alpha}}$ in (46). Now suppose that the two polyhedra are not equal. Then there is a point $\mathbf{p}$ in $K^{\vee}_{\underline{\alpha}}$ which is separated by an affine hyperplane from $0, \frac{1}{\alpha_1}\mathbf{a}_1, \ldots, \frac{1}{\alpha_N}\mathbf{a}_N$. That means that there is a vector $\mathbf{v} \in \mathbb{M}^{\vee}_{\mathbb{R}}$, perpendicular to the hyperplane, such that $\langle \mathbf{p}, \mathbf{v} \rangle < -1$ and $\langle \frac{1}{\alpha_j}\mathbf{a}_1, \mathbf{v} \rangle > -1$ for $j = 1, \ldots, N$. The last N inequalities imply according to (46) that $\mathbf{v}$ is in $K_{\underline{\alpha}}$, but then the first inequality contradicts $\mathbf{p} \in K^{\vee}_{\underline{\alpha}}$. So we conclude that the two polyhedra are equal. □

Next we use the projection from the point 0 to project $K^{\vee}_{\underline{\alpha}}$ into the hyperplane with equation $\langle \mathbf{p}, \mathbf{h} \rangle = 1$. This maps $K^{\vee}_{\underline{\alpha}}$ onto the polytope

$$\Delta_{\mathcal{A}} := \text{convex hull}\{\mathbf{a}_1, \ldots, \mathbf{a}_N\}. \tag{52}$$

The images of the codimension 1 faces of $K^{\vee}_{\underline{\alpha}}$ which do not contain the vertex 0 induce a subdivision of $\Delta_{\mathcal{A}}$ by the polytopes

$$\text{convex hull}\{\mathbf{a}_i\}_{i \in I} \qquad \text{for} \quad I \in T_{\mathrm{t}} , \tag{53}$$

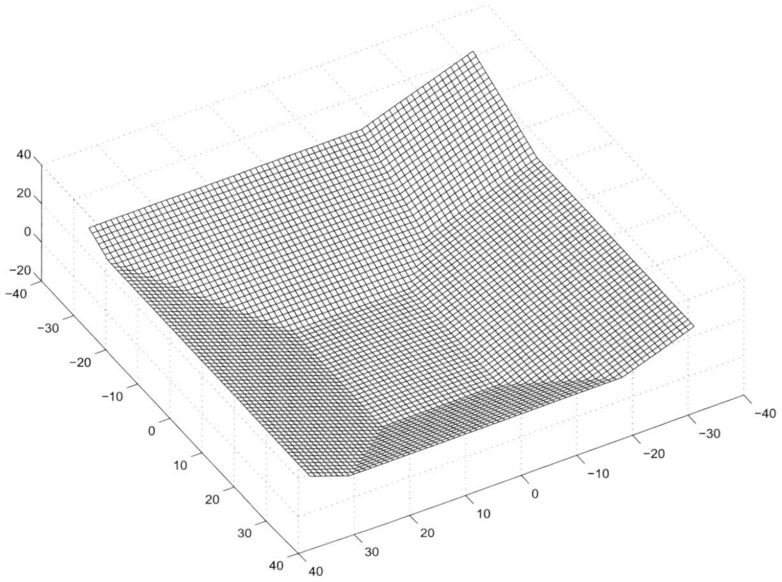

FIGURE 5. Example of a polyhedron $K_{\underline{\alpha}}$ for $\mathcal{A}$ as in (50).

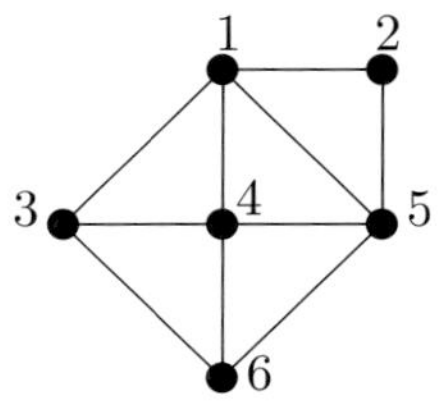

FIGURE 6. Triangulation corresponding with Figure 5

where $\mathsf{t} = \alpha_1 \mathbf{b}_1 + \ldots + \alpha_N \mathbf{b}_N$ as in Proposition 5 and T_t is the corresponding list of vertices of $\widehat{\pi}^{-1}(\mathsf{t})$ as in (43).

If the point t lies in some maximal cone $\mathcal{C}$ of the secondary fan, all members of the list $T_\mathsf{t} = T_\mathcal{C}$ have $N - d = k + 1$ elements. The polytopal subdivision of $\Delta_\mathcal{A}$ is then a *triangulation*; i.e., all polytopes in the subdivision (53) are k-dimensional simplices.

Definition 4. *The triangulations of* $\Delta_{\mathcal{A}}$ *obtained in this way are called* regular triangulations.

Definition 5. *One defines the* volume *of a* k*-dimensional simplex with vertex set* $\{\mathbf{a}_i\}_{i\in I}$ *to be*

$$\text{volume}(\{\mathbf{a}_i\}_{i\in I}) \;=\; |\det((\mathbf{a}_i)_{i\in I})| \,. \tag{54}$$

A regular triangulation of $\Delta_{\mathcal{A}}$ *is said to be* unimodular *if all* k*-dimensional simplices in the triangulation have volume equal to* 1.

By abuse of language we will just say "the triangulation $T_{\mathcal{C}}$" instead of "the triangulation corresponding with the maximal cone $\mathcal{C}$". Note that useful information about distances between vertices of $\widehat{\pi}^{-1}(\mathbf{t})$ gets lost in the passage to the (purely combinatorial) triangulation $T_{\mathcal{C}}$.

Remark. In general there can be triangulations of $\Delta_{\mathcal{A}}$ with vertices in $\{\mathbf{a}_1,\dots,\mathbf{a}_N\}$, which do not arise from the above construction and are therefore not regular.

4.3. The Secondary Polytope

The k-dimensional polytope $\Delta_{\mathcal{A}}$ defined in (52) is sometimes called the *primary polytope* associated with $\mathcal{A}$. By definition, the regular triangulations of $\Delta_{\mathcal{A}}$ correspond bijectively with the maximal cones of the secondary fan. To a regular triangulation $T_{\mathcal{C}}$ we assign the point $q_{\mathcal{C}} \in \mathbb{R}^N$ with

$$j^{\text{th}}\text{-coordinate of } q_{\mathcal{C}} \;=\; \sum_{I\in T_{\mathcal{C}} \text{ s.t. } j\in I} \text{volume}(\{\mathbf{a}_i\}_{i\in I})\,,$$

i.e., the sum of the volumes of the simplices in $T_{\mathcal{C}}$ of which $\mathbf{a}_j$ is a vertex.

Definition 6. *The* secondary polytope *associated with* $\mathcal{A}$ *is*

$$\text{Sec}(\mathcal{A}) \;=\; \textit{convex hull}\ \{q_{\mathcal{C}} \mid T_{\mathcal{C}} \textit{ regular triangulation of } \Delta_{\mathcal{A}}\,\}.$$

The map $\mathbb{R}^N \to \mathbb{M}_{\mathbb{R}}$ maps the j-th standard basis vector of $\mathbb{R}^N$ to $\mathbf{a}_j$. Thus the point $q_{\mathcal{C}}$ is mapped to

$$\begin{aligned}\sum_{j=1}^{N}\ \sum_{I\in T_{\mathcal{C}} \text{ s.t. } j\in I} \text{volume}(\{\mathbf{a}_i\}_{i\in I})\,\mathbf{a}_j \;&=\; \sum_{I\in T_{\mathcal{C}}} \text{volume}(\{\mathbf{a}_i\}_{i\in I}) \left(\sum_{j\in I}\mathbf{a}_j\right)\\ &=\; (k+1)\times\text{volume}(\Delta_{\mathcal{A}})\times\text{barycenter}(\Delta_{\mathcal{A}})\,.\end{aligned}$$

So the whole secondary polytope is mapped to one point. Therefore, after some translation in $\mathbb{R}^N$ we find the secondary polytope in $\mathbb{L}$:

$$\text{Sec}(\mathcal{A}) \subset \mathbb{L}\otimes\mathbb{R}\,.$$

As for the relation between secondary fan and secondary polytope we mention the following theorem, which is in a slightly different formulation proven in [13].

Theorem 2. ([13], p. 221, Thm. 1.7) *The secondary fan, which lies in* $\mathbb{L}^{\vee}_{\mathbb{R}}$*, is in fact the fan of outward pointing vectors perpendicular to the faces of* $\text{Sec}(\mathcal{A})$. $\square$

Example. In the example of $\mathbb{L} = \mathbb{Z}(-2, 1, 1) \subset \mathbb{Z}^3$ there are two maximal cones: $\mathbb{R}_{>0}$ and $\mathbb{R}_{<0}$. The corresponding triangulations are:

The secondary polytope is the line segment between the points $(0, 2, 2)$ and $(2, 1, 1)$ in $\mathbb{R}^3$.

Example. For Gauss's hypergeometric structures $\mathbb{L} = \mathbb{Z}(1, 1, -1, -1) \subset \mathbb{Z}^4$. From this one sees that $\mathbb{L}^\vee_\mathbb{R} \simeq \mathbb{R}$ and that there are two maximal cones: $\mathbb{R}_{>0}$ and $\mathbb{R}_{<0}$. The corresponding triangulations are:

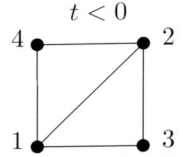

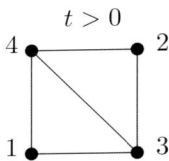

The secondary polytope is the line segment between the points $(2, 2, 1, 1)$ and $(1, 1, 2, 2)$ in $\mathbb{R}^4$.

Example: For $\mathbb{L} = \mathbb{Z}(-3\,1\,1\,1) \subset \mathbb{Z}^4$ we have $\mathbb{L}^\vee_\mathbb{R} \simeq \mathbb{R}$ and $\mathbf{b}_1 = -3$, $\mathbf{b}_2 = \mathbf{b}_3 = \mathbf{b}_4 = 1$. Corollary 1 shows for $t \in \mathbb{L}^\vee_\mathbb{R} \simeq \mathbb{R}$:

$$T_t = \begin{cases} \{\,\{1,3,4\}\,,\ \{1,2,4\}\,,\ \{1,2,3\}\,\} & \text{if} \quad t > 0 \\ \{\,\{1,2,3,4\}\,\} & \text{if} \quad t = 0 \\ \{\,\{2,3,4\}\,\} & \text{if} \quad t < 0 \end{cases}$$

So there are two maximal cones: $\mathbb{R}_{>0}$ and $\mathbb{R}_{<0}$. In terms of triangulations:

$$\mathcal{A} = \left\{ \begin{bmatrix} 1 \\ 0 \\ 0 \end{bmatrix}, \begin{bmatrix} 1 \\ 1 \\ 1 \end{bmatrix}, \begin{bmatrix} 1 \\ -1 \\ 0 \end{bmatrix}, \begin{bmatrix} 1 \\ 0 \\ -1 \end{bmatrix} \right\}$$

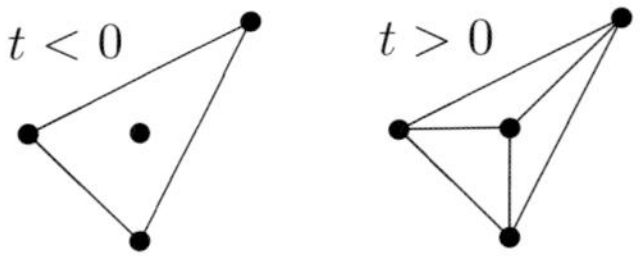

The secondary polytope is the line segment between the points $(0, 3, 3, 3)$ and $(3, 2, 2, 2)$ in $\mathbb{R}^4$.

Example: Figure 7 shows the secondary polytope with at each vertex the corresponding regular triangulation of $\Delta_\mathcal{A}$ for $\mathcal{A}$ as in (50).

5. The toric variety associated with the Secondary Fan

5.1. Construction of the toric variety for the secondary fan

The secondary fan is a complete fan of strongly convex polyhedral cones in $\mathbb{L}^\vee_\mathbb{R} := \mathrm{Hom}\,(\mathbb{L}, \mathbb{R})$ which are generated by vectors from the lattice $\mathbb{L}^\vee_\mathbb{Z} := \mathrm{Hom}\,(\mathbb{L}, \mathbb{Z})$. By the general theory of toric varieties [9] this lattice-fan pair gives rise to a toric variety. We are going to describe the construction of the toric variety for the case

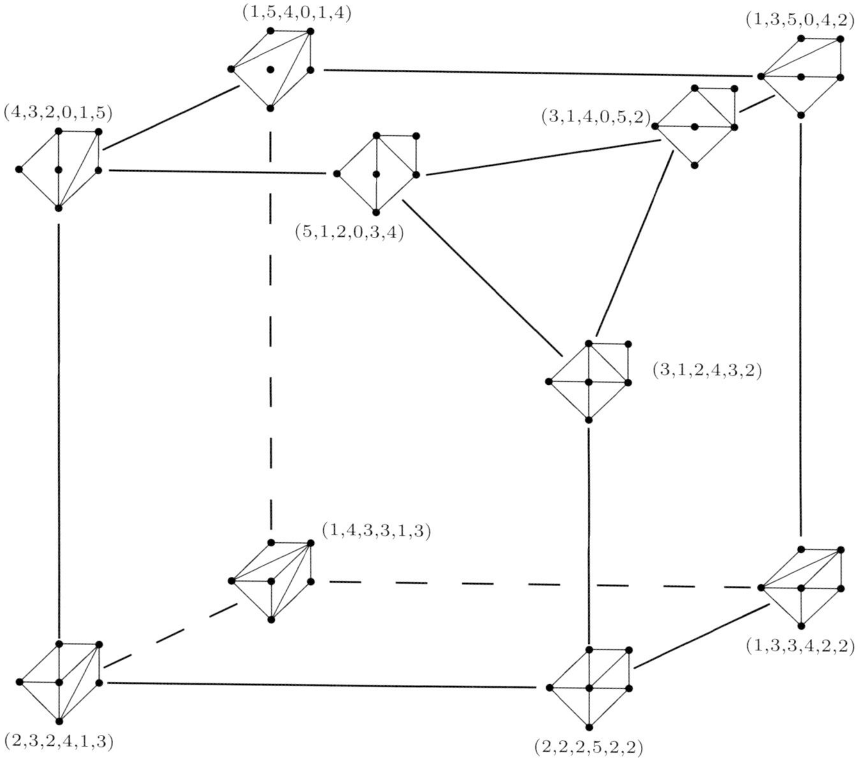

FIGURE 7. The secondary polytope and all regular triangulations for $\mathcal{A}$ as in (50).

of $\mathbb{L}_{\mathbb{Z}}^{\vee}$ and the secondary fan. Before starting we must point out that [9] works with a fan of *closed cones*, while the cones in our Definition 3 of the secondary fan are *not closed*; see also Formula (44). This difference, however, only affects a few minor subtleties in the formulation at intermediate stages. The monoids (55) and therefore also the resulting toric varieties are the same as in [9].

We denote the pairing between $\mathbb{L}_{\mathbb{R}} := \mathbb{L} \otimes_{\mathbb{Z}} \mathbb{R}$ and $\mathbb{L}_{\mathbb{R}}^{\vee}$ by $\langle , \rangle$. For each cone $\mathcal{C}$ in the secondary fan (see (44)) one considers the affine scheme $\mathcal{U}_{\mathcal{C}} :=$ Spec $\mathbb{Z}[\mathbb{L}_{\mathcal{C}}]$ associated with the monoid ring $\mathbb{Z}[\mathbb{L}_{\mathcal{C}}]$ of the monoid [1]

$$\mathbb{L}_{\mathcal{C}} := \{ \underline{\ell} \in \mathbb{L} \mid \langle \omega, \underline{\ell} \rangle \geq 0 \quad \text{for all} \quad \omega \in \mathcal{C} \} . \tag{55}$$

[1] Alternative terminology: monoid = semi-group

In down-to-earth terms, a complex point of $\mathcal{U}_\mathcal{C}$ is just a homomorphism from the additive monoid $\mathbb{L}_\mathcal{C}$ to the multiplicative monoid $\mathbb{C}$, sending $0 \in \mathbb{L}_\mathcal{C}$ to $1 \in \mathbb{C}$.

For cones $\mathcal{C}$ and $\mathcal{C}'$ in the secondary fan such that $\mathcal{C}'$ is contained in the closure $\overline{\mathcal{C}}$ of $\mathcal{C}$, there are inclusions

$$\mathcal{C}' \subset \overline{\mathcal{C}}, \qquad \mathbb{L}_{\mathcal{C}'} \supset \mathbb{L}_\mathcal{C}, \qquad \mathbb{Z}[\mathbb{L}_{\mathcal{C}'}] \supset \mathbb{Z}[\mathbb{L}_\mathcal{C}], \qquad \mathcal{U}_{\mathcal{C}'} \subset \mathcal{U}_\mathcal{C};$$

more precisely, the following lemma shows that the inclusion $\mathcal{U}_{\mathcal{C}'} \hookrightarrow \mathcal{U}_\mathcal{C}$ is an open immersion associated with the inversion of an element in the ring $\mathbb{Z}[\mathbb{L}_\mathcal{C}]$.

Lemma 5. *In the above situation* $\mathbb{L}_{\mathcal{C}'} = \mathbb{L}_\mathcal{C} + \mathbb{Z}\lambda$ *for some* $\lambda \in \mathbb{L}_\mathcal{C}$.

Proof. If $\mathcal{C}' = \mathcal{C}$, the result is trivial. So assume $\mathcal{C}' \neq \mathcal{C}$. Since $\mathcal{C}$ is a rational polyhedral cone it is spanned by finitely many $\omega_1, \ldots, \omega_p \in \mathbb{L}_\mathbb{Z}^\vee$, i.e., every point in $\mathcal{C}$ is a linear combination with non-negative real coefficients of $\omega_1, \ldots, \omega_p$. Moreover since $\mathcal{C}' \subset \overline{\mathcal{C}}$ there is a $\lambda \in \mathbb{L}_\mathcal{C}$ such that $\langle \omega', \lambda \rangle = 0$ for all $\omega' \in \mathcal{C}'$ and $\langle \omega, \lambda \rangle > 0$ for all $\omega \in \mathcal{C}$. Take $\mu \in \mathbb{R}_{>0}$ such that for $j = 1, \ldots, p$ we have $\langle \omega_j, \lambda \rangle > \mu$ if $\omega_j \notin \overline{\mathcal{C}'}$. For every $\underline{\ell} \in \mathbb{L}_{\mathcal{C}'}$ and every non-negative integer $r > -\frac{1}{\mu}\min_j \langle \omega_j, \underline{\ell} \rangle$, one now easily checks that $\langle \omega, \underline{\ell} + r\lambda \rangle \geq 0$ for every $\omega \in \mathcal{C}$, and hence $\underline{\ell} + r\lambda \in \mathbb{L}_\mathcal{C}$. □

Definition 7. *The* toric variety associated with the secondary fan *is the scheme that results from glueing the affine schemes* $\mathcal{U}_\mathcal{C}$, *where* $\mathcal{C}$ *ranges over all cones in the secondary fan, using the open immersions* $\mathcal{U}_{\mathcal{C}'} \hookrightarrow \mathcal{U}_\mathcal{C}$ *for* $\mathcal{C}' \subset \overline{\mathcal{C}}$. *We denote this toric variety by* $\mathcal{V}_\mathcal{A}$.

For every cone $\mathcal{C}$ of the secondary fan the monoid $\mathbb{L}_\mathcal{C}$ splits as a disjoint union $\mathbb{L}_\mathcal{C} = \mathbb{L}^0_\mathcal{C} \coprod \mathbb{L}^+_\mathcal{C}$ where $\mathbb{L}^0_\mathcal{C}$ (resp. $\mathbb{L}^+_\mathcal{C}$) is the set of elements which do (resp. do not) have an inverse in the additive monoid $\mathbb{L}_\mathcal{C}$. One easily checks that

$$\mathbb{L}^0_\mathcal{C} := \{\, \underline{\ell} \in \mathbb{L} \mid \langle \omega, \underline{\ell} \rangle = 0 \quad \forall \omega \in \mathcal{C} \,\}, \qquad \mathbb{L}^+_\mathcal{C} := \{\, \underline{\ell} \in \mathbb{L} \mid \langle \omega, \underline{\ell} \rangle > 0 \quad \forall \omega \in \mathcal{C} \,\}.$$

If $\mathcal{C} = \{0\}$, then $\mathbb{L}^0_\mathcal{C} = \mathbb{L}_\mathcal{C}$ and $\mathbb{L}^+_\mathcal{C} = \emptyset$. If $\mathcal{C} \neq \{0\}$, the elements of $\mathbb{L}^+_\mathcal{C}$ generate a proper ideal $I_\mathcal{C}$ in the ring $\mathbb{Z}[\mathbb{L}_\mathcal{C}]$ and one has in $\mathcal{U}_\mathcal{C}$ the closed subscheme

$$\mathcal{B}_\mathcal{C} := \operatorname{Spec} \mathbb{Z}[\mathbb{L}_\mathcal{C}] / I_\mathcal{C}.$$

Let us see what this amounts to for the complex points of $\mathcal{U}_\mathcal{C}$, viewed as homomorphisms from the additive monoid $\mathbb{L}_\mathcal{C}$ into the multiplicative monoid $\mathbb{C}$. Each such homomorphism has to send invertible elements to invertible elements, i.e., $\mathbb{L}^0_\mathcal{C}$ into $\mathbb{C}^*$. If $\mathcal{C} = \{0\}$, the set of complex points of $\mathcal{U}_\mathcal{C}$ can therefore be identified with the set (in fact, d-dimensional torus) of group homomorphisms from $\mathbb{L}$ into $\mathbb{C}^*$:

$$\mathcal{U}_{\{0\}}(\mathbb{C}) = \operatorname{Hom}(\mathbb{L}, \mathbb{C}^*) = \mathbb{L}^\vee_\mathbb{Z} \otimes_\mathbb{Z} \mathbb{C}^*. \tag{56}$$

If $\mathcal{C} \neq \{0\}$, the complex points of $\mathcal{B}_\mathcal{C}$ are those monoid homomorphisms that send all elements of $\mathbb{L}^+_\mathcal{C}$ to $\{0\}$. So, the set of complex points of $\mathcal{B}_\mathcal{C}$ can be identified with the set of group homomorphisms from $\mathbb{L}^0_\mathcal{C}$ into $\mathbb{C}^*$:

$$\mathcal{B}_\mathcal{C}(\mathbb{C}) = \operatorname{Hom}(\mathbb{L}^0_\mathcal{C}, \mathbb{C}^*).$$

This is a torus of dimension $d - \dim \mathcal{C}$.

If $\mathcal{C}$ is a maximal cone, then $\mathbb{L}^0_{\mathcal{C}} = \{0\}$ and $\mathcal{B}_{\mathcal{C}}(\mathbb{C})$ is only one point, which we denote as $\mathbf{p}_{\mathcal{C}}$. For every positive real number $r < 1$ the homomorphisms $\mathbb{L}_{\mathcal{C}} \to \mathbb{C}$ mapping $\mathbb{L}^+_{\mathcal{C}}$ into the disc of radius r centred at 0 in $\mathbb{C}$ form an open neighborhood of $\mathbf{p}_{\mathcal{C}}$, which we will also call the disc of radius r about $\mathbf{p}_{\mathcal{C}}$ in $\mathcal{U}_{\mathcal{C}}(\mathbb{C})$.

Example. For Gauss's hypergeometric structures $\mathbb{L} = \mathbb{Z}(1,1,-1,-1) \subset \mathbb{R}^4$. So, $\mathbb{L}^\vee_{\mathbb{R}} \simeq \mathbb{R}$ and the secondary fan has two maximal cones: $\mathbb{R}_{>0}$ and $\mathbb{R}_{<0}$. One can easily see that the associated toric variety is the projective line $\mathbb{P}^1$ (see [9], p. 6).

Example. For Appell's F_4 the secondary fan is shown in Figure 4. One can easily see that the associated toric variety is the projective plane $\mathbb{P}^2$ (see [9], p. 6–7).

Example. For Appell's F_1 the secondary fan is shown in Figure 3. One can easily see that the associated toric variety is the projective plane $\mathbb{P}^2$ with three points blown up (see [9]).

5.2. Convergence of Fourier Γ-series and the secondary fan

We want to use the toric variety associated with the secondary fan to put the domains of convergence of the Fourier Γ-series (37) in the proper perspective. Let us write $\widehat{\mathbf{w}} \in \mathbb{L}^\vee_{\mathbb{C}}$ for the image of $\mathbf{w} \in \mathbb{C}^N$ under the natural projection $\mathbb{C}^N \longrightarrow \mathbb{L}^\vee_{\mathbb{C}} := \mathrm{Hom}\,(\mathbb{L}, \mathbb{C})$ (linear dual; cf. (40)). Then $\mathbf{w} \cdot \underline{\ell} = \langle \widehat{\mathbf{w}}, \underline{\ell} \rangle$ for all $\underline{\ell} \in \mathbb{L}$ and (37) can be rewritten as

$$\Psi_{\mathbb{L},\underline{\gamma}}(\mathbf{w}) = e^{2\pi \mathrm{i} \mathbf{w} \cdot \underline{\gamma}} \sum_{\underline{\ell} \in \mathbb{L}} \frac{e^{2\pi \mathrm{i} \langle \widehat{\mathbf{w}}, \underline{\ell} \rangle}}{\prod_{j=1}^N \Gamma(\gamma_j + \ell_j + 1)} . \tag{57}$$

A vector $\widehat{\mathbf{w}} \in \mathbb{L}^\vee_{\mathbb{C}}$ defines a homomorphism

$$\mathbb{L} \to \mathbb{C}^* , \qquad \underline{\ell} \mapsto e^{2\pi \mathrm{i} \langle \widehat{\mathbf{w}}, \underline{\ell} \rangle}$$

and, hence, a complex point of the toric variety $\mathcal{V}_{\mathcal{A}}$. This point lies in the disc of radius $r < 1$ about the special point $\mathbf{p}_{\mathcal{C}}$ corresponding to a maximal cone $\mathcal{C}$ of the secondary fan if and only if $\langle \Im \widehat{\mathbf{w}}, \underline{\ell} \rangle > -\frac{\log r}{2\pi}$ for every non-zero $\underline{\ell} \in \mathbb{L}_{\mathcal{C}}$; this means that $\Im \widehat{\mathbf{w}}$ should lie 'sufficiently far' inside the cone $\mathcal{C}$.

Recall that a maximal cone $\mathcal{C}$ of the secondary fan corresponds to a regular triangulation of the polytope $\Delta_{\mathcal{A}}$. The index sets of the vertices of the maximal simplices in this triangulation constitute a list $T_{\mathcal{C}}$ of subsets of $\{1, \dots, N\}$ with $N - d$ elements, and according to (44)

$$\mathcal{C} = \bigcap_{J \in T_{\mathcal{C}}} (\text{positive span of } \{\mathbf{b}_j\}_{j \notin J}). \tag{58}$$

Now note that for $\underline{\ell} = (\ell_1, \dots, \ell_N) \in \mathbb{L}$ almost tautologically $\ell_j = \langle \underline{\ell}, \mathbf{b}_j \rangle$. This shows that for $\mathbb{L}_J$ as defined in (38)

$$\mathbb{L}_J \subset \mathbb{L}_{\mathcal{C}} \qquad \text{for every} \quad J \in T_{\mathcal{C}} . \tag{59}$$

The above arguments together with those in Section 3.5 show:

Proposition 7. *Let $\mathcal{C}$ be a maximal cone of the secondary fan. Let $J \in T_{\mathcal{C}}$. Let $\underline{\gamma} = (\gamma_1, \ldots, \gamma_N) \in \mathbb{C}^N$ be such that $\gamma_j \in \mathbb{Z}_{\leq 0}$ for $j \notin J$. Then there is a positive real constant $r < 1$ (depending on $\underline{\gamma}$) such that the Fourier Γ-series $\Psi_{\mathbb{L},\underline{\gamma}}(\mathbf{w})$ in (57) converges for every $\mathbf{w} \in \mathbb{C}^N$ for which $\widehat{\mathbf{w}}$ defines a point in the disc of radius r about the special point $\mathbf{p}_{\mathcal{C}}$ in the toric variety $\mathcal{V}_{\mathcal{A}}$.* □

5.3. Solutions of GKZ differential equations and the secondary fan

Let us look for local solutions to the GKZ differential equations (19)–(20) associated with an N-element subset $\mathcal{A} = \{\mathbf{a}_1, \ldots, \mathbf{a}_N\} \subset \mathbb{Z}^{k+1}$ and a vector $\mathbf{c} \in \mathbb{C}^{k+1}$. Let $\mathcal{C}$ be a maximal cone in the secondary fan of $\mathcal{A}$. According to Proposition 7, every vector $\underline{\gamma} = (\gamma_1, \ldots, \gamma_N) \in \mathbb{C}^N$ which satisfies

$$\gamma_1 \mathbf{a}_1 + \ldots + \gamma_N \mathbf{a}_N \quad = \quad \mathbf{c}\,, \tag{60}$$

$$\exists\, J \in T_{\mathcal{C}} \quad \text{such that} \qquad \gamma_j \in \mathbb{Z}_{\leq 0} \quad \text{for} \quad j \notin J\,, \tag{61}$$

yields a Fourier Γ-series $\Psi_{\mathbb{L},\underline{\gamma}}(\mathbf{w})$ converging for every $\mathbf{w} \in \mathbb{C}^N$ for which $\widehat{\mathbf{w}}$ defines a point in a sufficiently small disc about the point $\mathbf{p}_{\mathcal{C}}$ in $\mathcal{V}_{\mathcal{A}}$. According to Section 3.6 the corresponding Γ-series satisfies the GKZ differential equations (19)–(20) for $\mathcal{A}$ and $\mathbf{c}$. If $\underline{\gamma} \equiv \underline{\gamma}' \bmod \mathbb{L}$, then the two Fourier Γ-series are equal. Lemma 6 will imply that the number of $\mathbb{L}$-congruence classes of solutions to (60)–(61) is finite and, hence, *the Fourier Γ-series we obtain in this way have a common domain of convergence.*

Remark. Because of the factor $e^{2\pi i \mathbf{w}\cdot\underline{\gamma}}$ in (57) the Fourier Γ-series $\Psi_{\mathbb{L},\underline{\gamma}}(\mathbf{w})$ will in general not descend to a function on some disc about $\mathbf{p}_{\mathcal{C}}$ in $\mathcal{V}_{\mathcal{A}}$. On the other hand, if $\underline{\gamma}$ and $\underline{\gamma}'$ both satisfy (60)–(61), then $\underline{\gamma} - \underline{\gamma}' \in \mathbb{L}_{\mathbb{C}}$ and $\mathbf{w} \cdot (\underline{\gamma} - \underline{\gamma}') = \langle \widehat{\mathbf{w}}, \underline{\gamma} - \underline{\gamma}' \rangle$ for every $\mathbf{w} \in \mathbb{C}^N$. This means that the quotient $\Psi_{\mathbb{L},\underline{\gamma}}(\mathbf{w})\Psi^{-1}_{\mathbb{L},\underline{\gamma}'}(\mathbf{w})$ does descend to a function on some disc about $\mathbf{p}_{\mathcal{C}}$ in $\mathcal{V}_{\mathcal{A}}$.

Lemma 6. *Fix $\mathbf{c} \in \mathbb{C}^{k+1}$ and a $k+1$-element set $J \subset \{1, \ldots, N\}$ such that the vectors $\mathbf{a}_j$ with $j \in J$ are linearly independent. Then the number of classes modulo $\mathbb{L}$ of vectors $\underline{\gamma} = (\gamma_1, \ldots, \gamma_N) \in \mathbb{C}^N$ which satisfy Equation (60) and $\gamma_j \in \mathbb{Z}$ for $j \in J' := \{1, \ldots, N\} \setminus J$, is equal to $|\det((\mathbf{a}_j)_{j\in J})|$.*

Proof. Since the vectors $\mathbf{a}_j$ with $j \in J$ are linearly independent, the equation $\sum_{j=1}^N \gamma_j \mathbf{a}_j = \mathbf{c}$ can be solved in parametric form with the components γ_j for $j \in J'$ as free parameters. Every solution is the sum of one particular solution of the inhomogeneous system (e.g., the solution with $\gamma_j = 0$ for $j \in J'$) and a solution of the homogeneous system. So it suffices to determine the number of $\mathbb{L}$-equivalence classes of solutions of the equation $\sum_{j=1}^N \gamma_j \mathbf{a}_j = 0$ with $\gamma_j \in \mathbb{Z}$ for $j \in J'$. The solutions themselves lie in $\mathbb{L} \otimes \mathbb{Q}$.

Take any $d \times N$-matrix B whose rows form a $\mathbb{Z}$-basis of $\mathbb{L}$. This amounts to choosing an isomorphism $\mathbb{L} \simeq \mathbb{Z}^d$. Let $\mathsf{B}_{J'}$ (resp. B_J) denote the submatrix of B formed by the columns with index in J' (resp. in J). As in the proof of Lemma 2 one sees that the matrix $\mathsf{B}_{J'}$ is invertible over $\mathbb{Q}$ and that the set of solutions of $\sum_{j=1}^N \gamma_j \mathbf{a}_j = 0$ with $\gamma_j \in \mathbb{Z}$ for $j \in J'$ is $\mathbb{Z}^d(\mathsf{B}_{J'})^{-1} \subset \mathbb{Q}^d \simeq \mathbb{L} \otimes \mathbb{Q}$; the notation

$\mathbb{Z}^d(\mathsf{B}_{J'})^{-1}$ refers to the fact that here $\mathbb{Z}^d$ consists of row vectors. The number of classes modulo $\mathbb{L}$ of such solutions is therefore

$$\sharp\left(\mathbb{Z}^d(\mathsf{B}_{J'})^{-1}/\mathbb{Z}^d\right) = \sharp\left(\mathbb{Z}^d/\mathbb{Z}^d\mathsf{B}_{J'}\right) = |\det(\mathsf{B}_{J'})|\,.$$

Thus we must prove:

$$|\det(\mathsf{B}_{J'})| = |\det((\mathbf{a}_j)_{j\in J})|\,. \tag{62}$$

Proof of (62): Let A (resp. A_J resp. $\mathsf{A}_{J'}$) denote the matrix with columns $\mathbf{a}_j$ with $j \in \{1,\dots,N\}$ (resp. $j \in J$ resp. $j \in J'$). Then $\mathsf{A}\,\mathsf{B}^t = 0$ and hence

$$\mathsf{A}_J^{-1}\mathsf{A}_{J'} = -(\mathsf{B}_{J'}^{-1}\mathsf{B}_J)^t\,. \tag{63}$$

As in Cramer's rule one sees that the matrix entries on the left hand side of (63) are all of the form $\pm(\det\mathsf{A}_J)^{-1}(\det\mathsf{A}_I)$ with $I \subset \{1,\dots,N\}$ such that $\sharp I = k+1$ and $\sharp(I\cap J) = k$. The corresponding matrix entries on the right-hand side of (63) are $\pm(\det\mathsf{B}_{J'})^{-1}(\det\mathsf{B}_{I'})$ with $I' := \{1,\dots,N\}\setminus I$. Thus we see

$$|\det\mathsf{A}_J|^{-1}|\det\mathsf{A}_I| = |\det\mathsf{B}_{J'}|^{-1}|\det\mathsf{B}_{I'}|\,,$$

first for every $I \subset \{1,\dots,N\}$ such that $\sharp I = k+1$ and $\sharp(I\cap J) = k$ and then, by induction, for every $k+1$-element subset $I \subset \{1,\dots,N\}$. Consequently there are coprime positive integers a, b such that

$$b|\det\mathsf{A}_I| = a|\det\mathsf{B}_{I'}| \tag{64}$$

for every $k+1$-element subset $I \subset \{1,\dots,N\}$. Now recall that the columns $\mathbf{a}_1,\dots,\mathbf{a}_N$ of A generate $\mathbb{Z}^{k+1}$. This implies that the greatest common divisor of the numbers $\det\mathsf{A}_I$ is 1. So $a = 1$ in (64). On the other hand, the rows of B form a $\mathbb{Z}$-basis of $\mathbb{L}$. Therefore for every prime number p the rows the matrix $\mathsf{B} \bmod p\mathbb{Z}$ are linearly independent over the field $\mathbb{Z}/p\mathbb{Z}$ and at least one of the numbers $\det\mathsf{B}_{I'}$ must be not divisible by p. This shows $b = 1$ in (64) and finishes the proof of Formula (62). □

Lemma 7. *Let $\mathcal{C}$ be a maximal cone of the secondary fan. Let $\underline{\gamma}^1,\dots,\underline{\gamma}^p$ be solutions to the equations* (60)–(61) *such that $\underline{\gamma}^i \not\equiv \underline{\gamma}^j \bmod \mathbb{L}$ for $i \neq j$. Then the Fourier Γ-series $\Psi_{\mathbb{L},\underline{\gamma}^1}(\mathbf{w}),\dots,\Psi_{\mathbb{L},\underline{\gamma}^p}(\mathbf{w})$ are linearly independent over $\mathbb{C}$.*

Proof. Fix a positive real constant $r < 1$ such that all the given Fourier Γ-series converge for every $\mathbf{w} \in \mathbb{C}^N$ for which $\widehat{\mathbf{w}}$ defines a point in the disc of radius r about the special point $\mathbf{p}_\mathcal{C}$ in the toric variety $\mathcal{V}_\mathcal{A}$ (cf. Proposition 7). Next choose $\mathbf{w} \in \mathbb{C}^N$ such that $\widehat{\mathbf{w}}$ defines a point in that disc and such that no two of the numbers $\Im\mathbf{w}\cdot(\underline{\gamma}^j + \underline{\ell})$ with $1 \le j \le p$ and $\underline{\ell} \in \mathbb{L}_\mathcal{C}$ such that the $\underline{\ell}$-th term in the Fourier Γ-series $\Psi_{\mathbb{L},\underline{\gamma}^j}(\mathbf{w})$ is not 0, are equal. For this $\mathbf{w}$ the set of those numbers $\Im\mathbf{w}\cdot(\underline{\gamma}^j + \underline{\ell})$ assumes its minimum for a unique pair, say $(\underline{\gamma}^m, \underline{\ell}^m)$. This implies

$$\lim_{t\to\infty} e^{-2\pi i t\mathbf{w}\cdot(\underline{\gamma}^m+\underline{\ell}^m)}\Psi_{\mathbb{L},\underline{\gamma}^j}(t\mathbf{w}) = 0 \quad \text{if} \quad j \neq m\,, \qquad \text{resp.} \quad \neq 0 \quad \text{if} \quad j = m\,.$$

The linear independence claimed in the lemma now follows immediately. □

It follows from Lemma 6 that the number of $\mathbb{L}$-congruence classes of solutions to the Equations (60)–(61) is less than or equal to

$$\sum_{J \in T_{\mathcal{C}}} |\det((\mathbf{a}_j)_{j\in J})| = \text{volume}\,\Delta_{\mathcal{A}}\,.$$

Definition 8. *Let $\mathcal{C}$ be a maximal cone of the secondary fan of $\mathcal{A} = \{\mathbf{a}_1, \ldots, \mathbf{a}_N\}$ and let $\mathbf{c} \in \mathbb{C}^{k+1}$. One says that $\mathbf{c}$ is $\mathcal{C}$-*resonant *if the number of $\mathbb{L}$-congruence classes of solutions to the equations (60)-(61) is less than* volume $\Delta_{\mathcal{A}}$.

This means that $\mathbf{c}$ is $\mathcal{C}$-resonant if and only if there is a $\underline{\gamma} = (\gamma_1, \ldots, \gamma_N) \in \mathbb{C}^N$ which satisfies $\gamma_1\mathbf{a}_1 + \ldots + \gamma_N\mathbf{a}_N = \mathbf{c}$, and for which there are two different sets J_1 and J_2 on the list $T_{\mathcal{C}}$ such that $\gamma_j \in \mathbb{Z}$ for $j \in \{1, \ldots, N\} \setminus (J_1 \cap J_2)$.

Corollary 2. *If $\mathbf{c}$ is not $\mathcal{C}$-resonant, the Fourier Γ-series $\Psi_{\mathbb{L},\underline{\gamma}}(\mathbf{w})$ associated with solutions $\underline{\gamma}$ of the equations* (60)–(61) *are linearly independent and span a space of local solutions of the GKZ differential equations* (19)–(20) *of dimension equal to* volume $\Delta_{\mathcal{A}}$. *According to the discussion in Section* 2.7 *this is then the full space of local solutions if (for instance) the polytope $\Delta_{\mathcal{A}}$ admits a unimodular triangulation.* □

6. Extreme resonance in GKZ systems

In this section $\mathcal{C}$ is a maximal cone of the secondary fan of $\mathcal{A} = \{\mathbf{a}_1, \ldots, \mathbf{a}_N\}$ for which the corresponding regular triangulation of $\Delta_{\mathcal{A}}$ is unimodular, i.e.,

$$|\det((\mathbf{a}_j)_{j\in J})| = 1 \qquad \text{for every} \quad J \in T_{\mathcal{C}}.$$

This means that for every $J \in T_{\mathcal{C}}$ the set $\{\mathbf{a}_j\}_{j\in J}$ is a $\mathbb{Z}$-basis of $\mathbb{Z}^{k+1}$. Consequently, for every $\mathbf{c} \in \mathbb{Z}^{k+1}$ all solutions $\underline{\gamma} = (\gamma_1, \ldots, \gamma_N)$ of the equations (60)–(61) lie in $\mathbb{Z}^N$ and are therefore congruent modulo $\mathbb{L}$. So a vector $\mathbf{c} \in \mathbb{Z}^{k+1}$ is $\mathcal{C}$-resonant, in an extreme way: all Fourier Γ-series coming from solutions of (60)–(61) are equal! *In this section we will demonstrate how one can obtain, locally near the point $\mathbf{p}_{\mathcal{C}}$ on $\mathcal{V}_{\mathcal{A}}$, more solutions of the GKZ differential equations* (19)–(20) *from an 'infinitesimal deformation' of this Fourier Γ-series.*

Definition 9. *For $\mathcal{A}$ and $\mathcal{C}$ as above we define the ring*

$$\mathcal{R}_{\mathcal{A},\mathcal{C}} := \mathbb{Z}[E_1, \ldots, E_N] / (\mathcal{I}_{\mathcal{A}} + \mathcal{I}_{\mathcal{C}}) \tag{65}$$

where $\mathbb{Z}[E_1, \ldots, E_N]$ is just the polynomial ring over $\mathbb{Z}$ in N variables,
$\mathcal{I}_{\mathcal{A}}$ is the ideal generated by the linear forms which are the components of the vector

$$E_1\mathbf{a}_1 + \ldots + E_N\mathbf{a}_N\,, \tag{66}$$

and $\mathcal{I}_{\mathcal{C}}$ is the ideal generated by the monomials

$$E_{i_1} \cdots E_{i_s} \qquad \text{with} \quad \{i_1, \ldots, i_s\} \not\subset J \quad \text{for all} \quad J \in T_{\mathcal{C}}\,. \tag{67}$$

We write ε_j for the image of E_j in $\mathcal{R}_{\mathcal{A},\mathcal{C}}$.

So in $\mathcal{R}_{\mathcal{A},\mathcal{C}}$ we have the relations

$$\varepsilon_1\mathbf{a}_1 + \dots \varepsilon_N\mathbf{a}_N = 0\,, \tag{68}$$

$$\varepsilon_{i_1}\cdots\varepsilon_{i_s} = 0 \quad \text{if} \quad \{i_1,\dots,i_s\} \not\subset J \quad \text{for all} \quad J \in T_{\mathcal{C}}\,. \tag{69}$$

Relation (68) means that the vector $\underline{\varepsilon} = (\varepsilon_1,\dots,\varepsilon_N) \in \mathcal{R}^N_{\mathcal{A},\mathcal{C}}$ lies in $\mathbb{L}\otimes_{\mathbb{Z}}\mathcal{R}_{\mathcal{A},\mathcal{C}}$.

Remark. The ideal $\mathcal{I}_{\mathcal{C}}$ is well known in combinatorial algebra [23], where it is called the *Stanley–Reisner ideal* of the triangulation $T_{\mathcal{C}}$. The ring $\mathbb{Z}[E_1,\dots,E_N]/\mathcal{I}_{\mathcal{C}}$ is called the *Stanley–Reisner ring.*

The following facts about the ring $\mathcal{R}_{\mathcal{A},\mathcal{C}}$ are proven in [24], §2.

Proposition 8. 1. *$\mathcal{R}_{\mathcal{A},\mathcal{C}}$ is a free $\mathbb{Z}$-module of rank equal to* volume $\Delta_{\mathcal{A}}$.

2. *$\mathcal{R}_{\mathcal{A},\mathcal{C}}$ is a graded ring and each ε_j has degree* 1.
3. *Denoting the homogeneous part of degree i in $\mathcal{R}_{\mathcal{A},\mathcal{C}}$ by $\mathcal{R}^{(i)}_{\mathcal{A},\mathcal{C}}$ one has isomorphisms (see also §4.1)*

$$\mathcal{R}^{(0)}_{\mathcal{A},\mathcal{C}} = \mathbb{Z}\,, \qquad \mathcal{R}^{(1)}_{\mathcal{A},\mathcal{C}} \simeq \mathbb{L}^{\vee}_{\mathbb{Z}}\,, \quad \varepsilon_j \mapsto \mathbf{b}_j\,. \tag{70}$$

4. *The Poincaré series of the graded ring $\mathcal{R}_{\mathcal{A},\mathcal{C}}$ is*

$$\sum_{i\geq 0}\left(\operatorname{rank}\mathcal{R}^{(i)}_{\mathcal{A},\mathcal{C}}\right)T^i = \sum_{m=0}^{k+1} S_{\mathcal{C},m}\,T^m(1-T)^{k+1-m}\,, \tag{71}$$

where $S_{\mathcal{C},0}=1$ and $S_{\mathcal{C},m}$, for $m\geq 1$, is the number of simplices with m vertices in the triangulation of $\Delta_{\mathcal{A}}$ corresponding with $\mathcal{C}$. In particular $\mathcal{R}^{(i)}_{\mathcal{A},\mathcal{C}}=0$ for $i\geq k+1$ and the elements $\varepsilon_1,\dots,\varepsilon_N$ are nilpotent. □

For the examples at the end of Sections 4.3 and 4.1 (see also Figures 1, 3, 4) we find:

Example. For $\mathbb{L} = \mathbb{Z}(-2,1,1)\subset\mathbb{Z}^3$ there is only one unimodular triangulation, namely $T_{\mathcal{C}} = \{\{1,2\},\{1,3\}\}$. One easily checks that in this case

$$\mathcal{R}_{\mathcal{A},\mathcal{C}} = \mathbb{Z}\oplus\mathbb{Z}\varepsilon\,, \qquad \varepsilon^2 = 0\,, \qquad (\varepsilon_1,\varepsilon_2,\varepsilon_3) = (-2\varepsilon,\varepsilon,\varepsilon)\,.$$

Example. For Gauss, $\mathbb{L} = \mathbb{Z}(1,1,-1,-1)\subset\mathbb{Z}^4$ and there are two unimodular triangulations, which both lead to

$$\mathcal{R}_{\mathcal{A},\mathcal{C}} = \mathbb{Z}\oplus\mathbb{Z}\varepsilon\,, \qquad \varepsilon^2 = 0\,.$$

For one triangulation $(\varepsilon_1,\varepsilon_2,\varepsilon_3,\varepsilon_4)$ is $(-\varepsilon,-\varepsilon,\varepsilon,\varepsilon)$, for the other $(\varepsilon,\varepsilon,-\varepsilon,-\varepsilon)$.

Example: For $\mathbb{L} = \mathbb{Z}(-3\,1\,1\,1)\subset\mathbb{Z}^4$ there is only one unimodular triangulation, namely $T_{\mathcal{C}} = \{\,\{1,3,4\}\,,\{1,2,4\}\,,\{1,2,3\}\,\}$. One easily checks that in this case

$$\mathcal{R}_{\mathcal{A},\mathcal{C}} = \mathbb{Z}\oplus\mathbb{Z}\varepsilon\oplus\mathbb{Z}\varepsilon^2\,, \qquad \varepsilon^3 = 0\,, \qquad (\varepsilon_1,\varepsilon_2,\varepsilon_3,\varepsilon_4) = (-3\varepsilon,\varepsilon,\varepsilon,\varepsilon)\,.$$

Example. For Appell's F_1, $\mathbb{L} = \mathbb{Z}(1,-1,0,-1,1,0) \oplus \mathbb{Z}(1,0,-1,-1,0,1)$. There are six unimodular triangulations (see Figure 3). One can check that for the triangulation $T_{\mathcal{C}} = \{\{3,4,5,6\}, \{1,2,3,4\}, \{2,3,4,5\}\}$ the relations (66)–(67) yield

$$\begin{aligned}(\varepsilon_1,\varepsilon_2,\varepsilon_3,\varepsilon_4,\varepsilon_5,\varepsilon_6) &= \varepsilon(1,-1,0,-1,1,0) + \delta(1,0,-1,-1,0,1)\,,\\ \varepsilon_1\varepsilon_5 \;=\; \varepsilon_1\varepsilon_6 \;&=\; \varepsilon_2\varepsilon_6 \;=\; 0\,,\end{aligned}$$

and hence:

$$\mathcal{R}_{\mathcal{A},\mathcal{C}} \;=\; \mathbb{Z}\oplus\mathbb{Z}\varepsilon\oplus\mathbb{Z}\delta\,, \qquad \varepsilon^2 \;=\; \delta^2 \;=\; \varepsilon\delta \;=\; 0\,.$$

Example. For Appell's F_4, $\mathbb{L} = \mathbb{Z}(1,-1,1,-1,0,0) \oplus \mathbb{Z}(1,0,1,0,0,-1,-1)$. There are three unimodular triangulations (see Figure 4). One can check that for the triangulation $T_{\mathcal{C}} = \{\{1,3,4,6\}, \{1,3,4,5\}, \{1,2,3,5\}, \{1,2,3,6\}\}$ the relations (66)–(67) yield

$$\begin{aligned}(\varepsilon_1,\varepsilon_2,\varepsilon_3,\varepsilon_4,\varepsilon_5,\varepsilon_6) &= \varepsilon(1,-1,1,-1,0,0) + \delta(1,0,1,0,0,-1,-1)\,,\\ \varepsilon_2\varepsilon_4 \;=\; \varepsilon_5\varepsilon_6 \;&=\; 0\,,\end{aligned}$$

and hence:

$$\mathcal{R}_{\mathcal{A},\mathcal{C}} \;=\; \mathbb{Z}\oplus\mathbb{Z}\varepsilon\oplus\mathbb{Z}\delta\oplus\mathbb{Z}\varepsilon\delta\,, \qquad \varepsilon^2 \;=\; \delta^2 \;=\; 0\,.$$

For $z \in \mathbb{C}$ and nilpotent ε one can define $\frac{1}{\Gamma(z+\varepsilon)}$ as an element of $\mathbb{C}[\varepsilon]$ by using the Taylor expansion of the function $\frac{1}{\Gamma}$ at z:

$$\frac{1}{\Gamma(z+\varepsilon)} \;:=\; \frac{1}{\Gamma(z)} + \varepsilon\left(\frac{1}{\Gamma}\right)'(z) + \frac{\varepsilon^2}{2}\left(\frac{1}{\Gamma}\right)''(z) + \frac{\varepsilon^3}{3!}\left(\frac{1}{\Gamma}\right)'''(z) + \dots .$$

One defines similarly $\Gamma(1+\varepsilon)$. Thus for $z \in \mathbb{C}$ and nilpotent ε also $\frac{\Gamma(1+\varepsilon)}{\Gamma(z+1+\varepsilon)}$ has been defined. From (30) one sees that for $m \in \mathbb{Z}$:

$$\frac{\Gamma(1+\varepsilon)}{\Gamma(m+1+\varepsilon)} \;=\; \begin{cases} \dfrac{1}{(1+\varepsilon)(2+\varepsilon)\cdots(m+\varepsilon)} & \text{if} \quad m>0 \\ 1 & \text{if} \quad m=0 \\ \varepsilon(\varepsilon-1)(\varepsilon-2)\cdots(\varepsilon+m+1) & \text{if} \quad m<0. \end{cases} \tag{72}$$

Finally, for $z \in \mathbb{C}$, $u \in \mathbb{C}^*$ (with a choice of a branch of $\log u$) and nilpotent ε one has naturally

$$e^{\varepsilon\, z} \;:=\; \sum_{m\geq 0} \frac{1}{m!}\varepsilon^m\, z^m\,, \qquad u^{\varepsilon} \;:=\; e^{\varepsilon\,\log u}\,.$$

We are ready to present our deformation of the (Fourier) Γ-series:

Definition 10. *For* $\underline{\gamma} = (\gamma_1,\dots,\gamma_N) \in \mathbb{Z}^N$ *and* $\underline{\varepsilon} = (\varepsilon_1,\dots,\varepsilon_N) \in \mathcal{R}^N_{\mathcal{A},\mathcal{C}}$ *we define*

$$\Psi_{\mathbb{L},\underline{\gamma},\underline{\varepsilon}}(\mathbf{w}) \quad:=\quad \sum_{\underline{\ell}\in\mathbb{L}} \prod_{j=1}^{N} \frac{\Gamma(1+\varepsilon_j)}{\Gamma(\gamma_j+\ell_j+1+\varepsilon_j)}\; e^{2\pi \mathrm{i}\mathbf{w}\cdot(\underline{\gamma}+\underline{\ell}+\underline{\varepsilon})}\,, \tag{73}$$

$$\Phi_{\mathbb{L},\underline{\gamma},\underline{\varepsilon}}(\mathbf{u}) \quad:=\quad \sum_{\underline{\ell}\in\mathbb{L}} \prod_{j=1}^{N} \frac{\Gamma(1+\varepsilon_j)\, u_j^{\gamma_j+\ell_j+\varepsilon_j}}{\Gamma(\gamma_j+\ell_j+1+\varepsilon_j)}\,. \tag{74}$$

Remark. From the point of view of deforming $\underline{\gamma}$ it seems more natural to consider

$$\Psi_{\mathbb{L},\underline{\gamma}+\underline{\varepsilon}}(\mathbf{w}) \quad := \quad \sum_{\underline{\ell}\in\mathbb{L}} \prod_{j=1}^{N} \frac{1}{\Gamma(\gamma_j+\ell_j+1+\varepsilon_j)}\, e^{2\pi i \mathbf{w}\cdot(\underline{\gamma}+\underline{\ell}+\underline{\varepsilon})}\,, \tag{75}$$

$$\Phi_{\mathbb{L},\underline{\gamma}+\underline{\varepsilon}}(\mathbf{u}) \quad := \quad \sum_{\underline{\ell}\in\mathbb{L}} \prod_{j=1}^{N} \frac{u_j^{\gamma_j+\ell_j+\varepsilon_j}}{\Gamma(\gamma_j+\ell_j+1+\varepsilon_j)}\,; \tag{76}$$

i.e.,

$$\Psi_{\mathbb{L},\underline{\gamma}+\underline{\varepsilon}}(\mathbf{w}) = \frac{\Psi_{\mathbb{L},\underline{\gamma},\underline{\varepsilon}}(\mathbf{w})}{\prod_{j=1}^{N}\Gamma(1+\varepsilon_j)}\,, \qquad \Phi_{\mathbb{L},\underline{\gamma}+\underline{\varepsilon}}(\mathbf{u}) = \frac{\Phi_{\mathbb{L},\underline{\gamma},\underline{\varepsilon}}(\mathbf{u})}{\prod_{j=1}^{N}\Gamma(1+\varepsilon_j)}\,.$$

Indeed, expanding these functions in coordinates with respect to a basis of $\mathcal{R}_{\mathcal{A},\mathcal{C}}$ is for (75) and (76) essentially just Taylor expansion, if one views the expressions as functions of $\underline{\gamma}$, while the interpretation as (multi-valued) local solutions of GKZ differential equations with values in $\mathcal{R}_{\mathcal{A},\mathcal{C}} \otimes \mathbb{C}$ (see below) are equally true for (75)–(76) in place of (73)–(74). We prefer, however, the latter because their coordinates are series with rational coefficients, whereas the coefficients of the coordinate series of the former involve interesting, but mysterious non-rational numbers like the Euler–Masceroni constant and values of Riemann's zeta-function. We can be slightly more informative about the coefficients in (75)–(76): there is the well-known formula for the Γ-function due to Gauss

$$\Gamma(s) \;=\; \lim_{n\to\infty}\left[\frac{n!\,n^s}{s(s+1)\cdots(s+n)}\right]\,,$$

from which one easily derives the expansion

$$\log\Gamma(1+s) \;=\; -\curlyvee s + \sum_{m=2}^{\infty}(-1)^m\zeta(m)\frac{s^m}{m}$$

where $\curlyvee$ denotes the Euler–Masceroni constant and ζ is Riemann's zeta-function. By exponentiating and re-expanding one finds the Taylor expansion for $\Gamma(1+s)$ and then eventually the expansion of $\left[\prod_{j=1}^{N}\Gamma(1+\varepsilon_j)\right]^{-1}$.

Lemma 8. *There are finitely many* $\underline{\ell}^{(1)},\ldots,\underline{\ell}^{(r)} \in \mathbb{L}_{\mathcal{C}}$ *(with* $\mathbb{L}_{\mathcal{C}}$ *as in* (55)) *such that the series* $\Psi_{\mathbb{L},\underline{\gamma},\underline{\varepsilon}}(\mathbf{w})$ *and* $\Phi_{\mathbb{L},\underline{\gamma},\underline{\varepsilon}}(\mathbf{u})$ *involve only terms with*

$$\underline{\ell} \in \bigcup_{i=1}^{r}(-\underline{\ell}^{(i)} + \mathbb{L}_{\mathcal{C}})\,.$$

In particular for $\underline{\gamma} = 0$ *the series involve only terms with* $\underline{\ell} \in \mathbb{L}_{\mathcal{C}}$.

Proof. It follows immediately from (72) and (67) that for the terms which appear with non-zero coefficient, the set $\{j \mid \gamma_j + \ell_j < 0\}$ is contained in some J on the list $T_{\mathcal{C}}$. Suppose $\{j \mid \gamma_j + \ell_j < 0\} \subset J \in T_{\mathcal{C}}$. Then $\gamma_j + \ell_j \geq 0$ for every $j \in J' := \{1,\ldots,N\} \setminus J$. The vector $\sum_{j\in J'} \max(0,\gamma_j)\,\mathbf{a}_j$ is a $\mathbb{Z}$-linear combination of the vectors $\mathbf{a}_i$ with $i \in J$, because the triangulation is unimodular. Such a relation is an element of $\mathbb{L}$. Thus one sees that $\mathbb{L}$ contains an element $\underline{\ell}^J = (\ell_1^J, \ldots, \ell_N^J)$ with $\ell_j^J = \max(0, \gamma_j)$ for all $j \in J'$. So $\ell_j^J + \ell_j \geq 0$ for every $j \in J'$. In the notation introduced in (38) this can be written as $\underline{\ell}^J \in \mathbb{L}_J$ and $\underline{\ell}^J + \underline{\ell} \in \mathbb{L}_J$. The lemma now follows from (59). □

Partial sums (with finitely many terms) of the series (73) resp. (74) can be evaluated as elements in the ring $\mathcal{R}_{\mathcal{A},\mathcal{C}} \otimes \mathbb{C}$ and be written in coordinates with respect to a $\mathbb{Z}$-basis of the finite rank $\mathbb{Z}$-module $\mathcal{R}_{\mathcal{A},\mathcal{C}}$. These coordinates are again partial sums of series. In [24], §3 one finds estimates on the growth of the coefficients of these series and on a common domain of convergence. Thus $\Psi_{\mathbb{L},\underline{\gamma},\underline{\varepsilon}}(\mathbf{w})$ and $\Phi_{\mathbb{L},\underline{\gamma},\underline{\varepsilon}}(\mathbf{u})$ are functions with values in $\mathcal{R}_{\mathcal{A},\mathcal{C}} \otimes \mathbb{C}$. The function $\Psi_{\mathbb{L},\underline{\gamma},\underline{\varepsilon}}(\mathbf{w})$ is defined for $\mathbf{w} \in \mathbb{C}^N$ with $\Im\widehat{\mathbf{w}}$ 'sufficiently far' inside the cone $\mathcal{C}$ (cf. §5.2). Because of the appearance of logarithms $\Phi_{\mathbb{L},\underline{\gamma},\underline{\varepsilon}}(\mathbf{u})$ is actually a multi-valued function, defined on some open disc about $\mathsf{0}$ in $\mathbb{C}^{\mathcal{A}}$ with the divisor $u_1 \cdots u_N = 0$ removed. The multi-valuedness is easily described using the relation $u_j = e^{2\pi i w_j}$ which matches w_j with a choice of $\log u_j$. A different choice adds an integer to w_j. Now note that for $\mathsf{m} \in \mathbb{Z}^N$

$$\Psi_{\mathbb{L},\underline{\gamma},\underline{\varepsilon}}(\mathbf{w} + \mathsf{m}) = e^{2\pi i \mathsf{m}\cdot\underline{\varepsilon}}\, \Psi_{\mathbb{L},\underline{\gamma},\underline{\varepsilon}}(\mathbf{w})\,.$$

This formula can also be read as a precise expression for local monodromy. Since $\{\mathsf{m}\cdot\underline{\varepsilon} \mid \mathsf{m} \in \mathbb{Z}^N\} = \mathcal{R}^{(1)}_{\mathcal{A},\mathcal{C}}$, we can summarize our analysis of the multi-valuedness of $\Phi_{\mathbb{L},\underline{\gamma},\underline{\varepsilon}}(\mathbf{u})$ as follows:

Proposition 9. *$\Phi_{\mathbb{L},\underline{\gamma},\underline{\varepsilon}}(\mathbf{u})$ is a multi-valued function with values in $\mathcal{R}_{\mathcal{A},\mathcal{C}} \otimes \mathbb{C}$. Different branches of this function are related by multiplication with an element $e^{2\pi i\omega}$ with $\omega \in \mathcal{R}^{(1)}_{\mathcal{A},\mathcal{C}}$.* □

The same arguments as those used in Section 3.6 show immediately

Proposition 10. *The $\mathcal{R}_{\mathcal{A},\mathcal{C}} \otimes \mathbb{C}$-valued function $\Phi_{\mathbb{L},\underline{\gamma},\underline{\varepsilon}}(\mathbf{u})$ satisfies the GKZ system of differential equations* (19)–(20) *for $\mathcal{A}$ and $\mathbf{c} = \sum_{j=1}^N \gamma_j \mathbf{a}_j$.*

The $\mathbb{C}$-valued functions which arise as coordinates of $\Phi_{\mathbb{L},\underline{\gamma},\underline{\varepsilon}}(\mathbf{u})$ with respect to a basis of $\mathcal{R}_{\mathcal{A},\mathcal{C}}$ satisfy the same GKZ system of differential equations. □

Example. For $\mathbb{L} = \mathbb{Z}(-3,1,1,1)$ and $\mathcal{C}$ the cone corresponding to the unimodular triangulation $T_{\mathcal{C}} = \{\{1,2,3\},\{1,2,4\},\{1,3,4\}\}$,

$$\mathcal{R}_{\mathcal{A},\mathcal{C}} = \mathbb{Z} \oplus \mathbb{Z}\varepsilon \oplus \mathbb{Z}\varepsilon^2\,, \qquad \varepsilon^3 = 0\,, \qquad (\varepsilon_1,\varepsilon_2,\varepsilon_3,\varepsilon_4) = (-3\varepsilon,\varepsilon,\varepsilon,\varepsilon)\,.$$

For $\underline{0} = (0,0,0,0)$ one then finds, using (72) and setting $z = u_1^{-3}u_2u_3u_4$,

$$\begin{aligned}
\Phi_{\mathbb{L},\underline{0},\underline{\varepsilon}}(\mathbf{u}) &= \sum_{m\in\mathbb{Z}} \frac{\Gamma(1-3\varepsilon)}{\Gamma(1-3m-3\varepsilon)} \left(\frac{\Gamma(1+\varepsilon)}{\Gamma(1+m+\varepsilon)}\right)^3 u_1^{-3m-3\varepsilon} u_2^{m+\varepsilon} u_3^{m+\varepsilon} u_4^{m+\varepsilon} \\
&= z^{\varepsilon} \left(1 + \sum_{m\geq 1} \frac{(-3\varepsilon)(-3\varepsilon-1)\cdots(-3\varepsilon-3m+1)}{((1+\varepsilon)\cdots(m+\varepsilon))^3}\, z^m\right) \\
&= \left(1 + \varepsilon\log z + \frac{\varepsilon^2}{2}\log^2 z\right)\left(1 + \varepsilon G_1(z) + \varepsilon^2 G_2(z)\right) \\
&= 1 + (\log z + G_1(z))\varepsilon + (\tfrac{1}{2}\log^2 z + G_1(z)\log z + G_2(z))\varepsilon^2
\end{aligned}$$

with

$$\begin{aligned}
G_1(z) &= 3\sum_{m\geq 1} (-1)^m \frac{(3m-1)!}{(m!)^3}\, z^m \\
G_2(z) &= 9\sum_{m\geq 1} (-1)^m \frac{(3m-1)!}{(m!)^3} \left(\sum_{j=m+1}^{3m-1} \frac{1}{j}\right) z^m .
\end{aligned}$$

Similarly, for $\underline{\gamma} = (-1,0,0,0)$ we obtain

$$\begin{aligned}
\Phi_{\mathbb{L},\underline{\gamma},\underline{\varepsilon}}(\mathbf{u}) &= \sum_{m\in\mathbb{Z}} \frac{\Gamma(1-3\varepsilon)}{\Gamma(-3m-3\varepsilon)} \left(\frac{\Gamma(1+\varepsilon)}{\Gamma(m+1+\varepsilon)}\right)^3 u_1^{-1-3m-3\varepsilon} u_2^{m+\varepsilon} u_3^{m+\varepsilon} u_4^{m+\varepsilon} \\
&= u_1^{-1} \sum_{m\geq 0} \frac{(-3\varepsilon)(-3\varepsilon-1)\cdots(-3\varepsilon-3m)}{((1+\varepsilon)\cdots(m+\varepsilon))^3}\, (u_1^{-3}u_2u_3u_4)^{m+\varepsilon} \\
&= u_1^{-1}\left(1 + \varepsilon\log z + \frac{\varepsilon^2}{2}\log^2 z\right)\left(\varepsilon F_1(z) + \varepsilon^2 F_2(z)\right) \\
&= u_1^{-1} F_1(z)\varepsilon + u_1^{-1}(F_1(z)\log z + F_2(z))\varepsilon^2
\end{aligned}$$

with

$$\begin{aligned}
F_1(z) &= -3\sum_{m\geq 0} (-1)^m \frac{(3m)!}{(m!)^3} z^m , \\
F_2(z) &= -9\sum_{m\geq 1} (-1)^m \frac{(3m)!}{(m!)^3} \left(\sum_{j=m+1}^{3m} \frac{1}{j}\right) z^m .
\end{aligned}$$

Note that in agreement with Proposition 1

$$\Phi_{\mathbb{L},\underline{\gamma},\underline{\varepsilon}}(\mathbf{u}) = \frac{\partial}{\partial u_1}\Phi_{\mathbb{L},\underline{0},\underline{\varepsilon}}(\mathbf{u}) = -3u_1^{-1} z \frac{\partial}{\partial z}\Phi_{\mathbb{L},\underline{0},\underline{\varepsilon}}(\mathbf{u}) .$$

The components of $\Phi_{\mathbb{L},\underline{0},\underline{\varepsilon}}(\mathbf{u})$ are three linearly independent solutions of the GKZ system of differential equations with $\mathbf{c} = 0$, whereas the components of $\Phi_{\mathbb{L},\underline{\gamma},\underline{\varepsilon}}(\mathbf{u})$ yield only two linearly independent solutions of the GKZ system for $\mathbf{c} = -\mathbf{a}_1$.

Since in this case volume $\Delta_{\mathcal{A}} = 3$ we find enough solutions for $\mathbf{c} = 0$, but not enough for $\mathbf{c} = -\mathbf{a}_1$ (see Section 2.7).

The phenomenon observed at the end of the previous example — namely that our method yields enough solutions if $\mathbf{c} = 0$, but misses solutions if $\mathbf{c} \neq 0$ — occurs quite generally. Below, in Theorem 3, we quote [24] Theorem 5 and also recall some conclusions (e.g., Proposition 9) found earlier in the present notes:

Theorem 3. *Let $\mathcal{C}$ be a maximal cone of the secondary fan of $\mathcal{A}$ for which the corresponding regular triangulation of $\Delta_{\mathcal{A}}$ is unimodular. Let $\underline{0} = (0, \ldots, 0)$. Then the coordinates of the $\mathcal{R}_{\mathcal{A},\mathcal{C}} \otimes \mathbb{C}$-valued function $\Phi_{\mathbb{L},\underline{0},\underline{\varepsilon}}(\mathbf{u})$ with respect to a basis of the free $\mathbb{Z}$-module $\mathcal{R}_{\mathcal{A},\mathcal{C}}$ constitute a basis for the local solution space of the GKZ system of differential equations (19)–(20) for $\mathcal{A}$ and $\mathbf{c} = 0$. These multi-valued functions are invariant under the action (1) of the torus $\mathbb{T}^{k+1}$ and descend therefore to multi-valued functions on a disc minus a divisor centered at the point $\mathbf{p}_{\mathcal{C}}$ in the toric variety $\mathcal{V}_{\mathcal{A}}$. The multi-valuedness of these functions is given by multiplying $\Phi_{\mathbb{L},\underline{0},\underline{\varepsilon}}(\mathbf{u})$ with elements in the group $\{e^{2\pi i\omega} \mid \omega \in \mathcal{R}^{(1)}_{\mathcal{A},\mathcal{C}}\}$.* □

Remark. In [4] Anne de Boo carefully re-examined the preceding method and improved it by also taking $\underline{\gamma}$ into account. In this way he obtained full local solution spaces for GKZ systems of differential equations for many more instances of the triangulation of $\Delta_{\mathcal{A}}$ and of the parameter $\mathbf{c}$.

Very recently Borisov and Horja [5] found a way to obtain enough solutions for any $\mathbf{c} \in \mathbb{Z}^{k+1}$ and any triangulation. Their method is close in spirit to the method in this Section 6. We recommend [5] for further reading on this aspect of GKZ hypergeometric structures.

7. GKZ for Lauricella's F_D

Since Lauricella's F_D also plays an important role in other lectures in this School, we put details of the GKZ theory for Lauricella's F_D together in this section.

7.1. Series, $\mathbb{L}$, $\mathcal{A}$ and the primary polytope $\Delta_{\mathcal{A}}$

Recall that in Section 3.2.4 we found, starting from the power series expansion of Lauricella's F_D in $k-1$ variables

$$F_D(a, \mathbf{b}, c|\mathbf{z}) := \sum_{\mathbf{m}} \frac{(a)_{|\mathbf{m}|}(\mathbf{b})_{\mathbf{m}}}{(c)_{|\mathbf{m}|}\mathbf{m}!} \mathbf{z}^{\mathbf{m}},$$

that the lattice $\mathbb{L}$ is generated by the rows of the following $(k-1) \times (2k)$-matrix

$$\begin{pmatrix} 1 & -1 & 0 & \ldots & 0 & -1 & 1 & 0 & \ldots & 0 \\ 1 & 0 & -1 & \ddots & \vdots & -1 & 0 & 1 & \ddots & \vdots \\ \vdots & \vdots & \ddots & \ddots & 0 & \vdots & \vdots & \ddots & \ddots & 0 \\ 1 & 0 & \ldots & 0 & -1 & -1 & 0 & \ldots & 0 & 1 \end{pmatrix}. \tag{77}$$

So for $\mathcal{A}$ we can take the set of columns of the $(k+1)\times(2k)$-matrix

$$\begin{pmatrix} 0 & 0 & 0 & \dots & 0 & 1 & 1 & 1 & \dots & 1 \\ 1 & 0 & 0 & \dots & 0 & 1 & 0 & 0 & \dots & 0 \\ 0 & 1 & 0 & \dots & 0 & 0 & 1 & 0 & \dots & 0 \\ 0 & 0 & 1 & \ddots & \vdots & 0 & 0 & 1 & \ddots & \vdots \\ \vdots & \vdots & \ddots & \ddots & 0 & \vdots & \vdots & \ddots & \ddots & 0 \\ 0 & 0 & \dots & 0 & 1 & 0 & 0 & \dots & 0 & 1 \end{pmatrix}. \tag{78}$$

This notation is consistent with the main part of this text: $\mathcal{A}$ is a subset of $\mathbb{Z}^{k+1}$; moreover $N = 2k$ and $d = \operatorname{rank}\mathbb{L} = k-1$.

The primary polytope $\Delta_{\mathcal{A}}$ is the direct product of a $(k-1)$-simplex and a 1-simplex and, for $k=3$, looks like the prism in Figure 1. The vectors $\mathbf{a}_1,\dots,\mathbf{a}_k$ are in the bottom face of the prism; $\mathbf{a}_{k+1},\dots,\mathbf{a}_{2k}$ are in the top face. The numbering is such that the difference vectors $\mathbf{a}_{k+j}-\mathbf{a}_j$, for $j=1,\dots,k$ are all equal.

7.2. Integrals and differential equations for F_D

In [19] Lauricella's F_D in variables $z_0,\dots,z_n$ is introduced via the integrals

$$F_\alpha(z_0,\dots,z_n) := \int_\alpha (z_0-\zeta)^{-\mu_0}\cdots(z_n-\zeta)^{-\mu_n}\,d\zeta \tag{79}$$

over suitable intervals α, with endpoints in $\{z_0,\dots,z_n,\infty\}$. Note that because of the translation invariance property

$$F_\alpha(z_0+a,\dots,z_n+a) = F_\alpha(z_0,\dots,z_n) \tag{80}$$

the integral (79) is in fact a function of just n variables: $z_1-z_0,\dots,z_n-z_0$.

GKZ theory can deal efficiently with (multiplicative) torus actions on the variables, but it can not accommodate for translation invariance like (80). So we eliminate the translation invariance during the passage to GKZ and consider the integrals (with the same $\mu_0,\dots,\mu_n$)

$$I_\sigma(u_1,\dots,u_{2n}) = \int_\sigma (u_1+u_{n+1}\xi)^{-\mu_1}\cdots(u_n+u_{2n}\xi)^{-\mu_n}\xi^{-\mu_0}\,d\xi\,, \tag{81}$$

which are of the type considered in Section 2.4.

The GKZ differential equations satisfied by these integrals can be found with the methods used in Section 2.3. For instance, for $j=1,\dots,n$

$$\frac{\partial I_\sigma}{\partial u_j} = -\mu_j\int_\sigma (u_1+u_{n+1}\xi)^{-\mu_1}\cdots(u_n+u_{2n}\xi)^{-\mu_n}\xi^{-\mu_0}\,\frac{d\xi}{u_j+u_{j+n}\xi}$$

$$\frac{\partial I_\sigma}{\partial u_{j+n}} = -\mu_j\int_\sigma (u_1+u_{n+1}\xi)^{-\mu_1}\cdots(u_n+u_{2n}\xi)^{-\mu_n}\xi^{-\mu_0}\,\frac{\xi d\xi}{u_j+u_{j+n}\xi}$$

and, hence, for $i,\,j=1,\dots,n$

$$\frac{\partial^2 I_\sigma}{\partial u_i\partial u_{j+n}} = \frac{\partial^2 I_\sigma}{\partial u_j\partial u_{i+n}}\,,$$

i.e., I_σ satisfies the differential equations (19) with $\mathbb{L}$ as in (77) and $k=n$.

Similarly, for $s \in \mathbb{C}$ close to 1, we have

$$I_\sigma(u_1,\ldots,u_n,su_{n+1},\ldots,su_{2n}) = s^{\mu_0-1}I_\sigma(u_1,\ldots,u_{2n}),$$
$$I_\sigma(u_1,\ldots,u_{j-1},su_j,u_{j+1},\ldots,u_{j+n-1},su_{j+n},u_{j+n+1},\ldots,u_{2n})$$
$$= s^{-\mu_j}I_\sigma(u_1,\ldots,u_{2n}).$$

This leads to the differential equations (20) with $k=n$, $\mathcal{A}$ as in (78) and $\mathbf{c} = (\mu_0-1,-\mu_1,\ldots,-\mu_n)^t$.

As we have seen in Section 3.2.4 the power series $F_D(a,\mathbf{b},c|\mathbf{z})$ is, up to a constant factor, the Γ-series associated with the above $\mathbb{L}$ and with $\underline{\gamma} = (\gamma_1,\ldots,\gamma_N) = (c-1,-b_1,\ldots,-b_{k-1},-a,0,\ldots,0)$. The parameter $\mathbf{c}$ in the GKZ differential equations (20) is therefore

$$\mathbf{c} = \sum_{j=1}^{2k}\gamma_j\mathbf{a}_j = (-a,c-a-1,-b_1,\ldots,-b_{k-1})^t =: (c_0,c_1,c_2,\ldots,c_k)^t.$$

The system of differential equations (20) can now be written as

$$\frac{\partial\Phi}{\partial u_{j+k}} = -u_{j+k}^{-1}\left(u_j\frac{\partial\Phi}{\partial u_j} - c_j\Phi\right) \qquad \text{for} \quad j=1,\ldots,k, \tag{82}$$

$$u_1\frac{\partial\Phi}{\partial u_1} + \ldots + u_k\frac{\partial\Phi}{\partial u_k} = (-c_0+c_1+\ldots+c_k)\Phi. \tag{83}$$

The system (19) is equivalent with the following $\frac{1}{2}k(k-1)$ differential equations

$$\frac{\partial^2\Phi}{\partial u_i\partial u_{j+k}} = \frac{\partial^2\Phi}{\partial u_j\partial u_{i+k}} \qquad \text{for} \quad 1\le i<j\le k. \tag{84}$$

Next we substitute (82) into (84) and set

$$u_j = z_j \quad \text{if} \quad 1\le j\le k, \qquad u_j = 1 \quad \text{if} \quad k+1\le j\le 2k.$$

The result is the system of $\frac{1}{2}k(k-1)$ differential equations

$$(z_i-z_j)\frac{\partial^2\Phi}{\partial z_i\partial z_j} = c_i\frac{\partial\Phi}{\partial z_j} - c_j\frac{\partial\Phi}{\partial z_i} \qquad \text{for} \quad 1\le i<j\le k. \tag{85}$$

The above substitution turns (83) into

$$z_1\frac{\partial\Phi}{\partial z_1} + \ldots + z_k\frac{\partial\Phi}{\partial z_k} = (-c_0+c_1+\ldots+c_k)\Phi. \tag{86}$$

The system of differential equations (85)–(86) is then equivalent with the GKZ system (19)–(20) for Lauricella's F_D. The Equations (85) appear in this form also in [19], and (86) appears in loc.cit. in an 'integrated' form:

$$\Phi(e^tz_1,\ldots,e^tz_k) = e^{(-c_0+c_1+\ldots+c_k)t}\Phi(z_1,\ldots,z_k).$$

The match with [19] becomes exact, if one eliminates in loc. cit. the translation invariance by setting $z_0=0$ (like we did in passing from (79) to (81)).

7.3. Triangulations of $\Delta_{\mathcal{A}}$, secondary polytope and fan for F_D

Consider a triangulation $\mathcal{T}$ of the prism $\Delta_{\mathcal{A}}$ by k-dimensional simplices with vertices in the set $\mathcal{A}$. Then the bottom $(k-1)$-simplex $[\mathbf{a}_1, \ldots, \mathbf{a}_k]$ must be a face of exactly one k-simplex in the triangulation, say σ_1. Let $\mathbf{a}_{k+s_1}$ be the vertex of σ_1 opposite to the face $[\mathbf{a}_1, \ldots, \mathbf{a}_k]$. So $1 \leq s_1 \leq k$. The face of σ_1 opposite to the vertex s_1 has vertices $\mathbf{a}_{k+s_1}$ and $\mathbf{a}_i$ with $1 \leq i \leq k$, $i \neq s_1$. This must be a face of exactly one other k-dimensional simplex in the triangulation, say σ_2. Let $\mathbf{a}_{k+s_2}$ be the remaining vertex of σ_2. So $1 \leq s_2 \leq k$ and $s_2 \neq s_1$. The face of σ_2 opposite to the vertex $\mathbf{a}_{s_2}$ has vertices $\mathbf{a}_{k+s_1}$, $\mathbf{a}_{k+s_2}$ and $\mathbf{a}_i$ with $1 \leq i \leq k$, $i \neq s_1$, s_2. This must be a face of exactly one other k-dimensional simplex, say σ_3. Let $\mathbf{a}_{k+s_3}$ be the remaining vertex of σ_3. So $1 \leq s_3 \leq k$ and $s_3 \neq s_1$, s_2. And so on. Thus the triangulation $\mathcal{T}$ of $\Delta_{\mathcal{A}}$ determines a permutation τ of $\{1, \ldots, k\}$ with $\tau(i) = s_i$.

There is an obvious converse to this procedure associating to a permutation τ of $\{1, 2, 3, \ldots, k\}$ the triangulation with maximal simplices $\sigma_1^{(\tau)}, \ldots, \sigma_k^{(\tau)}$ where

$$\sigma_j^{(\tau)} := \text{convex hull } (\{\mathbf{a}_{\tau(i)} \mid j \leq i \leq k\} \cup \{\mathbf{a}_{k+\tau(i)} \mid 1 \leq i \leq j\}) . \tag{87}$$

These triangulations are unimodular; i.e., all k-simplices have volume 1. So when constructing the secondary polytope one only has to count for every triangulation how many simplices come together in the points $\mathbf{a}_1, \ldots, \mathbf{a}_N$. With the above formula for the simplex $\sigma_j^{(\tau)}$ one easily finds that the vector associated with the permutation τ is $(\tau^{-1}(1), \ldots, \tau^{-1}(k), k+1-\tau^{-1}(1), \ldots, k+1-\tau^{-1}(k))$.

The secondary polytope is the convex hull of these points as τ runs through all permutations of $\{1, 2, 3, \ldots, k\}$. By translating over the vector corresponding to the identity permutation the secondary polytope moves to the convex hull of the points $(\tau^{-1}(1)-1, \ldots, \tau^{-1}(k)-k, 1-\tau^{-1}(1), \ldots, k-\tau^{-1}(k))$ in the space $\mathbb{L}_{\mathbb{R}} := \mathbb{L} \otimes \mathbb{R}$.

Example/Exercise. The reader is invited to determine with the above algorithm the permutations corresponding to the maximal cones of the secondary fan of Appell's F_1 (= Lauricella's F_D with $k = 3$) shown in Figure 3.

Recall from Section 4.1 that the secondary fan is a partition of the real vector space $\mathbb{L}_{\mathbb{R}}^{\vee} = \mathrm{Hom}(\mathbb{L}, \mathbb{R})$ into rational cones, all with their apex in $\mathsf{0}$. Corollary 1 and Formula (44) describe these cones. The vectors $\mathbf{b}_1, \ldots, \mathbf{b}_N$ are the images of the standard basis vectors of $\mathbb{R}^N$ under the natural surjection $\mathbb{R}^N \longrightarrow \mathbb{L}_{\mathbb{R}}^{\vee}$. In the present situation we choose the rows of the matrix (77) as a basis for $\mathbb{L}_{\mathbb{R}} = \mathbb{L} \otimes \mathbb{R}$. On $\mathbb{L}_{\mathbb{R}}^{\vee}$ we use coordinates with respect to the dual basis. The columns of (77) then represent the vectors $\mathbf{b}_1, \ldots, \mathbf{b}_N$ in these coordinates.

Now consider a vector $\mathsf{t} = (t_2, \ldots, t_k)$ in $\mathbb{L}_{\mathbb{R}}^{\vee}$. Put $t_1 = 0$. Then t defines a partial ordering $<_{\mathsf{t}}$ on the set $\{1, 2, \ldots, k\}$ by

$$i <_{\mathsf{t}} j \qquad \Leftrightarrow \qquad t_i < t_j .$$

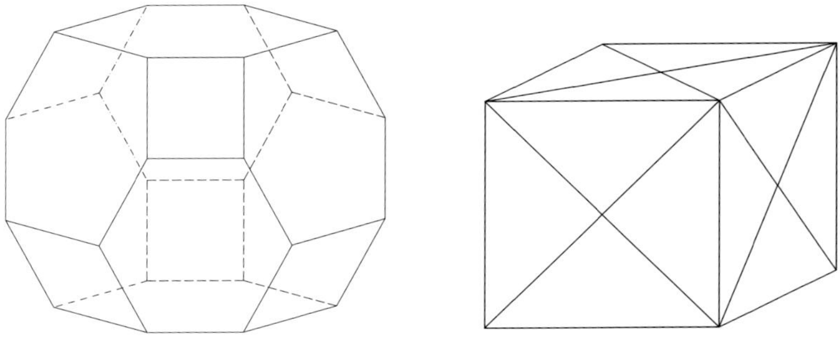

FIGURE 8. Secondary polytope (left) and Secondary fan (right) for F_D with $k = 4$. All cones in the fan have their apex at the centre of the cube. Shown are the intersections of the cones with some faces of the cube. The reader is invited to label the maximal cones with the permutations of $\{1, 2, 3, 4\}$.

The indexing and ordering is such that for $h = 1, \ldots, k$

$$\mathsf{t} = \sum_{i >_{\mathsf{t}} h} (t_i - t_h)\mathbf{b}_{i+k} + \sum_{i <_{\mathsf{t}} h} (t_h - t_i)\mathbf{b}_i . \tag{88}$$

One also easily checks that these are the only expresssions for t as positive linear combination of a linearly independent subset of $\{\mathbf{b}_1, \ldots, \mathbf{b}_N\}$. Corollary 1 and Formula (44) now tell exactly in which cone of the secondary fan t lies. In particular, t lies in the interior of a maximal cone if and only if no two of the numbers $t_1, t_2, \ldots, t_k$ are equal. In that case, the ordering $<_{\mathsf{t}}$ is a total ordering, or what amounts to the same a permutation of $\{1, 2, \ldots, k\}$. More precisely, if we associate with t the permutation τ defined by

$$\tau(1) <_{\mathsf{t}} \tau(2) <_{\mathsf{t}} \ldots <_{\mathsf{t}} \tau(k-1) <_{\mathsf{t}} \tau(k) ,$$

then the index set which effectively appears in (88) is the complement of the index set in (87) with $h = \tau(j)$.

Thus we have shown:

Corollary 3. *The maximal cones in the secondary fan for F_D are the connected components of the complement in $\mathbb{R}^{k-1}$ of the union of hyperplanes*

$$\bigcup_{1 \leq i < j \leq k} H_{ij} , \qquad \textit{with equation for} \quad H_{ij} : \quad t_i = t_j ,$$

where on the right $t_2, t_3, \ldots, t_k$ are coordinates on $\mathbb{R}^{k-1}$ and $t_1 = 0$. □

Remarks. Most GKZ systems do not have a secondary fan which is cut out by a hyperplane arrangement. F_D is something special.

Exercise. In Looijenga's lectures [19] the natural domain of definition for the *Schwarz map* is $\mathbb{P}(V_n^\circ)$. The notations are: $\mathbb{P}(V_n^\circ)$ is $(\mathbb{C}^{n+1})^\circ$ modulo the natural $\mathbb{C}^*$-action with weights $(1, 1, \ldots, 1)$ and modulo translations over $\mathbb{C}(1, 1, \ldots, 1)$,

$$(\mathbb{C}^{n+1})^\circ := \mathbb{C}^{n+1} \setminus \bigcup_{i<j} \{\text{hyperplane with equation } \; z_i = z_j\}.$$

How does $\mathbb{P}(V_n^\circ)$ relate to the toric variety $\mathcal{V}_\mathcal{A}$?

8. A glimpse of Mirror Symmetry

8.1. GKZ data from Calabi–Yau varieties

One of the manifestations of the Mirror Symmetry phenomenon is a relation between two families of 3-dimensional Calabi–Yau varieties, matching complex geometry on one family with symplectic geometry on the other. Our discussion of some aspects of Mirror Symmetry in connection with GKZ hypergeometric functions is, however, very asymmetric and biassed towards the complex geometry of one of the partners in the mirror pair. In the language of complex geometry a smooth Calabi–Yau variety is a compact smooth Kähler manifold X with trivial canonical bundle, i.e., $\Omega_X^{\dim X} \simeq \mathcal{O}_X$, which also satisfies $H^0(X, \Omega_X^i) = 0$ for $0 < i < \dim X$. Not all definitions in the literature require this second condition. Moreover, there are definitions which allow certain types of singularities. We will not deal with general Calabi–Yau varieties, but focus on concrete examples.

A Calabi–Yau variety of dimension 1 is an *elliptic curve*. A Calabi–Yau variety of dimension 2 is a *K3 surface*. A Calabi–Yau variety of dimension 3 is usually called a *Calabi–Yau threefold*. Standard examples of Calabi–Yau varieties, all given as complete intersections in a product of projective spaces, are shown in the second column of Table 1. From the homogeneous degrees of the defining equations and the coordinates of the ambient projective space one builds a lattice $\mathbb{L}$ for use in GKZ context. This is shown in the third column of Table 1. The lattice $\mathbb{L}$ comes naturally with an embedding into some $\mathbb{Z}^N$ and the quotient $\mathbb{M} := \mathbb{Z}^N/\mathbb{L}$ is torsion free, isomorphic to $\mathbb{Z}^{k+1}$, $k + 1 = N - d$. As in Section 4 we let $\mathbf{a}_1, \ldots, \mathbf{a}_N \in \mathbb{M}$ denote the images of the standard basis vectors of $\mathbb{Z}^N$ and $\mathcal{A} = \{\mathbf{a}_1, \ldots, \mathbf{a}_N\}$.

Behind the hypersurface cases in Table 1 one can see a pair of dual polytopes. For instance, a general cubic curve in $\mathbb{P}^2$ is given by a homogeneous equation of degree 3. Such an equation has 10 terms with exponents as shown in the left-hand picture in Figure 9. The polygon dual to the convex hull of this exponent set is the second from the left picture in Figure 9. This is the convex hull of the set $\mathcal{A}$. The other two pictures in Figure 9 show the exponents for the general homogeneous equation of degree $(2, 2)$ in $\mathbb{P}^1 \times \mathbb{P}^1$ and its dual. This dual polytope description works for all hypersurface cases in Table 1, but is not so easy to draw if the ambient space has dimension > 2; cf. [1]. For complete intersections of codimension > 1 there is not such a simple formulation with dual polytopes; cf. [3].

dim.	Calabi–Yau variety	B
1	cubic curve in $\mathbb{P}^2$	$(-3,1,1,1)$
1	$\bigcap$ two quadrics in $\mathbb{P}^3$	$(-2,-2,1,1,1,1)$
1	curve of degree $(2,2)$ in $\mathbb{P}^1\times\mathbb{P}^1$	$\begin{pmatrix} -2 & 1 & 0 & 1 & 0 \\ -2 & 0 & 1 & 0 & 1 \end{pmatrix}$
1	$\bigcap$ two surf. deg. $(1,1,1)$ in $(\mathbb{P}^1)^3$	$\begin{pmatrix} -1 & -1 & 1 & 0 & 0 & 1 & 0 & 0 \\ -1 & -1 & 0 & 1 & 0 & 0 & 1 & 0 \\ -1 & -1 & 0 & 0 & 1 & 0 & 0 & 1 \end{pmatrix}$
2	quartic surface in $\mathbb{P}^3$	$(-4,1,1,1,1)$
2	$\bigcap$ quadric and cubic in $\mathbb{P}^4$	$(-2,-3,1,1,1,1,1)$
2	$\bigcap$ three quadrics in $\mathbb{P}^5$	$(-2,-2,-2,1,1,1,1,1,1)$
2	surface of deg. $(2,2,2)$ in $(\mathbb{P}^1)^3$	$\begin{pmatrix} -2 & 1 & 0 & 0 & 1 & 0 & 0 \\ -2 & 0 & 1 & 0 & 0 & 1 & 0 \\ -2 & 0 & 0 & 1 & 0 & 0 & 1 \end{pmatrix}$
3	quintic hypersurface in $\mathbb{P}^4$	$(-5,1,1,1,1,1)$
3	$\bigcap$ two cubics in $\mathbb{P}^5$	$(-3,-3,1,1,1,1,1,1)$
3	3-fold of deg. $(3,3)$ in $\mathbb{P}^2\times\mathbb{P}^2$	$\begin{pmatrix} -3 & 1 & 0 & 1 & 0 & 1 & 0 \\ -3 & 0 & 1 & 0 & 1 & 0 & 1 \end{pmatrix}$
3	$\bigcap$ four quadrics in $\mathbb{P}^7$	$(-2,-2,-2,-2,1,1,1,1,1,1,1,1)$
3	3-fold of deg. $(2,2,2,2)$ in $(\mathbb{P}^1)^4$	$\begin{pmatrix} -2 & 1 & 0 & 0 & 0 & 1 & 0 & 0 & 0 \\ -2 & 0 & 1 & 0 & 0 & 0 & 1 & 0 & 0 \\ -2 & 0 & 0 & 1 & 0 & 0 & 0 & 1 & 0 \\ -2 & 0 & 0 & 0 & 1 & 0 & 0 & 0 & 1 \end{pmatrix}$
	$\bigcap$ means 'intersection of'.	$\mathbb{L}=\mathbb{Z}$-span of rows of B

TABLE 1. Standard examples of Calabi–Yau varieties.

cubic in $\mathbb{P}^2$ and its mirror (2,2)-curve in $\mathbb{P}^1\times\mathbb{P}^1$ and its mirror

FIGURE 9. Polygons of exponents.

Using Corollary 1 and Formula (44) one checks that in these examples the positive span of the last d columns of matrix B is a maximal cone $\mathcal{C}$ in the secondary fan of $\mathbb{L}$. Using (62) one checks that the triangulation of $\Delta_{\mathcal{A}}$ corresponding to $\mathcal{C}$ is unimodular. Next one computes the ring $\mathcal{R}_{\mathcal{A},\mathcal{C}}$ in Definition 65 and one finds that it is (isomorphic to) the cohomology ring of the ambient space in the second

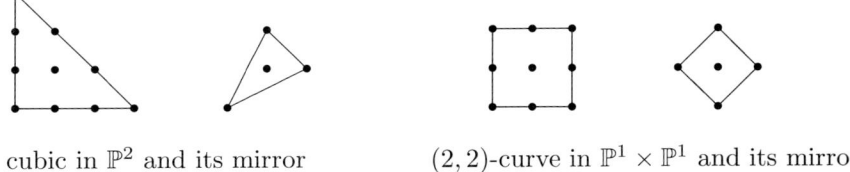

column of Table 1:

$$\mathcal{R}_{\mathcal{A},\mathcal{C}} = \begin{cases} \mathbb{Z}[\varepsilon]\,/\,(\varepsilon^{r+1}) & \text{if the ambient space is} \quad \mathbb{P}^r \\ \mathbb{Z}[\delta_1,\ldots,\delta_r]\,/\,(\delta_1^2,\ldots,\delta_r^2) & \text{if the ambient space is} \quad (\mathbb{P}^1)^r \\ \mathbb{Z}[\delta_1,\delta_2]\,/\,(\delta_1^3,\delta_2^3) & \text{if the ambient space is} \quad (\mathbb{P}^2)^2. \end{cases}$$

Choosing a $\mathbb{Z}$-basis for $\mathbb{M}$ we write $\mathbf{a}_j = (a_{1j},\ldots,a_{(k+1)\,j})$. With $\mathcal{A}$ one now associates the *Laurent polynomial* in the variables $x_1,\ldots,x_{k+1}$ with undetermined coefficients $u_j = \mathbf{u}_{\mathbf{a}_j}$ (cf.(11)):

$$P_{\mathcal{A}}(\mathbf{x}) = P_{\mathcal{A}}(x_1,\ldots,x_{k+1}) = \sum_{\mathbf{a}\in\mathcal{A}} \mathbf{u}_{\mathbf{a}}\mathbf{x}^{\mathbf{a}} = \sum_{j=1}^{N} u_j \prod_{i=1}^{k+1} x_i^{a_{ij}}\,. \tag{89}$$

Since for each $\mathbf{a}_j$ the coordinates sum to 1, this Laurent polynomial is homogeneous: $P_{\mathcal{A}}(t\mathbf{x}) = tP_{\mathcal{A}}(\mathbf{x})$ for every $t \in \mathbb{C}^*$. As the coefficients $\mathbf{u}$ vary the zero loci of $P_{\mathcal{A}}(\mathbf{x})$ sweep out a family of hypersurfaces in $(\mathbb{C}^*)^{k+1}/\mathbb{C}^* = (\mathbb{C}^*)^k$. Both $(\mathbb{C}^*)^k$ and the hypersurfaces can be suitably compactified. This family of compactified hypersurfaces is then the *mirror in the sense of* [3] *of the family of Calabi–Yau varieties* in the second column in Table 1. The members of this mirror family are Calabi–Yau varieties if the original Calabi–Yau varieties have codimension 1 in the ambient space. In case of codimension > 1 the mirror family consists of *generalized Calabi–Yau varieties* in the sense of [3].

We will now discuss details for the first three examples of Calabi–Yau threefolds. Other examples can be treated in the same way. Our text tells exactly how one obtains various tables of enumerative data in the literature, but it may read like a cook book. In Section 8.5 we will explain that what is actually being cooked is a kind of Schwarz map, which goes well with the menu of this summer school. In Section 8.6 we discuss in an informal way some features of our hypergeometric formulae which may be seen as manifestations of Mirror Symmetry.

8.2. The quintic in $\mathbb{P}^4$

This is the original example with which Mirror Symmetry entered the mathematical arena; see [7]. The matrix B in Table 1 is of the form $\mathsf{B} = (\widetilde{\mathsf{B}}\;\mathbb{I}_d)$, where $\mathbb{I}_d$ is the $d\times d$-identity matrix. The matrix $\mathsf{A} = (\mathbb{I}_{N-d}\;\; -\widetilde{\mathsf{B}}^t)$ then satisfies $\mathsf{B}\mathsf{A}^t = 0$ and its columns generate $\mathbb{Z}^{k+1}$. We apply row operations (i.e., a basis transformation in $\mathbb{Z}^{k+1}$) so that the Laurent polynomial $P_{\mathcal{A}}(\mathbf{x})$ in (89) assumes a pleasant form:

$$\mathsf{A} = \begin{pmatrix} 1&0&0&0&0&5\\ 0&1&0&0&0&-1\\ 0&0&1&0&0&-1\\ 0&0&0&1&0&-1\\ 0&0&0&0&1&-1 \end{pmatrix} \quad\rightsquigarrow\quad \begin{pmatrix} 1&1&1&1&1&1\\ 0&1&0&0&0&-1\\ 0&0&1&0&0&-1\\ 0&0&0&1&0&-1\\ 0&0&0&0&1&-1 \end{pmatrix}.$$

We let $\mathcal{A} = \{\mathbf{a}_1,\ldots,\mathbf{a}_N\}$ denote the columns of the right-hand matrix. Then

$$P_{\mathcal{A}}(\mathbf{x}) = x_1\left(u_1 + u_2x_2 + u_3x_3 + u_4x_4 + u_5x_5 + u_6(x_2x_3x_4x_5)^{-1}\right).$$

Remark. The Laurent polynomial $P_{\mathcal{A}}(\mathbf{x})$ can be dehomogenized by setting $x_1 = 1$ and subsequently be homogenized to the degree 5 polynomial in 5 variables

$$\begin{aligned}\widetilde{P}_{\mathcal{A}}(\mathsf{X}) &= u_1X_1X_2X_3X_4X_5 + u_2X_2^2X_3X_4X_5 + u_3X_2X_3^2X_4X_5 + \\ &\quad + u_4X_2X_3X_4^2X_5 + u_5X_2X_3X_4X_5^2 + u_6X_1^5\,.\end{aligned}$$

The polynomial $\widetilde{P}_{\mathcal{A}}(\mathsf{X})$ defines a family of special quintic hypersurfaces in $\mathbb{P}^4$, which is the mirror of the family of general quintic hypersurfaces in $\mathbb{P}^4$. Traditionally the mirror family is presented as a quotient of the hypersurface

$$Z_1^5 + Z_2^5 + Z_3^5 + Z_4^5 + Z_5^5 - 5\psi Z_1Z_2Z_3Z_4Z_5 = 0$$

by a specific action of the group $\left(\mathbb{Z}/5\mathbb{Z}\right)^3$ (see [7] §2). The hypergeometric integrals and series constructed from the periods of this 'Fermat-like quintic' are however the same as those coming from $\widetilde{P}_{\mathcal{A}}(\mathsf{X})$.

The periods of the mirror Calabi–Yau hypersurfaces are given by integrals

$$I_\sigma^-(\mathbf{u}) := \frac{1}{(2\pi \mathrm{i})^4}\int_\sigma P_{\mathcal{A}}(1,x_2,x_3,x_4,x_5)^{-1}\,\frac{dx_2}{x_2}\,\frac{dx_3}{x_3}\,\frac{dx_4}{x_4}\,\frac{dx_5}{x_5}\,. \tag{90}$$

As shown in Section 2.3, these integrals viewed as functions of $u_1,\ldots,u_6$, satisfy the GKZ system of differential equations (19)–(20) with $\mathcal{A}$ as above and $\mathbf{c} = -\mathbf{a}_1$.

If the numbers $\left|u_ju_1^{-1}\right|$ for $2 \le j \le 6$ are sufficiently small, the domain of integration for one of the above period integrals $I_\sigma^-(\mathbf{u})$ can be taken to be $\sigma = \{|x_2| = |x_3| = |x_4| = |x_5| = 1\}$. Using geometric series, the binomial and residue theorems, one obtains for this period integral the series expansion:

$$I_\sigma^-(\mathbf{u}) = u_1^{-1}\sum_{n\ge 0}(-1)^n\frac{(5n)!}{(n!)^5}z^n \qquad \text{with} \quad z = u_1^{-5}u_2u_3u_4u_5u_6\,. \tag{91}$$

Now look at the series $\Phi_{\mathbb{L},\underline{\gamma},\underline{\varepsilon}}(\mathbf{u})$ for $\mathbb{L} = \mathbb{Z}(-5,1,1,1,1,1)$ and $\underline{\gamma} = (-1,0,0,0,0,0)$ defined in (74). In this case $\underline{\varepsilon} = (\varepsilon_1,\varepsilon_2,\varepsilon_3,\varepsilon_4,\varepsilon_5,\varepsilon_6) = (-5\varepsilon,\varepsilon,\varepsilon,\varepsilon,\varepsilon,\varepsilon)$. The triangulation is given by the list $T_{\mathcal{C}}$ consisting of the five sets one gets by deleting from $\{1,2,3,4,5,6\}$ one number > 1. The minimal set not contained in a set on the list $T_{\mathcal{C}}$ is $\{2,3,4,5,6\}$. Thus we see

$$\mathcal{R}_{\mathcal{A},\mathcal{C}} = \mathbb{Z}\oplus\mathbb{Z}\varepsilon\oplus\mathbb{Z}\varepsilon^2\oplus\mathbb{Z}\varepsilon^3\oplus\mathbb{Z}\varepsilon^4\,, \qquad \varepsilon^5 = 0\,,$$

and, with Pochhammer symbol notation (30) and $z = u_1^{-5}u_2u_3u_4u_5u_6$,

$$\Phi_{\mathbb{L},\underline{\gamma},\underline{\varepsilon}}(\mathbf{u}) = -5\varepsilon u_1^{-1}\sum_{n\ge 0}(-1)^n\frac{(1+5\varepsilon)_{5n}}{((1+\varepsilon)_n)^5}z^{n+\varepsilon}\,. \tag{92}$$

So, the function $\Phi_{\mathbb{L},\underline{\gamma},\underline{\varepsilon}}(\mathbf{u})$ takes values in the vector space $\varepsilon\mathcal{R}_{\mathcal{A},\mathcal{C}}\otimes\mathbb{C}$. Now let $\mathrm{Ann}(\varepsilon) = \{x \in \mathcal{R}_{\mathcal{A},\mathcal{C}} \mid x\varepsilon = 0\}$ denote the annihilator ideal of ε and let $\overline{\mathcal{R}}_{\mathcal{A},\mathcal{C}} := \mathcal{R}_{\mathcal{A},\mathcal{C}}/\mathrm{Ann}(\varepsilon)$. Then as a vector space $\varepsilon\mathcal{R}_{\mathcal{A},\mathcal{C}}\otimes\mathbb{C}$ is isomorphic to $\overline{\mathcal{R}}_{\mathcal{A},\mathcal{C}}\otimes\mathbb{C}$. The latter has, however, the advantage of being a ring. Let $\overline{\varepsilon}$ denote the class of ε in $\overline{\mathcal{R}}_{\mathcal{A},\mathcal{C}}$. Then

$$\overline{\mathcal{R}}_{\mathcal{A},\mathcal{C}} = \mathbb{Z}\oplus\mathbb{Z}\overline{\varepsilon}\oplus\mathbb{Z}\overline{\varepsilon}^2\oplus\mathbb{Z}\overline{\varepsilon}^3\,, \qquad \overline{\varepsilon}^4 = 0\,.$$

Moreover we can write

$$\Phi_{\mathbb{L},\underline{\gamma},\underline{\varepsilon}}(\mathbf{u}) \;=\; -5u_1^{-1}z^{\overline{\varepsilon}}\sum_{n\geq 0}(-1)^n\frac{(1+5\overline{\varepsilon})_{5n}}{((1+\overline{\varepsilon})_n)^5}z^n\,, \tag{93}$$

and view it as a function with values in the ring $\overline{\mathcal{R}}_{\mathcal{A},\mathcal{C}}\otimes\mathbb{C}$. We expand this function with respect to the basis $\{1,\overline{\varepsilon},\overline{\varepsilon}^2,\overline{\varepsilon}^3\}$:

$$\Phi_{\mathbb{L},\underline{\gamma},\underline{\varepsilon}}(\mathbf{u}) \;=\; \Phi_0(\mathbf{u}) \;+\; \Phi_1(\mathbf{u})\overline{\varepsilon} \;+\; \Phi_2(\mathbf{u})\overline{\varepsilon}^2 \;+\; \Phi_3(\mathbf{u})\overline{\varepsilon}^3\,. \tag{94}$$

By-passing all motivations, justifications and interpretations from string theory and Hodge theory (see, however, Section 8.5 and [8], p. 263) we define the *canonical coordinate*

$$q \;:=\; -\exp\left(\frac{\Phi_1(\mathbf{u})}{\Phi_0(\mathbf{u})}\right) \tag{95}$$

and the *prepotential*

$$\mathcal{F}(q) \;=\; \frac{5}{2}\left(\frac{\Phi_1(\mathbf{u})}{\Phi_0(\mathbf{u})}\frac{\Phi_2(\mathbf{u})}{\Phi_0(\mathbf{u})} \;-\; \frac{\Phi_3(\mathbf{u})}{\Phi_0(\mathbf{u})}\right)\,. \tag{96}$$

Note that these are functions of z, because they are constructed from quotients of solutions to the same GKZ system. In fact $q = -z + O(z^2)$ and we can invert this relation so as to get z as a function $z(q)$ of q. We then want to view the prepotential as a function of q. The recipe for extracting results about the enumerative geometry of the general quintic threefold is to take

$$\mathcal{F}(q) \;=\; \tfrac{5}{6}\log^3 q \;+\; \sum_{j\geq 1} N_j\,\mathrm{Li}_3(q^j)\,, \tag{97}$$

where Li_3 is the *trilogarithm function* $\mathrm{Li}_3(x) := \sum_{n\geq 1}\frac{x^n}{n^3}$. Then one of the miracles of mirror symmetry is that all numbers N_j are positive integers and that in fact N_j equals the number of rational curves of degree j on a general quintic threefold [7, 8]. The first few of these N_j are shown in Table 2.

N_1	$=$	2875
N_2	$=$	609250
N_3	$=$	317206375
N_4	$=$	242467530000
N_5	$=$	229305888887625
N_6	$=$	248249742118022000
N_7	$=$	295091050570845659250
N_8	$=$	375632160937476603550000
N_9	$=$	503840510416985243645106250

TABLE 2. The numbers N_j for the quintic threefold.

Actual computations proceed as follows: compute $F_0,\ldots,f_3$ from

$$\sum_{n\geq 0}(-1)^n\frac{(1+5\overline{\varepsilon})_{5n}}{((1+\overline{\varepsilon})_n)^5}z^n = F_0(z) + F_1(z)\overline{\varepsilon} + F_2(z)\overline{\varepsilon}^2 + F_3(z)\overline{\varepsilon}^3$$

and $f_i(z) := \frac{F_i(z)}{F_0(z)}$. That implies

$$\Phi_{\mathbb{L},\underline{\gamma},\underline{\varepsilon}}(\mathbf{u}) = \Phi_0(\mathbf{u})z^{\overline{\varepsilon}}\left(1 + f_1(z)\overline{\varepsilon} + f_2(z)\overline{\varepsilon}^2 + f_3(z)\overline{\varepsilon}^3\right). \tag{98}$$

Comparing the expansions of $\log\Phi_{\mathbb{L},\underline{\gamma},\underline{\varepsilon}}(\mathbf{u})$ which result from (94) and (98), i.e.,

$$\begin{aligned}\log\Phi_{\mathbb{L},\underline{\gamma},\underline{\varepsilon}}(\mathbf{u}) &= \log(\Phi_0(z)) + \frac{\Phi_1(\mathbf{u})}{\Phi_0(\mathbf{u})}\overline{\varepsilon} + \left(-\frac{1}{2}\left[\frac{\Phi_1(\mathbf{u})}{\Phi_0(\mathbf{u})}\right]^2 + \frac{\Phi_2(\mathbf{u})}{\Phi_0(\mathbf{u})}\right)\overline{\varepsilon}^2\\ &\quad + \left(\frac{1}{3}\left[\frac{\Phi_1(\mathbf{u})}{\Phi_0(\mathbf{u})}\right]^3 - \frac{\Phi_1(\mathbf{u})}{\Phi_0(\mathbf{u})}\frac{\Phi_2(\mathbf{u})}{\Phi_0(\mathbf{u})} + \frac{\Phi_3(\mathbf{u})}{\Phi_0(\mathbf{u})}\right)\overline{\varepsilon}^3\\ &= \log(\Phi_0(z)) + (\log z + f_1(z))\overline{\varepsilon} + (-\tfrac{1}{2}f_1(z)^2 + f_2(z))\overline{\varepsilon}^2\\ &\quad + (\tfrac{1}{3}f_1(z)^3 - f_1(z)f_2(z) + f_3(z))\overline{\varepsilon}^3,\end{aligned}$$

we see that q and $\mathcal{F}(q)$ can easily be computed from the already known f_1, f_2, f_3:

$$\begin{aligned} q &= -z\exp(f_1(z)),\\ \mathcal{F}(q) &= \tfrac{5}{2}\left(\tfrac{1}{3}\log^3(-q) - \left(\tfrac{1}{3}f_1(z(q))^3 - f_1(z(q))f_2(z(q)) + f_3(z(q))\right)\right).\end{aligned}$$

8.3. The intersection of two cubics in $\mathbb{P}^5$

This is one of the examples discussed in [18]. Here we treat it as a highly resonant GKZ system. As in the case of the quintic the matrix B in Table 1 is of the form $\mathsf{B} = (\widetilde{\mathsf{B}}\ \mathbb{I}_d)$ and we apply row operations to the matrix $(\mathbb{I}_{N-d}\ -\widetilde{\mathsf{B}}^t)$ so that the Laurent polynomial $P_{\mathcal{A}}(\mathbf{x})$ in (89) assumes a pleasant form:

$$\begin{pmatrix} 1&0&0&0&0&0&0&3\\ 0&1&0&0&0&0&0&3\\ 0&0&1&0&0&0&0&-1\\ 0&0&0&1&0&0&0&-1\\ 0&0&0&0&1&0&0&-1\\ 0&0&0&0&0&1&0&-1\\ 0&0&0&0&0&0&1&-1\end{pmatrix} \rightsquigarrow \begin{pmatrix} 1&0&1&1&0&0&0&1\\ 0&1&0&0&1&1&1&0\\ 0&0&1&0&0&0&0&-1\\ 0&0&0&1&0&0&0&-1\\ 0&0&0&0&1&0&0&-1\\ 0&0&0&0&-1&1&0&0\\ 0&0&0&0&-1&0&1&0\end{pmatrix}.$$

We let $\mathcal{A} = \{\mathbf{a}_1,\ldots,\mathbf{a}_N\}$ denote the columns of the right-hand matrix. Then

$$P_{\mathcal{A}}(\mathbf{x}) = x_1\,P_{\mathcal{A},1}(x_3,x_4,x_5) + x_2\,P_{\mathcal{A},2}(x_5,x_6,x_7)$$

with

$$\begin{aligned} P_{\mathcal{A},1}(x_3,x_4,x_5) &= u_1 + u_3x_3 + u_4x_4 + u_8(x_3x_4x_5)^{-1}\\ P_{\mathcal{A},2}(x_5,x_6,x_7) &= u_2 + u_5x_5(x_6x_7)^{-1} + u_6x_6 + u_7x_7\,.\end{aligned}$$

This way of combining the two Laurent polynomials in five variables, $P_{\mathcal{A},1}$ and $P_{\mathcal{A},2}$, to one Laurent polynomial in seven variables $P_{\mathcal{A}}$ is known as *Cayley's trick* (see [12, 13, 14]). The two polynomials $P_{\mathcal{A},1}$ and $P_{\mathcal{A},2}$, suitably homogenized, define a family of Calabi–Yau complete intersection threefolds in $\mathbb{P}^1 \times \mathbb{P}^2 \times \mathbb{P}^2$. The corresponding period integrals are (cf. Section 2.4 and [12, 14])

$$I_\sigma^-(\mathbf{u}) := \frac{1}{(2\pi \mathsf{i})^5} \int_\sigma P_{\mathcal{A},1}(x_3, x_4, x_5)^{-1} P_{\mathcal{A},2}(x_5, x_6, x_7)^{-1} \frac{dx_3}{x_3} \frac{dx_4}{x_4} \frac{dx_5}{x_5} \frac{dx_6}{x_6} \frac{dx_7}{x_7} . \tag{99}$$

One can show as in Sections 2.3 and 2.4 that these integrals viewed as functions of $u_1, \ldots, u_8$, satisfy the GKZ system of differential equations (19)–(20) with $\mathcal{A}$ as above and $\mathbf{c} = -\mathbf{a}_1 - \mathbf{a}_2$.

If the numbers $|u_3 u_1^{-1}|$, $|u_4 u_1^{-1}|$, $|u_8 u_1^{-1}|$ and $|u_5 u_2^{-1}|$, $|u_6 u_2^{-1}|$, $|u_7 u_2^{-1}|$ are sufficiently small, the domain of integration for one of the above period integrals $I_\sigma^-(\mathbf{u})$ can be taken to be $\sigma = \{|x_3| = |x_4| = |x_5| = |x_6| = |x_7| = 1\}$. This period integral admits the series expansion, with $z = u_1^{-3} u_2^{-3} u_3 u_4 u_5 u_6 u_7 u_8$,

$$I_\sigma^-(\mathbf{u}) = u_1^{-1} u_2^{-1} \sum_{n \geq 0} \left(\frac{(3n)!}{(n!)^3} \right)^2 z^n . \tag{100}$$

The series $\Phi_{\mathbb{L},\underline{\gamma},\underline{\varepsilon}}(\mathbf{u})$ for $\mathbb{L} = \mathbb{Z}(-3, -3, 1, 1, 1, 1, 1, 1)$, $\underline{\gamma} = (-1, -1, 0, 0, 0, 0, 0, 0)$ and $\underline{\varepsilon} = (\varepsilon_1, \ldots, \varepsilon_8) = (-3\varepsilon, -3\varepsilon, \varepsilon, \varepsilon, \varepsilon, \varepsilon, \varepsilon, \varepsilon)$ reads

$$\Phi_{\mathbb{L},\underline{\gamma},\underline{\varepsilon}}(\mathbf{u}) = 9\varepsilon^2 u_1^{-1} u_2^{-1} \sum_{n \geq 0} \left(\frac{(1 + 3\varepsilon)_{3n}}{((1 + \varepsilon)_n)^3} \right)^2 z^{n+\varepsilon} , \tag{101}$$

and is evaluated in $\mathcal{R}_{\mathcal{A},\mathcal{C}} = \mathbb{Z} \oplus \mathbb{Z}\varepsilon \oplus \mathbb{Z}\varepsilon^2 \oplus \mathbb{Z}\varepsilon^3 \oplus \mathbb{Z}\varepsilon^4 \oplus \mathbb{Z}\varepsilon^5 , \qquad \varepsilon^6 = 0 .$
The function $\Phi_{\mathbb{L},\underline{\gamma},\underline{\varepsilon}}(\mathbf{u})$ actually takes values in the vector space $\varepsilon^2 \mathcal{R}_{\mathcal{A},\mathcal{C}} \otimes \mathbb{C}$. As in the case of the quintic, we replace this space by the isomorphic space $\overline{\mathcal{R}}_{\mathcal{A},\mathcal{C}} \otimes \mathbb{C}$, where $\mathrm{Ann}(\varepsilon^2) := \{x \in \mathcal{R}_{\mathcal{A},\mathcal{C}} \mid x\varepsilon^2 = 0\}$ and $\overline{\mathcal{R}}_{\mathcal{A},\mathcal{C}} := \mathcal{R}_{\mathcal{A},\mathcal{C}} / \mathrm{Ann}(\varepsilon^2)$. Let $\overline{\varepsilon}$ denote the class of ε in $\overline{\mathcal{R}}_{\mathcal{A},\mathcal{C}}$. Then

$$\overline{\mathcal{R}}_{\mathcal{A},\mathcal{C}} = \mathbb{Z} \oplus \mathbb{Z}\overline{\varepsilon} \oplus \mathbb{Z}\overline{\varepsilon}^2 \oplus \mathbb{Z}\overline{\varepsilon}^3 , \qquad \overline{\varepsilon}^4 = 0 .$$

Proceeding as in the case of the quintic we write

$$\begin{aligned} \Phi_{\mathbb{L},\underline{\gamma},\underline{\varepsilon}}(\mathbf{u}) &= 9 u_1^{-1} u_2^{-1} \sum_{n \geq 0} \left(\frac{(1 + 3\overline{\varepsilon})_{3n}}{((1 + \overline{\varepsilon})_n)^3} \right)^2 z^{n+\overline{\varepsilon}} \\ &= \Phi_0(\mathbf{u}) + \Phi_1(\mathbf{u})\overline{\varepsilon} + \Phi_2(\mathbf{u})\overline{\varepsilon}^2 + \Phi_3(\mathbf{u})\overline{\varepsilon}^3 \\ &= \Phi_0(\mathbf{u}) z^{\overline{\varepsilon}} \left(1 + f_1(z)\overline{\varepsilon} + f_2(z)\overline{\varepsilon}^2 + f_3(z)\overline{\varepsilon}^3\right) , \end{aligned}$$

and almost exactly as for the quintic we extract from $\log \Phi_{\mathbb{L},\underline{\gamma},\underline{\varepsilon}}(\mathbf{u})$ a *canonical coordinate* and a *prepotential*:

$$q := z \exp(f_1(z)) = z + O(z^2) , \tag{102}$$

$$\mathcal{F}(q) := \tfrac{9}{2} \left(\tfrac{1}{3} \log^3 q - \left(\tfrac{1}{3} f_1(z(q))^3 - f_1(z(q)) f_2(z(q)) + f_3(z(q)) \right) \right). \tag{103}$$

Finally we compute the numbers N_j from the expansion

$$\mathcal{F}(q) = \tfrac{3}{2} \log^3 q + \sum_{j \geq 1} N_j \,\mathrm{Li}_3(q^j) . \tag{104}$$

The first few of the numbers N_j are shown in Table 3 and agree with those in [18].

$$\begin{aligned}
N_1 &= 1053\\
N_2 &= 52812\\
N_3 &= 6424326\\
N_4 &= 1139448384\\
N_5 &= 249787892583\\
N_6 &= 62660964509532\\
N_7 &= 17256453900822009\\
N_8 &= 5088842568426162960\\
N_9 &= 1581250717976557887945
\end{aligned}$$

TABLE 3. The numbers N_j for the intersection of two cubics in $\mathbb{P}^5$.

8.4. The hypersurface of degree $(3,3)$ in $\mathbb{P}^2 \times \mathbb{P}^2$

Again the matrix B in Table 1 is of the form $\mathsf{B} = (\widetilde{\mathsf{B}}\ \mathbb{I}_d)$, and we apply row operations to the matrix $(\mathbb{I}_{N-d}\ -\widetilde{\mathsf{B}}^t)$:

$$\begin{pmatrix} 1&0&0&0&0&3&3\\ 0&1&0&0&0&-1&0\\ 0&0&1&0&0&0&-1\\ 0&0&0&1&0&-1&0\\ 0&0&0&0&1&0&-1 \end{pmatrix} \rightsquigarrow \begin{pmatrix} 1&1&1&1&1&1&1\\ 0&1&0&0&0&-1&0\\ 0&0&1&0&0&0&-1\\ 0&0&0&1&0&-1&0\\ 0&0&0&0&1&0&-1 \end{pmatrix}.$$

We let $\mathcal{A} = \{\mathbf{a}_1, \ldots, \mathbf{a}_N\}$ denote the columns of the right-hand matrix. Then

$$P_{\mathcal{A}}(\mathbf{x}) = x_1\left(u_1 + u_2x_2 + u_3x_3 + u_4x_4 + u_5x_5 + u_6(x_2x_4)^{-1} + u_7(x_3x_5)^{-1}\right).$$

The periods of the mirror Calabi–Yau hypersurfaces are given by integrals

$$I_\sigma^-(\mathbf{u}) := \frac{1}{(2\pi \mathsf{i})^4}\int_\sigma P_{\mathcal{A}}(1,x_2,x_3,x_4,x_5)^{-1}\,\frac{dx_2}{x_2}\frac{dx_3}{x_3}\frac{dx_4}{x_4}\frac{dx_5}{x_5}. \tag{105}$$

As shown in Section 2.3 these integrals viewed as functions of $u_1, \ldots, u_7$, satisfy the GKZ system of differential equations (19)–(20) with $\mathcal{A}$ as above and $\mathbf{c} = -\mathbf{a}_1$.

If the numbers $\left|u_ju_1^{-1}\right|$ for $2 \le j \le 7$ are sufficiently small, the domain of integration for one of the above period integrals $I_\sigma^-(\mathbf{u})$ can be taken to be $\sigma = \{|x_2| = |x_3| = |x_4| = |x_5| = 1\}$. This integral admits the expansion:

$$I_\sigma^-(\mathbf{u}) = u_1^{-1}\sum_{n_1,\,n_2\ge 0}(-1)^{n_1+n_2}\frac{(3n_1+3n_2)!}{(n_1!)^3(n_2!)^3}z_1^{n_1}z_2^{n_2} \tag{106}$$

with $z_1 = u_1^{-3}u_2u_4u_6$ and $z_2 = u_1^{-3}u_3u_5u_7$. Now look at the series $\Phi_{\mathbb{L},\underline{\gamma},\underline{\varepsilon}}(\mathbf{u})$ for $\mathbb{L} = \mathbb{Z}(-3,1,0,1,0,1,0) \oplus \mathbb{Z}(-3,0,1,0,1,0,1)$, $\underline{\gamma} = (-1,0,0,0,0,0,0)$ and $\underline{\varepsilon} = (\varepsilon_1,\varepsilon_2,\varepsilon_3,\varepsilon_4,\varepsilon_5,\varepsilon_6,\varepsilon_7) = \delta_1(-3,1,0,1,0,1,0) + \delta_2(-3,0,1,0,1,0,1)$. Using

Corollary 1 and the matrix B from Table 1 for this example one easily checks that the triangulation is given by the list $T_{\mathcal{C}}$ consisting of the nine sets one gets by deleting from $\{1,2,3,4,5,6,7\}$ one even and one odd number > 1. The minimal sets not contained in a set on the list $T_{\mathcal{C}}$ are $\{2,4,6\}$ and $\{3,5,7\}$. Thus we see

$$\begin{aligned}\mathcal{R}_{\mathcal{A},\mathcal{C}} &= \mathbb{Z}\oplus\mathbb{Z}\delta_1\oplus\mathbb{Z}\delta_2\oplus\mathbb{Z}\delta_1^2\oplus\mathbb{Z}\delta_1\delta_2\oplus\mathbb{Z}\delta_2^2\oplus\mathbb{Z}\delta_1^2\delta_2\oplus\mathbb{Z}\delta_1\delta_2^2\oplus\mathbb{Z}\delta_1^2\delta_2^2\,,\\ \delta_1^3 &= \delta_2^3 = 0\,.\end{aligned}$$

Thus, with $z_1 = u_1^{-3}u_2u_4u_6$ and $z_2 = u_1^{-3}u_3u_5u_7$,

$$\Phi_{\mathbb{L},\underline{\gamma},\underline{\varepsilon}}(\mathbf{u}) = -3(\delta_1+\delta_2)u_1^{-1}\sum_{n_1,\,n_2\geq 0}(-1)^{n_1+n_2}\frac{(1+3\delta_1+3\delta_2)_{3n_1+3n_2}}{((1+\delta_1)_{n_1}(1+\delta_2)_{n_2})^3}z_1^{n_1+\delta_1}z_2^{n_2+\delta_2}\,.$$

The function $\Phi_{\mathbb{L},\underline{\gamma},\underline{\varepsilon}}(\mathbf{u})$ takes values in the vector space $(\delta_1+\delta_2)\mathcal{R}_{\mathcal{A},\mathcal{C}}\otimes\mathbb{C}$. As in the previous cases, we replace this space by the isomorphic one $\overline{\mathcal{R}}_{\mathcal{A},\mathcal{C}}\otimes\mathbb{C}$, where $\mathrm{Ann}(\delta_1+\delta_2) := \{x\in\mathcal{R}_{\mathcal{A},\mathcal{C}}\mid x(\delta_1+\delta_2)=0\}$ and $\overline{\mathcal{R}}_{\mathcal{A},\mathcal{C}} := \mathcal{R}_{\mathcal{A},\mathcal{C}}/\,\mathrm{Ann}(\delta_1+\delta_2)$. Let $\overline{\delta}_1$ and $\overline{\delta}_2$ denote the classes of δ_1 and δ_2, respectively, in $\overline{\mathcal{R}}_{\mathcal{A},\mathcal{C}}$. Then

$$\begin{gathered}\overline{\mathcal{R}}_{\mathcal{A},\mathcal{C}} = \mathbb{Z}\oplus\mathbb{Z}\overline{\delta}_1\oplus\mathbb{Z}\overline{\delta}_2\oplus\mathbb{Z}\overline{\delta}_2^2\oplus\mathbb{Z}\overline{\delta}_1^2\oplus\mathbb{Z}\overline{\delta}_1^2\overline{\delta}_2\,,\\ \overline{\delta}_1\overline{\delta}_2 = \overline{\delta}_1^2+\overline{\delta}_2^2\,,\qquad \overline{\delta}_1^2\overline{\delta}_2 = \overline{\delta}_1\overline{\delta}_2^2\,,\qquad \overline{\delta}_1^3 = \overline{\delta}_2^3 = \overline{\delta}_1^2\overline{\delta}_2^2 = 0\,.\end{gathered}$$

Proceeding as in the previous examples we write

$$\begin{aligned}\Phi_{\mathbb{L},\underline{\gamma},\underline{\varepsilon}}(\mathbf{u}) &= -3u_1^{-1}\sum_{n_1,\,n_2\geq 0}(-1)^{n_1+n_2}\frac{(1+3\overline{\delta}_1+3\overline{\delta}_2)_{3n_1+3n_2}}{((1+\overline{\delta}_1)_{n_1}(1+\overline{\delta}_2)_{n_2})^3}z_1^{n_1+\overline{\delta}_1}z_2^{n_2+\overline{\delta}_2}\\ &= \Phi_0(\mathbf{u})+\Phi_{1,1}(\mathbf{u})\overline{\delta}_1+\Phi_{1,2}(\mathbf{u})\overline{\delta}_2+\Phi_{2,1}(\mathbf{u})\overline{\delta}_2^2+\Phi_{2,2}(\mathbf{u})\overline{\delta}_1^2+\Phi_3(\mathbf{u})\overline{\delta}_1^2\overline{\delta}_2\\ &= \Phi_0(\mathbf{u})z_1^{\overline{\delta}_1}z_2^{\overline{\delta}_2}\Big(1+f_{1,1}(\mathbf{z})\overline{\delta}_1+f_{1,2}(\mathbf{z})\overline{\delta}_2+f_{2,1}(\mathbf{z})\overline{\delta}_2^2+f_{2,2}(\mathbf{z})\overline{\delta}_1^2+f_3(\mathbf{z})\overline{\delta}_1^2\overline{\delta}_2\Big)\,.\end{aligned}$$

Here $\mathbf{z} = (z_1,z_2)$. From the $\overline{\delta}_1$ and $\overline{\delta}_2$ components we construct two *canonical coordinates* (cf. (111))

$$q_1 := -z_1\exp(f_{1,1}(\mathbf{z}))\,,\qquad q_2 := -z_2\exp(f_{1,2}(\mathbf{z}))\,. \tag{107}$$

We view z_1, z_2 as functions of q_1, q_2 via the inverse of relation (107). The *prepotential* in this case is (cf. (114))

$$\mathcal{F}(q_1,q_2) = \frac{3}{2}\left(\frac{\Phi_{1,1}(\mathbf{u})}{\Phi_0(\mathbf{u})}\frac{\Phi_{2,1}(\mathbf{u})}{\Phi_0(\mathbf{u})}+\frac{\Phi_{1,2}(\mathbf{u})}{\Phi_0(\mathbf{u})}\frac{\Phi_{2,2}(\mathbf{u})}{\Phi_0(\mathbf{u})}-\frac{\Phi_3(\mathbf{u})}{\Phi_0(\mathbf{u})}\right)\,. \tag{108}$$

The −-signs in (107) and the factor 3 in (108) are needed to match the calculations below with the results in [16] Appendix B2.

We expand $\log\Phi_{\mathbb{L},\underline{\gamma},\underline{\varepsilon}}(\mathbf{u})$ on the basis $\{1,\overline{\delta}_1,\overline{\delta}_2,\overline{\delta}_2^2,\overline{\delta}_1^2,\overline{\delta}_1^2\overline{\delta}_2\}$ of $\overline{\mathcal{R}}_{\mathcal{A},\mathcal{C}}$. The $\overline{\delta}_1^2\overline{\delta}_2$-coordinate is on the one hand

$$\log(-q_1)\log(-q_2)\log(q_1q_2)-\tfrac{2}{3}\mathcal{F}(q_1,q_2)$$

and on the other hand it is

$$f_{1,1}(\mathbf{z})^2f_{1,2}(\mathbf{z})+f_{1,1}(\mathbf{z})f_{1,2}(\mathbf{z})^2-f_{1,1}(\mathbf{z})f_{2,1}(\mathbf{z})-f_{1,2}(\mathbf{z})f_{2,2}(\mathbf{z})+f_3(\mathbf{z}).$$

Computing the coefficients N_{j_1,j_2} in the expansion

$$\mathcal{F}(q_1,q_2) = \tfrac{3}{2}\log(-q_1)\log(-q_2)\log(q_1q_2) + \textstyle\sum_{j_1,j_2\geq 0,\, j_1+j_2>0} N_{j_1,j_2}\mathrm{Li}_3(q_1^{j_1}q_2^{j_2})$$

is now somewhat more involved than in the previous examples. We leave it as an **exercise in Mathematica, Maple or PARI programming**.
A table of the numbers N_{j_1,j_2} for this example appears in [16] Appendix B2 under the name $X_{(3|3)}(1,1,1|1,1,1)$. In [16] one finds many more 2-parameter models.

8.5. The Schwarz map for some extended GKZ systems

In this section we briefly discuss how the $\overline{\mathcal{R}}_{\mathcal{A},\mathcal{C}}\otimes\mathbb{C}$-valued function $\Phi_{\mathbb{L},\underline{\gamma},\underline{\varepsilon}}(\mathbf{u})$ which we met in the preceding examples, can be viewed as a Schwarz map and what are some special features of the image.

First note that, since $\dim\overline{\mathcal{R}}_{\mathcal{A},\mathcal{C}}\otimes\mathbb{C} < \dim\mathcal{R}_{\mathcal{A},\mathcal{C}}\otimes\mathbb{C} = \text{volume}\,\Delta_{\mathcal{A}}$, the components of $\Phi_{\mathbb{L},\underline{\gamma},\underline{\varepsilon}}(\mathbf{u})$ with respect to some basis of $\overline{\mathcal{R}}_{\mathcal{A},\mathcal{C}}\otimes\mathbb{C}$ can not suffice as a basis for the solution space of the GKZ system. They do however constitute a basis for the solution space of some extension of the GKZ system (see [16]). So, strictly speaking we are not talking about the Schwarz map for the GKZ system, but for an extension thereof. Since we do explicitly have all these basis solutions for the extended system, we need not care about this system itself.

In the examples, coming from (families of) Calabi–Yau threefolds, the ring $\overline{\mathcal{R}}_{\mathcal{A},\mathcal{C}}$ is graded and splits in homogeneous pieces,

$$\overline{\mathcal{R}}_{\mathcal{A},\mathcal{C}} = \overline{\mathcal{R}}^{(0)}_{\mathcal{A},\mathcal{C}}\oplus\overline{\mathcal{R}}^{(1)}_{\mathcal{A},\mathcal{C}}\oplus\overline{\mathcal{R}}^{(2)}_{\mathcal{A},\mathcal{C}}\oplus\overline{\mathcal{R}}^{(3)}_{\mathcal{A},\mathcal{C}}\,,$$

with degrees 0, 1, 2, 3 and ranks 1, d, d, 1, respectively; recall $d = \operatorname{rank}\mathbb{L}$. We fix a basis for $\overline{\mathcal{R}}_{\mathcal{A},\mathcal{C}}$ by fixing bases for the homogeneous pieces

$$\overline{\mathcal{R}}^{(0)}_{\mathcal{A},\mathcal{C}}: e_0 = 1\,,\quad \overline{\mathcal{R}}^{(1)}_{\mathcal{A},\mathcal{C}}: e_{1,1},\ldots,e_{1,d}\,,\quad \overline{\mathcal{R}}^{(2)}_{\mathcal{A},\mathcal{C}}: e_{2,1},\ldots,e_{2,d}\,,\quad \overline{\mathcal{R}}^{(3)}_{\mathcal{A},\mathcal{C}}: e_3\,,$$

and expand $\Phi_{\mathbb{L},\underline{\gamma},\underline{\varepsilon}}(\mathbf{u})$ with respect to this basis

$$\Phi_{\mathbb{L},\underline{\gamma},\underline{\varepsilon}}(\mathbf{u}) = \Phi_0(\mathbf{u})e_0 + \sum_{i=1}^{d}\Phi_{1,i}(\mathbf{u})e_{1,i} + \sum_{i=1}^{d}\Phi_{2,i}(\mathbf{u})e_{2,i} + \Phi_3(\mathbf{u})e_3\,. \tag{109}$$

The Schwarz map lands in the projective space

$$\mathbb{P}\left(\overline{\mathcal{R}}_{\mathcal{A},\mathcal{C}}\otimes\mathbb{C}\right)$$

and $\Phi_0(\mathbf{u}),\ldots,\Phi_3(\mathbf{u})$ are homogeneous coordinates for the image points. Since these functions are solutions of the same GKZ system their quotients, and hence also the Schwarz map, are defined on some open set in $\mathcal{V}_{\mathcal{A}}$ near the special point $\mathbf{p}_{\mathcal{C}}$ corresponding to the maximal cone $\mathcal{C}$ in the secondary fan. The map is multi-valued and we do fully control the local monodromy.

The image of the Schwarz map has dimension equal to $\dim\mathcal{V}_{\mathcal{A}} = \operatorname{rank}\mathbb{L} = d$, whereas the projective space $\mathbb{P}\left(\overline{\mathcal{R}}_{\mathcal{A},\mathcal{C}}\otimes\mathbb{C}\right)$ has dimension $1+2d$. For the description of the image of the Schwarz map we want to profit from the description of the moduli of Calabi–Yau threefolds by Bryant and Griffiths [6]. In the theory of moduli of Calabi–Yau threefolds one writes the holomorphic 3-form in coordinates with respect to a basis of the third cohomology space given by topological 3-cycles. These coordinates are the period integrals of the 3-form. We know $\Phi_0(\mathbf{u})$ explicitly as a period integral (see (90), (99), (105)), but we still need an argument for the other coordinates in (109) to be periods. Such an argument may be that inspection of the local monodromy shows that the extended GKZ system of differential equations satisfied by the known period integral $\Phi_0(\mathbf{u})$ is irreducible, for then all periods must be linear combinations of $\Phi_0(\mathbf{u}),\ldots,\Phi_3(\mathbf{u})$.

Having matched (109) with the coordinates (= periods) of the holomorphic 3-form with respect to a basis of topological 3-cycles, we must check that the basis $e_0,\ldots,e_3$ satisfies the requirements for application of the Bryant–Griffiths theory, i.e., we need to know that with respect to the alternating bilinear form $\langle,\rangle$ on the third cohomology space of the Calabi–Yau threefold

$$\langle e_0, e_3\rangle = -\langle e_3, e_0\rangle = -\langle e_{1,i}, e_{2,i}\rangle = \langle e_{2,i}, e_{1,i}\rangle = 1 \quad \text{for } i = 1,\ldots,d, \tag{110}$$

and all other $\langle e_r, e_s\rangle = 0$. In an example at the end of this section we show how to derive (110) from the explicitly known local monodromy and logarithmic pieces of $\Phi_{\mathbb{L},\underline{\gamma},\underline{\varepsilon}}(\mathbf{u})$.

We are now all set for applying [6]. First define the *canonical coordinates*

$$q_i := \exp\left(\frac{\Phi_{1,i}(\mathbf{u})}{\Phi_0(\mathbf{u})}\right) \qquad \text{for } i = 1,\ldots d\,. \tag{111}$$

The derivations $q_i\frac{\partial}{\partial q_i}$ act on the cohomology spaces of the Calabi–Yau threefolds in the family. Griffiths' transversality and the Riemann bilinear relations imply

$$\left\langle \frac{\Phi_{\mathbb{L},\underline{\gamma},\underline{\varepsilon}}(\mathbf{u})}{\Phi_0(\mathbf{u})},\, q_i\frac{\partial}{\partial q_i}\left(\frac{\Phi_{\mathbb{L},\underline{\gamma},\underline{\varepsilon}}(\mathbf{u})}{\Phi_0(\mathbf{u})}\right)\right\rangle = 0\,. \tag{112}$$

Write $\varphi_3 := \frac{\Phi_3(\mathbf{u})}{\Phi_0(\mathbf{u})}$ and $\varphi_{a,j} := \frac{\Phi_{a,j}(\mathbf{u})}{\Phi_0(\mathbf{u})}$ for $a = 1, 2,\ j = 1,\ldots,d$. These are (multivalued) functions of $q_1,\ldots,q_d$, and in fact $\varphi_{1,j} = \log q_j$. Then the left hand side of (112) evaluates to

$$q_i\frac{\partial\varphi_3}{\partial q_i} - \sum_{j=1}^{d}\left(\varphi_{1,j}\,q_i\frac{\partial\varphi_{2,j}}{\partial q_i}\right) + \varphi_{2,i} = q_i\frac{\partial}{\partial q_i}\left(\varphi_3 - \sum_{j=1}^{d}\varphi_{1,j}\varphi_{2,j}\right) + 2\,\varphi_{2,i}\,.$$

According to (112) this equals 0 and thus

$$\varphi_{2,i} = q_i\frac{\partial\mathcal{F}}{\partial q_i} \tag{113}$$

where

$$\mathcal{F} := \frac{1}{2}\left(-\varphi_3 + \sum_{j=1}^{d}\varphi_{1,j}\varphi_{2,j}\right) \tag{114}$$

is the so-called *prepotential.*

Example. Thus we recover the canonical coordinate (95) for the quintic in $\mathbb{P}^4$ up to a $-$-sign and the prepotential (96) up to a factor 5 (which is the degree of the quintic). And similarly for the intersection of two cubics in $\mathbb{P}^5$ and the hypersurface of degree $(3,3)$ in $\mathbb{P}^2 \times \mathbb{P}^2$. The factors 'sign' and 'degree' are needed to match the results of our calculations with the tables of enumerative data in the literature. Moreover, if the wrong sign is used, the numbers $N_{j_1,\dots,j_d}$ are often even not integers.

Conclusion. *The above discussion shows that the canonical coordinates and the prepotential act like a parametrization for the image of the Schwarz map: the image points have coordinates* $(1, t_1, \dots, t_{2d+1})$ *with*

$$\begin{array}{rcll} t_j &=& \log q_j & \text{for} \quad j = 1, \dots, d\,, \\ t_{d+j} &=& q_j \frac{\partial \mathcal{F}}{\partial q_j} & \text{for} \quad j = 1, \dots, d\,, \\ t_{2d+1} &=& -2\,\mathcal{F} + \sum_{j=1}^{d} t_j t_{d+j}\,. & \end{array}$$

Remark. On the graded ring $\overline{\mathcal{R}}_{\mathcal{A},\mathcal{C}}$ there is an involution $.^*$ given for homogeneous elements by $x^* = (-1)^{\deg x} x$. We fix the linear map $\tau : \overline{\mathcal{R}}_{\mathcal{A},\mathcal{C}} \overset{\text{project}}{\longrightarrow} \overline{\mathcal{R}}^{(3)}_{\mathcal{A},\mathcal{C}} \overset{\simeq}{\longrightarrow} \mathbb{Z}$. Then, in the examples of Sections 8.2, 8.3, 8.4 the alternating bilinear form defined by the ordered basis and the relations (110), is in fact

$$\langle x, y \rangle = \tau(x^* y)\,. \tag{115}$$

Moreover, in those examples the trick of expanding $\log \Phi_{\mathbb{L},\underline{\gamma},\underline{\varepsilon}}(\mathbf{u})$ in two ways showed that the prepotential is a polynomial of degree 3 in $\log q_1, \dots, \log q_d$ plus a power series in $q_1, \dots, q_d$.

Example. Let us check that (110) holds for the ordered basis

$$\{e_0, e_{1,1}, e_{1,2}, e_{2,1}, e_{2,2}, e_3\} := \{1, \overline{\delta}_1, \overline{\delta}_2, \overline{\delta}_2^2, \overline{\delta}_1^2, \overline{\delta}_1^2\overline{\delta}_2\}$$

in the example of Section 8.4. The alternating bilinear form (on the third cohomology space in a family of Calabi–Yau threefolds) is invariant under the local monodromy operators, which in this case are given by multiplication by $\exp(2\pi \mathrm{i}\overline{\delta}_1)$ and $\exp(2\pi \mathrm{i}\overline{\delta}_2)$. This means that the matrices for multiplication with $\overline{\delta}_1$ and $\overline{\delta}_2$ and the Gram matrix of the alternating bilinear form, everything with respect to the above basis, must satisfy $\mathrm{Gram}^t = -\mathrm{Gram}$ and

$$\mathrm{mat}(\overline{\delta}_a) \cdot \mathrm{Gram} = -\mathrm{Gram} \cdot \mathrm{mat}(\overline{\delta}_a)^t\,, \qquad a = 1, 2\,. \tag{116}$$

One easily checks

$$\mathrm{mat}(\overline{\delta}_1) = \begin{pmatrix} 0&0&0&0&0&0 \\ 1&0&0&0&0&0 \\ 0&0&0&0&0&0 \\ 0&0&1&0&0&0 \\ 0&1&1&0&0&0 \\ 0&0&0&1&0&0 \end{pmatrix}, \qquad \mathrm{mat}(\overline{\delta}_2) = \begin{pmatrix} 0&0&0&0&0&0 \\ 0&0&0&0&0&0 \\ 1&0&0&0&0&0 \\ 0&1&1&0&0&0 \\ 0&1&0&0&0&0 \\ 0&0&0&0&1&0 \end{pmatrix}.$$

The general anti-symmetric matrix solution to (116) has, up to a non-zero scalar factor, the form

$$\begin{pmatrix} 0 & 0 & 0 & 0 & 0 & 1 \\ 0 & 0 & 0 & -1 & 0 & 0 \\ 0 & 0 & 0 & 0 & -1 & 0 \\ 0 & 1 & 0 & 0 & 0 & x \\ 0 & 0 & 1 & 0 & 0 & y \\ -1 & 0 & 0 & -x & -y & 0 \end{pmatrix}.$$

When we evaluate (112) using this Gram matrix, we find, for $i = 1, 2$,

$$\begin{aligned} & q_i \frac{\partial\varphi_3}{\partial q_i} - \varphi_{1,1} q_i \frac{\partial\varphi_{2,1}}{\partial q_i} - \varphi_{1,2} q_i \frac{\partial\varphi_{2,2}}{\partial q_i} + \varphi_{2,i} + \\ & + x\left(\varphi_{2,1} q_i \frac{\partial\varphi_3}{\partial q_i} - \varphi_3 q_i \frac{\partial\varphi_{2,1}}{\partial q_i}\right) + y\left(\varphi_{2,2} q_i \frac{\partial\varphi_3}{\partial q_i} - \varphi_3 q_i \frac{\partial\varphi_{2,2}}{\partial q_i}\right) = 0\,. \end{aligned} \tag{117}$$

We want to estimate the growth of the various terms in this expression by looking at the logarithmic pieces. So, recall that in this example

$$\Phi_{\mathbb{L},\underline{\gamma},\underline{\varepsilon}}(\mathbf{u}) = \Phi_0(\mathbf{u}) z_1^{\overline{\delta}_1} z_2^{\overline{\delta}_2} \times (\text{power series in } z_1, z_2)$$

and

$$\begin{aligned} z_1^{\overline{\delta}_1} z_2^{\overline{\delta}_2} &= (1 + (\log z_1)\overline{\delta}_1 + \tfrac{1}{2}(\log z_1)^2\overline{\delta}_1^2)(1 + (\log z_2)\overline{\delta}_2 + \tfrac{1}{2}(\log z_2)^2\overline{\delta}_2^2) \\ &= 1 + (\log z_1)\overline{\delta}_1 + (\log z_2)\overline{\delta}_2 + \left(\tfrac{1}{2}(\log z_2)^2 + (\log z_1)(\log z_2)\right)\overline{\delta}_2^2 \\ &\quad + \left(\tfrac{1}{2}(\log z_1)^2 + (\log z_1)(\log z_2)\right)\overline{\delta}_1^2 + \tfrac{1}{2}(\log z_1)(\log z_2)(\log z_1 z_2)\overline{\delta}_1^2\overline{\delta}_2\,. \end{aligned}$$

Moreover, up to addition of power series, $\log q_1 \asymp \log z_1$ and $\log q_2 \asymp \log z_2$. Thus we see that the highest order logarithmic contributions are

$$\begin{aligned} -\varphi_3 q_1 \frac{\partial\varphi_{2,1}}{\partial q_1} + \varphi_{2,1} q_1 \frac{\partial\varphi_3}{\partial q_1} &\asymp -\tfrac{1}{2}((\log q_1)^2(\log q_2) + (\log q_1)(\log q_2)^2)(\log q_2) \\ &\quad + \left(\tfrac{1}{2}(\log q_2)^2 + (\log q_1)(\log q_2)\right)\left((\log q_1)(\log q_2) + \tfrac{1}{2}(\log q_2)^2\right) \\ &= \tfrac{1}{2}(\log q_1)^2(\log q_2)^2 + \tfrac{1}{2}(\log q_1)(\log q_2)^3 + \tfrac{1}{4}(\log q_2)^4 \end{aligned}$$

and

$$\begin{aligned} & -\varphi_3 q_1 \frac{\partial\varphi_{2,2}}{\partial q_1} + \varphi_{2,2} q_1 \frac{\partial\varphi_3}{\partial q_1} \\ &\qquad \asymp -\tfrac{1}{2}((\log q_1)^2(\log q_2) + (\log q_1)(\log q_2)^2)(\log q_1 + \log q_2) \\ &\qquad\quad + \left(\tfrac{1}{2}(\log q_1)^2 + (\log q_1)(\log q_2)\right)\left((\log q_1)(\log q_2) + \tfrac{1}{2}(\log q_2)^2\right) \\ &\qquad = \tfrac{1}{4}(\log q_1)^2(\log q_2)^2. \end{aligned}$$

So if we consider (117) for $i = 1$ and $q_2 \downarrow 0$ the dominant $\frac{1}{4}(\log q_2)^4$ term forces $x = 0$. Having $x = 0$ we consider (117) for $i = 1$ and $q_1 = q_2 \downarrow 0$. Once again there is a dominant $\frac{1}{4}(\log q_2)^4$, forcing $y = 0$.

This finishes the proof for the fact that in the example of Section 8.4 *the ordered basis* $\{1, \overline{\delta}_1, \overline{\delta}_2, \overline{\delta}_2^2, \overline{\delta}_1^2, \overline{\delta}_1^2\overline{\delta}_2\}$ *satisfies* (110).

Exercise. Apply the techniques of the preceding example and show that the basis $\{1, \bar{\varepsilon}, \bar{\varepsilon}^2, \bar{\varepsilon}^3\}$ in Sections 8.2 and 8.3 satisfies (110).

8.6. Manifestations of Mirror Symmetry

In this section we discuss in an informal way some features of our hypergeometric formulas which can be seen as manifestations of Mirror Symmetry. The objects in the second and third columns of Table 1 should be considered as mirror partners.

The objects in the third column are given as zero loci of polynomials (89):

$$P_{\mathcal{A}} = \sum_{\mathbf{a}\in\mathcal{A}} \mathbf{u_a x^a} . \tag{118}$$

The coefficients $\mathbf{u_a}$ yield the complex structure of these varieties via period integrals like (90), (99), (105). These periods express in cohomology the position of the class of the holomorphic 3-form with respect to the classes of the topological 3-cycles. They also show how this position varies as a function of the parameters $\mathbf{u_a}$. These period functions satisfy an (extended) GKZ system of differential equations. From the series $\Phi_{\mathbb{L},\underline{\gamma},\underline{\varepsilon}}(\mathbf{u})$ (cf. (92), (101), 109)) one can derive all solutions to these (extended) GKZ differential equations as follows. The series $\Phi_{\mathbb{L},\underline{\gamma},\underline{\varepsilon}}(\mathbf{u})$ takes values in the vector space $\mathcal{R}_{\mathcal{A},\mathcal{C}} \otimes \mathbb{C}$. This vector space carries a non-degenerate bilinear form $\langle , \rangle$ (see (115)) and thus all solutions to the (extended) GKZ system can be obtained by pairing $\Phi_{\mathbb{L},\underline{\gamma},\underline{\varepsilon}}(\mathbf{u})$ with the elements of of $\mathcal{R}_{\mathcal{A},\mathcal{C}} \otimes \mathbb{C}$. This should hold in particular for the above period integrals. Thus given a non-trivial holomorphic 3-form Ω on the Calabi–Yau variety in the third column of Table 1 there should for every topological 3-cycle ρ on this CY variety be an element $\alpha_\rho \in \mathcal{R}_{\mathcal{A},\mathcal{C}} \otimes \mathbb{C}$ such that

$$\int_\rho \Omega = \left\langle \Phi_{\mathbb{L},\underline{\gamma},\underline{\varepsilon}}(\mathbf{u}),\, \alpha_\rho \right\rangle . \tag{119}$$

On the other hand, as we remarked before (see also [9], [24], Thm. 9) $\mathcal{R}_{\mathcal{A},\mathcal{C}}$ is the cohomology ring of the ambient space in the second column of Table 1; better even, it is the Chow ring of this ambient space [9]. The Chow ring is generated by rational equivalence classes of algebraic subvarieties of this space, with product structure coming from intersection theory. Thus evaluating the right-hand side of (119) becomes a computation in the intersection theory of algebraic subvarieties of the ambient space in the second column of Table 1.

This interpretation of the ring $\mathcal{R}_{\mathcal{A},\mathcal{C}}$ and of the right-hand side of (119) is however not the only possible one, and probably not even optimal. Via Chern characters one may want to lift things from cohomology or Chow ring to the Grothendieck group K_0 or the bounded derived category of coherent sheaves on the ambient space in the second column of Table 1. In the latter setting one then comes to Kontsevich's *Homological Mirror Symmetry Conjecture* [17]. The reader can find in [17] also the example of the quintic which we discussed in section 8.2.

In Section 3.5 we wrote the parameters $\mathbf{u_a}$ in exponential form:

$$\mathbf{u_a} = e^{2\pi i \mathbf{w_a}} .$$

In Section 5.2 this was reinterpreted as taking coordinates on the toric variety associated with the secondary fan. Tracing back through the interpretations of the secondary fan one finds in (46) that with $\mathbf{w} = (\mathbf{w_a})_{\mathbf{a}\in\mathcal{A}}$ is associated the polyhedron

$$K_{\mathbf{w}} := \{\mathbf{v} \in \mathbb{M}_{\mathbb{R}}^{\vee} \mid \langle \mathbf{a}, \mathbf{v} \rangle \geq -\Im \mathbf{w_a},\ \forall \mathbf{a} \in \mathcal{A}\}\,. \tag{120}$$

With [15] Appendix 2 one can then view $\Im\mathbf{w}$ as moduli parameters for the Kähler geometry on the ambient space in the second column of Table 1; there is here a slight subtlety in that our $K_{\mathbf{w}}$ is not compact and actually corresponds to a vector bundle over this ambient space, whereas in loc. cit. one has a compact polyhedron which corresponds to the ambient space itself. *This now raises the interesting problem of interpreting the right-hand side of* (119) *directly as a pairing of Kähler structure and algebraic cycles, or coherent sheaves, on the ambient space in the second column of Table* 1.

Let us now put also the variables $\mathbf{x}$ in (118) in exponential form:

$$\mathbf{x}^{\mathbf{a}} = e^{2\pi \mathrm{i}\langle \mathbf{a}, \mathbf{y}\rangle}\,.$$

This gives the polynomial $P_{\mathcal{A}}$ a new outfit:

$$P_{\mathcal{A}} = \sum_{\mathbf{a}\in\mathcal{A}} e^{2\pi \mathrm{i}(\mathbf{w_a} + \langle \mathbf{a}, \mathbf{y}\rangle)} \tag{121}$$

and reinterprets $K_{\mathbf{w}}$ as a requirement that the individual terms of the sum $P_{\mathcal{A}}$ should have absolute value ≤ 1. This also suggests possible relations with the developing new branch of geometry, called *Tropical Geometry* [22].

Since there is still a lot of activity in Mirror Symmetry and related fields, with constant emergence of new ideas, I feel it is too early to formulate a final conclusion. It may be clear, though, that hypergeometric objects like $\Phi_{\mathbb{L},\underline{\gamma},\underline{\varepsilon}}$ *do play a central role.*

References

[1] V. Batyrev, *Dual polyhedra and mirror symmetry for Calabi–Yau hypersurfaces in toric varieties,* J. Alg. Geom. **3** (1994), 493–535.

[2] V. Batyrev, *Variations of mixed Hodge structure of affine hypersurfaces in algebraic tori,* Duke Math. J. **69** (1993), 349–409.

[3] V. Batyrev, L. Borisov, *Dual Cones and Mirror Symmetry for Generalized Calabi–Yau Manifolds,* pp. 71–86 in *Mirror Symmetry II*, B. Greene, S.-T. Yau (eds.), Studies in Advanced Math., vol. 1, American Math. Soc. / International Press, 1997.

[4] A. de Boo, *Solving GKZ-hypergeometric systems using relative Stanley-Reisner ideals*, PhD thesis, Utrecht University, November 1999 (unpublished).

[5] L. Borisov, P. Horja, *Mellin-Barnes Integrals as Fourier-Mukai Transforms*, math.AG/0510486.

[6] R. Bryant, P. Griffiths, *Some observations on the infinitesimal period relations for regular threefolds with trivial canonical bundle*, in: Arithmetic and Geometry, vol. II, M. Artin and J. Tate (eds.), Progress in Math. vol. 36, Birkhäuser, Boston, 1983, pp. 77–102.

[7] Ph. Candelas, X. de la Ossa, P. Green, L. Parkes, *A Pair of Calabi–Yau Manifolds as an Exactly Soluble Superconformal Theory*, in: Essays on Mirror Manifolds, S-T. Yau (ed.), International Press, Hong Kong, 1992.

[8] D. Cox, S. Katz, *Mirror Symmetry and Algebraic Geometry*, Mathematical Surveys and Monographs, vol. 68, American Math. Soc., Providence, Rhode Island, 1999.

[9] W. Fulton, *Introduction to Toric Varieties,* Annals of Mathematics Studies 131, Princeton University Press, Princeton, New Jersey, 1993.

[10] I.M. Gelfand, A.V. Zelevinskii, M.M. Kapranov, *Hypergeometric functions and toral manifolds,* Functional Analysis and its Applications **23** (1989), 94–106.

[11] correction to [10], Funct. Analysis and its Appl. **27** (1993), 295.

[12] I.M. Gelfand, M.M. Kapranov, A.V. Zelevinsky, *Generalized Euler Integrals and A-Hypergeometric Functions,* Advances in Math. **84** (1990), 255–271.

[13] I.M. Gelfand, M.M. Kapranov, A.V. Zelevinsky, *Discriminants, Resultants and Multidimensional Determinants,* Birkhäuser Boston, 1994.

[14] I.M. Gelfand, M.M. Kapranov, A.V. Zelevinsky, *Hypergeometric Functions, Toric Varieties and Newton Polyhedra* in: Special Functions, M. Kashiwara, T. Miwa (eds.), ICM-90 Satellite Conference Proceedings, Springer-Verlag Tokyo, 1991.

[15] V. Guillemin, *Moment Maps and Combinatorial Invariants of Hamiltonian T^n-spaces,* Progress in Math. vol. 22, Birkhäuser Boston, 1994.

[16] S. Hosono, A. Klemm, S. Theisen, S.-T. Yau, *Mirror Symmetry, Mirror Map and Applications to Calabi–Yau Hypersurfaces*, Commun. Math. Phys. **167** (1995), 301–350; also hep-th/9308122.

[17] M. Kontsevich, *Homological Algebra of Mirror Symmetry*, in: Proceedings of the International Congress of Mathematicians Zürich 1994, Birkhäuser Basel, 1995.

[18] A. Libgober, J. Teitelbaum, *Lines on Calabi–Yau Complete Intersections, Mirror Symmetry, and Picard-Fuchs Equations*, Int. Math. Res. Notices **1** (1993), 29–39.

[19] E. Looijenga, *Uniformization by Lauricella Functions — An Overview of the Theory of Deligne-Mostow*, this volume; also math.CV/0507534.

[20] K. Mayr, *Über die Lösung algebraischer Gleichungssysteme durch hypergeometrische Funktionen*, Monatshefte für Math. **45** (1937), 280–313.

[21] M. Passare, A. Tsikh, *Algebraic Equations and Hypergeometric Series*, in: The Legacy of Niels Henrik Abel, O.A. Laudal and R. Piene (eds.), Springer Verlag, Berlin, 2004.

[22] J. Richter-Gebert, B. Sturmfels, T. Theobald, *First steps in tropical geometry*, math.AG/0306366.

[23] R.P. Stanley, *Combinatorics and Commutative Algebra (second edition),* Progress in Math. 41, Birkhäuser Boston, 1996.

[24] J. Stienstra, *Resonant Hypergeometric Systems and Mirror Symmetry*, in: Integrable Systems and Algebraic Geometry, Proceedings of the Taniguchi Symposium 1997, M.-H. Saito, Y. Shimizu, K. Ueno (eds.); World Scientific, Singapore, 1998; also alg-geom/9711002.

[25] B. Sturmfels, *Gröbner Bases and Convex Polytopes,* University Lecture Series vol. 8, American Math. Soc., 1996.

Jan Stienstra
Department of Mathematics
University of Utrecht
Budapestlaan 6
3584 CD Utrecht
The Netherlands
e-mail: stien@math.uu.nl

Progress in Mathematics, Vol. 260, 373–406

Orbifolds and Their Uniformization

A. Muhammed Uludağ

Abstract. This is an introduction to complex orbifolds with an emphasis on orbifolds in dimension 2 and covering relations between them.

Mathematics Subject Classification (2000). Primary 32Q30; Secondary 14J25.

Keywords. Orbifold, orbiface, uniformization, ball-quotient.

Contents

This work was supported by TÜBİTAK grant KARİYER-104T136 and achieved thanks to the support of the European Commission through its 6th Framework Program "Structuring the European Research Area" and the contract Nr. RITA-CT-2004-505493 for the provision of Transnational Access implemented as Specific Support Action.

Introduction

These notes aim to give an introduction to the theory of orbifolds and their uniformizations, along the lines settled in 1986 by M. Kato [13], with special emphasis on complex 2-dimensional orbifolds (orbifaces).

An orbifold is a space locally modeled on a smooth manifold modulo a finite group action, which is said to be uniformizable if it is a global quotient. They were first studied in the 1950s by Satake under the name "V-manifold" and renamed by Thurston in the 1970s. Orbifolds appear naturally in various fields of mathematics and physics and they are studied from several points of view. In these notes we focus on the uniformization problem and consider almost exclusively orbifolds with a smooth base space. In most cases this base will be a complex projective space. From this perspective, orbifolds can be viewed as a refinement of the double covering construction of algebraic varieties with special properties (for example with given Chern numbers, see [11]). The first steps in this refinement were taken by Hirzebruch [9], culminating in the monograph [1] devoted to Kummer coverings of $\mathbb{P}^2$ branched along line arrangements. Kobayashi [14] studied more general coverings with non-linear branch loci with non-nodal singularities.

Many basic topological invariants such as the fundamental group has an orbifold version, and the usual notion of Galois covering is extended to orbifolds in a straightforward way. It was observed by Yoshida that orbifold germs are related via covering maps. We elaborate on this observation and show that many interesting orbifolds (e.g., the ball-quotient orbifolds) are related via covering maps. Note that a covering relation between ball-quotient orbifolds is nothing but a commensurability among the corresponding lattices acting on the ball.

The plan of this paper is as follows: Section 1 gives some background on branched coverings. Section 2 includes fundamental facts and definitions about orbifolds. Section 3 is devoted to the local structure and singularities of orbifolds, especially in dimension 2. Section 4 sketches the solutions of the global uniformization problem for some special orbifolds. In particular, Section 4.1 includes a complete classification of abelian finite smooth branched coverings of $\mathbb{P}^2$. This amounts to the classification of algebraic surfaces with an abelian group action whose quotient is isomorphic to $\mathbb{P}^2$. There are also many examples of non-abelian coverings.

These notes were typeset during my stay in IHES in August 2005 and are based on talks delivered at the CIMPA Summer School *Arithmetic and Geometry around hypergeometric functions* (2005) held in Istanbul and the EMS Summer School *Braid Groups and Related Topics* (2005) held in Tiberias. I am grateful to

Professor Mina Teicher and Dr. Tzachi Ben-Itzhak for their hospitality. Diagrams of the paper were typeset by Paul Taylor's *Commutative Diagrams* package.

1. Branched Coverings

Here we collect some facts about branched coverings which can mostly be found in Namba's book [17]. In what follows a *variety* is always irreducible, defined over $\mathbb{C}$ and endowed with the strong topology.

A surjective finite holomorphic map $\varphi : M \to X$ of normal varieties is called a *branched covering*. A topological finite covering map is a very special kind of branched covering. Any non-constant map between compact Riemann surfaces is a finite branched covering. If $M \subset \mathbb{P}^n$ is an irreducible hypersurface, then the restriction onto M of a *generic* projection $\mathbb{P}^n \to \mathbb{P}^{n-1}$ is a finite branched covering. A blow-down is not a branched covering since it is not a finite map. An immersion into a higher dimensional space is not a branched covering since it is not surjective.

Example 1.1. (Model branched coverings) The map $\varphi_m : z \in \mathbb{C} \to z^m \in \mathbb{C}$ is a branched covering. More generally, the map

$$\varphi_m : (z_1, z_2, \ldots, z_n) \in \mathbb{C}^n \to (z_1^m, z_2, z_3, \ldots, z_n) \in \mathbb{C}^n$$

is a branched covering.

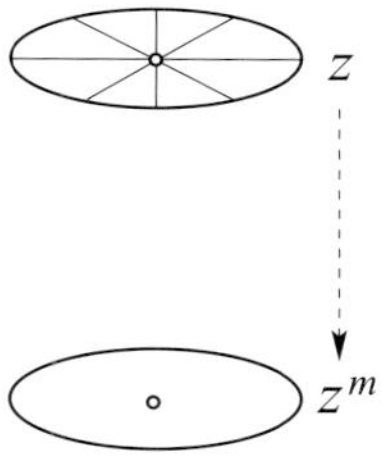

FIGURE 1.1. A model branched covering

A *morphism* between branched coverings $\varphi : M \to X$ and $\psi : N \to X$ is a surjective holomorphic map $\mu : M \to N$ such that the following diagram commutes:

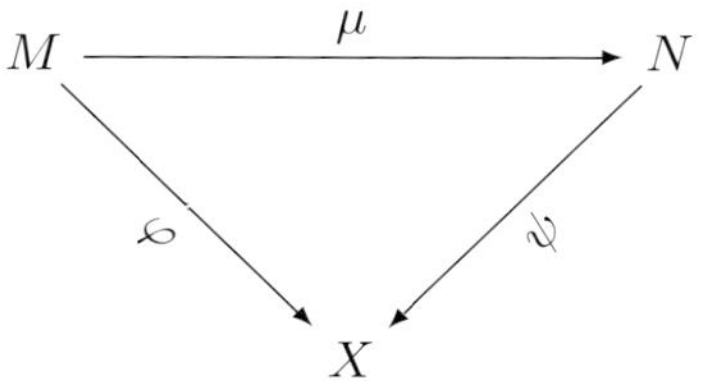

An isomorphism of branched covering spaces is a morphism that is a biholomorphism. The group G_φ of all automorphisms of φ is finite and acts on every fiber of φ. A finite branched covering $\varphi : M \to X$ is called a *finite branched Galois covering* if G_φ acts transitively on every fiber of φ. In this case the orbit space M/G_φ is biholomorphic to X (see [4]).

The *ramification locus* of a finite branched covering $\varphi : M \to X$ is defined by

$$R_\varphi := \{p \in M : \varphi \text{ is not biholomorphic around } p\}.$$

The image $B_\varphi := \varphi(R_\varphi)$ is called the *branch locus* of φ and the map φ is said to be branched *along* B_φ. In case φ is a topological covering then both R_φ and B_φ are empty, such φ is said to be *unbranched.* The restriction $\varphi : M\backslash R_\varphi \to X\backslash B_\varphi$ is an unbranched covering. Conversely, the Grauert–Remmert theorem states (see [19]):

Theorem 1.1. *Let X be a normal variety and B a finite union of proper subvarieties of codimension* 1. *Given a topological unbranched finite covering $\varphi' : M' \to X\backslash B$ with M' being connected, there exists an irreducible normal variety M with a finite branched covering $\varphi : M \to X$ and a homeomorphism $s : M' \to \varphi^{-1}(X\backslash B)$ such that $\varphi(x) = \varphi'(s(x))$ for all $x \in M'$.*

By this theorem, there is a correspondence between subgroups of $\pi_1(X\backslash B)$ of finite index and finite coverings of X branched along B. If φ' is Galois, then so is φ ([8], Proposition I.2.8) and therefore the covering φ is Galois if and only if the corresponding subgroup is normal.

Consider the model branched covering φ_m introduced in Example 1.1. Both R_φ and B_φ are hypersurfaces in $\mathbb{C}^n$ defined by the equation $z_1 = 0$. Let $\varphi : M \to X$ be a branched covering. If X is singular, then R_φ and B_φ can be of codimension > 1, even when φ is a non-trivial branched cover. If X is smooth, then by Zariski's "purity of the branch locus" theorem (see [26]), R_φ is a hypersurface in M and B_φ is a hypersurface in X.

The *ramification divisor* of a finite branched covering $\varphi : M \to X$ of smooth spaces is the divisor of its jacobian; for singular spaces it can be defined for the restriction of φ to smooth parts of M and X and then extended. (If φ is ramified only along a singular part then the ramification divisor is empty). The ramification divisor lives on M. If $\varphi : M \to X$ is Galois, it is possible to define the *branch divisor* on X as follows: let $H_1, \ldots , H_k$ be the irreducible components of the branch locus B_φ. Let $p \in H_i$ be a smooth point of B_φ. Let U be a small neighborhood of p and V be a connected component of $\varphi^{-1}(U)$. The degree m_i of $\varphi|_V$ does not depend on p and is called the *branching index* of φ along H_i. Then the branch divisor is defined as

$$D_\varphi := \sum_{i=1}^{k} m_i H_i.$$

Definition 1.2. Let X be a complex manifold and $D = \Sigma m_i H_i$ be a divisor with coefficients in $\mathbb{Z}_{>0}$. A Galois covering $\varphi : M \to X$ is said to be *branched at D* if $D_\varphi = D$.

Example 1.2. Let $H \subset \mathbb{C}^n$ be a hypersurface given by the reduced polynomial $f \in \mathbb{C}[x_1, \ldots, x_n]$ and let $M \subset \mathbb{C}^{n+1}$ be the hypersurface defined by the polynomial

$z^m - f \in \mathbb{C}[z, x_1, \ldots, x_n]$. Let π be the projection

$$\pi : (z, x_1, \ldots, x_n) \in \mathbb{C}^{n+1} \to (x_1, \ldots, x_n) \in \mathbb{C}^n$$

Then the restriction $\pi : M \to \mathbb{C}^n$ is a finite branched Galois covering with $\mathbb{Z}/(m)$ as the Galois group. The branch locus of π is precisely the hypersurface H, and the branch divisor is mH. Note that if the origin is a singular point of H then M also has a singularity at the origin.

Let X be a normal variety and $B = \cup H_i$ be a hypersurface with irreducible components H_i. Take a base point $\star \in X \backslash B$ and let $p \in H_i$ be a smooth point of B. A *meridian* of H_i in $X \backslash B$ is the homotopy class of a loop μ_p in $X \backslash B$ constructed as follows: Take a small disc Δ intersecting B transversally at p. Let ω be a path in $X \backslash B$ connecting $\star$ to a point of $\partial\Delta$. Then $\mu := \omega \cdot \delta \cdot \omega^{-1}$, where δ is the path obtained by following $\partial\Delta$ in the positive sense. It is well known that any two meridians of a fixed irreducible component H_i are conjugate elements in $\pi_1(X \backslash B)$.

Let $D = \Sigma_1^k m_i H_i$, where H_i are irreducible and take meridians μ_1 of H_1 $\ldots \mu_k$ of H_k in $X \backslash \cup H_i$. The *orbifold fundamental group* of the pair (X, D) is defined as

$$\pi_1^{orb}(X, D) := \pi_1(X \backslash D, \star)/\langle\langle \mu_1^{m_1}, \ldots, \mu_k^{m_k} \rangle\rangle,$$

where $\langle\langle\rangle\rangle$ denotes the normal closure. (Note that the definition of an *orbifold* will wait till the next section.)

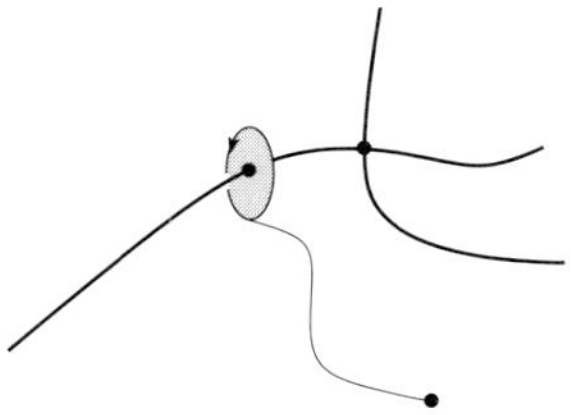

FIGURE 1.2. A meridian

Let $D = \Sigma_1^k m_i H_i$ and let K be a normal subgroup of finite index in $\pi_1(X \backslash D)$. The Galois covering corresponding to K is branched at D if and only if the following two conditions are satisfied:

Condition (i). K contains the elements $\mu_1^{m_1}, \ldots, \mu_k^{m_k}$.

Condition (ii). $\mu_i^m \notin K$ for $m < m_i$.

Condition (ii) will be called the *Branching Condition* in the sequel. Let $G := \pi_1(X \backslash D)/K$ be the corresponding Galois group. Then Condition (i) amounts to the existence of the factorization

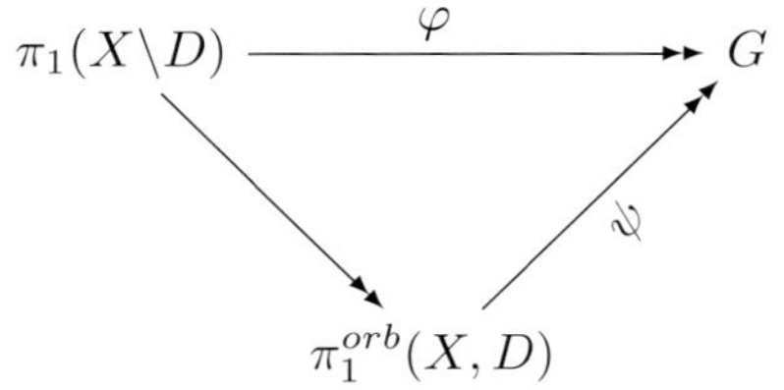

whereas the branching condition means that $\varphi(\mu_i) \in G$ is strictly of order m_i. We conclude that the coverings of X branched at D are really controlled by the group $\pi_1^{orb}(X, D)$, and there is a Galois correspondence between the Galois coverings of X branched at D and normal subgroups of $\pi_1^{orb}(X, D)$ satisfying the branching condition. In particular a covering of X branched at D is simply connected if and only if it is universal, i.e., the Galois group is the full group $\pi_1^{orb}(X, D)$. Observe that the group $\pi_1^{orb}(X, D)$ may fail to satisfy the branching condition. In this case there are no coverings of X branched at D (see Example 1.3 below).

The following lemma follows from ([7], §7).

Lemma 1.3. *Let $M \to X$ be a Galois covering branched at D and with Galois group G. We have the exact sequence*

$$0 \to \pi_1(M) \to \pi_1^{orb}(X, D) \to G \to 0.$$

Example 1.3. Let $X := \mathbb{P}^n$ where $n > 1$ and let $H_0, \dots, H_k$ be k hyperplanes in general position Let $m_0, m_1, \dots, m_k$ be $k+1$ distinct prime numbers and put $D := \Sigma_0^k m_i H_i$. By a result of Zariski the group $\pi_1(\mathbb{P}^n \backslash D)$ is abelian and it admits the presentation

$$\pi_1(\mathbb{P}^n \backslash D) \simeq \left\langle \mu_0, \dots, \mu_k \,\middle|\, \textstyle\sum_0^k m_i \mu_i = 0 \right\rangle,$$

where μ_i is a meridian of H_i for $i \in [0, k]$. Consequently, one has

$$\pi_1^{orb}(\mathbb{P}^n, D) \simeq \left\langle \mu_0, \dots, \mu_k \,\middle|\, m_0\mu_0 = \dots = m_k\mu_k = \textstyle\sum_0^k m_i \mu_i = 0 \right\rangle$$

It is easy to see that this latter group is trivial, hence there are no coverings of $\mathbb{P}^n$ branched at D. On the other hand, in case $m_0 = \dots = m_k = m$ there is a covering branched at D, since the group

$$\pi_1^{orb}(\mathbb{P}^n, D) \simeq \mathbb{Z}/(m) \oplus \ldots \oplus \mathbb{Z}/(m) \text{ (k copies)}$$

satisfies the Branching Condition. As we will see in the next section, the universal covering branched at D is smooth if $k \geq n$. In case $k = n$ we can show this immediately: The power map

$$\varphi_m : [z_0 : \dots : z_n] \in \mathbb{P}^n \to [z_0^m : \dots : z_n^m] \in \mathbb{P}^n$$

is a Galois covering map branched at the divisor $\Sigma_0^n m H_i$ where H_i is the hyperplane $\{z_i = 0\}$ (the arrangement $H_0 \cup \dots \cup H_n$ is unique up to projective transformations). Note that φ_m is the universal covering branched at D.

1.1. Branched coverings of $\mathbb{P}^1$

Example 1.3 concerns projective spaces of dimension > 1. The situation is very different in dimension 1. Let $X = \mathbb{P}^1$, take distinct points $p_0, \dots, p_k$ in $\mathbb{P}^1$ and let $m_0, \dots, m_k$ be integers > 1. Put $D := \Sigma_0^k m_i p_i$. (According to the definition that will be given in the next section, the pair $(\mathbb{P}^1, D)$ will be an orbifold). One has the presentation

$$\pi_1(\mathbb{P}^1 \backslash \{p_0, \dots, p_k\}) \simeq \left\langle \mu_0, \dots, \mu_k \,\middle|\, \mu_0 \dots \mu_k = 1 \right\rangle,$$

which is a free group of rank k. For the orbifold fundamental group one has

$$\pi_1^{orb}(\mathbb{P}^1, D) = \left\langle \mu_1, \dots, \mu_k \,\middle|\, \mu_0^{m_0} = \dots = \mu_k^{m_k} = \mu_0 \dots \mu_k = 1 \right\rangle$$

Let $M \to \mathbb{P}^1$ be a covering branched at D with G as the Galois group. By the Riemann–Hurwitz formula the euler number $e(M)$ of M equals

$$e(M) = |G| \left[e(\mathbb{P}^1 \backslash \{p_0, \dots, p_k\}) + \Sigma_0^k \frac{1}{m_i} \right] = |G| \left[1 - k + \Sigma_0^k \frac{1}{m_i} \right] \qquad (1.1)$$

Recall that by the Koebe–Poincaré theorem, up to biholomorphism there are only three simply connected Riemann surfaces: the Riemann sphere $\mathbb{P}^1$, the affine plane $\mathbb{C}$, and the Poincaré disc Δ. If M is a compact Riemann surface, either $e(M) > 0$ and $M \simeq \mathbb{P}^1$ (and therefore $e(M) = 2$), or $e(M) = 0$ and the universal cover of M is $\mathbb{C}$, or $e(M) < 0$ and the universal cover of M is Δ. Note that in (1.1) the signature of $e(M)$ is already determined by the data $(\mathbb{P}^1, D)$ and no information on G is needed. Accordingly, let us define the *orbifold euler number* of $(\mathbb{P}^1, D)$ as

$$e^{orb}(\mathbb{P}^1, D) = 1 - k + \sum_0^k \frac{1}{m_i} \qquad \Rightarrow \qquad e(M) = |G| e^{orb}(\mathbb{P}^1, D). \qquad (1.2)$$

In particular, if $M \to \mathbb{P}^1$ is a covering branched at D and with G as the Galois group, then

$$|G| = \frac{e(M)}{e^{orb}(\mathbb{P}^1, D)}. \qquad (1.3)$$

For $k = 0$ one has $e^{orb}(\mathbb{P}^1, D) = 1 + 1/m_0$, which is positive. Hence if $M \to \mathbb{P}^1$ is a covering branched at D, then $e(M) > 0 \Rightarrow M \simeq \mathbb{P}^1$. Suppose that this covering exists and let G be the Galois group. By (1.3) one has $|G| = 2/(1 + 1/m_0)$, which is not integral unless $m_0 = 1$. Hence for $k = 0$ there are no coverings branched at D, unless $m_0 = 1$. We could also deduce this result by looking at the group $\pi_1^{orb}(\mathbb{P}^1, D)$. Indeed, for $k = 0$ this group is trivial and the Branching Condition can not be satisfied.

In case $k = 1$ one has $e^{orb}(\mathbb{P}^1, D) = 1/m_0 + 1/m_1 > 0$. Hence, if a covering $M \to \mathbb{P}^1$ branched at D exists, then $M \simeq \mathbb{P}^1$. Suppose that it exists and let G be its Galois group. By (1.3) one should have $|G| = 2m_0 m_1/(m_0 + m_1) \in \mathbb{Z}_{>0}$. By the Branching Condition G must contain elements of order m_0 and m_1, in other words $|G|$ must be divisible by m_0 and m_1. This is possible only if $m := m_0 = m_1$. In this case a covering branched at D exists, it is the power map

Group	(m_0, m_1, m_2)	*order*
Cyclic	$(1, m, m)$	m
Dihedral	$(2, 2, m)$	$2m$
Tetrahedral	$(2, 3, 3)$	12
Octahedral	$(2, 3, 4)$	24
Icosahedral	$(2, 3, 5)$	60

TABLE 1. Finite subgroups of $\mathrm{PGL}(2, \mathbb{C})$.

$\varphi_m : [z_0 : z_1] \in \mathbb{P}^1 \to [z_0^m : z_1^m] \in \mathbb{P}^1$. We could also deduce this result by looking at the group $\pi_1^{orb}(\mathbb{P}^1, D)$ as in the example above.

Now let us consider the case $k = 2$. Observe that a three-point set in $\mathbb{P}^1$ is projectively rigid, i.e., any two such sets can be mapped onto each other by a projective transformation. Assume $m_0 \leq m_1 \leq m_2$ and put $\rho := 1/m_0 + 1/m_1 + 1/m_2$. The orbifold euler number is then $\rho - 1$.

If $\rho - 1 > 0$, then the covering must be $\mathbb{P}^1$. Hence, if a covering branched at D exists, the Galois group must be of order $2\rho^{-1}$. In this case, either $m_0 = m_1 = 2$ or (m_0, m_1, m_2) is one of $(2, 3, 3)$, $(2, 3, 4)$ or $(2, 3, 5)$, the corresponding Galois groups must be of orders $2n$, 12, 24 and 60 respectively. The group

$$\pi_1^{orb}(\mathbb{P}^1, m_0p_0 + m_1p_1 + m_2p_2) \simeq \left\langle \mu_0, \mu_1, \mu_2 \,|\, \mu_0^{m_0} = \mu_1^{m_1} = \mu_2^{m_2} = \mu_0\mu_1\mu_2 = 1 \right\rangle$$

is called a *triangle group*, it turns out that it is finite of (the right) order $2\rho^{-1}$ if $\rho > 1$ and satisfies the Branching Condition. Hence there exists Galois coverings $\mathbb{P}^1 \to \mathbb{P}^1$ branched at D. Historically this follows from Klein's classification of finite subgroups of $\mathrm{PGL}(2, \mathbb{C}) \simeq \mathrm{Aut}(\mathbb{P}^1)$. Each group is the symmetry group of one of the platonic solids inscribed in a sphere. An independent proof of this result will be given in Section 2.6, except in the icosahedral case.

If $\rho - 1 = 0$, then the orbifold Euler number of $(\mathbb{P}^1, m_0p_0 + m_1p_1 + m_2p_2)$ vanishes, and (m_0, m_1, m_2) is one of $(2, 3, 6)$, $(2, 4, 4)$ or $(3, 3, 3)$ (one may also add the triple $(2, 2, \infty)$). In these cases, the abelianizations $\mathcal{A}b\left(\pi_1^{orb}(X, D)\right)$ are finite and satisfy the Branching Condition. Hence, they are covered by Riemann surfaces of genus 1 (an elliptic curve), and their universal covering is $\mathbb{C}$. The groups $\pi_1^{orb}(X, D)$ are infinite solvable. Similary, the Galois coverings branched at the divisors $D := 2m_0 + 2m_1 + 2m_2 + 2m_3$ are also elliptic curves. Each one of these coverings corresponds to a regular tessellation of the plane.

Any pair $(\mathbb{P}^1, D)$ not considered above has negative orbifold euler characteristic. The question of existence of finite coverings branched at D in this case is known as Fenchel's problem. It amounts to finding finite quotients of $\pi_1^{orb}(\mathbb{P}^1, D)$ satisfying the Branching Condition and is of combinatorial group theoretical in nature. Fenchel's problem has been solved by Bundgaard–Nielsen [2] and was generalized by Fox [6] to pairs (R, D) where R is a Riemann surface. These pairs are covered by Riemann surfaces of genus > 1 and their universal covering is the Poincaré disc. Summing up, we have

Theorem 1.4 (Bundgaard–Nielsen, Fox). *Let $k \geq 2$ and let $D := \Sigma_0^k m_i p_i$ be a divisor on $\mathbb{P}^1$. Then there exists a finite Galois covering $M \to \mathbb{P}^1$ which is branched at D; and M is*

(i) **(elliptic case)** $\mathbb{P}^1$ *if $k = 1$ and $m_0 = m_1$ or $k = 2$ and $\Sigma_0^2 1/m_i > 1$,*
(ii) **(parabolic case)** *a Riemann surface of genus 1 if $k = 2$ and $\Sigma_0^2 1/m_i = 1$ or $k = 3$ and $m_0 = \cdots = m_3 = 2$, and*
(iii) **(hyperbolic case)** *a Riemann surface of genus > 1 otherwise.*

1.2. Fenchel's problem

In the last part of this section we present some results on branched coverings, which are of independent interest.

A natural generalization of Fenchel's problem to higher dimensions is: given a complex manifold X and a divisor with coefficients in $\mathbb{Z}_{>1}$ on X, decide whether there exists a Galois covering $M \to X$ branched at D, regardless of the question of desingularization. There is no hope for a complete solution of the generalized Fenchel's problem as in Theorem 1.4, since the group $\pi_1(X \backslash \text{supp}(D))$ does not admit a simple presentation, and it can be trivial, abelian, finite non-abelian, or infinite. However, there are some partial results obtained by several authors.

Theorem 1.5 (Kato). *Let $H := H_0 \cup \cdots \cup H_k$ be an arrangement of lines in $\mathbb{P}^2$ such that any line contains a point of multiplicity at least 3. Let $m_0, \ldots, m_k \in \mathbb{Z}_{>1}$ and put $D := \Sigma_0^k m_i H_i$. Then there exists a finite Galois covering of $\mathbb{P}^2$ branched at D.*

Kato also describes the resolution of singularities of the covering surfaces, and this resolution is compatible with the blowing-up of points of multiplicity > 2 of the branch locus. There is a generalization of the Kato theorem to conic arrangements given by Namba [17]. At the other extreme there is the following result concerning irreducible curves. Recall that for p, q coprime integers Oka [18] constructed an irreducible curve $C_{p,q}$ of degree pq and with $\pi_1(\mathbb{P}^2 \backslash C_{p,q}) \simeq \mathbb{Z}/(p) \star \mathbb{Z}/(q)$. For a proof of the following theorem see [24].

Theorem 1.6. *If $C_{p,q}$ is an Oka curve, then for any $m \geq 1$ there exists a finite Galois covering of $\mathbb{P}^2$ branched at $mC_{p,q}$.*

Given a projective manifold X, which groups can appear as the Galois group of a branched covering of X? This question has the following solution.

Theorem 1.7 (Namba [16]). (i) *For any projective manifold X and any finite group G there is a finite branched Galois covering $M \to X$ with G as the Galois group.*

(ii) *For $n \geq 2$ there exists a covering of the germ $(\mathbb{C}^n, 0)$ with a given finite Galois group.*

2. Orbifolds

In the previous section we studied branched Galois coverings of complex manifolds, which are possibly singular. Under which conditions is a finite branched covering of a complex manifold smooth? Loosely speaking, an orbifold is a pair (X, D) that admits locally a branched covering by a smooth manifold.

2.1. Transformation groups

A *transformation group* is a pair (G, M) where M is a connected complex manifold and G is a group of holomorphic automorphisms of M acting properly discontinuously, in particular for any $z \in M$ the isotropy group

$$G_z := \{g \in G : gz = z\}$$

is finite. The most important example of a transformation group is (G, M), where M is a symmetric space such as the polydisc Δ^n or the n-ball B_n. Let (G, M) be a transformation group and X its orbit space with the projection $\varphi : M \to X$. The orbit space X is an irreducible normal analytic space endowed with a *b-map* defined as

$$b_\varphi : x \in X \to |G_z| \in \mathbb{Z}_{>0}$$

where $z \in \varphi^{-1}(x)$. In dimension 1 the orbit space X is always smooth. In higher dimensions X may have singularities of quotient type.

The product of two transformation groups (G_1, M_1) and (G_2, M_2) is the transformation group $(G_1 \times G_2, M_1 \times M_2)$ where $G_1 \times G_2$ acts in the obvious way.

Example 2.1. (The power map) The model example of a transformation group is $(\mathbb{Z}/(m), \mathbb{C})$, where $m \in \mathbb{Z}_{>0}$ and the element $[j] \in \mathbb{Z}/(m)$ acts by

$$\psi_{[j]} : z \in \mathbb{C} \to \omega^j z \in \mathbb{C},$$

ω being a primitive m-th root of unity. The orbit space of $(\mathbb{Z}/(m), \mathbb{C})$ is $\mathbb{C}$. The orbit map is the power map $\varphi_m : z \in \mathbb{C} \to z^m \in \mathbb{C}$. The isotropy group of the origin is the full group $\mathbb{Z}/(m)$, whereas the isotropy group of any other point is trivial. Hence the b-map is

$$b_\varphi(x) = \begin{cases} m & x = 0 \\ 1 & x \neq 0. \end{cases} \tag{2.1}$$

More generally, consider the product transformation group $(\oplus_{i=1}^n \mathbb{Z}/(m_i), \mathbb{C}^n)$. Obviously $\mathbb{C}^n$ is the orbit space of $(\oplus_{i=1}^n \mathbb{Z}/(m_i), \mathbb{C}^n)$, and the orbit map is $\varphi_{\vec{m}} : (z_1, \ldots, z_n) \to (z_1^{m_1}, \ldots, z_n^{m_n})$. Let H_i be the hyperplane defined by $z_i = 0$. The b-map of $\varphi_{\vec{m}}$ is

$$b_{\varphi_{\vec{m}}}(p) = \prod_{p \in H_i} m_i$$

Example 2.2. (The projective power map) Let as above $(G, \mathbb{C}^{n+1})$ be the product of $n+1$ copies of the transformation group $(\mathbb{Z}/(m), \mathbb{C})$, where $G := \oplus_{i=0}^n \mathbb{Z}/(m)$. Let ω be a primitive mth root of unity, the element $([j_0], \ldots [j_n]) \in G$ acts by

$$\psi_{([j_0], \ldots, [j_n])} : (z_0, \ldots, z_n) \in \mathbb{C}^{n+1} \to (\omega^{j_0} z_0, \omega^{j_1} z_1 : \cdots : \omega^{j_n} z_n) \in \mathbb{C}^{n+1}.$$

Projectivizing $\mathbb{C}^{n+1}$, we get the projective space $\mathbb{P}^n$. The diagonal $\Delta := \{(g, \ldots, g) \mid g \in \mathbb{Z}/(m)\}$ of G acts trivially on $\mathbb{P}^n$. The quotient $G/\Delta \simeq (\mathbb{Z}/(m))^n$ acts faithfully on $\mathbb{P}^n$. The orbit space of $(G/\Delta, \mathbb{P}^n)$ is $\mathbb{P}^n$ itself. The orbit map is

$$\varphi_m : [z_0 : \cdots : z_n] \in \mathbb{P}^n \to [z_0^m : \cdots : z_n^m] \in \mathbb{P}^n$$

The map φ_m is called a polycyclic covering of $\mathbb{P}^n$. Let $H_i := \{z_i = 0\}$. For any point $p \in \mathbb{P}^n$ denote by $\alpha(p)$ the number of hyperplanes H_i through p. Then the b-map of φ_m is (see Figure 2.3 for the case $n = 2$).

$$b(p) = m^{\alpha(p)} \tag{2.2}$$

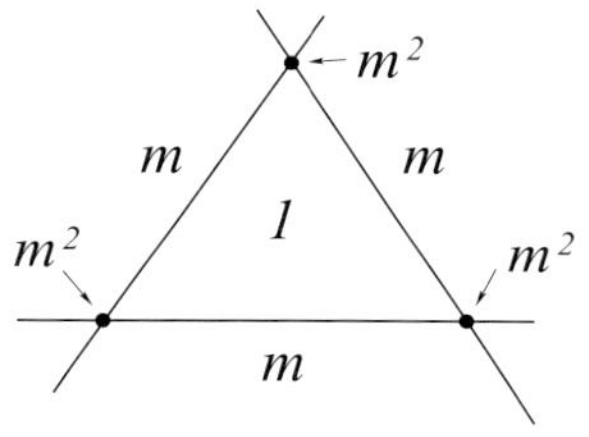

FIGURE 2.3. The b-map of the bicyclic covering $\varphi_2 : \mathbb{P}^2 \to \mathbb{P}^2$

Example 2.3. (A singular orbit space) Consider the action of $[j] \in \mathbb{Z}/(m)$ on $\mathbb{C}^2$ by

$$\psi_{[j]} : (x, y) \in \mathbb{C}^2 \to (\omega^j x, \omega^{-j} y) \in \mathbb{C}^2$$

The orbit space of $(\mathbb{Z}/(m), \mathbb{C}^2)$ is the hypersurface in $\mathbb{C}^3$ defined by $z^m = xy$, since the quotient map is $\psi : (x, y) \to (x^m, y^m, xy)$. This hypersurface has a cyclic quotient singularity at the origin.

2.2. Transformation groups and branched coverings

A transformation group is a locally finite branched Galois covering, as we now proceed to explain. Let (G, M) be transformation group with the orbit space X. Let $\varphi : M \to X$ be the orbit map and put

$$R_\varphi := \{z \in M \, : \, |G_z| > 1\} \text{ and } B_\varphi := \{x \in X \mid b_\varphi(x) > 1\} \; (= \varphi(R_\varphi)),$$

where G_z is the stabilizer of z. Let $\bar{X} := X \backslash \mathrm{Sing}(X)$ be the smooth part of X. (It can happen that a singularity of X lies on B_φ.) Let $x \in \bar{X}$ and $z \in \varphi^{-1}(x)$. Let M_z be the germ of M at z and X_x the germ of X at x. Then G_z acts on M_z, and the orbit space is X_x. Since $|G_z|$ is finite and X_x is smooth, the orbit map of germs

$$\varphi_z : M_z \to X_x$$

is a finite Galois covering branched along $B_{\varphi,x}$ Therefore locally one can define the branch divisor $D_{\varphi,x}$, and the local branch divisors patch and yield a global branch divisor D_φ supported by B_φ. Let $D_\varphi = \Sigma m_i H_i$, where $H_1, H_2 \ldots$ are the irreducible components of B_φ. The divisor D_φ is always locally finite and in all of the cases considered in this article, it is a finite sum. Thus D_φ is defined on the smooth part $\bar{X}$ of X — in what follows its closure in X will be denoted by D_φ again.

Let us turn our attention to the covering-germ $\varphi_z : M_z \to X_x$, which is a finite Galois covering branched *at* $D_{\varphi,x}$. Since M_z is a smooth germ, it is simply connected. Hence φ_z must be the universal covering branched at $D_{\varphi,x}$, in other words the Galois group of φ_z is

$$G_z \simeq \pi_1^{orb}(X, D_\varphi)_x,$$

where we denote the germ-pair (X_x, D_x) by $(X, D)_x$. In particular, one has

$$b(x) = |G_x| = |\pi_1^{orb}(X, D_\varphi)_x|$$

which also shows that the latter groups must be finite.

What is said above is in fact true for a singular point $x \in X$. For simplicity, assume that $x \notin B_\varphi$. Since M_z is a smooth germ it is simply connected and thus the covering germ $\varphi_z : M_z \to X_x$ must be universal. In other words the Galois group is $G_z \simeq \pi_1(X_x)$. For example, if $X \subset \mathbb{C}^3$ is defined by $z^m = xy$, then $\pi_1(X_O)$ is cyclic of order m, see Example 2.3.

2.3. b-spaces and orbifolds

Recall that a transformation group (G, M) induces a b-map on its orbit space X. Conversely, let X be a normal complex space and b a map $X \to \mathbb{Z}_{>0}$. The pair (X, b) is called a b-space. The basic question related to a b-space is the *uniformization problem:* Under which conditions does there exist a (finite) transformation group (G, M) with X as the orbit space and with the quotient map $\varphi : M \to X$ such that $b = b_\varphi$? In case such a transformation group exist, it is called a *uniformization* of (X, b) and (X, b) is said to be uniformized by (G, M). Observe that these definitions can be localized. Obviously, if (X, b) is uniformizable, then it is locally finitely uniformizable, that is the germs $(X, b)_x$ must admit finite uniformization.

Definition 2.1. A locally finite uniformizable b-space (X, b) is called an *orbifold.* The space X is said to be the *base space* of (X, b), and (X, b) is said to be an orbifold *over* X. The set $\{x \in X \mid b(x) > 1\}$ is called the *locus* of the orbifold. An orbifold with a two-dimensional base is called an *orbiface.*

Orbifolds (X, b) and (X', b') are said to be equivalent if there is a biholomorphism $\epsilon : X \to X'$ such that the following diagram commutes.

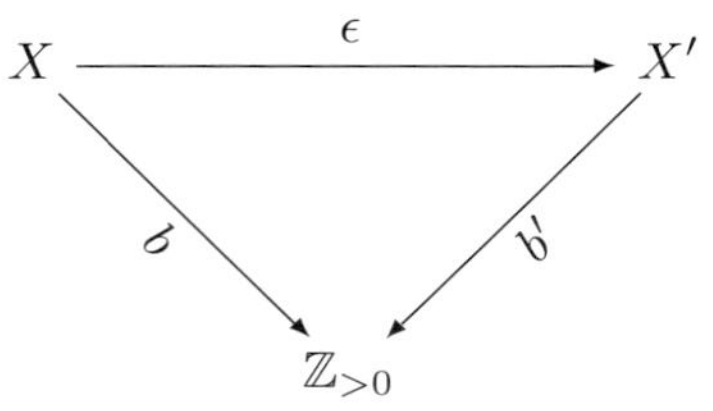

The product of two b-spaces (X_1, b_1) and (X_2, b_2) is the b-space $(X_1, b_1) \times (X_2, b_2)$ which is defined as $(X_1 \times X_2, b)$ where $b(x, y) := b_1(x)b_2(y)$. If (X_i, b_i) is an uniformized by (G_i, M_i) for $i = 1, 2$, then the product orbifold is uniformized by $(G_1, M_1) \times (G_2, M_2)$.

Let (X, b) be an orbifold. Then by locally finite uniformizability its locus B is a locally finite union of hypersurfaces $H_1, H_2 \ldots$, and b must be constant along $H_i \backslash (\mathrm{Sing}(B) \cup \mathrm{Sing}(X))$. Let m_i be this number, and put $D_b := \Sigma m_i H_i$ (in most cases of interest this will be a finite sum). The orbifold fundamental group of (X, b) is defined that of the pair (X, D_b).

Lemma 2.2. *If (X, b) is an orbifold, then $b(x) = |\pi_1^{orb}(X, b)_x|$ for any $x \in X$.*

Proof. Let $x \in X$. Since (X, b) is an orbifold, the germ $(X, b)_x$ admits a finite uniformization. Hence there is a (unique) transformation group (G_z, M_z) with $(X, b)_x$ as the orbit space, such that $b_{\varphi_z} = b_x$, where $\varphi_z : M_z \to (X, b)_x$ is the quotient map and $\varphi_z^{-1}(x) = \{z\}$ (in other words G_z stabilizes z). By Lemma 1.3 one has the exact sequence

$$0 \to \pi_1(M_z) \to \pi_1^{orb}(X, b)_x \to G_z \to 0$$

Since M_z is smooth, it is simply connected, so that $G_z \simeq \pi_1^{orb}(X, b)_x$. Hence $b(x) = |G_z| = |\pi_1^{orb}(X, b)|$. $\square$

Let (X, b) be an orbifold and let $D_b := \Sigma m_i H_i$ be the associated divisor. Since $\pi_1^{orb}(X, b)_x$ is defined as the group $\pi_1^{orb}(X, D_b)_x$, and since this latter group is determined by D_b, Lemma 2.2 implies that the b-function is completely determined by D_b. In other words the pair (X, D_b) determines the pair (X, b). On the other hand in dimensions ≥ 2 most pairs (X, D) do not come from an orbifold. The local uniformizability condition puts an important restriction on the possible pairs (X, D), in particular the local orbifold fundamental groups of (X, D) must be finite. In dimension 2 this latter condition is sufficient for local uniformizability, since by a theorem of Mumford a simply connected germ is smooth in dimension 2, see Theorem 3.1 below. This is no longer true in dimensions ≥ 3, see [3] for counterexamples.

Example 2.4. Consider the orbifold $(\mathbb{C}, b_m)$, where

$$b_m(z) = \begin{cases} m & z = 0 \\ 1 & z \neq 0. \end{cases}$$

This orbifold is uniformized by the transformation group $(\mathbb{Z}/(m), \mathbb{C})$, the uniformizing map is the power map. The (multivalued) inverse of a covering map is called a *developing map.* In this case the developing map is $\varphi_m^{-1} : [x : y] \in \mathbb{P}^1 \to [x^{1/m} : y^{1/m}] \in \mathbb{P}^1$.

Example 2.5. Let $p_0, \ldots, p_k$ be $k + 1$ distinct points in $\mathbb{P}^1$ and let $m_0, \ldots, m_k$ be positive integers. Let $b : \mathbb{P}^1 \to \mathbb{Z}_{>0}$ be the function with $b(p_i) = m_i$ for $i \in [0, k]$ and $b(p) = 1$ otherwise. Around the point p_i the b-space $(\mathbb{P}^1, b)$ is uniformized by the transformation group $(\mathbb{Z}/(m_i), \mathbb{C})$. Hence, $(\mathbb{P}^1, b)$ is an orbifold, which can also be denoted by $(\mathbb{P}^1, \Sigma_0^k m_i p_i)$. Theorem 1.4 completely answers the question of uniformizability of these orbifolds.

Example 2.6. (See Figure 2.4) Let p, q be two integers and consider the germ $(\mathbb{C}^2, b)_0$ where $b(0,0) = pq$, $b(x,0) = q$ for $x \neq 0$, $b(0,y) = p$ for $y \neq 0$ and $b(x,y) = 1$ for $xy \neq 0$. Put $H_1 := \{x = 0\}$ and $H_2 := \{y = 0\}$. The group $\pi_1(\mathbb{C}^2 \backslash (H_1 \cup H_2))_0$ is the free abelian group generated by the meridians of H_1, H_2 so that $\pi_1^{orb}(\mathbb{C}^2, b)_0 \simeq \mathbb{Z}/(p) \oplus \mathbb{Z}/(q)$ is finite. This is indeed an orbifold germ, the map $(x,y) \in \mathbb{C}^2 \to (x^p, y^q) \in \mathbb{C}$ is its uniformization. On the other hand, consider the germ of the pair $(\mathbb{C}^2, D)$ at the origin, where $D = pH_1 + qH_2 + rH_3$ with $H_1 := \{x = 0\}$, $H_2 := \{y = 0\}$ and $H_3 := \{x - y = 0\}$. One has

$$\pi_1(\mathbb{C}^2 \backslash (H_1 \cup H_2 \cup H_3)) \simeq \Big\langle \mu_1, \mu_2, \mu_3 \,|\, [\mu_i, \mu_1\mu_2\mu_3] = 1 \ (i \in [1,3]) \Big\rangle$$

where μ_i is a meridian of H_i for $i \in [1,3]$ (see [25]). The local orbifold fundamental group admits the presentation

$$\pi_1^{orb}(\mathbb{C}^2, D)_0 \simeq \Big\langle \mu_1, \mu_2, \mu_3 \,|\, [\mu_i, \mu_1\mu_2\mu_3] = \mu_1^p = \mu_2^q = \mu_3^r = 1 \ (i \in [1,3]) \Big\rangle.$$

Obviously, adding the relation $\delta = 1$ to this group gives a triangle group. Hence this group is a central extension of the triangle group and is finite of order $4\rho^{-2}$ if $\rho := 1/p + 1/q + 1/r - 1 > 0$, infinite solvable when $\rho = 0$ and "big" otherwise. (Here, "big" means that the group contains non-abelian free subgroups.) Hence $(\mathbb{C}^2, D)_0$ do not come from an orbifold germ if $\rho < 0$. For $\rho > 0$ it comes from an orbifold germ, its uniformization will be described explicitly in Section 3.2.

FIGURE 2.4. Some orbiface germs with a smooth base

2.4. A criterion for uniformizability

Let (X, b) be an orbifold and let D_b be the associated divisor. Recall that the group $\pi_1^{orb}(X, b)$ is by definition the group $\pi_1^{orb}(X, D_b)$. If

$$\rho : \pi_1^{orb}(X, b) \twoheadrightarrow G$$

is a surjection onto a finite group with $\mathrm{Ker}(\varphi)$ satisfying the branching condition, then there exists a Galois covering $\varphi : M \to X$ branched at D_b, where M is a possibly singular normal space.

Example 2.7. Let $(X, b) = (\mathbb{C}^2, b)$ with $b(0,0) = m^2$, $b(x,0) = m = b(0,y)$ $(x, y \neq 0)$ and $b(x,y) = 1$ otherwise, where $m \in \mathbb{Z}_{>1}$. Then $D_b = mH_1 + mH_2$, where $H_1 := \{x = 0\}$ and $H_2 := \{y = 0\}$. Consider the covering

$$\varphi : (x, y, z) \in \{z^m = xy\} \subset \mathbb{C}^3 \longrightarrow (x, y) \in \mathbb{C}^2.$$

This is a $\mathbb{Z}/(m)$-Galois covering branched at D with $b_\varphi(0,0) = b_\varphi(0,y) = b_\varphi(x,0) = m$ and $b_\varphi(x,y) = 1$ otherwise. The covering space is singular. Note that $b_\varphi \neq b$. On the other hand, the Galois covering $\psi : (x,y) \in \mathbb{C}^2 \to (x^m, y^m) \in \mathbb{C}^2$ satisfies $b_\psi = b$, and it is smooth.

Lemma 2.3. *Let (X, b) be an orbifold, and $\varphi : M \to X$ a Galois covering branched at D_b. Then M is smooth if and only if $b_\varphi \equiv b$.*

Proof. For any $x \in X$, there is the induced branched covering of germs $\varphi_x : M_z \to X_x$, where $z \in \varphi^{-1}(x)$. The stabilizer G_z is the Galois group of φ_x. The germ M_z is smooth only if φ_x is the uniformization map of the germ $(X, b)_x$, which is the universal branched covering and has $\pi_1^{orb}(X, b)_x$ as its Galois group. In other words, M_z is smooth if and only if $G_z \simeq \pi_1^{orb}(X, b)_x$, if and only if

$$b_\varphi(x) = |G(z)| = |\pi_1^{orb}(X, b)_x| = b(x). \qquad \square$$

For a point $x \in X$, there is a natural map

$$\pi_1(X \backslash D_b)_x \longrightarrow \pi_1(X \backslash D_b)$$

and therefore a map $\iota_x : \pi_1^{orb}(X, b)_x \to \pi_1^{orb}(X, b)$, induced by the inclusion. The group G_z is the image of the composition map

$$\rho \circ \iota_x : \pi_1^{orb}(X, b)_x \longrightarrow \pi_1^{orb}(X, b) \to G.$$

Theorem 2.4. *Let $\rho : \pi_1^{orb}(X, b) \twoheadrightarrow G$ be a surjection and let $\varphi : M \to X$ be the corresponding Galois covering of X branched along D_b. The pair (G, M) is a uniformization of the orbifold (X, b) if and only if for any $x \in X$, the map*

$$\rho \circ \iota_x : \pi_1^{orb}(X, b)_x \to G$$

is an injection.

Proof. One has $b_\varphi \equiv b$ if and only if for any $x \in X$ and $z \in \varphi^{-1}(x)$ the image G_z of $\rho \circ \iota_x$ is the full group $\pi_1^{orb}(X, b)_x$. The result then follows from Lemma 2.2. $\square$

2.5. Sub-orbifolds and orbifold coverings

Let (X, b) be an orbifold. An orbifold (X, b') is said to be a *sub-orbifold* of (X, b) if $b'(x)$ divides $b(x)$ for any $x \in X$. Let $\varphi : Y \to X$ be a uniformization of (X, b'). Define the function $c : Y \to \mathbb{Z}_{>0}$ by

$$c(y) := \frac{b(\varphi(y))}{b'(\varphi(y))}.$$

Then $\varphi : (Y, c) \to (X, b)$ is called an *orbifold covering*, and (Y, c) is called the *lifting of (X, b) to the uniformization Y of (X, b')*. The exact sequence of Lemma 1.3 can be generalized to the following commutative diagram:

$$\begin{array}{ccccccccc}
 & & 0 & & 0 & & & & \\
 & & \uparrow & & \uparrow & & & & \\
0 & \longrightarrow & \pi_1(Y) & \longrightarrow & \pi_1^{orb}(X,b') & \longrightarrow & G & \longrightarrow & 0 \\
 & & \uparrow & & \uparrow & & \uparrow \wr\!\simeq & & \\
0 & \longrightarrow & \pi_1^{orb}(Y,c) & \longrightarrow & \pi_1^{orb}(X,b) & \longrightarrow & G & \longrightarrow & 0
\end{array}$$

Example 2.8. Let $m, n \in \mathbb{Z}_{>0}$ and consider the orbifold $(\mathbb{C}, b_{mn})$ defined in Example 2.5. Then $(\mathbb{C}, b_m)$ is a suborbifold of $(\mathbb{C}, b_{mn})$, which is uniformized via $\varphi_m : z \in \mathbb{C} \to z^m \in \mathbb{C}$. Hence φ is an orbifold covering $(\mathbb{C}, b_n) \to (\mathbb{C}, b_{mn})$.

Remark 2.5. If $Y \subset X$ is an irreducible subvariety of positive codimension, then an orbifold structure (X, b) on X induces an orbifold structure on the normalization of Y as follows (note that Y may belong to the locus of (X, b)): Let $y \in Y$, and take an irreducible branch $\widetilde{Y}_y$ of the germ Y_y. Since (X, b) is an orbifold, there is a finite uniformization $\varphi_z : M_z \to X_y$. The germ $\varphi_z^{-1}(\widetilde{Y}_y)$ may or may not be irreducible. The restriction of φ_z to an irreducible component of $\varphi_z^{-1}(\widetilde{Y}_y)$ is a branched Galois covering onto $\widetilde{Y}_y$. Let b'_y be its b-map. The b-maps b'_y for varying y patch together and yield a b-map b' on Y. Then (Y, b') is the induced orbifold structure on Y, which might also be called a suborbifold of (X, b). If $\varphi : M \to X$ is a uniformization of (X, b), then its restriction to an irreducible component of $\varphi^{-1}(Y)$ is a uniformization of (Y, b'). The induced orbifold has a significance in relative proportionality, if $\dim(X) = 2$ and $\dim(Y) = 1$, then (Y, b') is relatively proportional only if and only if the natural map $\pi_1^{orb}(Y, b') \to \pi_1^{orb}(X, b)$ is an injection.

2.6. Covering relations among triangle orbifolds

Convention. In order to present an orbifold (X, b) one has to specify its b-map. However, since by Lemma 2.2 the pair (X, D_b) determines the orbifold (X, b), an orbifold can be presented by a pair (X, D). Since the latter presentation is sometimes more practical, we shall use it in the sequel. To be precise, in what follows the expression "the orbifold (X, D)" refers to the pair (X, b), where the b-map is defined by $b(p) := |\pi_1^{orb}(X, D)_p|$ (it is implicitly assumed that (X, b) is indeed an orbifold, i.e., it is locally finite uniformizable).

Let us illustrate the notion of orbifold coverings in the simplest, one-dimensional setting. In this subsection, we fix three points $p_0 = [1 : 0]$, $p_1 = [0 : 1]$, $p_2 := [1 : 1]$ in $\mathbb{P}^1$. Consider first the orbifold $(\mathbb{P}^1, rm_0p_0 + rm_1p_1)$. Then $(\mathbb{P}^1, rp_0 + rp_1)$

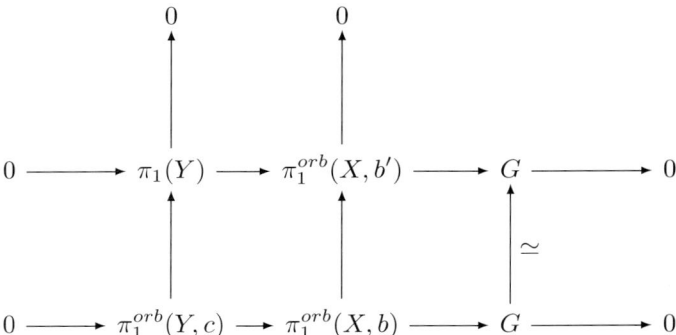

is a suborbifold, which is uniformized by $(\mathbb{Z}/(r), \mathbb{P}^1)$ via $\varphi_r : [x : y] \to [x^r : y^r]$. Hence, there is an orbifold covering

$$\varphi_r : (\mathbb{P}^1, m_0 p_0 + m_1 p_1) \to (\mathbb{P}^1, r m_0 p_0 + r m_1 p_1).$$

Coverings of triangle orbifolds, elliptic case. Now consider the orbifold $(\mathbb{P}^1, 2p_0 + 2p_1 + mp_2)$. Then $(\mathbb{P}^1, 2p_0 + 2p_1)$ is a suborbifold, which is uniformized by $\mathbb{P}^1$ via φ_2. Hence, there is a covering as in Figure 2.5, where $q_0 := [1 : 1]$, $q_1 := [1 : -1]$,

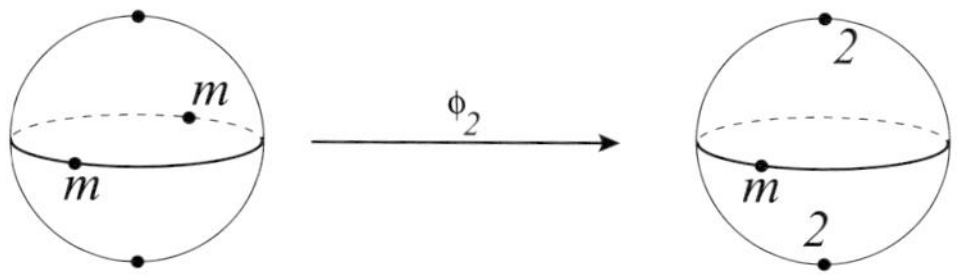

FIGURE 2.5. The covering $\varphi_2 : (\mathbb{P}^1, mq_0 + mq_1) \to (\mathbb{P}^1, 2p_0 + 2p_1 + mp_2)$

so that $\{q_0, q_1\} = \varphi_2^{-1}(p_2)$. One can map $(\mathbb{P}^1, mq_0 + mq_1)$ onto $(\mathbb{P}^1, mp_0 + mp_1)$ by a projective transformation. Since this latter orbifold is uniformized by φ_m, one has a chain of coverings

$$\mathbb{P}^1 \overset{\varphi_m}{\to} (\mathbb{P}^1, mq_0 + mq_1) \overset{\varphi_2}{\to} (\mathbb{P}^1, 2p_0 + 2p_1 + mp_2).$$

Then $\varphi_2 \circ \varphi_m$ is the uniformization of the dihedral orbifold $(\mathbb{P}^1, 2p_0 + 2p_1 + mp_2)$. (The covering $\varphi_2 \circ \varphi_m$ is Galois since it is universal). Now consider the octahedral orbifold $(\mathbb{P}^1, 2p_0 + 4p_1 + 3p_2)$. There is a covering

$$\varphi_2 : (\mathbb{P}^1, 2p_1 + 3q_0 + 3q_1) \to (\mathbb{P}^1, 2p_0 + 4p_1 + 3p_2).$$

Since any set of three points can be mapped to any set of three points on $\mathbb{P}^1$, one has $(\mathbb{P}^1, 2p_1 + 3q_0 + 3q_1) \simeq (\mathbb{P}^1, 3p_0 + 3p_1 + 2p_2)$. This latter orbifold admits the covering

$$\varphi_3 : (\mathbb{P}^1, 2r_0 + 2r_1 + 2r_2) \to (\mathbb{P}^1, 3p_0 + 3p_1 + 2p_2)$$

where $r_0 = [1 : 1]$, $r_1 := [1 : \omega]$, $r_2 := [1 : \omega^2]$ and ω being a primitive cubic root of unity, so that $\{r_0, r_1, r_2\} = \varphi_3^{-1}(p_2)$.

Exercise 2.1. Write down the uniformizing map of the octahedral orbifold explicitly.

Coverings of triangle orbifolds, parabolic case. Consider the orbifold $(\mathbb{P}^1, \Sigma_0^2 3p_i)$. The orbifold $(\mathbb{P}^1, 3p_0 + 3p_1)$ is a suborbifold uniformized by $\mathbb{P}^1$ via φ_3, and $\varphi_3^{-1}(p_2) = \{r_0, r_1, r_2\}$ as above. Hence, there is an orbifold covering as in Figure 2.6:

Since any two set of three points on $\mathbb{P}^1$ are projectively equivalent, we see that the orbifold $(\mathbb{P}^1, \Sigma_0^2 3p_i)$ admits a self-covering. This is not very surprising, since it is uniformized by the elliptic curve C which admits an automorphism of order 3, whose quotient is C.

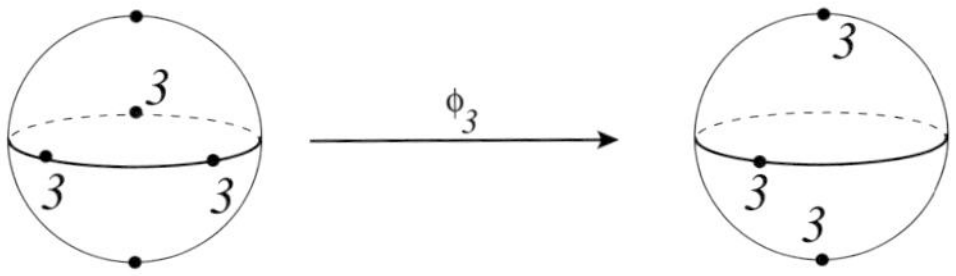

FIGURE 2.6. The covering $\varphi_3 : (\mathbb{P}^1, \Sigma_0^2 3r_i) \to (\mathbb{P}^1, \Sigma_0^2 3p_i)$

Exercise 2.2. Discover the coverings of the remaining parabolic orbifolds with parameters $(2, 4, 4)$, $(2, 3, 6)$ and $(2, 2, 2, 2)$ (one can also add the parameters (∞, ∞) and $(2, 2, \infty)$ to this list).

Coverings of triangle orbifolds, hyperbolic case. As an example, consider the orbifold $(\mathbb{P}^1, 5p_0 + 5p_1 + mp_2)$, which is hyperbolic for any $m \in \mathbb{Z}_{>1}$. The orbifold $(\mathbb{P}^1, 5p_0 + 5p_1)$ is a suborbifold uniformized by $\mathbb{P}^1$ via φ_5, and $\varphi_5^{-1}(p_2) = \{s_i := [1, \xi^i] \,|\, i \in [0, 4]\}$, where ξ is a primitive fifth root of unity. Hence, there is an orbifold covering as in Figure 2.7.

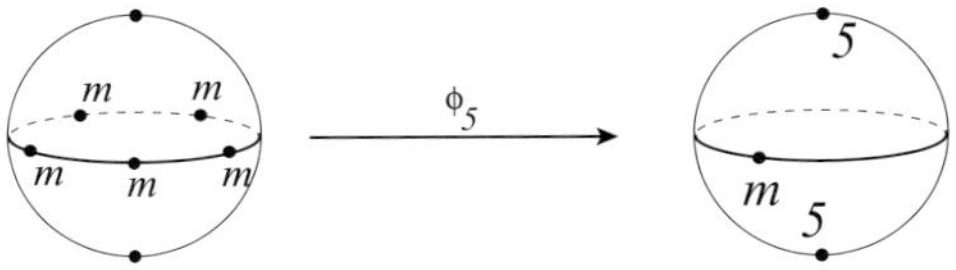

FIGURE 2.7. The covering $\varphi_5 : (\mathbb{P}^1, \Sigma_0^4 ms_i) \to (\mathbb{P}^1, 5p_0 + 5p_1 + mp_2)$

Now $(\mathbb{P}^1, ms_0 + ms_1)$ is a suborbifold of $(\mathbb{P}^1, \Sigma_0^4 ms_i)$, and it is clear how one can continue in this manner to get an infinite tower of hyperbolic orbifolds.

3. Orbifold Singularities

Recall that an orbifold germ $(X, b)_x$ is a germ that admits a finite uniformization by a transformation group (G_z, M_z), where M_z is a smooth germ and G_z is a finite group acting on M_z and fixes z. According to a classical result of Cartan [4], any orbifold germ $(X, b)_x$ is in fact equivalent to the quotient of the germ $\mathbb{C}_0^n$ by finite subgroup of $\mathrm{GL}(n, \mathbb{C})$. In other words, any orbifold germ $(X, b)_x$ admits a finite uniformization by $(G, \mathbb{C}^n)$ where G is a finite subgroup of $\mathrm{GL}(n, \mathbb{C})$. Observe that any finite group appears as a subgroup of $\mathrm{GL}(n, \mathbb{C})$ for sufficiently large n. For small n these subgroups can be effectively classified.

Any finite subgroup of $\mathrm{GL}(\mathbb{C}, 1) \simeq \mathbb{C}^*$ is cyclic and is generated by a root of unity, its orbit space is $\mathbb{C}$ and the quotient map is the power map. Hence in dimension one, any orbifold germ $(X, b)_x$ is of the form $(\mathbb{C}, mO)_O$, where $O \in \mathbb{C}$ is the origin. In higher dimensions, an orbifold germ $(X, b)_x$ may have singularities. Resolution graphs of all orbiface singularities can be found in the appendix to [14]. Let us first consider orbifolds (X, b) with a smooth base space X.

Let $G \subset \mathrm{GL}(n, \mathbb{C})$. Then G acts on the polynomial ring $\mathbb{C}[x_1, \ldots, x_n]$ by

$$M(P)(x) := P(M^{-1}x).$$

The ring of invariant polynomials under this action is denoted by $\mathbb{C}[x_1, \ldots, x_n]^G$. Recall that $M \in \mathrm{GL}(n, \mathbb{C})$ is called a *reflection* if one of its eigenvalues is a root of unity $\omega \neq 1$ and the remaining eigenvalues are all 1. A group $G \subset \mathrm{GL}(n, \mathbb{C})$ is called a *reflection group* if it is generated by reflections. By Chevalley's theorem [5] the ring $\mathbb{C}[x_1, \ldots, x_n]^G$ is generated by n algebraically independent homogeneous invariants if and only if G is a reflection group. In geometrical terms, the quotient $\mathbb{C}^n/G$ is isomorphic to $\mathbb{C}^n$ if and only if G is a reflection group. In other words, germs $(X, b)_x$ with a smooth base are in a one-to-one correspondence with finite reflection groups.

Irreducible finite reflection groups have been classified by Shepherd and Todd [20]. A group $G \subset \mathrm{GL}(n, \mathbb{C})$ is called *imprimitive* if $\mathbb{C}^n$ can be decomposed as a non-trivial direct sum of subspaces permuted by G, otherwise it is called *primitive*. Matrices permuting the coordinates of $\mathrm{GL}(n, \mathbb{C})$ generate the symmetric group S_n, which is primitive. Aside from S_n and an infinite family of imprimitive groups $G(m, p, n)$ there are only a finite number of primitive reflection groups, which are called exceptional reflection groups. There are no exceptional reflection groups in dimensions ≥ 9.

Observe that if G is a subgroup of $\mathrm{GL}(n, \mathbb{C})$, then its projectivization $\mathrm{P}G$ is a subgroup of $\mathrm{PGL}(n, \mathbb{C})$. The extension $G \to PG$ is central, since its kernel is generated by the multiples of the identity matrix I. If G is finite, then the kernel of $G \to \mathrm{P}G$ is generated by ωI, where ω is a root of unity.

3.1. Orbiface singularities

The following theorem gives a topological characterization of orbiface germs.

Theorem 3.1. *In dimension two, $(X, b)_x$ is an orbiface germ if and only its orbifold fundamental group $\pi_1^{orb}(X, b)_x$ is finite.*

Proof. We must show that $(X, b)_x$ admits a finite smooth uniformization. Since $\pi_1^{orb}(X, b)_x$ is finite, its universal covering is a finite covering by a simply connected germ. In dimension two, a simply connected germ is smooth by Mumford's theorem [15] (this is wrong in dimensions > 2, see [3] for a counterexample). The other direction is clear. □

We will mostly consider orbifaces with a smooth base. The following result characterizes their germs.

Theorem 3.2. *All orbiface germs with a smooth base are given in the table below.*

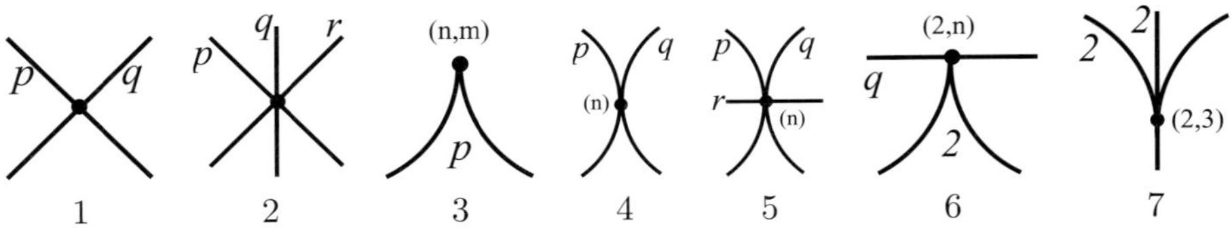

No.	*Equation*	*Condition*	*Order*
1	xy	$--$	pq
2	$xy(x+y)$	$0<\rho:=\frac{1}{p}+\frac{1}{q}+\frac{1}{r}-1$	$4\rho^{-2}$
3	x^n-y^m $(\gcd(n,m)=1)$	$0<\rho:=\frac{1}{n}+\frac{1}{m}+\frac{1}{p}-1$	$\frac{4}{nm}\rho^{-2}$
4	x^2-y^{2n} $(n\geq 2)$	$0<\rho:=\frac{1}{p}+\frac{1}{q}+\frac{1}{n}-1$	$\frac{4}{n}\rho^{-2}$
5	$y(x^2-y^{2n})$	$0<\rho:=\frac{1}{p}+\frac{1}{q}+\frac{1}{nr}-1$	$\frac{4}{n}\rho^{-2}$
6	$y(x^2-y^n)$ $(n\ \mathit{odd})$	$--$	$2nq^2$
7	$x(x^2-y^3)$	$--$	96

TABLE 2. Orbiface germs with a smooth base

In dimension 2, Yoshida observed the following facts (see [25]): If $H\subset \mathrm{GL}(2,\mathbb{C})$ is a reflection group with a non-abelian PG, then among the reflection groups with the same projectivization there is a maximal one G containing H. Every reflection group K with $PK=PG$ is a normal subgroup of this maximal reflection group. In other words, the germ $\mathbb{C}^2/K$ is a Galois covering of $\mathbb{C}^2/G$. If G is maximal reflection group, then the quotient $\mathbb{C}^2/G$ is familiar from Example 2.6; it is the orbiface $(\mathbb{C}^2, pX+qY+rZ)$ for some (p,q,r) with $1/p+1/q+1/r-1>0$, where X, Y, Z are three lines meeting at the origin (recall our convention in 2.6). Hence any orbiface germ $(X,b)_x$ with a smooth base X_x is a covering of the germ $(\mathbb{C}^2, pX+qY+rZ)_0$.

3.2. Covering relations among orbiface germs

Below we give some examples of covering relations among orbiface germs.

The abelian germs $(\mathbb{C}^2, pX+qY)_0$. Abelian reflection groups are always reducible, and therefore isomorphic to a $\mathbb{Z}/(p)\oplus\mathbb{Z}/(q)$ for some p,q. Let us study some coverings of the quotient orbiface germ, which is equivalent to $(\mathbb{C}^2, pX+qY)_0$ where $X:=\{x=0\}$ and $Y:=\{y=0\}$. Any smooth sub-orbiface of this orbiface is of the form $(\mathbb{C}^2, rX+sY)_0$ where $r|p$ and $s|q$ and $r,s\in\mathbb{Z}_{\geq 1}$. This latter orbiface germ is uniformized by $\mathbb{C}^2_0$ via the map $\varphi_{r,s}:(x,y)\in\mathbb{C}^2\to(x^r,y^s)\in\mathbb{C}^2$, with $\mathbb{Z}/(r)\oplus\mathbb{Z}/(s)$ as the Galois group. The lifting of $(\mathbb{C}^2, pX+qY)_0$ to this uniformization is the orbiface $(\mathbb{C}^2, \frac{p}{r}X, \frac{q}{s}Y)_0$. In other words, $\mathbb{Z}/(r)\oplus\mathbb{Z}/(s)$ acts on the orbiface germ $(\mathbb{C}^2, \frac{p}{r}X, \frac{q}{s}Y)_0$, and the quotient is $(\mathbb{C}^2, pX+qY)_0$.

The dihedral germ $(\mathbb{C}^2, 2X+2Y+mZ)_0$. Here we discuss the case where m is odd, the case of even m is left as an exercise. This orbiface has the sub-orbifaces $(\mathbb{C}^2,2X)_0$, $(\mathbb{C}^2,2Y)_0$, $(\mathbb{C}^2,mZ)_0$, $(\mathbb{C}^2,2X+2Y)_0$, $(\mathbb{C}^2,2Y+mZ)_0$ and

$(\mathbb{C}^2, mZ + 2X)_0$. Each one of these suborbifaces is uniformized by $\mathbb{C}_0^2$ via a bicyclic map $\varphi_{p,q}$, note that $\varphi_{r,s} \circ \varphi_{p,q} = \varphi_{rp,sq}$. The uniformizer of $(\mathbb{C}^2, 2X)_0$ is the map $\varphi_{2,1}$. Denote the branch $\varphi_{2,1}^{-1}(Y) = \{y = 0\}$ by Y and the branch $\varphi_{2,1}^{-1}(Z) = \{x^2 - y = 0\}$ by Z'. Hence $\varphi_{2,1}$ is an orbiface covering

$$(\mathbb{C}^2, 2Y + mZ')_0 \to (\mathbb{C}^2, 2X + 2Y + mZ)_0.$$

Now $\varphi_{1,2}$ is a covering of $(\mathbb{C}^2, 2Y + mZ')_0$ and one has $\varphi_{1,3}^{-1}(\mathbb{Z}') = \{x^2 - y^2 = 0\}$. Put $U := \{x + y = 0\}$ and $V := \{x - y = 0\}$. There is an orbiface covering

$$(\mathbb{C}^2, mU + mV)_0 \to (\mathbb{C}^2, 2Y + mZ')_0$$

which is related to the suborbiface $(\mathbb{C}^2, 2X+2Y)_0$ of the initial orbiface $(\mathbb{C}^2, 2X + 2Y + mZ)_0$. The germ $(\mathbb{C}^2, mU + mV)_0$ is uniformized by $\mathbb{C}^2$ via $\varphi_{m,m}$. Hence $\varphi_{2,1} \circ \varphi_{m,m}$ is the uniformization of the dihedral germ $(\mathbb{C}^2, 2X + 2Y + mZ)_0$.

The icosahedral germ $(\mathbb{C}^2, 2X + 3Y + 5Z)_0$**.** This orbiface has the suborbifaces $(\mathbb{C}^2, 2X)_0$, $(\mathbb{C}^2, 3Y)_0$, $(\mathbb{C}^2, 5Z)_0$, $(\mathbb{C}^2, 2X + 3Y)_0$, $(\mathbb{C}^2, 3Y + 5Z)_0$ and $(\mathbb{C}^2, 5Z + 2X)_0$. Keeping the notations of the preceding paragraph, there is an orbiface covering

$$\varphi_{1,2} : (\mathbb{C}^2, 3Y + 5Z')_0 \to (\mathbb{C}^2, 2X + 3Y + 5Z)_0.$$

Now $\varphi_{1,3}$ is a covering of $(\mathbb{C}^2, 3Y + 5Z')_0$, such that $\varphi_{1,3}^{-1}(\mathbb{Z}') = \{x^2 - y^3 = 0\}$, so that there is an orbiface covering

$$(\mathbb{C}^2, 5Z'')_0 \to (\mathbb{C}^2, 3Y + 5Z')_0$$

which is related to the suborbiface $(\mathbb{C}^2, 2X+3Y)_0$ of the initial orbiface $(\mathbb{C}^2, 2X + 3Y + 5Z)_0$. For coverings corresponding to other suborbifaces, see Figure 3.8.

The black dot on top of Figure 3.8 represents the isolated surface (Du Val) singularity of type E_8 given by the equation $S := \{(x, y, z) \in \mathbb{C}^3 \mid x^2 + y^3 + z^5 = 0\}$. It is clear how the projection $(x, y, z) \to (x, y)$ defines a $\mathbb{Z}/(5)$-orbiface covering by this singularity of the the orbiface $(\mathbb{C}^2, 5Z'')_0$. Other coordinate projections define respectively $\mathbb{Z}/(2)$ and $\mathbb{Z}/(3)$-coverings by the same singularity of the orbifaces $(\mathbb{C}^2, 2X'')_0$ $(\mathbb{C}^2, 3Y'')_0$, defined in the same way as $(\mathbb{C}^2, 5Z'')$. The germ of S at the origin is the universal homology covering (i.e., the maximal abelian covering) of the germ $(\mathbb{C}^2, 2X+3Y+5Z)_0$. Notice that S_0 is an orbiface germ with a singular base space and empty branch divisor.

Exercise 3.1. Study the covering relations among other orbiface germs with a smooth base. More generally, study the covering relation among orbiface germs with a singular base and possibly with branch loci.

3.3. Orbifaces with cusps

Many transformation groups (G, M) encountered in practice are not cocompact. In many cases, the orbit space M/G admits a "nice" compactification. It is possible to incorporate the compactifications into the orbifold theory by considering pairs (X, b) with extended b-functions with values in $\mathbb{N} \cup \{\infty\}$, and by declaring that the points with infinite b-value are added in the compactification process. Outside the points with an infinite b-value, the pair (X, b) remains an orbifold. Let us consider

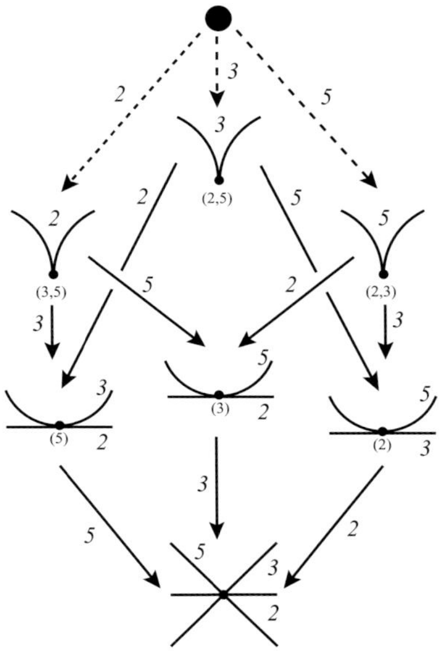

FIGURE 3.8. Coverings of the icosahedral orbiface germ

the case where M is the 2-ball B_2, and G is a finite volume discrete subgroup of $\mathrm{Aut}(B_2)$. Let (X,b) be the quotient orbifold. Then the germs $(X,b)_p$ with $b(p)=\infty$ are called *ball-cusp points*. For smooth X, a classification of ball-cusp points was given in [25]. It turns out that any such germ is a covering of one of the germs (i) $(\mathbb{C}^2, pH_1+qH_2+rH_3)_0$ with $\rho := 1/p+1/q+1/r = 1$ and (ii) $(\mathbb{C}^2, 2H_1+2H_2+2H_3+2H_4)_0$ where H_1, H_2, H_3 and H_4 are smooth branches meeting transversally at the origin. These germs are uniformized by a transformation group $(\Gamma, \mathbb{C}^2)$, where Γ is a parabolic subgroup of $\mathrm{Aut}(\mathbb{C}^2)$ generated by reflections. The orbifold fundamental groups of these germs are infinite solvable. Note that many ball-cusp points (with singular base and branch loci) are coverings of the germs (i) and (ii) above. For example the germ at the origin of the isolated surface singularity $z^3 = xy(x-y)$ is a triple covering of the germ $(\mathbb{C}^2, 3H_1+3H_2+3H_3)_0$ where H_1, H_2, H_3 are given by the polynomials x, y and $x-y$. This is called (somewhat paradoxically) an elliptic singularity, since it is resolved by a blow up which replace the origin by an elliptic curve.

In case M is the bidisc, the germs $(X,b)_p$ with $b(p)=\infty$ are called *cusp points*. In [14] it was shown that the only cusp point with a smooth base is the germ $(\mathbb{C}^2, 2H_1+2H_2+2H_3+2H_4)_0$ where H_1,H_2, H_3 and H_4 are given by the polynomials x, y and $x-y$ and $x-y^2$. This germ also has an infinite solvable orbifold fundamental group, and admits several coverings by germs with a singular base.

4. Orbifaces

Let M be an algebraic surface and let K be its canonical class. The number

$$c_1^2(M) := K \cdot K$$

is an important numerical invariant of M, and is called the *first Chern number of* M. Let $e(M)$ be the the Euler number of M (the Euler number is also called the *second Chern number of* M and denoted by $c_2(M)$). Hirzebruch proved in 1958 the celebrated proportionality theorem: If M is a quotient of the two-ball B_2, then one has

$$c_1(M)^2 = 3e(M).$$

Similarly, if M is a quotient of the bidisc $\Delta \times \Delta$, then the proportionality $c_1(M)^2 = 2e(M)$ holds. In 1977 Miyaoka and Yau proved the inequality $c_1(M)^2 \leq 3e(M)$ for an arbitrary algebraic surface and the following converse to Hirzebruch's proportionality theorem: if M satisfies the $c_1(M)^2 = 3e(M) > 0$, then either M is $\mathbb{P}^2$ or its universal covering is B_2. The analogue of this result for surfaces with $c_1(M)^2 = 2e(M) > 0$ is not correct.

Chern numbers are invariants of algebraic surfaces, but they have orbifold versions. Below we introduce the Chern numbers for orbifolds over the base $\mathbb{P}^2$ only.

Definition 4.1. Let $(\mathbb{P}^2, b)$ be an orbiface with the associated divisor $D_b = \Sigma_1^k m_i B_i$, the curves B_i being irreducible of degree d_i for $i \in [1, k]$. The *orbifold Chern numbers of* $(\mathbb{P}^2, b)$ are defined as

$$c_1^2(\mathbb{P}^2, b) := \Big[-3 + \sum_{i\in[1,k]} d_i\big(1 - \frac{1}{m_i}\big)\Big]^2$$

$$e(\mathbb{P}^2, b) := 3 - \sum_{i\in[1,k]} \big(1 - \frac{1}{m_i}\big)e(B_i \backslash \mathrm{Sing}(B)) - \sum_{p\in \mathrm{Sing}(B)} \big(1 - \frac{1}{b(p)}\big).$$

(If $(\mathbb{P}^2, b)$ is an orbiface with cusp points set $1/b(p) = 0$ whenever $b(p) = \infty$.)

The orbifold Chern numbers have the following property: if $M \to (X, b)$ is a finite uniformization with G as its Galois group, then

$$e(M) = |G|e(X, b) \text{ and } c_1^2(M) = |G|c_1^2(X, b). \tag{4.1}$$

The following orbiface analogue of the Miyaoka–Yau theorem was proved in 1989. We refer the reader to [14] for an introduction to metric uniformization theory of algebraic surfaces.

Theorem 4.2 (Kobayashi, Nakamura, Sakai). *Let* $(\mathbb{P}^2, b)$ *be an orbiface of general type, possibly with ball-cusp points. Then* $c_1^2(\mathbb{P}^2, b) \leq 3e(\mathbb{P}^2, b)$*, the equality holding if and only if* $(\mathbb{P}^2, b)$ *is uniformized by* B_2.

4.1. Orbifaces $(\mathbb{P}^2, D)$ with an abelian uniformization

Consider an orbifold $(\mathbb{P}^2, D)$ where $D = \Sigma_1^k m_i H_i$ where H_i are irreducible and let $B := \cup_1^k H_i$ be the support of D. Suppose $(\mathbb{P}^2, D)$ admits a uniformization with an abelian Galois group. Then for any point p, the local groups $\pi_1^{orb}(\mathbb{P}^2, D)_p$ must be abelian since these groups inject into the Galois group. Nodes are the only orbifold singularities with a smooth base and an abelian fundamental group. Hence B must be a nodal curve. Then by the Zariski conjecture proved by Deligne and Fulton, the group $\pi_1(\mathbb{P}^2\backslash B)$ is abelian and admits the presentation

$$\pi_1(\mathbb{P}^2\backslash B) \simeq \left\langle \mu_1, \ldots, \mu_k \,\middle|\, \sum_1^k d_i\mu_i = 0 \right\rangle$$

where $d_i := \deg(H_i)$. Therefore the group $\pi_1^{orb}(\mathbb{P}^2, D)$ is finite abelian and admits the presentation

$$\pi_1^{orb}(\mathbb{P}^2, D) \simeq \left\langle \mu_1, \ldots, \mu_k \,\middle|\, m_1\mu_1 = \cdots = m_k\mu_k = \sum_1^k d_i\mu_i = 0 \right\rangle.$$

Since the subgroup of the group $\langle \mu_1, \ldots, \mu_k \,|\, m_1\mu_1 = \cdots = m_k\mu_k = 0\rangle$ generated by $\langle \Sigma_1^k d_i\mu_i \rangle$ is of order $\operatorname{lcm}\{m_i/\gcd(m_i, d_i) \,|\, i \in [1, k]\}$, we find that

$$|\pi_1^{orb}(\mathbb{P}^2, D)| = \frac{\prod_1^k m_i}{\operatorname{lcm}\{b_i \,|\, i \in [1, k]\}} \tag{4.2}$$

where $b_i := m_i/\gcd(m_i, d_i)$.

We claim that if $(\mathbb{P}^2, D)$ admits a uniformization, then irreducible components of B must be smooth: Assume the contrary; e.g., suppose that H_i has a node at $p \in \mathbb{P}^2$. The local orbifold fundamental group of this node admits the presentation

$$\pi_1^{orb}(\mathbb{P}^2, D)_p \simeq \langle \mu_p, \mu_p' \,|\, m_i\mu_p = m_i\mu_p' = 0\rangle \simeq \mathbb{Z}/(m_i) \oplus \mathbb{Z}/(m_i)$$

where μ_p and μ_p' are meridians of the branches of H_i meeting at p. Since H_i is irreducible, μ_p and μ_p' are conjugate elements in $\pi_1^{orb}(\mathbb{P}^2, D)$. Since this latter group is abelian, one actually has $\mu_p = \mu_p'$. Hence, the subgroup of $\pi_1^{orb}(\mathbb{P}^2, D)$ generated by μ_p and μ_p' is at most $\mathbb{Z}/(m_i)$ and can not be isomorphic to the local orbifold fundamental group at p, which is $\mathbb{Z}/(m_i) \oplus \mathbb{Z}/(m_i)$.

Suppose that $(\mathbb{P}^2, D)$ is an orbiface with a nodal locus, whose irreducible components are all smooth. Since the group $\pi_1^{orb}(\mathbb{P}^2, D)$ is finite, either $(\mathbb{P}^2, D)$ is not uniformizable or there is a finite universal uniformization. Hence by Theorem 2.4, $(\mathbb{P}^2, D)$ is uniformizable if for every $p \in \mathbb{P}^2$, the image of the inclusion-induced map

$$\rho \circ \iota_* : \pi_1^{orb}(\mathbb{P}^2, D)_p \to \pi_1^{orb}(\mathbb{P}^2, D) \tag{4.3}$$

is an injection.

For a point in $\mathbb{P}^2\backslash B$ the local orbifold fundamental group is trivial, so that $\rho \circ \iota_*$ is always an injection. Now let $p \in H_i\backslash\operatorname{Sing}(B)$. Then $\pi_1^{orb}(\mathbb{P}^2, D)_p \simeq \mathbb{Z}/(m_i)$, and $\rho \circ \iota_*$ is an injection only if the condition below is satisfied:

Condition 1. For any $i \in [1,k]$, the subgroup $\langle \mu_i \rangle$ is of order m_i in $\pi_1^{orb}(\mathbb{P}^2, D)$.

(The notation $\langle A \rangle$ means the subgroup generated by the subset A). Finally, if p is a point of intersection of H_i and H_j, $i \neq j$, then

$$\pi_1^{orb}(\mathbb{P}^2, D)_p = \pi_1^{orb}(\mathbb{P}^2, m_iH_i + m_jH_j)_p \simeq \mathbb{Z}/(m_i) \oplus \mathbb{Z}/(m_j),$$

and $\rho \circ \iota_*$ is injective only if the following condition is satisfied:

Condition 2. For any pair of distinct integers $i, j \in [1,k]$, the subgroup $\langle \mu_i, \mu_j \rangle$ is of order $m_i m_j$ in $\pi_1^{orb}(\mathbb{P}^2, D)$.

Obviously, Condition 2 implies Condition 1 (since any two curves intersects in $\mathbb{P}^2$). Let $D - (m_iH_i + m_jH_j)$ be the divisor obtained from D by removing H_i and H_j. Then Condition 2 is equivalent to

$$|\pi_1^{orb}(\mathbb{P}^2, D)| = |\langle \mu_i, \mu_j \rangle||\pi_1^{orb}(\mathbb{P}^2, D - (m_iH_i + m_jH_j))| \quad \forall i, j \in [1,k],\ i \neq j.$$

By (4.2), this is equivalent to the condition

$$\frac{\prod_1^k m_i}{\operatorname{lcm}\{b_i \mid i \in [1,k]\}} = \frac{\prod_1^k m_i}{\operatorname{lcm}\{b_i \mid i \in [1,k]\backslash\{i,j\}\}} \quad \forall i, j \in [1,k],\ i \neq j$$

$$\Leftrightarrow \operatorname{lcm}\{b_i \mid i \in [1,k]\} = \operatorname{lcm}\{b_i \mid i \in [1,k]\backslash\{i,j\}\} \quad \forall i, j \in [1,k],\ i \neq j. \tag{4.4}$$

Finally, one has the following condition, equivalent to Condition 2:

Condition 3. Any prime power dividing one of $b_1, \dots, b_k$ must divide at least two others.

We have proved the following theorem:

Theorem 4.3. *Let $D = \Sigma_1^k m_iH_i$ where H_i are irreducible of degree d_i and let $B := \cup_1^k H_i$. Then $(\mathbb{P}^2, D)$ is an orbiface with an abelian uniformization if and only if B is a nodal curve whose irreducible components are all smooth, and the numbers $b_1, \dots, b_k$ satisfy Condition 3, where $b_i := m_i/\gcd(m_i, d_i)$.*

Let p be a prime, $\alpha \in \mathbb{Z}_{>0}$ and take numbers $\alpha_i \in [0, \alpha]$ for $i \in [4,k]$. Then the vector

$$[p^\alpha, p^\alpha, p^\alpha, p^{\alpha_4}, p^{\alpha_5}, \dots, p^{\alpha_k}],$$

as well as any of its permutations, satisfies Condition 3. Any vector $[b_1, \dots, b_k]$ satisfying Condition 3 admits a unique factorization into a product of such vectors with distinct p (where the product is taken component-wise).

For $k \leq 2$, Condition 3 is satisfied only if $b_1 = b_2 = 1$, that is, when m_i divides d_i $(i = 1,2)$. For $k = 3$, it is satisfied only if $b_1 = b_2 = b_3$. Some solutions for $k = 4$ can be given as

$$[b_1, b_2, b_3, b_4] = [p^3, p^3, p^3, p] \circledast [q^2, q^6, q^6, q^6] \circledast [r, r, r, r] \dots$$

where p, q, r are distinct primes and $\circledast$ is the operation of component-wise multiplication. In general, Condition 3 is always satisfied if $k \geq 2$ and $b_1 = \dots = b_k$.

Exercise 4.1. The study of algebraic surfaces from the point of view of possible values of $(c_1^2, e) \in \mathbb{Z}_{>0} \times \mathbb{Z}_{>0}$ is called the *surface geography*. Study the geography of abelian uniformizations of $\mathbb{P}^2$.

Orbifaces with a linear locus. Now suppose that $B = \cup_1^k H_i$ is a line arrangement. By Theorem 4.3 the lines $H_1, \ldots, H_k$ must be in general position. Then $d_i = 1$ for $i \in [1, k]$, so that $b_i = m_i$. Obviously Condition 3 is not satisfied unless $k \geq 3$, except the trivial case $b_1 = b_2 = 1$. As we have already seen, in case $k = 3$ and $m_1 = m_2 = m_3 =: m$ the uniformizing surface is $\mathbb{P}^2$ itself, with the polycyclic map

$$\varphi_m : [z_1 : z_2 : z_3] \in \mathbb{P}^2 \to [z_1^m : z_2^m : z_3^m] \in \mathbb{P}^2$$

as the uniformizing map, where we assumed $H_i = \{z_i = 0\}$ for $i = 1, 2, 3$.

The orbifold $(\mathbb{P}^2, \Sigma_1^4 2H_i)$ is uniformized by $\mathbb{P}^1 \times \mathbb{P}^1$. Indeed, $(\mathbb{P}^2, \Sigma_1^3 2H_i)$ is a suborbifold which is uniformized by $\mathbb{P}^2$ via φ_2, and the lifting of $(\mathbb{P}^2, \Sigma_1^4 2H_i)$ to this uniformization is the orbifold $(\mathbb{P}^2, 2Q)$, where $Q \simeq \varphi_2^{-1}(H_4)$ is a smooth quadric. This latter orbifold is uniformized by $\mathbb{P}^1 \times \mathbb{P}^1$ as we shall show below (see Theorem 4.5).

Note that the orbifaces $(\mathbb{P}^2, D)$ may admit intermediate uniformizations (e.g., uniformizations which are not universal). For example, consider the case $D = \Sigma_1^6 2H_i$. There is a surjection of degree 2

$$\pi_1^{orb}(\mathbb{P}^2, D) \simeq \left\langle \mu_1, \ldots, \mu_6 \,\middle|\, 2\mu_1 = \cdots = 2\mu_6 = \sum_1^6 2\mu_i = 0 \right\rangle \twoheadrightarrow$$

$$\left\langle \mu_0, \ldots, \mu_5 \,\middle|\, 2\mu_1 = \cdots = 2\mu_6 = \mu_1 + \mu_2 + \mu_3 = \mu_4 + \mu_5 + \mu_6 = 0 \right\rangle$$

Then the latter group G satisfies Condition 2, hence there is a uniformization with G as the Galois group. The uniformizing surface is an Enriques surface N. As we shall see below, the universal uniformization of $(\mathbb{P}^2, \Sigma_1^6 2H_i)$ is a K3 surface, which is a double covering of N. Observe that the arrangement of hyperplanes $\cup_1^6 H_i$ is not projectively rigid, so that $(\mathbb{P}^2, \Sigma_1^6 2H_i)$ is in fact an orbiface family.

4.1.1. K3 orbifaces. A simply connected algebraic surface M with $c_1^2(M) = 0$ is called a $K3$ surface. It is known that all K3 surfaces have the same Euler number, which is 24. An orbiface uniformized by a K3 surface M is called a *K3 orbiface*. Since M is simply connected, this uniformization must be universal. Let $(\mathbb{P}^2, D)$ be a K3 orbiface uniformized by the K3 surface M, where $D = \Sigma_1^k m_i H_i$ and H_i is an irreducible and reduced curve of degree d_i. Put $B = \cup_1^k H_i$, then B is of degree $d = \Sigma_1^k d_i$. Then

$$c_1^2(\mathbb{P}^2, D) = \frac{c_1^2(M)}{|\pi_1^{orb}(\mathbb{P}^2, D)|} = 0 \tag{4.5}$$

and

$$e(\mathbb{P}^2, D) = \frac{e(M)}{|\pi_1^{orb}(\mathbb{P}^2, D)|} = \frac{24}{|\pi_1^{orb}(\mathbb{P}^2, D)|}. \tag{4.6}$$

Equation 4.5 implies that

$$\sum_{i\in[1,k]} d_i\big(1-\frac{1}{m_i}\big)=3 \Leftrightarrow \sum_{i\in[1,k]} \frac{d_i}{m_i}=d-3 \tag{4.7}$$

which in turn implies that $4\leq d\leq 6$. Equation 4.6 implies that $24/e(\mathbb{P}^2,D)$ must be an integer, which equals the order of the orbifold fundamental group. Under the assumption that $(\mathbb{P}^2,D)$ admits an abelian uniformization, this group order can be computed easily. It is possible to classify all "abelian" K3 orbifaces in this way, see [21] for details. Let us carry out this program for K3 orbifaces with a linear support.

Abelian K3 orbifaces with a linear locus. Suppose $k=6$. Equation (4.7) implies $\Sigma_1^6 1/m_i=3$, which forces $m_1=\cdots=m_6=2$. This orbifold satisfies the conditions of Theorem 4.3 and is uniformizable. Hence, the universal uniformization is a K3 surface M_1. The orbifold fundamental group

$$\pi_1^{orb}(\mathbb{P}^2,D)\simeq\Big\langle \mu_1,\ldots,\mu_6 \;\Big|\; 2\mu_1=\cdots=2\mu_6=\Sigma_0^6\mu_i=0\Big\rangle$$

is of order 32. Let us verify that $e(M_1)=24$. For any H_i, there are 5 singular points of B lying on $H_i\simeq\mathbb{P}^1$, so that $e(H_i\backslash\mathrm{Sing}(B))=e(H_i)-e(\mathrm{Sing}(B))=2-5=-3$. Since the local orbifold fundamental group at the point $H_i\cap H_j$ is of order m_im_j, one has

$$e(\mathbb{P}^2,D)=3+3\sum_1^6(1-\frac{1}{m_i})-\sum_{1\leq i\neq j\leq 6}(1-\frac{1}{m_im_j})=\frac{3}{4}$$

so that $e(M_1)=32e(\mathbb{P}^2,D)=24$.

For $k=5$ there are no abelian K3 orbifaces with a linear support, this can be proved by a case by case analysis. Suppose $k=4$. Equation (4.7) implies

$$\frac{1}{m_1}+\frac{1}{m_2}+\frac{1}{m_3}+\frac{1}{m_4}=1. \tag{4.8}$$

For any H_i, there are 3 singular points of B lying on $H_i\simeq\mathbb{P}^1$, so that $e(H_i\backslash\mathrm{Sing}(B))=e(H_i)-e(\mathrm{Sing}(B))=2-3=-1$. Suppose without loss of generality that $m_1\leq m_2\leq m_3\leq m_4$. There are finitely many 4-tuples satisfying (4.8). It can be shown by case-by-case analysis that the only 4-tuples satisfying Condition 3 are $[4,4,4,4]$ and $[2,6,6,6]$. Hence, the universal uniformizations of these orbifolds are K3 surfaces, say M_2 and M_3 respectively. On the other hand, assumption (4.8) gives

$$e(\mathbb{P}^2,D)=\sum_{1\leq i\neq j\leq 4}\frac{1}{m_im_j}.$$

By using the formula $|\pi_1^{orb}(\mathbb{P}^2,D)|=\prod_1^4 m_i/\mathrm{lcm}(m_1,\ldots,m_4)$ one can verify that $e(M_2)=e(M_3)=24$.

Let us now prove that the surface M_2 is the Fermat quartic surface, the hypersurface in $\mathbb{P}^3$ defined by the equation $M_2: z_4^4=z_1^4+z_2^4+z_3^4$. Since any two 4-line arrangements are projectively equivalent, one can assume that $H_i=\{z_i=0\}$

for $i \in [1,3]$, and $H_4 = \{z_1 + z_2 + z_3 = 0\}$. The suborbifold $(\mathbb{P}^2, \Sigma_1^3 4H_i)$ is uniformized by $\mathbb{P}^2$ via φ_4. Lifting the initial orbifold yields the orbifold $(\mathbb{P}^2, 4K)$, where K is the Fermat quartic curve $z^4 + z_2^4 + z_3^4 = 0$. Now it is easy to see that the restriction of the projection $[z_1 : z_2 : z_3 : z_4] \in \mathbb{P}^3 \to [z_1 : z_2 : z_3] \in \mathbb{P}^2$ to M_2 is a Galois covering branched at $4K$.

Exercise 4.2. Classify the abelian K3 orbifaces and study the covering relations between them.

4.2. Covering relations among orbifaces $(\mathbb{P}^2, D)$ uniformized by $\mathbb{P}^2$

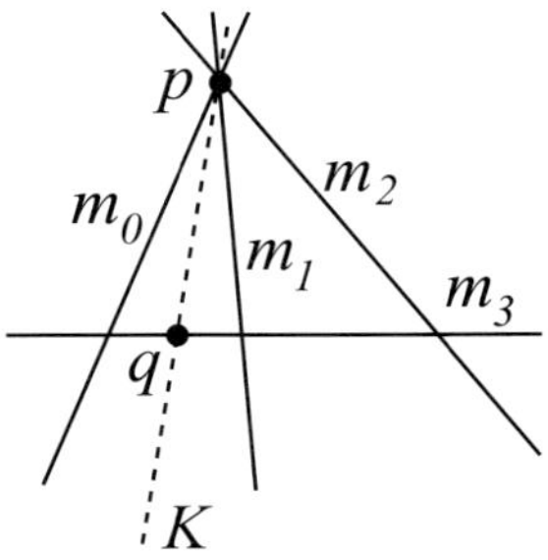

FIGURE 4.9. The orbiface $(\mathbb{P}^2, \Sigma_0^3 m_i H_i)$

Now let us consider the simplest orbiface with a non-abelian fundamental group. Let $H_0 := \{x = 0\}$, $H_1 := \{y = 0\}$, $H_2 := \{x = y\}$ and $H_3 := \{z = 0\}$ be four lines in $\mathbb{P}^2$. Observe that the arrangement $\cup_0^3 H_i$ is projectively rigid. Consider the orbiface $(\mathbb{P}^2, D)$ where $D = \Sigma_0^3 m_i H_i$ The point $p := [0 : 0 : 1]$ is an orbiface germ only if $\Sigma_0^2 1/m_i > 1$. Assume this is the case. Consider another line K through p. Take a base point $\star \in K$, a meridian $\mu_p \subset$ of $p \in K$ and a meridian $\mu_3 \subset K$ of the point $q := K \cap H_3 \in K$ (see Figure 4.9). Since K is topologically a sphere, the loop $\mu_p \mu_3$ is contractible in $K \backslash \{p, q\}$ and hence in $\mathbb{P}^2 \backslash (\cup_0^3 H_i)$. Hence, $\mu_p = \mu_3^{-1}$ in the group $\pi_1^{orb}(\mathbb{P}^2, D)$. In particular, these two meridians are of the same order. Now let m_p be the order of μ_p, considered as an element of $\pi_1^{orb}(\mathbb{P}^2, D)_p$. If the orbiface is uniformizable, this latter group injects into the global orbifold fundamental group. Hence, if $(\mathbb{P}^2, D)$ is uniformizable, the element μ_p, and therefore the element μ_3 must be of order m_p $\pi_1^{orb}(\mathbb{P}^2, D)$. In other words, $m_3 = m_p = 2(\Sigma_0^2 m_i - 1)^{-1}$. Hence, (m_0, m_1, m_2, m_3) must be one of $(2, 2, r, 2r)$, $(3, 3, 2, 12)$, $(2, 4, 3, 24)$ or $(2, 3, 5, 60)$.

Exercise 4.3. Compute the Chern numbers of these orbifolds and check that $c_1^2(\mathbb{P}^2, D) = 3e(\mathbb{P}^2, D)$.

The case $(2, 2, r, 2r)$**:** Observe that $(\mathbb{P}^2, 2H_0 + 2H_1 + 2H_3)$ is a suborbiface of $(\mathbb{P}^2, 2H_0 + 2H_1 + rH_2 + 2rH_3)$. This suborbiface is uniformized by $\mathbb{P}^2$ via the bicyclic covering

$$\varphi_2 : [x : y : z] \in \mathbb{P}^2 \longrightarrow [x^2 : y^2 : z^2] \in \mathbb{P}^2.$$

The lifting $\varphi_2^{-1}(H_2)$ consists of two lines given by the equation $x^2 = y^2$, which we denote by H_2^1 and H_2^2. Denote $\varphi_2^{-1}(H_3)$ by H_3 again. Hence φ_2 is an orbiface covering

$$\varphi_2 : (\mathbb{P}^2, rH_2^1 + rH_2^2 + rH_3) \longrightarrow (\mathbb{P}^2, 2H_0 + 2H_1 + rH_2 + 2rH_3).$$

Obviously, the covering orbiface is uniformized by $\mathbb{P}^2$ via φ_r.

The case $(2, 4, 3, 24)$**:** Observe that $(\mathbb{P}^2, 2H_0 + 2H_1 + 2H_3)$ is a suborbiface of $(\mathbb{P}^2, 2H_0 + 4H_1 + 3H_2 + 24H_3)$. Let φ_2 be its uniformization, denote $\varphi_2^{-1}(H_1)$ by H_1 and $\varphi_2^{-1}(H_3)$ by H_3. As in the previous case, denote the lines $\varphi_2^{-1}(H_2)$ by H_2^1 and H_2^2. Hence there is an orbiface covering

$$\varphi_2 : (\mathbb{P}^2, 3H_1^1 + 3H_1^2 + 2H_2 + 12H_3) \longrightarrow (\mathbb{P}^2, 2H_0 + 4H_1 + 3H_2 + 24H_3).$$

Observe that the covering orbiface is equivalent to the orbiface $(\mathbb{P}^2, 3H_0 + 3H_1 + 2H_2 + 12H_3)$.

The case $(3, 3, 2, 12)$**:** Observe that $(\mathbb{P}^2, 3H_0 + 3H_1 + 3H_3)$ is a suborbiface of $(\mathbb{P}^2, 3H_0 + 3H_1 + 2H_2 + 12H_3)$. This suborbiface is uniformized by $\mathbb{P}^2$ via the bicyclic covering

$$\varphi_3 : [x : y : z] \in \mathbb{P}^2 \longrightarrow [x^3 : y^3 : z^3] \in \mathbb{P}^2.$$

The lifting $\varphi_3^{-1}(H_2)$ consists of two lines given by the equation $x^3 = y^3$, which we denote by H_2^1, H_2^2 and H_2^3. Denote $\varphi_2^{-1}(H_3)$ by H_3 again. Hence φ_3 is an orbiface covering

$$\varphi_3 : (\mathbb{P}^2, 2H_2^1 + 2H_2^2 + 2H_2^3 + 4H_3) \longrightarrow (\mathbb{P}^2, 3H_0 + 3H_1 + 2H_2 + 12H_3).$$

The covering orbiface appeared in the first case with $r = 2$ and is uniformized by $\mathbb{P}^2$.

4.3. Orbifaces $(\mathbb{P}^2, D)$ uniformized by $\mathbb{P}^1 \times \mathbb{P}^1$, $\mathbb{C} \times \mathbb{C}$ and $\Delta \times \Delta$

It is well known that the quotient of $\mathbb{P}^1 \times \mathbb{P}^1$ under the obvious action of the symmetric group Σ_2 is the projective plane. To put in another way, one has the following fact:

Lemma 4.4. *Let $Q \subset \mathbb{P}^2$ be a smooth quadric. Then there is a uniformization $\psi : Q \times Q \to (\mathbb{P}^2, 2Q)$. Let $p \in Q$ and put $T_p^v := \{p\} \times Q$, $T_p^h := Q \times \{p\}$. Then $T_p := \psi(T_p^h) = \psi(T_p^v) \subset \mathbb{P}^2$ is a line tangent to Q at the point $p \in Q$.*

Proof. Since any two smooth quadrics are projectively equivalent, it suffices to prove this for a special quadric. Consider the $\mathbb{Z}/(2)$-action defined by $(x, y) \in \mathbb{P}^1 \times \mathbb{P}^1 \to (y, x) \in \mathbb{P}^1 \times \mathbb{P}^1$. The diagonal $Q = \{(x, x) : x \in \mathbb{P}^1\}$ is fixed under this action. Let $x = [a : b] \in \mathbb{P}^1$ and $y = [c : d]$, then the symmetric polynomials

$\sigma_1([a:b],[c:d]) := ad+bc$, $\sigma_2([a:b],[c,d]) := bd$, $\sigma_3([a:b],[c:d]) := ac$ are invariant under this action, and the Viéte map

$$\psi : (x,y) \in \mathbb{P}^1 \times \mathbb{P}^1 \longrightarrow [\sigma_1(x,y) : \sigma_2(x,y) : \sigma_3(x,y)] \in \mathbb{P}^2$$

is a branched covering map of degree 2. The branching locus $\subset \mathbb{P}^2$ can be found as the image of Q. Note that the restriction of ψ to the diagonal Q is one-to-one, so that one can denote $\psi(Q)$ by the letter Q again. One has $\psi(Q) = [2ab : b^2 : a^2]$ ($[a:b] \in \mathbb{P}^1$), so that Q is a quadric given by the equation $4yz = x^2$. One can identify the surface $\mathbb{P}^1 \times \mathbb{P}^1$ with $Q \times Q$, via the projections of the diagonal $Q \subset \mathbb{P}^1 \times \mathbb{P}^1$. Let $p \in Q$, and put $T_p^h := Q \times \{p\}$, $T_p^v : \{p\} \times Q$. Then $T_p := \psi(T_p^h) = \psi(T_p^v) \subset \mathbb{P}^2$ is a line tangent to Q. Indeed, if $p = [a:b]$, then $\psi(T_p^h)$ is parametrized as $[cb+da : db : ca]$ ($[c:d] \in \mathbb{P}^1$), and can be given by the equation $b^2z + a^2y - abx = 0$, which shows that T_p is tangent to Q at the point $[2ab : b^2 : a^2]$. □

Now let $Q \subset \mathbb{P}^2$ be a smooth quadric and $T_0, \ldots, T_n$ tangents to Q at distinct points $p_i := Q \cap T_i$, $i \in [0,n]$. The configuration $Q \cup T_0 \cup T_1 \cup T_2$ is called the *Apollonius configuration*. Consider the orbiface $(\mathbb{P}^2, aQ + \Sigma_0^n m_i T_i)$. By Theorem 3.2 this is an orbiface provided $1/a + 1/m_i \geq 1/2$. An immediate consequence of Lemma 4.4 is the following result.

Proposition 4.5. *There is an orbiface covering*

$$(\mathbb{P}^1 \times \mathbb{P}^1, aQ + \Sigma_0^n m_i(T_i^v + T_i^h)) \longrightarrow (\mathbb{P}^2, 2aQ + \Sigma_0^n m_i T_i)$$

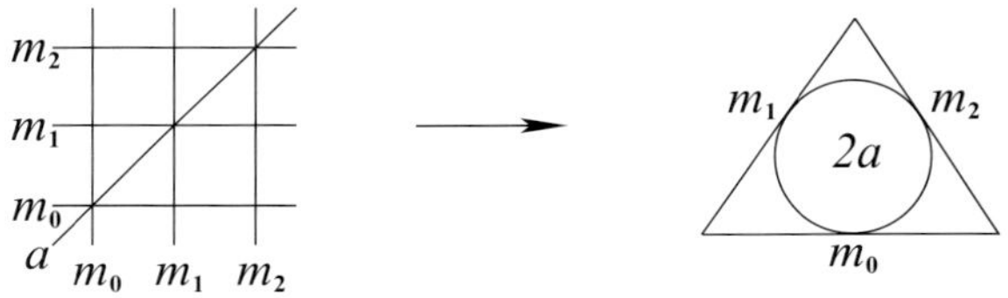

FIGURE 4.10. The covering $(\mathbb{P}^1 \times \mathbb{P}^1, aQ + \Sigma_0^n m_i(T_i^v + T_i^h)) \longrightarrow (\mathbb{P}^2, 2aQ + \Sigma_0^n m_i T_i)$

In particular, when $a = 1$, there is an orbiface covering

$$(\mathbb{P}^1, \Sigma_0^n m_i p_i) \times (\mathbb{P}^1, \Sigma_0^n m_i p_i) \longrightarrow (\mathbb{P}^2, 2Q + \Sigma_0^n m_i T_i)$$

By Theorem 1.4, the covering orbiface above is uniformized by $\mathbb{P}^1 \times \mathbb{P}^1$ if $n = 1$ and $m_0 = m_1$, or if $n = 2$ and $\Sigma_0^2 1/m_i > 1$. Hence the following orbifaces are uniformized by $\mathbb{P}^1 \times \mathbb{P}^1$.

Similarly, the orbifaces of the following Figure 4.3 are uniformized by $\mathbb{C} \times \mathbb{C}$.

Otherwise, the orbifolds $(\mathbb{P}^2, 2Q + \Sigma m_i T_i)$ are uniformized by the bidisc $\Delta \times \Delta$. The orbifaces in Figure 4.11 were first discovered in 1982 by Kaneko, Tokunaga and Yoshida who also gave a complete classification of the orbifaces $(\mathbb{P}^2, D)$ uniformized

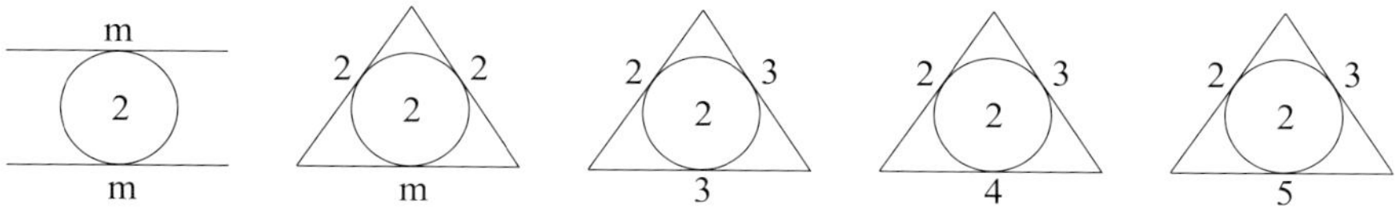

FIGURE 4.11. Orbifaces uniformized by $\mathbb{P}^1 \times \mathbb{P}^1$

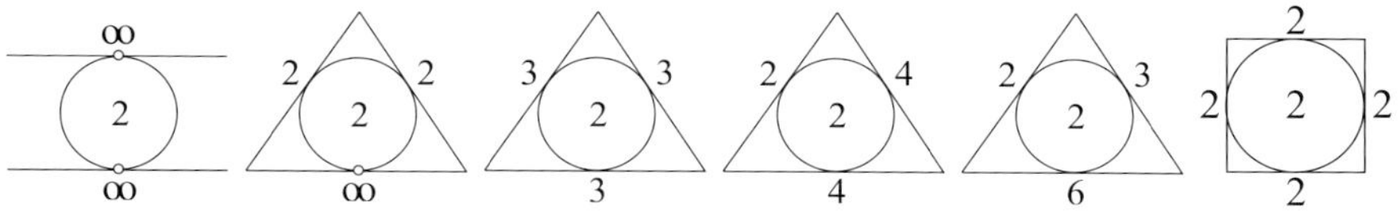

FIGURE 4.12. Orbifaces A uniformized by $\mathbb{C} \times \mathbb{C}$

by $\mathbb{C} \times \mathbb{C}$ (see [12]). Note that the Apollonius configuration is projectively rigid. Except the first and the fifth orbifolds in Figure 4.11 these orbifolds admit liftings to $\mathbb{P}^2$ and give rise to new orbifaces uniformized by $\mathbb{C} \times \mathbb{C}$. Below we shall study the coverings of the fourth orbiface in detail.

Coverings of the orbiface $(\mathbb{P}^2, 2Q + 2T_0 + 4T_1 + 4T_2)$. This orbiface has the suborbifold $(\mathbb{P}^2, 2T_0 + 2T_1 + 2T_2)$, which is uniformized by $\mathbb{P}^2$ via φ_2. We can assume that in projective coordinates the tangent lines are given by $T_i := \{z_i = 0\}$, in these coordinates φ_2 is the map $[z_0 : z_1 : z_2] \to [z_0^2 : z_1^2 : z_2^2]$. A quadric tangent to both the lines $z_0 z_1 z_2 = 0$ is given by the equation $a\sqrt{z_0} + b\sqrt{z_1} + c\sqrt{z_2} = 0$. Hence $\varphi_2^{-1}(Q)$ is given by $\pm a z_0 \pm b z_1 \pm c z_2 = 0$, in other words the lifting of Q consists of four lines $\varphi_2^{-1}(Q) := Q_1 \cup Q_2 \cup Q_3 \cup Q_4$ which meet two by two on the lines $z_0 z_1 z_2 = 0$. The arrangement $T_1 \cup T_2 \cup_1^4 Q_i$ is known as the complete quadrilateral, since it is the set of lines through two points among four points in general position in $\mathbb{P}^2$ (see Figure 4.13).

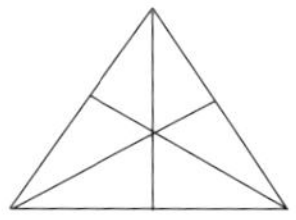

FIGURE 4.13. The complete quadrilateral

Hence the lifting of $(\mathbb{P}^2, 2Q + 2T_0 + 4T_1 + 4T_2)$ is the orbifold $(\mathbb{P}^2, 2T_1 + 2T_2 + \Sigma_1^4 2Q_i)$. Since any two sets of 4 points in general position in $\mathbb{P}^2$ are projectively equivalent, the complete quadrilateral is projectively rigid. Hence there are projective coordinates in which the locus of $(\mathbb{P}^2, 2T_1 + 2T_2 + \Sigma_1^4 2Q_i)$ is given by the equation $z_0 z_1 z_2 (z_0 - z_1)(z_1 - z_2)(z_2 - z_3) = 0$, which is another equation for the complete quadrilateral. Let us name these lines $L_1, \ldots, L_6$ respectively. Now $(\mathbb{P}^2, \Sigma_1^3 2L_i)$ is a suborbifold of $(\mathbb{P}^2, \Sigma_1^6 2L_i)$. This orbifold is uniformized by $\mathbb{P}^2$ via φ_2. The liftings of L_4, L_5, L_6 are given by the equation $(z_0^2 - z_1^2)(z_1^2 - z_2^2)(z_2^2 - z_3^2) = 0$. But this is another equation for the complete quadrilateral. This shows that the orbiface $(\mathbb{P}^2, 2T_1 + 2T_2 + \Sigma_1^4 2Q_i)$ admits self coverings and proves the following result.

Lemma 4.6. *The orbiface* $(\mathbb{P}^2, 2Q + 2T_0 + 4T_1 + 4T_2)$ *has an infinite tower of coverings.*

Observe the analogy with the one-dimensional case: The orbifold $(\mathbb{P}^1, 2p_0 + 3p_1 + 6p_2)$ is covered by $(\mathbb{P}^1, 3_0 + 3p_1 + 3_2)$, which admits self-coverings.

For a higher dimensional version of the results in this subsection, see [23].

4.4. Covering relations among ball-quotient orbifolds

The orbifaces $(\mathbb{P}^2, aQ + \Sigma_0^2 m_i T_i)$ supported by the Apollonius configuration were throughly studied in [10] and [22]. The Chern numbers of $(\mathbb{P}^2, aQ + \Sigma_0^2 m_i T_i)$ are given by

$$c_1^2 = \left[2 - \frac{2}{a} - \sum_1^3 \frac{1}{m_i}\right]^2$$

$$e = 1 - \frac{1}{a} - \sum_1^3 \frac{1}{m_i} + \sum_{1 \le i \ne j \le 3} \frac{1}{m_i m_j} + \frac{1}{2}\sum_1^3 \left[\frac{1}{m_i} + \frac{1}{a} - \frac{1}{2}\right]^2 .$$

One has

$$(3e - c_1^2)(\mathbb{P}^2, aQ + \Sigma_0^2 m_i T_i) = \frac{1}{2}\left[\sum_1^3 \frac{1}{m_i} - \frac{1}{a} - \frac{1}{2}\right]^2 \tag{4.9}$$

which vanishes for the following orbifaces:

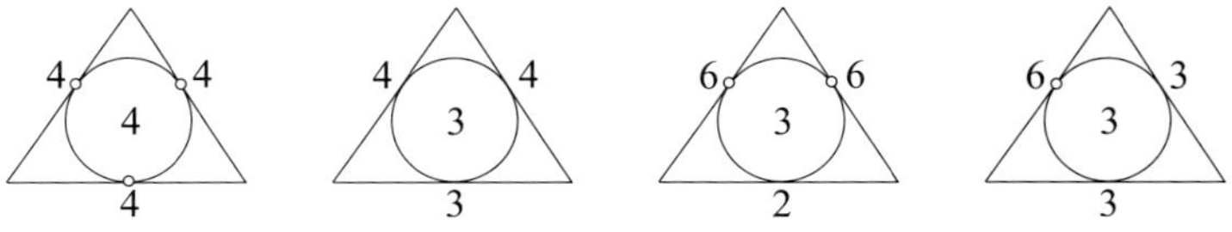

FIGURE 4.14. Orbifolds A uniformized by B_2

By Theorem 4.2 these orbifolds are uniformized by the 2-ball. Observe that the orbiface $(\mathbb{P}^2, 2Q + 2T_0 + 4T_1 + 4T_2)$ is a suborbifold of the first orbiface in Figure 4.14. By Lemma 4.6 this suborbifold admits an infinite tower of coverings. The orbiface $(\mathbb{P}^2, 4Q + 4T_0 + 4T_1 + 4T_2)$ can be lifted to these coverings, and these liftings give an infinite tower of orbifaces uniformized by the 2-ball. Since the group $\pi_1^{orb}(\mathbb{P}^2, 4Q + 4T_0 + 4T_1 + 4T_2)$ is Picard modular, this tower is called a *Picard modular tower.*

Exercise 4.4. Find the first three steps of the Picard modular tower.

Exercise 4.5. Study the coverings of the ball-quotient orbifolds in Figure 4.14.

Question 4.1. The orbifaces $(\mathbb{P}^2, 3Q+3T_0+4T_1+2T_2)$ and $(\mathbb{P}^2, 6Q+2T_0+3T_1+3T_2)$ satisfy $2e - c_1^2 = 0$. What is their universal uniformization?

References

[1] G. Barthel, F. Hirzebruch, Th. Höfer, Th., Geradenconfigurationen und algebrische Flächen. Aspects of Mathematics D 4, Vieweg, Braunschweig-Wiesbaden, 1986.

[2] S. Bundgaard, J. Nielsen, *On normal subgroups of finite index in F-groups*, Math. Tidsskrift B (1951), 56–58.

[3] E.V. Brieskorn, *Examples of singular normal complex spaces which are topological manifolds*, Proc. Natl. Acad. Sci. USA **55** (1966), 1395–1397.

[4] H. Cartan, *Quotient d'un espace analytique par un groupe d'automorphismes*, Algebraic Geometry and Topology, A sympos. in honor of S. Lefschetz, Princeton University Press, 1957, 90–102.

[5] C. Chevalley, *Invariants of finite groups generated by reflections*, Am. J. Math. **77** (1955), 778–782.

[6] R.H. Fox, *On Fenchel's conjecture about F-groups,* Math. Tidsskrift B (1952), 61–65.

[7] R.H. Fox, *Covering spaces with singularities*, Princeton Math. Ser. **12** (1957), 243–257.

[8] E. Hironaka, *Abelian coverings of the complex projective plane branched along configurations of real lines*, Mem. Am. Math. Soc. **502** (1993).

[9] F. Hirzebruch, *Arrangements of lines and algebraic surfaces*, Progress in Mathematics **36**, Birkhäuser, Boston, 1983, 113–140.

[10] R.-P. Holzapfel, V. Vladov, *Quadric-line configurations degenerating plane Picard-Einstein metrics I–II.* Sitzungsber. d. Berliner Math. Ges. 1997–2000, Berlin, 2001, 79–142.

[11] B. Hunt, *Complex manifold geography in dimension* 2 *and* 3. J. Differ. Geom. **30** no. 1 (1989), 51–153.

[12] J. Kaneko, S. Tokunaga, M. Yoshida, *Complex crystallographic groups II*, J. Math. Soc. Japan **34** no. 4 (1982), 581–593.

[13] M. Kato, *On uniformizations of orbifolds.* Adv. Stud. Pure Math. **9** (1987), 149–172.

[14] R. Kobayashi, *Uniformization of complex surfaces*, in: Kähler metric and moduli spaces, Adv. Stud. Pure Math. **18** no. 2 (1990), 313–394.

[15] D. Mumford, *The topology of normal singularities of an algebraic surface and a criterion for simplicity*, Inst. Hautes Études Sci. Publ. Math. **9** (1961), 5–22.

[16] M. Namba, *On finite Galois covering germs*, Osaka J. Math. **28** no. 1 (1991), 27–35.

[17] M. Namba, Branched coverings and algebraic functions, Pitman Research Notes in Mathematics Series 161, 1987.

[18] M. Oka, *Some plane curves whose complements have non-abelian fundamental groups*, Math. Ann. **218** (1975), 55–65.

[19] J.-P. Serre, *Revêtements ramifiés du plan projectif* Séminaire Bourbaki **204**, 1960.

[20] G.C. Shephard, J.A. Todd, *Finite unitary reflection groups*, Can. J. Math. **6** (1954), 274–304.

[21] A.M. Uludağ, *Coverings of the plane by K3 surfaces,* Kyushu J. Math. **59** no. 2 (2005), 393–419.

[22] A.M. Uludağ, *Covering relations between ball-quotient orbifolds*, Math. An. **308** no. 3 (2004), 503–523.

[23] A.M. Uludağ, *On Branched Galois Coverings of* $\mathbb{P}^n$ *by products of discs*, International J. Math. **4** no. 10 (2003), 1025–1037.

[24] A.M. Uludağ, *Fundamental groups of a family of rational cuspidal plane curves,* Ph.D. thesis, Institut Fourier 2000.

[25] M. Yoshida, Fuchsian Differential Equations, Vieweg Aspekte der Math., 1987.

[26] O. Zarsiki, *On the Purity of the Branch Locus of Algebraic Functions* Proc. Natl. Acad. Sci. USA. **44** no. 8 (1958), 791–796.

A. Muhammed Uludağ
Galatasaray University
Mathematics Department
34357 Ortaköy/ Istanbul
Turkey
e-mail: muludag@gsu.edu.tr

Progress in Mathematics, Vol. 260, 407–429

From the Power Function to the Hypergeometric Function

Masaaki Yoshida

Abstract. The hypergeometric function is a slight generalization of the power fucntion. We will see this by the Schwarz map of the hypergeometric equation focussing on the behavior of this map when the local exponent-differences are purely imaginary.

Mathematics Subject Classification (2000). Primary 33C05; Secondary 34M35.

Keywords. Hypergeometric functions, Schwarz s-function, Schottky groups.

Contents

The author is grateful to the MPIM in Bonn and the JSPS.

1. Introduction

The hypergeometric function is a slight generalization of the power fucntion. We will see this in several aspects.

1.1. Power series

The binomial theorem tells that the power function admits a power series expansion as

$$(1-x)^{-a} = \sum_{n=0}^{\infty} \frac{(a,n)}{(1,n)} x^n,$$

where $(a,n) = a(a+1)\cdots(a+n-1)$, in particular, $(1,n) = n!$ By putting the terms like $(*,n)$ to the denominators and the numerators, one defines the hypergeometric function by the power series

$$F(a,b,c;x) = \sum_{n=0}^{\infty} \frac{(a,n)(b,n)}{(c,n)(1,n)} x^n.$$

1.2. Differential equations

If we set the coefficients of the binomial series as

$$A_n = \frac{(a,n)}{(1,n)},$$

then they satisfy

$$A_{n+1} = A_n \frac{n+a}{n+1}.$$

This shows that the power function $(1-x)^{-a}$ solves the linear differential equation

$$(D+a)u = (D+1)\frac{1}{x}u,$$

where $D = x\frac{d}{dx}$. In fact, since $Dx^n = nx^n$, we have

$$(D+a)\sum A_n x^n = \sum A_n(D+a)x^n = \sum A_n(n+a)x^n = \sum A_{n+1}(n+1)x^n$$

and

$$(D+1)\frac{1}{x}\sum A_n x^n = \sum A_n(D+1)x^{n-1} = \sum A_n n x^{n-1}.$$

Note that the above equation is equivalent to

$$(1-x)u' - au = 0, \quad \text{where} \quad u' = \frac{du}{dx}.$$

This equation has singularity at $x = 1$ and ∞. By exactly the same way, we see that the hypergeometric series $F(a,b,c;x)$ solves the differential equation

$$(D+a)(D+b)u = (D+c)(D+1)\frac{1}{x}u,$$

which is equivalent to

$$x(1-x)u'' + \{c - (a+b+1)x\}u' - abu = 0;$$

either of them is called the hypergeometric differential equation, and is denoted by $E(a,b,c)$. This equation has singularity at $x = 0, 1$ and ∞. It is linear and of second order; these imply that

- at any point $x_0 \neq 0, 1, \infty$, the local solutions at x_0 (they are holomorphic) form a 2-dimensional linear space over $\mathbb{C}$, and
- any solution at x_0 can be continued holomorphically along any curve starting x_0 not passing through $0, 1$ or ∞.

1.3. The aim of this lecture

In this lecture, I would like to show that the map (often called the Schwarz map)

$$s : x \longmapsto u_0(x)/u_1(x)$$

defined by linearly independent two solutions of $E(a,b,c)$ is a slightly enriched version of the map $x \mapsto x^a$ defined by the power function x^a.

2. Power functions

Since the power function is the prototype of the hypergeometric function, we study it carefully. The reader will find that this seemingly simple function is in fact a fairly complicated function. Recall first that the power function x^a is by definition, the composition of the three maps:

$$x \overset{\log}{\longrightarrow} \log x \overset{\times a}{\longrightarrow} a\log x \overset{\exp}{\longrightarrow} \exp(a\log x) = x^a.$$

We here recall the exponential function and the logarithmic function; see Figure 1. The upper half-plane $\mathbb{H}_x = \{x \in \mathbb{C} \mid \Im x > 0\}$ in x-space is mapped under the logarithmic function $x \mapsto \exp x$ onto a belt. The logarighmic function is the inverse of the exponential function. So, in a sense, the map $x \mapsto x^a$ is just *conjugate* to

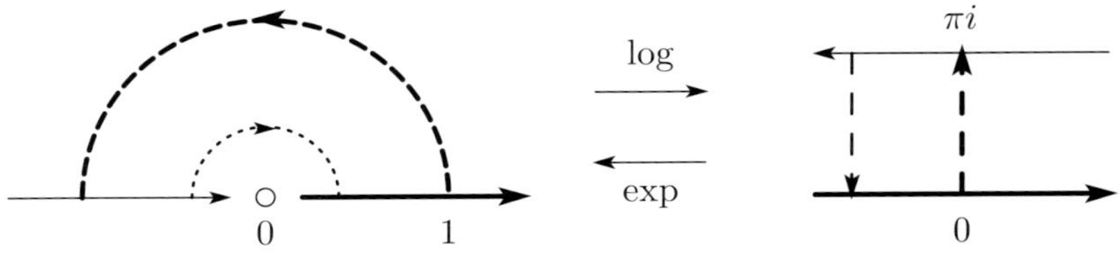

FIGURE 1. The logarithmic function and the exponential function

the multiplication $\times a$ by a, which rotates and enlarges/shrinks the belt. Note that the inverse function of a power function is again a power function.

2.1. When a is real

If the exponent a is real — for simplicity we assume $a > 0$ — the map $x \mapsto z = x^a$ takes the upper half-plane conformally onto the fan

$$\{z \in \mathbb{C} \mid 0 < \arg z < \pi a\},$$

with angle πa. If $|a| \leq 1$ then this is also univalent. Note that a fan is considered as a diangle; see Figure 2. The whole behavior of this map can be understood via

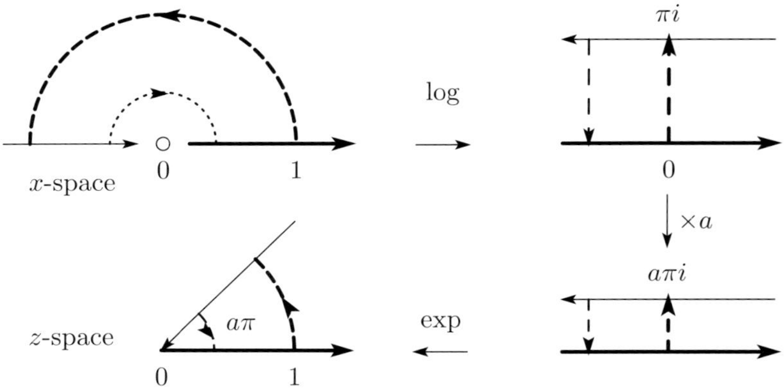

FIGURE 2. Power function with real exponent

repeated use of the Schwarz reflection principle along the intervals $(-\infty, 0)$ and $(0, +\infty)$ in the x-space, and the two sides of the diangle in the z-space.

If one travels around the origin in the x-space, the value changes from z to $e^{2\pi i a}z$ since $\exp(a\log x)$ changes into $\exp(a(\log x+2\pi i))$. The group $M(a)$ of transformations

$$z\mapsto e^{2\pi i a n}z,\quad n\in\mathbb{Z}$$

acts properly discontinuously on the z-space if and only if a is rational.

2.2. When a is purely imaginary

If the exponent a is purely imaginary, set $a=i\theta$ and assume for simplicity $\theta>0$. Then the upper half-plane $\mathbb{H}_x$ covers the annulus $\{z\in\mathbb{C}\mid e^{-\pi\theta}<|z|<1\}$ infinitely many times. If we restrict the map to the hemi-annulus $\{x\in\mathbb{H}_x\mid e^{-\pi}<|x|<1\}$, it is mapped univalently onto the hemi-annulus $\{z\in\mathbb{H}_z\mid e^{-\pi\theta}<|z|<1\}$; see Figure 3.

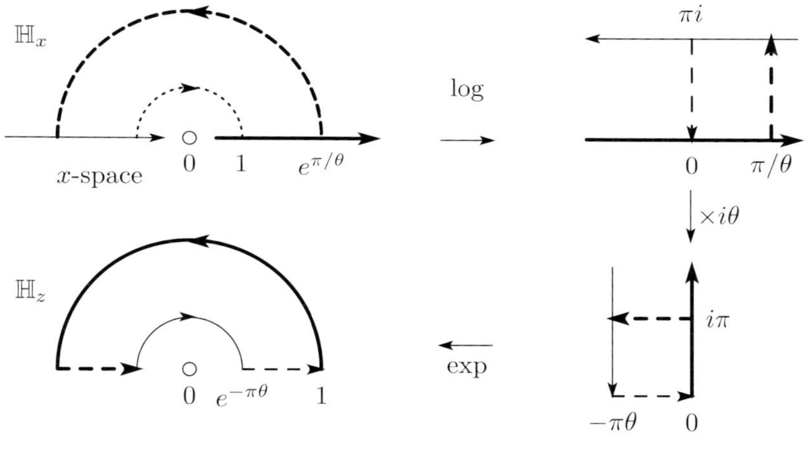

FIGURE 3. Power function with purely imaginary exponent $a=i\theta,\ \theta>0$

The whole behavior of this map can be understood via repeated use of the Schwarz reflection principle along the corresponding edges of the hemi-annulus. Note that the arcs correspond to the intervals, and vice versa. So if one turns around zero in the x-space then one approaches 0 or ∞ in the z-space, and vice versa.

The group $M(a)$ of transformations $z\mapsto e^{2\pi\theta n}z,\ n\in\mathbb{Z}$ acts properly discontinuously on the punctured z-space $\mathbb{C}_z^\times$. Note that the quotient space $\mathbb{C}_z^\times/M(a)$ is an elliptic curve, of which moduli is a bit special; I leave it to the reader.

2.3. Otherwise

If the exponent a is not real or purely imaginary, the image will be a spiral-ribbon. It may be interesting to draw the images of the images of the hemi-annulus $\{x \in \mathbb{H}_x \mid e^{-\pi} < |x| < 1\}$ for $a = e^{it}$, $0 \le t \le 1/2$. Figure 4 shows the half-way model (when $t = 1/4$); one sees the 2-fold symmetry.

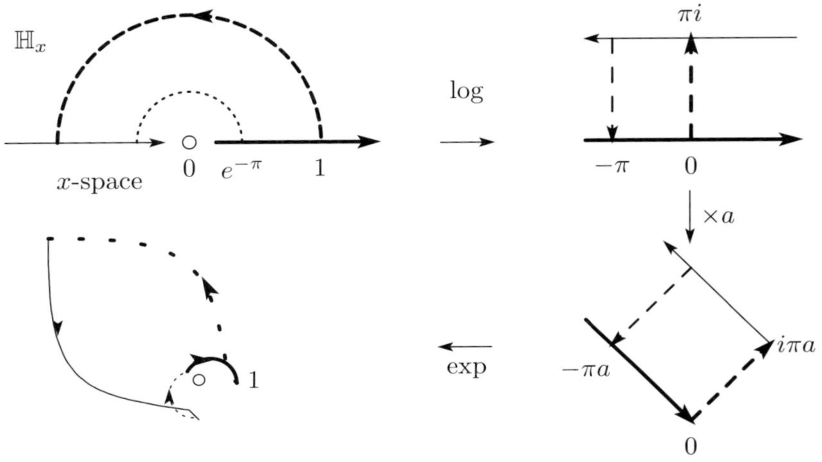

FIGURE 4. Power function with complex exponent

The group $M(a)$ of transformations $z \mapsto e^{2\pi i a n} z$, $n \in \mathbb{Z}$ acts properly discontinuously on the punctured z-space $\mathbb{C}_z^\times$. Note that the quotient space $\mathbb{C}_z^\times / M(a)$ is a (general) elliptic curve.

3. Some local properties of the hypergeometric differential equation

3.1. At a regular point

At a regular point $\xi \neq 0, 1, \infty$, we have a unique solution of $E(a, b, c)$ in the form

$$u_0 = 1 + O((x - \xi)^2),$$

and a unique solution in the form

$$u_1 = x + O((x - \xi)^2);$$

they form a basis of the space of the local solutions at x_0. Here $O((x - \xi)^2)$ stands for the product of $(x - \xi)^2$ and a convergent power series in $x - \xi$.

3.2. Around $x = 0$

Since we have $D(x^\alpha u) = \alpha x^\alpha u + x^\alpha Du = x^\alpha(\alpha + D)u$, regarding D and x^α as operators acting on functions, we have

$$Dx^\alpha = x^\alpha(\alpha + D).$$

This can be remembered as 'when x^α passes through D from right to left, D gets α' or 'when x^α passes through D from left to right, D loses α.'

Recall that the series $F(a, b, c; x)$ solves

$$(D + a)(D + b)u = (D + c)(D + 1)\frac{1}{x}u.$$

Multiply x^{c-1} to the both sides of this equation from the left, and let it move from left to right. Then every D loses $c - 1$, and it changes into

$$(D + a - c + 1)(D + b - c + 1)(x^{c-1}u) = (D + 1)(D - c + 2)\frac{1}{x}(x^{c-1}u).$$

Since $v = F(a - c + 1, b - c + 1, -c + 2; x)$ solves

$$(D + a - c + 1)(D + b - c + 1)v = (D + 1)(D - c + 2)\frac{1}{x}v,$$

we conclude that $x^{1-c}F(a-c+1, b-c+1, -c+2; x)$ is another solution to $E(a, b, c)$ at $x = 0$. But if $c = 1$, this gives just the series $F(a, b, c; x)$; in such a case we must work a little more.

Since the power function x^{1-c} is not single-valued (in general), when we concern seriously its value, we must assign the argument of x.

Anyway, we call $\{0, 1 - c\}$ the set of local exponents of $E(a, b, c)$ at $x = 0$.

3.3. Around $x = 1$ and ∞

Around other singular points $x = 1$ and ∞, one can use the so-called symmetry of the hypergeometric equation. By the change $x = 1 - y$, the equation

$$x(1 - x)u'' + \{c - (a + b + 1)x\}u' - abu = 0$$

changes into

$$y(1 - y)u'' + \{-c + a + b + 1 - (a + b + 1)y - abu = 0,$$

which is just the hypergeometric equation $E(a, b, -c + a + b + 1)$. So

$F(a, b, -c + a + b + 1; 1 - x)$ and $(1 - x)^{c-a-b}F(c - a, c - b, c + 1 - a - b; 1 - x)$

solve $E(a, b, c)$. So $\{0, c - a - b\}$ is the set of local exponents of $E(a, b, c)$ at $x = 1$.

At infinity, one better makes use of the form

$$(D + a)(D + b)u = (D + c)(D + 1)\frac{1}{x}u.$$

The change $x = 1/y$ will take this equation again to a hypergeometric equation, and one will get $\{a, b\}$ as the set of local exponents of $E(a, b, c)$ at $x = \infty$; here I omit the details.

Set the exponent differences as

$$\lambda = 1 - c, \quad \mu = c - a - b, \quad \nu = a - b.$$

4. Some global properties of the hypergeometric differential equation

4.1. Connection matrices

If we have two sets of bases of local solutions at a regular point (I mean the point is not equal to $0, 1, \infty$), they are linearly related. The matrices representing these linear relations are called connection matrices. For the hypergeometric equation the connection matrices are known explicitly. Here we present one example. We consider two sets of solutions

$$f_0(0,x) = F(a,b,c;x), \quad f_0(\lambda,x) = x^{\lambda}F(a-c+1, b-c+1, 2-c; x),$$

and

$$f_1(0,x) = F(a,b,a+b-c+1; 1-x),$$
$$f_1(\mu,x) = (1-x)^{\mu}F(c-a, c-b, c+1-a-b; 1-x),$$

which we found in the previous section. We compare these solutions on the interval $(0,1)$, where the real positive number x and $1-x$ are assigned to have zero argument. Then they are related as

$$(f_0(0,x), f_0(\lambda,x)) = (f_1(0,x), f_1(\mu,x))P,$$

where

$$P = \begin{pmatrix} D & C \\ B & A \end{pmatrix}$$

for

$$D = \frac{\Gamma(c)\Gamma(c-a-b)}{\Gamma(c-a)\Gamma(c-b)}, \quad C = \frac{\Gamma(2-c)\Gamma(c-a-b)}{\Gamma(1-a)\Gamma(1-b)},$$
$$B = \frac{\Gamma(c)\Gamma(a+b-c)}{\Gamma(a)\Gamma(b)}, \quad A = \frac{\Gamma(2-c)\Gamma(a+b-c)}{\Gamma(a-c+1)\Gamma(b-c+1)},$$

and Γ denotes the Gamma function. This has been found by Gauss. A proof can be found in any textbook treating the hypergeometric function; see for example [IKSY], where find many different proofs can be found.

4.2. A set of generators of the monodromy group

The fundamental group of $\mathbb{C} - \{0,1\}$ with a base point, say $x = 1/2$, can be generated by a loop ρ_0 around 0, and a loop ρ_1 around 1 as are shown in Figure 5. If one continues analytically the pair $(f_0(0,x), f_0(\lambda,x))$ along the loop ρ_0, then it changes into

$$(f_0(0,x), f_0(\lambda,x)) \begin{pmatrix} 1 & 0 \\ 0 & e^{2\pi i\lambda} \end{pmatrix},$$

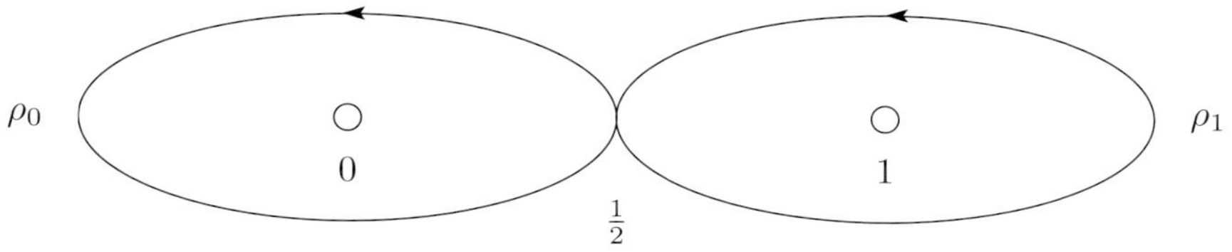

FIGURE 5. Loops ρ_0 and ρ_1

and if one continues analytically the pair $(f_1(0,x), f_1(\mu,x))$ along the loop ρ_1, then it changes into

$$(f_1(0,x), f_1(\mu,x)) \begin{pmatrix} 1 & 0 \\ 0 & e^{2\pi i\mu} \end{pmatrix};$$

we here used the behavior of the power function, which we studied in §2. Since the pairs $(f_0(0,x), f_0(\lambda,x))$ and $(f_1(0,x), f_1(\mu,x))$ are related as in the previous subsection, if one continues analytically the pair $(f_0(0,x), f_0(\mu,x))$ along the loop ρ_1, then it changes into

$$(f_0(0,x), f_0(\mu,x))P^{-1} \begin{pmatrix} 1 & 0 \\ 0 & e^{2\pi i\mu} \end{pmatrix} P.$$

The monodromy group of the differential equation $E(a,b,c)$ (with respect to a pair of linearly independent solutions $(f_0(0,x), f_0(\lambda,x))$ with base $x = 1/2$) is by definition, the group generated by

$$\begin{pmatrix} 1 & 0 \\ 0 & e^{2\pi i\lambda} \end{pmatrix}, \qquad P^{-1} \begin{pmatrix} 1 & 0 \\ 0 & e^{2\pi i\mu} \end{pmatrix} P.$$

If one makes use of another pair of solutions at another base point, the resulting group is conjugate in $GL_2(\mathbb{C})$ to this group. So the differential equation determines a conjugacy class represented by this group. Any representative, or its conjugacy class, is called the monodromy group of the equation.

Note that these two matrices are generators of the monodromy group of the hypergeometric equation $E(a,b,c)$. Well, there are many ways to define a group; among others, giving just a set of generators by matrices is the worst way, I am afraid. In most cases, these generators give almost no information about the group.

5. The Schwarz map of the hypergeometric differential equation with real exponents

We are interested in the Schwarz map $s : x \to z = u_0(x)/u_1(x)$, where u_0 and u_1 are linearly independent solutions of the hypergeometric equation $E(a,b,c)$. In this section, we assume that the parameters a, b and c are real; so the coefficients of the equation $E(a, b, c)$ are real, if x is real. Along each of the three intervals

$$(-\infty, 0), \quad (0, 1), \quad (1, +\infty),$$

there are two linearly independent solutions which are real-valued on the interval. Note that a real-valued solution along an interval may not be real valued along another interval. Since the Schwarz map is in general multi-valued, we restrict it on the upper half-plane $\mathbb{H}_x$, and study the shape of its image in the target space $\mathbb{P}^1_z$ $(\cong \mathbb{C} \cup \{\infty\})$, the projective line with coordinate z. The Schwarz map depends on the choice of the two linearly independent solutions; if we choose other two such solutions then the new and the old Schwarz maps relate with a linear fractional transformation, which is an automorphism of $\mathbb{P}^1_z$.

Recall the following fundamental fact: A linear fractional transformation takes a circle to a circle, here a line is considered to be a circle which passes through ∞.

Around the singular points $x = 0, 1$ and ∞, the Schwarz map s is, after performing suitable linear fractional transformations, near to the power functions $x \to x^{|\lambda|}, (1-x)^{|\mu|}$ and $x^{|\nu|}$. Thus small hemi-disks with center $x = 0, 1$ and ∞ are mapped to horns with angle $\pi|\lambda|, \pi|\mu|$ and $\pi|\nu|$. See Figure 6.

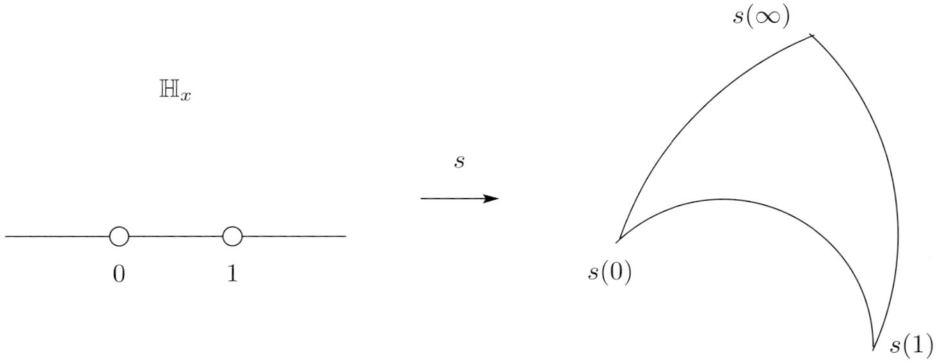

FIGURE 6. Schwarz triangle

Summing up, the image of $\mathbb{H}_x$ under s is bounded by three arcs (part of circles), and the three arcs meet with angle $\pi|\lambda|, \pi|\mu|$ and $\pi|\nu|$. If

$$|\lambda| < 1, \quad |\mu| < 1, \quad |\nu| < 1,$$

then the image is an arc-triangle, which is called the *Schwarz triangle*. In this case, the Schwarz map gives a conformal equivalence between $\mathbb{H}_x$ and the Schwarz triangle.

5.1. Real but general exponents

If the condition above is not satisfied, though the image is indeed bounded by three arcs, it might cover the target $\mathbb{P}^1_z$ many times. I give some examples (for a complete description, see [Yo3]). Suppose the three circles generated by the three curves cut sphere $\mathbb{P}^1_s$ into eight triangles $A, B, C, D, \bar{A}, \ldots, \bar{D}$; see Figure 7. We

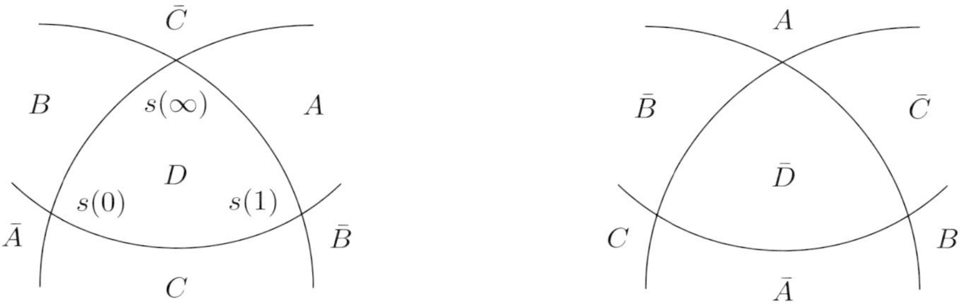

FIGURE 7. Eight triangles $A, \ldots, \bar{D}$ tessellating the s-sphere

draw in $\mathbb{H}_x$ the inverse images under s of these eight triangles; see Figure 8. To make the pictures look nice, I transformed the half plane $\mathbb{H}_x$ into a disc. For the condition $|\lambda|, |\mu|, |\nu| < 1$, there is only one triangle $\mathbb{H}_x$ itself colored in D.

5.2. Schwarz map

The global behavior of the Schwarz map can be seen by applying repeatedly the Schwarz reflection principle to the three sides of the Schwarz triangle. For generic parameters, the picture of the reflected triangles gradualy become caotic. For special parameters, it can remain neat. For example, if

$$|\lambda| = \frac{1}{p}, \quad |\mu| = \frac{1}{q}, \quad |\nu| = \frac{1}{r}, \quad p, q, r \in \{2, 3, \ldots, \infty\},$$

then the whole image is nice; note that this is not a *necessary* condition. Paint the original triangle black (since it is called the Schwarz triangle), adjacent ones white, and so on. The picture of these black and white triangles thus obtained can be roughly classified into three cases depending on whether

$$\frac{1}{p} + \frac{1}{q} + \frac{1}{r}$$

is bigger than or equal to or less than 1.

- In the first case, the number of the triangles is finite, and those cover the whole sphere. The possible triples $(p \leq q \leq r)$ are

$$(2, 2, r), \quad (2, 3, 3), \quad (2, 3, 4), \quad (2, 3, 5).$$

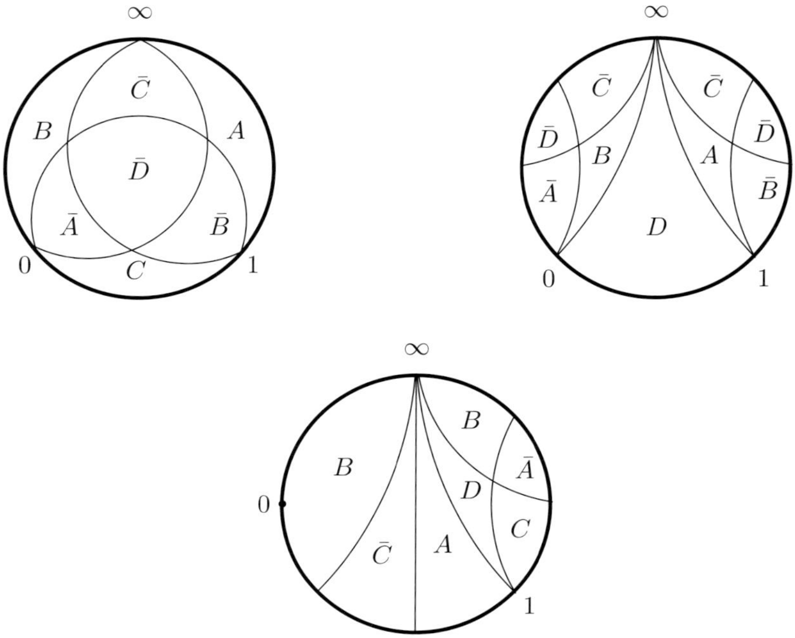

FIGURE 8. Inverse images of the eight triangles under different s

The monodromy groups are the dihedral, tetrahedral, octahedral and icosahedral groups, respectively.

- In the second case, the three circles generated by the three sides of the triangles pass through a common point. If you send this point to infinity, then the triangles are bounded by (straight) lines, and they cover the whole plane. The possible triples $(p \le q \le r)$ are

$$(2,2,\infty), \quad (3,3,3), \quad (2,4,4), \quad (2,3,6).$$

- In the third case, there is a unique circle which is perpendicular to all of the circles generated by the sides of the triangles, and these triangles fill the disc bounded by this circle. In particular, when

$$(p,q,r) = (2,3,\infty), \quad (\infty,\infty,\infty), \quad (2,3,7),$$

one can find in many books nice pictures of the tessellation by these triangles; we do not repeat them here.

6. The Schwarz map of the hypergeometric differential equation with purely imaginary exponent-differences

In this section, our exponent-differences will be purely imaginary. Before restricting the exponents in this way, we recall an elementary but important transformation.

6.1. A projective change of the unknown

Since the Schwarz map $s : x \to z = u_0(x)/u_1(x)$ is defined by the ratio of the solutions, the change of the unknown u by multiplying any (multivalued OK, with any singularity or zero or pole OK, but not identically zero of course) function to u does not affect the Schwarz map. Following is a very famous (considered to be the origin of the Schwarzian derivative; if you do not know this derivative, just forget it for a moment, we do not use this in this lecture) change: Consider in general an equation of the form

$$u'' + pu' + qu = 0.$$

If we make a change from u to v by $u = fv$, then we have

$$v'' + \left(p + 2\frac{f'}{f}\right) v' + \left(q + p\frac{f'}{f} + \frac{f''}{f}\right) v = 0.$$

Choose f solving $p + 2f'/f = 0$. Then the coefficient of v is given as

$$q - \frac{1}{2}p' - \frac{1}{4}p^2.$$

6.2. The image of each interval is a circle

If you make the transformation in the previous subsection for our $E(a, b, c)$, we get (we write u in place of v)

$$u'' + \frac{1}{4}\left(\frac{1-\lambda^2}{x^2} + \frac{1-\mu^2}{(1-x)^2} + \frac{1+\nu^2-\lambda^2-\mu^2}{x(1-x)}\right) u = 0.$$

So if the exponent-differences λ, μ, ν are purely imaginary, the coefficient of the above equation is real, if x is real. As in the previous section, along each of the three intervals

$$(-\infty, 0), \quad (0, 1), \quad (1, +\infty),$$

there are two linearly independent solutions which are real-valued on the interval. Therefore the image of the three intervals under the Schwarz map are circles.

We will study the behavior of the Schwarz map on the upper half x-plane. Since this map takes the three real intervals to three circles, it is natural to ask for the arrangement of the three circles. Note that, if the three circles do not intersect, there are two ways topologically to put three disjoint circles on the sphere: like a dartboard and like a pig-nose. See Figure 9. Please guess which is the case.

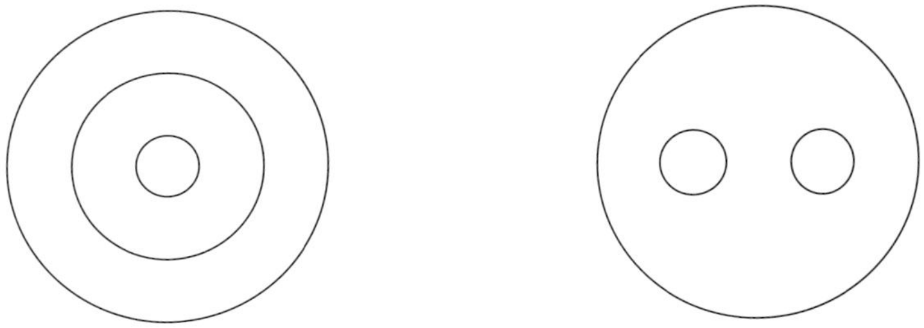

FIGURE 9. A dartboard and a pig-nose

6.3. Around the origin, this map is very near to $x \mapsto x^\lambda$

Put

$$\lambda = i\theta_0, \quad \mu = i\theta_1, \quad \nu = i\theta_2,$$

$(\theta_0, \theta_1, \theta_2 \in \mathbb{R})$, that is,

$$a = \frac{1}{2} - i\frac{\theta_0 + \theta_1 + \theta_2}{2}, \quad b = \frac{1}{2} - i\frac{\theta_0 + \theta_1 - \theta_2}{2}, \quad c = 1 - i\theta_0.$$

We assume that the three purely imaginary exponents have positive imaginary parts. (This assumption is made only to make figure-drawing easier.) We define two Schwarz maps as

$$s_0 = \frac{f_0(x;\lambda)}{f_0(x;0)} \quad \text{and} \quad s_1 = \frac{f_1(x;\mu)}{f_1(x;0)},$$

which are related as

$$s_0 = \frac{As_1 + C}{Bs_1 + D} \quad \text{and} \quad s_1 = \frac{Ds_0 - C}{-Bs_0 + A}.$$

Note their local behavior:

$$\begin{aligned} s_0 &\approx x^\lambda \quad \text{around } x = 0, \\ s_1 &\approx (1-x)^\mu \quad \text{around } x = 1, \end{aligned}$$

where $\approx$ stands for 'is very near to'.

Recall that the images of the intervals $(-\infty, 0)$ and $(0, 1)$ are circles.

In a sufficiently small neighborhood of $x = 0$ in the upper half-plane, the map s_0 can be approximated by the power function x^λ as closely as we like. Recall the local behavior of the power function studied in §2.2: For any $n \in N$, it gives conformal equivalence between the hemi-annulus $\{x \in \mathbb{H}_x \mid e^{-2n-1} < |x| < e^{-2n}\}$ and the hemi-annulus $\{z \in \mathbb{H}_z \mid e^{-\pi\theta_0} < |z| < 1\}$. In a very small neighborhood (say, hemi-disc of radius ϵ) of $x = 0$, s is very near to the power function x^λ, which takes the interval $(-\epsilon, 0)$ and $(0, \epsilon)$ to the circle of radii 1 and $e^{-\pi\theta_0}$. On the other hand we know in advance that s_0 takes these intervals to circles. So we conclude that s takes these interval exactly to these circles, and any small hemi-disk to the annulus $\{z \in \mathbb{C}_z \mid e^{-\pi\theta_0} < |z| < 1\}$.

So we conclude that any sufficiently small neighborhood of $x = 0$ in the upper half-plane is mapped under s_0 onto the ring of radii $\exp(-\theta_0\pi)$ and 1; the unit circle is the image of the interval $(0, 1)$, and the smaller one is the image of the interval $(-\infty, 0)$.

In the same way, any sufficiently small neighborhood of $x = 1$ of the upper half-plane is mapped under s_1 to the ring of radii 1 and $\exp(\theta_1\pi)$; the unit circle is the image of the interval $(0, 1)$. (This is because the variable $1 - x$ is real positive on the interval $(0, 1)$.) See Figure 10.

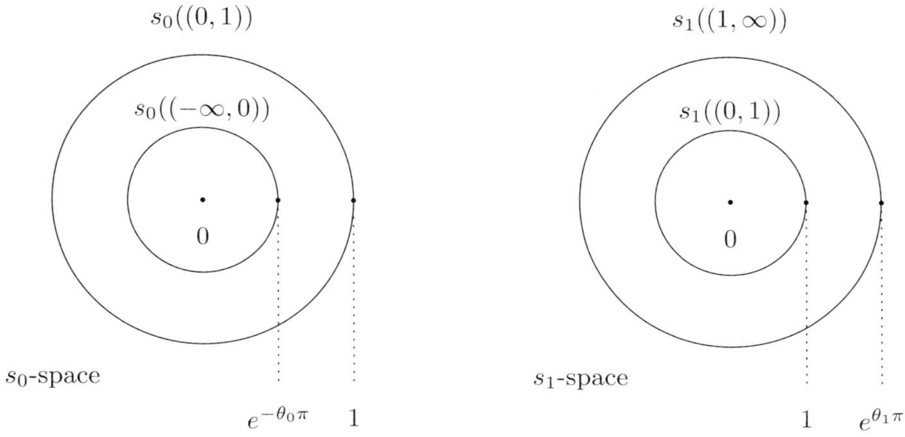

FIGURE 10. A ring in s_0-space and a ring in s_1-space

6.4. Arrangement of the three circles

We would like to draw these two rings in the same plane, say in the s_1-plane. The connection matrix presented in §4.2 would make it possible. Then we will know whether the three circles intersects, and if they do not intersects, whether they form a dartbord or a pig nose. The correct answer is a pig nose; the three circles in the s_1-plane (a bit rotated, see below) are drawn in Figure 11. Precisely speaking, it becomes

Proposition 6.1. *Let the three exponent-differences be* $\lambda = i\theta_0$, $\mu = i\theta_1$, $\nu = i\theta_2$, $(\theta_0, \theta_1, \theta_2 > 0)$. *Then the image of the upper half* x*-plane under the Schwarz map* $s_1 = f_1(x; \mu)/f_1(x; 0)$ *is bounded by the three disjoint circles:*

- *the unit circle with center* 0*, the image of the interval* $(0, 1)$*,*
- *the circle of radius* $\exp(\theta_1\pi)$ *with center* 0*, the image of the interval* $(1, \infty)$*, and*
- *the circle of radius* R *with center* K*, the image of the interval* $(-\infty, 0)$*, where the precise values of* R *and* K *are given in the next subsection.*

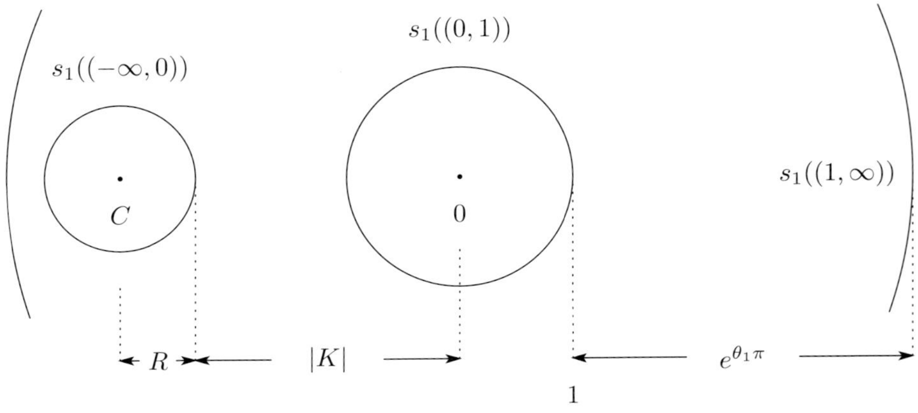

FIGURE 11. Three circles showing the pig-nose $\mathcal{PN}$ in the s-space

The second one encircles the other two, making a pig-nose.

In Figure 11 we rotated the s_1-plane so that the second circle has its center on the negative real axis, that is, at $-|K|$. Or, we better define a new Schwarz map s as

$$s : x \longmapsto \frac{\bar{\xi}}{|\xi|} s_1(x),$$

so that the three circles are exactly as in Figure 11. Let us denote by $\mathcal{PN}$ the complement of the two small discs in the big disc, and call it the pig-nose.

Do not hastily conclude that the image of the upper half-plane under s is $\mathcal{PN}$; wait till §6.6.2.

The next subsection is devoted to a proof of this fact.

6.5. Proof

Note that, since $a - 1/2$, $b - 1/2$ and $c - 1$ are purely imaginary, we have

$$\bar{a} = 1 - a, \quad \bar{b} = 1 - b, \quad \bar{c} = 2 - c$$

and that, since the Gamma function is a real function (I mean $\Gamma(\bar{t}) = \bar{\Gamma}(t)$), we have

$$A = \bar{D}, \quad B = \bar{C}.$$

Since s_0 and s_1 is related by a linear fractional way as in §6.3, a straightforward computation leads to

Lemma 6.2. *If* s_0 *moves along the circle of radius* r, *then* s_1 *moves along the circle with center* K *and radius* R, *where*

$$K = -\frac{\xi(1-r^2)}{|\xi|^2 - r^2}, \qquad R = \frac{r(1-|\xi|^2)}{|\xi|^2 - r^2}, \qquad \textit{where} \quad \xi = \frac{\bar{A}}{B}.$$

Thus the unit circle is mapped to the unit circle, and the circle in the s_0-plane of radius

$$r = \exp(-\theta_0 \pi)$$

is mapped to the circle in the s_1-plane with the center and the radius given just above. The following lemma will garantee that the circle in question lies gently between the two other circles.

Lemma 6.3. $|K| - R - 1 > 0, \quad \exp(\theta_1 \pi) - |K| - R > 0.$

To prove this, we first recall the well-known formula

$$\Gamma(z)\Gamma(1-z) = \frac{\pi}{\sin \pi z},$$

(for a geometric meaning of this formula, see [MY]) which implies in particular

$$|\Gamma(\frac{1}{2} + iy)|^2 = \frac{\pi}{\cosh \pi y}, \qquad y: \text{ real}.$$

On the other hand, by the definition of ξ, we have

$$|\xi| = \left| \frac{\Gamma(a)}{\Gamma(a + i\theta_0)} \frac{\Gamma(b)}{\Gamma(b + i\theta_0)} \right|.$$

Since the real parts of a and b are $1/2$, we can apply the formula above to get

$$\begin{aligned} |\xi|^2 &= \frac{\cosh((-\theta_0 + \theta_1 + \theta_2)\pi/2)\cosh((\theta_0 - \theta_1 + \theta_2)\pi/2)}{\cosh((\theta_0 + \theta_1 + \theta_2)\pi/2)\cosh((\theta_0 + \theta_1 - \theta_2)\pi/2)} \\ &= \frac{\cosh \theta_2 \pi + \cosh(\theta_0 - \theta_1)\pi}{\cosh \theta_2 \pi + \cosh(\theta_0 + \theta_1)\pi}. \end{aligned}$$

So we can conclude that $|\xi|$, as a function of $\theta_2 \geq 0$, increases monotonically to 1 and that

$$1 > |\xi|_{\theta_2 = 0} = \frac{\exp(\theta_1 - \theta_0)\pi + 1}{\exp(-\theta_0 \pi) + \exp(\theta_1 \pi)} > r \; (= e^{-\theta_0 \pi}).$$

Now we are ready to prove the lemma by the identities:

$$\begin{aligned}|K|-R-1 &= \{(1-r^2)|\xi|-r|1-|\xi|^2|-||\xi|^2-r^2|\}/||\xi|^2-r^2| \\ &= \{(1-r^2)|\xi|-r(1-|\xi|^2)-(|\xi|^2-r^2)\}/(|\xi|^2-r^2) \\ &= (1-r)(|\xi|-r)(1-|\xi|)/(|\xi|^2-r^2),\end{aligned}$$

$$\begin{aligned}e^{\theta_1\pi}-|K|-R &= e^{\theta_1\pi}-\frac{|\xi|(1-r^2)}{|\xi|^2-r^2}-\frac{r(1-|\xi|^2)}{|\xi|^2-r^2} \\ &= e^{\theta_1\pi}-\frac{(1-r|\xi|)(|\xi|+r)}{|\xi|^2-r^2} \\ &= \frac{e^{\theta_1\pi}(|\xi|-r)-1+r|\xi|}{|\xi|-r} \\ &= \frac{(e^{\theta_1\pi}+e^{-\theta_0\pi})|\xi|-(e^{(\theta_1-\theta_0)\pi}+1)}{|\xi|-r}.\end{aligned}$$

6.6. Global study of the Schwarz map

Our Schwarz map s, if we continue freely in $\mathbb{C}-\{0,1\}$, is doubly multivalued, that is, ∞-to-∞. So we restricted this map on the upper half x-plane. But it is still ∞-to-one.

We define domains F_x in $\mathbb{H}_x$ and F_s in $\mathbb{H}_s$ so that the restriction of the Schwarz map s on F_x gives a one-to-one correspondence between F_x and F_s, and that the restricted map reproduces the whole Schwarz map by applying the reflection principle. They are made as follows: Cut the pig-nose $\mathcal{PN}$ along the real axis. The upper half part with two-arched bridge shape will be denoted by F_s, and its inverse image will be denoted by F_x.

Though the domain F_s also has the shape of a two-arched bridge (see Figure 12), it is a little distorted — bounded by three real intervals and three curves, which are not arcs (part of circles). This fact is not surprising, because the inverse function of a Schwarz map is not a Schwarz map of a hypergeometric equation.

Anyway, these two two-arched-bridge-shaped domains F_x and F_s will be called *fundamental domains* for s. Recall that in §2.2.2 we got two single-arched bridges for the purely imaginary exponent.

6.6.1. Analytic continuation I. We see what happens if we apply repeatedly Schwarz's reflection principle to the restricted map

$$s|_{F_x} : F_x \longrightarrow F_s$$

through the three real intervals (the intersection of the real axis and the closure of F_x) to the (complex conjugate) mirror image $\bar{F}_x$ of F_x. Since the images of the three intervals are hemi-circles, each image of $\bar{F}_x$ can be known by the reflection with respect to one of these circles. (See Figure 13.) Next we apply reflection principle again through the three intervals, and so on. Eventually we get

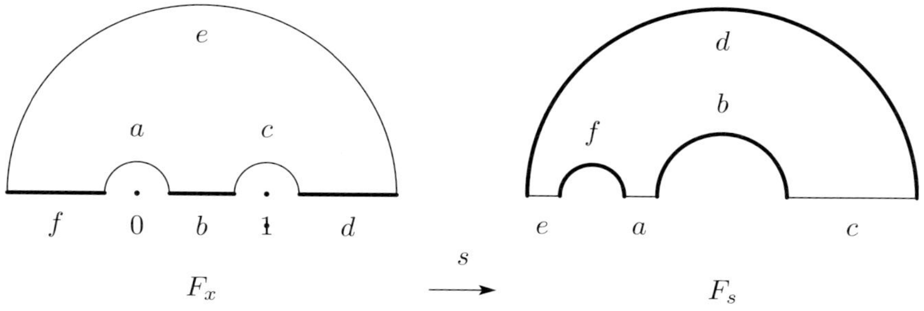

FIGURE 12. The fundamental domains F_x and F_s

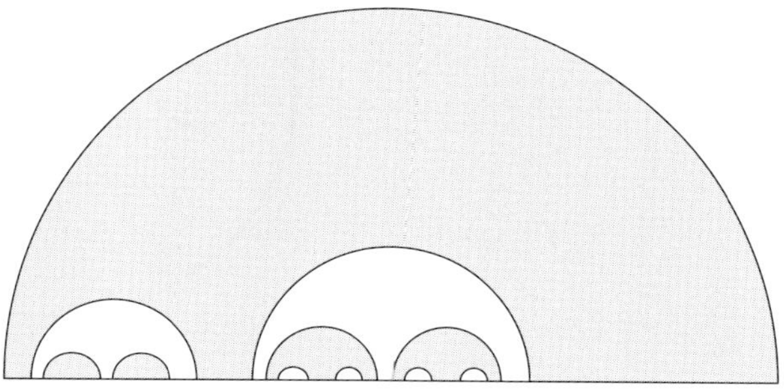

FIGURE 13. Action of a Schottky group on $\mathbb{H}_s$

Proposition 6.4. *The whole image of the* 2*-connected domain* $F_x \cup \bar{F}_x \cup \{$*the three intervals*$\}$ *under* s *will cover the upper half* s*-plane* $\mathbb{H}_s$*. The inverse map, defined on* $\mathbb{H}_s$*, is single-valued, covers infinitely many times the domain* $F_x \cup \bar{F}_x \cup \{$*the three intervals*$\}$*, and gives the isomorphism*

$$\mathbb{H}_s/\Lambda \cong F_x \cup \bar{F}_x \cup \{\textit{the three intervals}\},$$

where Λ *is the monodromy group of the hypergeometric equation.*

The group Λ is a so-called Schottky group (of genus 2). Automorphic functions with respect to this Schottky group is constructed in [IY1].

6.6.2. Analytic continuation II. We next continue analytically the restricted map $s|_{F_x} : F_x \mapsto F_s$ to the upper half x-plane $\mathbb{H}_x$.

Proposition 6.5. *If we apply the reflection principle to the inverse map of* $s|_{F_x}$ *through the real intervals (the intersection of the real axis and the closure of* F_s *) to the (complex conjugate) mirror image* $\bar{F}_s$ *of* F_s*, back to* F_s *through the three intervals, and continue as we did in the* x*-space in the previous section, then the* *whole image under the inverse map, i.e.* $s^{-1}(\mathcal{PN})$, *does not cover the entire half-plane* $\mathbb{H}_x$ *but is the complement of infinitely many disjoint (topological) discs in* $\mathbb{H}_x$.

Sketch of proof. Let us apply the reflection principle for s along the curve c bounding F_x and the line segment c bounding F_s. See Figure 12 and its zoomed one (focussed on $x = 1$) Figure 14. The line segment (resp. hemi-circle) b and the line

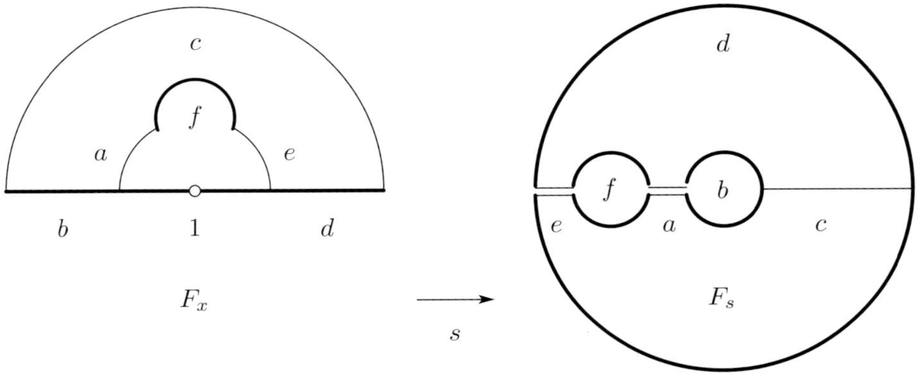

FIGURE 14. Analytic continuation through the curve/segment c

segment (resp. hemi-circle) d are prolonged toward 1 (resp. to the lower s-space). The inverse image under s of the line segment e has one end on the line segment d and the other end *in* H_x, *not* on the real axis. This is because the intervals $(0, 1)$ and $(1, +\infty)$ are mapped to the circles b and d, respectively, and so no part of these interval can be mapped to the other circle f. Similarly the inverse image of the line segment a has one end on the line segment a and the other end in H_x. (This is the essential difference from the Schottky case appeared in the previous subsection.) The completion of the proof is now immediate. □

The proposition above can be paraphrased as follows: the upper half-plane $\mathbb{H}_x$ covers under the map s the whole s-sphere infinitely many times, and the three real intervals bounding $\mathbb{H}_x$ are mapped to the three circles. (See Figure 15.) Maybe Figure 16 is more impressive; the upper half plane $\mathbb{H}_x$ is transformed into a disc, in which bubbles colored in b, d, f form a foam.

7. Towards further study

Since a Schottky group is stable under small deformation, if we move the parameter (λ, μ, ν) a little from a purely imaginary point, the monodromy group is still a

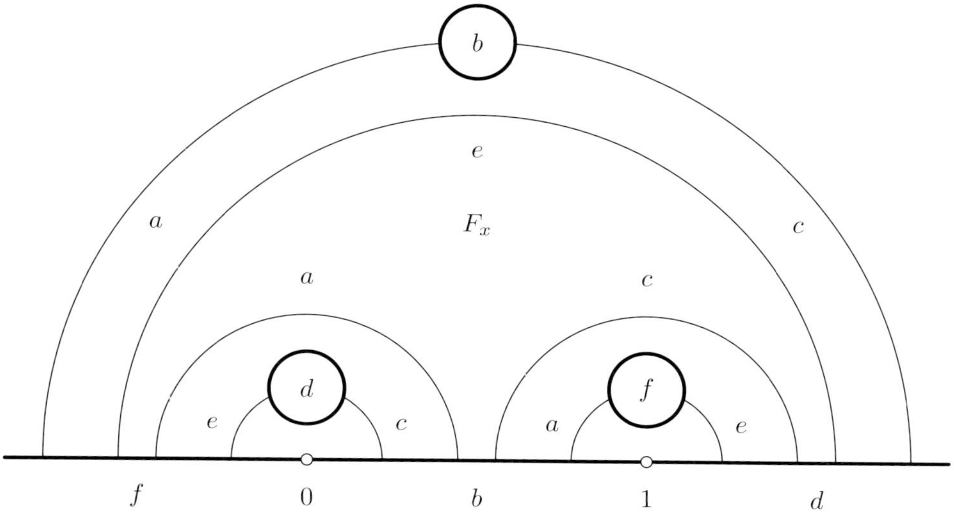

FIGURE 15. The inverse image of $\mathcal{PN}$

Schottky group. Thus the set of points (a, b, c) such that the monodromy group of $E(a, b, c)$ is a Schottky group (of genus 2) form an open subset of the (complex 3-dimensional) space of parameters (a, b, c). We started to study this space in [IY2].

8. Closing

I showed in this lecture, I hope, that though the hypergeometric function is a slightly enriched version of the power function, there are many interesting problems about this function still waiting to be studied; most of them can be stated in a quite elementary language.

I expect that young researchers will find a new aspect of this function, pose a new problem, and (perhaps) solve it.

Acknowledgement. I am grateful to Jurgen Wolfart, who kindly showed me the reference [Zap], which gave me a key to start the study of the Schwarz map of the hypergeometric function with not necessarily real exponents. I also would like to thank Takeshi Sasaki, who is tolerant for my quoting some part of our joint paper. My thanks goes also to the CIMPA, the Galatasary University and the local organizers.

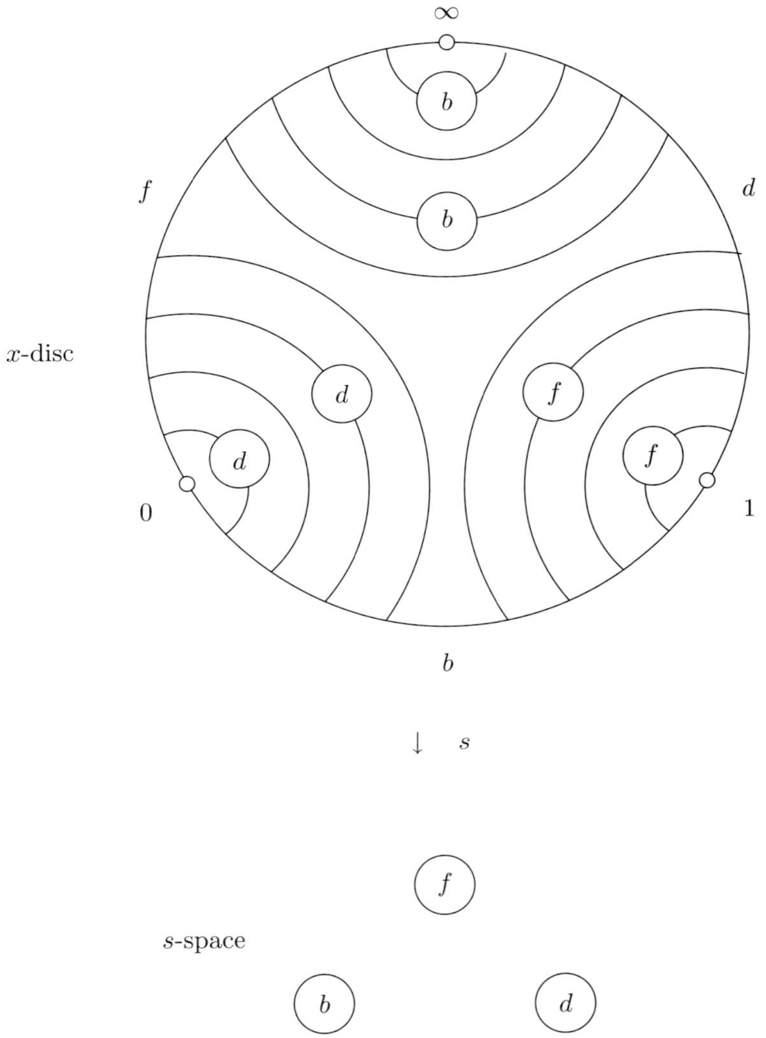

FIGURE 16. More impressive picture: bubbles colored in b, d, f forming a foam

References

[IY1] T. Ichikawa and M. Yoshida, On Schottky groups arising from the hypergeometric equation with imaginary exponents, *Proc. AMS* **132** (2003), 447–454.

[IY2] T. Ichikawa and M. Yoshida, A family of Schottky groups arising from the hypergeometric equation, to appear in *Proc. AMS.*

[IKSY] K. Iwasaki, H. Kimura, S. Shimomura and M. Yoshida, *From Gauss to Painlevé — A modern theory of special functions,* Vieweg Verlag, Wiesbaden, 1991.

[KY] M. Kaneko and M. Yoshida, The kappa function, *Internat. J. Math.* **14** (2003), 1003–1013.

[MY] K. Matsumoto and M. Yoshida, Recent progress of interesection theory for twisted (co)homology groups, *Advanced Studies in Pure Math.* **27** (2000), 217–237.

[SY] T. Sasaki and M. Yoshida, A geometric study of the hypergeometric fucntion with imaginary exponents, *Experimental Math.* **10** (2000), 321–330.

[Sch1] F. Schilling, Beiträge sur geometrischen Theorie der Schwarzschen s-Funktion, *Math. Annalen* **44** (1894), 161–260.

[Sch2] F. Schilling, Die Geometrischen Theorie der Schwarzschen s-Funktion für komplexe Exponenten, *Math. Annalen* **46** (1895), 62–76, 529–538.

[Sch] F. Schottky, Über eine specielle Function, welche bei einer bestimmten linearen Transformation ihres Arguments univerändert bleibt, *J. Reine Angew. Math.* **101** (1887) 227–272.

[Yo1] M. Yoshida, *Fuchsian Differential Equations,* Vieweg Verlag, 1987.

[Yo2] M. Yoshida, *Hypergeometric Functions, My Love,* Vieweg Verlag, 1997.

[Yo3] M. Yoshida, A naive-topological study of the contiguity relations of the hypergeometric function. To appear in Bedlewo Proceedings.

[Zap] M. Zapf Abbildungseigenschaften allgemeiner Dreiecksfunktionen, Diplomarbeit der Universität Frankfurt, 1994.

Masaaki Yoshida
Department of Mathematics
Kyushu University
Fukuoka 810-8560
Japan
e-mail: `myoshida@math.kyuhsu-u.ac.jp`

Progress in Mathematics, Vol. 260, 431–437

Problem Session

Edited by Celal Cem Sarıoğlu

Abstract. This article contains the open problems discussed during the problem session of the CIMPA summer school "Arithmetic and Geometry Around Hypergeometric Functions" held at Galatasaray University, İstanbul, 2005.

1. Introduction

The Problem Session of the summer school *Arithmetic and Geometry Around the Hypergeometric Functions* (AGAHF) was held on June 24, 2005 in İstanbul. Nine open problems were presented during the problem session by Professors Rolf Peter Holzapfel, Jürgen Wolfart, Shiguyeki Kondo, Igor Dolgachev, Daniel Allcock and Hironori Shiga. Professors Jan Stienstra and A. Muhammed Uludağ added their own problems after the problem session.

The aim of this article is to gather the open problems of AGAHF summer school. Some notes taken at the problem session were first brought together and then sent to lecturers to take their opinion. After their corrections and additional background information, this problem session article took its final form.

The following is a list of the 11 problems presented on June 24, 2005 at the problem session of AGAHF summer school.

2. Open Problems

Problem 1 (R.P. Holzapfel). *Determine the Heegner series* $Heeg_{\mathbf{D}_\Gamma}(q)$ *of a Picard line* D_Γ *on the most classical orbital Picard modular plane* $\hat{X}_\Gamma$ *as linear combination of a basis of the space* $\mathcal{M}_3(D_{K/\mathbb{Q}}, \chi_K)$ *of modular forms of weight* 3, *of Nebentypus* χ_K, K *the field of Eisenstein numbers, as it was explained in* [4] *for an orbital curve on the orbital Picard modular plane of Gauß numbers with orbital Apollonius cycle by a* (Q)*-linear combination of Jacobi and Hecke theta series.*

Thereby, the most classical Picard modular surface (plane) is $\mathbb{P}^2$ together with orbital cycle consisting of 6 lines through 4 points (complete quadrilateral) with orbital weights 1 (in general), 3, 9, ∞, respectively, as drawn in the following picture:

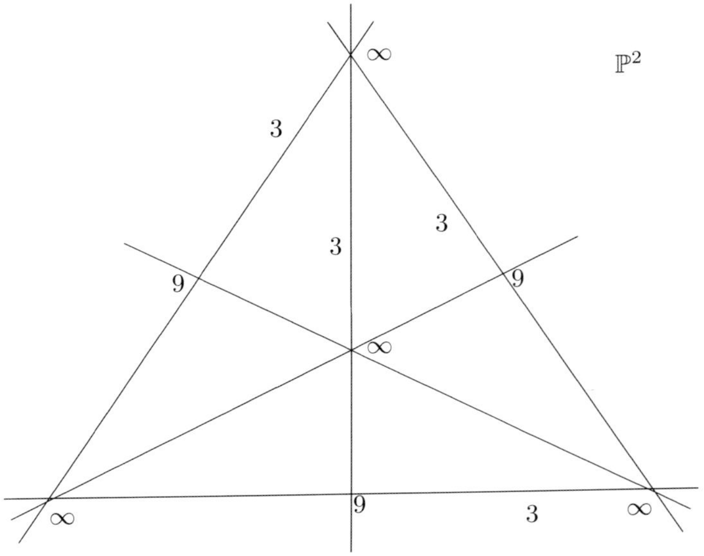

Interpretation as ball quotient: Let $\Gamma_1 = \mathbb{U}((2,1), \mathbb{Z} + \mathbb{Z}\rho)$, ρ a primitive 3-rd unit root, be the full Picard modular group of the field $K = \mathbb{Q}(\sqrt{-3})$ of Eisenstein numbers acting on the complex two-dimensional unit ball $\mathbb{B}$. The corresponding Picard modular plane is the Baily–Borel compactification (add 4 points with weight ∞) of the quotient surface $\Gamma_1(\sqrt{-3})\backslash\mathbb{B}$, where $\Gamma_1(\sqrt{-3})$ is the principal congruence subgroup of Γ_1 with respect to the ideal $(\sqrt{-3})$ in the ring of Eisenstein integers. This was proved in [5]. The quotient map $\mathbb{B} \to \mathbb{P}^2$ has the complete quadrihedral (minus the 4 cusp points) as branch locus. The ramification index of each of the 6 lines is 3, as indiceted in the picture. For D_Γ you can take one of this lines. This is the quotient of a complete linear subdisc $\mathbb{D}$ of $\mathbb{B}$. For more details, e.g., the definition of the orbital line $\mathbf{D}_\Gamma$, see [4].

Problem 2 (R.P. Holzapfel).

a) *Determine (by explicit equations) all Shimura curves of fixed small genus (e.g. the rational ones) of the three orbital Shimura planes described in* [4] *and in Problem* 1.

Two of them are Picard modular and one is Hilbert modular. The Picard planes support the above quadrilateral orbital cycle or the Appolonius cycle, respectively. The former belong to a Picard modular congruence subgroup of Eisenstein numbers, the latter to Gauß numbers. The Hilbert modular plane with orbital Cartesius cycle belongs to the real quadratic number field $\mathbb{Q}(\sqrt{2})$. Use the Relative Proportionality Theorem in [4], which can be written as explicit genus formula for curves in the projective plane joining genus, curve degree and singularity data like Plücker's formula.

b) *Determine their Heegner series as in Problem* 1.

Problem 3 (I. Dolgachev). *Are there any strata in the space of configurations of hyperplanes in higher-dimensional projective space which can be isomorphic to a ball quotient?*

Take $\mathbb{P}^1$. $z_1, z_2, \ldots, z_m \in \mathbb{C}$ and $\mu_1, \mu_2, \ldots, \mu_m \in (0,1)$ such that $\mu_1+\mu_2+\ldots+\mu_m = 2$. Deligne and Mostow gives a list of possible $\underline{\mu}$ such that $(\mathbb{P}^1)^m//_{\underline{\mu}} \mathrm{SL}(2)$ is a ball quotient. A natural question is whether $\mathbb{P}^1$ can be replaced with arbitrary $\mathbb{P}^n$. Let us take $\mathbb{P}^n$, $z_1, z_2, \ldots, z_m \in \mathbb{C}$ and $\mu_1, \mu_2, \ldots, \mu_m \in (0,1)$ such that $\mu_1 + \mu_2 + \ldots + \mu_m = 1+n$. Then one considers the GIT-quotient $(\mathbb{P}^n)^m//_{\underline{\mu}} \mathrm{SL}(n+1)$. Unfortunately, even under some arithmetical conditions on the μ_i's similar to Deligne–Mostow conditions (see [12]), it seems essentially no new examples of ball quotients arise in this way if $n > 1$. Here essentially means up to the association isomorphism

$$(\mathbb{P}^{m-n-2})^m//_{\underline{\mu}} \mathrm{SL}(m-n-1) \simeq (\mathbb{P}^n)^m//_{\underline{\mu}} \mathrm{SL}(n+1).$$

To remedy this unfortunate situation, one has to consider some special strata in the space of configurations defined by the following degeneracy conditions. Pick a collection I of subsets $I_1, I_2, \ldots, I_k$ of $\{1, 2, \ldots, m\}$ of cardinality $n+1$ and consider the subvariety X_I of $(\mathbb{P}^n)^m$ defined by the conditions that the points defined by a subset I_j are linearly dependent. The problem is now whether $X_I \subset (\mathbb{P}^n)^m//_{\underline{\mu}} \mathrm{SL}(n+1)$ is a ball quotient for some values of the parameters μ_i and I.

Problem 4 (D. Allcock). *Are there reflection groups* Γ *acting discretely on the complex ball* B^n*, with finite-volume fundamental domains, for arbitrarily large* n*?*

This question arose during discussion of Prof. Dolgachev's question about ball quotients. A reflection group means a group generated by complex reflections, and a complex reflection is an automorphism of B^n of finite order that fixes a hyperplane pointwise (For example, a complex reflection of the Poincaré disk B^1 is a finite-order rotation around a point.).

B^n is also known as complex hyperbolic space, and the answer to the real-hyperbolic analogue of this problem is 'no': the quotient of H^n by a reflection group cannot be compact if $n \geq 30$ (see [13]) and cannot have finite volume if $n \geq 996$ (see [11]). The techniques that prove these results do not extend to the complex case, where I suspect (not strongly enough to call it a conjecture) that the answer is 'yes'. This suspicion is based on a curious phenomenon expressed as theorems 3.1–3.3 of [1]; a good candidate for a discrete reflection group with finite-volume fundamental domain in B^{4n+1} is the reflection group of the lattice $\left(E_8^{\mathcal{E}}\right)^n \oplus \left(\begin{smallmatrix} 0 & \bar{\theta} \\ \theta & 0 \end{smallmatrix}\right)$. Here $E_8^{\mathcal{E}}$ is the E_8 lattice regarded as a 4-dimensional lattice over the Eisenstein integers $\mathcal{E} = \mathbb{Z}[\sqrt[3]{1}]$ and $\theta = \sqrt{-3}$. See [1] for a discussion of the cases $n = 1$, 2 and 3.

Here is some background. In all the examples we have seen in this summer school, and in several more, the groups Γ acting on B^n such that B^n/Γ is a moduli space are all generated by complex reflections. This is natural because in each case B^n is a ramified cover of a moduli space of some type of object, ramified along the divisor representing the singular objects. This ramification implies that the group of deck transformations contains elements of finite order with fixed-point sets of codimension 1; the only such transformations are complex reflections. Therefore reflection groups are especially interesting in algebraic geometry. The largest n for which a finite-covolume reflection group is known to act on B^n is $n = 13$; I constructed it in [1] by using the Leech lattice. The largest n for which the quotient of B^n by such a group has a known interesting moduli interpretation is $n = 10$; this is the moduli space of cubic three folds in $\mathbb{C}P^4$, a work in progress of J. Carlson, D. Toledo and myself. The possibility of a moduli interpretation for the $n = 13$ example is part of a circle of conjectures I have formulated involving the monster simple group; details will appear elsewhere.

Problem 5 (H. Shiga). *Can a K3 surface having toric structure be obtained from orbifold? If answer is 'yes', what are the its period map and differential equation?*

Problem 6 (J. Wolfart). *Determine the transcendence degree of the field generated by the hypergeometric functions $F(a, b, c; z)$ (say all $a, b, c \in \mathbb{Q}$ with some fixed denominator) over the field $\overline{\mathbb{C}(z)}$ of all algebraic functions, or even better a complete list of algebraic equations over the field $\overline{\mathbb{C}(z)}$ among these $F(a, b, c; z)$.*

Examples of such relations are Gauss's relations between contiguous hypergeometric functions and classical results about algebraicity of $F(a, b, c; z)$. To include also Kummer's relations, quadratic and higher transformations, it is perhaps reasonable to include in the problem somehow also rational transformations in the variable z .

Problem 7 (J. Wolfart). *Does the Jacobian (respectively Prym) of the hypergeometric curve $y^p = u^r(u-1)^s(u-z)^t$ have complex multiplication type only for finitely many z if z runs over a fixed number field $\mathbb{K}$? Stronger version: do we get CM type for only finitely many z if we restrict to algebraic z of bounded degree $[\mathbb{Q}(z) : \mathbb{Q}] < M$?*

The elliptic curve $y^2 = u(u-1)(u-\lambda)$ has in fact complex multiplication only for finitely many λ of degree $< M$: in CM points, the absolute invariant generates class fields, and their degree grows with the discriminant according to a famous result by Siegel–Gross–Zagier.

Problem 8 (S. Kondo). *Are the arithmetic subgroups $\Gamma_{(\mu_i)}$ in the Deligne–Mostow's list related to K3 surfaces?*

Let μ_i be a positive rational number ($0 \le i \le d+1$ or $i = \infty$) satisfying $\Sigma_i \mu_i = 2$. Set

$$F_{gh}(x_2, \cdots, x_{d+1}) = \int_g^h u^{-\mu_0}(u-1)^{-\mu_1} \cdot \prod_{i=2}^{d+1} (u - x_i)^{-\mu_i} du$$

where $g, h \in \{\infty, 0, 1, x_2, \cdots, x_{d+1}\}$. Then F_{gh} is a multi-valued function on

$$M = \{(x_i) \in (\mathbb{P}^1)^{d+3} : x_i \neq \infty, 0, 1 \text{ and } x_i \neq x_j \text{ when } i \neq j\}.$$

These functions generate a $(d+1)$-dimensional vector space which is invariant under monodromy. Let $\Gamma_{(\mu_i)}$ be the image of $\pi_1(M)$ in $PGL(d+1, \mathbb{C})$ under monodromy action. In [2] and [9], Deligne and Mostow gave a sufficient condition for which $\Gamma_{(\mu_i)}$ is lattice in the projective unitary group $PU(d+1)$, that is, $\Gamma_{(\mu_i)}$ is discrete and of finite covolume, and gave a list of such (μ_i) (see [10] for the correction of their list).

Denote $\mu_i = \dfrac{\bar{\mu}_i}{D}$, where D is a common denominator. In case of $D = 3$, 4, 5 or 6, $\Gamma_{(\mu_i)}$ is related to K3 surfaces (see [3], [6], [7] and [8]). In these cases, the corresponding K3 surfaces have an isotrivial pencil whose general fiber is an elliptic curve with an automorphism of an order 4 or 6 (the case D =3, 4 or 6) or a smooth curve of genus 2 with an automorphism of order 5 (the case D =5). We also remark that $12 \sum \mu_i = 24 =$ *Euler characteristic of K3.* All of these examples related to $K3$ surfaces are unirational.

Problem 9 (S. Kondo). *Find a non-unirational moduli which is a ball quotient by using K3 surfaces.*

Problem 10 (A. M. Uludağ). *Families of K3 surfaces with a non-symplectic symmetry as ball quotients.*

In a recent article *Galois coverings of the plane by K3 surfaces*, I classified all smooth K3 surfaces with a group action such that the quotient is isomorphic to the projective plane. For example, the universal orbifold covering of the plane branched along an arrangement of a smooth conic with four lines in general position is a K3 surface (the double covering branched along the same sextic curve has some singularities, its resolution is also a K3 surface). These sextics — or the corresponding K3s — form a 5-dimensional family and are related to 8-point configurations on the projective line as follows: Take the smooth conic as the projective line and 8 intersection points with the remaining four lines as the point configuration. Obviously, points belonging to the same line can be permuted, whereas the points belonging to different lines can not be permuted. Hence the configuration space of these points is a certain "coloured-braid space", which is isogeneous to the usual braid space. The corresponding K3 family must be isogeneous to the one associated by Kondo to 8-point configurations. But in the location cited above there are many other families of K3 surfaces "with symmetry" which can be related to point configurations, for example an arrangement of a conic with 3 lines in general position-related to 6 points on the projective line exactly in the same way. (In this case the standart double-covering argument will not work, since the orbifold structure is more complicated.) It seems promising to apply the machinery developed by Dolgachev and Kondo in this volume to these K3 families. In many cases, it is easy to find elliptic fibrations on these K3 surfaces. Obviously, this story is connected with two-dimensional non-linear hypergeometric integrals associated to the corresponding branch loci (e.g. a conic with four lines).

Problem 11 (J. Stienstra). *Compare the approaches to Lauricella's F_D in the lectures by Looijenga (i.e., Deligne–Mostow), Varchenko (i.e., Knizhnik–Zamolodchikov), Stienstra (i.e., Gel'fand–Kapranov–Zelevinsky).*

Deligne–Mostow theory gives global monodromy information. Gel'fand–Kapranov–Zelevinsky theory is strong in its local analysis, but weak on getting global monodromy. Knizhnik–Zamolodchikov theory gives the differential equations as a connection (i.e., first order, vector valued) using a combination of F_D's with different exponents.

References

[1] Allcock, D., *The Leech lattice and complex hyperbolic reflections*, Invent. Math. **140** (2000), 283–301.

[2] Deligne, P., Mostow,G.W., *Monodromy of hypergeometric functions and non-lattice integral monodromy*, Publ. Math. IHES **63** (1972), 543–560.

[3] Dolgachev, I., van Geemen, B., Kondo, S., *A complex ball uniformization of the moduli space of cubic surfaces via periods of K3 surfaces*, math.AG/0310342, J.reine angew. Math., To appear.

[4] Holzapfel, R.P., *Relative Propertionality on Picard and Hilbert Modular Surfaces*, To appear in the proceedings of the AGAHF summer school.

[5] Holzapfel, R.P., *Geometry and Arithmetic around Euler partial differential equations*, Dt. Verlag d. Wiss., Berlin / Reidel Publ. Comp., Dordrecht, (1986)

[6] Kondo, S., *The moduli space of curves of genus 4 and Deligne-Mostow's complex reflection groups*, Adv. Studies Pure Math. **36** (2002), Algebraic Geometry (2000), Azumino, 383–400.

[7] Kondo, S., *The moduli space of 5 points on $\mathbb{P}^1$ and K3 surfaces*, math.AG/0507006.

[8] Kondo, S., *The moduli space of 8 points on $\mathbb{P}^1$ and automorphic forms*, math.AG/0504233, To appear in the proceedings of the conference "Algebraic Geometry in the honor of Igor Dolgachev".

[9] Mostow, G.W., *Generalized Picard lattices arising from half-integral conditions*, Publ. Math. IHES **63** (1986), 91–106.

[10] Thurston, W.P., *Shape of polyhedra and triangulations of the sphere*, Geometry & Topology Monograph **1** (1998), 511–549.

[11] Prokhorov, M., *Absence of discrete reflection groups with a non-compact polyhedron of finite-volume in Lobachevsky spaces of large dimension*, Math. USSR Izv. **28** (1987), 401–411.

[12] Varchenko, A., *Hodge filtration of hypergeometric integrals associated with an affine configuration of general position and a local Torelli theorem*, I. M. Gelfand Seminar, Adv. Soviet Math., 16, Part 2, Amer. Math. Soc., Providence, RI, (1993), 167–177.

[13] Vinberg, È. B., *The nonexistence of crystallographic reflection groups in Lobachevskii spaces of large dimension*, Funct. Anal. Appl. **15** (1981), 207–216.

Edited by Celal Cem Sarıoğlu
Department of Mathematics
Faculty of Arts and Sciences
Dokuz Eylül University
35160 Buca, İzmir
Turkey
e-mail: `celalcem.sarioglu@deu.edu.tr`